Human Genetics

Concepts and Applications

Library

Seventh Edition

Ricki Lewis

CareNet Medical Group
Schenectady, New York

Mc Graw Hill **Higher Education**

Boston   Burr Ridge, IL   Dubuque, IA   Madison, WI   New York   San Francisco   St. Louis
Bangkok   Bogotá   Caracas   Kuala Lumpur   Lisbon   London   Madrid   Mexico
Milan   Montreal   New Delhi   Santiago   Seoul   Singapore   Sydney   Taipei   Toronto

Higher Education

HUMAN GENETICS: CONCEPTS AND APPLICATIONS, SEVENTH EDITION

Published by McGraw-Hill, a business unit of The McGraw-Hill Companies, Inc., 1221 Avenue of the Americas, New York, NY 10020.

Some ancillaries, including electronic and print components, may not be available to customers outside the United States.

This book is printed on recycled, acid-free paper containing 10% postconsumer waste.

1 2 3 4 5 6 7 8 9 0 DOW/DOW 0 9 8 7 6 5

ISBN-13 978–0–07–110689–4
ISBN-10 0–07–110689–8

The credits section for this book begins on page C- 1 and is considered an extension of the copyright page.

About the Author

Ricki Lewis has built a multifaceted career around communicating the excitement of life science, especially genetics and biotechnology. She earned her Ph.D. in genetics in 1980 from Indiana University, working with homeotic mutations in *Drosophila melanogaster.*

Ricki is the original author of *Life,* an introductory biology text; co-author of two human anatomy and physiology textbooks; and author of *Discovery: Windows on the Life Sciences,* an essay collection about research and the nature of scientific investigation. She writes frequently on research and news in genetics, biotechnology, and neuroscience for several magazines. Since 1980, Ricki has published widely, including one of the first stories on DNA fingerprinting in *Discover* magazine. She has taught a variety of life science courses at Miami University, the University at Albany, Empire State College, and community colleges. Ricki has been a genetic counselor for a large private medical practice in Schenectady, NY, since 1984, and recently became a hospice volunteer. She enjoys travel, often to give talks on genomes, stem cell biology, and media coverage of biology and health.

Ricki lives in upstate New York with chemist husband Larry, three daughters, and many cats.

Dedicated to Ray,
who opened my
eyes to human
genetics in a new
way.

Brief Contents

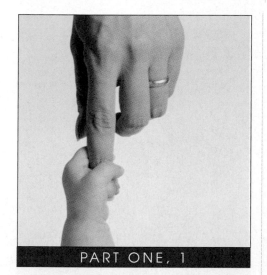

PART ONE, 1

PART TWO, 75

PART THREE, 171

PART FOUR, 267

PART FIVE, 331

PART SIX, 377

List of Boxes

READINGS

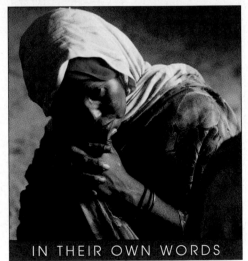

IN THEIR OWN WORDS

BIOETHICS

Clinical Coverage

Chapter Opening Case Studies

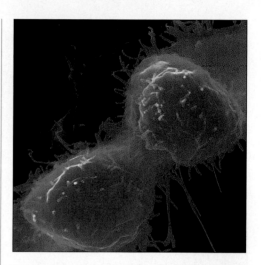

Solving a Problem

Contents

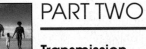

PART THREE

DNA and
Chromosomes   171

PART FOUR

Population Genetics 267

PART FIVE

Immunity and Cancer 331

PART SIX

Genetic Technologies 377

Preface

Genetics on the Personal Level

Genetics, like other sciences, is built around units: a DNA nucleotide, a gene, a chromosome, a cell, a family, and a population. But the truly important unit of inheritance is the person.

The person has been the focus of this book through its seven editions, even as research has shifted from a single gene approach to considering entire genomes. With this revision, however, came an experience that has opened my eyes in a new way to the connections among the various levels of genetics. I have come to know a young man who has Huntington disease. My new friend is unable to speak or move, but he communicates with what remains of his facial expression, especially his eyes. I am astonished at how an extra bit of DNA out on a tip of the fourth largest chromosome can slowly drown a small part of the brain in protein, diminishing the body while leaving the person intact. The mutation reverberates through my friend's family, striking many through the generations.

Knowing someone with an inherited disease renews my wonder at how, most of the time, our 24,000 or so genes work just fine, and their variants create the distinctions that make us individuals. New to this edition, glimpses of personal impacts of genetics open each chapter. They span the familiar (infections and lactose intolerance), the exceedingly rare (inborn errors of metabolism), the practical (forensics) to the just weird (improving pig manure). All capture attention and catapult the reader into the chapter material. Within the chapters, a feature called Relating the Concepts refocuses the discussion on the initial vignette. Other personal touches from editions past remain, such as "In Their Own Words" boxed essays, in-text examples, and questions, problems, and cases at the end of each chapter.

When we step back from individual views of genetics, a bigger picture emerges. Reading 16.1 encapsulates much recent research in *What Makes Us Human?* It continues ideas from the opening essay, which evokes a time when more than one type of hominid roamed the earth. The final chapter considers the human genome in the context of other species with whom we share the planet and includes Figure 22.9, a periodic table-like chart comparing selected organisms whose genomes have been sequenced. I look forward to updating it as we learn more.

What Sets This Book Apart

Unique Writing Style

From the beginning, the clarity, flavor, and immediacy of *Human Genetics: Concepts and Applications* has been uniquely interesting and accessible to non-scientist readers. Into the book flows my 25 years of experience as a journalist, 20 years as a genetic counselor, and my PhD background in genetics. As a frequent contributor to the magazine The Scientist for 15 years, I have had access to scientific meetings and researchers that are not available to everyone. From my network of sources, I learn of new research results before they are published. This allows me to keep *Human Genetics: Concepts and Applications* current—and that's vital for a scientific field that is racing ahead as fast as genetics/genomics.

My approach is straightforward: present the essential concepts in clear language, then demonstrate them with the very best, sometimes quirky, examples. Students appreciate this approach to writing and learn from it.

Pedagogical Tools

Pedagogical aids ensure that students can identify and understand basic concepts and relate the many examples and cases. Chapters open with an outline and an essay. At the end of each major section, **Key Concepts** reinforce important core material. **Relating the Concepts** questions are embedded in the narrative, where they connect chapter content to the opening essay. **Chapter Summaries** review content, calling attention to important vocabulary.

Each chapter ends with a great variety of **Review Questions** to measure content knowledge and **Applied Questions** to practice using that knowledge. Both sets of questions are written to engage students in understanding the mechanisms of genetics and enable them to master content from a basic to a more advanced level—often with humor. Following the Applied Questions are **Cases and Research Results**, many of them new to this edition. Answers to all questions are at the end of the book.

Web Activities encourage students to dig deeper. They provide an opportunity to find the newest genetic information and to use some of the latest tools and databases in genetic analysis.

Dynamic Art

Art Presented in Steps   Illustrating an entire biological process in one figure can overwhelm students and hinder understanding. Therefore, in this edition, appropriate figures have been reorganized, and summary tables and step-by-step figures have been added where appropriate to help students focus on the important concepts.

For example, building a polypeptide from genetic instructions lies at the crux of understanding genetics, and is a difficult topic more easily learned in pieces. The figures in Chapter 10, *Gene Action: From DNA to Protein*, clarify concepts by gradually explaining the various components of the concept.

- Orientations of DNA and mRNA are uniform in all figures in chapter 10.

- New figure 10.6 is small but important: it depicts the physical relationships among tRNA, mRNA, and amino acid.

- In previous editions, figure 10.7 showed 3 stages of transcription in part a and simultaneous transcription in part b. It is now 2 figures. Similarly, translation elongation and termination, once a single figure, are now two.

- New figure 10.19 summarizes the entire process: From DNA to RNA to protein.

- The final section, on protein folding, now includes a table of disorders—kuru might be unfamiliar, but Alzheimer's disease and Parkinson's disease are not.

Chapter 14, *When Allele Frequencies Stay Constant*, has two new figures (14.4 and 14.5) that calculate carrier frequencies for two disorders. Figure 15.9 in chapter 15, *Changing Allele Frequencies*, uses colored shapes to trace how nonrandom mating, migration, genetic drift, mutations, and natural selection change allele frequencies, driving evolution. This new edition retains the entire figure, but also breaks it down, introducing each component separately before the culmination.

Comparisons Used to Aid Understanding   In contrast to art presented in steps, some concepts in genetics reflect dichotomies, and so presenting them together can provide clarity. Figure 5.13, for example, contrasts independent assortment and linkage. Figure 7.10 and the accompanying discussion distinguish a linkage study from an association study. Table 14.5 compares VNTRs and STRs, two classes of DNA repeats used in DNA profiling.

New to This Edition

- Each chapter now opens with a **case study** that provides students with a glimpse of the personal impact of genetics.

- **Relating the Concepts** sections pose questions related to the opening case study, so students apply concepts to real-life situations.

- **New and updated information has been integrated throughout the text,** giving students a modern perspective of human genetics.

Chapter 1, *Overview of Genetics*

- Boxed reading introduces DNA structure and function, so instructors can easily cover DNA or Mendel first.

- New section on metagenomics discusses sampling DNA from ecosystems.

Chapter 4, *Mendelian Inheritance*

- Illustrations reworked so the short and tall pea plants are easier to distinguish.

- Several new end-of-chapter questions.

Chapter 5, *Exceptions and Extensions to Mendel's Laws*

- New figure compares independent assortment to linkage.

Chapter 8, *The Genetics of Behavior*

- Fuller explanation of neuron function begins the chapter.

Chapter 14, *When Allele Frequencies Stay Constant*

- Section on forensics is rewritten for clarity and currency.

- Chapter 15, *Changing Allele Frequencies*

  - *Reading, Antibiotic Resistance: Genomics to the Rescue?* revised to include information on methicillin-resistant staph aureus (MRSA) and microbial genome sequences that determine how resistance arises.

  - Figure 15.9 from the sixth edition, *Forces that change allele frequencies*, is broken down into its parts at the appropriate section of text, and the parts reunited into a complete figure at the end of the chapter.

Chapter 18, *Genetics of Cancer*

- New section, table, and figures on Origins of Cancer Cells.

Chapter 19, *Genetic Technologies*

- **Reorganized** to cover patenting, amplifying, modifying, and monitoring DNA.

Chapter 20, *Genetic Testing, Genetic Counseling, and Gene Therapy*

- Emphasizes genetic testing, counseling, and privacy.

Chapter 22, *The Age of Genomics*

- Offers more information on comparative genomics including a new figure : "A periodic table of genomes" that highlights 9 sequenced (non-human) genomes.

- **Unparalleled coverage of stem cells** Dr. Lewis' specialty is stem cells, and coverage in this text reflects her frequent conversations with leading researchers in this field. The information and illustrations on stem cells are unparalleled.

  - In Chapter 2, *Cells*, four illustrations accompany the clear narrative, progressing from basic to applied views of stem cells. Figures 2.23 and 2.24 stress the defining characteristic of self-renewal, as well as depicting the route to differentiation of daughter cells. Figure 2.25 walks through somatic cell nuclear transfer, and figure 2.26 considers how stem cells can heal a heart.

  - The stem cell theme continues in figure 11.4, which depicts development of the pancreas into a uniquely dual structure—from a single type of progenitor cell.

  - A new section in Chapter 18, *Genetics of Cancer*, explores the role of stem cells in causing cancer, with several new figures (18.8, 18.9, 18.10).

- **New problems, cases, and research questions** are scattered throughout the text. The popular **Solving a Problem** feature introduced in the last edition includes new problems.

- **Focus Placed on the Important Concepts** Fewer bold terms and a more concise glossary emphasize the important terms—if a term isn't directly related to genetics, or if it is a disorder name, it isn't emphasized for memorization. At the same time, all disorders are identified by their Online Mendelian Inheritance in Man number, to ease acquiring further information. Suggested Readings have been moved to the Online Learning Center, providing space for more end-of-chapter exercises.

Teaching and Learning Supplements

McGraw-Hill offers various tools and teaching products to support the seventh edition of *Human Genetics: Concepts and Applications.* Students can order supplemental study materials by contacting their local bookstore. Instructors can obtain teaching aids by calling the Customer Service Department at 800-338-3987, visiting the text website at www.mhhe.com/lewisgenetics7, or contacting your local McGraw-Hill sales representative.

Digital Content Manager

This multimedia collection of visual resources allows instructors to utilize artwork from the text in multiple formats to create customized classroom presentations, visually based tests and quizzes, dynamic course website content, or attractive printed support materials. The digital assets on this cross-platform CD-ROM are grouped by chapter within the following easy-to-use folders:

- **Art Library**—All textbook art in a format compatible with presentation or word processing software.
- **Photo Library**—All photos from the textbook are available in color.
- **PowerPoint Presentations**—Ready-made presentations cover each chapter of the textbook.
- **Active Art Library**—Key figures from the textbook are saved in manipulable layers that can be isolated and customized to meet the needs of the lecture environment. Build images from simple to complex to suit your lecture style.
- **Animations Library**—Numerous full-color animations of key processes are provided. Harness the visual impact of processes in motion by importing these files into classroom presentations or course websites.

Instructor Testing and Resource CD-ROM (ITRCD)

The ITRCD is a cross-platform CD-ROM providing a wealth of resources for the instructor. Supplements featured on this CD-ROM include a computerized test bank utilizing McGraw-Hill's EZ Test. EZ Test is a flexible and easy-to-use electronic testing program. The program allows instructors to create tests from book-specific items. It accommodates a wide range of question types, and instructors may add their own questions. Multiple versions of the test can be created and any test can be exported for use with course management systems such as WebCT, BlackBoard or PageOut. The program is available for Windows and Macintosh environments.

Other assets on the ITRCD are grouped within easy-to-use folders. The Instructor's Manual is available, and Word files of the test bank are included for those instructors who prefer to work outside of the test generator software.

Instructor's Manual

The Instructor's Manual, prepared by William Perry Baker of Midwestern University, is available on the ITRCD or through the Instructor Resources of the Online Learning Center, (www.mhhe.com/lewisgenetics7). The manual includes chapter outlines and overviews, a chapter-by-chapter resource guide to use of visual supplements, answers to questions in the textbook, additional questions and answers for each chapter, and Internet resources and activities.

Overhead Transparencies

A set of 100 full-color transparencies showing key illustrations from the textbook is available for adopters.

For the Student
Genetics: From Genes to Genomes CD-ROM

This easy-to-use CD covers the most challenging concepts in the course and makes them more understandable through presentation of full-color animations and interactive exercises.

Online Learning Center (OLC)

Get online at www.mhhe.com/lewisgenetics7. The OLC offers an extensive array of learning and teaching tools. Explore this dynamic site designed to help you get ahead and stay ahead in your study of human genetics. Some of the activities you will find on the website include:

- Self-quizzes to help you master material in each chapter
- Flash cards to ease learning of new vocabulary
- Case studies to practice application of your knowledge of human genetics
- Links to resource articles, popular press coverage, and support groups

Case Workbook to accompany *Human Genetics*, seventh edition, by Ricki Lewis

This workbook is specifically designed to support the concepts presented in *Human Genetics* through real cases adapted from recent scientific and medical journals, interviews, and meetings. The workbook has been extensively updated and provides practice for constructing and interpreting pedigrees; applying Mendel's laws; reviewing the relationships of DNA, RNA, and proteins; analyzing the effects of mutations; evaluating phenomena that distort Mendelian ratios; designing gene therapies; and applying new genomic approaches to understanding inherited disease. A special set of exercises at the end of the workbook link concepts across chapters. An answer key is available for the instructor.

Acknowledgements

Human Genetics: Concepts and Applications, seventh edition, would not have been possible without the editorial and production dream team: many thanks to Rose Koos, Developmental Editor; Toni Michaels, Carrie Burger and Chris Hammond, photo editors; Anne Cody, copyeditor; Sheila Frank, project manager, and Laurie Janssen, designer. Many thanks also to Deborah Allen, who guided the book through previous editions. Special thanks to Don Watson, dedicated reader, who pointed out errors. Thanks also to Jim McGivern of Gannon University for insightful comments on every edition.

I also thank my wonderful family: Larry, daughters Heather, Sarah, and Carly, and our legions of felines.

Reviewers

Michael A. Abruzzo
California State University, Chico

Joel Adams-Stryker
Evergreen Valley College

William P. Baker
Midwestern University

Jay L. Brewster
Pepperdine University

Barry Chess
Pasadena City College

Thomas R. Danford
West Virginia Northern Community College

David W. Essar
Winona State University

Lori Estes
Georgetown University

Anne M. Galbraith
University of Wisconsin, La Crosse

Gail E. Gasparich
Towson University

Dr. Burt Goldberg, Ph.D.
New York University

Gary Gussin
University of Iowa

Cheryl L. Jorcyk
Boise State University

Trace Jordan
New York University

Mary King Kananen
Penn State Altoona

Arthur Koch
Indiana University

Karen Kurvink
Moravian College

Patricia Matthews
Grand Valley State University

Gerard P. McNeil
York College, City University of New York

Philip Meneely
Haverford College

Joanne Odden
University of Oregon

Frank Potter
Fort Hays State University

James V. Price
Utah Valley State College

Regina Rector
William Rainey Harper College

Laura S. Rhoads
State University of New York at Potsdam

Wendy Rothwell
University of California, Santa Cruz

Monica M. Skinner
Oregon State University

Patricia M. Walsh
University of Delaware

Yunqiu Wang
University of Miami

Robert J. Wiggers
Stephen F. Austin State University

Visual Preview

Instructional Art Program

Art program puts molecular and cellular information into a familiar context.

Figures of complex processes focus on essentials and are presented in easy-to-follow steps

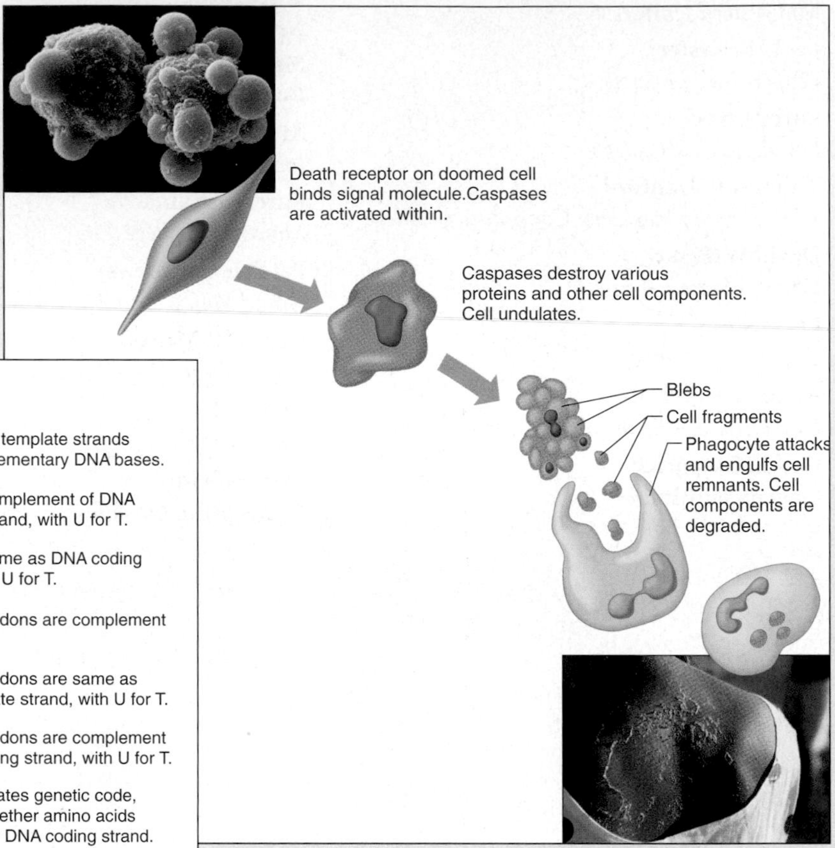

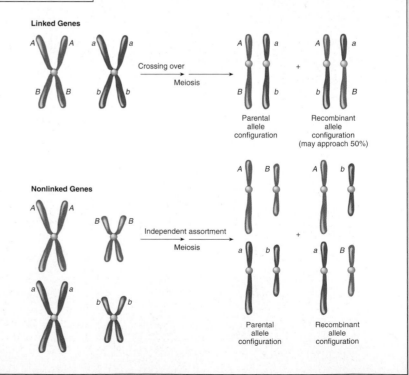

Death receptor on doomed cell binds signal molecule. Caspases are activated within.

Caspases destroy various proteins and other cell components. Cell undulates.

- Blebs
- Cell fragments
- Phagocyte attacks and engulfs cell remnants. Cell components are degraded.

Figure 2.20

Type of molecule	Sample sequence	Rules
DNA coding strand	GCT GTC AAA TGC GGT	**1** Coding and template strands have complementary DNA bases.
DNA template strand	CGA CAG TTT ACG CCA	**2** mRNA is complement of DNA template strand, with U for T. **3** mRNA is same as DNA coding strand, with U for T.
mRNA	GCU GUC AAA UGC GGU	**4** tRNA anticodons are complement of mRNA. **5** tRNA anticodons are same as DNA template strand, with U for T.
tRNA anticodons	CGA CAG UUU ACG CCA	**6** tRNA anticodons are complement of DNA coding strand, with U for T. **7** tRNA translates genetic code, bringing together amino acids specified by DNA coding strand.
Amino acids (protein)	alanine-valine-lysine-cysteine-glycine	**8** Amino acids bond to form a protein.

Figure 10.19

Comparative figures provide clarity and aid understanding

Linked Genes

A A   a a

B B   b b

Crossing over → Meiosis →

A a   A a

B b   b B

+

Parental allele configuration

Recombinant allele configuration (may approach 50%)

Nonlinked Genes

A A   B B

a a   b b

Independent assortment → Meiosis →

A B   A b

a b   a B

+

Parental allele configuration

Recombinant allele configuration

Figure 5.13

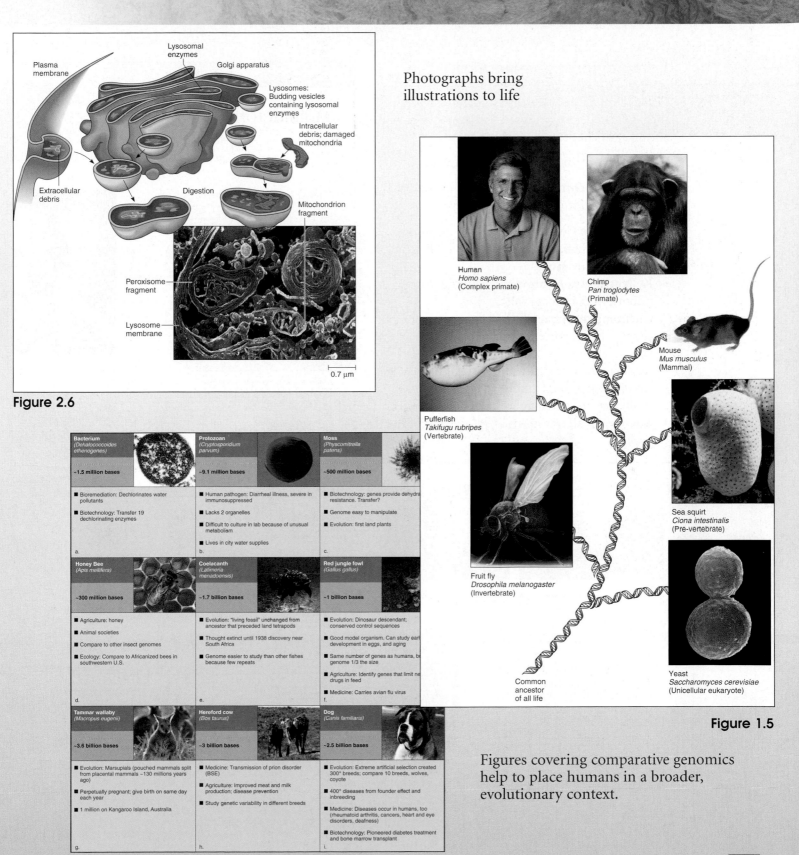

Figure 2.6

Photographs bring
illustrations to life

Figure 1.5

Figures covering comparative genomics
help to place humans in a broader,
evolutionary context.

Figure 22.9

Extraordinary Learning Aids

Pedagogical aids ensure that students can identify the basic concepts presented and exemplified within each chapter.

- **Chapter outline** previews the chapter contents.

- **Chapter opening case study** provides a real-life story related to the chapter concepts.

- **Relating the Concepts** questions encourage students to consider the concepts in the context of the real-life story.

- **Key concepts** are summarized to reinforce important core material.

- **Chapter summaries** review the contents of the chapter, calling attention to important new vocabulary.

- **Review questions** measure content knowledge.

- **Applied questions** guide students in solving challenges that genetic information presents.

- **Answers** to all questions are at the end of the book.

CHAPTER 7

Multifactorial Traits

CHAPTER CONTENTS

7.1 Genes and the Environment Mold Most Traits
    Polygenic Traits Are Continuously Varying
    Fingerprint Patterns
    Height
    Eye Color
    Skin Color

7.2 Methods Used to Investigate Multifactorial Traits
    Empiric Risk
    Heritability
    Adopted Individuals
    Twins
    Association Studies

7.3 Two Multifactorial Traits
    Heart Health
    Weight

CLEFT LIP AND PALATE

The young couple was shocked when they first saw their daughter. She had a cleft lip and palate—a hole between her nose and upper lip. The parents soon discovered that feeding Emily was difficult, because she could not maintain suction. Special nipples on her bottles helped.

Today, Emily is 14, and has a glorious smile. The defect occurred between weeks 4 and 12 of prenatal development, when the nose and jaw failed to meet and close. Emily endured many surgeries. Her first procedure, at 4 months, repaired her lip; the second, at a year, connected the edges of her palate (the roof of the mouth). The doctor also repositioned tissue at the back of her throat to correct her nasal speech. A speech and language therapist helped with Emily's early feeding problems and assisted her when frequent ear infections, due to openings at the back of her throat, caused hearing loss. At age seven Emily had orthodontia to make room for her permanent teeth, and at age 10, bone from her hip was used to bolster her palate so that it could support the teeth. At age 16, Emily can have surgery on her nose to build it up and straighten it.

Cleft lip, with or without cleft palate, is very variable in severity and has genetic and environmental components. Known causes include at least three

Cleft lip is more likely to occur in a person who has a relative with the

researchers to search human genome data to identify nearby genes whose protein products could be absent or altered in a way that explains symptoms. This was the case, for example, for Crohn disease, in which the intestines become severely inflamed in response to the presence of certain bacteria.

Two different research groups identified a gene that increases the risk of developing Crohn disease. One group consulted older linkage data that indicated a causative gene on chromosome 16 in some families. These researchers considered all the genes identi-

Individuals also differ in "large-scale copy variations," which are very long DNA base sequences present in some people but not in others. A 200,000 base sequence might be absent in one individual, present in one copy in another, and present in two copies in yet another. So far researchers have discovered 255 places in the human genome that harbor these mysterious sequences.

Figure 7.10 compares linkage analysis to association studies. **Table 7.6** reviews terms used in the study of multifactorial traits.

RELATING THE CONCEPTS

Researchers have identified a gene, *IRF6*, that causes isolated cleft lip/cleft palate in some families. An ongoing study is investigating 150 SNPs located among more than 140,000 DNA bases surrounding this gene. Explain how haplotypes based on these SNPs might be used to predict risk for the condition in the general population.

Figure 7.13 The environment influences gene expression. Comparison of average body weights among the Arizona population of Pima Indians (a) and the Mexican population (b) reveals the effects of the environment.

Arizona. By the 1970s, the Arizona Indians no longer farmed nor ate a low-calorie, low-fat diet, but instead consumed 40 percent of their calories from fat. With this extreme change in lifestyle, they developed the highest prevalence of obesity of any population on earth. (Prevalence is the total number of individuals with a certain condition in a particular population at a given time.) Half of the Arizona group had diabetes by age 35, weighing, on average, 57 pounds (26 kilograms) more than their southern relatives, who still eat a low-fat diet and are very active.

The Pima Indians demonstrate that future obesity is not sealed in the genes at conception, but instead is much more likely to occur if the environment provides too many calories and too much fat. They illustrate what geneticist James Neel termed the

"thrifty gene hypothesis" in 1962. He suggested that long ago, the hunter-gatherers who survived famine had genes that enabled them to efficiently conserve fat. Today, with food plentiful, the genetic tendency to retain fat is no longer healthful, but harmful. Unfortunately, for many of us, our genomes hold an energy-conserving legacy that works too well—it is much easier to gain weight than to lose it, for sound evolutionary reasons.

Interactions and contributions of genes and the environment provide some of the greatest challenges in studying human genetics. Why does one smoker develop lung cancer, but another does not? Why can one person consistently overeat and never gain weight, while another gains easily? Because we exist in an environment, no gene functions in a vacuum. Subtle interac-

tions of nature and nurture mold our lives and make us all—each of us—unique individuals.

Key Concepts

1. Genes that affect lipid metabolism, blood clotting, leukocyte adhesion, and blood pressure influence cardiovascular health.
2. Genes that encode leptin, the leptin receptor, and proteins that transmit or counter leptin's signals affect body weight.
3. Studies on adopted individuals and twins indicate a heritability of 80 percent for obesity.
4. Populations that suddenly become sedentary and switch to a fatty diet reflect environmental influences on body weight.

Summary

7.1 Genes and the Environment Mold Most Traits

1. **Multifactorial traits** reflect the environment and genes. A **polygenic trait** is determined by more than one gene and varies continuously in expression. The frequency distribution of phenotypes for a polygenic trait forms a bell curve.

7.2 Methods Used to Investigate Multifactorial Traits

2. **Empiric risk** measures the likelihood that a multifactorial trait will recur based on its prevalence in a population. The risk rises as genetic closeness to an affected individual increases, as the severity of the phenotype increases, and as the number of affected relatives rises.

3. **Heritability** estimates the proportion of variation in a multifactorial trait attributable to genetic variation. Heritability describes a trait in a particular population at a particular time. Heritability equals 1 when the actual incidence of a shared trait among people related in a certain way exceeds the expected incidence. Rare dominant alleles can contribute to heritability.

---

4. Characteristics shared by adopted people and their biological parents are mostly inherited, whereas similarities between adopted people and their adoptive parents reflect environmental influences.

5. **Concordance** measures the frequency of expression of a trait in both members of MZ or DZ twin pairs. The more influence genes exert over a trait, the higher the

differences in concordance between MZ and DZ twins.

6. **Association studies** correlate SNP patterns to increased risk of developing a disorder.

7.3 Two Multifactorial Traits

7. Genes that control lipid metabolism and blood clotting contribute to cardiovascular health.

8. Leptin, its receptor, its transporter, neuropeptide Y, and the melanocortin-4 receptor are proteins that affect body weight. Fat cells secrete leptin in response to starvation, and the protein acts in the hypothalamus. Populations that switch to a fatty, high-calorie diet and a less-active lifestyle reveal the effects of the environment on weight.

Review Questions

1. Consider the traits of eye color and body weight. Which is more likely to be inherited as a Mendelian trait, and which is multifactorial? Cite reasons for your answer.

2. Cite two examples from the chapter of a rare illness that helped researchers understand a process that could be applied to treat or help more people.

3. What is the difference between a Mendelian multifactorial trait and a polygenic multifactorial trait?

4. Which has a greater heritability—eye color or height? State a reason for your answer.

5. How can skin color have a different heritability at different times of the year?

6. In a large, diverse population, why are medium brown skin colors more common than very white or very black skin?

7. Using the information in figure 7.6, what percentage of genes does Tim share with
    a. Joan?
    b. Hailey?
    c. Eliot?
    d. Ricki?

8. Describe the type of information in a(n)
    a. empiric calculation.
    b. twin study.
    c. adoption study.
    d. association study.

9. Why does SNP mapping require extensive data?

10. Name three types of proteins that affect cardiovascular functioning and three that affect body weight.

11. Describe or sketch the circuitry for appetite control during starvation, based on figure 7.12.

Applied Questions

1. Rebecca breeds Maine coon cats. The partial pedigree to the right describes how her current cats are related—the umbrellalike lines indicate littermates, which are the equivalent of fraternal (DZ) twins in humans.

Cat lover Sam wishes to purchase a pair of Rebecca's cats to breed, but wants them to share as few genes as possible to minimize the risk that their kittens will inherit certain multifactorial disorders. Sam is quite taken with Farfel, but can't decide among Marbles, Juice, or Angie for Farfel's mate.

Calculate the percentage of genes that Farfel shares with each of these female relatives. With which partner would the likelihood of healthy kittens be greatest?

Maine coon cats

"Solving a Problem" sections appear throughout the book where students can perform a genetic analysis. Each new section presents a step-by-step sample computation.

Case studies and Research Results found after each chapter apply, and sometimes extend, concepts. These case studies supplement those in the *Case Workbook to accompany Human Genetics* by Ricki Lewis.

Web activities encourage students to find information about human genetics that particularly interests them. They also provide an opportunity to find the latest genetic information and to use some of the latest tools and databases in genetic analysis.

Reading 7.1
Solving a Problem: Connecting Cousins

With more genetic tests becoming available as the human genome sequence is analyzed, more people are learning that relatives beyond their immediate families have certain gene variants that might affect their health. Because the genetic closeness of the relationship impacts the risk of developing certain conditions, it may be helpful to be able to calculate the percentage of the genome that two relatives share.

The pedigree in **figure 1** displays an extended family, with "YOU" as the starting point. Calculate the percent of the genome shared for your first cousins once and twice removed (that is, removed from you by one or two generations, respectively)—in the figure, in generations III and II, while YOU are in generation IV. A second, third, or fourth cousin, by contrast, is in the same generation on a pedigree as the individual in question; see, for example, individual V-1 in figure 1.) **Table 1** summarizes the genetic relationships between cousins.

SOLUTION
The rules: Every step between parent and child, or sibling and sibling, has a value of 1/2, because these types of pairs share approximately 1/2 of their genes, according to Mendel's first law (chromosome segregation).

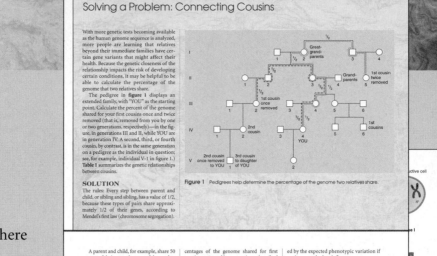

Figure 1 Pedigrees help determine the percentage of the genome two relatives share.

A parent and child, for example, share 50 percent of their genes, because of the mechanism of meiosis. Siblings share on average 50 percent of their genes, because they have a 50 percent chance of inheriting each allele for a gene from each parent. Genetic counselors use the designations of primary (1°), secondary (2°), and tertiary (3°) relatives when calculating risks (table 7.4 and **figure 7.6**). For extended or complicated pedigrees, the value of 1/2 or 50 percent between siblings and between parent-child pairs can be used to trace and calculate the percentage of genes shared between people related in other ways. **Reading 7.1** discusses how to calculate per-

centages of the genome shared for first cousins separated by generations, described as "removed" by one or more generations.

If the heritability of a trait is very high, then of a group of 100 sibling pairs, nearly 50 would be expected to have the same phenotype, because siblings share on average 50 percent of their genes. Height is a trait for which heritability reflects the environmental influence of nutrition. Of 100 sibling pairs in a population, for example, 40 might be the same number of inches tall. Heritability for height among this group of sibling pairs is .40/.50, or 80 percent, which is the observed phenotypic variation divid-

ed by the expected phenotypic variation if environment had no influence.

Genetic variance for a polygenic trait is mostly due to the additive effects of recessive alleles of different genes. For some traits, a few dominant alleles can greatly influence the phenotype, but because they are rare, they do not contribute greatly to heritability. This is the case for heart disease caused by a faulty LDL receptor, a rare dominant condition that is also influenced by many other genes. Epistasis (interaction between alleles of different genes) can also influence heritability. To account for the fact that different genes affect a phenotype

142 PART TWO Transmission Genetics

---

the other (genotypes *RRYy*, *RrYY*, *rrYy*, or *Rryy*), or heterozygous for both genes (genotype *RrYy*). Mendel could explain the 9:3:3:1 proportion of progeny classes only if one gene does not influence transmission of the other. Each parent would produce equal numbers of four different types of gametes: *RY*, *Ry*, *rY*, and *ry*. Note that each of these combinations has one gene for each trait. A Punnett square for this cross shows that the four types of seeds:

1. round, yellow (*RRYY*, *RrYY*, *RRYy*, and *RrYy*)
2. round, green (*RRyy* and *Rryy*)
3. wrinkled, yellow (*rrYY* and *rrYy*)
4. wrinkled, green (*rryy*)

are present in the ratio 9:3:3:1, just as Mendel found (**figure 4.10**).

Solving a Problem: Following More Than One Segregating Gene

A Punnett square for three genes has 64 boxes; for four genes, 256 boxes. An easier way to predict genotypes and phenotypes in multi-gene crosses is to use the mathematical laws of probability on which Punnett squares are based. Probability predicts the likelihood of an event.

An application of probability theory called the product rule can predict the chance that parents with known genotypes can produce offspring of a particular genotype. The product rule states that the chance that two independent events will both occur equals the product of the chance that either event will occur alone. Consider the probability of obtaining a plant with

wrinkled, green peas (genotype *rryy*) from dihybrid (*RrYy*) parents. Do the reasoning for one gene at a time, then multiply the results (**figure 4.11**).

A Punnett square for *Rr* crossed to *Rr* shows that the probability of *Rr* plants producing *rr* progeny is 25 percent, or 1/4. Similarly, the chance of two *Yy* plants producing a *yy* plant is 1/4. Therefore, the chance of dihybrid parents (*RrYy*) producing homozygous recessive (*rryy*) offspring is 1/4 multiplied by 1/4, or 1/16. Now consult the 16-box Punnett square for Mendel's dihybrid cross again (figure 4.10). Only one of the 16 boxes is *rryy*, just as the product rule predicts. **Figure 4.12** shows how probability and Punnett squares can be used to predict offspring genotypes and phenotypes for three human traits simultaneously.

86 PART TWO Transmission Genetics

Web Activities

Visit the Online Learning Center (OLC) at www.mhhe.com/lewisgenetics7. Select **Student Edition, chapter 4** and **Web Activities** to find the website links needed to complete the following activities.

19. Go to the website for the National Organization for Rare Disorders. Identify an autosomal recessive disorder and an autosomal dominant disorder. Create a family for each one, and describe transmission of the disease over three generations.

20. Go to the website for Gene Gateway—Exploring Genes and Genetic Disorders. Select two disorders or traits that would demonstrate independent assortment if present in the same family, and two that would not.

Case Studies and Research Results

21. On the daytime drama "The Young and the Restless," several individuals suffer from SORAS, which stands for "soap opera rapid aging syndrome." It is not listed in OMIM. In SORAS, a young child is sent off to boarding school and returns three months later an angry teenager. In the Newman family, siblings Nicholas and Victoria aged from ages six and eight years,

respectively, to 16 and 18 years within a few months. Their parents, Victor and Nikki, are curiously not affected; in fact, they never seem to age at all.
    a. What is the mode of inheritance of the rapid aging disorder affecting Nicholas and Victoria?
    b. How do you know what the mode of inheritance is?
    c. Draw a pedigree to depict this portion of the Newman family.

22. Sam Fitzgerald is a carpet salesman who, at age 46, begins to slur his speech and stagger slightly when he walks. His speech worsens, he develops a shuffling gait to avoid falling, and he loses his job, because customers complain that he is intoxicated. His children, who know he does not drink alcohol, urge him to seek medical care. Eventually, after much counseling, he is tested and learns that he has HD. As his symptoms worsen, his sister Pam, is tested and is free of the mutation, cares for him—something she also did as a teen when their parents, Ruth and Alan, died in a car crash in their thirties. Another sister, Sue, refuses to be tested.
    a. Draw a pedigree for this family.
    b. What is the risk that Sam's daughter has inherited HD?

c. What is the risk that Sue's son has inherited HD?
    d. When Sue hears that Pam was tested and is free of the mutation, she assumes that this raises the risk that she has inherited the disease. Is she correct? Explain your answer in terms of Mendel's first law.

23. Recall Mackenzie, the young woman from chapter 1 who underwent genetic testing. The tests revealed that she has a 1 in 10 chance of developing lung cancer, and a 2 in 1,000 chance of developing colon cancer. What is the probability that she will develop both cancers?

Learn to apply the skills of a genetic counselor with these additional cases found in the *Case Workbook in Human Genetics*:

Acrocephalosyndactyly
Carnosinemia
Huntington-like disorder
Restless leg syndrome
Schneckenbecken dysplasia

VISIT YOUR ONLINE LEARNING CENTER
Visit your online learning center for additional resources and tools to help you master this chapter. See us at **www.mhhe.com/lewisgenetics7.**

94 PART TWO Transmission Genetics

CHAPTER

1

Overview of Genetics

SUPERBOY

The German boy appeared different from birth—his prominent arm and thigh muscles suggested he'd been lifting weights while in the womb. His exceptional muscular endowment continued, and by five years of age, his muscles twice normal size, he could lift heavier weights than many adults. He also had half the normal amount of body fat.

Relatives share his prowess. The boy's mother was a professional sprinter and is unusually strong, as are three close male relatives. One is a construction worker who regularly and effortlessly lifts very heavy stones. Genes from both parents cause the boy's unusual body composition. His cells cannot produce a protein called myostatin, which normally stops stem cells from making a muscle too large. A mutation turns off this genetic brake, and the muscles bulge. The boy is healthy, but since myostatin is also made in the heart, he may develop heart problems.

Other species have myostatin mutations. "Double muscling" cattle are valued for their high weights early in life—these animals occur naturally. Chicken breeders dampen myostatin production to yield meatier birds, and researchers have created "mighty mice" with blocked myostatin genes to study muscle overgrowth.

Understanding myostatin will have clinical applications. Blocking myostatin activity to stimulate muscle growth could reverse the ravages of muscular dystrophy and muscle-wasting from AIDS and cancer. But blocking myostatin levels also has the potential for bodybuilding abuse. Performance enhancement isn't the only ethically questionable application of this genetic knowledge. Theoretically, infants could be tested to identify those with myostatin gene variants that predict athletic prowess, given the right training.

As in many matters in human genetics, understanding how this one gene functions has great potential for improving the quality of life for many people—but also presents an opportunity for abuse.

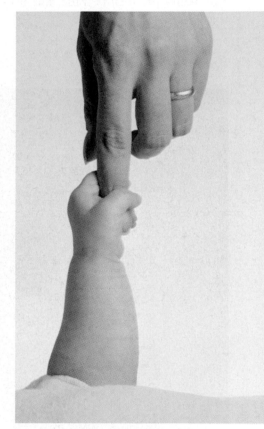

An infant with myostatin deficiency is an overly muscled "superbaby."

Genetics is the study of inherited traits and their variation. Sometimes people confuse genetics with genealogy, which considers relationships but not traits. With the advent of tests that can predict genetic illness, some people have even compared genetics to fortunetelling! But genetics is neither genealogy nor fortunetelling—it is a life science.

Genes are the units of heredity. They are biochemical instructions that tell **cells,** the basic units of life, how to manufacture certain proteins. These proteins control the characteristics that create much of our individuality, from our hair and eye color, to the shapes of our body parts, to our talents, personality traits, and health **(figure 1.1).**

Figure 1.1 Inherited traits. This young lady is the proud possessor of an unusual gene variant that confers her red hair, fair skin, and freckles. About 80 percent of individuals like her have a variant of a gene that encodes a protein called the melanocortin 1 receptor—it controls the balance of pigments in the skin. The little girl is, therefore, a mutant.

A gene is composed of the molecule **deoxyribonucleic acid (DNA).** Some traits are determined nearly entirely by genes; most traits, however, also have environmental components. The complete set of genetic information characteristic of an organism, including protein-encoding genes and other DNA sequences, constitutes a **genome.** We have known the sequence of the human genome since 2000, but researchers are still determining what all the information does—an effort that will take many years.

Genetics directly affects our lives, as well as those of our descendants. Principles of genetics also touch history, politics, economics, sociology, art, and psychology, and they force us to wrestle with concepts of benefit and risk, even tapping our deepest feelings about right and wrong. A field of study called bioethics was founded in the 1970s to address personal issues that arise in applying medical technology. Bioethicists today address concerns that new genetic knowledge raises, such as privacy, confidentiality, and discrimination.

An even newer field is **genomics,** which considers many genes at a time, in contrast to the emphasis on single-gene traits that pervaded genetics during the twentieth century. Genomics addresses the more common illnesses influenced by many genes that interact with each other and the environment. Considering genomes also enables us to compare ourselves to other species, as in the myostatin mutation seen in humans, cattle, chickens, and mice, discussed in the chapter opening case study. The similarities in genomes can be astonishing and quite humbling.

Many of the basic principles of genetics were discovered before DNA was recognized as the genetic material, from experiments and observations on patterns of trait transmission in families. For many years, genetics textbooks (such as this one) presented concepts in the order that they were understood, discussing pea plant experiments before delving into the structure of DNA. Now, since even grade-schoolers know what DNA is, a "sneak preview" of DNA structure and function is appropriate **(Reading 1.1)** for those who wish to consider the early discoveries

in genetics (Chapter 4) from a modern perspective.

1.1 Genetic Testing

Past editions of this textbook began with a scenario of two college students taking genetic tests sometime in the near future. The future is now. Although entire human genomes can be sequenced, it is more cost-effective to detect only health-related gene variants most likely to be present in a particular individual, based on clues such as personal health, family history, and ethnic background.

Young people might take genetic tests to prevent, delay, control, or treat symptoms that are likely to develop, or to gain information, perhaps to help decide whether to have children. Consider two 19-year-old college roommates, Mackenzie and Laurel, who have tailored genetic tests. Some of the results illustrate the type of information found in DNA.

Mackenzie requests three panels of tests, based on her family background. An older brother and her father smoke cigarettes and drink too much alcohol, and her father's mother, also a smoker, died of lung cancer. Two relatives on her mother's side had colon cancer. Older relatives on both sides have Alzheimer disease. Mackenzie has tests to detect genes that predispose her to developing addictions, certain cancers, and inherited forms of Alzheimer disease.

Laurel requests different tests. She, her sister, and her mother frequently have bronchitis and pneumonia, so she has a test for cystic fibrosis (CF), which can increase susceptibility to respiratory infections.

She also has tests for type 2 (non-insulin-dependent) diabetes mellitus, because several relatives have it, and she knows that diet and exercise can help control symptoms. Laurel refuses a test for inherited susceptibility to Alzheimer disease, even though a grandfather died of it. She does not want to know if this condition is likely to lie in her future, because, at least for now, it can't be treated. Finally, because past blood tests revealed elevated cholesterol, Laurel seeks information about her risk of developing heart and blood vessel (cardiovascular) disease.

Introducing DNA

We have probably wondered about heredity since our beginnings, when our distant ancestors noticed family traits such as a beaked nose or unusual talent. Awareness of heredity appears in ancient Jewish law that excuses a boy from circumcision if his brothers or cousins bled to death following the ritual. Nineteenth-century biologists thought that body parts controlled traits, and gave the hypothetical units of inheritance such colorful names as "pangens," "ideoblasts," "gemules," and simply "characters."

When Gregor Mendel meticulously bred pea plants to follow trait transmission, establishing the basic laws of inheritance, he inferred that units of inheritance of some kind were at play. His work is all the more amazing because he had no knowledge of cells, chromosomes, or DNA. This short reading is meant to recount, very briefly, what Mendel did not know—that is, the nature of DNA, and how it confers inherited traits.

DNA resembles a spiral staircase or double helix in which the "rails" or backbone are the same from molecule to molecule, but the "steps" are pairs of four types of building blocks whose sequence varies (**figure 1**). The chemical groups that form the steps are adenine (A) and thymine (T), which attract, and cytosine (C) and guanine (G), which attract. The two strands are in opposite directions. A, T, C, and G are called bases, for short. DNA functions as the genetic material because it holds information in the sequences of A, T, C, and G.

DNA uses its information in two ways. If the sides of the helix part, each half can reassemble its other side by pulling in free building blocks—A and T attracting and G and C attracting. This process, called DNA replication, is essential to maintain the information when the cell divides. DNA also directs the production of specific proteins. In a process called transcription, the sequence of part of one strand of a DNA molecule is copied into a related molecule, messenger RNA. Each three such RNA bases

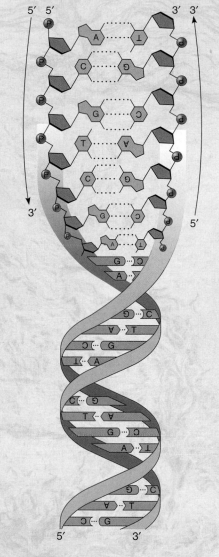

Figure 1   The DNA double helix.

in a row attracts another type of RNA that functions as a connector, bringing with it a particular amino acid, which is a building block of protein. As the two types of RNA temporarily bond, the amino acids align and join, forming a protein that is then released. DNA, RNA, and proteins can be thought of as three languages (**figure 2**).

Knowing the nature of a protein can explain how it confers a trait or illness.

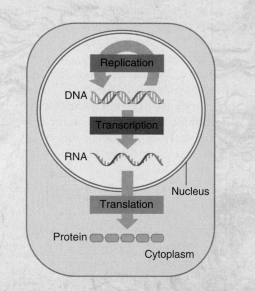

Figure 2   The language of life: DNA to RNA to protein.

Consider sickle cell disease, in which red blood cells bend into crescent shapes that lodge in tiny blood vessels, causing extreme pain as body parts are robbed of oxygen. The altered part of the responsible protein, called beta globin, has a single "wrong" DNA base. The replaced amino acid causes the globin molecules to attach to one another differently, forming sticky sheets in places where the oxygen level is low. This action, in turn, distorts the shapes of the red blood cells containing the abnormal proteins. The result of the blocked circulation: strokes, blindness, kidney damage, lung pain, and excruciating pain in the hands and feet.

Identifying the exact alteration in a gene and understanding how the affected protein disrupts normal functions provides valuable information that medical researchers can use to develop new treatments. But understanding how a gene functions is only the beginning of understanding how a trait arises. Genes and their protein products interact with each other, and with signals from the environment, in highly complex ways.

Each student proceeds through the steps outlined in **figure 1.2.** The first step is to record a complete family history. Next, each young woman swishes a cotton swab on the inside of her cheek to obtain cells, which are sent to a laboratory for analysis. There, DNA is extracted and cut into pieces, then tagged with molecules that fluoresce under certain types of light. The tagged DNA pieces are then applied to "DNA chips," postage-stamp-sized pieces of glass or nylon with particular sequences of DNA attached. Because the genes on the chip are aligned in fixed positions, this device is technically called a **microarray.** It shows, at a glance, which genes the students have.

A typical microarray bears hundreds or thousands of DNA pieces. One of Mackenzie's microarrays includes genes that regulate her circadian (daily) rhythms and encode proteins on nerve cells that receive chemical messages. Certain variants of these genes may increase her risk of developing addictive behaviors. Another microarray screens for gene variants that greatly increase the risk for lung cancer, and a third targets colon cancer genes. Her fourth microarray is smaller, bearing genes that cause Alzheimer disease and several DNA sequences associated with increased risk of developing other types of dementia.

Laurel's microarrays suit her background and requests, including several dozen of the most common versions of the CF gene associated with milder symptoms, plus a few variants of other genes that interact with the CF gene. The microarray for diabetes bears gene variants that reflect how Laurel's circulation transports glucose and how efficiently her cells take it up, factors that contribute to the development of diabetes. Largest and most diverse is the microarray for cardiovascular disease. Its thousands of genes influence blood pressure, blood clotting, and how cells use cholesterol and other lipids.

A few days later, a **genetic counselor** explains the findings. Mackenzie learns that she is indeed predisposed to develop addictive behaviors and lung cancer—a dangerous combination. But she does not face increased risk for inherited forms of colon cancer or Alzheimer disease.

Laurel has mild CF, which explains her frequent respiratory infections. The microarray indicates which types of infections she is most susceptible to, and which antibiotics will most effectively treat them. The diabetes test panel reveals her risk is lower than that for the general population, but she does have several gene variants that raise her blood cholesterol level. The cardiovascular disease microarray panel indicates which cholesterol-lowering drug she will respond to best, should diet and exercise habits not sufficiently counter her inherited tendency to accumulate lipids in the bloodstream.

Mackenzie and Laurel can take additional genetic tests as their interests and health status change. For example, shortly before each young woman

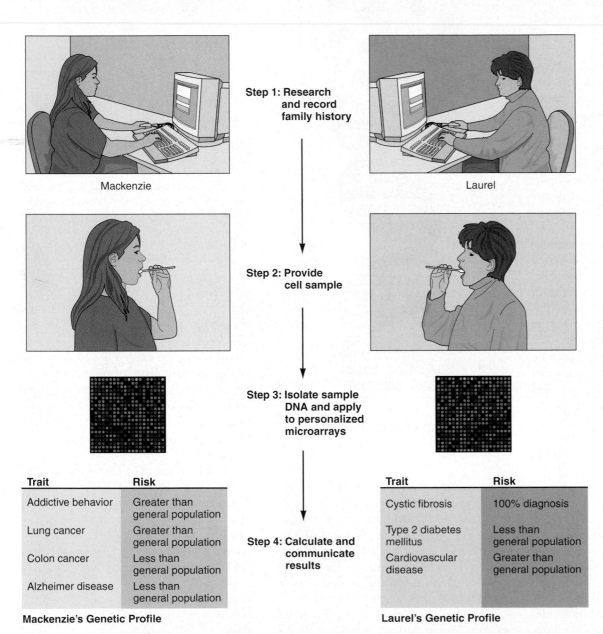

Mackenzie

Laurel

Step 1: Research and record family history

Step 2: Provide cell sample

Step 3: Isolate sample DNA and apply to personalized microarrays

Step 4: Calculate and communicate results

Mackenzie's Genetic Profile

Trait	Risk
Addictive behavior	Greater than general population
Lung cancer	Greater than general population
Colon cancer	Less than general population
Alzheimer disease	Less than general population

Laurel's Genetic Profile

Trait	Risk
Cystic fibrosis	100% diagnosis
Type 2 diabetes mellitus	Less than general population
Cardiovascular disease	Greater than general population

Figure 1.2 Genetic testing. Genetic tests are slowly becoming part of health care, revealing probabilities of developing certain conditions and refining medical diagnoses.

tries to become pregnant, she and her partner might take tests to detect whether they are carriers for any of several hundred illnesses, because two carriers of the same condition can pass it to offspring even when they are not themselves affected. If either Laurel or Mackenzie is in this situation, then tests on DNA from a fetus can determine whether it has inherited the illness.

Illness may prompt Laurel or Mackenzie to seek further testing. If either young woman suspects she may have cancer, for example, a type of microarray called an expression panel can determine which genes are turned on or off in the affected cells sampled from the tumor or from blood. **"Gene expression"** refers to the cell's use of the information in the DNA sequence to synthesize a particular protein. In contrast, DNA from cheek lining cells reveals specific gene variants and DNA sequences that are present in *all* cells of the body.

DNA expression microarrays are very useful in diagnosing and treating cancer. They can identify cancer cells very early, when treatment is more likely to work, and estimate if and how quickly the disease will progress. Microarrays can also show how tumor cells and the individual's immune system are likely to respond to particular drugs, and which drugs will likely produce intolerable side effects.

Laurel and Mackenzie's genetic test results will be kept confidential. Laws prevent employers and insurers from discriminating based on genetic information. In general, insurance companies decide whom to insure and at what rates based on symptoms present before or at the time of request for coverage. The results of genetic tests are not clinical diagnoses or even predictions, but probability statements about how likely certain symptoms are to arise in an individual.

Key Concepts

1. Genetics investigates inherited traits and their variations.
2. Genes, composed of DNA, are the units of inheritance, and they specify particular proteins.
3. A genome is the complete set of genetic instructions for an organism.
4. Human genome information will personalize medicine.

1.2 Levels of Genetics

Genetics considers the transmission of information at several levels, from the molecular level to populations to the evolution of species (**figure 1.3**).

DNA

Genes consist of sequences of four types of DNA building blocks, or bases—adenine, guanine, cytosine, and thymine, abbreviated A, G, C, and T. Each base bonds to a sugar and a phosphate group to form a unit called a nucleotide, and nucleotides are linked into huge DNA molecules. In genes, DNA bases provide an alphabet of sorts. Each consecutive three DNA bases is a code for a particular amino acid, and amino acids are the building blocks of proteins. Another type of molecule, **ribonucleic acid (RNA)**, uses the information in certain DNA sequences to construct specific proteins. These proteins confer the trait.

Proteomics is a new field that considers the types of proteins made in a particular type of cell. A muscle cell, for example, requires abundant contractile proteins. A skin cell, in contrast, contains mostly scaly proteins called keratins. DNA remains in the nucleus to be passed on when a cell divides.

The human genome includes about 24,000 protein-encoding genes. Those known to cause disorders or traits are catalogued and described in a database called Online Mendelian Inheritance in Man. It can be accessed through the National Center for Biotechnology Information (http://www.ncbi.nlm.nih.gov/) or directly at http://www.ncbi.nlm.nih.gov/entrez/query.fcgi?db=OMIM. Throughout the text, first mention of a disease includes its OMIM number.

Despite the fact that we know the sequence of DNA bases that comprise the human genome, there is much we still do not know. For example, only about 1.5 percent of the DNA in the human genome encodes protein. The rest includes many highly repeated sequences with unknown functions, sequences that turn protein-encoding genes on or off, and other sequences whose roles are yet to be discovered. Only recently, for example, have researchers discovered that RNA actually controls which proteins a cell manufactures. This process, called **RNA interference,** is discussed in chapter 11. Its discovery illustrates how we are constantly learning about new aspects of gene function.

Genes, Chromosomes, and Genomes

The same protein-encoding gene may vary slightly from person to person. These variants of a gene are called **alleles,** and the changes in DNA sequence that distinguish alleles arise by a process called **mutation.**

Once a gene mutates, the change is passed on when the cell that contains it divides. If the change is in a sperm or egg cell that becomes a fertilized egg, it is passed to the next generation.

Some mutations cause disease; others provide variation, such as freckled skin; and some mutations may help. For example, one mutation makes a person's cells unable to manufacture a surface protein that binds HIV. These people are resistant to HIV infection. The myostatin mutation in the German family described in the chapter opener is an advantage to an athlete. Many mutations have no visible effect because they do not change the encoded protein in a way that affects its function, just as a minor spelling errror does not obscure the meaning of a sentence.

Parts of the DNA sequence can vary among individuals, yet not change a person's appearance or health. Such a variant in sequence that is present in at least 1 percent of a population is called a **polymorphism,** which means "many forms." Researchers have identified millions of **single nucleotide polymorphisms** (SNPs, pronounced "snips"), which are single base sites that differ among individuals. Microarrays can include both disease-causing mutations and SNPs that just mark places where people differ. Researchers can identify combinations of SNPs that are found almost exclusively among people with a particular disorder and use the SNP patterns to estimate disease risks.

DNA wraps around proteins and winds tightly into structures called **chromosomes.** A human cell has 23 pairs of chromosomes. Twenty-two pairs are **autosomes,** which do not differ between the sexes. The autosomes

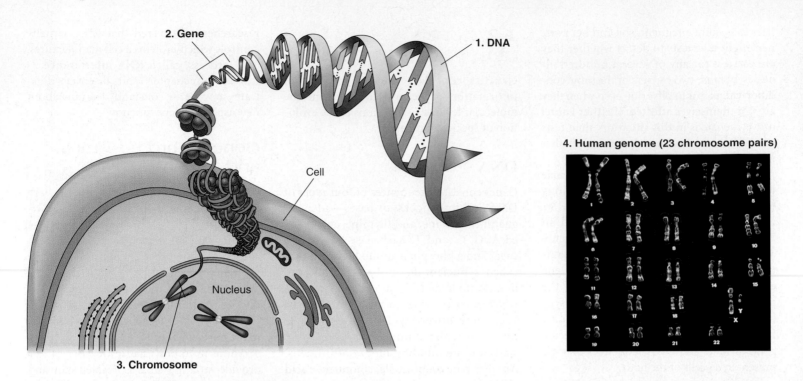

Figure 1.3   From molecule to population.   Genetics can be considered at several levels, from DNA, to genes, to chromosomes, to genomes, to individuals, families, and populations.

are numbered from 1 to 22, with 1 the largest. The other two chromosomes, the X and the Y, are **sex chromosomes.** The Y chromosome bears genes that determine maleness. In humans, a female has two X chromosomes and a male has one X and one Y. Charts called **karyotypes** order the chromosome pairs from largest to smallest.

A human cell has two complete sets of genetic information. The 24,000 or more protein-encoding genes are scattered among 3 billion DNA bases among each set of 23 chromosomes.

Cells, Tissues, and Organs

A human body consists of trillions of cells. All cells except red blood cells contain all of the genetic instructions, but cells differ in appearance and function because they use only some of their genes, a process called **differentiation.** For example, a muscle cell manufactures its abundant contractile protein fibers, but not the scaly keratins that fill skin cells, or the collagen and elastin proteins characteristic of connective tissue cells. All three cell types, however, have complete genomes. Specialized cells aggregate and

interact, forming tissues, which in turn form organs and organ systems (**figure 1.4).**

Parts of organs are made up of rare, unspecialized **stem cells** that can divide to yield another stem cell and a cell that differentiates. Thanks to stem cells, organs can maintain a reserve supply of cells to grow and repair damage. Yet stem cells are controlled—lifting of this control in the German boy described in the chapter-opening case study led to overgrowth of his muscles.

Individual

Two terms distinguish the alleles that are *present* in an individual from the alleles that are *expressed.* The **genotype** refers to the underlying instructions (alleles present), while the **phenotype** is the visible trait, biochemical change, or effect on health (alleles expressed). Alleles are further distinguished by how many copies it takes to affect the phenotype. A **dominant** allele has an effect when present in just one copy (on one chromosome), whereas a **recessive** allele must be present on both chromosomes to be expressed.

Family

Individuals are genetically connected into families. A person has half his or her genes in

common with each parent and each sibling, and one-quarter with each grandparent. First cousins share one-eighth of their genes.

The study of traits in families is called transmission or Mendelian genetics, for Gregor Mendel, who pioneered the study of single genes using pea plants. Molecular genetics, which considers DNA, RNA, and proteins, often begins with transmission genetics, when an interesting family trait or illness comes to a researcher's attention. Charts called **pedigrees** represent the members of a family and indicate which individuals have particular inherited traits. Figure 1.3 includes a pedigree for a mother, father, and son.

Population

Above the family level of genetic organization is the population. In a strict biological sense, a population is a group of interbreeding individuals. In a genetic sense, a population is a large collection of alleles, distinguished by their frequencies. People from a Swedish population, for example, would have a greater frequency of alleles that specify light hair and skin than people from a population in Ethiopia who tend to have dark hair and skin. The fact that

5. Individual

6. Family (pedigree)

Mother — Father

Son

7. Population

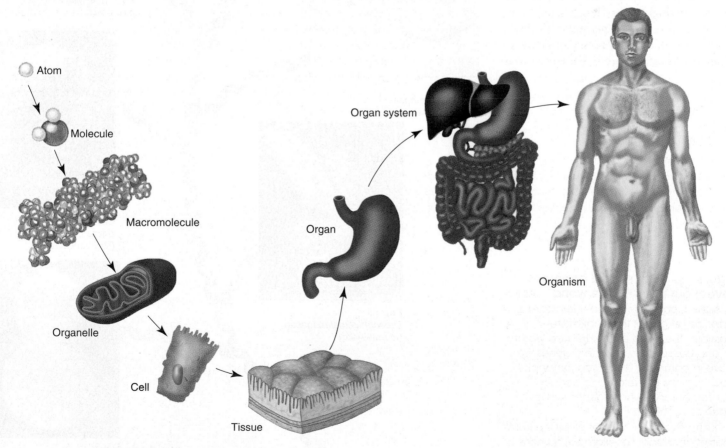

Atom

Molecule

Macromolecule

Organelle

Cell

Tissue

Organ

Organ system

Organism

Figure 1.4 Levels of biological organization.

groups of people look different and may suffer from different health problems reflects the frequencies of their distinctive sets of alleles. All the alleles in a population constitute the gene pool. (An individual does not have a gene pool.)

Population genetics is applied in health care and forensics. It is also the very basis of evolution, which is defined as changing allele frequencies in populations. These small-scale genetic changes foster the more obvious species distinctions we most often associate with evolution.

Evolution

Comparing DNA sequences for individual genes, or the amino acid sequences of the proteins that the genes encode, can reveal how closely related different types of organisms are (**figure 1.5**). The underlying assumption is that the more similar the sequences are, the more recently two species diverged from a shared ancestor.

Genome sequence comparisons reveal more about evolutionary relationships than comparing single genes. Humans, for example, share more than 98 percent of the DNA sequence with chimpanzees. Our genomes differ from theirs more in gene organiza-

tion and in the number of copies of genes than in the overall sequence. Still, learning the functions of the human-specific genes may explain the differences between us and them—such as our lack of hair and use of spoken language. Genome comparisons can clarify our kinship with other species, too. Consider the aardvark, a mammal with a

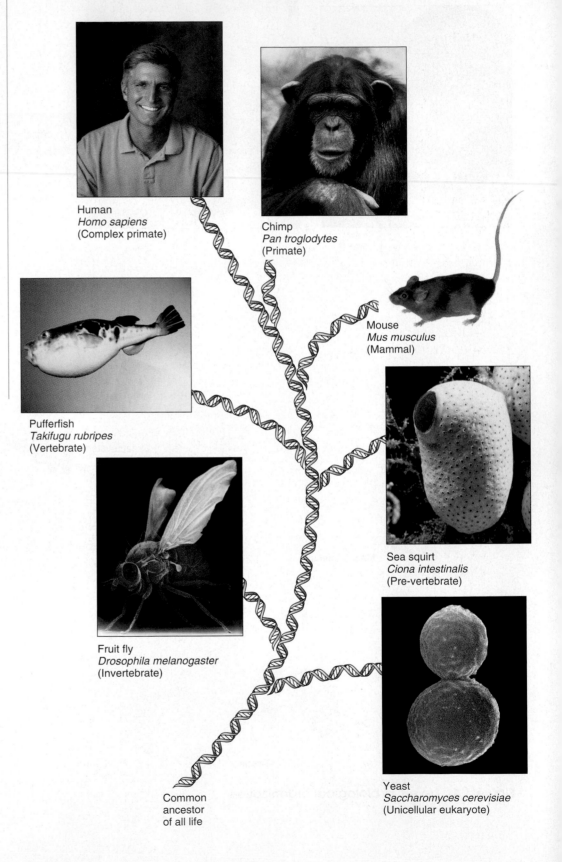

Figure 1.5 Genes and genomes reveal our place in the world. All life is related, and different species share a basic set of genes that makes life possible. The more closely related we are to another species, the more genes we have in common. This illustration depicts how humans are related to certain contemporaries whose genomes have been sequenced.

During evolution, species diverged from shared ancestors. For example, humans diverged more recently from chimps, our closest relative, than from mice, pufferfish, sea squirts, flies, or yeast.

Human
Homo sapiens
(Complex primate)

Chimp
Pan troglodytes
(Primate)

Mouse
Mus musculus
(Mammal)

Pufferfish
Takifugu rubripes
(Vertebrate)

Sea squirt
Ciona intestinalis
(Pre-vertebrate)

Fruit fly
Drosophila melanogaster
(Invertebrate)

Yeast
Saccharomyces cerevisiae
(Unicellular eukaryote)

Common
ancestor
of all life

distinctive long snout, also known as the "earth pig." A study that compared specific genes on the chromosomes of various placental mammals found that humans differ the most from the aardvark. This suggests that the aardvark is the most primitive placental mammal (a mammal that nurtures its unborn young through a maternal organ called a placenta).

Humans also share many DNA sequences with mice, pufferfish, and fruit flies. Dogs get many of the same genetic diseases that we do! At the level of genetic instructions for building a body, we are not very different from other organisms. We even share some genes necessary for life with single-celled organisms such as yeast and bacteria.

Comparisons of person to person at the genome level reveal that we are incredibly like one another. Studies of polymorphisms among different modern ethnic groups reveal that their gene pools are all subsets of the modern African gene pool, which indicates that modern humans arose in Africa. For example, comparing 377 highly variable genome regions in 52 populations from Africa, Eurasia, East Asia, Oceania, and the Americas found 99.9 percent identity of the DNA sequence.

Genome analyses also confirm that race, as defined by skin color, is a social concept, not a biological one. "Race" is actually defined by fewer than 0.01 percent of our genes. Put another way, two members of different races may have more alleles in common than two members of the same race. Very few, if any, gene variants are unique to any one racial or ethnic group. Imagine if we defined race by a different gene, such as the ability to taste bitter substances!

Table 1.1 presents the basic vocabulary of genetics.

1.3 Most Genes Do Not Function Alone

Until recently, genetics dealt almost exclusively with traits and illnesses that are clearly determined by single genes, called Mendelian traits. Genetics is much more complicated, however, than a one-gene-one-disease paradigm. Most genes do not

Table 1.1

A Mini-Glossary of Genetic Terms

Term	Definition
Allele	An alternate form of a gene; a gene variant.
Autosome	A chromosome that does not include a gene that determines sex.
Chromosome	A structure, consisting of DNA and protein, that carries the genes.
DNA	Deoxyribonucleic acid; the molecule whose building block sequence encodes the information that a cell uses to construct a particular protein.
Dominant	An allele that exerts an effect when present in just one copy.
Gene	A sequence of DNA that has a known function, such as encoding protein or controlling gene expression.
Gene expression	A cell's use of DNA information to manufacture specific proteins.
Gene pool	All of the genes in a population.
Genome	A complete set of genetic instructions in a cell, including DNA that encodes protein as well as other DNA.
Genomics	The field of investigating how genes interact and comparing genomes.
Genotype	The allele combination in an individual.
Karyotype	A size-order display of chromosomes.
Mendelian trait	A trait completely determined by a single gene.
Multifactorial trait	A trait determined by one or more genes and by the environment. Also called a complex trait.
Mutation	A change in a gene that affects the individual's health, appearance, or biochemistry.
Pedigree	A diagram used to follow inheritance of a trait in a family.
Phenotype	The observable expression of an allele combination.
Polymorphism	A site in a genome that varies in 1 percent or more of a population.
Recessive	An allele that exerts an effect only when present in two copies.
RNA	Ribonucleic acid; the molecule that enables a cell to synthesize proteins using the information in DNA sequences.
Sex chromosome	A chromosome that carries genes whose presence or absence determines sex.

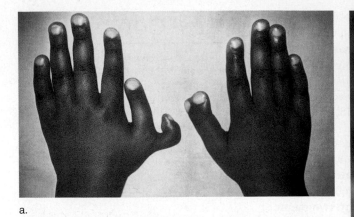

a.

Figure 1.6 Mendelian versus multifactorial traits.
(a) Polydactyly—extra fingers and/or toes—is a Mendelian trait, determined by a single gene. **(b)** Hair color is multifactorial, controlled by at least three genes plus environmental factors such as the bleaching effects of sun exposure.

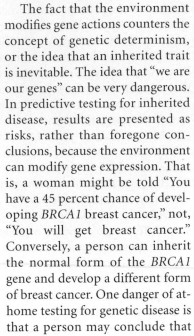

b.

function alone, but are influenced by the actions of other genes, and often by factors in the environment as well. Traits determined by one or more genes and the environment are called **multifactorial,** or complex, traits **(figure 1.6).** (The term *complex traits* has different meanings in a scientific and a popular sense, so this book uses the more precise term *multifactorial*.)

Confusing matters further is the fact that some illnesses occur in different forms—some inherited, some not, some Mendelian, some multifactorial. Usually the inherited forms are rarer, as is the case for Alzheimer disease, breast cancer, and Parkinson disease. Researchers can develop treatments based on the easier-to-study inherited form of an illness that physicians can then use to treat the more common, multifactorial forms. For example, drugs that millions of people take to lower cholesterol were developed from work on the one-in-a-million children with familial hypercholesterolemia (OMIM 144010) (see figure 5.2).

Knowing whether a trait or illness is Mendelian or multifactorial is important for predicting the risk of recurrence. The probability that a Mendelian trait will occur in another family member is simple to calculate using the laws that Mendel derived, discussed in chapter 4. In contrast, predicting the recurrence of a multifactorial trait is difficult because several contributing factors are at play. One form of inherited breast cancer illustrates how the fact that

genes rarely act alone can complicate calculation of risk.

Mutations in a gene called *BRCA1* cause fewer than 5 percent of all cases of breast cancer (OMIM 113705). But studies of the disease incidence in different populations have confusing results. In Jewish families of eastern European descent (Ashkenazim) with many members affected at a young age, inheriting the most common *BRCA1* mutation confers an 86 percent chance of developing the disease over a lifetime. But women from other ethnic groups who inherit this allele may have only a 45 percent chance. Perhaps different alleles of other genes interact with *BRCA1* genes in different populations.

Environmental factors may also affect the gene's expression. For example, exposure to pesticides that mimic the effects of the hormone estrogen may be an environmental contributor to breast cancer. It can be difficult to tease apart genetic and environmental contributions to disease. *BRCA1* breast cancer, for example, is especially prevalent in Long Island, New York. This population includes both many Ashkenazim and many people exposed to pesticides.

Increasingly, predictions of inherited disease are considered in terms of "modified genetic risk," which takes into account single genes as well as environmental and family background information. A modified genetic risk is necessary to predict *BRCA1* breast cancer occurrence in a family.

The fact that the environment modifies gene actions counters the concept of genetic determinism, or the idea that an inherited trait is inevitable. The idea that "we are our genes" can be very dangerous. In predictive testing for inherited disease, results are presented as risks, rather than foregone conclusions, because the environment can modify gene expression. That is, a woman might be told "You have a 45 percent chance of developing *BRCA1* breast cancer," not, "You will get breast cancer." Conversely, a person can inherit the normal form of the *BRCA1* gene and develop a different form of breast cancer. One danger of at-home testing for genetic disease is that a person may conclude that the detection of a mutation means unavoidable disease. Such test results only predict risk, and they must be considered with other factors.

Genetic determinism as part of social policy can be particularly harmful. In the past, for example, the assumption that one ethnic group is genetically less intelligent than another led to lowered expectations and fewer educational opportunities for those perceived as biologically inferior. Environment, in fact, has a huge impact on intellectual development. The bioethics essay in chapter 8 considers genetic determinism further.

Key Concepts

1. Inherited traits are determined by one gene (Mendelian) or by one or more genes and the environment (multifactorial).
2. Even the expression of single genes is affected to some extent by the actions of other genes.
3. Genetic determinism is the false idea that an inherited trait cannot be modified.

1.4 Statistics Represent Risks

Predicting the inheritance of traits in individuals is not a precise science, largely because of the many influences on gene function and the uncertainties of analyzing multiple factors. Genetic counselors calculate risks for clients who want to know the chance that a new family member will inherit a particular disease—or has inherited it, but does not yet have symptoms.

In general, risk assessment estimates the degree to which a particular event or situation endangers a population. In genetics, that event is the likelihood of inheriting a particular gene or gene combination. The genetic counselor can infer that information from a detailed family history, or from the results of tests that identify a gene variant or an absent or abnormal protein.

Risks are expressed as absolute or relative figures. **Absolute risk** is the probability that an individual will develop a particular condition. **Relative risk** is the likelihood that an individual from a particular population will develop a condition compared with individuals from another group, usually the general population. Relative risk is expressed as a ratio of the probability in one group compared to another. In genetics, relative risks might be calculated by evaluating any situation that might elevate the risk of developing a particular condition, such as one's ethnic group, age, or exposure to a certain danger. The threatening situation is called a risk factor. For example, "advanced" maternal age is a risk factor because chromosome abnormalities are more common in the offspring of older mothers. Pregnant women who are tested for Down syndrome caused by an extra chromosome 21 are compared by age to the general population of pregnant women to derive the relative risk that they are carrying a fetus that has the syndrome.

Determining a relative risk may seem unnecessary, because an absolute risk applies to an individual. However, relative risks help health care providers identify patients most likely to have the conditions for which absolute risks can be calculated, and patients most likely to benefit from particular medical tests. A frequent problem in assessing genetic risk is that statistics tend to lose their meaning in a one-on-one situation. To a couple learning that their fetus has Down syndrome, it is immaterial that the relative risk was low based on the woman's age.

Absolute and relative risk are represented mathematically in different ways. Odds and percentages are used to depict absolute risk.

For example, Mackenzie's absolute risk of developing inherited Alzheimer disease over her lifetime is 4 in 100 (the odds), or 4 percent. Determining her relative risk requires knowing the risk to the general population. If that risk is 10 in 100, then Mackenzie's relative risk is 4 percent divided by 10 percent, or 0.4. A relative risk less than 1 indicates the chance that an individual will develop a particular illness is less than that for the general population; a value greater than 1 indicates risk greater than that for the general population. Mackenzie's 0.4 relative risk means she has 40 percent as much risk of inheriting Alzheimer disease as the average person in the general population; a relative risk of 8.4, by contrast, would indicate a greater-than-8-fold risk compared to an individual in the general population.

Determining the risks for Alzheimer disease is actually more complicated than depicted in this hypothetical case. Several genes are involved, the percentage of inherited cases isn't known, and prevalence is highly associated with age. Elevated risk is linked to having more than one affected relative and to early age of onset.

Risk estimates can change depending upon how the groups being compared are defined. For a couple who has a child with an extra chromosome, such as a child with Down syndrome, the risk of recurrence is 1 in 100, a figure derived from looking at many families who have at least one such child. Therefore, the next time the couple has a child, two risk estimates are possible for Down syndrome—1 in 100, because they already have an affected child, and the age-associated risk. A genetic counselor presents the highest risk, to describe a worst-case scenario. Consider a 23-year-old and a 42-year-old woman who have each had one child with the extra chromosome of Down syndrome. Each faces a recurrence risk of 1 in 100 based on medical history, but different age-associated risks—the 23-year-old's is 1 in 500, but the 42-year-old's is 1 in 63. The counselor provides the 1 in 100 figure to the younger woman, but the age-associated 1 in 63 figure to the older woman (**figure 1.7**).

Geneticists derive risk figures in several ways. Empiric risk comes from population-level observations, such as the 1 in 100 risk of having another child with an extra chromosome. Risk estimates also derive from Mendel's laws. A child whose parents are

Patient 1: Rebecca	
Age:	23
Age-dependent risk for Down syndrome:	1 in 500
Risk based on previous child with Down syndrome:	1 in 100

Patient 2: Diane	
Age:	42
Age-dependent risk for Down syndrome:	1 in 63
Risk based on previous child with Down syndrome:	1 in 100

Figure 1.7 Relative risk. Risk may differ depending on how the population group is defined.

both carriers of sickle cell disease, for example, faces a 1 in 4, or 25 percent, chance of inheriting the disease. This child also has a 1 in 2, or 50 percent, chance of being a carrier, like the parents. The risk is the same for each offspring. It is a common error to conclude that if two carrier parents have a child with an inherited disorder, the next three children are guaranteed to be healthy. This isn't so, because each conception is an independent event.

Key Concepts

1. In genetics, risk estimates the likelihood that a particular individual will develop a particular condition.
2. Absolute risk is the probability that an individual will develop a certain condition.
3. Relative risk is based on the person's population group compared to another population group.

1.5 Applications of Genetics

Barely a day goes by without some mention of genetics in the news. Genetics is impacting many areas of our lives, from health care choices, to what we eat and wear, to unraveling our pasts and controlling our futures. Thinking about genetics evokes fear, hope, anger, and wonder, depending on context and circumstance. **Figure 1.8** shows an artistic view of genetics. Following are glimpses of applications of genetics that we will explore more fully in subsequent chapters.

Establishing Identity and Origins

Comparing DNA sequences to establish or rule out identity, relationships, or ancestry is becoming routine. This approach, called DNA profiling, has many applications.

Forensics

Before September 11, 2001, the media reported on DNA profiling rarely, usually in the wake of plane crashes where victims needed to be identified or in spectacular criminal cases. After the 2001 terrorist attacks, investigators compared DNA sequences in bone and teeth collected from the scenes to hair and skin samples from hairbrushes, toothbrushes, and clothing of missing people, and to DNA samples from relatives. DNA profiling was used on a larger scale to identify victims of the tsunami in Asia in 2004 and hurricane Katrina in the U.S. in the fall, 2005.

A more conventional forensic application matches a rare DNA sequence in tissue left at a crime scene to that of a sample from a suspect. This is statistically strong evidence that the accused person was at the crime scene (or that someone planted evidence). DNA databases of convicted felons often get "cold hits"—when DNA at a crime scene matches a criminal's DNA in the database.

The United Kingdom, where DNA profiling was pioneered in the middle 1980s, has for years collected DNA from all convicts. In the United States, Virginia was the first state to establish such a database. Since 1989, law enforcement officials in Virginia have scored cold hits in hundreds of crimes.

DNA profiling has been very successful in overturning convictions. Illinois has led the way; there, in 1996, DNA tests exonerated the Ford Heights Four, men convicted of a gang rape and double murder who had spent eighteen years in prison, two of them on death row. In 1999, the men received compensation of $36 million for their wrongful convictions. A journalism class at Northwestern University initiated the investigation that gained the men their freedom. The case led to new state laws granting death row inmates new DNA tests if their convictions could have arisen from mistaken identity, or if DNA tests were performed when they were far less accurate. In 2003, Governor George Ryan was so disturbed by the number of overturned convictions based on DNA evidence that shortly before he left office, he commuted the sentences of everyone on death row to life imprisonment.

DNA profiling also helps adopted individuals locate blood relatives. The Kinsearch Registry maintains a database of DNA information on people adopted in the United States from China, Russia, Guatemala, and South Korea, the sources of most foreign adoptions. Adopted individuals can provide a DNA sample and search the database by country of origin to find siblings.

Maintaining DNA databases on convicted felons is generally accepted because criminals give up certain civil rights, and it is of great value to adopted individuals because participation is voluntary. Establishing such databases on the general public is another story. *Bioethics: Choices for the Future* describes some of the first general population databases.

Rewriting History

DNA can help to flesh out details of history, and sometimes springs surprises. Consider the offspring of Thomas Jefferson's slave, Sally Hemings (**figure 1.9**). Rumor at the time placed Jefferson near Hemings nine

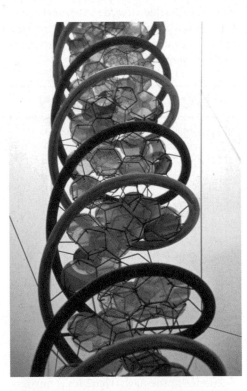

Figure 1.8 Genetic science inspires art. This sleek, symmetrical depiction of the DNA double helix adorns the four-story spiral staircase in the Life Sciences building at the University of California at Davis.

Bread lists for 1815. 148

Monticello	farm	Tufton	Lego.
Betty Brown	Abram	Bagwell	Charles
Edwin	Doll	Minerva	Aggy
Robert	Shepherd	Willis	Polly
Mary.	Barnaby	Archy	James B.
Sally	Stannard	Jordan.	Rachael
Beverly	Bartlet.	Mary	Joe
Harriet	Davy senr.	Washington	Lania.
Madison	Isabel	John B.	Gloster
Eston.	Indridge	Virginia.	Washington
Billy B.	Thrimston	Robert	Edmund
Burwell	Lovilo.	Esther.	Eve.

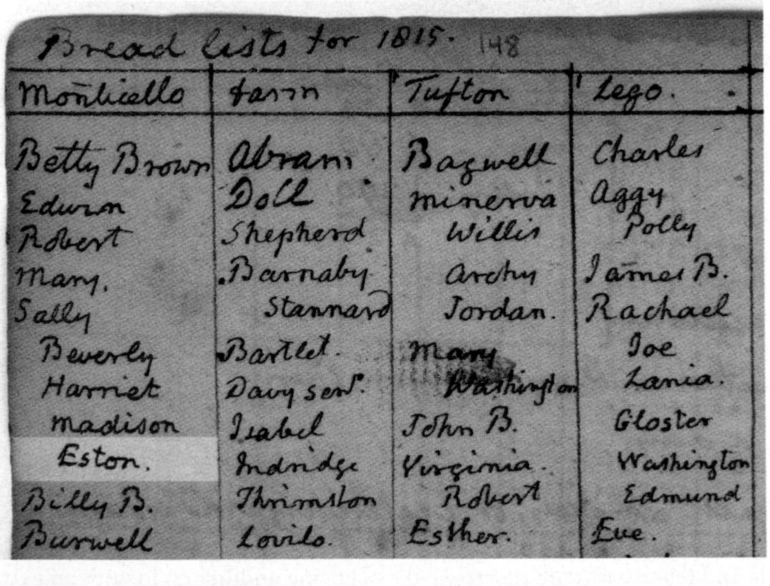

Figure 1.9 DNA clarifies history. Analysis of DNA sequences on the Y chromosomes of some of Thomas Jefferson's descendants indicate that either the president, his brother, or one of his nephews fathered Eston Hemings, a son of slave Sally Hemings.

Bioethics: Choices for the Future

Population Biobanks

More than a dozen nations are recording and scrutinizing genetic, genealogical, lifestyle, and health information on their citizens to discover and archive the inherited and environmental influences on common disorders. These "biobank" projects vary in how people participate, but they raise similar concerns: Who will have access to the information? How can people benefit from providing it? How might it be abused?

The first country to make headlines for collecting genetic information on a population level was Iceland. In 1998, a company called deCODE Genetics received government permission to collect existing health and genealogy records and to add DNA sequence data. Many Icelandic citizens can trace their families back more than a thousand years and have family tree diagrams etched in blood on old leather. Participation in the database is presumed—citizens must file a special form to opt out of the project. Despite initial concerns (mostly from outside Iceland) that the populace would feel pressured to participate, and perhaps do so against their will, that hasn't been the case. The head of the project claims that 95 percent of everyone who has lived in the nation since 1703, when the first census was conducted, are represented in the database.

DeCODE has used the information to identify genes that contribute to more than 25 common disorders, and is pursuing new treatments for many. These include Alzheimer disease, anxiety disorder, hypertension, arthritis, heart disease, schizophrenia, stroke, and diabetes. Their strategy groups people by clinical condition and identifies parts of the genome that they uniquely share, then finds genes in these regions whose functions could explain the symptoms.

The Estonian Genome Foundation uses registries for patients with cancer, Parkinson disease, diabetes mellitus, and osteoporosis. When patients show up for appointments, they learn about the project and are asked for details of their health histories and to donate DNA. Researchers then match variations in the DNA sequence to particular medical conditions.

Researchers in the United Kingdom are recruiting half a million individuals between the ages of 45 and 69, when many common illnesses begin, to donate DNA to a biobank. Investigators are searching for connections among DNA sequence variants, health, and lifestyle characteristics as the population ages. Unlike deCODE, the UK Biobank is run by the government. Another effort, called GenomeEUtwin, has collected data on more than 600,000 pairs of twins from eight European nations for decades. The Estonian Foundation, UK Biobank, GenomeEUtwin, and a Canadian project called CARTaGENE, have joined to form the Public Population Project in Genomics. Like deCODE, the other population genetic databases are using genetic information to develop diagnostic tests and treatments.

Ideally, a biobank must meet several criteria. It should

- have data and tissue samples from at least 500,000 people.

- draw conclusions based on a population that is representative of the nation.

- have clinical information collected over many years.

- include family trees that link generations.

- compare results to those of other populations to validate DNA-disease associations.

Bioethicists have suggested strategies to ensure that individuals benefit from such projects, such as:

- Preserving choice in seeking genetic tests.

- Protecting privacy by legally restricting access to genome information.

- Tailoring genetic tests to genes that are most relevant to an individual.

- Refusing to screen for trivial traits in embryos or fetuses.

Global cooperation in establishing biobanks can tap the unique resources of different countries. For example, India and Pakistan fulfill the requirements for a useful biobank, but in addition, their cultures support consanguineous (blood relative) marriages, which has led to higher incidences of certain single-gene disorders compared to other populations. The people are splintered into many ethnic groups that suffer from unique combinations of health problems. Biobanks in India and Pakistan may hold information about diseases that could ultimately help many people.

Anyone can participate in a biobank effort. The Genographic Project tests DNA samples, obtained with a cheek brush, from anyone, and reports back on personal ancestries. For further information, see http://www3.nationalgeographic.com/genographic/

Table 1

The First Biobanks

Biobank	Population	Website
CARTaGENE	Canada	http://www.cartagene.qc.ca/en/
DeCODE Genetics	Iceland	http://www.decode.com/
Estonian Genome Foundation	Estonia	http://www.geenivaramu.ee/index.php?show=main&lang=engl
GenomeEUtwin	Europe	http://www.genomeutwin.helsinki.fi/
UK Biobank	UK	http://www.ukbiobank.ac.uk/

months before each of her seven children was born, and the children themselves claimed to be presidential offspring. A Y chromosome analysis revealed that Thomas Jefferson could have fathered Heming's youngest son, Eston—but so could any of 26 other Jefferson family members. The Y chromosome, because it is only in males, passes from father to son. Researchers identified very unusual DNA sequences on the Y chromosomes of descendants of Thomas Jefferson's paternal uncle, Field Jefferson. (These men were checked because the president's only son with wife Martha died in infancy, so he had no direct descendants.) The Jefferson family's unusual Y chromosome matched that of descendants of Eston Hemings, supporting the talk of the time.

Reaching farther back, DNA profiling can clarify relationships from Biblical times. Consider a small group of Jewish people, the cohanim, who share distinctive Y chromosome DNA sequences and enjoy special status as priests. By considering the number of DNA differences between cohanim and other Jewish people, how long it takes DNA to mutate, and the average generation time of 25 years, researchers extrapolated that the cohanim Y chromosome pattern originated 2,100 to 3,250 years ago—which includes the time when Moses lived. According to religious documents, Moses' brother Aaron was the first priest.

The Jewish priest DNA signature also appears today among the Lemba, a population of South Africans with black skin. Researchers looked at them for the telltale gene variants because their customs suggest a Jewish origin—they do not eat pork (or hippopotamus), they circumcise their newborn sons, and they celebrate a weekly day of rest. Today, the Lemba clearly practice Judaism (**figure 1.10**).

DNA profiling can also trace origins for organisms other than humans. For example, researchers analyzed DNA from the leaves of 300 varieties of wine grapes, in search of the two parental strains that gave rise to the sixteen major types of wine grapes existing today (**figure 1.11**). One parent was already known—the bluish-purple Pinot grape. But the second parent, revealed in the DNA, was a surprise—a variety of white grape called Gouais blanc that was so unpopular it hadn't been cultivated for years and was actually banned

Figure 1.10 Y chromosome DNA sequences reveal origins.
The Lemba, a modern people with dark skin, have the same Y chromosome DNA sequences as the cohanim, a group of Jewish priests. The Lemba practiced Judaism long before DNA analysis became available.

during the Middle Ages. Thanks to DNA analysis, vintners now know to maintain both parental stocks, to preserve the gene pool from which all wines descend.

Health Care

The new genetics/genomics is changing the face of health care. New as well as experienced practitioners are learning how to integrate new genetic knowledge and technology into patient care. In the past, physicians typically encountered genetics only as extremely rare disorders caused by single genes. Today, medical science is increasingly recognizing the role that genes play not only in many common conditions, but also in how people react to medications. Disease is beginning to be seen as the consequence of complex interactions among genes and environmental factors.

Inherited illness caused by a single gene variant differs from other types of illnesses in several ways (**table 1.2**). First, we can predict the recurrence risk for single-gene

a.

b.

Figure 1.11 Surprising wine origins.
(a) Gouais blanc and **(b)** Pinot (noir) grapes gave rise to nineteen modern popular wines, including Chardonnay.

Table 1.2
How Genetic Diseases Differ from Other Diseases
1. One can predict recurrence risk in other family members.
2. Predictive testing is possible.
3. Different populations may have different characteristic frequencies.
4. Correction of the underlying genetic abnormality may be possible.

disorders using the laws of inheritance chapter 4 describes. In contrast, an infectious disease requires that a pathogen pass from one person to another—a much less predictable circumstance.

A second key distinction of inherited illness is that the risk of developing symptoms can be predicted. This is because all genes are present in all cells, even if they are not expressed in every cell. The use of genetic testing to foretell disease is termed predictive medicine. For example, some women who have lost several relatives at young ages to *BRCA1* breast cancer and who know they have inherited the gene variant that causes the illness call themselves "previvors," in contrast to survivors. Some have their breasts removed to prevent the cancer. A medical diagnosis, however, is still based on symptoms. This is because some people who inherit gene variants associated with particular symptoms never develop them, because of interactions with other genes or environmental factors.

A third feature of genetic disease is that an inherited disorder may be much more common in some populations than others. Genes do not "like" or "dislike" certain types of people, but we tend to pick partners in nonrandom ways that can cause particular gene variants to cluster in certain groups. This phenomenon has economic consequences. While it might not be "politically correct" to offer a "Jewish genetic disease screen," as several companies do, it makes biological and economic sense—a dozen disorders are much more common among Ashkenazim.

So far, tests can identify about 1,000 single-gene disorders, but each year, only about 250,000 people in the United States take these tests. Many people fear that employers or insurers will discriminate based on the results of genetic tests—or even for taking the tests. Yet millions of people regularly have their cholesterol checked! Studies from Canada after more than a decade of offering predictive genetic testing for Huntington disease indicate that it may not be fear of health insurance discrimination that deters people from taking a test—Canada has national health care. More older people took the test, to guide financial decisions and future plans, than did younger people to help make decisions about having children.

Despite the slow start to predictive genetic testing in some nations, in the United States legislation to prevent the misuse of genetic information in the insurance industry has been in development since 1993. The 1996 Health Insurance Portability and Accountability Act passed by the U.S. Congress stated that genetic information, without symptoms, does not constitute a preexisting condition, and that individuals could not be excluded from group coverage on the basis of a genetic predisposition. But the law did not cover individual insurance polices, nor did it stop insurers from asking people to have genetic tests. In 2000, U.S. President Bill Clinton issued an executive order prohibiting the federal government from obtaining genetic information for employees or job applicants and from using such information in promotion decisions. Since then, more than a dozen bills have been introduced in Congress to prevent genetic discrimination, and most states have enacted antidiscrimination legislation. Yet because the legislation is still in flux, and because the media reports cases of health insurance denial or higher premiums following a genetic test, many people continue to fear the misuse of genetic information.

Genetic tests may actually, eventually, lower health care costs. If people know their inherited risks, they can forestall or ease symptoms that environmental factors might trigger—for example, by eating healthy foods, not smoking, exercising regularly, avoiding risky behaviors, having frequent medical exams, and beginning treatment earlier. Genetic tests can also enable people to make more informed reproductive decisions. People who know that they can transmit an inherited illness may elect not to have children, or to use one of the assisted reproductive technologies chapter 21 discusses.

A few genetic diseases can be treated. Supplying a missing protein can prevent some symptoms, such as providing a clotting factor to a person who has a bleeding disorder. Gene therapy theoretically provides a more lasting cure by replacing the instructions for producing the protein in the cells that are affected in the illness. Chapter 20 discusses gene therapy.

Because it is unethical to experiment on humans to investigate genetic disease, researchers take advantage of the fact that animal species have many genes in common. This is the case for lissencephaly (OMIM 607432). Like the Greek "smooth brain" from which the name derives, the brains of affected children lack the characteristic coils of the cortex region (see figure 8.1). This causes severe mental retardation, seizures, and shortened lifespan. To study how this rare disorder unfolds during development, which cannot be done on human embryos and fetuses, Guy Caldwell, a researcher at the University of Alabama, uses the roundworm *Caenorhabditis elegans*. It has a gene very similar in DNA sequence to the human lissencephaly gene. When mutant, the gene causes worms to have seizures! Although the worm's 302-celled brain is much too simple to have coils, it lacks a key "motor molecule" that normally shuttles cell contents to appropriate places. Researchers are now focusing on this molecule to discover how similarly misguided nerve cells in a human embryo's forming brain lead to lissencephaly.

Agriculture

The field of genetics arose from agriculture. Traditional agriculture is the controlled breeding of plants and animals to select individuals with certain combinations of useful inherited traits, such as seedless fruits or lean meat. **Biotechnology,** the use of organisms to produce goods (including foods and drugs) or services, is an ancient art as well as a modern science. One ancient example of biotechnology is using microorganisms to ferment fruits to manufacture alcoholic beverages, a technique the Babylonians used by 6000 B.C.

Traditional agriculture is imprecise, because it shuffles many genes—and, therefore, many traits—at a time. The application of DNA-based techniques, part of modern biotechnology, enables researchers to manipulate one gene at a time, adding control and precision to agriculture. Organisms altered to have new genes or to over- or underexpress their own genes are termed genetically modified (GM). If the organism has genes from another species, it is termed transgenic. Golden rice, for example, manufactures beta carotene (a vitamin A precursor) using "transgenes" from plants and bacteria. It stores twice as much iron as unaltered rice because one of its own genes is overexpressed (**figure 1.12**). These nutritional boosts bred into edible rice strains may help prevent vitamin A and iron deficiencies in people who eat them.

People in the United States have been safely eating GM foods for more than a decade. In Europe, many people object to GM foods, seemingly on ethical grounds or based on fear. Officials in France and Austria have called such crops "not natural," "corrupt," and "heretical." **Figure 1.13** shows an artist's rendition of these fears. Food labels in Europe, and some in the United States, indicate whether a product is "GM-free." Some objections to GM foods arise from lack of knowledge. A public opinion poll in the United Kingdom discovered, for example, that a major reason citizens avoid GM foods is that they do not want to eat DNA! One British geneticist wryly observed that the average meal provides about 150,000 kilometers (about 93,000 miles) of DNA. Ironically, the British ate GM foods for years. For example, tomatoes with a gene added to delay ripening vastly outsold regular tomatoes in England, because they were cheaper.

Other concerns about GM organisms may be better founded. Labeling foods can prevent allergic reaction to an ingredient in a food that wouldn't naturally be there, such as a peanut protein in corn. Field tests may not adequately predict the effects of GM crops on ecosystems. GM plants have been found far beyond where they were planted, thanks to wind pollination. Some GM organisms, such as fish that grow to twice normal size or can survive at temperature extremes, may be so unusual that they disrupt ecosystems. GM crops may also lead to extreme genetic uniformity, which could be disastrous.

Some researchers are combining traditional breeding techniques that tap natural trait diversity with biotechnology. Consider a strategy to increase yield of a tomato crop. Researchers in Israel crossed wild plants with cultivated plants that differ from each other by known, small regions of the tomato genome. After growing the plants for several generations, the researchers exposed them to various environmental stresses to select plant lines that remained robust and plentiful. The next step: searching among the genome sections propagated in the surviving strains for specific gene vari-

ants that confer high yield under stressful conditions.

Ecology

Genomics extends beyond the controlled ecosystems of agriculture. In a subdiscipline called metagenomics, researchers sample and sequence genomes of all residents of a targeted habitat. Such areas might range from the microorganisms in a drop of seawater to those in a bit of soil or an insect's gut. This information is revealing how organisms and species interact, and may even yield new drugs and reveal novel energy sources.

Instead of collecting organisms that may be difficult or impossible to keep alive in a laboratory, metagenomics researchers collect DNA and consult databases of known genes and genomes to imagine what the organisms might be like. Consider the metagenomics approach to discover and describe life in the Sargasso Sea. This 2-million-square-mile oval area off the coast of Bermuda has long been thought to lack life beneath its thick cover of seaweed, which is so abundant that Christopher Columbus thought he'd reached land when his ships came upon it. Many a vessel has been lost in the

Figure 1.12 Nutrient-boosted crops, courtesy of biotechnology: Golden rice, a transgenic plant, will be widely available. Its developers claim the extra nutrients can help combat vitamin A and iron deficiencies. But critics called the "golden-rice-will-save-the-developing-world" campaign more an effort to improve the image of gene modification technology than a public health effort.

Figure 1.13 An artist's view of biotechnology. Artist Alexis Rockman vividly captures some fears of biotechnology, including a pig used to incubate spare parts for sick humans, a muscle-boosted, boxy cow, a featherless chicken with extra wings, a mini-warthog, and a mouse with a human ear growing out of its back.

Sargasso Sea, which includes the area known as the Bermuda Triangle. When researchers sampled the depths, they collected more than a billion DNA bases. Database searches revealed that this catch represents about 1,800 microbial species, including at least 148 not seen before. More than a million new genes were discovered. Another metagenomics project is collecting DNA from air samples taken in lower Manhattan.

Yet another ecosystem that metagenomics researchers are probing is the human mouth, home to some 500 species of bacteria. Only about 150 of the bacteria can grow in the laboratory, but that is not a barrier to genome scientists. One oral resident of the human mouth, *Treponema denticola*, holds a place in medical history as the first microorganism that the father of microscopy, Antonie van Leeuwenhoek, sketched in the 1670s. Its genome revealed how it survives amid the films formed by other bacteria in the mouth, and how it causes gum disease. Researchers were surprised to find that this microorganism is genetically very different from other spiral-shaped bacteria that were thought to be close relatives—those that cause syphilis and Lyme disease. Therefore, genomics showed that appearance (a spiral shape) does not necessarily reflect the closeness of the evolutionary relationship between two types of organisms.

Genetics from a Global Perspective

Because genetics so intimately affects us, it cannot be considered solely as a branch of life science. Equal access to testing, misuse of information, and abuse of genetics to intentionally cause harm are compelling issues that parallel scientific progress.

Genetics and genomics are spawning technologies that may vastly improve quality of life. But at first, tests and treatments will be costly and not widely available. While people in economically and politically stable nations may look forward to genome-based individualized health care—what some have called "Cadillac medicine"—those in other nations just try to survive, often lacking basic vaccines and medicines. In an African nation where two out of five children suffer from AIDS and many die from other infectious diseases, newborn screening for rare single-gene defects hardly seems practical. However, genetic disorders weaken people so that they become more susceptible to infectious diseases, which they can pass to others.

Human genome information can ultimately benefit everyone. Consider drug development. Today, there are fewer than 500 types of drugs. Genome information from humans and our pathogens and parasites is revealing new drug targets. For example, malaria is an infectious disease caused by a single-celled parasite transmitted through the bite of a female mosquito. The genomes of the parasite, mosquito, and human have been sequenced, and within this vast amount of information likely lie clues that researchers can use to develop new types of anti-malarial drugs.

Global organizations, including the United Nations, World Health Organization, and the World Bank, are discussing how nations can share new diagnostic tests and therapeutics that arise from genome information.

Key Concepts

1. Genetics has applications in diverse areas. Matching DNA sequences can clarify relationships, which is useful in forensics, establishing identity, and understanding historical events.
2. Inherited disease differs from other disorders in its predictability; predictive testing; characteristic frequencies in different populations; and the potential of gene therapy.
3. Agriculture, both traditional and biotechnological, applies genetic principles.
4. Collecting DNA from habitats and identifying the sequences in databases is a new way to analyze ecosystems.
5. Human genome information has tremendous potential but must be carefully managed.

Summary

1.1 Genetic Testing

1. **Genes** are the instructions to manufacture proteins, which determine inherited traits.
2. A **genome** is a complete set of genetic information. A **cell** contains two genomes of **DNA. Genomics** is the study of many genes and their interactions.
3. People can choose specific gene tests, based on family and health history, to detect or estimate risk of developing certain conditions. DNA **microarrays** detect many genes at once. Expression arrays indicate which proteins a cell makes.

1.2 Levels of Genetics

4. Genes encode proteins and the **RNA** molecules that carry out protein synthesis. RNA carries the gene sequence information so that it can be utilized, while the DNA is transmitted when the cell divides. Much of the genome does not encode protein.
5. Variants of a gene, called **alleles,** arise by **mutation.** They may differ slightly from one another, but encode the same product. A **polymorphism** is a site or sequence of DNA that varies in one percent or more of a population. The **phenotype** is the gene's expression. An allele combination constitutes the **genotype.** Alleles may be **dominant** (exerting an effect in a single copy) or **recessive** (requiring two copies for expression).
6. **Chromosomes** consist of DNA and protein. The 22 types of **autosomes** do not include genes that specify sex. The X and Y **sex chromosomes** bear genes that determine sex.
7. The human genome contains about 3 billion DNA bases. Cells **differentiate** by

expressing subsets of genes. **Stem cells** divide to yield other stem cells and cells that differentiate.

8. Pedigrees are diagrams used to study traits in families.

9. Genetic populations are defined by their collections of alleles, termed the gene pool.

10. Genome comparisons among species reveal evolutionary relationships.

1.3 Most Genes Do Not Function Alone

11. Single genes determine Mendelian traits.

12. **Multifactorial** traits reflect the influence of one or more genes and the environment. Recurrence of a Mendelian trait is predicted based on Mendel's laws; predicting recurrence of a multifactorial trait is more difficult.

13. Genetic determinism is the idea that expression of an inherited trait cannot be changed.

1.4 Statistics Represent Risks

14. Risk assessment estimates the probability of inheriting a particular gene. **Absolute risk,** expressed as odds or a percentage, is the probability that an individual will develop a particular trait or illness over a lifetime.

15. **Relative risk** is a ratio that estimates how likely a person is to develop a particular phenotype compared to another group.

16. Risk estimates are empiric, based on Mendel's laws, or modified to account for environmental influences.

1.5 Applications of Genetics

17. DNA profiling can establish identity, relationships, and origins.

18. In inherited diseases, recurrence risks are predictable and a mutation may be detected before symptoms arise. Some inherited disorders are more common among certain population groups. Gene therapy attempts to correct mutations. Studying genes and genomes of nonhuman animals can help us understand causes of diseases in humans.

19. Genetic information can be misused.

20. Agriculture is selective breeding. **Biotechnology** is the use of organisms or their parts for human purposes. A transgenic organism harbors a gene or genes from a different species.

21. In metagenomics, DNA collected from habitats is used to reconstruct ecosystems.

Review Questions

1. Place the following terms in size order, from largest to smallest, based on the structures or concepts they represent:
   a. chromosome
   b. gene pool
   c. gene
   d. DNA
   e. genome

2. Distinguish between:
   a. an autosome and a sex chromosome
   b. genotype and phenotype
   c. DNA and RNA
   d. recessive and dominant traits
   e. absolute and relative risks
   f. pedigrees and karyotypes
   g. gene and genome

3. List four ways that inherited disease differs from other types of illnesses.

4. Cystic fibrosis is a Mendelian trait; height is a multifactorial trait. How do the causes of these characteristics differ?

5. Mutants are often depicted in the media as being abnormal, ugly, or evil. Why is this not necessarily true?

6. Health insurance forms typically ask for applicants to list existing or preexisting symptoms. How do the results of a genetic test differ from this?

Applied Questions

1. At the same time the media reported the story of the giant-muscled German boy, another young man, an 8-year-old poet, died of a muscle-wasting condition. How might a mutation in the myostatin gene cause an effect opposite the one seen in the German boy?

2. Should researchers publish genome sequences of pathogens, or should such information be restricted to prevent the development of bioweapons? Cite a reason for your answer.

3. Breast cancer caused by the *BRCA1* gene affects 1 in 800 women in the general U.S. population. Among Jewish people of eastern European descent, it affects 2 in 100. What is the relative risk for this form of breast cancer among eastern European Jewish women in the United States?

4. In a search for a bone marrow transplant donor, why would a patient's siblings be considered before first cousins?

5. DNA databases of convicted felons have solved many crimes and exonerated many innocent people. What might be the benefits and dangers of establishing databases on everyone? How should such a program be instituted?

Web Activities

Visit the Online Learning Center (OLC) at www.mhhe.com/lewisgenetics7. Select **Student Edition, chapter 1** and **Web Activities** to find the website links needed to complete the following activities.

6. Many artists have been inspired by aspects of genetics, from the symmetry of DNA to common fears of genetic technologies. Visit the websites provided on the OLC, select a work of art, and describe what it represents.

7. Consult the website for the Council for Responsible Genetics. Select a controversy covered and present both sides of the issue. This may require some additional research!

8. Consult the Combined DNA Index System (CODIS) of the Federal Bureau of Investigation on the Web. This database utilizes forensic DNA information and computer technology so that the local, state, and federal governments can easily exchange information about suspected criminals. Do you think this information is useful, or an invasion of privacy?

9. Genetics inspires cartoonists, too. Visit the website provided on the OLC and search under "DNA." Select a cartoon that misrepresents genetics, and explain how it is inaccurate, misleading, or sensationalized.

10. The website from GeneLink Inc. announces "the world's first family-centered DNA bank and hereditary genetic information services." A client sends a sample of his or her DNA, obtained with a cheekbrush, to the company, which then examines certain genes. Explore the website, and discuss the pros and cons of using this type of service to learn about your DNA.

Case Studies and Research Results

11. Morris has a DNA microarray test for several genes that predispose to developing prostate cancer. He learns that his overall relative risk is 1.5, compared to the risk in the general population. Overjoyed, he tells his wife that his risk of developing prostate cancer is only 1.5 percent. She says no, his risk is 50 percent greater than that of the average individual in the general population. Who is correct? Why?

12. Benjamin undergoes a genetic screening test and receives the following relative risks:

– addictive behaviors	0.6
– coronary artery disease	2.3
– kidney cancer	1.4
– lung cancer	5.8
– diabetes	0.3
– depression	1.2

Which conditions is he more likely to develop than someone in the general population, and which conditions is he less likely to develop?

13. In the mid-1980s, a project in a county in northern Sweden sought to lower the incidence of heart disease and early death by amassing medical data—specifically, lifestyle habits, blood pressure, lipid profiles, and blood samples. Genetic testing was not part of the original program. In the mid-1990s, with the human genome project progressing, the university that ran the program and the local government formed a company to develop commercial products using the information collected on the population. Mostly because of arguments over intellectual property rights, the proposed biotechnology company never became established.

   a. How does the Swedish biobank differ from deCODE Genetics?

   b. How does the Swedish biobank differ from the UK Biobank?

   c. What roles do you think governments, companies, universities, and consumer groups should play in the establishment and running of biobanks?

14. The Larsons have a child who has inherited cystic fibrosis. Their physician tells them that if they have other children, each faces a 1 in 4 chance of also inheriting the illness. The Larsons tell their friends, the Espositos, of their visit with the doctor. Mr. and Mrs. Esposito are expecting a child, so they ask their physician to predict whether he or she will one day develop multiple sclerosis—Mr. Esposito is just beginning to show symptoms. They are surprised to learn that, unlike the situation for cystic fibrosis, recurrence risk for multiple sclerosis cannot be easily predicted. Why not?

15. Burlington Northern Santa Fe Railroad asked its workers for a blood sample, and then supposedly tested for a gene variant that predisposes a person for carpal tunnel syndrome, a disorder of the wrists caused by repetitive motion. The company threatened to fire a worker who refused to be tested; the worker sued the company. The Equal Employment Opportunity Commission ruled in the worker's favor, agreeing that the company's action violated the Americans with Disabilities Act.

   a. Do you agree with the company or the worker? What additional information would be helpful in taking sides?

   b. How is the company's genetic testing not based on sound science?

   c. How can tests such as those described for the two students earlier in this chapter be instituted in a way that does not violate a person's right to privacy, as the worker in the railroad case contended?

Learn to apply the skills of a genetic counselor with this additional case found in the *Case Workbook in Human Genetics.*
   Genetics in the news

CHAPTER 2

Cells

CHAPTER CONTENTS

STEM CELLS RESTORE SIGHT, BUT NOT VISION

In 1960, three-year-old Michael M. lost his left eye in an accident. Because much of the vision in his right eye was already impaired from scars on the cornea, the transparent outer layer, he could see only distant dim light. Several corneal transplants failed, only adding scar tissue. At age 39, Michael received donor stem cells from a cornea, rather than the entire structure. When the bandages were removed, Michael saw his wife and two sons for the first time. But he quickly learned that vision is more than seeing—his brain had to interpret the images his eyes captured. Because the development of his visual system had stalled, and he had only one eye, he could discern shapes and colors, but not three-dimensional objects, such as facial details. In fact, he had been more comfortable skiing blind, using verbal cues, than he was with sight—the looming trees were terrifying. For a while, post-surgery, he skied blindfolded. Michael's brain had to catch up to his rejuvenated eye.

Michael's doctors used stem cells to repair an injury. Stem cells may also be used to correct inherited disorders, such as retinitis pigmentosa (RP). The nerve cells or blood vessels of the retina degenerate, resulting in blindness. (The retina is a layered structure at the back of the eyeball that collects and sends visual information to the brain.) Investigators injected stem cells from human bone marrow into one eye of mice with RP. In each animal, the treated eye developed normally, but the retina of the untreated eye degenerated. The injected stem cells divided to yield some cells that specialized to form the linings of blood vessels, and some stem cells. The injected stem cells seemed to "know" to settle near cells that guide the blood vessel building blocks towards each other, and they also saved the nerve cells in the retina.

Stem cells are the body's way of growing and healing. Medical science is trying to harness them to treat a variety of types of disease.

When Michael M. received stem cells to heal his eyes, his sight (sensation of light) was restored, but not his vision (his brain's perception of the images). Slowly, his brain caught up with his senses, and he was able to see his family for the first time.

The activities and abnormalities of cells underlie inherited traits, quirks, and illnesses. The muscles of a boy with muscular dystrophy, for example, weaken because they lack a protein that normally supports the cells' shape during forceful contractions (**figure 2.1**).

Understanding what goes wrong in certain cells when a disease occurs can suggest ways to treat the condition—we learn what must be repaired or replaced. Understanding cell function also reveals how a healthy body works, and how it develops from one cell to trillions. Our bodies include many variations on the cellular theme, with such specialized cell types as bone and blood, nerve and muscle, and even subtypes of those. Equally important are unspecialized cells that are nestled into organs. These **stem cells,** able to replicate themselves as well as to generate specialized cells, enable a body to develop, grow, and repair damage. Stem cells healed the eyes of Michael M., described in the chapter-opening case study.

Cells interact. They send, receive, and respond to information. Some cells aggregate with others of like function, forming tissues, which in turn interact to form organs and organ systems. Other cells move about the body. Cell numbers are important, too—they are critical to development, growth, and healing. These processes reflect a precise balance between cell division, which adds cells, and cell death, which takes them away.

2.1 The Components of Cells

All cells share certain features that enable them to perform the basic life functions of reproduction, growth, response to stimuli, and energy use. Body cells also have specialized features, such as the contractile proteins in a muscle cell. The more than 260 specialized or differentiated cell types in a human body arise because the cells express different parts of the genome.

The human body's cells fall into four broad categories, or tissue types: epithelium (lining cells), muscle, nerve, and connective tissues (including blood, bone, cartilage, and adipose cells). Other multicellular organisms, including other animals, fungi, and plants, also have differentiated cells. Some single-celled organisms, such as the familiar paramecium and ameba, have very distinctive cells as complex as our own. The most abundant organisms on the planet, however, are simpler and single-celled. These microorganisms are nonetheless successful life forms, because they have occupied earth longer than we have.

Biologists recognize three broad varieties of cells that define three major "domains" of life: the Archaea, the Bacteria, and the Eukarya. A domain is a broader classification than the familiar kingdom.

The archaea and bacteria are both single-celled, but they differ from each other in the sequences of many of their genetic molecules and in the types of molecules in their membranes. Archaea and bacteria are, however, both **prokaryotes,** which means that they lack a **nucleus,** the structure that contains DNA in the cells of other types of organisms.

The third domain of life, the Eukarya or **eukaryotes,** includes single-celled organisms that have nuclei, as well as all multicellular organisms such as ourselves. Eukaryotic cells are also distinguished from prokaryotic cells by structures called **organelles,** which perform specific functions. The cells of all three domains contain globular assemblies of RNA and protein called **ribosomes** that are essential for protein synthesis. The eukaryotes may have arisen from an ancient fusion of a bacterium with an archaean.

Chemical Constituents of Cells

Cells are composed of molecules. Some of the chemicals of life (biochemicals) are so large that they are called macromolecules.

The major macromolecules that make up and fuel cells are **carbohydrates** (sugars and starches), **lipids** (fats and oils), **proteins,** and **nucleic acids.** Cells require vitamins and minerals in much smaller amounts.

Carbohydrates provide energy and contribute to cell structure. Lipids form the basis of several types of hormones, provide insulation, and store energy. Proteins have many diverse functions in the human body. They participate in blood clotting, nerve transmission, and muscle contraction and form the bulk of the body's connective tissue. **Enzymes** are especially important proteins because they facilitate, or catalyze, biochemical reactions so that they occur swiftly enough to sustain life. Most important to the study of genetics are the nucleic acids DNA and RNA, which translate information from past generations into specific collections of proteins that give a cell its individual characteristics.

Macromolecules often combine to form larger structures within cells. For example, the membranes that surround cells and compartmentalize their interiors consist of double layers (bilayers) of lipids embedded with carbohydrates, proteins, and other lipids.

Life is based on the chemical principles that govern all matter; genetics is based on a highly organized subset of the chemical reactions of life. **Reading 2.1** describes some drastic effects that result from major biochemical abnormalities.

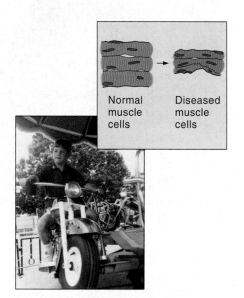

Figure 2.1 Genetic disease at the whole-person and cellular levels. This young man has Duchenne muscular dystrophy. The condition has not yet severely limited his activities, but he shows an early sign of the illness—overdeveloped calf muscles that result from his inability to rise from a sitting position the usual way. Lack of the protein dystrophin causes his skeletal muscle cells to collapse when they contract.

Normal muscle cells → Diseased muscle cells

Inborn Errors of Metabolism Affect the Major Biomolecules

Enzymes are proteins that catalyze (speed or facilitate) specific chemical reactions, and therefore control a cell's production of all types of macromolecules. When the gene that encodes an enzyme mutates so that the enzyme is not made or cannot function, the result can be too much or too little of the product of the biochemical reaction that the enzyme catalyzes. Genetic disorders that result from deficient or absent enzymes are called inborn errors of metabolism. Following are some examples.

Carbohydrates

The newborn yelled and pulled up her chubby legs in pain a few hours after each feeding. She developed watery diarrhea, even though she was breastfed. Finally, a doctor identified *lactase deficiency*—lack of the enzyme lactase, which enables the digestive system to break down the carbohydrate lactose. Bacteria multiplied in the undigested lactose in the child's intestines, producing gas, cramps, and bloating. Switching to a soybean-based, lactose-free infant formula helped. (A different disorder with milder symptoms is lactose intolerance [OMIM 150200], common in adults.)

Lipids

A sudden sharp pain began in the man's arm and spread to his chest—the first sign of a heart attack. At age 36, he was younger than most people who suffer heart attacks, but he had inherited a gene that halved the number of protein receptors for cholesterol on his liver cells. Because cholesterol could not enter the liver cells efficiently, it built up in his arteries, constricting blood flow in his heart and eventually causing a mild heart attack. A fatty diet had accelerated his *familial hypercholesterolemia* (OMIM 144010), an inherited form of heart disease, but taking a cholesterol-lowering drug helps.

Proteins

The first sign that the infant was ill was also the most innocuous—his urine smelled like maple syrup. Tim slept most of the time, and he vomited so often that he hardly grew. A blood test revealed that Tim had inherited *maple syrup urine disease* (OMIM 248600). He could not digest three types of amino acids (protein building blocks), so these amino acids accumulated in his bloodstream. A diet very low in these amino acids has helped Tim.

Nucleic Acids

From birth, Michael's wet diapers contained orange, sandlike particles, but otherwise he seemed healthy. By six months of age, he was in pain when urinating. A physician noted that Michael's writhing movements were involuntary rather than normal crawling.

The orange particles in Michael's diaper indicated *Lesch-Nyhan syndrome* (OMIM 300322), caused by deficiency of an enzyme called HGPRT. Michael's body could not recycle two of the four types of DNA building blocks, instead converting them into uric acid, which crystallizes in urine. Other symptoms that appeared later were not as easy to explain—severe mental retardation, seizures, and aggressive and self-destructive behavior. By age three or so, Michael responded to stress by uncontrollably biting his fingers, lips, and shoulders. His parents had his teeth removed to keep him from harming himself, and he was kept in restraints. He would probably die before the age of 30 of kidney failure or infection.

Vitamins

Vitamins enable the body to use the carbohydrates, lipids, and proteins we eat. Julie inherited *biotinidase deficiency* (OMIM 253260), which greatly slows the rate at which her body can use the vitamin biotin. If Julie hadn't been diagnosed as a newborn and quickly started on biotin supplements, by early childhood she would have shown biotin deficiency symptoms: mental retardation, seizures, skin rash, and loss of hearing, vision, and hair. Her slow growth, caused by her body's inability to extract energy from nutrients, would have eventually proved lethal.

Minerals

Ingrid, in her thirties, lives in the geriatric ward of a mental hospital, unable to talk or walk. She grins and drools, but she is alert and communicates using a computer. When she was a healthy high-school senior, symptoms of *Wilson disease* (OMIM 277900) began to appear, as her weakened liver could no longer control the excess copper her digestive tract absorbed from food. The initial symptoms were stomachaches, headaches, and an inflamed liver (hepatitis). Then odd changes began—slurred speech; loss of balance; a gravelly, low-pitched voice; and altered handwriting. A psychiatrist noted the telltale greenish rings around her irises, caused by copper buildup, and diagnosed Wilson disease (**figure 1**). Finally Ingrid received penicillamine, which enabled her to excrete the excess copper in her urine. The treatment halted the course of the illness, saving her life.

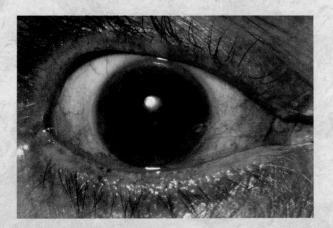

Figure 1 **Wilson disease.** A greenish ring around the iris is one sign of the copper buildup of Wilson disease.

Organelles

A eukaryotic cell holds a thousand times the volume of a bacterial or archaeal cell (**figure 2.2**). In order to carry out the activities of life in such a large cell, **organelles** divide the labor, partitioning off certain areas or serving specific functions. Saclike organelles sequester biochemicals that might harm other cellular constituents. Some organelles consist of membranes studded with enzymes embedded in the order in which they participate in the chemical reactions that produce a particular molecule. In general, organelles keep related biochemicals and structures close enough to one another to interact efficiently. This eliminates the need to maintain a high concentration of a particular biochemical throughout the cell.

Organelles enable a cell to retain as well as use its genetic instructions, acquire energy, secrete substances, and dismantle debris. The coordinated functioning of the organelles in a eukaryotic cell is much like the organization of departments in a department store (**figure 2.3**).

The most prominent organelle, the nucleus, is enclosed in a layer called the nuclear envelope. Nuclear pores are rings of proteins that allow certain biochemicals to exit or enter the nucleus (**figure 2.4**). Within the nucleus, an area that appears darkened under a microscope, the nucleolus ("little nucleus"), is the site of ribosome production. The nucleus is filled with DNA complexed with many proteins to form chromosomes. Other proteins form fibers that give the nucleus a roughly spherical shape. RNA is abundant too, as are enzymes and proteins required to synthesize RNA from DNA. The material in the nucleus, minus these contents, is called nucleoplasm.

The remainder of the cell—that is, everything but the nucleus, organelles, and the outer boundary, or **plasma membrane**—is the **cytoplasm.** Other cellular components include stored proteins, carbohydrates, and lipids; pigment molecules; and various other small chemicals.

Secretion—The Eukaryotic Production Line

Organelles interact to coordinate basic life functions and sculpt the characteristics of specialized cell types. Secretion, which is the release of a substance from a cell, illustrates how organelles function together.

Secretion begins when the body sends a biochemical message to a cell to begin producing a particular substance. For example, when a newborn first suckles the mother's breast, the stimulation causes her brain to release hormones that signal cells in her breast, called lactocytes, to rapidly increase the production of the complex mixture that makes up milk, which began with hormonal changes at the birth (**figure 2.5**). In response, information in certain genes is copied into molecules of **messenger RNA** (mRNA), which then exit the nucleus (see steps 1 and 2 in figure 2.5). In the cytoplasm, the mRNAs, with the help of ribosomes and another type of RNA called **transfer RNA,** direct the manufacture of milk proteins. These include nutritive proteins called caseins, antibody proteins that protect against infection, and various enzymes.

Most protein synthesis occurs on a maze of interconnected membranous tubules and sacs called the **endoplasmic reticulum** (ER) (see step 3 in figure 2.5). The ER winds from the nuclear envelope outward to the plasma membrane. The portion of the ER nearest the nucleus, which is flattened and studded with ribosomes, is called the rough ER because the ribosomes make it appear fuzzy when viewed under an electron microscope. Messenger RNA attaches to the ribosomes on the rough ER. Amino acids from the cytoplasm are then linked, following the instructions in the mRNA's sequence, to form particular proteins that will either exit the cell or become part of membranes (step 3, figure 2.5). Proteins are also synthesized on ribosomes not associated with the ER. These proteins remain in the cytoplasm.

The ER acts as a quality control center for the cell. Its chemical environment enables the forming protein to start folding into the three-dimensional shape necessary for its specific function. Misfolded proteins are pulled out of the ER and degraded, much as an obviously defective toy might be pulled from an assembly line at a toy factory and discarded. Misfolded proteins that are not destroyed can cause disease, as discussed further in chapter 10.

As the rough ER winds out toward the plasma membrane, the ribosomes become fewer, and the diameters of the tubules widen, forming a section called smooth ER. Here, lipids are made and added to the

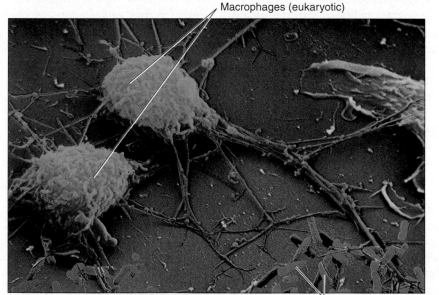

Macrophages (eukaryotic)

Bacteria (prokaryotic)

Figure 2.2 Eukaryotic and prokaryotic cells. A human cell is eukaryotic and much more complex than a bacterial cell, while an archaean cell looks much like a bacterial cell. Here, human macrophages (blue) capture bacteria (yellow). Note how much larger the human cells are.

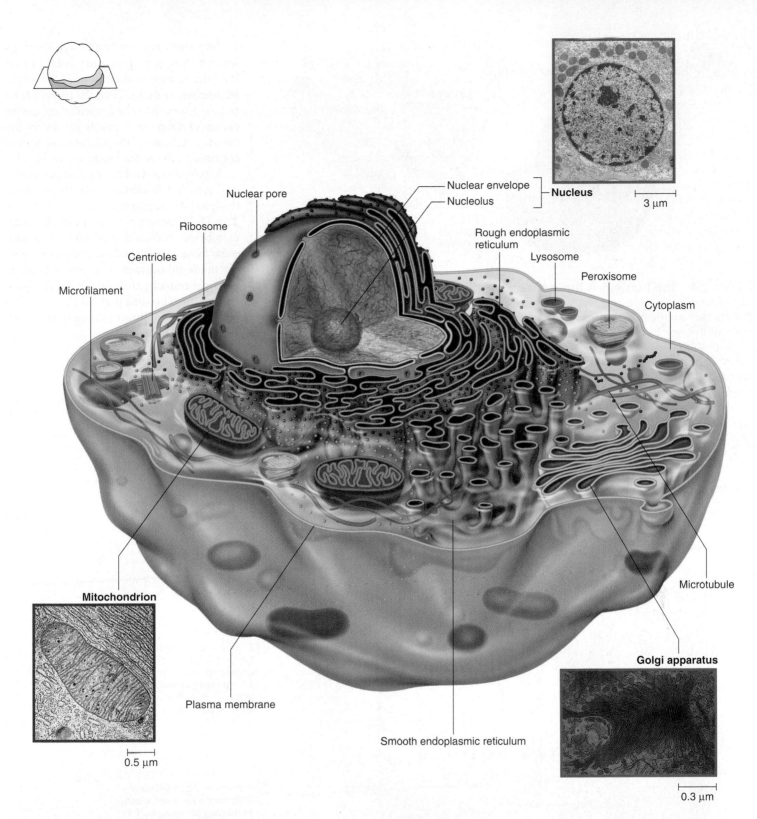

Nuclear pore

Ribosome

Centrioles

Microfilament

Nuclear envelope

Nucleolus

Nucleus

3 μm

Rough endoplasmic reticulum

Lysosome

Peroxisome

Cytoplasm

Microtubule

Mitochondrion

0.5 μm

Plasma membrane

Smooth endoplasmic reticulum

Golgi apparatus

0.3 μm

Figure 2.3 Generalized animal cell. Organelles provide specialized functions for the cell. Most of these structures are transparent; colors are used here to distinguish them.

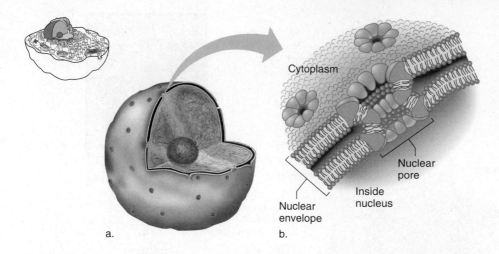

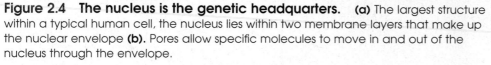

a.

b.

Figure 2.4 The nucleus is the genetic headquarters. **(a)** The largest structure within a typical human cell, the nucleus lies within two membrane layers that make up the nuclear envelope **(b).** Pores allow specific molecules to move in and out of the nucleus through the envelope.

proteins arriving from the rough ER (step 4, figure 2.5). The lipids and proteins travel until the tubules of the smooth ER eventually narrow and end. Then the proteins exit in membrane-bounded, saclike organelles called **vesicles** that pinch off from the tubular endings of the membrane. Lipids continue without the need for a vesicle.

A loaded vesicle takes its contents to the next stop in the secretory production line, the nearby **Golgi apparatus** (step 5, figure 2.5). This processing center is a stack of flat, membrane-enclosed sacs. Here, the milk sugar lactose is synthesized and other sugars are made that attach to proteins to form **glycoproteins** or to lipids to form **glyco-lipids,** which become parts of plasma membranes. Proteins finish folding in the Golgi apparatus (step 6, figure 2.5).

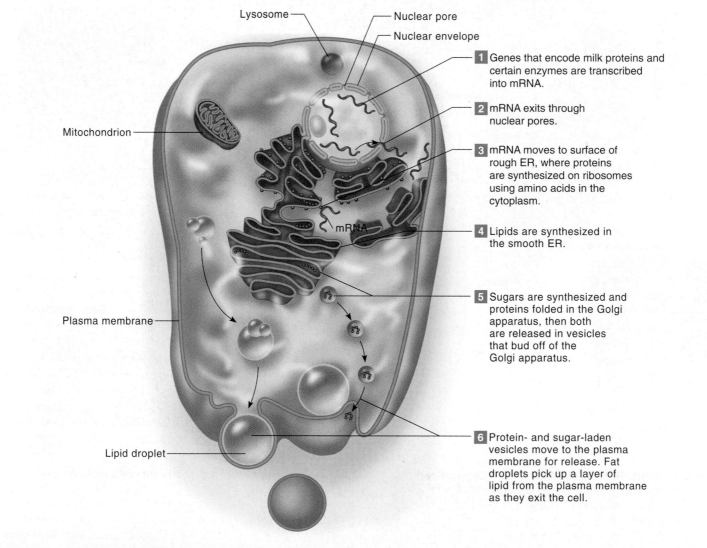

1 Genes that encode milk proteins and certain enzymes are transcribed into mRNA.

2 mRNA exits through nuclear pores.

3 mRNA moves to surface of rough ER, where proteins are synthesized on ribosomes using amino acids in the cytoplasm.

4 Lipids are synthesized in the smooth ER.

5 Sugars are synthesized and proteins folded in the Golgi apparatus, then both are released in vesicles that bud off of the Golgi apparatus.

6 Protein- and sugar-laden vesicles move to the plasma membrane for release. Fat droplets pick up a layer of lipid from the plasma membrane as they exit the cell.

Figure 2.5 Secretion: Making milk. Milk production and secretion illustrate organelle functions and interactions in a cell from a mammary gland: (1) through (6) indicate the order in which organelles participate in this process. Lipids are secreted in separate droplets from proteins and their attached sugars.

The components of complex secretions, such as milk, are temporarily stored in the Golgi apparatus. Droplets of proteins and sugars then bud off in vesicles that move outward to the plasma membrane, fleetingly becoming part of the membrane until they are secreted to the cell's exterior. Lipids exit the plasma membrane directly, taking bits of it with them (step 6, figure 2.5).

Within the breast, epithelial cells called lactocytes form tubules, into which they secrete the components of milk. When the baby suckles, contractile cells squeeze the milk through the tubules and out of holes in the nipples. This "ejection reflex" is so powerful that the milk can actually shoot across a room!

Intracellular Digestion— Lysosomes and Peroxisomes

Just as clutter and garbage tend to accumulate in an apartment or dorm room, debris builds up in cells—and must be discarded. Eukaryotic cells use organelles called **lysosomes** to handle the garbage. Lysosomes are membrane-bounded sacs that contain enzymes that dismantle bacterial remnants, worn-out organelles, and other debris (**figure 2.6**). The enzymes also break down some digested nutrients into forms that the cell can use.

Lysosomes fuse with vesicles carrying debris from outside or within the cell, and the lysosomal enzymes then degrade the contents. A loaded lysosome moves toward the plasma membrane and fuses with it, releasing its contents to the outside. The word *lysosome* means "body that lyses"; *lyse* means "to cut." Lysosomes maintain the very acidic environment that their enzymes require to function, without harming other cellular constituents.

Cells differ in the number of lysosomes they contain. Certain white blood cells and macrophages that move about and engulf bacteria are loaded with lysosomes. Liver cells require many lysosomes to break down cholesterol and toxins.

All lysosomes contain more than 40 types of digestive enzymes, which must maintain a correct balance. Absence or malfunction of an enzyme causes a lysosomal storage disease. In these inherited disorders, the molecule that the missing or abnormal enzyme

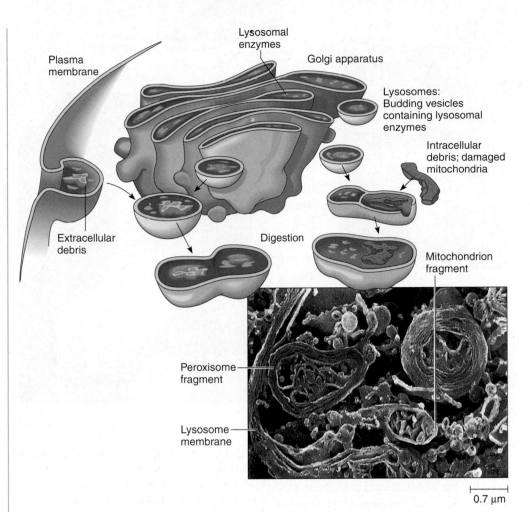

Figure 2.6 Lysosomes are trash centers. Lysosomes fuse with vesicles or damaged organelles, activating the enzymes within to recycle the molecules for the cell's use. Lysosomal enzymes also dismantle bacterial remnants. These enzymes require a very acidic environment to function.

normally degrades accumulates. The lysosome swells, crowding organelles and interfering with the cell's functions. In Tay-Sachs disease (OMIM 272800), for example, an enzyme that normally breaks down lipids in the cells that surround nerve cells is deficient, burying the nervous system in lipid. An affected infant begins to lose skills at about six months of age, then gradually loses sight, hearing, and the ability to move, typically dying within three years. Even before birth, the lysosomes of affected cells swell.

Peroxisomes are sacs with outer membranes that are studded with several types of enzymes. These enzymes perform a variety of functions, including breaking down certain lipids and rare biochemicals, synthesizing bile acids used in fat digestion, and detoxifying compounds that result from exposure to damaging oxygen free radicals. Peroxisomes are large and abundant in liver and kidney cells (**figure 2.7**).

The 1992 film *Lorenzo's Oil* recounted the true story of a child with an inborn error of metabolism caused by an absent peroxisomal enzyme. Lorenzo Odone had adrenoleukodystrophy (OMIM 202370). His peroxisomes lacked a protein that normally transports an enzyme into peroxisomes. The enzyme catalyzes a reaction that helps break down a type of lipid called a very-long-chain fatty acid. Without the transporter protein, the cells of the brain and spinal cord accumulate the fatty acid. Early symptoms include low blood sugar, skin darkening, muscle weakness, and heartbeat irregularities. The patient eventually loses control over the limbs and usually dies within a few years. Eating a type of lipid in rapeseed (canola) oil—the oil in the

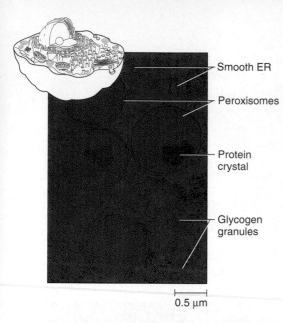

- Smooth ER
- Peroxisomes
- Protein crystal
- Glycogen granules

0.5 µm

Figure 2.7 Peroxisomes are detox centers. The high concentration of enzymes within a peroxisome crystallizes proteins. Glycogen is a storage form of carbohydrate. Peroxisomes are abundant in liver cells, where they assist in detoxification.

film's title—slows buildup of the very-long-chain fatty acids in blood plasma and the liver. But the rapeseed lipid cannot enter the brain, where it is required to combat the symptoms. It slows progression of the disease, but isn't a cure.

Energy Production— Mitochondria

The activities of secretion, as well as the many chemical reactions taking place in the cytoplasm, require enormous and continual energy. Organelles called **mitochondria** provide energy by breaking down nutrients from foods.

A mitochondrion has an outer membrane similar to those in the ER and Golgi apparatus and an inner membrane that is folded into structures called cristae (**figure 2.8**). These folds hold enzymes that catalyze the biochemical reactions that release energy from the chemical bonds of nutrient molecules. This liberated energy from food is captured and stored in the bonds that hold together a molecule called adenosine triphosphate (ATP). Therefore, ATP serves as a cellular energy currency.

The number of mitochondria in a cell varies from a few hundred to tens of thou-

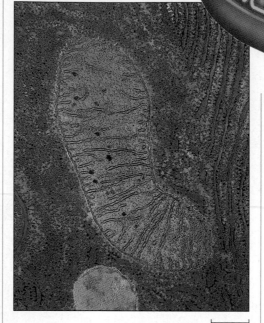

- Cristae
- Outer membrane
- Inner membrane

0.5 µm

Figure 2.8 A mitochondrion extracts energy. Cristae, infoldings of the inner membrane, increase the available surface area containing enzymes for energy reactions in a mitochondrion.

sands, depending upon the cell's activity level. A typical liver cell, for example, has about 1,700 mitochondria, but a muscle cell, with its very high energy requirements, has many more.

Mitochondria are especially interesting because, like the nucleus, they contain DNA, although a very small amount. Another unusual characteristic of mitochondria is that they are almost always inherited from the mother only, because mitochondria are located in the middle regions of sperm cells, but usually not in the head regions that enter eggs. Moreover, mitochondria that do enter with a sperm are usually destroyed in the very early embryo. A class of inherited diseases whose symptoms result from abnormal mitochondria are always passed from mother to offspring. These illnesses usually produce extreme muscle weakness, because muscle cells have so many mitochondria. Chapter 5 discusses mitochondrial inheritance. Evolutionary biologists study mitochondrial genes to trace the beginnings of humankind, as discussed in chapter 15.

Table 2.1 summarizes the structures and functions of organelles.

Table 2.1

Structures and Functions of Organelles

Organelle	Structure	Function
Endoplasmic reticulum	Membrane network; rough ER has ribosomes, smooth ER does not	Site of protein synthesis and folding; lipid synthesis
Golgi apparatus	Stacks of membrane-enclosed sacs	Site where sugars are made and linked into starches or joined to lipids or proteins; proteins finish folding; secretions stored
Lysosome	Sac containing digestive enzymes	Degrades debris, recycles cell contents
Mitochondrion	Two membranes; inner membrane enzyme-studded	Releases energy from nutrients, participates in cell death
Nucleus	Porous sac containing DNA	Separates DNA from rest of cell
Peroxisome	Sac containing enzymes	Breaks down and detoxifies various molecules
Ribosome	Two associated globular subunits of RNA and protein	Scaffold and catalyst for protein synthesis
Vesicle	Membrane-bounded sac	Temporarily stores or transports substances

The Plasma Membrane

Just as the character of a community is molded by the people who enter and leave it, the special characteristics of different cell types are shaped in part by the substances that enter and leave. The plasma membrane controls this process. It forms a selective barrier that completely surrounds the cell and monitors the movements of molecules in and out of the cell. How the chemicals that comprise the plasma membrane associate with each other determines which substances can enter or leave the cell. Similar membranes form the outer boundaries of several organelles, and some organelles consist entirely of membranes. A cell's membranes are more than mere coverings. Some of their constituent or associated molecules carry out specific functions.

A biological membrane is built of a double layer (bilayer) of molecules called phospholipids. A phospholipid is a fat molecule with attached phosphate groups, and is often depicted as a head with two parallel tails. (A phosphate group (PO_4) is a phosphorus atom bonded to four oxygen atoms.) Phospholipid molecules can organize into sheetlike structures, which makes membrane formation possible (**figure 2.9**). The molecules do this because their ends have opposite reactions to water: The phosphate end of a phospholipid is attracted to water, and thus is hydrophilic (water-loving); the other end, which consists of two chains of fatty acids, moves away from water, and is therefore hydrophobic (water-fearing). Because of these water preferences, phospholipid molecules in water spontaneously form bilayers, with the hydrophilic surfaces exposed to the watery exterior and interior of the cell, and the hydrophobic surfaces facing each other on the inside of the bilayer, away from the water.

A phospholipid bilayer forms the structural backbone of a biological membrane. Proteins are embedded in the bilayer with some traversing the entire structure, while others extend from either face (**figure 2.10**).

Proteins, glycoproteins, and glycolipids jut from a plasma membrane, creating surface topographies that are important in a cell's interactions with other cells. The surfaces of your cells indicate that they are part of your body, and also that they have differentiated in a particular way.

Many molecules that extend from the plasma membrane function as **receptors,** which are structures that have indentations or other shapes that fit and hold molecules outside the cell. The molecule that binds to the receptor, called the **ligand,** sets into motion a cascade of chemical reactions that carries out a particular cellular activity.

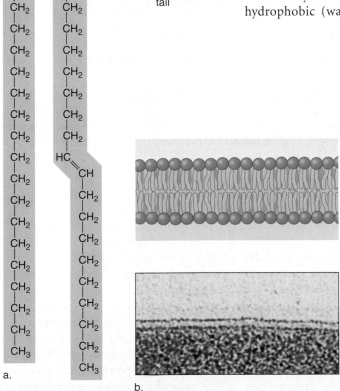

a.

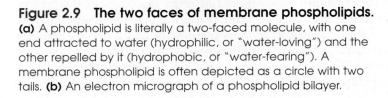

b.

Figure 2.9   The two faces of membrane phospholipids.
(a) A phospholipid is literally a two-faced molecule, with one end attracted to water (hydrophilic, or "water-loving") and the other repelled by it (hydrophobic, or "water-fearing"). A membrane phospholipid is often depicted as a circle with two tails. **(b)** An electron micrograph of a phospholipid bilayer.

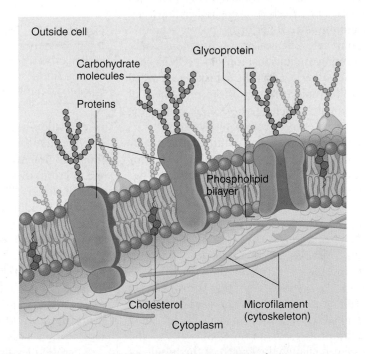

Figure 2.10   Anatomy of a plasma membrane.   In a plasma membrane, mobile proteins are embedded throughout a phospholipid bilayer. Other types of lipids aggregate to form "rafts," and an underlying mesh of protein fibers provides support. Carbohydrates jut from the membrane's outer face.

The phospholipid bilayer is oily, and some proteins move within it like ships on a sea. Proteins with related functions may cluster on "lipid rafts" that float on the phospholipid bilayer. The rafts are rich in cholesterol and other types of lipids. This clustering of proteins eases their interaction.

Proteins aboard lipid rafts have several functions. They contribute to the cell's identity; act as transport shuttles into the cell; serve as gatekeepers; and can let in certain toxins and pathogens. HIV, for example, enters a cell by breaking a lipid raft.

The inner hydrophobic region of the phospholipid bilayer blocks entry and exit to most substances that dissolve in water. However, certain molecules can cross the membrane through proteins that form passageways, or when they are escorted by a "carrier" protein. Some membrane proteins form channels for ions (atoms or molecules with an electrical charge). **Reading 2.2** describes how faulty ion channels can cause disease.

The Cytoskeleton

The **cytoskeleton** is a meshwork of protein rods and tubules that molds the distinctive structures of a cell, positioning organelles and providing three-dimensional shape. The proteins of the cytoskeleton are broken down and built up as a cell performs specific activities. Some cytoskeletal elements function as rails, forming conduits that transport cellular contents; other parts of the cytoskeleton, called motor molecules, power the movement of organelles along these rails by converting chemical energy to mechanical energy.

The cytoskeleton includes three major types of elements—**microtubules, microfilaments,** and **intermediate filaments (figure 2.11).** They are distinguished by protein type, diameter, and how they aggregate into larger structures. Other proteins connect these components, creating the meshwork that provides the cell's strength and ability to resist force and maintain shape.

Long, hollow microtubules provide many cellular movements. A microtubule is composed of pairs (dimers) of a protein, called tubulin, assembled into a hollow tube. The cell can change the length of the tubule by adding or removing tubulin molecules.

Cells contain both formed microtubules and individual tubulin molecules. When the cell requires microtubules to carry out a spe-

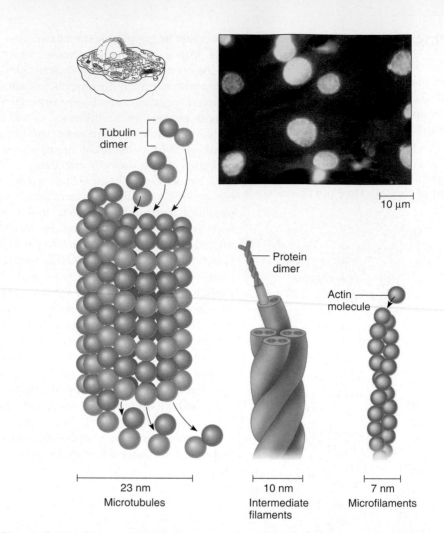

Figure 2.11 The cytoskeleton is made of protein rods and tubules. The three major components of the cytoskeleton are microtubules, intermediate filaments, and microfilaments. Through special staining, the cytoskeletons in these cells appear orange under the microscope. (The abbreviation nm stands for nanometer, which is a billionth of a meter.)

cific function—cell division, for example—the free tubulin dimers self-assemble into more tubules. After the cell divides, some of the microtubules fall apart into individual tubulin dimers. This replenishes the cell's supply of building blocks. Cells are in a perpetual state of flux, building up and breaking down microtubules. Some drugs used to treat cancer affect the microtubules that pull a cell's duplicated chromosomes apart, either by preventing tubulin from assembling into microtubules, or by preventing microtubules from breaking down into free tubulin dimers. In each case, cell division stops.

Microtubules also form structures called cilia, which are hairlike structures that move in a coordinated fashion, producing a wavelike motion. An individual cilium is constructed of nine microtubule pairs that sur-

round a central, separated pair. A type of motor protein called dynein connects the outer microtubule pairs and also links them to the central pair. Dynein supplies the energy to slide adjacent microtubules against each other, bending the cilium. Coordinated movement of cilia generates a wave that moves the cell or propels substances along its surface. For example, cilia beat particles up and out of respiratory tubules, and move egg cells in the female reproductive tract.

Another component of the cytoskeleton, the microfilament, is a long, thin rod composed of many molecules of the protein actin. In contrast to microtubules, microfilaments are solid and narrower. Microfilaments provide strength for cells to survive stretching and compression. They also help to anchor one cell to another, and provide

Faulty Ion Channels Cause Inherited Disease

What do collapsing horses, irregular heartbeats in teenagers, and cystic fibrosis have in common? All result from abnormal ion channels in plasma membranes.

Ion channels are protein-lined tunnels in the phospholipid bilayer of a biological membrane. These passageways permit electrical signals in the form of ions (charged particles) to pass through membranes.

Ion channels are specific for calcium (Ca^{+2}), sodium (Na^+), potassium (K^+), or chloride (Cl^-). A plasma membrane may have a few thousand ion channels for each of these ions. Ten million ions can pass through an ion channel in one second! The following disorders result from abnormal ion channels.

Hyperkalemic Periodic Paralysis and Sodium Channels

The quarterhorse was originally bred in the 1600s to run the quarter mile, but one of the four very fast stallions used to establish much of today's population of 3 million animals inherited *hyperkalemic periodic paralysis* (HPP) (OMIM 170500). The horse, a champion, collapsed from sudden attacks of weakness and paralysis.

HPP results from abnormal sodium channels in the plasma membranes of muscle cells. But the trigger for the temporary paralysis is another ion: potassium. A rising blood potassium level, which may follow intense exercise, slightly alters the electrical charge in the plasma membranes of muscle cells. Normally, this slight change would have no effect, but in horses with HPP, sodium channels open too widely, allowing too much sodium into muscle cells. The cells cannot respond to nervous stimulation for awhile, and the racehorse falls.

People can inherit HPP, too. In one family, several members collapsed suddenly after eating bananas! These fruits are very high in potassium, which caused the symptoms.

Long-QT Syndrome and Potassium Channels

Four children in a Norwegian family were born deaf, and three of them died at ages four, five, and nine. All of the children had inherited from unaffected carrier parents *long-QT syndrome associated with deafness* (OMIM 176261). ("QT" refers to part of a normal heart rhythm.) These children had abnormal potassium channels in the cells of the heart muscle and in the inner ear. In the heart cells, the malfunctioning ion channels disrupted electrical activity, causing a fatal disturbance to the heart rhythm. In the cells of the inner ear, the abnormal ion channels increased the extracellular concentration of potassium ions, impairing hearing. Some cases of long-QT syndrome are caused not by faulty ion channels, but by the proteins, called ankyrins, that hold the channels in place within the plasma membrane.

Cystic Fibrosis and Chloride Channels

A seventeenth-century English saying, "A child that is salty to taste will die shortly after birth," described the consequence of abnormal chloride channels in CF (OMIM 219700). The chloride channel is called CFTR, for cystic fibrosis transductance regulator. In most cases of CF, CFTR protein remains in the cytoplasm, unable to reach the plasma membrane, where it would normally function (**figure 1**). CF is inherited from carrier parents. The major symptoms of difficulty breathing, frequent severe respiratory infections, and a clogged pancreas that disrupts digestion all result from buildup of extremely thick mucous secretions.

Abnormal chloride channels in cells lining the lung passageways and ducts of the pancreas cause the symptoms of CF. The primary defect in the chloride channels also causes sodium channels to malfunction. The result: salt trapped inside cells draws moisture in and thickens surrounding mucus. **Figure 2** shows a parent helping a child cough up mucus. This is done twice a day.

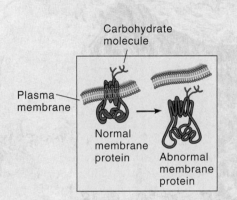

Figure 1 CFTR protein remains in the cytoplasm, rather than anchoring in the plasma membrane. This prevents normal chloride channel function.

Figure 2 Treating cystic fibrosis. In cystic fibrosis, the thick, sticky mucus that clogs airways must be coughed up at least twice every day.

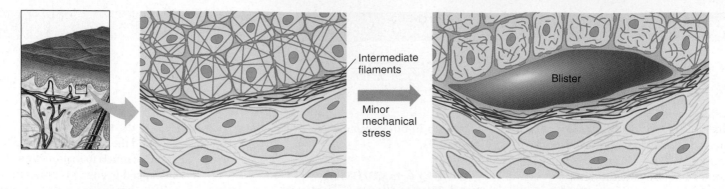

Figure 2.12 Intermediate filaments in skin. Keratin intermediate filaments internally support cells in the basal (bottom) layer of the epidermis. Abnormal intermediate filaments in the skin cause epidermolysis bullosa, in which skin easily blisters.

many other functions within the cell through proteins that interact with actin. When any of these proteins is absent or abnormal, a genetic disease results.

Intermediate filaments have diameters intermediate between those of microtubules and microfilaments, and are made of different proteins in different cell types. However, all intermediate filaments share a common overall organization of dimers entwined into nested coiled rods. Intermediate filaments are scarce in many cell types, but are very abundant in cells of the skin.

The intermediate filaments in actively dividing skin cells in the bottommost layer of the epidermis (the upper skin layer) form a strong inner framework that firmly attaches the cells to each other and to the underlying tissue. These cellular attachments are crucial to the skin's barrier function. In a group of inherited conditions called epidermolysis bullosa (OMIM 226500, 226650, 131750), intermediate filaments are abnormal. The skin blisters easily as tissue layers separate (**figure 2.12**).

Disruption in the structures of cytoskeletal proteins, or in how they interact, can be devastating. Consider hereditary spherocytosis, which disturbs the interface between the plasma membrane and the cytoskeleton in red blood cells.

The doughnut shape of normal red blood cells enables them to squeeze through the narrowest blood vessels. Their cytoskeletons provide the ability to deform. Rods of a protein called spectrin form a meshwork beneath the plasma membrane, strengthening the red blood cell, and ankyrin proteins attach the spectrin rods to the plasma membrane (**figure 2.13**). Spectrin molecules also attach to microfilaments and microtubules. Spectrin

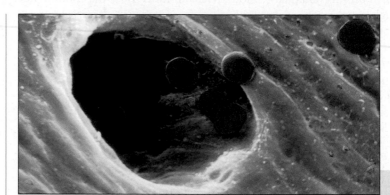

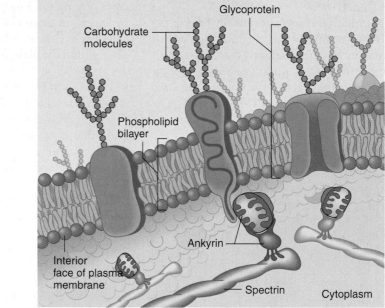

Figure 2.13 The red blood cell plasma membrane. The cytoskeleton that supports the plasma membrane of a red blood cell is specialized to withstand the great turbulent force of circulation. Proteins called ankyrins bind molecules of spectrin from the cytoskeleton to the inner membrane surface. On its other end, ankyrin binds proteins that help ferry molecules across the plasma membrane. In hereditary spherocytosis, abnormal ankyrin causes the plasma membrane to collapse, and the cell then balloons out—a problem for a cell whose function depends upon its shape. In the inset, red blood cells move from a large blood vessel into a much smaller capillary. A red blood cell travels about 900 miles during its four-month existence. Falsely colored scanning electron micrograph (1,400×).

molecules are like steel girders, and ankyrins are like nuts and bolts. If either molecule is absent, the red blood cell cannot maintain its shape and collapses.

In hereditary spherocytosis, the ankyrins are abnormal, and parts of the red blood cell plasma membrane disintegrate, causing the cell to balloon out. The bloated cells obstruct narrow blood vessels—especially in the spleen, the organ that normally disposes of aged red blood cells. Anemia develops as the spleen destroys red blood cells more rapidly than the bone marrow can replace them, producing great fatigue and weakness. Removing the spleen can treat the condition.

Key Concepts

1. Cells are the units of life. They consist mostly of carbohydrates, lipids, proteins, and nucleic acids.
2. Organelles subdivide specific cell functions. They include the nucleus, the endoplasmic reticulum (ER), Golgi apparatus, mitochondria, lysosomes, and peroxisomes.
3. The plasma membrane is a flexible, selective phospholipid bilayer with embedded proteins and lipid rafts.
4. The cytoskeleton is an inner structural framework made of protein rods and tubules, connectors and motor molecules.

2.2 Cell Division and Death

The cell numbers in a human body must be in balance to promote normal growth and development. The process of mitotic cell division, or **mitosis,** provides new cells by forming two cells from one. Mitosis occurs in **somatic cells** (all cells but the sperm and eggs). Some cells must die as a body forms, just as a sculptor must take away some clay to shape the desired object. A foot, for example, starts out as a webbed triangle of tissue; toes emerge as certain cells die. This type of cell death, which is a normal part of development, is termed **apoptosis,** from the Greek for leaves falling from a tree. Apoptosis is a precise, genetically programmed sequence of events, as is mitosis (**figure 2.14**).

Another form of cell death, called necrosis, is a response to injury. It is not part of normal development. Yet another form of cell death occurs in the breasts of a preg-

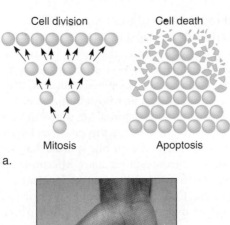

a.

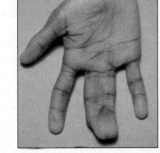

b.

Figure 2.14 Mitosis and apoptosis mold a body. Biological structures in animal bodies enlarge, allowing organisms to grow, as opposing processes regulate cell number. **(a)** Cell numbers increase from mitosis and decrease from apoptosis. **(b)** In the embryo, fingers and toes are carved from webbed structures. In syndactyly, normal apoptosis fails to carve digits, and webbing persists.

nant woman, when fatty tissue shrinks and milk-secreting, glandular tissue grows.

The Cell Cycle

Many cell divisions transform a fertilized egg into a many-trillion-celled person. A series of events called the **cell cycle** describes the sequence of activities as a cell prepares for division and then divides.

Cell cycle rate varies in different tissues at different times. A cell lining the small intestine's inner wall may divide throughout life; a cell in the brain may never divide; a cell in the deepest skin layer of a 90-year-old may divide more if the person lives long enough. Frequent mitosis enables the embryo and fetus to grow rapidly. By birth, the mitotic rate slows dramatically. Later, mitosis must maintain the numbers and positions of specialized cells in tissues and organs.

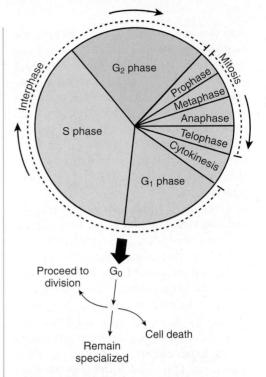

Figure 2.15 The cell cycle. The cell cycle is divided into interphase, when cellular components are replicated to prepare for division, and mitosis, when the cell splits, distributing its contents into two daughter cells. Interphase is divided into G_1 and G_2, when the cell duplicates specific molecules and structures, and a phase S, when it replicates DNA. Mitosis is divided into four stages plus cytokinesis, when the cells separate. G_0, is a "time-out" when a cell "decides" which course of action to follow.

The cell cycle is continual, but we divide it into stages based on what we see. The two major stages are **interphase** (not dividing) and mitosis (dividing) (**figure 2.15**). In mitosis, a cell duplicates its chromosomes, then apportions one set into each of two resulting cells, called daughter cells. This maintains the set of 23 chromosome pairs characteristic of a human somatic cell. Another form of cell division, meiosis, produces sperm or eggs, which have half the usual amount of genetic material, or 23 single chromosomes. Chapter 3 discusses meiosis.

Interphase—A Time of Great Activity

Interphase is a very active time, when the cell continues the basic biochemical functions of life and also replicates its DNA and other subcellular structures. It is divided into two gap (**G₁** and **G₂**) **phases** and one synthesis (**S**) **phase.** In addition, a cell can exit the cell cycle at G₁ to enter a quiescent phase called **G₀.** A cell in G₀ maintains its specialized characteristics but does not replicate its DNA or divide. From G₀, a cell may also proceed to mitosis and divide, or die. Apoptosis may ensue if the cell's DNA is so damaged that cancer might result. G₀ then, is when a cell's fate is either decided or put on hold.

During G₁, which follows mitosis, the cell resumes synthesis of proteins, lipids, and carbohydrates. These molecules will contribute to building the extra plasma membrane required to surround the two new cells that form from the original one. G₁ is the period of the cell cycle that varies the most in duration among different cell types. Slowly dividing cells, such as those in the liver, may exit at G₁ and enter G₀, where they remain for years. In contrast, the rapidly dividing cells in bone marrow speed through G₁ in 16 to 24 hours. Cells of the early embryo may skip G₁ entirely.

During S phase, the cell replicates its entire genome, so that each chromosome consists of two copies joined at an area called the **centromere.** In most human cells, S phase takes 8 to 10 hours. Many proteins are also synthesized during this phase, including those that form the mitotic **spindle** that will pull the chromosomes apart. Microtubules form structures called **centrioles** near the nucleus. Centriole microtubules are oriented at right angles to each other, forming paired oblong structures that organize other microtubules into the spindle.

G₂ occurs after the DNA has been replicated but before mitosis begins. More proteins are synthesized during this phase. Membranes are assembled from molecules made during G₁ and are stored as small, empty vesicles beneath the plasma membrane. These vesicles will merge with the plasma membrane to enclose the two daughter cells.

Mitosis—The Cell Divides

As mitosis begins, the replicated chromosomes are condensed enough to be visible, when stained, under a microscope. The two long strands of identical chromosomal material in a replicated chromosome are called **chromatids (figure 2.16).** At a certain point during mitosis, a replicated chromosome's centromere splits, allowing its chromatid pair to separate into two individual chromosomes. (Although the centromere of a replicated chromosome appears as a constriction, its DNA is replicated.)

During **prophase,** the first stage of mitosis, DNA coils tightly, shortening and thickening the chromosomes and enabling them to more easily separate **(figure 2.17).** Microtubules assemble from tubulin building blocks in the cytoplasm to form the spindles. Toward the end of prophase, the nuclear membrane breaks down. The nucleolus is no longer visible.

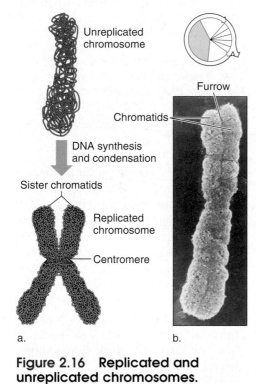

Unreplicated chromosome

DNA synthesis and condensation

Sister chromatids

Replicated chromosome

Centromere

Furrow

Chromatids

a.

b.

Figure 2.16 Replicated and unreplicated chromosomes.

Chromosomes are replicated during S phase, before mitosis begins. Two genetically identical chromatids of a replicated chromosome join at the centromere **(a).** In the photograph **(b),** a human chromosome is forming two chromatids.

Figure 2.17 Mitosis in a human cell. In a separate process, cytokinesis, the cytoplasm and other cellular structures distribute and pinch off into two daughter cells. (For simplicity, not all 23 chromosome pairs are depicted.)

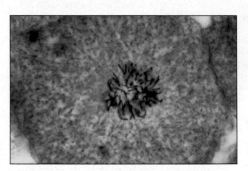

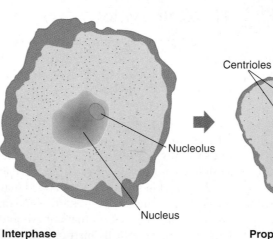

Nucleolus

Nucleus

Interphase
Chromosomes are uncondensed.

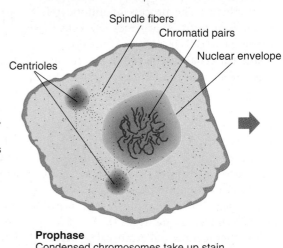

Spindle fibers

Chromatid pairs

Nuclear envelope

Centrioles

Prophase
Condensed chromosomes take up stain. The spindle assembles, centrioles appear, and the nuclear envelope breaks down.

Metaphase follows prophase. Chromosomes attach to the spindle at their centromeres and align along the center of the cell, which is called the equator. Metaphase chromosomes are under great tension, but they appear motionless because they are pulled with equal force on both sides, like a tug-of-war rope pulled taut.

Next, during **anaphase,** the plasma membrane indents at the center, where the metaphase chromosomes line up. A band of microfilaments forms on the inside face of the plasma membrane, constricting the cell down the middle. Then the centromeres part, which relieves the tension and releases one chromatid from each pair to move to opposite ends of the cell—like a tug-of-war rope breaking in the middle and the participants falling into two groups. Microtubule movements stretch the dividing cell. During the very brief anaphase stage, a cell temporarily contains twice the normal number of chromosomes because each chromatid becomes an independently moving chromosome, but the cell has not yet physically divided.

In **telophase,** the final stage of mitosis, the cell looks like a dumbbell with a set of chromosomes at each end. The spindle falls apart, and nucleoli and the membranes around the nuclei re-form at each end of the elongated cell. Division of the genetic material is now complete. Next, during **cytokinesis,** organelles and macromolecules are distributed between the two daughter cells. Finally, the microfilament band contracts like a drawstring, separating the newly formed cells.

Control of the Cell Cycle

When and where a somatic cell divides is crucial to health, and regulation of mitosis is a daunting task. Quadrillions of mitoses occur in a lifetime, and not at random. Too little mitosis, and an injury goes unrepaired; too much, and an abnormal growth forms.

Groups of interacting proteins function at times in the cell cycle called checkpoints to ensure that chromosomes are faithfully replicated and apportioned into daughter cells (**figure 2.18**). A "DNA damage checkpoint," for example, temporarily pauses the cell cycle while special proteins repair damaged DNA. An "apoptosis checkpoint" turns on as mitosis begins. During this checkpoint, proteins called survivins override signals telling the cell to die, ensuring that mitosis (division) rather than apoptosis (death) occurs. Later during mitosis, the "spindle assembly checkpoint" oversees construction of the spindle and the binding of chromosomes to it.

Cells obey an internal "clock" that tells them approximately how many times to divide. Mammalian cells grown (cultured) in a dish divide about 40 to 60 times. A connective tissue cell from a fetus, for example, will ultimately divide about 50 times. But a similar cell from an adult divides only 14 to 29 more times. That is, the number of divisions left declines with age.

How can a cell "know" how many divisions remain? The answer lies in the chromosome

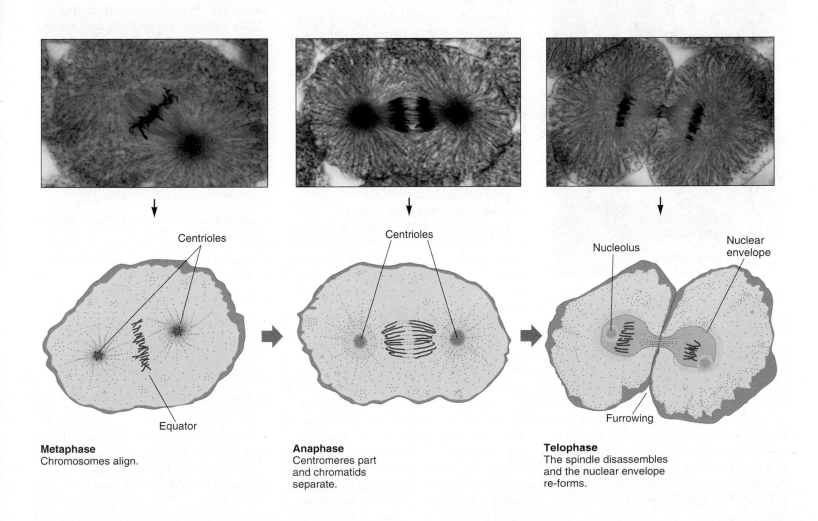

Metaphase
Chromosomes align.

Anaphase
Centromeres part
and chromatids
separate.

Telophase
The spindle disassembles
and the nuclear envelope
re-forms.

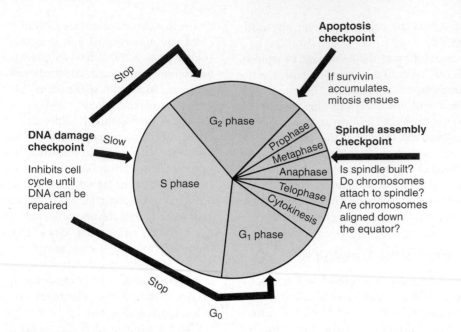

Figure 2.18 **Cell cycle checkpoints.** Checkpoints ensure that mitotic events occur in the correct sequence. Many types of cancer result from faulty checkpoints.

Figure 2.19 **Telomeres.** Fluorescent tags mark the telomeres in this human cell.

tips, called **telomeres (figure 2.19).** Telomeres function like a cellular fuse that burns down as pieces are lost from the ends. Telomeres consist of hundreds to thousands of repeats of a specific six DNA-base sequence. At each mitosis, the telomeres lose 50 to 200 endmost bases, gradually shortening the chromosome. After about 50 divisions, a critical length of telomere DNA is lost, which signals mitosis to stop. The

cell may remain alive but not divide again, or it may die.

Not all cells have shortening telomeres. In eggs and sperm, in cancer cells, and in a few types of normal cells that must continually supply new cells (such as bone marrow cells), an enzyme called telomerase keeps chromosome tips long (see figure 18.3). However, most cells do not produce telomerase, and their chromosomes gradually shrink.

Outside factors also affect a cell's mitotic clock. Crowding can slow or halt mitosis. Normal cells growing in culture stop dividing when they form a one-cell-thick layer lining the container. This limitation to division is called contact inhibition. If the layer tears, the cells that border the tear grow and divide, filling in the gap, but stop dividing once it is filled. Perhaps a similar mechanism in the body limits mitosis.

Chemical signals control the cell cycle from outside as well as from inside the cell. Hormones and growth factors are biochemicals from outside the cell that influence mitotic rate. A hormone is a substance synthesized in a gland and transported in the bloodstream to another part of the body, where it exerts a specific effect. Hormones secreted in the brain, for example, signal the cells lining a woman's uterus to build up each month by mitosis in preparation for possible pregnancy. Growth factors act

more locally. Epidermal growth factor, for example, stimulates cell division beneath a scab.

Two types of proteins, cyclins and kinases, interact inside cells to activate the genes whose products carry out mitosis. The two types of proteins form pairs. Cyclin levels fluctuate regularly throughout the cell cycle, while kinase levels stay the same. A certain number of cyclin-kinase pairs turn on the genes that trigger mitosis. Then, as mitosis begins, enzymes degrade the cyclin. The cycle starts again as cyclin begins to build up during the next interphase.

Apoptosis

Apoptosis rapidly and neatly dismantles a cell into membrane-enclosed pieces that a phagocyte (a cell that engulfs and destroys another) can mop up. It is a little like taking the contents of a messy room and packaging them into garbage bags—then disposing of it all. In contrast is necrosis, a form of cell death associated with inflammation, rather than an orderly, contained destruction.

Like mitosis, apoptosis is a continuous, stepwise process. It begins when a "death receptor" on the doomed cell's plasma membrane receives a signal to die. Within seconds, enzymes called caspases are activated inside the cell, stimulating each other and snipping apart various cell components. These killer enzymes:

- Destroy the cytoskeletal threads that support the nucleus so that it collapses, condensing the DNA within.

- Demolish enzymes that replicate and repair DNA.

- Activate enzymes that cut DNA into similarly sized small pieces.

- Tear apart the cytoskeleton.

- Cause mitochondria to release molecules that trigger further caspase activity, end the cell's energy supply, and destroy these organelles.

- Abolish the cell's ability to adhere to other cells.

- Send a certain phospholipid from the plasma membrane's inner face to its outer surface, where it attracts phagocytes that dismantle the cell remnants.

A dying cell has a characteristic appearance (**figure 2.20**). It rounds up as contacts with other cells are cut off, and the plasma membrane undulates and forms bulges called blebs. The nucleus bursts, releasing same-sized DNA pieces. Mitochondria decompose. Then the cell shatters. Almost instantly, pieces of membrane encapsulate the cell fragments, which prevents inflammation. Within an hour, the cell is gone.

From the embryo onward through development, mitosis and apoptosis are synchronized, so that tissue neither overgrows nor shrinks. In this way, a child's liver retains much the same shape as she grows into adulthood. During early development, mitosis and apoptosis orchestrate the ebb and flow of cell number as new structures form. Later, these processes protect. Mitosis produces new skin to heal a scraped knee; apoptosis peels away sunburnt skin cells that might otherwise become cancerous. Cancer is a profound derangement of the balance between cell division and death, with mitosis occurring too frequently or too many times, or apoptosis too infrequently. Chapter 18 discusses cancer in detail.

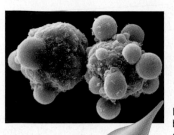

Death receptor on doomed cell binds signal molecule. Caspases are activated within.

Caspases destroy various proteins and other cell components. Cell undulates.

Blebs
Cell fragments
Phagocyte attacks and engulfs cell remnants. Cell components are degraded.

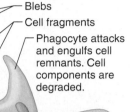

Figure 2.20 Death of a cell. A cell undergoing apoptosis loses its characteristic shape, forms blebs, and finally falls apart. Caspases destroy the cell's insides. Phagocytes digest the remains. Note the blebs on the dying liver cells in the first photograph. Sunburn peeling is one example of apoptosis.

Key Concepts

1. Mitosis and apoptosis regulate cell numbers during development, growth, and repair.
2. The cell cycle includes interphase and mitosis. During G_0, the cell "decides" to divide, die, or stay differentiated. Interphase includes two gap (G) phases and a synthesis (S) phase that prepares the cell for mitosis. During S phase, DNA is replicated. Proteins, carbohydrates, and lipids are synthesized during G_1 and more proteins are synthesized in G_2. During prophase, metaphase, anaphase, and telophase, replicated chromosomes condense, align, split, and distribute into daughter cells.
3. The cell cycle is controlled by checkpoints; telomeres; hormones and growth factors from outside the cell; and cyclins and kinases from within.
4. During apoptosis, cells receive a death signal, activate caspases, and break apart in an orderly fashion.

2.3 Cell-Cell Interactions

Precisely coordinated biochemical steps orchestrate the cell-cell interactions that make multicellular life possible. Defects in cell communication and interaction cause certain inherited illnesses. Two broad types of interactions among cells are signal transduction and cellular adhesion.

Signal Transduction

In **signal transduction,** molecules on the plasma membrane assess, transmit, and amplify incoming messages to the cell's interior. *Transduce* means to change one form of something (such as energy or information) into another. In signal transduction, the cell changes various types of stimuli into specific biochemical reactions. Some signal molecules must bind receptors for the cell to function normally; others, such as a signal to divide when cell division is not warranted, must be ignored.

Signal transduction is carried out by the interaction between cytoplasmic proteins and proteins embedded in the plasma membrane that extend from one or both faces. The process begins at the cell surface. First, a receptor binds an incoming molecule, called the first messenger, and contorts in a way that touches a nearby protein called a regulator (**figure 2.21**). Next, the regulator activates a nearby enzyme, which catalyzes (speeds) a specific chemical reaction. The product of this reaction is called the second messenger, and it lies at the crux of the entire process; it elicits the cell's response, typically by activating certain enzymes. A single stimulus can trigger the production of many second messenger molecules, and this is how signal transduction amplifies incoming information. Because cascades of proteins carry out signal transduction, it is a genetically controlled process.

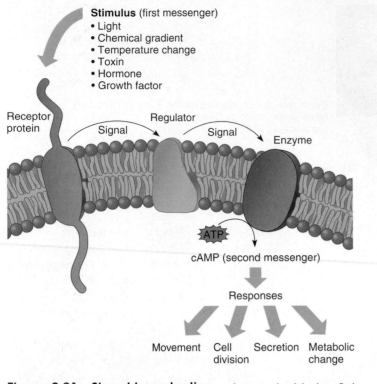

Stimulus (first messenger)
• Light
• Chemical gradient
• Temperature change
• Toxin
• Hormone
• Growth factor

Receptor protein

Signal   Regulator   Signal

Enzyme

ATP

cAMP (second messenger)

Responses

Movement   Cell division   Secretion   Metabolic change

Figure 2.21   Signal transduction.   A receptor binds a first messenger, triggering a cascade of biochemical activity at the cell's surface. An enzyme catalyzes a reaction inside the cell that circularizes ATP to cyclic AMP, the second messenger. cAMP then stimulates various responses, such as cell division, metabolic changes, and muscle contraction. Splitting ATP also releases energy.

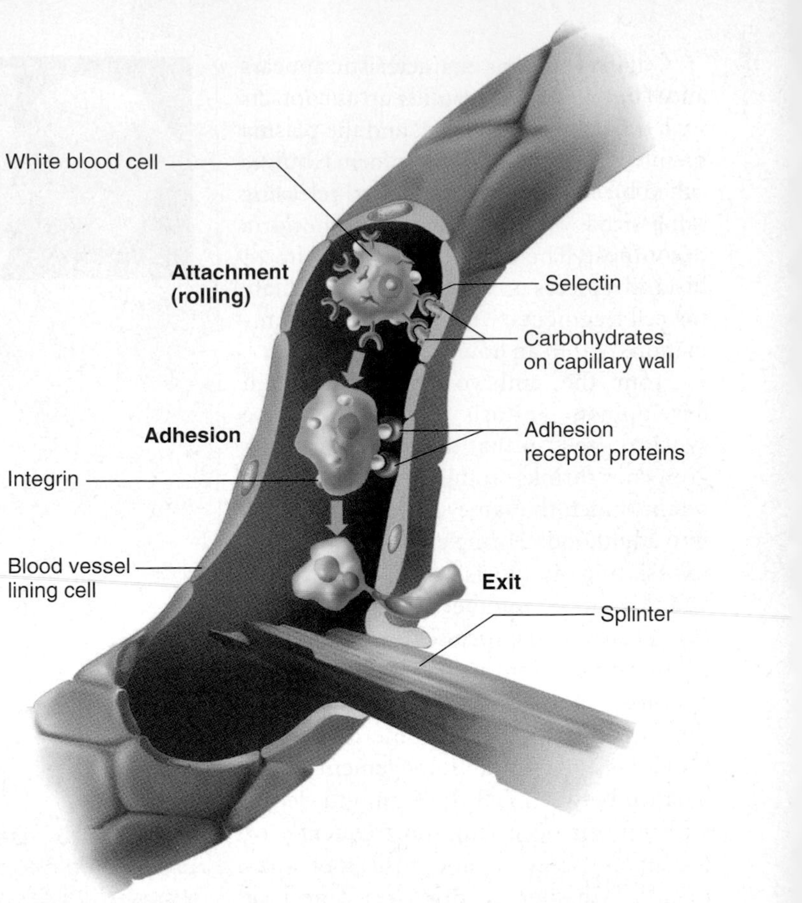

White blood cell

Attachment (rolling)

Adhesion

Integrin

Blood vessel lining cell

Selectin

Carbohydrates on capillary wall

Adhesion receptor proteins

Exit

Splinter

Figure 2.22   Cellular adhesion.   Cellular adhesion molecules (CAMs), including selectins, integrins, and adhesion receptor proteins, direct white blood cells to injury sites.

Defects in signal transduction underlie many inherited disorders. In neurofibromatosis type 1 (NF1) (OMIM 162200), for example, tumors (usually benign) grow in nervous tissue, particularly under the skin. At the cellular level, NF1 occurs when cells fail to block transmission of a growth factor signal that triggers cell division. Affected cells misinterpret the signal and divide when it is inappropriate.

RELATING THE CONCEPTS

How is signal transduction part of the healing that occurred in Michael M.'s eye and in the mice with retinitis pigmentosa?

Cellular Adhesion

Cellular adhesion is a precise sequence of interactions among the proteins that join cells. Inflammation—the painful, red swelling at a site of injury or infection— illustrates one type of cellular adhesion. Inflammation occurs when white blood cells (leukocytes) move in the circulation to the injured or infected body part, where they squeeze between cells of the blood vessel walls to reach the site. Cellular adhesion molecules, or CAMs, help guide white blood cells to the injured area.

Three types of CAMs carry out the inflammatory response (**figure 2.22**). First, selectins attach to the white blood cells, and slow them to a roll by also binding to carbohydrates on the capillary wall. (This is a little like putting out your arms to slow your ride down a slide.) Next, clotting blood, bacteria, or decaying tissues release chemical attractants that signal white blood cells to stop. The chemical attractants activate CAMs called integrins, which latch onto the white blood cells, and CAMs called adhesion receptor proteins, which extend from the capillary wall at the injury site. The integrins and adhesion receptor proteins then guide the white blood cells between the tile-like lining cells to the injury site.

If the signals that direct white blood cells to injury sites fail, a condition called leukocyte-adhesion deficiency (OMIM 116920) results. The first symptom is often teething sores that do not heal. These and other small wounds never accumulate the pus (bacteria, cellular debris, and white blood cells) that indicates the body is fighting infection. The person lacks the CAMs that enable white blood cells to stick to blood vessel walls, and so blood cells zip right past wounds. An affected individual must avoid injury and infection, and receive anti-infective treatments for even the slightest wound.

More common disorders may also reflect abnormal cellular adhesion. Cancer cells journey easily from one part of the body to another thanks to impaired cellular adhesion. Arthritis may occur when the wrong adhesion molecules rein in white blood cells, inflaming a joint where no injury exists.

Cellular adhesion is critical to many other functions. CAMs guide cells surrounding an embryo to grow toward maternal cells and form the placenta, the supportive organ linking a pregnant woman to the fetus. Sequences of CAMs also help establish connections among the nerve cells that underlie learning and memory.

Key Concepts

1. In signal transduction, cell surface receptors receive information from first messengers (stimuli) and pass them to second messengers, which then trigger a cellular response.
2. Cellular adhesion molecules (CAMs) guide white blood cells to injury sites using a sequence of cell-protein interactions.

2.4 Stem Cells and Cell Specialization

Bodies grow and heal thanks to cells that retain the ability to divide, generating both new cells like themselves and cells that go on to specialize. **Stem cells** and **progenitor cells** renew tissues so that as the body grows, or loses cells to apoptosis, injury, and disease, other cells arise to take their places.

Cell Lineages

A stem cell divides by mitosis to yield either two daughter cells that are stem cells like itself, or one that is a stem cell and one that is a partially specialized progenitor cell (**figure 2.23**). The characteristic of self-renewal is what makes a stem cell a stem cell—its ability to continue the lineage of cells that can divide to give rise to another cell like itself. A progenitor cell's daughters usually specialize as any of a restricted number of cell types. A fully differentiated cell, such as a mature blood cell, descends from a sequence of increasingly specialized progenitor cell intermediates, each one less like a stem cell and more like a blood cell. Our 260 or so differentiated cell types develop from lineages of stem and progenitor cells. **Figure 2.24** shows parts of a few lineages.

Stem cells and progenitor cells are described in terms of developmental poten-

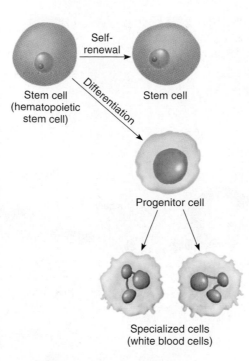

Figure 2.23 Stem cells and progenitor cells. A stem cell is less specialized than the progenitor cell that descends from it by mitosis. Various types of stem cells provide the raw material for producing the specialized cells that comprise tissues, while retaining the ability to generate new cells. A hematopoietic stem cell resides in the bone marrow and can produce progenitors whose daughter cells may specialize as certain blood cell types. Hematopoietic stem cells seem to have the ability to travel and produce daughter cells that differentiate in a number of ways.

tial—that is, according to the number of possible fates of their daughter cells. A fertilized ovum and the cells of the very early embryo, when it is just a small ball of identical-appearing cells, are totipotent, which means that they can give rise to every cell type. In contrast, stem cells that persist until later in development and progenitor cells are pluripotent: Their daughter cells have fewer possible fates. This is a little like a freshman's consideration of many majors, compared to a junior's more narrowed focus in selecting courses.

As cells specialize, they express some genes and ignore others. An immature bone cell forms from a progenitor cell by manufacturing mineral-binding proteins and enzymes. In contrast, an immature

muscle cell forms from a muscle progenitor cell that accumulates contractile proteins. The bone cell does not produce muscle proteins, nor does the muscle cell produce bone proteins. All cells, however, synthesize proteins for basic "housekeeping" functions, such as energy acquisition and protein synthesis.

Many, if not all, of the organs in an adult human body harbor stem or progenitor cells that can begin to divide when injury or illness occurs and new cells are needed to replace damaged ones. Stem cells in the adult may have been set aside in the embryo or fetus in particular organs as repositories of future healing. Alternatively, or perhaps also, stem cells or progenitor cells may travel from the bone marrow to replace damaged or dead cells in response to signals that are released when injury or disease occurs. Some stem and progenitor cells are actually more "plastic" than researchers had first thought—for example, hematopoietic stem cells in bone marrow can form not only blood cells, but also nerve, muscle, liver, and blood vessel lining cells, under certain conditions. Because every cell contains all of an individual's genetic material, it is theoretically possible that, given appropriate signals, any cell type can become any other. But this may not occur naturally, or may only happen under unusual conditions, such as catastrophic injury.

Using Embryos

Physicians are beginning to use stem cells to treat particular disorders or injuries. Using stem cells to heal is one type of "regenerative medicine," which replaces damaged tissue with materials that include cells that can divide.

Stem cells that may be used in regenerative medicine have several sources. Although rare, stem cells probably exist in all organs. They can be derived from the earliest embryos through the elderly, and even from corpses and medical waste, such as the fatty material discarded after liposuction and surgically removed organs. The most promising cells for therapy, according to many biologists, are embryonic stem (ES) cells. These are obtained and cultured from a 5-day embryo, called a blastocyst. It is a hollow ball of cells with a few cells, comprising

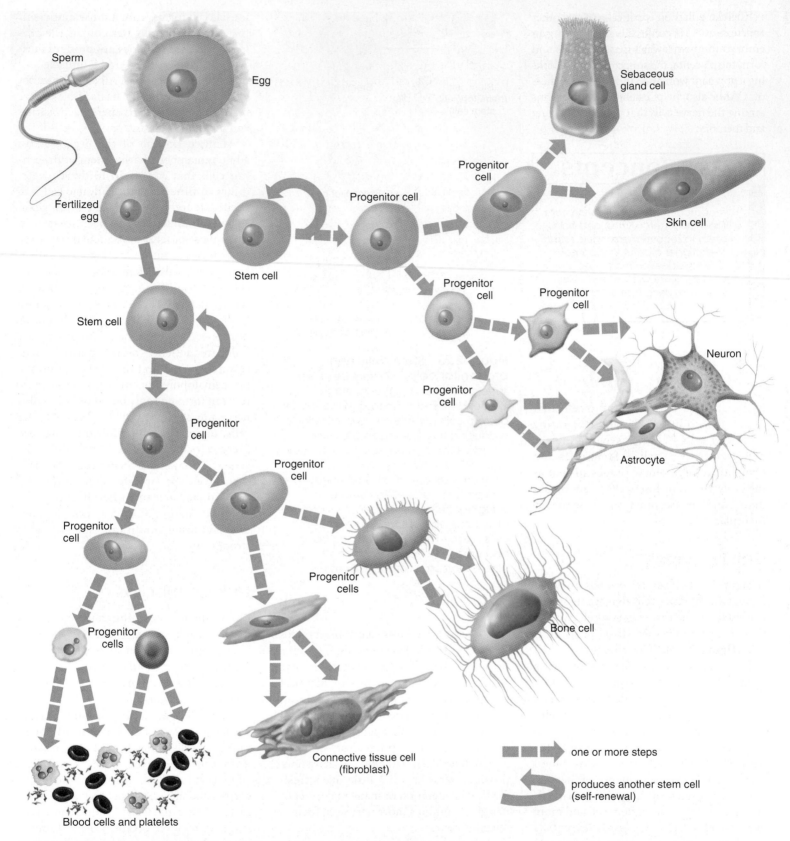

Figure 2.24 Pathways to cell specialization. All cells in the human body descend from stem cells, through the processes of mitosis and differentiation. The differentiated cells on the left are all connective tissues (blood, connective tissue, and bone), but the blood cells are more closely related to each other than they are to the other two cell types. On the right, the skin and sebaceous gland cells share a recent progenitor, and both share a more distant progenitor with neurons and supportive astrocytes. Imagine how complex the illustration would be if it embraced all 260-plus types of cells in a human body!

a structure called the inner cell mass (ICM), on the inside. ICM cells, given appropriate biochemical signals, can divide to give rise to totipotent ES cells, which can generate any cell type. In contrast is the more limited repertoire of possibilities for stem or progenitor cells in tissues. ES cell-derived cells are also less likely to provoke rejection by the recipient's immune system.

ES cells come from two sources. One is to use embryos from fertility clinics where couples undergoing *in vitro* (test tube) fertilization have frozen the extras. This approach could create banks of cell types not precisely matched to a particular individual. "Typing" would have to be done, as it is for transfusions and transplants. Because the embryos already exist and are destined for discard, many people argue that they can ethically be used to develop treatments for conditions such as Parkinson disease and spinal cord injury.

A second source of ES cells is to create an embryo using the nucleus from a somatic cell from a patient, such as a person who has suffered a spinal cord injury (**figure 2.25**). The nucleus is injected into or fused with a donated egg cell whose nucleus has been removed. The resulting cell—not a fertilized egg because no sperm is involved—develops for 5 days. ICM cells are then removed and cultured to yield ES cells, then given growth factors to differentiate as needed—such as to patch a spinal cord injury.

Researchers in Korea have performed the early part of this procedure, which is called **somatic cell nuclear transfer** (SCNT). They introduced nuclei from somatic cells from people with spinal cord injuries, type 1 diabetes, or an inherited immune deficiency into oocytes whose nuclei had been removed, then chemically activated the cells to divide. They obtained ICM cells and cultured them to form ES cells. Ultimately, if researchers can learn how to guide the cells

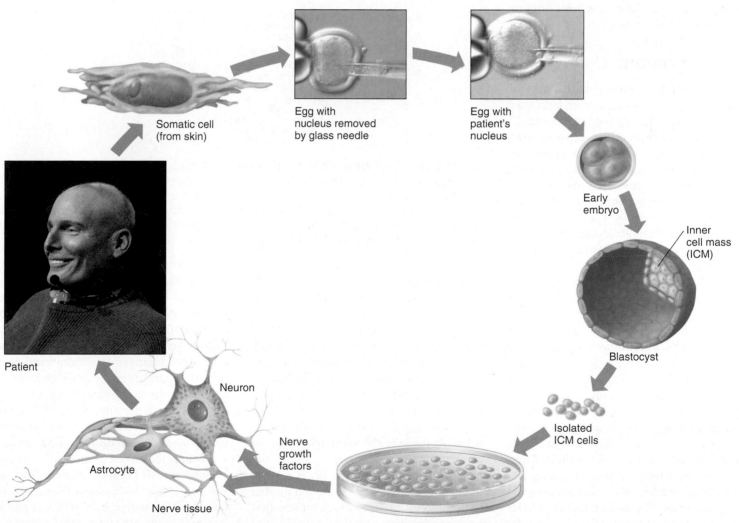

Figure 2.25   Somatic cell nuclear transfer yields embryonic stem cells genetically matched to a patient.

A new way to possibly treat degenerative diseases and injuries is to culture cells whose nuclei come from a patient's own cells, and use the new cells to replace diseased or damaged cells. The immune system would not reject these cells, because they contain the patient's genome. An alternative approach to treat nervous system problems is to use neural stem cells taken from cadavers, but these would not match cells from the patient. The prospective patient in this illustration is the late actor Christopher Reeve, who had a spinal cord injury.

to replace diseased or injured tissues without overgrowing, the person's body will theoretically accept the cells because they are a genetic match. SCNT is sometimes called "therapeutic cloning" because clones are genetically identical cells or individuals. In contrast, "reproductive cloning" would create a new individual (see Bioethics: Choices for the Future 3.1).

Using stem cells from fertility clinic "leftovers" or SCNT is controversial. Some people object to using existing embryos because they were created to become children. Either approach prevents a blastocyst from developing further. Nations vary in their policies: Some permit both ways to obtain ES cells, some allow one only, and others ban or restrict government funding for either or both approaches.

Using Somatic Cells

Researchers are developing treatments that use cells that can be taken from an individual without harm. (It is not quite accurate to call these cells "adult stem cells" because such tissue-based stem cells are also present in embryos and fetuses.) Using somatic stem cell implants is not new. Bone marrow transplants have delivered hematopoietic stem cells for half a century. Today, using stem cells from stored umbilical cord blood is routine in treating a variety of blood disorders.

Continuing basic research on stem cells from adults may lead to novel clinical applications. For example, new drugs to treat breast cancer seek the rare stem cells tucked between the lining and muscular layers of the milk ducts, where cancer begins.

Hearts were once thought to be unrepairable, but recent stem cell discoveries have changed that view (**figure 2.26**). When several men who had received heart transplants from women died, autopsies showed that their hearts had differentiated cells with Y chromosomes, which indicated that the cells came from the male recipient. The new "heart patches" also included progenitor cells marked with the telltale Y. The men's bodies had recruited their own progenitor cells to become part of the new heart. These cells came either from the bone marrow, or from stem cells in the bit of their own tissue to which the new heart was stitched. Perhaps a smaller female heart, stressed in a sick man's chest cavity, released

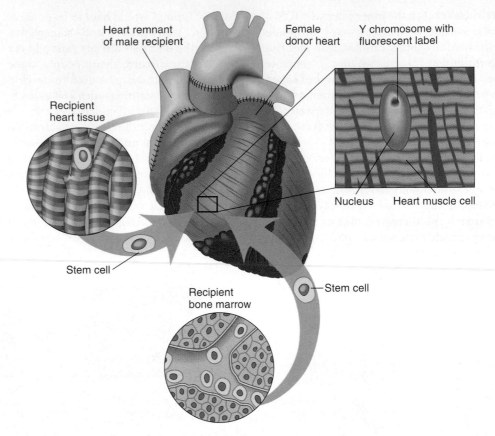

Figure 2.26 Can stem cells heal hearts? A study of the hearts of eight men who had received heart transplants from women revealed that stem cells from the recipients' bodies helped to accept the new organs. The stem cells came from the bone marrow or from remnants of the men's hearts—or both. The new hearts contained progenitor cells and special differentiated cells that had Y chromosomes—meaning that they must have come from the male recipients.

growth factors and other signals that activated the recipient's stem and progenitor cells.

The heart transplant study inspired experimental treatment for a 16-year-old who was shot in the chest with a nail gun. Physicians gave him a drug to coax his bone marrow to produce stem cells that could travel to the heart. The young man improved and did not require the transplant. Apparently the stem cells induced blood vessel growth in and around the heart, rather than replacing heart muscle. Similarly, researchers have coaxed human ES cells to give rise to heart muscle that beats and has been genetically modified to induce blood vessel formation—in a laboratory dish. These "heart patches" are being tested in pigs to aid recovery from heart attack.

It was also once thought that brains could not add cells. Then researchers discovered that new neurons appear in the brains of birds when they learn songs. Looking at animals more closely related to humans, researchers used a stain called BrdU to reveal small clusters of neural stem cells in the brains of tree shrews and marmosets. These stem cells can give rise to daughter cells that differentiate into neurons or the cells that support them. Demonstrating the existence of stem cells in human brains proved more difficult, since no one was willing to provide a sample of what one researcher calls "brain marrow." Then, in the late 1990s, researchers found volunteers—people with cancer of the tongue or larynx being treated with BrdU, which is a cancer drug. After death, their brains revealed pockets of neural stem cells stained with BrdU. Researchers may one day be able to treat spinal cord injuries and conditions such as Parkinson disease or multiple sclerosis by coaxing a person's own neural stem cells to heal the damage. These stem cells give rise to neurons as well as the glial cells that also form nervous tissue.

Stem cell technology may be used to treat less serious conditions, too. For example, a single stem cell in the skin gives rise to skin cells, hair follicle cells, and sebaceous (oil) gland cells. Manipulating these stem cells might lead to treatments for baldness, acne, and hair removal.

We still have much to learn about stem cells. For example, what makes a stem cell a stem cell? Researchers have identified a set of genes that must be expressed to impart a state of "stemness" to a cell, many of which are involved in signal transduction. Because most of the genes are also expressed in non-stem cells, it appears that "stemness"

arises from a combination of genes expressed at a particular time, as well as from the actions of a few distinctive genes. As analysis of the human genome continues, researchers will more precisely define the genetic functions that enable a cell to retain developmental potential—essential to building and maintaining bodies, the subject of the next chapter.

RELATING THE CONCEPTS

Why was the stem cell treatment that restored Michael M.'s vision a more lasting cure than the treatment for retinitis pigmentosa in mice?

Key Concepts

1. All cells descend from progenitor and stem cells, most of which are pluripotent. The fertilized egg and cells of the early embryo are totipotent.
2. Differential gene expression underlies cell specialization.
3. Stem cells exist at all stages of development and throughout the body.
4. Embryonic stem cells are the most promising for regenerative medicine. They derive from fertilized ova stored at fertility clinics and from somatic cell nuclear transfer.
5. Stem cells from adults may have a variety of medical applications.

Summary

2.1 The Components of Cells

1. Cells are the fundamental units of life and comprise the human body. Inherited traits and illnesses can be understood at the cellular and molecular levels.

2. All cells share certain features, but they are also specialized because they express different subsets of genes. Cells consist primarily of water and several types of macromolecules: **carbohydrates, lipids, proteins,** and **nucleic acids.**

3. The three domains of life—Archaea, Bacteria, and Eukarya—have characteristic cells. The archaea and bacteria are simple, small, and lack **nuclei** and other **organelles. Eukaryotic** cells have organelles, and their genetic material is contained in a nucleus.

4. Organelles sequester related biochemical reactions, improving efficiency of life functions and protecting the cell. The cell also consists of **cytoplasm** and other chemicals.

5. The nucleus contains DNA and a nucleolus, which is a site of ribosome synthesis. **Ribosomes** provide scaffolds for protein synthesis; they exist free in the cytoplasm or complexed with the **rough endoplasmic reticulum** (ER).

6. In secretion, the rough ER is the site of protein synthesis and folding, the smooth ER is the site of lipid synthesis, transport, and packaging, and the **Golgi apparatus** packages secretions into vesicles, which

exit through the **plasma membrane. Lysosomes** contain enzymes that dismantle debris, and **peroxisomes** house enzymes that perform a variety of functions. Enzymes in **mitochondria** extract energy from nutrients.

7. The plasma membrane is a protein-studded phospholipid bilayer. It controls which substances exit and enter the cell, and how the cell interacts with other cells.

8. The **cytoskeleton** is a protein framework of hollow microtubules, made of tubulin, and solid microfilaments, which consist of actin. Intermediate filaments are made of more than one protein type and are abundant in skin. The cytoskeleton and the plasma membrane distinguish different types of cells.

2.2 Cell Division and Death

9. Coordination of cell division (**mitosis**) and cell death (**apoptosis**) maintains cell numbers, enabling structures to enlarge during growth and development but preventing abnormal growth.

10. The **cell cycle** describes whether a cell is dividing (mitosis) or not (**interphase**). Interphase consists of two gap phases, when proteins and lipids are produced, and a synthesis phase, when DNA is replicated.

11. Mitosis proceeds in four stages. In **prophase,** replicated chromosomes consisting of two **chromatids** condense,

the **spindle** assembles, the nuclear membrane breaks down, and the nucleolus is no longer visible. In **metaphase,** replicated chromosomes align along the center of the cell. In **anaphase,** the **centromeres** part, equally dividing the now unreplicated chromosomes into two daughter cells. In **telophase,** the new cells separate. Cytokinesis apportions other components into daughter cells.

12. Internal and external factors control the cell cycle. Checkpoints are times when proteins regulate the cell cycle. **Telomere** (chromosome tip) length determines how many more mitoses will occur. Crowding, hormones, and growth factors signal cells from the outside; the interactions of cyclins and kinases trigger mitosis from inside.

13. In apoptosis, a receptor on the plasma membrane receives a death signal, then activates caspases that tear apart the cell in an orderly fashion. Membrane surrounds the pieces, preventing inflammation.

2.3 Cell-Cell Interactions

14. In **signal transduction,** a stimulus (first messenger) activates a cascade of action among membrane proteins, culminating in production of a second messenger that turns on enzymes that provide the response.

15. Cellular adhesion molecules enable cells to interact. Selectins slow the movement of

leukocytes, and integrins and adhesion receptor proteins guide the blood cell through a capillary wall to an injury site.

2.4 Stem Cells and Cell Specialization

16. **Stem cells** produce daughter cells that retain the ability to divide and that specialize in particular ways.

17. Totipotent stem cells can become anything. Pluripotent stem cells can differentiate as any of a variety of cell types. **Progenitor cells** can specialize as any of a restricted number of cell types.

18. Embryonic stem (ES) cells have more medical applications and are less likely to be rejected than stem cells from somatic tissues.

19. ES cells can be obtained from existing embryos (IVF "leftovers") or be tailor-made (through **somatic cell nuclear transfer**).

20. Researchers are developing ways to use the body's stem and progenitor cells to heal.

Review Questions

1. Match each organelle to its function.

Organelle	Function
a. lysosome	1. lipid synthesis
b. rough ER	2. houses DNA
c. nucleus	3. energy extraction
d. smooth ER	4. dismantles debris
e. Golgi apparatus	5. detoxification
f. mitochondrion	6. protein synthesis
g. peroxisome	7. processes secretions

2. Explain the functions of the following proteins:
   a. tubulin and actin
   b. caspases
   c. cyclins and kinases
   d. checkpoint proteins
   e. cellular adhesion molecules

3. List four types of controls on cell cycle rate.

4. How can all of a person's cells contain exactly the same genetic material, yet specialize as bone cells, nerve cells, muscle cells, and connective tissue cells?

5. Distinguish between
   a. a bacterial cell and a eukaryotic cell.
   b. interphase and mitosis.
   c. mitosis and apoptosis.
   d. rough ER and smooth ER.
   e. microtubules and microfilaments.
   f. a stem cell and a progenitor cell.
   g. totipotent and pluripotent.

6. Select a process described in the chapter (such as signal transduction or apoptosis). List the steps and state why the cell could not survive without this ability.

7. How are intermediate filaments similar to microtubules and microfilaments, and how are they different?

8. What advantage does compartmentalization provide to a large and complex cell?

9. What role does the plasma membrane play in signal transduction?

10. Explain how stem cells obtained from IVF leftovers and somatic cell nuclear transfer differ in terms of their genomes.

Applied Questions

1. How might abnormalities in each of the following contribute to cancer?
   a. cellular adhesion
   b. signal transduction
   c. balance between mitosis and apoptosis
   d. cell cycle control
   e. telomerase activity

2. Why do many inherited conditions result from defective enzymes?

3. In neuronal ceroid lipofuscinosis, a child experiences seizures, loss of vision, and lack of coordination, and dies. The body lacks an enzyme that normally breaks down certain proteins, causing them to accumulate and destroying the nervous system. Name two organelles that could be affected in this illness.

4. How do stem cells maintain their populations within tissues that consist of mostly differentiated cells?

5. Explain why mitosis that is too frequent or too infrequent, or apoptosis that is too frequent or too infrequent, can endanger health.

6. Why wouldn't a cell in an embryo likely be in phase G_0?

7. A defect in which organelle would cause fatigue?

8. Describe three ways that drugs can be used to treat cancer, based on disrupting microtubule function, telomere length, and signal transduction.

9. How can signal transduction, the plasma membrane, and the cytoskeleton function together?

10. What abnormality at the cellular or molecular level lies behind each of the following disorders?
    a. cystic fibrosis
    b. adrenoleukodystrophy
    c. neurofibromatosis type 1
    d. leukocyte adhesion deficiency
    e. syndactyly

11. A child with sickle cell disease endures periods of crisis, when circulation becomes painfully poor, starving parts of the body of oxygen. The blood of a child in crisis contains many more stem cells, sent from the bone marrow, than does the blood of a child not in crisis. What does this suggest about stem cell function?

12. The thymus gland in the chest manufactures white blood cells that protect against infection. It begins to shrink in adolescence. Researchers have discovered that a single variety of stem cell can, in a dish, be stimulated to regrow a thymus. List the steps to use somatic cell nuclear transfer to create a thymus gland to help a person suffering from AIDS.

Web Activities

Visit the Online Learning Center (OLC) at www.mhhe.com/lewisgenetics7. Select **Student Edition, chapter 2** and **Web Activities** to find the website links needed to complete the following activities.

13. The Coalition for the Advancement of Medical Research includes scientists, foundations, and patients advocating stem cell research for regenerative medicine. Consult the website provided on the OLC to learn the latest news on legislative efforts to either ban or spare stem cell research, and explain which types of research the bills would and would not allow.

14. Select ten nations and, using a web search engine, research whether they allow use of IVF leftovers to obtain human ES cells, somatic cell nuclear transfer to obtain the cells, neither, or both.

Case Studies and Research Results

15. Anthony Wright and Julia Green are 32-year-old participants in a clinical trial to evaluate stem cell therapy for multiple myeloma, a cancer of certain bone marrow cells. They are assigned to different treatment groups. Both receive standard therapy, which is alpha interferon. Anthony's group also receives conventional doses of four chemotherapeutic drugs, which kill any rapidly dividing cells. Julia's group's treatment is more drastic. She receives extremely high doses of the four drugs, enough to kill her bone marrow cells, then she has an umbilical cord stem cell transplant, which is an infusion of the material in a vein in her arm. Julia survives a year longer (5 years total) than Anthony, and the other people in the study have similar outcomes—those receiving stem cells live longer. The conclusion: Stem cell therapy is more effective for treating multiple myeloma than combination chemotherapy alone.

   a. It takes several weeks for Julia and Anthony to recover from the effects of the treatments. Which would have been more likely to have suffered serious side effects? Cite a reason for your answer.

   b. Julia's father reads about the experimental treatment in the newspaper and is outraged. "It is unethical to use embryos as spare parts," he tells her. How has he misunderstood the procedure?

   c. Why might Julia have eventually relapsed?

Learn to apply the skills of a genetic counselor with these additional cases found in the *Case Workbook for Human Genetics*:

   Carnitine-acylcarnitine translocase deficiency

   Combined factors V and VIII deficiency

Development

CYCLOPS AND CHICKENS

In Greek mythology, a Cyclops was a one-eyed creature that was the first blacksmith. It triggered volcanoes and gave the god Zeus thunderbolts on Mt. Olympus. A real Cyclops can result from a severe birth defect called holoprosencephaly or HPE (OMIM 236100).

HPE begins in the embryo, when the front of the brain (the forebrain) fails to grow forward and divide into two lobes, or hemispheres. Instead, various brain and facial structures remain fused or never form. Severity depends upon when normal forebrain development halts. The most severe form, cyclopia, arises earliest. The forebrain remains one large mass, with a single large eye in the middle of the face and a tubelike, closed nose on the forehead. Severe HPE affects 1 in 16,000 live births, but appears in 1 in 200 spontaneous abortions, indicating that most affected individuals are never born. In the moderate form, the forebrain partially divides, and facial abnormalities include missing teeth, cleft lip and/or palate, close-set eyes, and a small, flat nose with a single nostril. The severe and moderate forms also cause mental retardation, seizures, and hormonal problems. A mild variant is rarest, with only a few brain structures fused, slightly unusual facial features, and learning disabilities.

It has been difficult to study HPE because it is so rare after birth, but researchers discovered an unlikely helper: chickens. Facial development in chickens is similar to that of humans, and is controlled by the same genes, particularly one called "sonic hedgehog." Researchers added a chemical, cyclopamine, to chicken embryos at various points in development, to interfere with sonic hedgehog functioning. Adding cyclopamine early produced a Cyclops chick! The later the treatment, the less severe the outcome. The researchers conclude that HPE is caused by a faulty gene plus an environmental trigger that may occur at various times during development—explaining the differing degrees of severity. The chicken embryo system will allow investigators not only to understand how HPE arises, but to test ways to detect the disorder in pregnant women and possibly intervene.

A mythological cyclops.

Genes orchestrate our physiology from shortly after conception through adulthood. As a result, disorders caused by the malfunction of single genes, or genetic predispositions, affect people of all ages. Certain single-gene mutations act before birth, causing broken bones, dwarfism, or even cancer. Many other mutant genes exert their effects during childhood, and it may take parents months or even years to realize their child has a health problem. Duchenne muscular dystrophy (see figure 2.1), for example, usually begins as clumsiness in early childhood. Inherited forms of heart disease and breast cancer can appear in early or middle adulthood, earlier than multifactorial forms of these conditions. Pattern baldness is an inherited trait that may not become obvious until well into adulthood.

This chapter explores the stages of the human life cycle. Genes function against this developmental backdrop.

3.1 The Reproductive System

The formation of a new individual begins with a **sperm** from a male and an ovum (more precisely, an **oocyte**) from a female. Sperm and oocytes are **gametes,** or sex cells. They provide a mechanism for forming a new individual and mix genetic contributions from past generations. As a result, because there are so many genes and so many variants of them, each person (except for identical multiples) has a unique combination of inherited traits.

Sperm and oocytes are produced in the reproductive system, which is organized similarly in the male and female. Each system has paired structures, called **gonads,** where the sperm and oocytes are manufactured; tubules to transport these cells; and hormones and secretions that control the process.

The Male

Sperm cells develop within a 125-meter-long network of seminiferous tubules, which are packed into paired, oval organs called testes (sometimes called testicles) (**figure 3.1**). The testes are the male gonads.

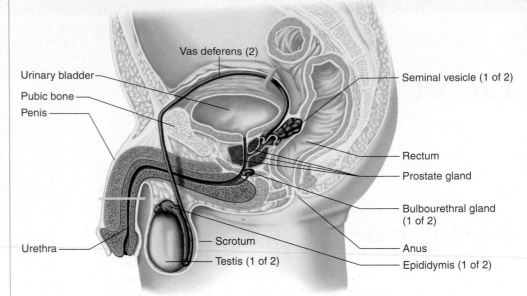

Figure 3.1   The human male reproductive system.   Sperm cells are manufactured within the seminiferous tubules, tightly wound within the testes, which descend into the scrotum. The prostate gland, seminal vesicles, and bulbourethral glands add secretions to the sperm cells to form seminal fluid. Sperm mature and are stored in the epididymis and exit through the vas deferens. The paired vasa deferentia join in the urethra, through which seminal fluid exits the body.

They lie outside the abdomen within a sac called the scrotum. This location keeps the testes cooler than the rest of the body, which is necessary for sperm to develop. Leading from each testis is a tightly coiled tube, the epididymis, in which sperm cells mature and are stored; each epididymis continues into another tube, the vas deferens. Each vas deferens bends behind the bladder and joins the urethra, the tube that carries sperm and urine out through the penis.

Along the sperm's path, three glands add secretions. The vasa deferentia pass through the prostate gland, which produces a thin, milky, alkaline fluid that activates the sperm to swim. Opening into the vas deferens is a duct from the seminal vesicles, which secrete fructose (a sugar that supplies energy), plus hormonelike prostaglandins, which may stimulate contractions in the female that help sperm and oocyte meet. The bulbourethral glands, each about the size of a pea, join the urethra where it passes through the body wall. They secrete an alkaline mucus that coats the urethra before sperm are released. All of these secretions combine to form the seminal fluid that carries sperm.

During sexual arousal, the penis becomes erect so that it can penetrate and deposit sperm in the female reproductive tract. At the peak of sexual stimulation, a pleasurable sensation called orgasm occurs, accompanied by rhythmic muscular contractions that eject the sperm from each vas deferens through the urethra and out the penis. The discharge of sperm from the penis, called ejaculation, delivers about 200 to 600 million sperm cells.

The Female

The female sex cells develop within paired organs in the abdomen called ovaries (**figure 3.2**), which are the female gonads. Within each ovary of a newborn female are about a million immature oocytes. Each individual oocyte is surrounded by nourishing follicle cells, and each ovary houses oocytes in different stages of development. After puberty, about once a month, one ovary releases the most mature oocyte. Beating cilia sweep the mature oocyte into

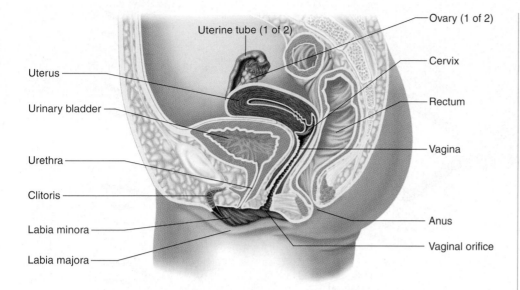

Uterine tube (1 of 2)

Uterus

Urinary bladder

Urethra

Clitoris

Labia minora

Labia majora

Ovary (1 of 2)

Cervix

Rectum

Vagina

Anus

Vaginal orifice

Figure 3.2　The human female reproductive system.　Oocytes are packed into the paired ovaries. Once a month after puberty, an ovary releases one oocyte, which is drawn into a nearby uterine tube. If a sperm fertilizes the oocyte in the uterine tube, the fertilized ovum continues into the uterus, where for nine months it develops into a new individual. If the oocyte is not fertilized, the body expels it, along with the built-up uterine lining.

the fingerlike projections of one of two uterine tubes. The tube carries the oocyte into a muscular, saclike organ called the uterus, or womb.

The released oocyte may encounter a sperm, usually in a uterine tube. If the sperm enters the oocyte so that the DNA of the two cells merges into a new nucleus, the result is a fertilized ovum. After about a day, this cell undergoes a series of rapid cell divisions while moving through the tube and nestles into the lining of the uterus. Here, it may continue to develop. If fertilization does not occur, the oocyte, along with much of the uterine lining, is shed as the menstrual flow. Hormones coordinate the monthly menstrual cycle.

The lower end of the uterus narrows and leads to the cervix, which opens into the tubelike vagina that exits from the body. The vaginal opening is protected on the outside by two pairs of fleshy folds. At the upper juncture of both pairs is a 2-centimeter-long structure called the clitoris, which is anatomically similar to the penis. Rubbing the clitoris triggers female orgasm. Hormones control the cycle of oocyte maturation and the preparation of the uterus to nurture a fertilized ovum.

Key Concepts

1. Sperm develop in the seminiferous tubules, mature and collect in each epididymis, enter the vasa deferentia, and move through the urethra in the penis. The prostate gland adds an alkaline fluid, seminal vesicles add fructose and prostaglandins, and bulbourethral glands secrete mucus to form seminal fluid.

2. In the female, ovaries contain oocytes. Each month, an ovary releases an oocyte, which enters a uterine tube leading to the uterus. If the oocyte is fertilized, it begins rapid cell division and nestles into the uterine lining to develop. Otherwise, the oocyte exits the body with the menstrual flow. Hormones control the monthly cycle of oocyte development.

3.2 Meiosis

Gametes form from special cells, called germline cells, in a type of cell division called **meiosis** that halves the chromosome number. A further process, maturation, sculpts the distinctive characteristics of

sperm and oocyte. The organelle-packed oocyte has 90,000 times the volume of the streamlined sperm, which is little more than a genetic package atop a propulsion system.

Unlike other cells in the human body, gametes contain 23 different chromosomes—half the usual amount of genetic material, but still a complete genome. Somatic (nonsex) cells contain 23 pairs, or 46 chromosomes. The chromosome pairs are called **homologous pairs,** or *homologs* for short. Homologs have the same genes in the same order but may carry different alleles, or forms, of the same gene. Gametes are **haploid** ($1n$), which means that they have only one of each type of chromosome and therefore one copy of the human genome. Somatic cells are **diploid** ($2n$), signifying that they have two copies of the genome.

Halving the number of chromosomes during gamete formation makes sense. If the sperm and oocyte each contained 46 chromosomes, the fertilized ovum would contain twice the normal number of chromosomes, or 92. Such a genetically overloaded cell, called a polyploid, usually does not develop. About one in a million newborns is polyploid, but these infants have abnormalities in all organ systems and usually only live a few days. However, studies on spontaneously aborted embryos indicate that about 1 percent of conceptions have three chromosome sets instead of the normal two. Therefore, most of these abnormal embryos do not survive to be born.

Meiosis mixes up trait combinations. For example, a person might produce one gamete containing alleles encoding green eyes and freckles, yet another encoding brown eyes and no freckles. Meiosis explains why siblings differ genetically from each other and from their parents.

In a much broader sense, meiosis, as the mechanism of sexual reproduction, provides genetic diversity, which can help a population to survive a challenging environment. A population of sexually reproducing organisms is made up of individuals with different genotypes and phenotypes. In contrast, a population of asexually reproducing organisms consists of identical individuals. Should a new threat arise, such as an infectious disease that kills only individuals with a certain genotype, then the entire asexual population could be wiped

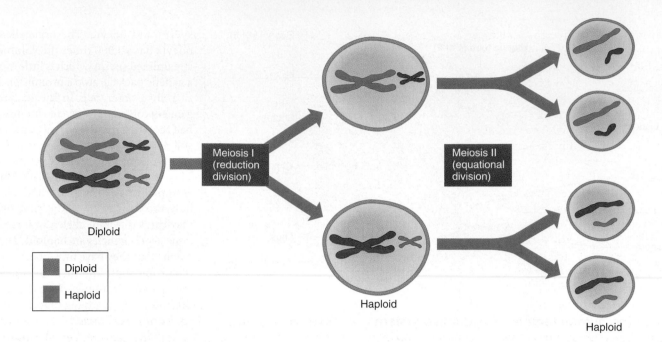

Figure 3.3 Overview of meiosis. Meiosis is a form of cell division in which certain cells are set aside, and give rise to haploid gametes. The first meiotic division reduces the number of chromosomes to 23, all in the replicated form. In the second meiotic division, the cells essentially undergo mitosis. The result of the two meiotic divisions is four haploid cells. In this illustration, homologous pairs of chromosomes are indicated by size, and parental origin of chromosomes by color.

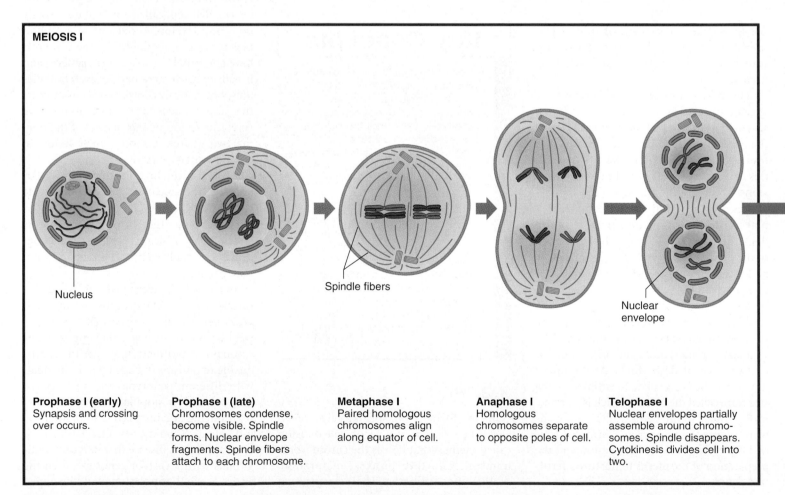

MEIOSIS I

Prophase I (early)
Synapsis and crossing over occurs.

Prophase I (late)
Chromosomes condense, become visible. Spindle forms. Nuclear envelope fragments. Spindle fibers attach to each chromosome.

Metaphase I
Paired homologous chromosomes align along equator of cell.

Anaphase I
Homologous chromosomes separate to opposite poles of cell.

Telophase I
Nuclear envelopes partially assemble around chromosomes. Spindle disappears. Cytokinesis divides cell into two.

Figure 3.4 Meiosis.

out. However, in a sexually reproducing population, individuals that inherited a certain combination of genes might survive. This differential survival of certain genotypes is the basis of evolution, discussed in chapter 16.

Meiosis entails two divisions of the genetic material. The first division is called **reduction division** (or meiosis I) because it reduces the number of replicated chromosomes from 46 to 23. The second division, called the **equational division** (or meiosis II), produces four cells from the two cells formed in the first division by splitting the replicated chromosomes. **Figure 3.3** shows an overview of the process, and **figure 3.4** depicts the major events of each stage.

As in mitosis, meiosis occurs after an interphase period when DNA is replicated (doubled) **(table 3.1).** For each chromosome pair in the cell undergoing meiosis, one homolog comes from the person's mother, and one from the father. In figures

3.3 and 3.4, the colors represent the contributions of the two parents, whereas size indicates different chromosomes.

After interphase, prophase I (so called because it is the prophase of meiosis I) begins as the replicated chromosomes condense and become visible when stained. A spindle forms. Toward the middle of prophase I, the homologs line up next to one another, gene by gene, in an event

Table 3.1

Comparison of Mitosis and Meiosis

Mitosis	Meiosis
One division	Two divisions
Two daughter cells per cycle	Four daughter cells per cycle
Daughter cells genetically identical	Daughter cells genetically different
Chromosome number of daughter cells same as that of parent cell (2n)	Chromosome number of daughter cells half that of parent cell (1n)
Occurs in somatic cells	Occurs in germline cells
Occurs throughout life cycle	In humans, completes after sexual maturity
Used for growth, repair, and asexual reproduction	Used for sexual reproduction, producing new gene combinations

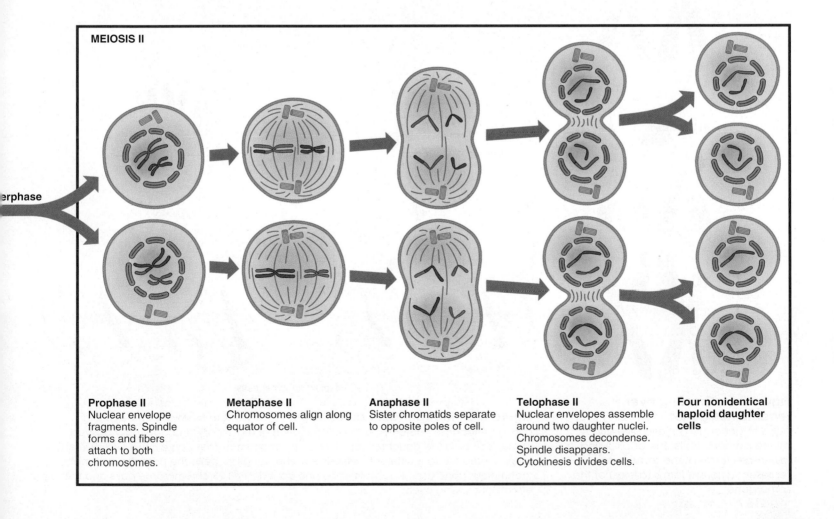

MEIOSIS II

Interphase

Prophase II
Nuclear envelope fragments. Spindle forms and fibers attach to both chromosomes.

Metaphase II
Chromosomes align along equator of cell.

Anaphase II
Sister chromatids separate to opposite poles of cell.

Telophase II
Nuclear envelopes assemble around two daughter nuclei. Chromosomes decondense. Spindle disappears. Cytokinesis divides cells.

Four nonidentical haploid daughter cells

called synapsis. A mixture of RNA and protein holds the chromosome pairs together. At this time, the homologs exchange parts in a process called **crossing over (figure 3.5).** All four chromatids that comprise each homologous chromosome pair are pressed together as exchanges occur. After crossing over, each homolog bears genes from both parents. (Prior to this, all of the genes on a homolog were derived from one parent.) New gene combinations arise from crossing over when the parents carry different alleles. Toward the end of prophase I, the synapsed chromosomes separate but remain attached at a few points along their lengths.

To understand how crossing over mixes trait combinations, consider a simplified example. Suppose that homologs carry genes for hair color, eye color, and finger length. One of the chromosomes carries alleles for blond hair, blue eyes, and short fingers. Its homolog carries alleles for black hair, brown eyes, and long fingers. After crossing over, one of the chromosomes might bear alleles for blond hair, brown eyes, and long fingers, and the other might bear alleles for black hair, blue eyes, and short fingers.

Meiosis continues in metaphase I, when the homologs align down the center of the cell. Each member of a homolog pair attaches to a spindle fiber at opposite poles. The pattern in which the chromosomes align during metaphase I is important in generating genetic diversity. For each homolog pair, the pole the maternally or paternally derived member goes to is random. The sit-uation is analogous to the number of different ways that 23 boys and 23 girls could line up in boy-girl pairs. The greater the number of chromosomes, the greater the genetic diversity generated at this stage.

For two pairs of homologs, four (2^2) different metaphase configurations are possible. For three pairs of homologs, eight (2^3) different combinations can occur. Our 23 chromosome pairs can line up in 8,388,608 (2^{23}) different ways. This random arrangement of the members of homolog pairs in metaphase is called **independent assortment (figure 3.6).** It accounts for a basic law of inheritance discussed in the next chapter.

Homologs separate in anaphase I and finish moving to opposite poles by telophase I, establishing a haploid set of still-replicated chromosomes at each end of the stretched-out cell. Unlike in mitosis, the centromeres of each homolog in meiosis I remain together. During a second interphase, chromosomes unfold into very thin

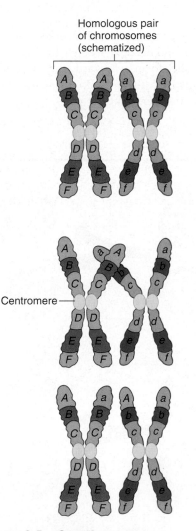

Homologous pair
of chromosomes
(schematized)

Centromere

Figure 3.5 Crossing over recombines genes. Crossing over helps to generate genetic diversity by mixing parental traits. The capital and lowercase forms of the same letter represent different forms (alleles) of the same gene.

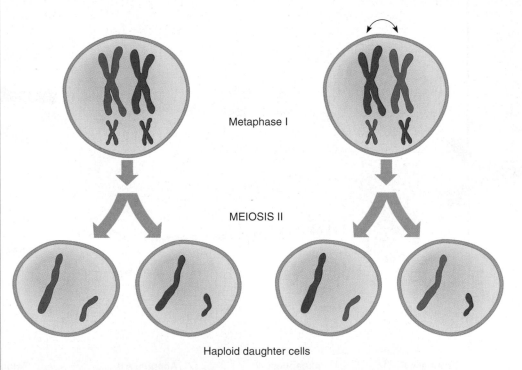

Metaphase I

MEIOSIS II

Haploid daughter cells

Figure 3.6 Independent assortment. The pattern in which homologs align during metaphase I determines the combination of maternally and paternally derived chromosomes in the daughter cells. Two pairs of chromosomes can align in two different ways to produce four different possibilities in the daughter cells. The potential variability that meiosis generates skyrockets when one considers all 23 chromosome pairs and the effects of crossing over.

threads. Proteins are manufactured, but the genetic material is not replicated a second time. The single DNA replication, followed by the double division of meiosis, halves the chromosome number.

Prophase II marks the start of the second meiotic division. The chromosomes are again condensed and visible. In metaphase II, the replicated chromosomes align down the center of the cell. In anaphase II, the centromeres part, and the newly formed chromosomes, each now in the unreplicated form, move to opposite poles. In telophase II, nuclear envelopes form around the four nuclei, which then separate into individual cells. The net result of meiosis is four haploid cells, each carrying a new assortment of genes and chromosomes that represent a single copy of the genome.

Meiosis generates astounding genetic variety. Any one of a person's more than 8 million possible combinations of chromosomes can meet with any one of the more than 8 million combinations of a partner, raising potential variability to more than 70 trillion $(8,388,608^2)$ genetically unique individuals! Crossing over contributes even more genetic variability.

Key Concepts

1. The haploid sperm and oocyte are derived from diploid germline cells by meiosis and maturation.
2. Meiosis maintains the chromosome number over generations and mixes gene combinations.
3. In the first meiotic (or reduction) division, the number of replicated chromosomes is halved.
4. In the second meiotic (or equational) division, each of two cells from the first division divides again, yielding four cells from the original one.
5. Chromosome number is halved because the DNA replicates once, but the cell divides twice.
6. Crossing over and independent assortment generate further genotypic diversity by creating new combinations of alleles.

3.3 Gamete Maturation

Meiosis occurs in both sexes, but further steps elaborate the very different-looking sperm and oocyte. Each type of gamete is haploid, but different distributions of other cell components create their distinctions. The cells of the maturing male and female proceed through similar stages, but with sex-specific terminology and different timetables. A male begins manufacturing sperm at puberty and continues throughout life, whereas a female begins meiosis when she is a fetus. Meiosis in the female completes only if a sperm fertilizes the oocyte.

Sperm Formation

Spermatogenesis, the formation of sperm cells, begins in a diploid stem cell called a **spermatogonium (figure 3.7).** This cell divides mitotically, yielding two daughter cells. One continues to specialize into a mature sperm, and the other remains a stem cell.

Bridges of cytoplasm join several spermatogonia, and their daughter cells enter meiosis together. As they mature, these spermatogonia accumulate cytoplasm and replicate their DNA, becoming primary spermatocytes.

During reduction division (meiosis I), each primary spermatocyte divides, forming two equal-sized haploid cells called secondary spermatocytes. In meiosis II, each

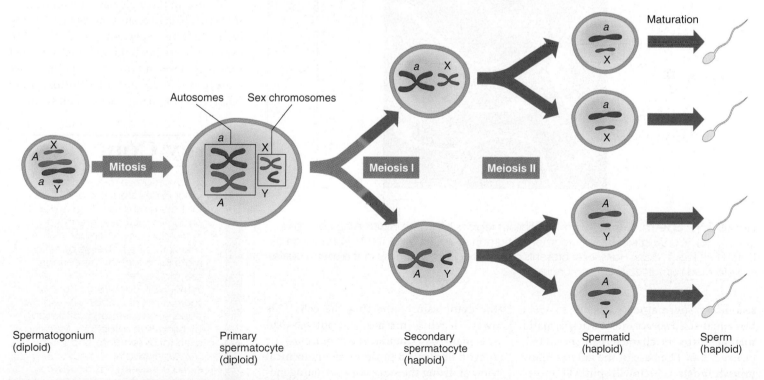

Spermatogonium (diploid)

Primary spermatocyte (diploid)

Secondary spermatocyte (haploid)

Spermatid (haploid)

Sperm (haploid)

Figure 3.7   Sperm formation (spermatogenesis).   Primary spermatocytes have the normal diploid number of 23 chromosome pairs. The large pair of chromosomes represents autosomes (non-sex chromosomes). The X and Y chromosomes are sex chromosomes.

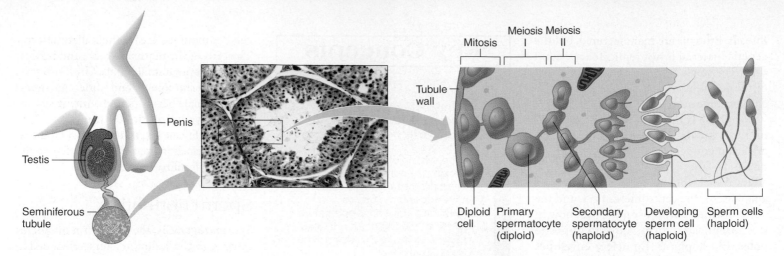

Figure 3.8   Meiosis produces sperm cells.   Diploid cells divide through mitosis in the linings of the seminiferous tubules. Some of the daughter cells then undergo meiosis, producing haploid spermatocytes, which differentiate into mature sperm cells.

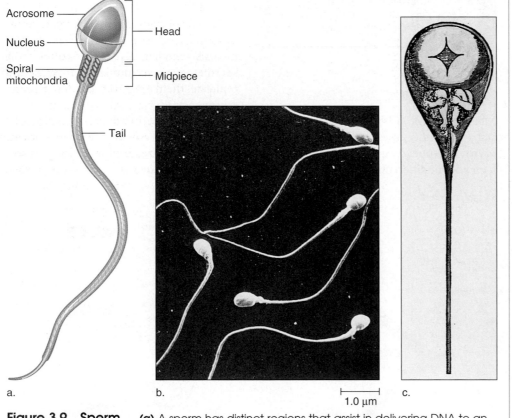

a.   b.   c.

1.0 µm

Figure 3.9   Sperm.   **(a)** A sperm has distinct regions that assist in delivering DNA to an oocyte. **(b)** Scanning electron micrograph of human sperm cells. **(c)** This 1694 illustration by Dutch histologist Niklass Hartsoeker presents a once-popular hypothesis that a sperm carries a preformed human called a homunculus.

timeters (7 inches) to reach an oocyte. Each sperm cell consists of a tail, body or midpiece, and a head region (**figure 3.9**). A membrane-covered area on the front end, the acrosome, contains enzymes that help the cell penetrate the protective layers around the oocyte. Within the bulbous sperm head, DNA is wrapped around proteins. The sperm's DNA at this time is genetically inactive. A male manufactures trillions of sperm in his lifetime. Although many of these will come close to an oocyte, very few will actually touch one.

Meiosis in the male has built-in protections that help prevent sperm from causing birth defects. Spermatogonia that are exposed to toxins tend to be so damaged that they never mature into sperm. More mature sperm cells exposed to toxins are often so damaged that they cannot swim.

secondary spermatocyte divides to yield two equal-sized spermatids. Each spermatid then develops the characteristic sperm tail, or flagellum. The base of the tail has many mitochondria, which will split ATP molecules to release energy that will propel the sperm inside the female reproductive tract. After spermatid differentiation, some of the cytoplasm connecting the cells falls away, leaving mature, tadpole-shaped spermatozoa (singular *spermatozoon*), or sperm. **Figure 3.8** presents an anatomical view showing the stages of spermatogenesis within the seminiferous tubules.

A sperm, which is a mere 0.006 centimeter (0.0023 inch) long, must travel about 18 cen-

Key Concepts

1. Spermatogonia divide mitotically, yielding one stem cell and one cell that accumulates cytoplasm and becomes a primary spermatocyte.
2. In meiosis I, each primary spermatocyte halves its genetic material to form two secondary spermatocytes.
3. In meiosis II, each secondary spermatocyte divides, yielding two equal-sized spermatids attached by bridges of cytoplasm. Maturing spermatids separate and shed some cytoplasm.
4. A mature sperm has a tail, body, and head, with an enzyme-containing acrosome covering the head.

Oocyte Formation

Meiosis in the female, called **oogenesis** (egg making), begins, as does spermatogenesis, with a diploid cell, an oogonium. Unlike the male cells, oogonia are not attached, but follicle cells surround each one. Each oogonium grows, accumulates cytoplasm, and replicates its DNA, becoming a primary oocyte. The ensuing meiotic division in oogenesis, unlike that in spermatogenesis, produces cells of different sizes.

In meiosis I, the primary oocyte divides into two cells: a small cell with very little cytoplasm, called a first **polar body,** and a much larger cell called a secondary oocyte (**figure 3.10**). Each cell is haploid, with the chromosomes in replicated form. In meiosis II, the tiny first polar body may divide to yield two polar bodies of equal size, with unreplicated chromosomes; or it may simply decompose. The secondary oocyte, however, divides unequally in meiosis II to produce another small polar body, with unreplicated chromosomes, and the mature egg cell, or ovum, which contains a large volume of cytoplasm. **Figure 3.11** summarizes meiosis in the female, and **figure 3.12** provides an anatomical view of the process.

Most of the cytoplasm among the four meiotic products in the female is concentrated in only one cell, the ovum. The woman's body absorbs the polar bodies, and they normally play no further role in development. Rarely, a sperm fertilizes a polar body. The woman's hormones respond as if she is pregnant, but a disorganized clump of cells that is not an embryo grows for a few weeks, and then leaves the woman's body. This event is a type of miscarriage called a "blighted ovum."

Before birth, a female's million or so oocytes arrest in prophase I. By puberty, about 400,000 oocytes remain. After puberty, meiosis I continues in one or several oocytes each month, but halts again at metaphase II. In response to specific hormonal cues each month, one ovary releases a secondary oocyte; this event is ovulation. If a sperm penetrates the oocyte membrane, then female

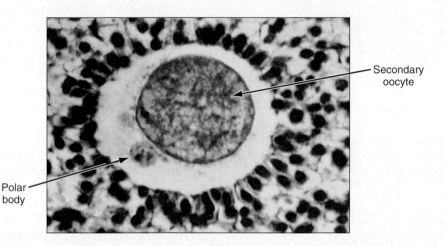

Figure 3.10 Meiosis in a female produces a secondary oocyte and a polar body. Unequal division enables the cell destined to become a fertilized ovum to accumulate the bulk of the cytoplasm and organelles from the primary oocyte, but with only one genome's worth of DNA. (×700)

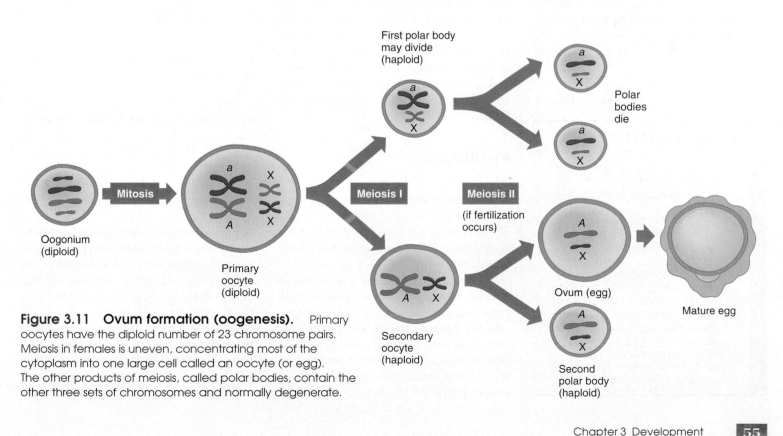

Figure 3.11 Ovum formation (oogenesis). Primary oocytes have the diploid number of 23 chromosome pairs. Meiosis in females is uneven, concentrating most of the cytoplasm into one large cell called an oocyte (or egg). The other products of meiosis, called polar bodies, contain the other three sets of chromosomes and normally degenerate.

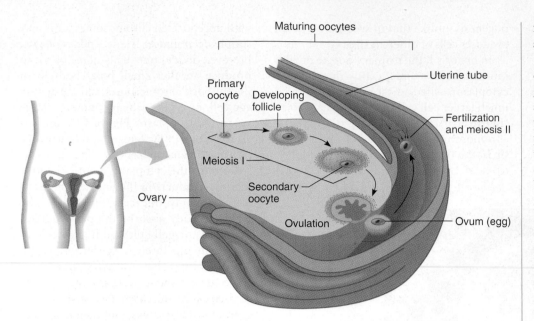

Maturing oocytes

Primary oocyte

Developing follicle

Meiosis I

Secondary oocyte

Ovary

Ovulation

Uterine tube

Fertilization and meiosis II

Ovum (egg)

Figure 3.12 The making of oocytes. Oocytes develop within the ovary in protective follicles. An ovary contains many oocytes in various stages of maturation. After puberty, the most mature oocyte in one ovary bursts out each month, an event called ovulation.

meiosis completes, and a fertilized ovum forms. If the secondary oocyte is not fertilized, it degenerates and leaves the body in the menstrual flow, meiosis never completed.

A female ovulates about 400 oocytes between puberty and menopause. (However, experiments in mice suggest that stem cells may produce oocytes even past menopause.) Most oocytes are destined to degrade, because fertilization is very rare.Only one in three of the oocytes that do meet and merge with a sperm cell will continue to grow, divide, and specialize to eventually form a new individual.

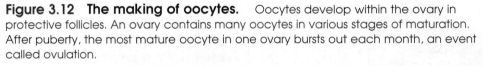

Key Concepts

1. An oogonium accumulates cytoplasm and replicates its chromosomes, becoming a primary oocyte.
2. In meiosis I, the primary oocyte divides, forming a small polar body and a large, haploid secondary oocyte.
3. In meiosis II, the secondary oocyte divides, yielding another small polar body and a mature haploid ovum.
4. Oocytes arrest at prophase I until puberty, after which one or several oocytes complete the first meiotic division during ovulation each month. The second meiotic division completes at fertilization.

3.4 Prenatal Development

A prenatal human is considered an **embryo** for the first eight weeks. During this time, rudiments of all body parts form. The embryo in the first week is considered to be in a "preimplantation" stage because it has not yet settled into the uterine lining. Prenatal development after the eighth week is the fetal period, when structures grow and specialize. From the start of the ninth week until birth, the human organism is a **fetus.**

Fertilization

Hundreds of millions of sperm cells are deposited in the vagina during sexual intercourse. A sperm cell can survive in the woman's body for up to six days, but the oocyte can only be fertilized in the 12 to 24 hours after ovulation.

The woman's body helps sperm reach an oocyte. A process in the female called capacitation chemically activates sperm, and the oocyte secretes a chemical that attracts sperm. Sperm are also assisted by contractions of the female's muscles, by their moving tails, and by upwardly moving mucus propelled by cilia on cells of the female reproductive tract. Still, only 200 or so sperm come near the oocyte.

A sperm first contacts a covering of follicle cells, called the corona radiata, that guards a secondary oocyte. The sperm's acrosome then bursts, releasing enzymes that bore through a protective layer of glycoprotein, called the zona pellucida, beneath the corona radiata. Fertilization, or conception, begins when the outer membranes of the sperm and secondary oocyte meet (**figure 3.13**). The encounter is dramatic. A wave of electricity spreads physical and chemical changes across the entire oocyte surface—changes that keep other sperm out. More than one sperm can enter an oocyte, but the resulting cell has too much genetic material for development to follow.

Usually only the sperm's head enters the oocyte. Within 12 hours of the sperm's penetration, the ovum's nuclear membrane disappears, and the two sets of chromosomes, called pronuclei, approach one another. Within each pronucleus, DNA replicates. Fertilization completes when the two genetic packages meet and merge, forming the genetic instructions for a new individual. The fertilized ovum is called a **zygote.** The Bioethics: Choices for the Future reading on cloning and stem cell technology describes cloning, one way to start development without a fertilized egg.

Cleavage and Implantation

About a day after fertilization, the zygote divides by mitosis, beginning a period of frequent cell division called **cleavage (figure 3.14).** The resulting early cells are called **blastomeres.** When the blastomeres form a solid ball of sixteen or more cells, the embryo is called a **morula** (Latin for "mulberry," which it resembles).

During cleavage, organelles and molecules from the secondary oocyte's cytoplasm still control cellular activities, but some of the embryo's genes begin to function. The ball of cells hollows out, and its center fills with fluid, creating a **blastocyst**—the "cyst" referring to the fluid-filled center. Some of the cells form a clump. This is the **inner cell mass** that is used to derive embryonic stem cells (see figure 2.25). Formation of the inner cell mass is the first event that distinguishes

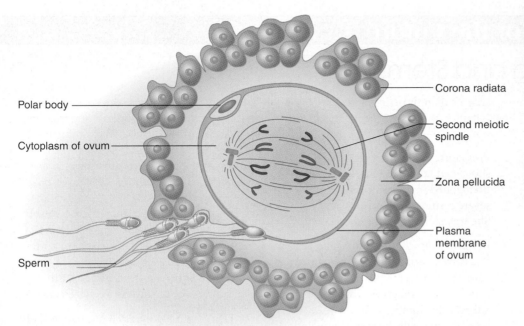

Polar body

Cytoplasm of ovum

Sperm

Corona radiata

Second meiotic spindle

Zona pellucida

Plasma membrane of ovum

a.

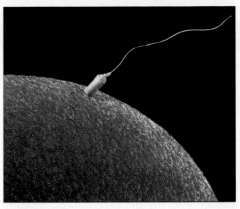

b.

Figure 3.13   Fertilization.
(a) Fertilization by a sperm cell induces the oocyte (arrested in metaphase II) to complete meiosis. Before fertilization occurs, the sperm's acrosome bursts, spilling enzymes that help the sperm's nucleus enter the oocyte. **(b)** A series of chemical reactions ensues that helps to ensure that only one sperm nucleus enters an oocyte.

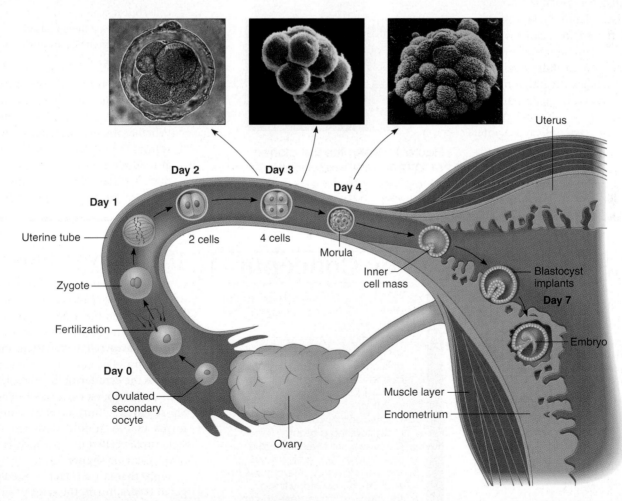

Day 2

Day 1

Uterine tube

Zygote

Fertilization

Day 0

Ovulated secondary oocyte

Ovary

2 cells

Day 3

4 cells

Day 4

Morula

Inner cell mass

Muscle layer

Endometrium

Uterus

Blastocyst implants

Day 7

Embryo

Figure 3.14   Cleavage: From ovulation to implantation.   The zygote forms in the uterine tube when a sperm nucleus fuses with the nucleus of an oocyte. The first divisions proceed while the zygote moves toward the uterus. By day 7, the zygote, now called a blastocyst, begins to implant in the uterine lining.

Considering Cloning and Stem Cell Technology

Cloning is the creation of a genetic replica of an individual. In fiction, renegade scientists have cloned Nazis, politicians, dinosaurs, and children at a boarding school in England to be organ donors. Real scientists have cloned sheep, mice, cats, and taken cloned human embryos to the blastocyst stage. Cloning transfers a nucleus from a somatic cell into an oocyte whose nucleus has been removed, and then develops new cells or a new individual from the manipulated cell. Figure 2.25 illustrates cloning taken to the 5-day stage of the embryo to supply stem cells tailored to a sick or injured individual. In contrast, reproductive cloning seeks to create a baby using the nucleus from the cell of a particular individual who will then, supposedly, be duplicated.

The premises behind reproductive cloning may be flawed, for a clone is not an exact replica of an individual. Many of the distinctions between an individual and a clone arise from epigenetic phenomena—effects that do not change genes, but alter their expression. There are other distinctions, too (parentheses indicate chapters that discuss these subjects further):

- In some species, telomeres of chromosomes in the donor nucleus are shorter than those in the recipient cell (chapter 2). This is true for sheep, but not cattle—no one yet knows why.

- Premature aging, as evidenced in shortened telomeres, may be why the first cloned mammal, Dolly, contracted a severe respiratory infection at six years. She was euthanized (**figure 1**).

- In normal development, for some genes, one copy is turned off, depending upon which parent transmits it. That is, some genes must be inherited from either the father or the mother to be active, a phenomenon called genomic imprinting. In cloning, genes in a donor nucleus do not pass through a germline cell (sperm or oocyte) before they go back to the beginning of development, and thus are not imprinted. Lack of imprinting in clones may cause the "large offspring syndrome" that affects many of the few cloned animals that actually develop far enough to be born. Not only is the first cell of an embryo derived from a somatic cell nucleus not equivalent to a fertilized ovum, but experiments in nonhuman animals indicate that regulation of gene expression is abnormal at many times during prenatal development (chapter 5).

- DNA from a donor cell has had years to accumulate mutations. Such a somatic mutation might not be noticeable if it occurs in one of millions of somatic cells, but it could be devastating if that somatic cell nucleus is used to program the development of an entire new individual (chapter 11).

- At a certain time in early prenatal development in all female mammals, one X chromosome is inactivated. Whether the inactivated X chromosome is from the mother or the father occurs at random in each cell, creating an overall mosaic pattern of expression for genes on the X chromosome. The

Figure 1   Dolly, the first cloned mammal.   She was put to sleep when she contracted a respiratory illness in 2003. Dolly lived six years and had several healthy lambs.

cells from each other in terms of their relative positions, other than the inside and outside of the morula. The cells of the inner cell mass will continue developing to form the embryo.

A week after conception, the blastocyst begins to nestle into the rich lining of the woman's uterus. This event, called implantation, takes about a week. As it starts, the outermost cells of the embryo, called the trophoblast, secrete the "pregnancy hormone," human chorionic gonadotropin (hCG), which prevents menstruation. hCG detected in a woman's urine or blood is one sign of pregnancy.

Key Concepts

1. Following sexual intercourse, sperm are capacitated and drawn to the secondary oocyte.
2. Acrosomal enzymes assist the sperm's penetration of the oocyte, and chemical and electrical changes in the oocyte's surface block additional sperm from entering.
3. The two sets of chromosomes meet, forming a zygote.
4. Cleavage cell divisions form a morula and then a blastocyst.
5. The outer layer of cells invades and implants in the uterine lining.
6. The inner cell mass develops into the embryo.
7. Certain blastocyst cells secrete hCG.

The Embryo Forms

During the second week of prenatal development, a space called the amniotic cavity forms between the inner cell mass and the outer cells anchored to the uterine lining. Then the inner cell mass flattens into a two-layered disc. The layer nearest the amniotic cavity is the **ectoderm;** the inner layer, closer to the blastocyst cavity, is the **endoderm.** Shortly after, a third layer, the **mesoderm,** forms in the middle. This three-layered structure is called the primordial embryo, or the **gastrula (figure 3.15).**

Once these three layers, called **primary germ layers,** form, the fates of many cells are determined, which means that they are destined to develop as a specific cell type. Each layer gives rise to certain structures.

pattern of X inactivation of a female clone would most likely not match that of her nucleus donor (chapter 6).

- Mitochondria contain DNA. A clone's mitochondria descend from the recipient oocyte, not from the donor cell.

The environment is another powerful factor in why a clone isn't an identical copy. For example, coat color patterns differ in cloned calves. This is because when the calves were embryos, cells destined to produce pigment moved in a unique way in each calf, producing different color patterns. In humans, such factors as experience, nutrition, stress, and exposure to infectious disease join our genes in molding who we are. Identical twins, although they have the same DNA sequence (except for somatic mutations), are not exact replicas of each other. Similarly, cloning a deceased child would probably disappoint parents seeking to recapture their lost loved one.

A compelling argument against reproductive cloning that embraces ethics, biology, and the results of experiments on other animals, is that it would likely create a child who would suffer.

Most cloning attempts fail. The reasons may lie in the fact that, as one researcher puts it, "The whole natural order is broken," referring to meiosis, which in the female completes at fertilization. In cloning, a diploid nucleus is introduced into oocyte cytoplasm, where signals direct it to do what a female secondary oocyte tends to do—shed half of itself as a polar body. If the out-of-place donor nucleus does this, the new cell jettisons half its chromosomes—one genome copy—and becomes haploid. It cannot develop.

Cloned cats reveal vividly that animals are more than the products of genes. Consider Rainbow and her clone, Carbon Copy, or "Cc." Rainbow is shy; Cc is aggressive and playful. The company that created Cc, Genetic Savings and Clone, began to clone pet cats for $50,000 in late 2003 (**figure 2**). They use a variation of cloning called chromatin transfer, in which factors associated with differentiation are removed from the donor material. Pigs have been successfully cloned too, and like cats, are no more alike than siblings.

On a more serious note, the essence of the ethical objection to cloning is that we are dissecting and defining our very individuality, reducing it to a biochemistry so supposedly simple that we can duplicate it. We probably can't.

Figure 2 Cloned cats are now a reality.

Cells in the ectoderm become skin, nervous tissue, or parts of certain glands. Endoderm cells form parts of the liver and pancreas and the linings of many organs. The middle layer of the embryo, the mesoderm, forms many structures, including muscle, connective tissues, the reproductive organs, and the kidneys.

The idea that cell fates are set in the gastrula, however, is changing. Certain progenitor cells can transdifferentiate, which means that they can divide to yield specialized cells characteristic of a different germ layer. For example, stem cells in bone marrow can migrate to the brain and give rise to neurons. Researchers had thought that once a cell was in a particular primary germ layer, it didn't stray. Yet in this example, bone marrow stem cells arise from mesoderm, while neurons normally derive from ectoderm. The word *normally* may be important here—transdifferentiation may occur naturally only in response to a certain type of injury. Perhaps no one noticed transdifferentiation in the past because nobody thought to look for it!

Table 3.2 summarizes the stages of early prenatal development.

Supportive Structures Form

As an embryo develops, structures form that support and protect it. These include chorionic villi, the placenta, the yolk sac, the allantois, the umbilical cord, and the amniotic sac.

By the third week after conception, finger-like projections called chorionic villi extend from the area of the embryonic disc close to the uterine wall, projecting into pools of the woman's blood. Her blood system and the embryo's are separate, but nutrients and oxygen diffuse across the chorionic villi from her circulation to the embryo, and wastes leave the embryo's circulation and enter the woman's circulation to be excreted.

By 10 weeks, the placenta is fully formed. It links woman and fetus for the rest of the pregnancy. The placenta secretes hormones that maintain pregnancy and alter the woman's metabolism to send nutrients to the fetus.

Other structures nurture the developing embryo. The yolk sac manufactures blood

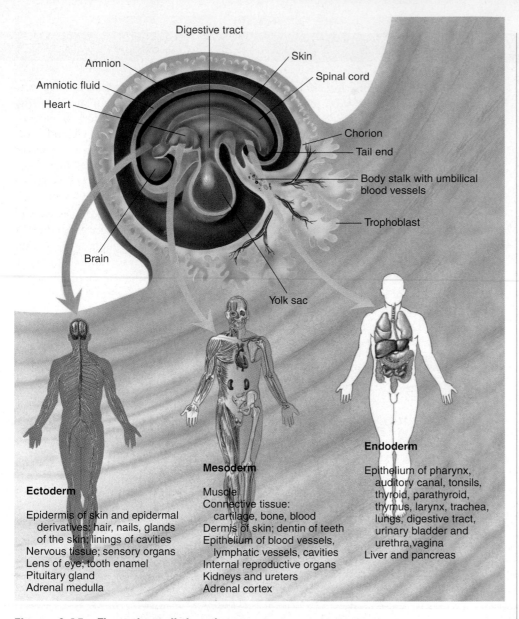

Digestive tract
Skin
Spinal cord
Amnion
Amniotic fluid
Heart
Chorion
Tail end
Body stalk with umbilical blood vessels
Trophoblast
Brain
Yolk sac

Ectoderm

Epidermis of skin and epidermal
  derivatives: hair, nails, glands
  of the skin; linings of cavities
Nervous tissue; sensory organs
Lens of eye; tooth enamel
Pituitary gland
Adrenal medulla

Mesoderm

Muscle
Connective tissue:
  cartilage, bone, blood
Dermis of skin; dentin of teeth
Epithelium of blood vessels,
  lymphatic vessels, cavities
Internal reproductive organs
Kidneys and ureters
Adrenal cortex

Endoderm

Epithelium of pharynx,
  auditory canal, tonsils,
  thyroid, parathyroid,
  thymus, larynx, trachea,
  lungs, digestive tract,
  urinary bladder and
  urethra,vagina
Liver and pancreas

Figure 3.15 The primordial embryo.   When the three basic layers of the embryo form at gastrulation, many cells become "determined" to follow a specific developmental pathway. However, each layer retains stem cells as the organism develops, and these cells may be capable of producing daughter cells that can specialize as many cell types, including some not associated with the layer of origin.

Table 3.2

Stages and Events of Early Human Prenatal Development

Stage	Time Period	Principal Events
Fertilized ovum	12–24 hours following ovulation	Oocyte fertilized; zygote has 23 pairs of chromosomes and is genetically distinct
Cleavage	30 hours to third day	Mitosis increases cell number
Morula	Third to fourth day	Solid ball of cells
Blastocyst	Fifth day through second week	Hollowed ball forms trophoblast (outside) and inner cell mass, which implants and flattens to form embryonic disc
Gastrula	End of second week	Primary germ layers form

cells, as does the allantois, a membrane surrounding the embryo that gives rise to the umbilical blood vessels. The umbilical cord forms around these vessels and attaches to the center of the placenta. Toward the end of the embryonic period, the yolk sac shrinks, and the amniotic sac swells with fluid that cushions the embryo and maintains a constant temperature and pressure. The amniotic fluid contains fetal urine and cells.

Two of the supportive structures that develop during pregnancy provide the material for prenatal tests (see figure 13.6), discussed in chapter 13. Chorionic villus sampling examines chromosomes from cells snipped off the chorionic villi at 10 weeks. Because the villi cells and the embryo's cells come from the same fertilized ovum, an abnormal chromosome detected in villi cells should also be in the embryo. In amniocentesis, a sample of amniotic fluid is taken after the fourteenth week of pregnancy, and fetal cells in the fluid are examined for biochemical, genetic, and chromosomal anomalies.

Key Concepts

1. Germ layers form in the second week. Cells in a specific germ layer later become parts of particular organ systems as a result of differential gene expression.
2. During week 3, chorionic villi extend toward the maternal circulation, and the placenta begins to form.
3. Nutrients and oxygen enter the embryo, and wastes pass from the embryo into the maternal circulation.
4. The yolk sac and allantois manufacture blood cells, the umbilical cord forms, and the amniotic sac expands with fluid.

Multiples

Twins and other multiples arise early in development. Twins are either fraternal or identical. Fraternal, or **dizygotic** (DZ), twins result when two sperm fertilize two oocytes. This can happen if ovulation occurs in two ovaries in the same month, or if two oocytes leave the same ovary and are both fertilized. DZ twins are no more alike than any two siblings, although they share a very early environment in the uterus. The tendency to have DZ twins may run in families if the women tend to ovulate two oocytes a month.

Identical, or **monozygotic** (MZ), twins descend from a single fertilized ovum and therefore are genetically identical. They are natural clones. Three types of MZ twins can form, depending upon when the fertilized ovum or very early embryo splits (**figure 3.16**). This difference in timing determines which supportive structures the twins share. About a third of all MZ twins have completely separate chorions and amnions, and about two-thirds share a chorion but have separate amnions. Slightly fewer than 1 percent of MZ twins share both amnion and chorion. (The amnion is the sac that contains fluid that surrounds the fetus. The chorion develops into the placenta.) These differences may expose the different types of MZ twins to slightly different uterine environments. For example, if one chorion develops more attachment sites to the maternal circulation, one twin may receive more nutrients and gain more weight than the other.

In 1 in 50,000 to 100,000 pregnancies, an embryo divides into twins after the point at which the two groups of cells can develop as two individuals, between days 13 and 15. The result is conjoined or "Siamese" twins. The latter name comes from Chang and Eng, who were born in Thailand, then called Siam, in 1811. They were joined by a band of tissue from the navel to the breastbone, and could easily have been separated today. Chang and Eng lived for 63 years, attached. They fathered 22 children and divided each week between their wives.

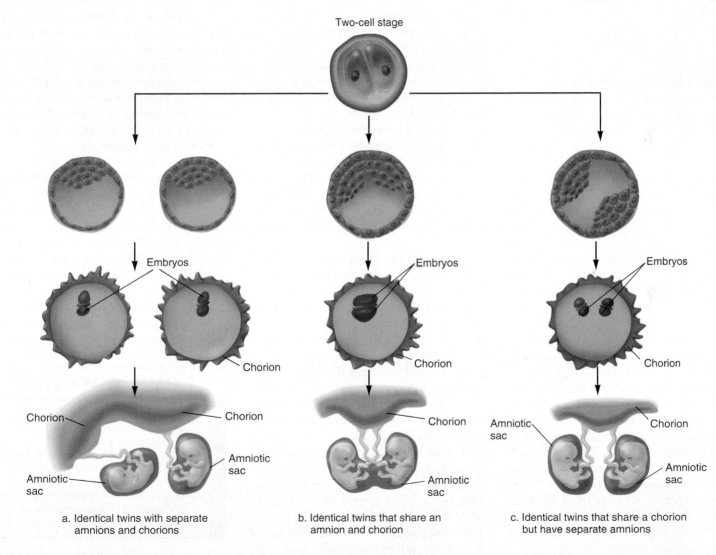

Figure 3.16 Facts about twins. Identical twins originate at three points in development. **(a)** In about one-third of identical twins, separation of cells into two groups occurs before the trophoblast forms on day 5. These twins have separate chorions and amnions. **(b)** About 1 percent of identical twins share a single amnion and chorion, because the tissue splits into two groups after these structures have already formed. **(c)** In about two-thirds of identical twins, the split occurs after day 5 but before day 9. These twins share a chorion but have separate amnions. Fraternal twins result from two sperm fertilizing two secondary oocytes. These twins develop their own amniotic sacs, yolk sacs, allantois, placentae, and umbilical cords.

Figure 3.17 Conjoined twins.
Abby and Britty Hensel are the result of incomplete twinning during the first two weeks of prenatal development.

In the case of Abigail and Brittany Hensel, shown in **figure 3.17,** the separation occurred after day 9 of development, but before day 14. Biologists determined this because the girls' shared organs contain representatives of ectoderm, mesoderm, and endoderm; that is, when the lump of cells divided incompletely, the three primary germ layers had not yet completely sorted themselves out. The Hensel girls are extremely rare "incomplete twins." Each girl has her own neck, head, heart, stomach, and gallbladder. Each has one leg and one arm, and a third arm between their heads was surgically removed. Each girl also has her own nervous system! The twins share a large liver, a single bloodstream, and all organs below the navel. They have three lungs and three kidneys. Because Abby and Britty were strong and healthy, doctors suggested surgery to separate them. But their parents, aware from other cases that only one child would likely survive a separation, chose to let their daughters be.

MZ twins occur in 3 to 4 pregnancies per 1,000 births worldwide. In North America, twins occur in about 1 in 81 pregnancies, which means that 1 in 40 of us is a twin. However, not all twins survive to be born.

One study of twins detected early in pregnancy showed that up to 70 percent of the eventual births are of a single child. This is called the "vanishing twin" phenomenon.

Key Concepts

1. Monozygotic twins arise from a single fertilized ovum and may share supportive structures.
2. Dizygotic twins arise from two fertilized ova.

The Embryo Develops

As the days and weeks of prenatal development proceed, different rates of cell division in different parts of the embryo fold the forming tissues into intricate patterns. In a process called embryonic induction, the specialization of one group of cells causes adjacent groups of cells to specialize. Gradually, these changes mold the three primary germ layers into organs and organ systems. Organogenesis is the transformation of the simple three layers of the embryo into distinct organs. During the weeks of organogenesis, the developing embryo is particularly sensitive to environmental influences such as chemicals and viruses.

During the third week of prenatal development, a band called the primitive streak appears along the back of the embryo. The primitive streak gradually elongates to form an axis that other structures organize around as they develop. The primitive streak eventually gives rise to connective tissue precursor cells and the notochord, a structure that forms the basic framework of the skeleton. The notochord induces overlying ectoderm to specialize into a hollow **neural tube,** which develops into the brain and spinal cord (central nervous system). Some nations designate day 14 of prenatal development and primitive streak formation as the point beyond which they ban research on the human embryo. The reason is that the primitive streak is the first sign of a nervous system, and day 14 is also the time at which implantation is complete.

Appearance of the neural tube marks the beginning of organ development. Shortly after, a reddish bulge containing the heart appears. The heart begins to beat around day 18, and this is easily detectable by day 22. Soon the central nervous system starts to form.

The fourth week of embryonic existence is one of spectacularly rapid growth and differentiation (**figure 3.18**). Arms and legs begin to extend from small buds on the torso. Blood cells form and fill primitive blood vessels. Immature lungs and kidneys appear.

If the neural tube does not close normally by about day 28, a neural tube defect results, leaving an area of the spine open and allowing parts of the brain or spinal cord to protrude (see Reading 16.1). If this happens, a substance from the fetus's liver called alpha fetoprotein (AFP) leaks at an abnormally rapid rate into the pregnant woman's circulation. A maternal blood test at the fifteenth week of pregnancy measures AFP level. If it is elevated, further tests measure AFP in the amniotic fluid, and ultrasound is used to visualize a defect. (Ultrasound scanning bounces sound waves off the fetus, creating an image; see figure 13.7.)

By the fifth and sixth weeks, the embryo's head appears to be too large for the rest of its body. Limbs end in platelike structures with tiny ridges, and gradually apoptosis sculpts the fingers and toes. The eyes are open, but they do not yet have lids or irises. By the seventh and eighth weeks, a skeleton composed of cartilage forms. The embryo is now about the length and weight of a paper clip. At eight weeks of gestation, the prenatal human has rudiments of all of the structures that will be present at birth. It is now a fetus.

Key Concepts

1. During week 3, the primitive streak appears, followed rapidly by the notochord, neural tube, heart, central nervous system, limbs, digits, facial features, and other organ rudiments.
2. By week 8, all of the organs that will be present in the newborn have begun to develop.

The Fetus Grows

During the fetal period, body proportions approach those of a newborn (**figure 3.19**). Initially, the ears lie low, and the eyes are

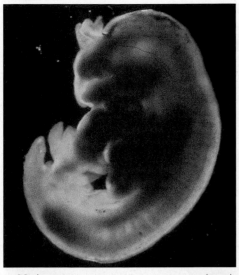

a. 28 days

4–6 mm

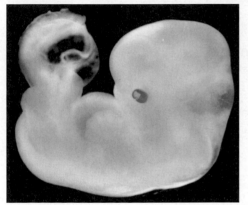

b. 42 days

12–15 mm

c. 56 days

23–32 mm

Figure 3.18 Human embryos.
Embryos at **(a)** 28 days, **(b)** 42 days, and **(c)** 56 days.

Figure 3.19 A fetus at 24 weeks.
At this stage and beyond, a fetus can survive outside of the uterus—but many do not.

widely spaced. Bone begins to replace the softer cartilage. As nerve and muscle functions become coordinated, the fetus moves.

Sex is determined at conception, when a sperm bearing an X or Y chromosome meets an oocyte, which always carries an X chromosome. An individual with two X chromosomes is a female, and one with an X and a Y is a male. A gene on the Y chromosome, called SRY (for "sex-determining region of the Y"), determines maleness. Differences between the sexes do not appear until week 6, after the SRY gene is activated in males. Male hormones then stimulate male reproductive organs and glands to differentiate from existing, indifferent structures. In a female, the indifferent structures of the early embryo develop as female organs and glands. Differences may begin to be noticeable on ultrasound scans by 12 to 15 weeks. Sexuality is discussed further in chapter 6.

By week 12, the fetus sucks its thumb, kicks, makes fists and faces, and has the beginnings of teeth. It breathes amniotic fluid in and out, and urinates and defecates into it. The first trimester (three months) of pregnancy ends.

By the fourth month, the fetus has hair, eyebrows, lashes, nipples, and nails. By 18 weeks, the vocal cords have formed, but the fetus makes no sound because it doesn't breathe air. By the end of the fifth month, the fetus curls into a head-to-knees position. It weighs about 454 grams (1 pound). During the sixth month, the skin appears wrinkled because there isn't much fat beneath it (figure 3.19). The skin turns pink as capillaries fill with blood. By the end of the second trimester, the woman feels distinct kicks and jabs and may even detect a fetal hiccup. The fetus is now about 23 centimeters (9 inches) long.

In the final trimester, fetal brain cells rapidly form networks as organs elaborate and grow. A layer of fat forms beneath the skin. The digestive and respiratory systems mature last, which is why infants born prematurely often have difficulty digesting milk and breathing. Approximately 266 days after a single sperm burrowed its way into an oocyte, a baby is ready to be born.

The birth of a live, healthy baby is against the odds. Of every 100 secondary oocytes exposed to sperm, 84 are fertilized. Of these 84, 69 implant in the uterus, 42 survive one week or longer, 37 survive six weeks or longer, and only 31 are born alive. Of the fertilized ova that do not survive, about half have chromosomal abnormalities that cause problems too severe for development to proceed.

RELATING THE CONCEPTS

During which stage of prenatal development do the defects of holoprosencephaly occur?

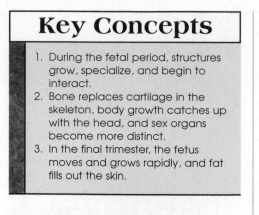

3.5 Birth Defects

When genetic abnormalities or toxic exposures affect an embryo or fetus, developmental problems occur, resulting in birth defects. Only a genetically caused birth defect can be passed to future generations. Although development can be derailed in many ways, about 97 percent of newborns appear healthy at birth.

The Critical Period

The specific nature of a birth defect usually depends on which structures are developing when the damage occurs. The time when genetic abnormalities, toxic substances, or viruses can alter a specific structure is its **critical period (figure 3.20).** Some body parts, such as fingers and toes, are sensitive for short periods of time. In contrast, the brain is sensitive throughout prenatal development, and connections between nerve cells continue to change throughout life. Because of the brain's continuous critical period, many birth defect syndromes include mental retardation.

About two-thirds of all birth defects arise from a disruption during the embryonic period. More subtle defects, such as learning disabilities, that become noticeable only after infancy are often caused by interventions during the fetal period. A disruption in the first trimester might cause mental retardation; in the seventh month of pregnancy, it might cause difficulty in learning to read.

Some birth defects can be attributed to an abnormal gene that acts at a specific point in prenatal development. In a rare inherited condition called phocomelia, for example, an abnormal gene halts limb development from the third to the fifth

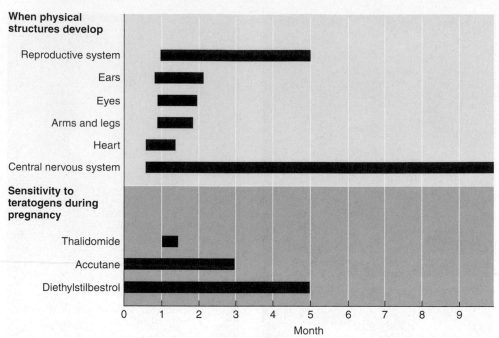

When physical structures develop

Reproductive system
Ears
Eyes
Arms and legs
Heart
Central nervous system

Sensitivity to teratogens during pregnancy

Thalidomide
Accutane
Diethylstilbestrol

0 1 2 3 4 5 6 7 8 9

Month

Figure 3.20 Critical periods of development. The nature of a birth defect resulting from drug exposure depends upon which structures were developing at the time of exposure. The time when a particular structure is vulnerable is called its critical period. Accutane is an acne medication. Diethylstilbestrol (DES) was used in the 1950s to prevent miscarriage. Thalidomide was used to prevent morning sickness.

week of the embryonic period, causing "flippers" to develop in place of arms and legs. The risk that a genetically caused birth defect will affect a particular family member can be calculated.

Many birth defects are caused not by genes but by toxic substances the pregnant woman encounters. These environmentally caused problems will not affect another family member unless the exposure occurs again. Chemicals or other agents that cause birth defects are called **teratogens** (Greek for "monster-causing"). While it is best to avoid teratogens while pregnant, some women may need to remain on potentially teratogenic drugs to maintain their own health.

RELATING THE CONCEPTS

Holoprosencephaly can be caused by a faulty gene or by exposure to a chemical. How do the experiments using chickens to investigate the facial defects of the disorder reveal the critical period for this part of the body?

Teratogens

Most drugs are not teratogens. **Table 3.3** lists some that are.

Thalidomide

The idea that the placenta protects the embryo and fetus from harmful substances was tragically disproven between 1957 and 1961, when 10,000 children were born in Europe with what seemed, at first, to be phocomelia. Because doctors realized that this genetic disorder is very rare, they began to look for another cause. They soon discovered that the mothers had all taken a mild tranquilizer, thalidomide, early in pregnancy, during the time an embryo's limbs form, to alleviate the nausea of morning sickness. Many "thalidomide babies" were born with incomplete or missing legs and arms.

The United States was spared from the thalidomide disaster because an astute government physician noted the drug's adverse effects on laboratory monkeys. Still, several "thalidomide babies" were born in South America in 1994, where

Table 3.3

Teratogenic Drugs

Drug	Medical Use	Risk to Fetus
Alkylating agents	Cancer chemotherapy	Growth retardation
Aminopterin, methotrexate	Cancer chemotherapy	Skeletal and brain malformations
Coumadin derivatives	Seizure disorders	Tiny nose Hearing loss Bone defects Blindness
Diethylstilbestrol (DES)	Repeat miscarriage	Vaginal cancer, vaginal adenosis Small penis
Diphenylhydantoin (Dilantin)	Seizures	Cleft lip, palate Heart defects Small head
Isotretinoin (Accutane)	Severe acne	Cleft palate Heart defects Abnormal thymus Eye defects Brain malformation
Lithium	Bipolar disorder	Heart and blood vessel defects
Penicillamine	Rheumatoid arthritis	Connective tissue abnormalities
Progesterone in birth control pills	Contraception	Heart and blood vessel defects Masculinization of female structures
Tetracycline	Antibiotic	Stained teeth
Thalidomide	Leprosy, AIDS, multiple myeloma	Limb defects

pregnant women were given the drug. In spite of its teratogenic effects, thalidomide is still a valuable drug—it is used to treat leprosy, AIDS, and certain blood and bone marrow cancers.

Cocaine

Cocaine is very dangerous to the unborn. It can cause spontaneous abortion by inducing a stroke in the fetus. Cocaine-exposed infants who do survive are more distracted and unable to concentrate on their surroundings than unexposed infants. Other health and behavioral problems arise as these children grow.

One problem in evaluating the prenatal effects of cocaine is that affected children are often exposed to other environmental influences that could account for their symptoms. Cocaine use by a father can affect an embryo because the cocaine binds to sperm.

Cigarettes

Chemicals in cigarette smoke stress a fetus. Carbon monoxide crosses the placenta and prevents the fetus's hemoglobin molecules from adequately binding oxygen. Other chemicals in smoke prevent nutrients from reaching the fetus. Smoke-exposed placentas lack important growth factors, causing poor growth before and after birth. Cigarette smoking during pregnancy increases the risk of spontaneous abortion, stillbirth, prematurity, and low birth weight.

Alcohol

A pregnant woman who has just one or two alcoholic drinks a day, or perhaps a large amount at a single crucial time, risks fetal alcohol syndrome (FAS) in her unborn child. In the future, tests for gene variants that encode proteins that regulate alcohol metabolism will be able to predict which women and fetuses are at elevated risk for developing FAS.

A child with FAS has a characteristic small head and a flat face (**figure 3.21**). Growth is slow before and after birth. Intellectual impairment ranges from minor learning disabilities to mental retardation. Teens and young adults who have FAS are short and have small heads. More than 80 percent of them retain the facial characteristics of a young child with FAS.

The long-term mental effects of prenatal alcohol exposure are more severe than the physical vestiges. Many adults with FAS function at early grade-school level. They often lack social and communication skills and find it difficult to understand the consequences of actions, form friendships, take initiative, and interpret social cues.

Aristotle noticed problems in children of alcoholic mothers more than 23 centuries ago. In the United States today, 1 to 3 of every 1,000 infants has the syndrome, meaning 2,000 to 6,000 affected children are born each year. Many more children have milder "alcohol-related effects." A fetus of a woman with active alcoholism has a 30 to 45 percent chance of harm from prenatal alcohol exposure.

Nutrients

Certain nutrients ingested in large amounts, particularly vitamins, act as drugs. The acne medicine isotretinoin (Accutane) is a vitamin A derivative that causes spontaneous abortion and defects of the heart, nervous system, and face in exposed embryos. Physicians first noted the tragic effects of this drug nine months after dermatologists began prescribing it to young women in the early 1980s. Another vitamin A-based drug, used to treat psoriasis, as well as excesses of vitamin A itself, also cause birth defects. Some forms of vitamin A are stored in body fat for up to three years.

Excessive exposure to vitamin C can also harm a fetus. The fetus becomes accustomed to the large amounts the woman takes. After birth, when the vitamin supply suddenly plummets, the baby may develop symptoms of vitamin C deficiency (scurvy), bruising easily and being prone to infection.

Malnutrition threatens the fetus as well. A woman must consume extra calories while she is pregnant or breastfeeding.

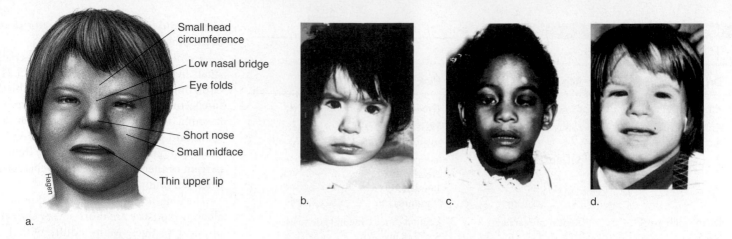

Figure 3.21 Fetal alcohol syndrome. Some children whose mothers drank alcohol during pregnancy have characteristic flat faces **(a)** that are strikingly similar in children of different races **(b, c,** and **d).**

Obstetrical records of pregnant women before, during, and after World War II link inadequate nutrition in early pregnancy to an increase in the incidence of spontaneous abortion. The aborted fetuses had very little brain tissue. Poor nutrition later in pregnancy affects the development of the placenta and can cause low birth weight, short stature, tooth decay, delayed sexual development, and learning disabilities.

Occupational Hazards

Teratogens are present in some workplaces. Researchers note increased rates of spontaneous abortion and children born with birth defects among women who work with textile dyes, lead, certain photographic chemicals, semiconductor materials, mercury, and cadmium. Men whose jobs expose them to sustained heat, such as smelter workers, glass manufacturers, and bakers, may produce sperm that can fertilize an oocyte and then cause spontaneous abortion or a birth defect. A virus or a toxic chemical carried in semen may also cause a birth defect.

Viral Infection

Viruses are small enough to cross the placenta and reach a fetus. Some viruses that cause mild symptoms in an adult, such as the virus that causes chickenpox, may devastate a fetus. Men can transmit viral infections to an embryo or fetus during sexual intercourse.

HIV can reach a fetus through the placenta or infect a newborn via blood contact during birth. Fifteen to 30 percent of infants born to HIV-positive women are HIV positive themselves. The risk of transmission is significantly reduced if a pregnant woman takes anti-HIV drugs. All fetuses of HIV-infected women are at higher risk for low birth weight, prematurity, and stillbirth if the woman's health is failing.

German measles (rubella) is a well-known viral teratogen. Australian physicians first noted its effects in 1941. In the United States, rubella did not gain public attention until the early 1960s, when an epidemic of the usually mild illness caused 20,000 birth defects and 30,000 stillbirths. Women who contract the virus during the first trimester of pregnancy run a high risk of bearing children with cataracts, deafness, and heart defects. In fetuses exposed during the second or third trimesters of pregnancy, rubella may cause, much later, learning disabilities, speech and hearing problems, and type 1 diabetes mellitus.

The incidence of these problems, called congenital rubella syndrome, has dropped markedly thanks to widespread vaccination that has eliminated the disease in the U.S. However, the syndrome resurfaces in unvaccinated populations. A resurgence in 1991 was attributed to a cluster of unvaccinated Amish women in rural Pennsylvania. In that isolated group, 14 of every 1,000 newborns had congenital rubella syndrome, compared to an incidence then of 0.006 per 1,000 in the general U.S. population.

Herpes simplex virus can harm a fetus or newborn whose immune system is not yet completely functional. Forty percent of babies exposed to active vaginal herpes lesions become infected, and half die. Of the survivors, 25 percent sustain severe nervous system damage, and another 25 percent have widespread skin sores. A woman who has sores at the time of birth can have a surgical delivery to protect the child.

Pregnant women are routinely checked for hepatitis B infection, which in adults causes liver inflammation, great fatigue, and other symptoms. Each year in the United States, 22,000 infants are infected with this virus during birth. These babies are healthy, but at high risk for developing serious liver problems as adults. When infected women are identified, a vaccine can be given to their newborns to help prevent complications.

Key Concepts

1. The critical period is the time during prenatal development when a structure is sensitive to damage from a faulty gene or environmental insult.
2. Most birth defects develop during the embryonic period and are more severe than problems that arise during fetal development.
3. Teratogens are agents that cause birth defects.

3.6 Maturation and Aging

"Aging" means moving through the life cycle, and it begins at conception. In adulthood, as we age, the limited life spans of cells are reflected in the waxing and waning of biological structures and functions. Although some aspects of our anatomy and physiology peak very early—such as the number of brain cells or hearing acuity, which do so in childhood—age 30 seems to be a turning point for decline. Some researchers estimate that, after this age, the human body becomes functionally less efficient by about 0.8 percent each year.

Many diseases that begin in adulthood, or are associated with aging, have genetic components. Often these disorders are multifactorial, because it takes many years for environmental exposures to alter gene expression in ways that noticeably affect health. Following is a closer look at how genes may impact health throughout life.

Adult-Onset Inherited Disorders

Human prenatal development is a highly regulated program of genetic switches that are turned on at specific places and times. Environmental factors can affect how certain genes are expressed before birth, creating risks that appear much later. Specifically, adaptations that enable a fetus to grow despite near-starvation become risk factors for certain common illnesses of adulthood, such as coronary artery disease, stroke, hypertension, and type 2 diabetes mellitus. A fetus that does not receive adequate nutrition has intrauterine growth retardation (IUGR), and though born on time, is very small. Premature infants, in contrast, are small but are born early, and are not predisposed to conditions resulting from IUGR.

More than one hundred studies clearly correlate low birth weight due to IUGR with increased incidence of cardiovascular disease. Much of the data come from war records—enough time has elapsed to study the effects of prenatal malnutrition as people age. A study of nearly 15,000 people born in Sweden from 1915 to 1929 correlates IUGR to heightened cardiovascular disease risk after age 65. Similarly, an analysis of individuals who were fetuses during a seven-month famine in the Netherlands in 1943 indicates a high rate of diabetes today. Experiments on intentionally starved sheep and rat fetuses support these historical findings.

How can poor nutrition before birth reverberate as disease many decades later? Perhaps to survive, the starving fetus redirects its circulation to protect vital organs such as the brain, as muscle mass and hormone production change to conserve energy. Growth-retarded babies have too little muscle tissue, and since muscle is the primary site of insulin action, glucose metabolism is altered. Thinness at birth, and the accelerated weight gain in childhood that often occurs to compensate, sets the stage for coronary heart disease and type 2 diabetes much later.

In contrast to the delay before health declines in people who were malnourished as fetuses, symptoms of single gene disorders can begin at any time (**Table 3.4**). In general, conditions that affect children are recessive. A fetus who has inherited osteogenesis imperfecta ("brittle bone disease"), for example, may already have broken bones (**figure 3.22a**). Most dominantly inherited conditions more often start to affect health in early to middle adulthood. This is the case for polycystic kidney disease (OMIM 173900). Cysts that may have been present in the kidneys during one's twenties begin causing bloody urine, high blood pressure, and abdominal pain in the thirties. The joint destruction of osteoarthritis may begin in one's thirties, but not become painful for twenty years. The uncontrollable movements, unsteady gait, and diminishing mental faculties of Huntington disease typically begin near age 40.

Five to 10 percent of Alzheimer disease cases are inherited and produce initial symptoms in the forties and fifties. German neurologist Alois Alzheimer first identified the condition in 1907 as affecting people in mid-adulthood. Noninherited Alzheimer disease typically begins later in life (figure 3.22b).

Whatever the age of onset, Alzheimer disease starts gradually. Mental function declines steadily for three to ten years after the first symptoms appear. Confused and forgetful, Alzheimer patients often wander away from family and friends. Finally, the patient cannot perform basic functions such as speaking or eating and usually must be cared for in a hospital or nursing home.

On autopsy, the brains of Alzheimer disease patients are found to contain deposits of a protein called beta amyloid in learning and memory centers. Alzheimer brains also contain structures called neurofibrillary tangles, which consist of a protein called tau. Tau binds to and disrupts microtubules in nerve cell branches, destroying the shape of the cell.

Disorders That Resemble Accelerated Aging

Genes control aging both passively (as structures break down) and actively (by initiating new activities). A group of inherited disorders whose symptoms are those that accompany aging may hold clues to how genes control the process. It isn't clear whether these conditions actually speed aging, or produce symptoms that resemble those more common in older people.

The most severe rapid aging disorders are the segmental progeroid syndromes. They were once called progerias, but the newer terminology reflects the fact that they do not hasten all aspects of aging. Most of these disorders, and possibly all of them, are caused by cells' inability to adequately repair DNA. This enables mutations that would ordinarily be corrected to persist. Over time, the accumulation of mutations destabilizes the entire genome, and even more mutations occur in somatic cells. The various changes that we associate with aging occur.

Table 3.5 lists the more common segmental progeroid syndromes. They vary in severity. People with Rothmund-Thomson syndrome, for example, may lead a normal life span, but develop gray hair or baldness, cataracts, cancers, and osteoporosis at young ages. The child in **figure 3.23,** in contrast, shows the extremely rapid aging of Hutchinson-Gilford syndrome. An affected child appears normal at birth but slows in growth by the first birthday. Within just a few years, the child becomes wrinkled and bald, with the facial features characteristic of advanced age. The body ages on the inside as well, as arteries clog with fatty deposits. The child usually dies of a heart attack or a stroke by age 13, although some patients live into their twenties. Only a few

Table 3.4

Time of Onset of Genetic Disorders

Prenatal Period	Birth	10 Years	20 Years	30 Years	40 Years	50 Years
Osteogenesis imperfecta	Adrenoleukodystrophy	Familial hypertrophic cardiomyopathy	Multiple endocrine neoplasia	Hemochromatosis	Gout	Fatal familial insomnia
Pituitary dwarfism	Chronic granulomatous disease	Wilson disease	Marfan syndrome	Breast cancer	Huntington disease	Alzheimer disease
Lissencephaly	von Willebrand disease			Polycystic kidney disease	Pattern baldness	Porphyria
Wilms' tumor	Xeroderma pigmentosum					Amyotrophic lateral sclerosis
Polydactyly	Diabetes insipidus					
	Colorblindness					
	Familial hypercholesterolemia					
	Albinism					
	Duchenne muscular dystrophy					
	Menkes disease					
	Sickle cell disease					
	Rickets					
	Cystic fibrosis					
	Hemophilia					
	Tay-Sachs disease					
	Phenylketonuria					
	Progeria					

Table 3.5

Rapid Aging Syndromes

Disorder	Incidence	Average Life Span	OMIM Number
Ataxia telangiectasia	1/60,000	20	208900
Cockayne syndrome	1/100,000	20	216400
Hutchinson-Gilford syndrome	<1/1,000,000	13	176670
Rothmund-Thomson syndrome	<1/100,000	normal	268400
Trichothiodystrophy	<1/100,000	10	601675
Werner syndrome	<1/100,000	50	277700

dozen cases of this syndrome have ever been reported.

Werner syndrome becomes apparent before age 20, causing death before age 50 from diseases associated with aging. Young adults with Werner syndrome develop atherosclerosis, type 2 diabetes mellitus, hair graying and loss, osteoporosis, cataracts, and wrinkled skin. They are short because they skip the growth spurt of adolescence.

Not surprisingly, the cells of segmental progeroid syndrome patients show aging-related changes. Recall that normal cells growing in culture divide about 50 times before dying. Cells from progeroid syndrome patients die in culture after only 10 to 30 divisions. Understanding how and why these cells race through the aging process may help us to understand genetic control of normal aging.

Is Longevity Inherited?

Aging reflects genetic activity plus a lifetime of environmental influences. Families with many very aged members have a fortuitous collection of genes plus shared environmental influences such as good nutrition, excellent health care, devoted relatives, and other advantages. A genome-level approach

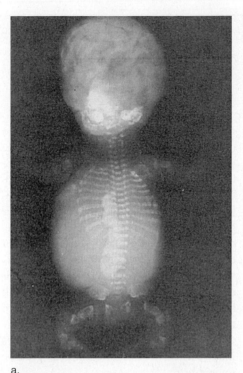

a.

b.

Figure 3.22   Genes act at various stages of development and life.
(a) Osteogenesis imperfecta breaks bones, even before birth. This fetus has broken limb bones, a beaded appearance of the ribs due to fractures, and a poorly mineralized skull. **(b)** At the funeral of former president Richard M. Nixon in April 1994, all was not right with former president Ronald Reagan. He was forgetful and responded inappropriately to questions. Six months later he penned a moving letter confirming that he had Alzheimer disease. He wrote,"My fellow Americans, I have recently been told that I am one of the millions of Americans who will be afflicted with Alzheimer's disease." By 1997, Reagan no longer knew the names of his closest relatives. By 1999, he didn't remember anyone, and by 2001 he no longer recalled being president. He died in June 2004. Because of the late onset of symptoms, Ronald Reagan's Alzheimer disease is probably not due to the malfunction of a single gene, but is multifactorial.

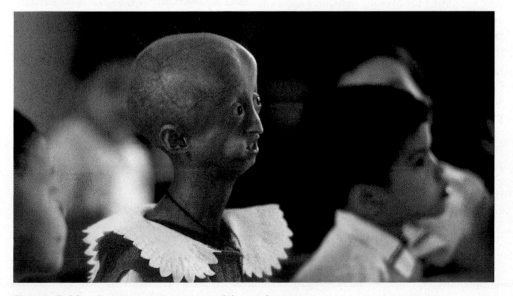

Figure 3.23   Segmental progeroid syndromes.   This child has Hutchinson-Gilford syndrome, which is extremely rare.

to identifying causes of longevity identified a region of chromosome 4 that houses gene variants associated with long life. Genome comparisons among people who've passed their 100th birthdays and those who have died of the common illnesses of older age will reveal other genes that influence longevity (**Reading 3.1**).

It is difficult to tease apart inborn from environmental influences on life span. One approach compares adopted individuals to both their biological and adoptive parents. In a study from Denmark, adopted individuals with one biological parent who died of natural causes before age 50 were more than twice as likely to die before age 50 as were adoptees whose biological parents lived beyond this age, suggesting an inherited component to longevity. Interestingly, adopted individuals whose natural parents died early due to infection were more than five times as likely to also die early of infection, perhaps because of inherited immune system deficiencies. The adoptive parents' ages at death had no influence on that of their adopted children. Chapter 7 explores the "nature versus nurture" phenomenon more closely.

Key Concepts

1. Starvation before birth can set the stage for later disease by affecting gene expression in certain ways.
2. Most single-gene disorders are recessive and strike early in life. Some single-gene disorders have an adult onset.
3. The segmental progeroid syndromes are single-gene disorders that speed aging.
4. Families with many aged members can probably thank their genes as well as the environment. Chromosome 4 houses longevity genes, and genomewide screens are identifying others.
5. Adoption studies compare the effects of genes versus environmental influences on longevity.

The Centenarian Genome

The human genome is like a vast library that holds the clues to good health. One way to identify those clues is to probe the genomes of those who have lived the longest, past 100 years. These fortunate people are called centenarians. Usually they enjoy excellent health, remaining active and interested in community affairs, then succumb rapidly to a disease that usually claims people decades earlier.

Centenarians fall into three broad groups—about 20 percent of them never

Figure 1 This woman has enjoyed living for more than a century. Researchers are discovering clues to good health by probing the genomes of centenarians.

get the diseases that kill most people; 40 percent get these diseases, but at a much older age than average; and the other 40 percent live with and survive the more common disorders of aging. Researchers hope that learning which gene variants offer this protection will lead to better understanding of the disorders that strike most of us in later adulthood—including heart disease, stroke, cancers, type 2 diabetes mellitus, and Alzheimer disease and other dementias.

While the environment seems to play an important role in the deaths of people ages 60 to 85, past that age, genes predominate. That is, someone who dies at age 68 of lung cancer can probably blame a lifetime of cigarette smoking. But a smoker who dies at age 101 of the same disease probably had gene variants that protected against lung cancer. Centenarians have higher levels of large lipoproteins that carry cholesterol (HDL) than other people, which researchers estimate adds 20 years of life.

Evidence that longevity is largely inherited is that children and siblings of centenarians tend to be long-lived as well. Brothers of centenarians are 17 times as likely to live past age 100 as the average man, and sisters are 8.5 times as likely. The fact that some people more than 100 years of age have less-than-healthful habits suggests that genes are protecting them. One researcher suggests that the saying, "The older you get, the sicker you get" be replaced with "The older you get, the healthier you've been."

Centenarians have luckily inherited two types of gene variants—those that directly protect them, and variants of genes that,

when mutant, cause disease. Research focuses on individual genes as well as genomewide scans to identify variants that make it more likely that a person will live past age 100. **Table 1** lists some "candidate" gene types that may control longevity. To find other gene variants that promote long life, researchers are comparing the genomes of centenarians with those of people with particular conditions associated with aging. For example, a group of these very old people all have certain gene variants that differ from those in people with type 2 diabetes mellitus, suggesting that these DNA sequences may protect against deranged glucose metabolism.

The New England Centenarian Study, headed at Boston University, began in 1988 to amass information on families of the oldest citizens in the United States. By studying those who have lived past the century mark, and have lived well, researchers will be able to identify a "healthy standard genome." Perhaps it will provide information that will help the majority of us who do not live as long.

Table 1

Single genes important in aging affect:

- control of insulin secretion and glucose metabolism.
- immune system functioning.
- control of the cell cycle.
- lipid (cholesterol) metabolism.
- response to stress.
- production of antioxidant enzymes.

Summary

3.1 The Reproductive System

1. The male and female reproductive systems include paired **gonads** and networks of tubes in which **sperm** and **oocytes** are manufactured.

2. Male **gametes** originate in seminiferous tubules within the paired testes. They then pass through the epididymis and vasa deferentia, where they mature before exiting the body through the urethra during sexual intercourse. The prostate gland, the seminal vesicles, and the bulbourethral glands add secretions.

3. Female gametes originate in the ovaries. Each month after puberty, one ovary releases an oocyte into a uterine tube. The oocyte then moves to the uterus for implantation (if fertilized) or expulsion.

3.2 Meiosis

4. **Meiosis** reduces the chromosome number in gametes to one genome, from **diploid** to **haploid.** This maintains the chromosome number from generation to generation. Meiosis ensures genetic variability by partitioning different combinations of genes into gametes as a result of **crossing over** and **independent assortment** of chromosomes.

5. Meiosis I, **reduction division,** halves the number of chromosomes. Meiosis II, **equational division,** produces four cells from the two that result from meiosis I, without another DNA replication.

6. Crossing over occurs during prophase I. It mixes up paternally and maternally derived genes on **homologous pairs of chromosomes.**

7. Chromosomes segregate and independently assort in metaphase I, which determines the distribution of genes from each parent in the gamete.

3.3 Gamete Maturation

8. **Spermatogenesis** begins with spermatogonia, which accumulate cytoplasm and replicate their DNA to become primary spermatocytes. After meiosis I, the cells become haploid secondary spermatocytes. In meiosis II, the secondary spermatocytes divide to each yield two spermatids, which then differentiate into spermatozoa.

9. In **oogenesis,** some oogonia grow and replicate their DNA, becoming primary oocytes. In meiosis I, the primary oocyte divides to yield one large secondary oocyte and a much smaller **polar body.** In meiosis II, the secondary oocyte divides to yield the large ovum and another small polar body. Female meiosis is completed at fertilization.

3.4 Prenatal Development

10. In the female, sperm are capacitated and drawn toward a secondary oocyte. One sperm burrows through the oocyte's protective layers with acrosomal enzymes. Fertilization occurs when the sperm and oocyte fuse and their DNA combines in one nucleus, forming the **zygote.** Electrochemical changes in the egg surface block additional sperm from entering. **Cleavage** begins and a 16-celled **morula** forms. Between days 3 and 6, the morula arrives at the uterus and hollows, forming a **blastocyst** made up of **blastomeres.** The trophoblast and inner cell mass form. Around day 6 or 7, the blastocyst implants, and trophoblast cells secrete hCG, which prevents menstruation.

11. During the second week, the amniotic cavity forms as the inner cell mass flattens. **Ectoderm** and **endoderm** form, and then **mesoderm** appears, establishing the **primary germ layers.** Cells in each germ layer begin to develop into specific organs. During the third week, the placenta, yolk sac, allantois, and umbilical cord begin to form as the amniotic cavity swells with fluid. **Monozygotic** twins result when one fertilized ovum splits. **Dizygotic** twins result from two fertilized ova. Organs form throughout the embryonic period. Structures including the primitive streak, the notochord and **neural tube,** arm and leg buds, the heart, facial features, and the skeleton gradually appear.

12. At the eighth week, the **embryo** becomes a **fetus,** with all structures present but not fully grown. Organ rudiments laid down in the embryo grow and specialize. The developing organism moves and reacts, and gradually, its body proportions resemble those of a newborn baby. In the last trimester, the brain develops rapidly, and fat is deposited beneath the skin. The digestive and respiratory systems mature last.

3.5 Birth Defects

13. Birth defects can result from a malfunctioning gene or an environmental intervention.

14. A substance that causes birth defects is a **teratogen.** Environmentally caused birth defects are not transmitted to future generations.

15. The time when a structure is sensitive to damage from an abnormal gene or environmental intervention is its **critical period.**

3.6 Maturation and Aging

16. Genes cause or predispose us to illness throughout life. Single-gene disorders that strike early tend to be recessive, whereas most adult-onset single-gene conditions are dominant.

17. Malnutrition before birth can alter gene expression to increase the risk of type 2 diabetes mellitus and cardiovascular disease much later in life.

18. The segmental progeroid syndromes are single-gene disorders that increase the rate of aging-associated changes.

19. Long life is due to genetics and environmental influences.

Review Questions

1. How many sets of human chromosomes are present in each of the following cell types?

   a. an oogonium

   b. a primary spermatocyte

   c. a spermatid

   d. a cell from either sex during anaphase of meiosis I

   e. a cell from either sex during anaphase of meiosis II

   f. a secondary oocyte

   g. a polar body derived from a primary oocyte

2. List the structures and functions of the male and female reproductive systems.

3. A dog has 39 pairs of chromosomes. Considering only the independent assortment of chromosomes, how many genetically different puppies are possible when two dogs mate? Is this number an underestimate or overestimate of the actual total? Why?

4. How does meiosis differ from mitosis?

5. What do oogenesis and spermatogenesis have in common, and how do they differ?

6. How does gamete maturation differ in the male and female?

7. Describe the events of fertilization.

8. Write the time sequence in which the following structures begin to develop: notochord, gastrula, inner cell mass, fetus, zygote, morula.

9. Give an example of transdifferentiation (you can make one up).

10. Exposure to teratogens tends to produce more severe health effects in an embryo than in a fetus. Why?

11. The same birth defect syndrome can be caused by a mutant gene or exposure to a teratogen. How do the consequences of each cause differ for future generations?

12. List four teratogens, and explain how they disrupt prenatal development.

13. Why is an "anti-aging" pill, diet, or contraption impossible?

14. Cite two pieces of evidence that genes control aging.

Applied Questions

1. Based on your knowledge of human prenatal development, at what stage do you think it is ethical to ban experimentation? Cite reasons for your answer. (The options of banning the research altogether, or of allowing research at any stage, are as valid an option as pinpointing a particular stage.)

2. Under a microscope, a first and second polar body look alike. What structure would a researcher have to look at to distinguish them?

3. Armadillos always give birth to identical quadruplets. Are the offspring clones?

4. Some Vietnam War veterans who were exposed to the herbicide Agent Orange claim that their children—born years after the exposure—have birth defects caused by dioxin, a contaminant in the herbicide. What types of cells would the chemical have to have affected in these men to cause birth defects years later?

5. In about 1 in 200 pregnancies, a sperm fertilizes a polar body instead of an oocyte. A mass of tissue that is not an embryo develops. Why can't a polar body support development of an entire embryo?

6. Should a woman be held legally responsible if she drinks alcohol, smokes, or abuses drugs during pregnancy and it harms her child? Should liability apply to all substances that can harm a fetus, or only to those that are illegal?

7. Would you want to one day have your genome scanned to estimate how long you are likely to live? Why or why not?

8. What types of evidence have led researchers to hypothesize that a poor prenatal environment can raise the risk for certain adult illnesses? How are genes part of this picture?

Web Activities

Visit the Online Learning Center (OLC) at www.mhhe.com/lewisgenetics7. Select **Student Edition, chapter 3** and **Web Activities** to find the website links needed to complete the following activities.

9. Look over the "Living to 100 life expectancy calculator" at the website provided on the OLC and list ten ways that you can change your behavior to possibly live longer. What does this quiz suggest about the relative role of genes and the environment in determining longevity?

10. Go to the website provided on the OLC and look at the photographs of Lily, Daffodil, Crocus, Forsythia, and Rose. What is the evidence that these calf clones are not identical?

Case Studies and Research Results

11. Miguel and Maria know that they are each carriers of cystic fibrosis, and that the condition is severe in their families. They have a procedure called preimplantation genetic diagnosis (see fig. 21.6) to ensure that their child does not inherit the condition. Maria's oocyte is fertilized with Miguel's sperm in a laboratory dish, and it develops to the 8-cell stage. One cell is removed and tested for the mutant CF allele that is on both sides of the family. Only the wild type allele is detected.

    Anna and Peter are also carriers of a genetic disorder that can affect either sex. They cannot get into a preimplantation genetic diagnosis clinical trial, which would be free, their insurance will not cover the procedure, and they cannot afford it. So, they choose chorionic villus sampling (CVS), in which a cell from the developing placenta is tested for the presence of a mutant allele at the tenth week of gestation. Their fetus is found to be a carrier, like them.

    A third couple, Vivian and Max, are not willing to take the higher risk of miscarriage associated with CVS, so they wait until the sixteenth week, and Vivian has amniocentesis. Vivian knows from her family history that she may be a carrier for hemophilia A. If she is, a son would face a

50 percent chance of inheriting the disorder. The amniocentesis indicates a daughter.

a. Why can a gene test performed on one cell of an 8-celled embryo, the developing placenta, or shed from a fetus into the fluid surrounding it, predict the future health of the individual?

b. At the time of preimplantation genetic diagnosis, is the embryo a cleavage embryo, an inner cell mass, or a gastrula?

c. What structures are present in Vivian and Max's fetus that have not yet developed in Anna and Peter's at the time of their prenatal tests?

12. Surgical separation of conjoined twins is more likely to succeed if fewer body parts are shared or attached. This was the case for Maria de Jesus and Maria Teresa, born in Guatemala in 2001 and separated before they celebrated their first birthday. They were joined at the head, but facing opposite directions, so they could not do much more than roll around. The surgery took 23 hours! Today they are well. The outcome wasn't good for Landan and Laleh Bijani, 29-year-old Iranian conjoined twins who could no longer stand being joined along their heads, with their brains fused. They died shortly after 50 hours of surgery in 2003.

If you had conjoined twins, what would you do? Would you attempt surgical separation?

Learn to apply the skills of a genetic counselor with this additional case found in the *Case Workbook in Human Genetics:*

Embryos and "former embryos" in research

Mendelian Inheritance

CHAPTER CONTENTS

BATTLING BATTEN DISEASE

Inheriting a single-gene disorder is like rolling dice—the chance of each outcome is the same for each roll. For the Milto family of Indiana, two of their four sons were unlucky in the game of genetics, inheriting Batten disease.

Nathan Milto was born in 1994. His mother, Tricia, noticed the first symptom when Nathan was 2 1/2 years old and became disturbed at the change in light in a movie theater. A few months later, Nathan's father Phil noticed that the boy couldn't track a baseball. Soon signs of developmental delay were clear, and his vision worsened, symptoms suggesting Batten disease (OMIM 204200). A DNA test confirmed the diagnosis. Each parent was an unaffected carrier, and gave him a gene combination that caused the disease. Nathan would eventually develop jerky movements, poor coordination, and seizures. Within certain brain cells, lack of a lysosomal enzyme causes a substance called lipofuscin to build up. It eventually buries parts of the brain in lipid, causing symptoms.

Soon after Nathan was diagnosed, Phil launched an organization to seek treatment. Then in April, 2001, disaster struck again—Nathan's brother P. J. developed symptoms. Brothers Nicholas and Joey are unaffected. The organization, Nathan's Battle, continued with renewed energy, bringing together academic researchers and pharmaceutical and biotech companies. Three approaches are being pursued—novel drugs, adult stem cell therapy, and gene therapy.

In summer 2004, Nathan and P. J. received experimental gene therapy that delivers a healing gene into the brain. Both boys are doing well. Says their father, "We have done all the testing that we can do on animals, and the only way to know if it will work is to try it on humans. But so far, so good. The boys smile and laugh often and truly love life."

The pattern of inheritance seen so vividly in the Milto family was first discovered in pea plants, many years ago.

A child inherits half of his or her genes from each parent. It is interesting to observe how traits reassort with each generation.

Figure 4.1 Inherited similarities.
Facial similarities are not always as obvious as those between rocker Steven Tyler of Aerosmith and his actress daughter, Liv. Liv's son Milo inherited her family's famous facial features.

Inherited similarities can be startling. When Aerosmith singer Steven Tyler first met his daughter, Liv, when she was nine years old, he knew with one glance that she was his child. He burst into tears, so compelling was the resemblance. Father and daughter have strikingly similar facial features (figure 4.1).

4.1 Following the Inheritance of One Gene—Segregation

Noting resemblances among blood relatives is one way to recognize heredity. A specific disorder not due to infection, present in more than one family member, is another—as was the case for the Milto family and their two boys with Batten disease. Yet another way to analyze heredity is to identify specific DNA sequences, decipher the proteins that they encode, and discover the functions of the proteins.

The beginnings of the field of genetics bridged these macroscopic and microscopic views of heredity. Gregor Mendel used a series of clever breeding experiments in pea plants to describe units of inheritance that pass traits from generation to generation. He called these units "elementen." He could not see them, but inferred their existence from the appearances of his plants. Although Mendel knew nothing of DNA, or even cells or chromosomes, his "laws" of inheritance

have not only stood the test of time, but explain trait transmission in any diploid species—as the problems at the end of the chapter attest.

Mendel the Man

Mendel spent his early childhood in a small village in what is now the Czech Republic, near the Polish border. His father was a farmer, and his mother was the daughter of a gardener, so Mendel learned early how to tend fruit trees. At age 10 he left home to attend a special school for bright students, supporting himself by tutoring. After a few years at a preparatory school, Mendel became a priest at the Augustinian monastery of St. Thomas in Brnö. At this atypical monastery, the priests were also teachers, and they did research in natural science. From them, Mendel learned how to artificially pollinate crop plants to control their breeding.

Mendel wanted to teach natural history, but had difficulty passing the necessary exams due to test anxiety. At age 29, he was such an effective substitute teacher that he was sent to earn a college degree. At the University of Vienna, courses in the sciences and statistics fueled his interest in plant breeding and got him thinking about experiments to address a question that had confounded other plant breeders—why did certain traits disappear in one generation, yet reappear in the next? To solve this puzzle, Mendel bred hybrids and applied the statistics he had learned in college.

From 1857 to 1863, Mendel crossed and cataloged traits in 24,034 plants, through several generations. He deduced that consistent ratios of traits in the offspring indicated that the plants transmitted distinct units, or "elementen." He derived two hypotheses to explain how inherited traits are transmitted. Mendel described his work to the Brnö Medical Society in 1865 and published it in the organization's journal the next year. The remarkably clear paper discusses plant hybridization, the reappearance of traits in the third generation, and the joys of working with peas, plus data.

Mendel's hypotheses eventually became laws because they apply to all diploid species. But it took years for his findings to be recognized. His treatise was published in English in 1901. Then three botanists (Hugo DeVries, Karl Franz Joseph Erich Correns,

and Seysenegg Tschermak) independently rediscovered the laws of inheritance, then found Mendel's paper. All credited Mendel, who came to be regarded as the "father of genetics."

In the twentieth century, researchers discovered the molecular basis of some of the traits that Mendel studied. The "short" and "tall" plants reflected the expression of a gene that enables a plant to produce the hormone gibberellin, which elongates the stem. One tiny change to the DNA, and a short plant results. Likewise, "round" and "wrinkled" peas arise from the R gene, whose encoded protein connects sugars into branching polysaccharide molecules. Seeds with a mutant gene cannot attach the sugars. As a result, water exits the cells, and the peas wrinkle.

Mendel's Experiments

Peas are ideal for probing heredity because they are easy to grow, develop quickly, and have many traits that take one of two easily distinguishable forms. **Figure 4.2** illustrates the seven traits that Mendel followed. When analyzing genetic crosses, the first generation is the parental generation, or P_1; the second generation is the first filial generation, or F_1; the next generation is the second filial generation, or F_2, and so on.

Mendel's first experiments dealt with single traits with two expressions, such as "short" and "tall." He set up all combinations of possible artificial pollinations, manipulating fertilizations to cross tall with tall, short with short, and tall with short. This last combination, plants with one trait variant crossed to plants with the alternate, produces hybrids, or offspring with one of each. Mendel noted that short plants crossed to other short plants were "true-breeding," always producing short plants.

The crosses of tall plants to each other were more confusing. Some tall plants were true-breeding, but others crossed with each other yielded short plants in about one-quarter of the next generation. It appeared as if in some tall plants, tallness could mask shortness. One trait that masks another is said to be **dominant;** the masked trait is **recessive.** Wrote Mendel,

> **In the case of each of the seven crosses the hybrid character resembles that of one of the parental forms so closely**

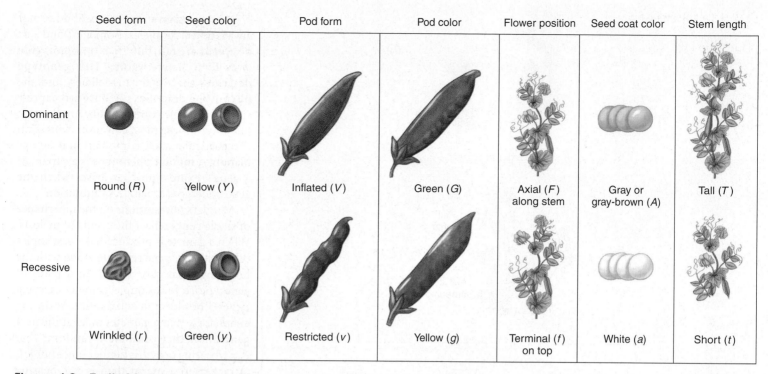

	Seed form	Seed color	Pod form	Pod color	Flower position	Seed coat color	Stem length
Dominant	Round (*R*)	Yellow (*Y*)	Inflated (*V*)	Green (*G*)	Axial (*F*) along stem	Gray or gray-brown (*A*)	Tall (*T*)
Recessive	Wrinkled (*r*)	Green (*y*)	Restricted (*v*)	Yellow (*g*)	Terminal (*f*) on top	White (*a*)	Short (*t*)

Figure 4.2 Traits Mendel studied. Gregor Mendel studied the transmission of seven traits in the pea plant. Each trait has two easily distinguished expressions, or phenotypes.

that the other either escapes observation completely or cannot be detected with certainty. . . . The expression "recessive" has been chosen because the characters thereby designated withdraw or entirely disappear in the hybrids, but nevertheless reappear unchanged in their progeny.

Mendel conducted up to 70 hybrid crosses for each of the seven traits. Because one trait is followed and the parents are hybrids, this is called a **monohybrid cross.**

When Mendel allowed the non-true-breeding tall plants—monohybrids—to self-fertilize, the progeny were in the ratio of one-quarter short to three-quarters tall plants (**figure 4.3**). In further crosses, he found that two-thirds of the tall plants from the monohybrid F_1 cross were non-true-breeding, and the remaining third were true-breeding.

In these experiments, Mendel confirmed that hybrids hide one expression of a trait—short, in this case—which reappears when hybrids are self-crossed. He tried to explain how this happened: Gametes distribute "elementen" because these cells physically link generations. Paired sets of elementen separate

as gametes form. When gametes join at fertilization, the elementen combine anew. Mendel reasoned that each element was packaged in a separate gamete. If opposite-sex gametes combine at random, he could mathematically explain the different ratios of traits produced from his pea plant crossings. Mendel's idea that elementen separate in the gametes would later be called the **law of segregation.** Frustrated at the lack of recognition he received as a scientist, Mendel eventually turned his energies to monastery administration.

When Mendel's ratios were demonstrated in several species in the early 1900s, just when chromosomes were being described for the first time, it became apparent that elementen and chromosomes had much in common. Both paired elementen and pairs of chromosomes separate at each generation and are transmitted—one from each parent—to offspring. Both elementen and chromosomes are inherited in random combinations. Therefore, chromosomes provided a physical mechanism for Mendel's hypotheses. In 1909, English embryologist William Bateson renamed Mendel's elementen *genes* (Greek for "give birth to"). It wasn't until the 1940s, however, that scientists began investigating the gene's chemical basis. We

pick up the historical trail at this point in chapter 9.

Terms and Tools to Follow Segregating Genes

We can describe the law of segregation in terms of the behavior of chromosomes and genes during meiosis. Because a gene is a long sequence of DNA, it can vary in many ways. An individual with two identical alleles for a gene is **homozygous** for that gene. An individual with two different alleles is **heterozygous**—what Mendel called "non-true-breeding" or "hybrid."

When considering two alleles, it is common to symbolize the dominant one with a capital letter and the recessive with the corresponding small letter. If both alleles are recessive, the individual is homozygous recessive. Two small letters, such as *tt* for short plants, symbolize this. An individual with two dominant alleles is homozygous dominant. Two capital letters, such as *TT* for tall pea plants, represent this. Another possible allele combination is one dominant and one recessive allele—*Tt* for non-true-breeding tall pea plants, or heterozygotes.

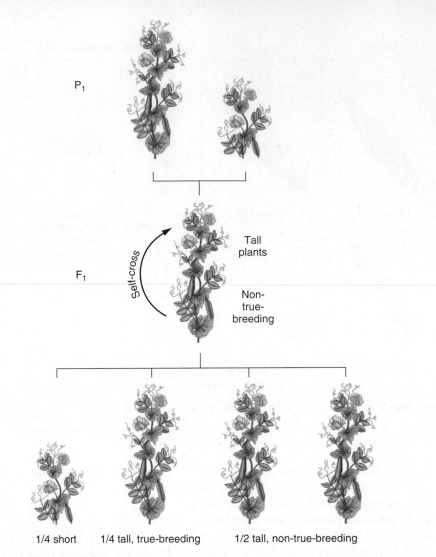

P₁

Self-cross

F₁

Tall plants

Non-true-breeding

F₂

1/4 short   1/4 tall, true-breeding   1/2 tall, non-true-breeding

Figure 4.3   A monohybrid cross. When Mendel crossed true-breeding tall plants with short plants, the next generation plants were all tall. When he self-crossed the F₁ plants, one-quarter of the plants in the next generation, the F₂, were short, and three-quarters were tall. Of the tall plants in the F₂, one-third were true-breeding, and the other two-thirds were not true-breeding. He could tell this by conducting further crosses of the tall plants to short plants, to see which bred true.

Table 4.1
Mendel's Law of Segregation

Experiment	Total	Dominant	Recessive	F₂ Phenotypic Ratios
1. Seed form	7,324	5,474	1,850	2.96:1
2. Seed color	8,023	6,022	2,001	3.01:1
3. Seed coat color	929	705	224	3.15:1
4. Pod form	1,181	882	299	2.95:1
5. Pod color	580	428	152	2.82:1
6. Flower position	858	651	207	3.14:1
7. Stem length	1,064	787	277	2.84:1
				Average = 2.98:1

An organism's appearance does not always reveal its alleles. Both a *TT* and a *Tt* pea plant are tall, but *TT* is a homozygote and *Tt* a heterozygote. The **genotype** describes the organism's alleles, and the **phenotype** describes the outward expression of an allele combination. A **wild type** phenotype is the most common expression of a particular allele combination in a population. A **mutant** phenotype is a variant of a gene's expression that arises when the gene undergoes a change, or **mutation.**

Mendel's observations on the inheritance of single genes reflect the events of meiosis. When a gamete is produced, the two copies of a particular gene separate along with the homologs that carry them. In a plant of genotype *Tt*, for example, gametes carrying either *T* or *t* form in equal numbers during anaphase I. When gametes meet at the next generation, they combine at random. That is, a *t*-bearing oocyte is neither more nor less attractive to a sperm than is a *T*-bearing oocyte. These two factors—equal allele distribution into gametes and random combinations of gametes—underlie Mendel's law of segregation (**figure 4.4).**

Meiosis explains what Mendel saw when he crossed short and tall plants. When he crossed short plants (*tt*) with true-breeding tall plants (*TT*), the seeds grew into F₁ plants that were all tall (genotype *Tt*). Next, he self-crossed the F₁ plants. The three possible genotypic outcomes were *TT, tt,* and *Tt*. A *TT* individual resulted when a *T* sperm fertilized a *T* oocyte; a *tt* plant resulted when a *t* oocyte met a *t* sperm; and a *Tt* individual resulted when either a *t* sperm fertilized a *T* oocyte, or a *T* sperm fertilized a *t* oocyte.

Because two of the four possible gamete combinations produce a heterozygote, and each of the others produces a homozygote, the genotypic ratio expected of a monohybrid cross is 1 *TT*: 2 *Tt*: 1 *tt*. The corresponding phenotypic ratio is three tall plants to one short plant, a 3:1 ratio. Mendel saw these results for all seven traits that he studied, although, as **table 4.1** shows, the ratios were not exact. Today we use a diagram called a **Punnett square** to derive these ratios (**figure 4.5).** A Punnett square represents how particular genes in gametes come together, assuming they are carried on different chromosomes. Experiments yield numbers of offspring that approximate these ratios.

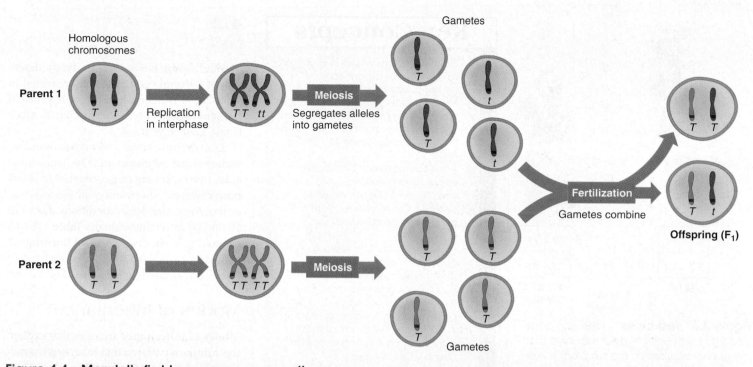

Figure 4.4 **Mendel's first law—gene segregation.** During meiosis, homologous pairs of chromosomes (and the genes that compose them) separate from one another and are packaged into separate gametes. At fertilization, gametes combine at random to form the individuals of a new generation. Green and blue denote different parental origins of the chromosomes. In this example, offspring of genotype *TT* are also generated, along with genotype *Tt* .

RELATING THE CONCEPTS

Phil and Tricia Milto are heterozygotes for the Batten disease gene. Nathan, Joey, and P. J. are homozygotes. Nicholas has not been tested, but is healthy.

1. Use the letters *B* and *b* to represent the genotypes of Nathan and P. J., the affected boys.
2. What are two possible genotypes for Nicholas?
3. Construct a Punnett square to represent the Milto family.

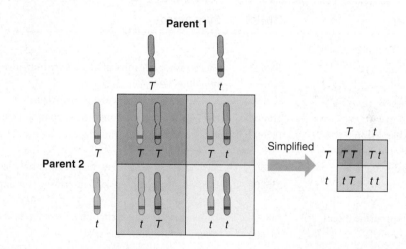

Figure 4.5 **A Punnett square.** A Punnett square is a diagram of how two alleles of a gene combine in a cross between two individuals. The different types of gametes of one parent are listed along the top of the square, with those of the other parent listed on the left-hand side. Each compartment within the square contains the genotype that results when gametes that correspond to that compartment join. The Punnett square here describes a monohybrid cross of two tall pea plants. Among the progeny, tall plants outnumber short plants 3:1. Can you determine the genotypic ratio? Punnett squares usually indicate only the alleles, as shown at the right.

Mendel distinguished the two genotypes resulting in tall progeny—*TT* from *Tt*—with additional crosses (**figure 4.6**). He bred tall plants of unknown genotype with short (*tt*) plants. If a tall plant crossed with a *tt* plant produced both tall and short progeny, Mendel knew it was genotype *Tt;* if it produced only tall plants, he knew it must be *TT.* Crossing an individual of unknown genotype with a homozygous recessive individual is called a test cross. It is based on the logic that the homozygous recessive is the only genotype that can be identified by its phenotype—that is, a short plant is always *tt.* The homozygous recessive is therefore a "known" that can reveal the unknown genotype of another individual when the two are crossed.

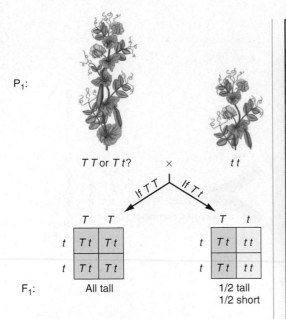

P₁:

$T\,T$ or $T\,t$? × $t\,t$

If $T\,T$ If $T\,t$

	T	T
t	Tt	Tt
t	Tt	Tt

	T	t
t	Tt	tt
t	Tt	tt

F₁: All tall   1/2 tall
                          1/2 short

Figure 4.6 Test cross. Breeding a tall pea plant with homozygous recessive short plants reveals whether the tall plant is true-breeding (TT) or non-true-breeding (Tt).

Key Concepts

1. From observing crosses in which two tall pea plants produced short offspring, along with other crosses, Mendel deduced that "elementen" for height segregate during meiosis, then combine at random with those from the opposite gamete at fertilization.
2. A homozygote has two identical alleles, and a heterozygote has two different alleles. The allele expressed in a heterozygote is dominant; the allele not expressed is recessive.
3. A monohybrid cross yields a genotypic ratio of 1:2:1 and a phenotypic ratio of 3:1.
4. Punnett squares help calculate expected genotypic and phenotypic ratios among progeny.
5. A test cross uses a homozygous recessive individual to reveal an unknown genotype.

4.2 Single-Gene Inheritance in Humans

Mendel's first law addresses traits determined by single genes, as demonstrated in pea plants. Transmission of single genes in humans is called Mendelian, unifactorial, or single-gene inheritance.

Even the most familiar Mendelian disorders, such as sickle cell disease and Duchenne muscular dystrophy, are rare compared to infectious diseases, cancer, and multifactorial disorders. Most Mendelian conditions affect 1 in 10,000 or fewer individuals. **Table 4.2** lists some Mendelian disorders, and **Reading 4.1** considers some interesting traits described in *Online Mendelian Inheritance in Man*.

Modes of Inheritance

Modes of inheritance are rules that explain the common patterns that inherited characteristics follow as they are passed through

Table 4.2

Some Mendelian Disorders in Humans

Disorder	OMIM	Symptoms
Autosomal Recessive		
Ataxia telangiectasis	208900	Facial rash, poor muscular coordination, involuntary eye movements, high risk for cancer, sinus and lung infections
Batten disease	204500	Visual loss, developmental delay, seizures
Cystic fibrosis	219700	Lung infections and congestion, poor fat digestion, male infertility, poor weight gain, salty sweat
Familial hypertrophic cardiomyopathy	192600	Overgrowth of heart muscle, causing sudden death in young adults
Gaucher disease	230800	Swollen liver and spleen, anemia, internal bleeding, poor balance
Hemochromatosis	235200	Iron retention; high risk of infection, liver damage, excess skin pigmentation, heart and pancreas damage
Maple syrup urine disease	248600	Lethargy, vomiting, irritability, mental retardation, coma, and death in infancy
Phenylketonuria	261600	Mental retardation, fair skin
Sickle cell disease	603903	Joint pain, spleen damage, high risk of infection
Tay-Sachs disease	272800	Nervous system degeneration
Autosomal Dominant		
Achondroplasia	100800	Dwarfism with short limbs, normal-size head and trunk
Familial hypercholesterolemia	144010	Very high serum cholesterol, heart disease
Huntington disease	143100	Progressive uncontrollable movements and personality changes, beginning in middle age
Lactose intolerance	150220	Inability to digest lactose, causing cramps after ingestion
Marfan syndrome	154700	Long limbs, sunken chest, lens dislocation, spindly fingers, weakened aorta
Myotonic dystrophy	160900	Progressive muscle wasting
Neurofibromatosis (1)	162200	Brown skin marks, benign tumors beneath skin
Polycystic kidney disease	173900	Cysts in kidneys, bloody urine, high blood pressure, abdominal pain
Polydactyly	174200	Extra fingers and/or toes
Porphyria variegata	176200	Red urine, fever, abdominal pain, headache, coma, death

Reading 4.1

It's All in the Genes

Do you have uncombable hair, misshapen toes or teeth, or a pigmented tongue tip? Are you unable to smell a squashed skunk, or do you sneeze repeatedly in bright sunlight? Do you lack teeth, eyebrows, eyelashes, nasal bones, thumbnails, or fingerprints? If so, your unusual trait may be one of thousands described in *Online Mendelian Inheritance in Man* (OMIM, at www.ncbi.nlm.nih.gov/entrez/query.fcgi?db=OMIM), a database that is updated daily. Entering a disease name retrieves family histories, clinical descriptions, mode of inheritance, and molecular information. Amidst the medical terminology and genetic jargon are the stories behind some fascinating inherited traits.

Genes control whether hair is blond, brown, or black, has red highlights, and is straight, curly, or kinky. Widow's peaks, cowlicks, a whorl in the eyebrow, and white forelocks run in families; so do hairs with triangular cross-sections. Some people have multicolored hairs, like cats; others have hair in odd places, such as on the elbows, nose tip, knuckles, palms, or soles. Teeth can be missing or extra, protuberant or fused, present at birth, shovel-shaped, or "snow-capped." A person can have a grooved tongue, duckbill lips, flared ears, egg-shaped pupils, three rows of eyelashes, spotted nails, or "broad thumbs and great toes." Extra breasts are known in humans and guinea pigs, and one family's claim to genetic fame is a double nail on the littlest toe.

Unusual genetic variants can affect metabolism, producing either disease or harmless, yet noticeable, effects. Members of some families experience "urinary excretion of odoriferous component of asparagus" or "urinary excretion of beet pigment," producing a strange odor or dark pink urine after consuming the offending vegetable. In blue diaper syndrome, an infant's urine turns blue on contact with air, thanks to an inherited inability to break down an amino acid.

One bizarre inherited illness is the Jumping Frenchmen of Maine syndrome (OMIM 244100). This exaggerated startle reflex was first noted among French-Canadian lumberjacks from the Moosehead Lake area of Maine, whose ancestors were from the Beauce region of Quebec. Physicians first reported the condition at a medical conference in 1878. Geneticists videotaped the startle response in 1980, and the condition continues to appear in genetics journals. OMIM offers a most vivid description:

> If given a short, sudden, quick command, the affected person would respond with the appropriate action, often echoing the words of command.... For example, if one of them was abruptly asked to strike another, he would do so without hesitation, even if it was his mother and he had an ax in his hand.

The Jumping Frenchmen of Maine syndrome may be an extreme variant of the more common Tourette syndrome, which causes tics and other uncontrollable movements. **Figure 1** illustrates some other genetic variants.

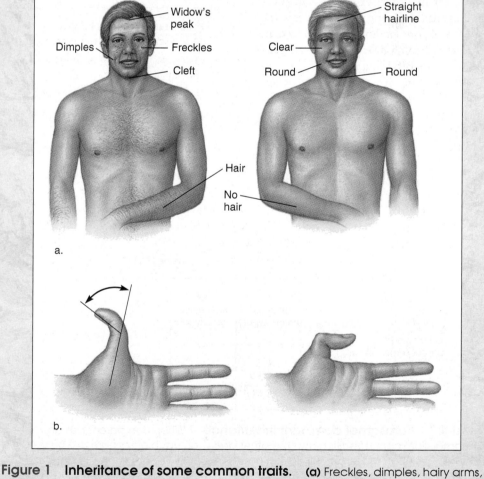

a.

b.

Figure 1   Inheritance of some common traits.   (a) Freckles, dimples, hairy arms, widow's peak, and a cleft chin are examples of dominant traits. **(b)** The ability to bend the thumb backward or forward is inherited.

families. Knowing the mode of inheritance makes it possible to calculate the probability that a particular couple will have a child who inherits a particular condition. For example, after Nathan Milto was diagnosed with Batten disease, his parents learned that each of their children faced a 1 in 4 chance of inheriting the illness.

Mendel derived his laws by studying traits carried on autosomes (non-sex chromosomes). The way those laws affect the modes of inheritance depends on whether a trait is transmitted on an autosome or a sex chromosome, and whether an allele is recessive or dominant. **Autosomal dominant** and **autosomal recessive** are the two modes of inheritance directly derived from Mendel's laws.

Autosomal Dominant Inheritance

In autosomal dominant inheritance, a trait can appear in either sex because an autosome carries the gene. If a child has the trait, at least one parent must also have it. Autosomal dominant traits do not skip generations. If no offspring inherit the trait in one generation, its transmission stops because the offspring can pass on only the recessive form of the gene. **Figure 4.7** uses a Punnett square to predict the genotypes and phenotypes of offspring of a mother

who has an autosomal dominant trait and a father who does not.

A young man, James Poush, discovered an autosomal dominant trait in his family while in high school and published the first full report on distal symphalangism. James and certain relatives had stiff fingers and toes with tiny nails. When he studied genetics, he realized this might be a Mendelian trait. James identified 27 affected individuals among 156 relatives, and concluded that the trait is autosomal dominant. Of 63 relatives with an affected parent, 27 (43 percent) were affected—close to the 50 percent expected for autosomal dominant inheritance (**table 4.3**). Bioethics: Choices for the Future (page 84) considers the dilemma of detecting autosomal dominant mutant

---

Table 4.3

Criteria for an Autosomal Dominant Trait

1. Males and females can be affected. Male-to-male transmission can occur.

2. Males and females transmit the trait with equal frequency.

3. Successive generations are affected.

4. Transmission stops after a generation in which no one is affected.

---

genes before symptoms arise for an untreatable illness.

Autosomal Recessive Inheritance

An autosomal recessive trait can appear in either sex. Affected individuals have a homozygous recessive genotype, whereas in heterozygotes—also called "carriers"—the wild type allele masks expression of the mutant allele. Consider the Deford family. Sportswriter Frank Deford and his wife, Carol, had their daughter Alex in 1972. She died at age eight of cystic fibrosis. Frank Deford wrote about his feelings at the time of diagnosis in a book, *Alex, the Life of a Child:*

> **I went to the encyclopedia and read about this cystic fibrosis. To me, at that point, it was one of those vague diseases you hear about now and then. . . . One out of every 20 whites carries the defective gene, as I do, as Carol does, as perhaps 10 million other Americans do—a population about the size of Illinois or Ohio. . . . If Carol and I had been lucky, if our second-born had not been cursed with cystic fibrosis, then we probably never would have had another child, and our bad genes would merely have been passed on, blissfully unknown to us.**

Frank Deford described the very essence of recessive inheritance: skipped generations in the expression of a trait. Because the human generation time is so long—about 25 to 30 years—we can usually trace a trait or illness for only two or three generations. For this reason, the Defords did not know of any other affected relatives; neither did the Miltos.

Alex Deford inherited a mutant allele on chromosome 7 from each of her carrier parents. They were unaffected because they each also had a dominant allele that encodes enough functional protein for health. **Figure 4.8** shows another monohybrid cross in humans.

Mendel's first law can be used to calculate the probability that an individual will have either of two phenotypes. The probabilities of each possible genotype are added. For example, the chance that a child whose parents are both carriers of cystic fibrosis

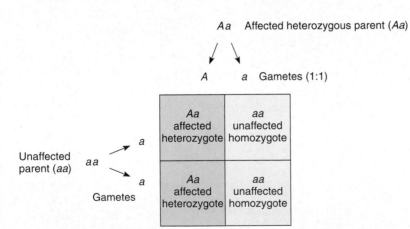

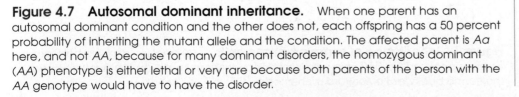

Figure 4.7 Autosomal dominant inheritance. When one parent has an autosomal dominant condition and the other does not, each offspring has a 50 percent probability of inheriting the mutant allele and the condition. The affected parent is *Aa* here, and not *AA*, because for many dominant disorders, the homozygous dominant (*AA*) phenotype is either lethal or very rare because both parents of the person with the *AA* genotype would have to have the disorder.

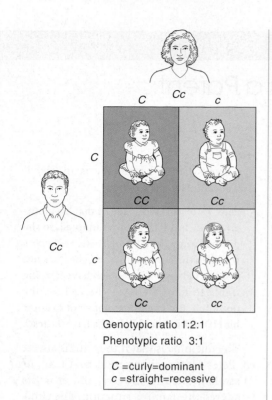

Genotypic ratio 1:2:1
Phenotypic ratio 3:1

C =curly=dominant
c =straight=recessive

Figure 4.8 Autosomal recessive inheritance. A 1:2:1 genotypic ratio results from a monohybrid cross, whether in peas or people. Curly hair (*C*) is dominant to straight hair (*c*). This pedigree depicts a monohybrid cross for hair curliness.

will *not* have the condition is the sum of the probability that she has inherited two normal alleles (1/4) plus the chance that she herself is a heterozygote (1/2), or 3/4. Note that this also equals 1 minus the probability that she is homozygous recessive and has the condition.

The ratios that Mendel's first law predicts for autosomal recessive inheritance apply to each offspring anew. Misunderstanding this concept leads to a common problem in genetic counseling. Many people conclude that if they have already had a child affected by an autosomal recessive illness, then their next three children are guaranteed to escape it. This isn't true. Each child faces the same 25 percent risk of inheriting the condition, as the Miltos experienced.

Most autosomal recessive conditions occur unexpectedly in families. However, blood relatives who have children together have a much higher risk of having a child with an autosomal recessive condition.

Marriage between relatives introduces **consanguinity**, which means "shared blood"—a figurative description, since genes are not passed in blood. Relatives can trace their families back to a common ancestor. For example, an unrelated man and woman have eight different grandparents, but first cousins have only six, because they share one pair through their parents, who are siblings (see figure 4.14*c*). Consanguinity increases the likelihood of inherited disease because the parents may have inherited the same mutant recessive allele from a shared grandparent. That is, the probability of two relatives inheriting the same disease-causing recessive allele is greater than that of two unrelated people having the same allele by chance. Chapter 7 discusses different types of cousins.

The nature of the phenotype is important when evaluating transmission of Mendelian traits. For example, each adult sibling of a person who is a known carrier of Tay-Sachs disease has a two-thirds chance of being a carrier. The probability is two-thirds, and not one-half, because there are only three genotypic sources for an adult—homozygous for the normal allele, or a carrier who inherits the mutant allele from either mother or father. A homozygous recessive individual would not have survived childhood.

Geneticists who study human traits and illnesses can hardly set up crosses as Mendel did, but they can pool information from families whose members have the same trait or illness. Consider a simplified example of 50 couples in whom both partners are carriers of sickle cell disease. If 100 children are born, about 25 of them would be expected to have sickle cell disease. Of the remaining 75, theoretically 50 would be carriers like their parents, and the remaining 25 would have two wild type alleles. **Table 4.4** lists criteria for an autosomal recessive trait.

Solving a Problem: Segregation

Using Mendel's laws to predict phenotypes and genotypes requires a careful reading of the problem to identify and organize relevant information. Sometimes common sense is useful, too. The following general

steps can help to solve a problem that addresses Mendel's first law, the inheritance of a single-gene trait.

1. List all possible genotypes and phenotypes for the trait.
2. Determine the genotypes of the individuals in the first (P_1) generation. This may require deductive reasoning based on information about those people's parents.
3. After determining the genotypes, determine the possible alleles in gametes produced by each individual in the cross.
4. Unite these gametes in all combinations, using a Punnett square if necessary, to reveal all possible genotypes. Calculate ratios for the first generation of offspring (F_1).
5. To extend predictions to the second offspring (F_2) generation, use the genotypes of the specified F_1 individuals and repeat steps 3 and 4.

As an example, consider curly hair, depicted in figure 4.8. If *C* is the dominant allele, conferring curliness, and *c* is the recessive allele, then both *CC* and *Cc* genotypes result in curly hair. A person with *cc* genotype has straight hair.

Wendy has beautiful curls, and her husband Rick has straight hair. Wendy's father is bald, but once had curly hair, and her mother has stick-straight hair. What is the probability that Wendy and Rick's child will have straight hair? Steps 1 through 5 solve the problem:

1. State possible genotypes:
   CC, Cc = curly   *cc* = straight
2. Determine genotypes: Rick must be *cc*, because his hair is straight. Wendy must

When Diagnosing a Fetus Also Diagnoses a Parent: Huntington Disease

Mendel's laws apply to all diploid organisms. They can have profound effects on families in which some members inherit a disease caused by a mutation in a single gene. The case of Huntington disease (HD) is particularly complex, because symptoms usually do not appear until about 38 to 40 years of age—often after people have had children and made career and other choices (see Tables 12.4 and 12.8). In HD, initial symptoms of clumsiness and slurred speech progress to a near-constant writhing and repetitive, dance-like movements—ceasing only during sleep. People typically live with the worsening disease for fifteen to twenty years, usually dying of infection. Some become demented; others remain aware. Drugs can dampen symptoms, but there is no cure.

HD is autosomal dominant, so each offspring of an affected individual has a 50 percent (1 in 2) chance of inheriting the condition. Before diagnosis, the child of an affected parent, or the sibling of an affected individual, is said to be "at-risk." (They aren't really carriers because this is not a recessive disorder. *HH* and *Hh* genotypes cause the same symptoms.) In the United States, about 150,000 people are at-risk. Unlike the case in many other disorders with a genetic component, inheriting the mutant HD gene means that the disease will develop with close to 100 percent certainty. A genetic marker test, requiring that several family members be tested to establish a pattern, became available to some families in 1983, offering a prediction accuracy of 96 to 99 percent. In 1993, direct testing for the mutation became possible with the discovery of the HD gene on one tip of chromosome 4. It encodes a protein called huntingtin that, when mutant, contains a stretch of extra amino acids that causes it to misfold and form clumps in certain brain cells. This causes the symptoms.

The ability to predict HD posed psychological difficulties for at-risk individuals, who usually know what the disease is like. Who should be tested? One study found that from 10 to 20 percent of at-risk individuals offered the gene test actually take it, generally after months of genetic counseling. Clinicians feared that receiving bad news would cause some people to commit suicide, but studies have alleviated this concern. A study of at-risk individuals offered HD testing in Canada, where there is no fear of health insurance discrimination because of national health care, found that older people were most likely to be tested. They used the information primarily to make financial, rather than reproductive, decisions.

HD testing brought other surprises. Often individuals given a genetic reprieve feel depressed. This "survivor guilt," rather than euphoria, is common, particularly if one's siblings inherit the disease. Some people who had made life decisions based on a worst-case scenario—that they would one day have HD—found they did not have the mutation after all, but had already made irreversible decisions, such as not having children. One young man, convinced he faced a future with HD, took up very risky behaviors—skydiving, bungee jumping, hang gliding, and gambling. He also had a vasectomy. What he didn't have was the mutation.

Yet predictive testing for HD is a relief for some people. Said one woman who did not have the mutation, "The devil you know is far better than the one you don't know." Her siblings refused to be tested. Researchers have found that those who psychologically could not handle the test results tend to refuse testing, which may explain why the suicide rate is lower than expected. A dilemma arises with testing, however, when a pregnancy is involved. Consider the following case:

> Peter M. discovered at age 24 that his mother, age 45, has HD. She was adopted, so she did not know of a family history. Peter's wife Martha is pregnant, and she does not want to have a child who will have HD. She wants to have the fetus checked for the mutation. But Peter does not want to know his HD status. Should Martha have the test?

The dilemma is that if the fetus is affected, Peter's fate will be revealed. A team of Australian doctors, lawyers, and ethicists pondered this family's situation. The clinicians decided it was right to offer testing to Peter first, and then test the fetus only if Peter had the mutation. This would be medically safer and would leave the choice with Peter. They discouraged Martha from having the test and keeping the results secret, because this would place great stress on all involved. The physicians acknowledged that Martha could insist and have the test, because it would be her body that would be tested, but they discouraged it because of the late-onset nature of the disorder—that is, even an affected child would have many good years. In contrast, the legal team concluded that only Martha's consent would be needed, but left the final recommendations to the doctors.

The ethicists advised providing support for the individual with the most compelling need. If Peter would become suicidal with a test result indicating HD, then his right to refuse to know should be paramount. However, if he could handle such information, then Martha should have the test, because not doing so would compromise her ability to make an informed choice about the pregnancy.

What would you do?

be *Cc*, because her mother has straight hair and therefore gave her a *c* allele.

3. Determine gametes: Rick's sperm carry only *c*. Half of Wendy's oocytes carry *C*, and half carry *c*.

4. Unite the gametes:

		Wendy	
		C	*c*
Rick	*c*	*Cc*	*cc*
	c	*Cc*	*cc*

5. Conclusion: Each child of Wendy and Rick has a 50 percent chance of having curly hair (*Cc*) and a 50 percent chance of having straight hair (*cc*).

On the Meaning of Dominance and Recessiveness

Determining whether an allele is dominant or recessive is critical in medical genetics because it helps predict which individuals are at high risk of inheriting a particular condition. Dominance and recessiveness reflect the characteristics or abundance of a protein.

Mendel based his definitions of dominance and recessiveness on what he could see—one allele masking the other. Today we can often add a cellular or molecular explanation. Consider inborn errors of metabolism, which are caused by the absence of an enzyme. These disorders tend to be recessive; although cells of a carrier make half the normal amount of the enzyme, this is usually sufficient to maintain health. The one normal allele, therefore, compensates for the mutant one, to which it is dominant. The situation is similar in pea plants. Short stem length results from deficiency of an enzyme that activates a growth hormone, but the *Tt* plants produce enough hormone to attain the same height as *TT* plants.

A recessive trait is sometimes called a "loss of function" because the recessive allele usually causes the loss of normal protein production and function, so there is too little of a particular protein. In contrast, some dominantly inherited disorders result from the action of an abnormal protein that interferes with the function of the normal protein. Huntington disease is an example of such a "gain of function" disorder. The dominant mutant allele encodes an abnormally elongated protein that prevents the normal protein from functioning in the brain. Researchers determined that Huntington

disease represents a gain of function because individuals who are missing one copy of the gene do not have the illness. That is, the protein encoded by the mutant HD allele must be abnormal, not absent, to cause the disease.

Recessive disorders tend to be more severe, and produce symptoms at much earlier ages, than dominant disorders. Disease-causing recessive alleles can remain, and even flourish, in populations because heterozygotes carry them without becoming ill and pass them to future generations. In contrast, if a dominant mutation arises that causes severe illness early in life, people who have the allele are either too ill or do not live long enough to reproduce, and the allele eventually becomes rare in the population unless it is replaced by mutation. Dominant disorders whose symptoms do not appear until adulthood, or that do not drastically disrupt health, tend to remain in a population because they do not affect health until after a person has reproduced. Therefore, the dominant conditions that persist tend to be those that first cause symptoms in middle adulthood—such as Huntington disease.

Key Concepts

1. A Mendelian trait is caused by a single gene.
2. Modes of inheritance reveal whether a Mendelian trait is dominant or recessive and whether the gene that controls it is carried on an autosome or a sex chromosome.
3. Autosomal dominant traits do not skip generations and can affect both sexes; autosomal recessive traits can skip generations and can affect both sexes.
4. Rare autosomal recessive disorders sometimes recur in families when blood relatives have children together.
5. Mendel's first law, which can predict the probability that a child will inherit a Mendelian trait, applies anew to each child.
6. Genetic problems are solved with logic and by applying Mendel's laws to follow gametes.
7. At the biochemical level, dominance refers to the ability of a protein encoded by one allele to compensate for a missing or abnormal protein encoded by another allele.

4.3 Following the Inheritance of Two Genes—Independent Assortment

The law of segregation follows the inheritance of two alleles for a single gene. In a second set of experiments, Mendel examined the inheritance of two different traits, each attributable to a gene with two different alleles.

Mendel's Second Law

The second law, the **law of independent assortment,** states that for two genes on different chromosomes, the inheritance of one does not influence the chance of inheriting the other. The two genes thus "independently assort" because they are packaged into gametes at random (**figure 4.9**). Two genes that are far apart on the same chromosome also appear to independently assort, because so many crossovers occur between them that it is as if they are carried on separate chromosomes (see figure 3.5.)

Mendel looked at seed shape, which was either round or wrinkled (determined by the *R* gene), and seed color, which was either yellow or green (determined by the *Y* gene). When he crossed true-breeding plants that had round, yellow seeds to true-breeding plants that had wrinkled, green seeds, all the progeny had round, yellow seeds. These offspring were double heterozygotes, or dihybrids, of genotype *RrYy*. From their appearance, Mendel deduced that round is dominant to wrinkled, and yellow to green.

Next, he self-crossed the dihybrid plants in a **dihybrid cross,** so named because two genes and traits are followed. Mendel found four types of seeds in the next, third generation: 315 plants with round, yellow seeds; 108 plants with round, green seeds; 101 plants with wrinkled, yellow seeds; and 32 plants with wrinkled, green seeds. These classes occurred in a ratio of 9:3:3:1.

Mendel then took each plant from the third generation and crossed it to plants with wrinkled, green seeds (genotype *rryy*). These test crosses established whether each plant in the third generation was true-breeding for both genes (genotypes *RRYY* or *rryy*), true-breeding for one gene but heterozygous for

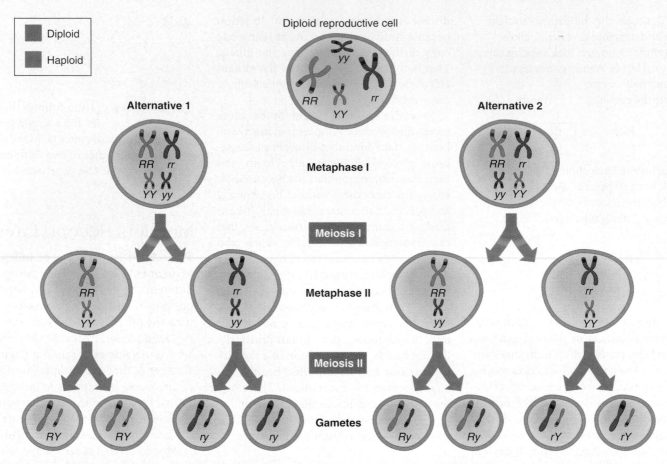

Figure 4.9   Mendel's second law—independent assortment.   The independent assortment of genes carried on different chromosomes results from the random alignment of chromosome pairs during metaphase of meiosis I. An individual of genotype *RrYy*, for example, manufactures four types of gametes, containing the dominant alleles of both genes (*RY*), the recessive alleles of both genes (*ry*), and a dominant allele of one with a recessive allele of the other (*Ry* or *rY*). The allele combination depends upon which chromosomes are packaged together in a gamete—and this happens at random.

the other (genotypes *RRYy, RrYY, rrYy,* or *Rryy*), or heterozygous for both genes (genotype *RrYy*). Mendel could explain the 9:3:3:1 proportion of progeny classes only if one gene does not influence transmission of the other. Each parent would produce equal numbers of four different types of gametes: *RY, Ry, rY,* and *ry.* Note that each of these combinations has one gene for each trait. A Punnett square for this cross shows that the four types of seeds:

1. round, yellow (*RRYY, RrYY, RRYy,* and *RrYy*)

2. round, green (*RRyy* and *Rryy*)

3. wrinkled, yellow (*rrYY* and *rrYy*) and

4. wrinkled, green (*rryy*)

are present in the ratio 9:3:3:1, just as Mendel found (**figure 4.10**).

Solving a Problem: Following More Than One Segregating Gene

A Punnett square for three genes has 64 boxes; for four genes, 256 boxes. An easier way to predict genotypes and phenotypes in multi-gene crosses is to use the mathematical laws of probability on which Punnett squares are based. Probability predicts the likelihood of an event.

An application of probability theory called the product rule can predict the chance that parents with known genotypes can produce offspring of a particular genotype. The product rule states that the chance that two independent events will both occur equals the product of the chance that either event will occur alone. Consider the probability of obtaining a plant with

wrinkled, green peas (genotype *rryy*) from dihybrid (*RrYy*) parents. Do the reasoning for one gene at a time, then multiply the results (**figure 4.11**).

A Punnett square for *Rr* crossed to *Rr* shows that the probability of *Rr* plants producing *rr* progeny is 25 percent, or 1/4. Similarly, the chance of two *Yy* plants producing a *yy* plant is 1/4. Therefore, the chance of dihybrid parents (*RrYy*) producing homozygous recessive (*rryy*) offspring is 1/4 multiplied by 1/4, or 1/16. Now consult the 16-box Punnett square for Mendel's dihybrid cross again (figure 4.10). Only one of the 16 boxes is *rryy*, just as the product rule predicts. **Figure 4.12** shows how probability and Punnett squares can be used to predict offspring genotypes and phenotypes for three human traits simultaneously.

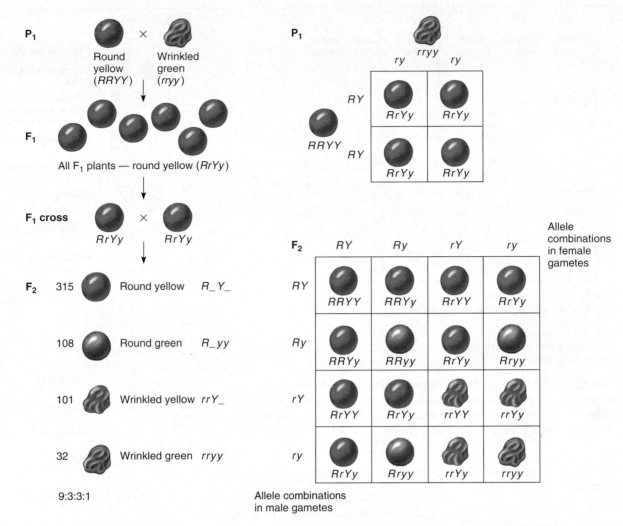

P₁ Round yellow (RRYY) × Wrinkled green (rryy)

F₁ All F₁ plants — round yellow (RrYy)

F₁ cross RrYy × RrYy

F₂
315 Round yellow R_Y_
108 Round green R_yy
101 Wrinkled yellow rrY_
32 Wrinkled green rryy

9:3:3:1

Allele combinations in female gametes

Allele combinations in male gametes

Figure 4.10 Plotting a dihybrid cross. A Punnett square can represent the random combinations of gametes produced by dihybrid individuals. An underline in a genotype (in the F₂ generation) indicates that either a dominant or recessive allele is possible. The numbers in the F₂ generation are Mendel's experimental data.

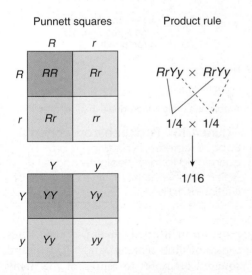

Punnett squares

	R	r
R	RR	Rr
r	Rr	rr

	Y	y
Y	YY	Yy
y	Yy	yy

Product rule

RrYy × RrYy

1/4 × 1/4

1/16

Figure 4.11 The product rule.

Until recently, Mendel's second law has not been nearly as useful in medical genetics as the first law, because not enough genes were known to follow the transmission of two or more traits at a time. But human genome information and DNA microarray technology are changing that practice. It is common now to screen for hundreds or thousands of gene variants or expressed genes at once. The increasingly computational nature of genetics in the twenty-first century has produced an entirely new field called bioinformatics. So, in this sense, genetics is continuing the theme of mathematical analysis that Gregor Mendel began more than a century ago.

Key Concepts

1. Mendel's law of independent assortment considers genes transmitted on different chromosomes.
2. In a dihybrid cross of heterozygotes for seed color and shape, Mendel saw a phenotypic ratio of 9:3:3:1. He concluded that transmission of one gene does not influence that of another.
3. Meiotic events explain independent assortment.
4. Punnett squares and probability can be used to follow independent assortment.
5. Knowing the human genome sequence has made it possible to analyze more than one gene at a time.

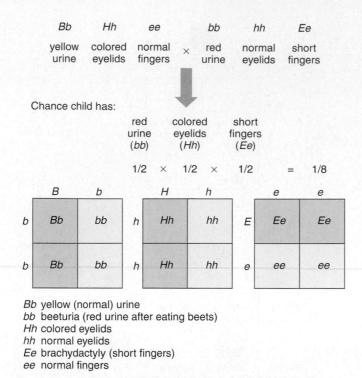

Chance child has:

red urine (*bb*)	colored eyelids (*Hh*)	short fingers (*Ee*)
1/2 ×	1/2 ×	1/2 =	1/8

Bb yellow (normal) urine
bb beeturia (red urine after eating beets)
Hh colored eyelids
hh normal eyelids
Ee brachydactyly (short fingers)
ee normal fingers

Figure 4.12 Using probability to track three traits. A man with normal urine, colored eyelids, and normal fingers wants to have children with a woman who has red urine after she eats beets, normal eyelids, and short fingers. The chance that a child of theirs will have red urine after eating beets, colored eyelids, and short fingers is 1/8.

4.4 Pedigree Analysis

For researchers, families are tools, and the bigger the family the better—the more children in a generation, the easier it is to discern modes of inheritance. Geneticists use charts called **pedigrees** to display family relationships and to depict which relatives have specific phenotypes and, sometimes, genotypes. A human pedigree serves the same purpose as one for purebred dogs or cats or thoroughbred horses—it represents relationships and traits.

A pedigree consists of lines that connect shapes. Vertical lines represent generations; horizontal lines that connect two shapes at their centers depict partners; shapes connected by vertical lines that are joined horizontally represent siblings. Squares indicate males; circles, females; and diamonds, individuals of unspecified sex. Roman numerals designate generations. Arabic numerals or names indicate individuals. **Figure 4.13** shows these and other commonly used pedigree symbols. Colored or shaded shapes indicate individuals who express the trait under study, and half-filled shapes represent known carriers. A genetic counselor

will often sketch out a pedigree while interviewing a client, then use a computer program and add test results that indicate genotypes.

Pedigrees Then and Now

The earliest pedigrees were genealogical, indicating family relationships but not traits. **Figure 4.14** shows such a pedigree for a highly inbred part of the ancient Egyptian royal family. The term *pedigree* arose in the fifteenth century, from the French *pie de grue,* which means "crane's foot." Pedigrees at that time, typically depicting large families, showed parents linked by curved lines to their offspring. The overall diagram often resembled a bird's foot.

One of the first pedigrees to trace an inherited illness was an extensive family tree of several European royal families, indicating which members had the clotting disorder hemophilia (see figure 6.8). The mutant gene probably originated in Queen Victoria of England in the nineteenth century. In 1845, a genealogist named Pliny Earle constructed a pedigree of a family with colorblindness using musical notation—half

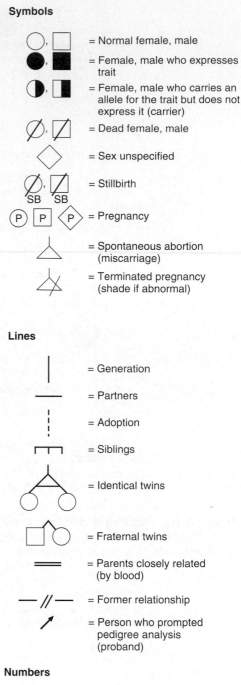

Figure 4.13 Pedigree components. Symbols representing individuals are connected to form pedigree charts, which display the inheritance patterns of particular traits.

notes for unaffected females, quarter notes for colorblind females, and filled-in and squared-off notes to represent the many colorblind males. In the early twentieth century, eugenicists tried to use pedigrees to

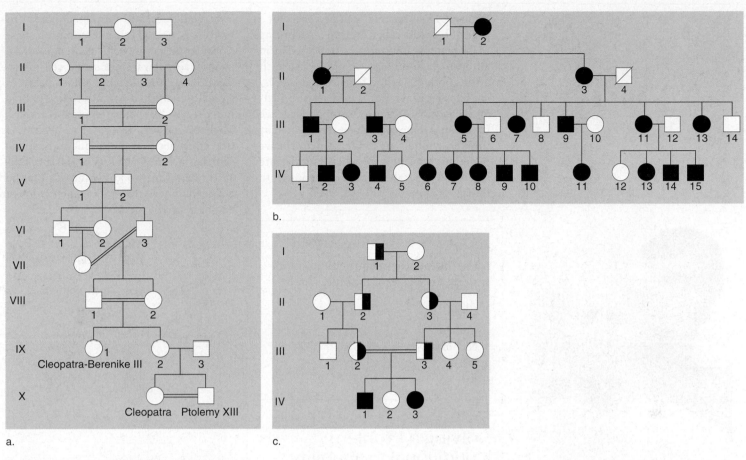

Figure 4.14 Some unusual pedigrees. **(a)** A partial pedigree of Egypt's Ptolemy dynasty shows only genealogy, not traits. It appears almost ladderlike because of the extensive inbreeding. From 323 B.C. to Cleopatra's death in 30 B.C., the family experienced one pairing between cousins related through half-brothers (generation III), four brother-sister pairings (generations IV, VI, VIII, and X), and an uncle-niece relationship (generations VI and VII). Cleopatra married her brother, Ptolemy XIII, when he was 10 years old! These marriage patterns were an attempt to preserve the royal blood. **(b)** In contrast to the Egyptian pedigree, a family with polydactyly (extra fingers and toes) extends laterally, with many children. **(c)** The most common form of consanguinity is marriage of first cousins. They share one set of grandparents, and therefore risk passing on the same recessive alleles to offspring.

show that traits such as criminality, feeblemindedness, and promiscuity were the consequence of faulty genes.

Today, pedigrees are important both for helping families identify the risk of transmitting an inherited illness and as starting points for identifying a gene from the human genome sequence. People who have kept meticulous family records are invaluable in helping researchers follow the inheritance of particular genes in groups such as the Mormons and the Amish. Very large pedigrees provide information on many individuals with a particular disorder. The researchers can then search these individuals' DNA to identify a particular sequence they have all inherited that is not found in healthy family members. Discovery of the gene that causes Huntington disease, for example, took researchers to a remote vil-

lage in Venezuela to study an enormous family. The gene was eventually traced to a sailor who introduced the mutation in the nineteenth century.

Pedigrees Display Mendel's Laws

A person familiar with Mendel's laws can often tell a mode of inheritance just by looking carefully at a pedigree. Consider an autosomal recessive trait, albinism. The homozygous recessive individual lacks an enzyme necessary to manufacture the pigment melanin and, as a result, has very pale hair and skin. Recall that an autosomal recessive trait can affect both sexes and can (but doesn't necessarily) skip generations. **Figure 4.15** shows a pedigree for albinism. If a condition is known to be inherited as an

autosomal recessive trait, carrier status can be inferred for individuals who have affected (homozygous recessive) children. In figure 4.15, because individuals III-1 and III-3 are affected, individuals II-2 and II-3 must be carriers. One partner from each pair of grandparents must also be a carrier, which can sometimes be determined using a carrier test, inferred from family history, or deduced from the DNA sequence.

An autosomal dominant trait does not skip generations and can affect both sexes. A typical pedigree for an autosomal dominant trait has some squares and circles filled in to indicate affected individuals in each generation (**figure 4.16**).

A pedigree may be inconclusive, which means that either autosomal recessive or autosomal dominant inheritance can explain the pattern of filled-in symbols.

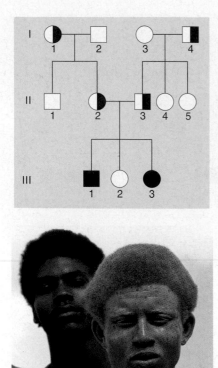

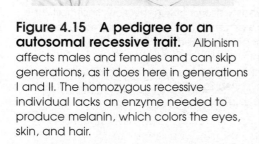

Figure 4.15 A pedigree for an autosomal recessive trait.
Albinism affects males and females and can skip generations, as it does here in generations I and II. The homozygous recessive individual lacks an enzyme needed to produce melanin, which colors the eyes, skin, and hair.

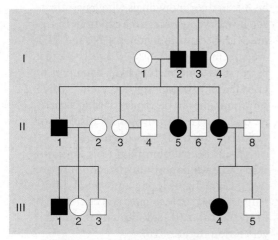

Figure 4.16 A pedigree for an autosomal dominant trait.
Autosomal dominant traits do not skip generations. This trait is brachydactyly, or short fingers.

Figure 4.17 shows one such pedigree, for a type of hair loss called alopecia. According to the pedigree, this trait can be passed in an autosomal dominant mode because it affects both males and females and is present in every generation. However, the pedigree can also depict autosomal recessive inheritance if the individuals represented by unfilled symbols are carriers. Inconclusive pedigrees tend to arise when families are small and the trait is not severe enough to impair fertility.

RELATING THE CONCEPTS

Use the following information to construct a pedigree for the Milto family:

Child	Birth Year
Nathan	1994
Nicholas	1996
P. J.	1997
Joey	2003

Solving a Problem: Conditional Probability

Often genetic counselors are asked to predict the probability that a condition will occur in a particular individual. Mendel's

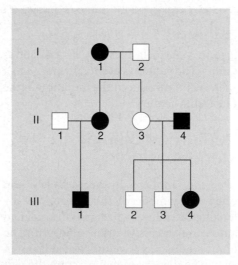

Figure 4.17 An inconclusive pedigree.
This pedigree could account for an autosomal dominant trait or an autosomal recessive trait that does not prevent affected individuals from having children. (Unfilled symbols could represent carriers.)

laws, pedigrees, and Punnett squares provide clues, as do logic and common sense. Consider the family depicted in **figure 4.18.**

Michael Stewart has sickle cell disease (OMIM 603903), which is autosomal recessive. His unaffected parents, Kate and Brad, must each be heterozygotes (carriers). Michael's sister, Ellen, also healthy, is expecting her first child. Ellen's husband, Tim, has no family history of sickle cell disease. Ellen wants to know the risk that her child will inherit the mutant allele from her and be a carrier.

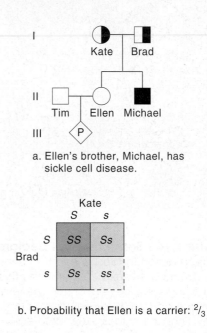

a. Ellen's brother, Michael, has sickle cell disease.

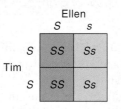

b. Probability that Ellen is a carrier: $2/3$

c. If Ellen is a carrier, chance that fetus is a carrier: $1/2$

Total probability = $2/3 \times 1/2 = 1/3$

Figure 4.18 Making predictions.
Ellen's brother, Michael, has sickle cell disease, as depicted in this pedigree **(a)**. Ellen wonders what the chance is that her fetus has inherited the sickle cell allele from her. First, she must calculate the chance that she is a carrier. The Punnett square in **(b)** shows that this risk is 2 in 3. (She must be genotype *SS* or *Ss*, but cannot be *ss* because she does not have the disease.) The risk that the fetus is a carrier, assuming that the father is not a carrier, is half Ellen's risk of being a carrier, or 1 in 3 **(c)**.

Ellen's request raises two questions. First, what is the risk that she herself is a carrier? Because Ellen is the product of a monohybrid cross, and we know that she is not homozygous recessive, she has a 2 in 3 chance of being a carrier, as the Punnett square indicates. If Ellen is a carrier, what is the chance that she will pass the mutant allele to an offspring? It is 1 in 2, because she has two copies of the gene, and according to Mendel's first law, only one goes into each gamete.

To calculate the overall risk to Ellen's child, we can apply the product rule and multiply the probability that Ellen is a carrier by the chance that, if she is, she will pass the mutant allele on. This result, following two events, is a conditional probability, because the likelihood of the second event—the child being a carrier—depends upon the first event—that Ellen is a carrier. If we assume Tim is not a carrier, Ellen's chance of giving birth to a child who carries the mutant allele is therefore 2/3 times 1/2, which equals 2/6, or 1/3. Ellen thus has a theoretical 1 in 3 chance of giving birth to a child who is a carrier for sickle cell disease.

Pedigrees can be difficult to construct and interpret for several reasons. People sometimes hesitate to supply information because they are embarrassed by symptoms affecting behavior or mental stability. Family relationships can be complicated by adoption, children born out of wedlock, serial relationships, blended families, and assisted reproductive technologies such as surrogate mothers and intrauterine insemination by donor (chapter 21). Moreover, many people cannot trace their families back more than three or four generations, so they lack sufficient evidence to reveal a mode of inheritance. Still, the pedigree remains a powerful way to see, at a glance, how a trait passes from generation to generation—just as Gregor Mendel did with peas.

Key Concepts

1. Pedigrees are charts that depict family relationships and the transmission of inherited traits. . Squares represent males, and circles, females; horizontal lines link two partners, vertical lines show generations, and elevated horizontal lines depict siblings. Symbols for heterozygotes are half-shaded, and symbols for individuals who express the trait under study are completely shaded.
2. Pedigrees can reveal modes of inheritance. Along with Punnett squares, they are tools that apply Mendel's first law to predict the recurrence risks of inherited disorders or traits.

Summary

4.1 Following the Inheritance of One Gene—Segregation

1. Gregor Mendel described the two basic laws of inheritance using pea plant crosses. The laws, which derive from the actions of chromosomes during meiosis, apply to all diploid organisms.

2. Mendel used a statistical approach to investigate why some traits seem to disappear in the hybrid generation. The **law of segregation** states that alleles of a gene are distributed into separate gametes during meiosis. Mendel demonstrated this using seven traits in pea plants.

3. A diploid individual with two identical alleles of a gene is **homozygous.** A **heterozygote** has two different alleles of a gene. A gene may have many alleles.

4. A **dominant** allele masks the expression of a **recessive** allele. An individual may be homozygous dominant, homozygous recessive, or heterozygous.

5. Mendel repeatedly found that when he crossed two true-breeding types, then bred the resulting hybrids to each other, the two variants of the trait appeared in a 3:1 phenotypic ratio. Crossing these progeny further revealed a genotypic ratio of 1:2:1.

6. A **Punnett square** is a chart used to follow the transmission of alleles. It is based on probability.

4.2 Single-Gene Inheritance in Humans

7. Traits or disorders caused by single genes are called Mendelian or unifactorial traits.

8. **Modes of inheritance** enable geneticists to predict phenotypes. In **autosomal dominant** inheritance, males and females may be affected, and the trait does not skip generations. Inheritance of an **autosomal recessive** trait may affect either males or females and may skip generations. Autosomal recessive conditions are more likely to occur in families with **consanguinity.** Recessive disorders tend to be more severe and cause symptoms earlier than dominant disorders.

9. Genetic problems can be solved by tracing alleles as gametes form and then combine in a new individual.

10. Dominance and recessiveness reflect how alleles affect the abundance or activity of the gene's protein product.

4.3 Following the Inheritance of Two Genes—Independent Assortment

11. Mendel's second law, the **law of independent assortment,** follows the transmission of two or more genes on different chromosomes. It states that a random assortment of maternally and paternally derived chromosomes during meiosis results in gametes that have different combinations of these genes.

12. The chance that two independent genetic events will both occur is equal to the product of the probabilities that each event will occur on its own. This principle, called the product rule, is useful in calculating the risk that certain individuals will inherit a particular genotype and in following the inheritance of two genes on different chromosomes.

4.4 Pedigree Analysis

13. A **pedigree** is a chart that depicts family relationships and patterns of inheritance for particular traits. A pedigree can be inconclusive.

Review Questions

1. How does meiosis explain Mendel's laws of segregation and independent assortment?

2. How was Mendel able to derive the two laws of inheritance without knowing about chromosomes?

3. Distinguish between
   a. autosomal recessive and autosomal dominant inheritance.
   b. Mendel's first and second laws.
   c. a homozygote and a heterozygote.
   d. a monohybrid and a dihybrid cross.
   e. a Punnett square and a pedigree.

4. Why would Mendel's results for the dihybrid cross have been different if the genes for the traits he followed were located near each other on the same chromosome?

5. Why are extremely rare autosomal recessive disorders more likely to appear in families in which blood relatives have children together?

6. How does the pedigree of the ancient Egyptian royal family in figure 4.14a differ from a pedigree a genetic counselor might use today?

7. People who have Huntington disease inherit one mutant and one normal allele. How would a person who is homozygous dominant for the condition arise?

8. What is the probability that two individuals with an autosomal recessive trait, such as albinism, will have a child with the same genotype and phenotype as they do?

Applied Questions

1. Predict the phenotypic and genotypic ratios for crossing the following pea plants:
   a. short × short
   b. short × true-breeding tall
   c. true-breeding tall × true-breeding tall

2. What are the genotypes of the pea plants that would have to be bred to yield one plant with restricted pods for every three plants with inflated pods?

3. If pea plants with all white seed coats are crossed, what are the possible phenotypes of their progeny?

4. Pea plants with restricted yellow pods are crossed to plants that are true-breeding for inflated green pods. The F_1 are then crossed. Derive the phenotypic and genotypic ratios for the F_2 generation.

5. A gene that the media described as the "bad hair day" gene is technically called frizzled (OMIM 601723), named for its well-studied counterparts in fruit flies and mice. Frizzled is inherited as an autosomal recessive trait, and the gene is on chromosome 2. A gene for "strikingly red hair," also autosomal recessive, is on chromosome 4 (OMIM 266300).

    Gina Rollins and Spencer Davis have boring, straight brown hair. They are amazed to discover that each has a mother (Inga and Magda) with strikingly red, frizzled hair. Their fathers, Ralph and Fred, both have boring, straight brown hair. If Gina and Spencer have children, what is the probability that each will have either trait, both traits, or neither?

6. What is the probability that Joey Milto can have a child with Batten disease if his partner does not have the mutant allele?

7. If Nathan Milto's gene therapy is successful and he has children, can he pass on the allele for Batten disease? Why or why not?

8. Below are four pedigrees depicting families with achondroplasia, a common form of hereditary dwarfism that causes very short limbs, stubby hands, and an enlarged forehead. What is the most likely mode of inheritance? Cite a reason for your answer.

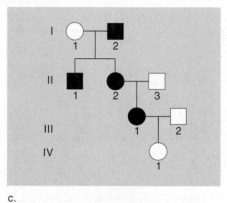

c.

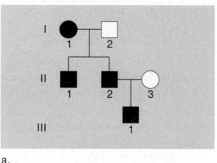

a.

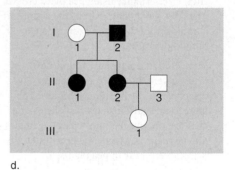

d.

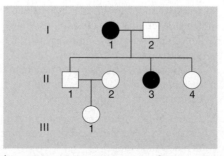

b.

9. Draw a pedigree to depict the following family:
   One couple has a son and a daughter with normal skin pigmentation. Another couple has one son and two daughters with normal skin pigmentation. The daughter from the first couple has three children with the son of the second couple. Their son and one daughter have albinism (OMIM 203100); their other daughter has normal skin pigmentation.

10. Chands syndrome (OMIM 214350) is an autosomal recessive condition characterized by very curly hair, underdeveloped nails, and

abnormally shaped eyelids. In the following pedigree, which individuals must be carriers?

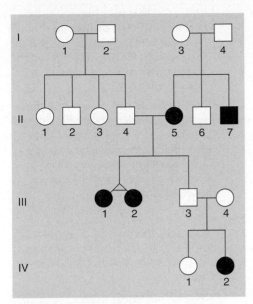

Chands syndrome

11. Caleb has a double row of eyelashes (OMIM 126300), which he inherited from his mother as a dominant trait. His maternal grandfather is the only other relative to have the trait. Veronica, a woman with normal eyelashes, falls madly in love with Caleb, and they marry. Their first child, Polly, has normal eyelashes. Now Veronica is pregnant again and hopes they will have a child who has double eyelashes. What chance does a child of Veronica and Caleb have of inheriting double eyelashes? Draw a pedigree of this family.

12. Congenital insensitivity to pain with anhidrosis (OMIM 256800) is an extremely rare autosomal recessive condition that causes fever, inability to sweat (anhidrosis), mental retardation, inability to feel pain, and self-mutilating behavior. Researchers compared the following three families with this condition. What do these families have in common that might explain the appearance of this rare illness?

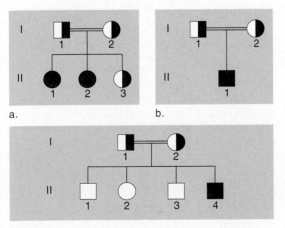

13. The child in figure 4.12 who has red urine after eating beets, colored eyelids, and short fingers, is of genotype *bbHhEe*. The genes for these traits are on different chromosomes. If he has children with a woman who is a trihybrid for each of these genes, what are the expected genotypic and phenotypic ratios for their offspring?

14. In this pedigree, individual III–1 died at age two of Tay-Sachs disease, an autosomal recessive disorder. Which other family members must be carriers, and which could be?

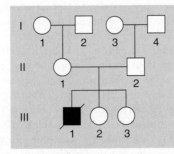

Tay-Sachs disease

15. Sclerosteosis (OMIM 269500) causes overgrowth of the skull and jaws that produces a characteristic face, gigantism, facial paralysis, and hearing loss. The overgrowth of skull bones can cause severe headaches and even sudden death. In this pedigree for a family with sclerosteosis:

   a. What is the relationship between the individuals who are connected by slanted double lines?

   b. Which individuals in the pedigree must be carriers of this autosomal recessive condition?

16. According to this pedigree from the soap opera "All My Children," is Charlie the product of a consanguineous relationship? The trait being tracked is freckles.

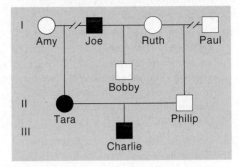

Freckles

17. On "General Hospital," six-year-old Maxi suffered from Kawasaki syndrome, an inflammation of the heart. She desperately needed a transplant, and received one from BJ, who died in a bus accident. Maxi and BJ had the same unusual blood type, which is inherited. According to this pedigree, how are Maxi and BJ related?

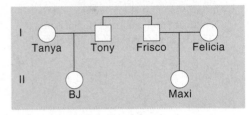

18. A man has a blood test for Tay-Sachs disease and learns that he is a carrier. His body produces half the normal amount of an enzyme. Why doesn't he have symptoms?

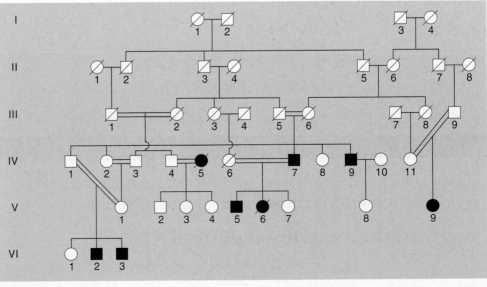

Sclerosteosis

Web Activities

Visit the Online Learning Center (OLC) at www.mhhe.com/lewisgenetics7. Select **Student Edition, chapter 4** and **Web Activities** to find the website links needed to complete the following activities.

19. Go to the website for the National Organization for Rare Disorders. Identify an autosomal recessive disorder and an autosomal dominant disorder. Create a family for each one, and describe transmission of the disease over three generations.

20. Go to the website for Gene Gateway—Exploring Genes and Genetic Disorders. Select two disorders or traits that would demonstrate independent assortment if present in the same family, and two that would not.

Case Studies and Research Results

21. On the daytime drama "The Young and the Restless," several individuals suffer from SORAS, which stands for "soap opera rapid aging syndrome." It is not listed in OMIM. In SORAS, a young child is sent off to boarding school and returns three months later an angry teenager. In the Newman family, siblings Nicholas and Victoria aged from ages six and eight years, respectively, to 16 and 18 years within a few months. Their parents, Victor and Nikki, are curiously not affected; in fact, they never seem to age at all.

a. What is the mode of inheritance of the rapid aging disorder affecting Nicholas and Victoria?

b. How do you know what the mode of inheritance is?

c. Draw a pedigree to depict this portion of the Newman family.

22. Sam Fitzgerald is a carpet salesman who, at age 46, begins to slur his speech and stagger slightly when he walks. His speech worsens, he develops a shuffling gait to avoid falling, and he loses his job, because customers complain that he is intoxicated. His children, who know he does not drink alcohol, urge him to seek medical care. Eventually, after much counseling, he is tested and learns that he has HD. As his symptoms worsen, his sister Pam, who is tested and is free of the mutation, cares for him—something she also did as a teen when their parents, Ruth and Alan, died in a car crash in their thirties. Another sister, Sue, refuses to be tested.

a. Draw a pedigree for this family.

b. What is the risk that Sam's daughter has inherited HD?

c. What is the risk that Sue's son has inherited HD?

d. When Sue hears that Pam was tested and is free of the mutation, she assumes that this raises the risk that she has inherited the disease. Is she correct? Explain your answer in terms of Mendel's first law.

23. Recall Mackenzie, the young woman from chapter 1 who underwent genetic testing. The tests revealed that she has a 1 in 10 chance of developing lung cancer, and a 2 in 1,000 chance of developing colon cancer. What is the probability that she will develop both cancers?

Learn to apply the skills of a genetic counselor with these additional cases found in the *Case Workbook in Human Genetics:*

- Acrocephalosyndactyly
- Carnosinemia
- Huntington-like disorder
- Restless leg syndrome
- Schneckenbecken dysplasia

VISIT YOUR ONLINE LEARNING CENTER

Visit your online learning center for additional resources and tools to help you master this chapter. See us at

www.mhhe.com/lewisgenetics7.

CHAPTER

5

Extensions and Exceptions to Mendel's Laws

THE MANY FACES OF ALKAPTONURIA

Pat Wright became aware that she had alkaptonuria (OMIM 203500) at age 15, when she suffered back spasms. As she got older, her spine continued to degenerate, and the cartilage in her left knee also broke down. She managed to have five children and to teach for 26 years, but retired on disability at age 57.

In Wright's medical records was a comment that she had a "harmless" metabolic disorder. Wright's parents had taken her to the doctor after noticing dark-stained diapers. The pediatrician, suspecting alkaptonuria, sent a blackened diaper to geneticists, who diagnosed the disorder—but it was never explained well to the anxious parents, who were both carriers. Alkaptonuria was one of four disorders described by English physician Sir Archibald Garrod in 1902, to introduce the idea of an "inborn error of metabolism."

Wright didn't hear a name for her condition until 1997, when the surgeon who was replacing her knee was amazed to find blackened cartilage around the joint. This explained the pains, stained diapers, and even the dark blue-gray color of Wright's ears, he told her, as well as other problems, including hearing loss, gallstones, and heart valve damage.

Alkaptonuria is pleiotropic, which means that a single mutation causes multiple symptoms that reflect the varied distribution of the encoded protein. Deficiency of the enzyme homogentisic acid oxidase leads to buildup of melanin pigment in urine, nails, skin, and cartilage. When urine is exposed to oxygen, it turns black. A form of alkaptonuria is seen in laboratory mice—the wood shavings in their cages, if soaked in urine and left too long, turn blue-black!

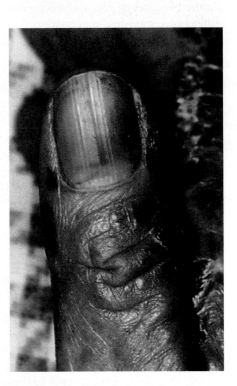

Blackened nails are just one sign of alkaptonuria, an inborn error of metabolism.

The transmission of inherited traits is not always as straightforward as Mendel's pea experiments indicated. This chapter examines extensions and exceptions to Mendel's laws.

5.1 When Gene Expression Appears to Alter Mendelian Ratios

Mendel's crosses yielded offspring that were easily distinguished from each other. A pea is either yellow or green; a plant tall or short. For some characteristics, though, offspring classes do not occur in the proportions that Punnett squares or probabilities predict. In other cases, transmission patterns of a visible trait are not consistent with a mode of inheritance, such as autosomal recessive or autosomal dominant. In these instances, Mendel's laws operate, and the underlying genotypic ratios persist, but either the nature of the phenotype or influences from other genes or the environment alter phenotypic ratios—that is, what is actually seen. Following are several circumstances in which phenotypic ratios appear to contradict Mendel's laws—although the laws actually still apply.

Lethal Allele Combinations

A genotype (allele combination) that causes death is, by definition, lethal. In a population and evolutionary sense, though, a lethal genotype has a more specific meaning—it causes death before the individual can reproduce, which prevents passage of his or her genes to the next generation.

In organisms used in experiments, such as fruit flies, pea plants, or mice, lethal allele combinations remove an expected progeny class following a specific cross. For example, a cross of two heterozygous flies, in which the homozygous recessive progeny die as embryos, would leave only heterozygous and homozygous dominant flies.

In humans, early-acting lethal alleles cause spontaneous abortion (technically called "miscarriages" if they occur after the embryonic period). When a man and woman each carries a recessive lethal allele for the same gene, each pregnancy has a 25 percent chance of spontaneously aborting—

a proportion representing the homozygous recessive class. Sometimes a double dose of a dominant allele is lethal, as is the case for Mexican hairless dogs (**figure 5.1**). Inheriting one dominant allele confers the coveted hairlessness trait, but inheriting two dominant alleles is lethal to the unlucky embryo. Breeders cross hairless to hairy ("powderpuff") dogs, rather than hairless to hairless, to avoid losing the lethal homozygous dominant class—a quarter of the pups.

Multiple Alleles

A person has two alleles for any autosomal gene—one allele on each homolog. However, a gene can exist in more than two allelic

forms in a population because it can mutate in many ways. Different allele combinations can produce variations in phenotype.

It would be very useful if testing for a particular genotype could always enable physicians to predict the course of an illness. This is often difficult because other genes and environmental effects can modify the phenotype, a point we return to soon. Sometimes knowing the genotype can predict the phenotype. This is the case for phenylketonuria (PKU) (OMIM 261600), an inborn error of metabolism in which an enzyme is deficient or absent, causing the amino acid phenylalanine to build up in brain cells. More than 300 mutant alleles combine to form four basic phenotypes:

Figure 5.1 Lethal alleles.

(a) This Mexican hairless dog has inherited a dominant allele that makes it hairless. Inheriting two such dominant alleles is lethal during embryonic development. **(b)** Breeders cross Mexican hairless dogs to hairy ("powderpuff") dogs to avoid dead embryos and stillbirths that represent the *HH* genotypic class.

a.

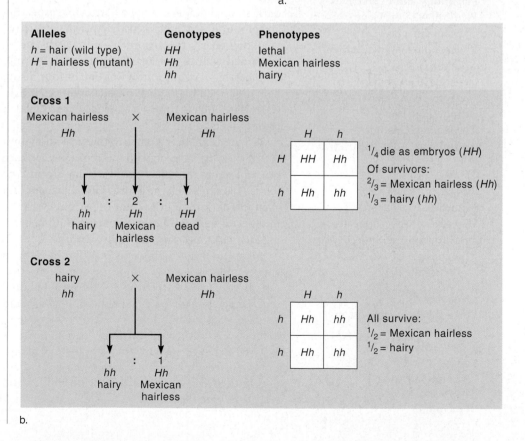

b.

- classic PKU with profound mental retardation

- moderate PKU

- mild PKU

- asymptomatic PKU, with excretion of excess phenylalanine in urine

Eating a special diet extremely low in phenylalanine from birth to at least eight years of age allows normal brain development. Knowing the allele combination can guide how strict the diet need be, and how long it must continue.

Multiple alleles are considered in carrier testing for cystic fibrosis (CF). When the CF gene was discovered in 1989, researchers identified one mutant allele, called ΔF508, that causes about 70 percent of cases in many populations. As the allele list grew, researchers discovered that not all allele combinations cause the exact same symptoms. People homozygous for ΔF508 have severe symptoms, including frequent serious respiratory infections, very sticky mucus in the lungs, and poor weight gain due to insufficient pancreatic function. Another genotype increases susceptibility to bronchitis and pneumonia, and another causes only absence of the vas deferens. Genetic tests probe panels of CF mutations that are the most common in a patient's ethnic group, maximizing the likelihood of detecting carriers and avoiding the cost of testing for 1,000⁺ alleles. CF testing is standard for women during pregnancy. If a woman has a disease-causing allele, then the father-to-be is tested, and if he has a mutant allele too, then the fetus may be tested.

RELATING THE CONCEPTS

Several mutations are known in the gene that encodes the enzyme that is deficient in alkaptonuria. How is this possible if a person only has two alleles?

Different Dominance Relationships

In complete dominance, one allele is expressed, while the other isn't. In **incomplete dominance,** the heterozygous phenotype is intermediate between that of either homozygote.

In a sense, enzyme deficiencies in which a threshold level is necessary for health illustrate both complete and incomplete dominance—depending upon how one evaluates the phenotype. For example, on a whole-body level, Tay-Sachs disease displays complete dominance because the heterozygote (carrier) is as healthy as a homozygous dominant individual. However, if phenotype is based on enzyme level, then the heterozygote is intermediate between the homozygous dominant (full enzyme level) and homozygous recessive (no enzyme). Half the normal amount of enzyme is sufficient for health, which is why at the whole-person level, the wild type allele is completely dominant.

A more obvious example of incomplete dominance occurs in the snapdragon plant. A red-flowered plant of genotype *RR* crossed to a white-flowered *rr* plant can give rise to an *Rr* plant—which has pink flowers. This intermediate color is presumably due to an intermediate amount of pigment.

Familial hypercholesterolemia (FH) is an example of incomplete dominance in humans that can be observed on the molec-

ular and whole-body levels. A person with two disease-causing alleles lacks receptors on liver cells that take up the low density lipoprotein (LDL) form of cholesterol from the bloodstream. A person with one disease-causing allele has half the normal number of receptors. Someone with two wild type (the most common) alleles has the normal number of receptors. **Figure 5.2** shows how measurement of plasma cholesterol reflects these three genotypes. The phenotypes parallel the number of receptors—those with two mutant alleles die as children of heart attacks, those with one mutant allele may suffer heart attacks in young adulthood, and those with two wild type alleles do not develop this inherited form of heart disease.

Different alleles that are both expressed in a heterozygote are **codominant.** The ABO blood group is based on the expression of codominant alleles.

Blood types are determined by the patterns of cell surface molecules on red blood cells. Most of these molecules are proteins embedded in the plasma membrane with attached sugars that extend from the cell

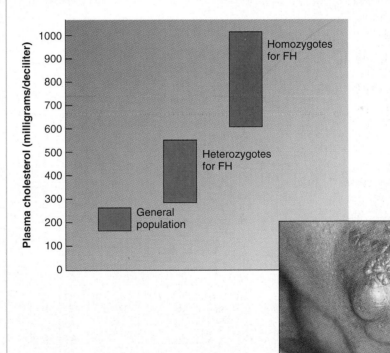

Figure 5.2 Incomplete dominance. A heterozygote for familial hypercholesterolemia (FH) has approximately half the normal number of cell surface receptors in the liver for LDL cholesterol. An individual with two mutant alleles has the severe form of FH, with liver cells that totally lack the receptors. Serum cholesterol level is very high. The photograph shows cholesterol deposits on the elbow of an affected young man.

surface. The sugar is the antigen, which is the molecule that the immune system recognizes. People who belong to blood group A have an allele that encodes an enzyme that adds a final piece to a certain sugar called antigen A. In people with blood type B, the allele and its encoded enzyme are slightly different, which causes a different piece to attach to the sugar, producing antigen B. People in blood group AB have both antigen types. Blood group O reflects yet a third allele of this gene. It is missing just one DNA nucleotide, but this drastically changes the encoded enzyme in a way that robs the sugar chain of its final piece (**figure 5.3**).

In the past, ABO blood types have been described as variants of a gene called "I,"

although OMIM now abbreviates the designations. The older I system is easier to understand (**table 5.1**). ("I" stands for isoagglutinin.) The three alleles are I^A, I^B, and i. People with blood type A have antigen A on the surfaces of their red blood cells, and may be of genotype $I^A I^A$ or $I^A i$. People with blood type B have antigen B on their red blood cell surfaces, and may be of genotype $I^B I^B$ or $I^B i$. People with the rare blood type AB have both antigens A and B on their cell surfaces, and are genotype $I^A I^B$. People with blood type O have neither antigen, and are genotype ii.

Television program plots often misuse ABO blood type terminology, assuming that a child's ABO type must match that of

one parent. This is not true, because a person with type A or B blood can be heterozygous. A person who is genotype $I^A i$ and a person who is $I^B i$ can jointly produce offspring of any ABO genotype or phenotype, as **figure 5.4** illustrates.

Epistasis—One Gene Affects Another's Expression

Mendel's laws can appear to not operate when one gene masks or otherwise affects the phenotype associated with another. This phenomenon is called **epistasis.** (Do not confuse this with dominance relationships between alleles of the *same* gene.) The Bombay phenotype, for example, is a result of two interacting genes: the I and H genes. Their relationship affects ABO blood type. In epistasis, the blocked gene is expressed (copied into RNA), but something about the interaction of the encoded proteins inactivates or impairs one of them.

The normal H allele encodes an enzyme that inserts a sugar molecule, called antigen H, onto a particular glycoprotein on the surface of an immature red blood cell. The recessive h allele produces an inactive form of the enzyme that cannot insert the sugar. (The H gene's product is fucosyltransferase 1, and in OMIM the H gene is called *FUT*1.) The A and B antigens attach to the H antigen. As long as at least one H allele is present, the ABO genotype dictates the ABO blood type. However, in a person with genotype hh, no H antigen binds to the A and B antigens, and they fall away. The person has blood type O based on phenotype (a blood test), but may have any ABO genotype. For example, Mendel's laws predict that each child of a man who has type B blood and genotype $I^B I^B$ and a woman who has type A blood with genotype $I^A I^A$ (like the upper left Punnett square in **figure 5.4**) must be type AB. But if the parents are each Hh, then each offspring has a 25 percent chance of being hh, and having a phenotype of type O blood, because the A and B antigens cannot bind to the red blood cells. The child's genotype, however, would be $I^A I^B$, hh. The predicted Mendelian phenotypic ratios change because of the action of the second gene.

Individuals with the hh genotype are very rare, with one notable exception—residents

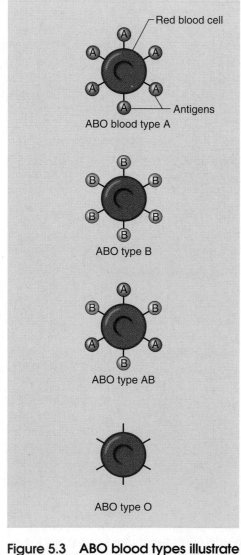

Figure 5.3 ABO blood types illustrate codominance. ABO blood types are based on antigens on red blood cell surfaces. The size of the A and B antigens is greatly exaggerated in this drawing.

Table 5.1		
The ABO Blood Group		
Genotypes	Phenotypes	
	Antigens on Surface	ABO Blood Type
$I^A I^A$	A	Type A
$I^A i$	A	
$I^B I^B$	B	Type B
$I^B i$	B	
$I^A I^B$	AB	Type AB
ii	None	Type O

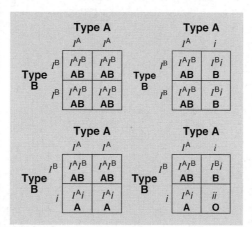

Figure 5.4 Codominance.
The I^A and I^B alleles of the I gene are codominant, but they follow Mendel's law of segregation. These Punnett squares follow the genotypes that could result when a person with type A blood produces offspring with a person with type B blood.

of Reunion Island, in the Indian Ocean east of Madagascar. Apparently the settlers of this isolated island included at least one *hh* or *Hh* individual. With time, large families, and some consanguinity (blood relatives having children together), the allele multiplied in the population.

Sometimes epistasis is just a matter of common sense. For example, a hairless gene in dogs and mice, and a spineless gene in cucumbers, prevent the actions of genes that color the hairs and spines. This is similar to the Bombay phenotype in that the two genes affect related structures.

Penetrance and Expressivity

The same allele combination can produce different degrees of a phenotype in different individuals because a gene does not act alone. Nutrition, exposure to toxins, other illnesses, and actions of other genes may influence the expression of most genes. For example, two individuals who have the most severe genotype for CF may nonetheless have different clinical experiences. One may be much sicker than the other because she also inherited genes predisposing her to develop asthma and respiratory allergies.

Many Mendelian traits and illnesses have distinctive phenotypes, despite all of these influences. The terms *penetrance* and *expressivity* describe degrees of expression of a single gene. **Penetrance** refers to the all-or-none expression of a genotype; **expressivity** refers to severity or extent.

An allele combination that produces a phenotype in everyone who inherits it is considered completely penetrant. Huntington disease (see Bioethics: Choices for the Future in chapter 4) is completely penetrant—all who inherit the mutant allele will develop symptoms if they live long enough, although symptoms may not begin until late in life. A genotype is incompletely penetrant if some individuals do not express the phenotype (have no symptoms). Polydactyly (see figure 1.6) is incompletely penetrant. Some people who inherit the dominant allele have more than five digits on a hand or foot, yet others who must have the allele (because they have an affected parent and child) have the normal number of fingers and toes. The penetrance of a gene is described numerically. If 80 of 100 people who have inherited the dominant polydactyly allele have extra digits, the genotype is 80 percent penetrant.

A phenotype is variably expressive if symptoms vary in intensity in different people. One person with polydactyly might have an extra digit on both hands and a foot, but another might have just one extra fingertip. Therefore, polydactyly is both incompletely penetrant and variably expressive.

It is hard to imagine how other genes or the environment can influence the numbers of fingers or toes. For familial hypercholesterolemia, variable expressivity reflects greater influence of other genes and the environment (see figure 5.2). FH heterozygotes develop heart disease due to high serum cholesterol in middle adulthood. Healthful diet and exercise habits can delay symptom onset.

RELATING THE CONCEPTS

Symptoms of alkaptonuria vary in the number of joints that turn black and become painful. Does this indicate incomplete penetrance or variable expressivity?

Pleiotropy—One Gene, Many Effects

A Mendelian disorder with many symptoms, or a gene that controls several functions or has more than one effect, is termed **pleiotropic.** Such conditions can be difficult to trace through families because people with different subsets of symptoms may appear to have different disorders. This is the case for porphyria variegata (OMIM 176200), an autosomal dominant, pleiotropic, inborn error of metabolism. The disease affected several members of the royal families of Europe (**figure 5.5**).

King George III ruled England during the American Revolution. At age 50, he first experienced abdominal pain and constipation, followed by weak limbs, fever, a fast pulse, hoarseness, and dark red urine. Next, nervous system signs and symptoms began, including insomnia, headaches, visual problems, restlessness, delirium, convulsions, and stupor. His confused and racing thoughts, combined with actions such as ripping off his wig and running about naked while at the peak of a fever, convinced court observers that the king was mad. Just as Parliament was debating his ability to rule, he recovered.

But the king's ordeal was far from over. He relapsed thirteen years later, then again three years after that. Always the symptoms appeared in the same order, beginning with abdominal pain, fever, and weakness, and progressing to nervous system symptoms. Finally, an attack in 1811 placed George in a prolonged stupor, and the Prince of Wales dethroned him. George III lived for several more years, experiencing further episodes.

In George III's time, doctors were permitted to do very little to the royal body, and their diagnoses were based on what the king told them. Twentieth-century researchers found that porphyria variegata caused George's red urine. Lacking a particular enzyme, a part of the blood pigment hemoglobin called a porphyrin ring is routed into the urine instead of being broken down and metabolized in cells. Porphyrin builds up and attacks the nervous system, causing symptoms. An examination of physicians' reports on George's relatives—easy to obtain for a royal family—showed that the disorder appeared to be several different illnesses. Today, porphyria variegata remains rare, and people who have it are often misdiagnosed with a seizure disorder. Unfortunately, some seizure medications and anesthetics worsen symptoms.

On a molecular level, pleiotropy occurs when a single protein affects different body parts or participates in more than one biochemical reaction. Consider Marfan syndrome (OMIM 154700), an autosomal dominant defect in an elastic connective tissue protein called fibrillin. The protein is abundant in the lens of the eye, in the aorta (the largest artery in the body, leading from the heart), and in the bones of the limbs, fingers, and ribs. Once researchers knew this, the Marfan syndrome symptoms of lens dislocation, long limbs, spindly fingers, and a caved-in chest made sense. The most serious symptom is a life-threatening weakening in the aorta, which can suddenly burst. If the weakening is detected early, a synthetic graft can replace the section of artery wall.

Pleiotropy can be confusing even to medical doctors. *The New England Journal of Medicine* ran a contest to see if its physician

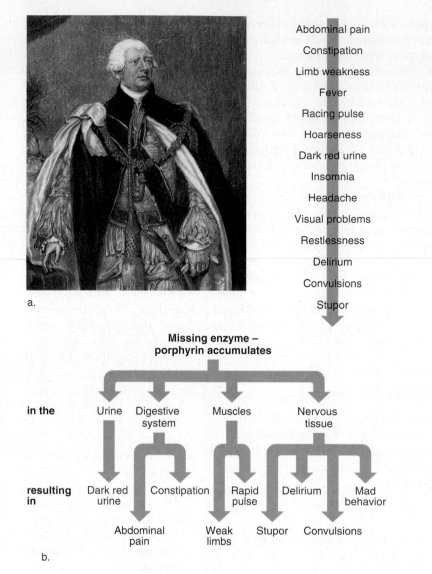

Abdominal pain

Constipation

Limb weakness

Fever

Racing pulse

Hoarseness

Dark red urine

Insomnia

Headache

Visual problems

Restlessness

Delirium

Convulsions

Stupor

a.

**Missing enzyme –
porphyrin accumulates**

in the Urine Digestive system Muscles Nervous tissue

resulting in Dark red urine Constipation Rapid pulse Delirium Mad behavior

Abdominal pain Weak limbs Stupor Convulsions

b.

Figure 5.5 Pleiotropy. King George III **(a)** suffered from the autosomal dominant disorder porphyria variegata—and so did several other family members. Because of pleiotropy, the family's varied illnesses and quirks appeared to be different, unrelated disorders. In King George, symptoms appeared every few years in a particular order **(b)**.

readers could identify the cause of three symptoms of alkaptonuria, discussed in the chapter opening essay. Many couldn't.

RELATING THE CONCEPTS

Explain the basis of the pleiotropy in alkaptonuria.

Phenocopies—When It's Not in the Genes

An environmentally caused trait that appears to be inherited is a **phenocopy.** Such a trait can either produce symptoms that resemble those of a Mendelian disorder or mimic inheritance patterns by occurring in certain relatives. For example, the limb birth defect caused by the drug thalidomide, discussed in chapter 3, is a phenocopy of the inherited illness phocomelia. Physicians recognized the environmental disaster when they began seeing many children born with what looked like the very rare phocomelia. A birth defect caused by exposure to a teratogen was more likely than a sudden increase in incidence of a rare inherited disease.

A phenocopy of alkaptonuria occurred in some women with dark brown skin who used a bleaching cream that contained a chemical called hydroquinone. It caused darkening of the fingers and ears, just like alkaptonuria.

An infection can be a phenocopy. Children who have AIDS may have parents who also have the disease, but these children acquired AIDS by viral infection, not by inheriting a gene. A phenocopy caused by a highly contagious infection can seem to be inherited if it affects more than one family member.

Sometimes, common symptoms may resemble those of an inherited condition until medical tests rule heredity out. For example, an underweight child who has frequent colds may show some signs of cystic fibrosis, but may instead suffer from malnutrition. A negative test for several CF alleles would alert a physician to look for another cause.

Genetic Heterogeneity— More Than One Way to Inherit a Trait

Different genes can produce the same phenotype, a phenomenon called **genetic heterogeneity.** This redundancy of function can make it appear that Mendel's laws are not operating. For example, 132 forms of hearing loss are transmitted as autosomal recessive traits. If a man who is homozygous for a hearing loss gene on one chromosome has a child with a woman who is homozygous for another hearing loss gene on a different chromosome, then the child would not be deaf, because he or she would be heterozygous for both hearing-related genes. Cleft palate (see the opening photograph in chapter 7) and albinism (see figure 4.15) are other traits that are genetically heterogeneic—that is, different genes can cause them.

Genetic heterogeneity can occur when genes encode different enzymes that catalyze the same biochemical pathway, or different proteins that are part of the pathway. For example, eleven biochemical reactions lead to blood clot formation. Clotting disorders may result from mutations in the genes that specify any of the enzymes that catalyze these reactions, leading to several types of bleeding disorders. Similarly, the fatal heart rhythm disorder long-QT syndrome (see Reading 2.2) can be caused by an abnormality in the potassium channel protein or in the ankyrin protein that holds the channel open in a heart muscle cell's plasma membrane.

The Human Genome Sequence Adds Perspective

Sequencing of the human genome has modified and in some cases clarified the extensions to Mendel's laws. Knowing the DNA sequence of a protein-encoding gene, for example, greatly increases the number of known alleles—previously, we only knew of those that obviously affect the phenotype.

Epistasis may not be as rare as previously thought—many genes influence each other. Gene interactions also underlie penetrance and expressivity, once thought to be strictly a characteristic of a particular gene. Similarly, DNA microarrays that reveal gene expression patterns in different tissues are painting detailed portraits of pleiotropy—which, like epistasis, may not be unusual after all. That is, inherited disorders may affect more tissues or organs than we realize. Finally, more cases of genetic heterogeneity are being discovered as genes with redundant or overlapping functions are identified. **Table 5.2** summarizes several of the phenomena that appear to alter Mendelian inheritance.

Gregor Mendel derived the two laws of inheritance working with traits conferred by genes located on different chromosomes in the nucleus. When genes do not conform to these conditions, however, the associated traits may not appear in Mendelian ratios. The remainder of this chapter considers two types of gene transmission that do not fulfill the requirements for Mendelian inheritance.

Key Concepts

1. A lethal allele combination is never seen as a progeny class.
2. In incomplete dominance, the heterozygote phenotype is intermediate between those of the homozygotes; in codominance, two different alleles for the same gene are each expressed.
3. In epistasis, one gene influences expression of another.
4. Genotypes vary in penetrance and expressivity of the phenotype.
5. A gene with more than one expression is pleiotropic.
6. A trait caused by the environment but resembling a known genetic trait or occurring in certain family members is a phenocopy.
7. Genetic heterogeneity occurs when different genes cause the same phenotype.
8. The human genome sequence reveals that several "exceptions" to Mendel's laws are actually more common than thought.

5.2 Maternal Inheritance and Mitochondrial Genes

The basis of the law of segregation is that both parents contribute genes equally to offspring. This is not the case for genes in mitochondria, the organelles that house the biochemical reactions that provide energy. Mitochondria in human cells contain several copies of a "mini-chromosome" that carries just 37 genes.

The inheritance patterns and mutation rates for mitochondrial genes differ from those for genes in the nucleus. Mitochondrial genes are maternally inherited. They are passed only from an individual's mother because sperm almost never contribute mitochondria when they fertilize an oocyte. In the rare instances when mitochondria from sperm enter an oocyte, they are usually selectively destroyed early in development. Pedigrees that follow mitochondrial genes show a woman passing the trait to all her children, while a male cannot pass the trait to any of his. The pedigree in **figure 5.6** illustrates maternal transmission of a mutation in a mitochondrial gene.

Unlike DNA in the nucleus, mitochondrial DNA (mtDNA) does not cross over. Mitochondrial DNA also mutates faster than nuclear DNA for two reasons: It

Table 5.2

Factors That Alter Mendelian Phenotypic Ratios

Phenomenon	Effect on Phenotype	Example
Lethal alleles	A phenotypic class does not survive to reproduce.	Spontaneous abortion
Multiple alleles	Many variants or degrees of a phenotype occur.	Cystic fibrosis
Incomplete dominance	A heterozygote's phenotype is intermediate between those of two homozygotes.	Familial hypercholesterolemia
Codominance	A heterozygote's phenotype is distinct from and not intermediate between those of the two homozygotes.	ABO blood types
Epistasis	One gene masks or otherwise affects another's phenotype.	Bombay phenotype
Penetrance	Some individuals with a particular genotype do not have the associated phenotype.	Polydactyly
Expressivity	A genotype is associated with a phenotype of varying intensity.	Polydactyly
Pleiotropy	The phenotype includes many symptoms, with different subsets in different individuals.	Porphyria variegata
Phenocopy	An environmentally caused condition has symptoms and a recurrence pattern similar to those of a known inherited trait.	Infection
Genetic heterogeneity	Different genotypes are associated with the same phenotype.	Hearing impairment

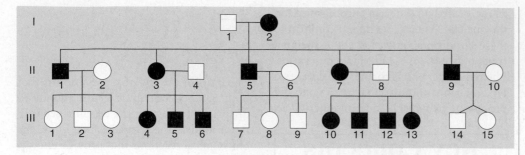

Figure 5.6   Inheritance of mitochondrial genes. Mothers pass mitochondrial genes to all offspring. Fathers do not transmit mitochondrial genes because sperm only very rarely contribute mitochondria to fertilized ova.

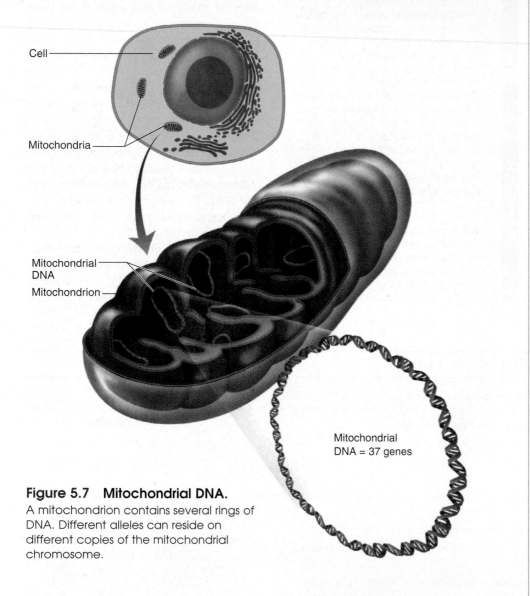

Figure 5.7   Mitochondrial DNA.
A mitochondrion contains several rings of DNA. Different alleles can reside on different copies of the mitochondrial chromosome.

harbors several copies of its chromosome (**figure 5.7** and **table 5.3**). Mitochondria with different alleles for the same gene can reside in the same cell.

Mitochondrial Disorders

Mitochondrial genes encode proteins that participate in protein synthesis and energy production. Twenty-four of the 37 genes encode RNA molecules (22 transfer RNAs and 2 ribosomal RNAs) that help assemble proteins. The other 13 mitochondrial genes encode proteins that function in cellular respiration. These are the biochemical reactions that use energy from digested nutrients to synthesize ATP, the biological energy molecule.

In diseases resulting from mutations in mitochondrial genes, symptoms arise from tissues whose cells have many mitochondria, such as skeletal muscle. It isn't surprising that a major symptom is often great fatigue. Inherited illnesses called mitochondrial myopathies, for example, produce weak and flaccid muscles and intolerance to exercise. Skeletal muscle fibers appear red and ragged when stained and viewed under a light microscope, their abundant abnormal mitochondria visible beneath the plasma membrane. When the first mitochondrial illness was recognized in 1962, it was viewed as a rarity. Today researchers suspect that mitochondrial disorders could be common.

A defect in an energy-related gene can produce symptoms other than fatigue. This is the case for Leber optic atrophy (OMIM 535000), which impairs vision. First described in 1871 and its maternal transmission noted, this disorder was not associ-

lacks DNA repair enzymes (discussed in chapter 9), and the mitochondrion is the site of the energy reactions that produce oxygen free radicals that damage DNA. Also unlike nuclear DNA, mtDNA is not wrapped in proteins, nor are genes "interrupted" by DNA sequences that do not encode protein. Finally, inheritance of mitochondrial genes differs from inheritance of nuclear genes simply because a human cell has one nucleus but many mitochondria—and each mitochondrion

Table 5.3
Features of Mitochondrial DNA
No crossing over
No DNA repair
Maternal inheritance
Many copies per mitochondrion and per cell
High exposure to oxygen free radicals
No histones (DNA-associated proteins)
No introns

ated with a mitochondrial mutation that impairs cellular energy reactions until 1988. Symptoms usually begin in early adulthood with a loss of central vision. Eyesight worsens and color vision vanishes as the central portion of the optic nerve degenerates.

A mutation in a mitochondrial gene that encodes a tRNA or rRNA can be devastating because it impairs the cell's ability to manufacture proteins. Consider what happened to Linda S., a once active and articulate dental hygienist and travel agent. In her forties, Linda gradually began to slow down at work. She heard a buzzing in her ears and developed difficulty talking and walking. Then her memory began to fade in and out, she became lost easily in familiar places, and her conversation made no sense. Her condition worsened, and she developed diabetes, seizures, and pneumonia and became deaf and demented. She was finally diagnosed with MELAS, which stands for "mitochondrial myopathy encephalopathy lactic acidosis syndrome" (OMIM 540000). Linda died. Her son and daughter will likely develop the condition because they inherited her mitochondria.

A new technique called ooplasmic transfer can enable a woman to avoid transmitting a mitochondrial disorder. Mitochondria from a healthy woman's oocyte are injected into the oocyte of a woman who is infertile. Then, the bolstered oocyte is fertilized in a laboratory dish by the partner's sperm, and the zygote is implanted in her uterus. Several dozen children, apparently free of mitochondrial disease, have been born from this technique.

Heteroplasmy Complicates Mitochondrial Inheritance

The fact that a cell contains many mitochondria makes possible a rare condition called **heteroplasmy,** in which a particular mutation may be present in some mitochondrial chromosomes, but not others. At each cell division, the mitochondria are distributed at random into daughter cells. Over time, the chromosomes within a mitochondrion tend to be all wild type or all mutant for any particular gene (a condition called homoplasmy).

Heteroplasmy has several consequences for the inheritance of mitochondrial phenotypes. Expressivity may vary widely among siblings, depending upon how many mutation-bearing mitochondria were in the oocyte that became each brother or sister. Severity of symptoms is also affected by which tissues have cells whose mitochondria bear the mutation. This is the case for a family with Leigh syndrome (OMIM 256000), which affects the enzyme that directly produces ATP. Two boys died of the severe form of the disorder because the brain regions that control movement rapidly degenerated. Another sibling was blind and had central nervous system degeneration. Several relatives, however, suffered only mild impairment of their peripheral vision. The more severely affected family members had more brain cells that received the mutation-bearing mitochondria.

The most severe mitochondrial illnesses are heteroplasmic. This is because homoplasmy—when all mitochondria bear the mutant allele—too severely impairs protein synthesis or energy production for embryonic development to complete. Often, severe heteroplasmic mitochondrial disorders do not produce symptoms until adulthood because it takes many cell divisions, and therefore much time, for a cell to receive enough mitochondria bearing mutant alleles to cause symptoms. For this reason, Leber optic atrophy usually does not affect vision until adulthood. Reading 9.1 relates how investigators used detection of heteroplasmy to solve a crime of historic import.

Mitochondrial DNA Studies Clarify the Past

Interest in mitochondrial DNA extends beyond the medical. Mitochondrial DNA provides a powerful forensic tool used to link suspects to crimes, identify war dead, and support or challenge historical records. The technology, for example, identified the son of Marie Antoinette and Louis XVI, who supposedly died in prison at age 10. In 1845, the boy was given a royal burial, but some people thought the buried child was an imposter. The boy's heart had been stolen at the autopsy, and through a series of bizarre events, wound up, dried out, in the possession of the royal family. Recently, researchers compared mitochondrial DNA sequences from cells in the boy's heart to corresponding sequences in heart and hair cells from Marie Antoinette (her decapitated body identified by her fancy underwear), two of her sisters, and still-living relatives Queen Anne of Romania and her brother. The genetic evidence showed that the unfortunate boy was indeed the prince, Louis XVII.

5.3 Linkage

Most of the traits that Mendel studied in pea plants were conferred by genes on different chromosomes. (Two were actually at opposite ends of the same chromosome.) When genes are located close to each other on the same chromosome, they usually do not separate during meiosis. Instead, they are packaged into the same gametes (**figure 5.8**). **Linkage** refers to the transmission of genes on the same chromosome. Linked genes do not assort independently and do not produce Mendelian ratios for crosses tracking two or more genes. Understanding linkage has been critical in identifying disease-causing genes, and helped pave the way for sequencing the human genome.

Linkage Was Discovered in Pea Plants

William Bateson and R. C. Punnett first observed the unexpected ratios indicating linkage in the early 1900s, again in pea plants. Bateson and Punnett crossed true-breeding plants with purple flowers and long pollen grains (genotype *PPLL*) to true-breeding plants with red flowers and round pollen grains (genotype *ppll*). The plants in the next generation, of genotype *PpLl*, were then self-crossed. But this dihybrid cross did not yield the expected 9:3:3:1

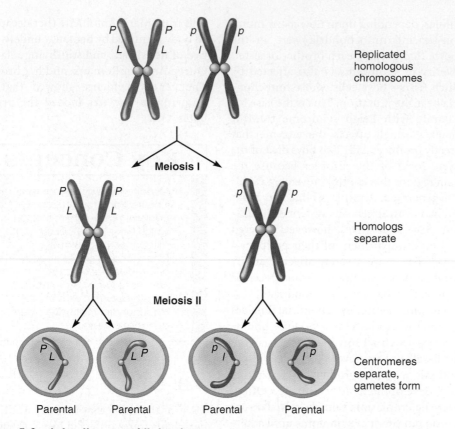

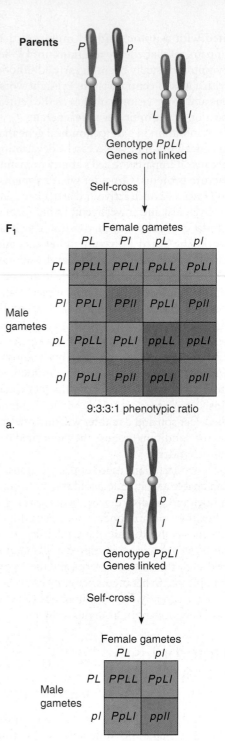

Figure 5.8 Inheritance of linked genes. Genes linked closely to one another on the same chromosome are usually inherited together when that chromosome is packaged into a gamete.

phenotypic ratio that Mendel's second law predicts (**figure 5.9**).

Bateson and Punnett noticed that two types of third-generation peas, those with the parental phenotypes *P_L_* and *ppll*, were more abundant than predicted, while the other two progeny classes, *ppL_* and *P_ll*, were less common (the blank indicates that the allele can be dominant or recessive). The more prevalent parental allele combinations, Bateson and Punnett hypothesized, could reflect genes that are transmitted on the same chromosome and that therefore do not separate during meiosis. The two less common offspring classes could also be explained by a meiotic event—crossing over. Recall that this is an exchange between homologs that mixes up maternal and paternal gene combinations without disturbing the sequence of genes on the chromosome (**figure 5.10**).

Progeny that exhibit this mixing of maternal and paternal alleles on a single chromosome are called **recombinant.** *Parental* and *recombinant* are relative terms. Had the parents in Bateson and Punnett's crosses been of genotypes *ppL_* and *P_ll*,

then *P_L_* and *ppll* would be recombinant rather than parental classes.

Two other terms describe the configurations of linked genes in dihybrids. Consider a pea plant with genotype *PpLl*. These alleles can be part of the chromosomes in either of two ways. If the two dominant alleles are on one chromosome and the two recessive alleles on the other, the genes are in "cis." In the opposite configuration, with one dominant and one recessive allele on each chromosome, the genes are in "trans" (**figure 5.11**). Whether alleles in a dihybrid are in cis or trans is important in distinguishing recombinant from parental progeny classes in specific crosses.

Linkage Maps

As Bateson and Punnett were discovering linkage in peas, geneticist Thomas Hunt Morgan and his coworkers at Columbia University were doing the same using the fruit fly *Drosophila melanogaster*, taking the concept of linkage further by assigning genes to relative positions on chromo-

Figure 5.9 Expected results of a dihybrid cross. **(a)** When genes are not linked, they assort independently. The gametes then represent all possible allele combinations. The expected phenotypic ratio of a dihybrid cross is 9:3:3:1. **(b)** If genes are linked on the same chromosome, only two allele combinations are expected in the gametes. The phenotypic ratio is 3:1, the same as for a monohybrid cross.

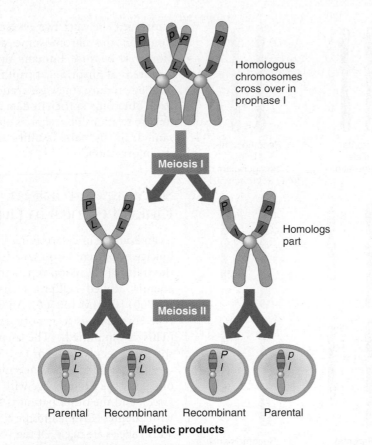

Homologous chromosomes cross over in prophase I

Meiosis I

Homologs part

Meiosis II

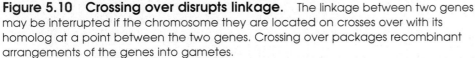

Parental   Recombinant   Recombinant   Parental

Meiotic products

Figure 5.10 Crossing over disrupts linkage. The linkage between two genes may be interrupted if the chromosome they are located on crosses over with its homolog at a point between the two genes. Crossing over packages recombinant arrangements of the genes into gametes.

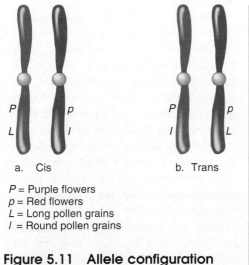

a. Cis   b. Trans

P = Purple flowers
p = Red flowers
L = Long pollen grains
l = Round pollen grains

Figure 5.11 Allele configuration is important. Parental chromosomes can be distinguished from recombinant chromosomes only if the allele configuration of the two genes is known—they are either in cis **(a)** or in trans **(b)**.

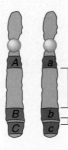

Genes A and B far apart; crossing over more likely

Genes B and C close together; crossing over less likely

Figure 5.12 Breaking linkage. Crossing over is more likely to occur between the widely spaced linked genes A and B, or between A and C, than between the more closely spaced linked genes B and C, because there is more room for an exchange to occur.

somes. These researchers compared progeny class sizes to assess whether traits were linked. They soon realized that the pairs of traits fell into four groups. Within each group, crossed dihybrids did not produce offspring according to the proportions Mendel's second law predicts. The number of these linkage groups—four—is exactly the number of chromosome pairs in the fly. The traits fell into four groups based on progeny class proportions because the genes controlling traits that are inherited together are transmitted on the same chromosome.

Morgan wondered why the size of the recombinant classes varied depending upon which genes were studied. Might the differences reflect the physical relationship of the genes on the chromosome? Exploration of this idea fell to an undergraduate, Alfred Sturtevant. In 1911, Sturtevant developed a theory and technique that would profoundly affect genetics. He proposed that the farther apart two genes are on a chromosome, the more likely they are to cross over simply

because more physical distance separates them (**figure 5.12**).

The correlation between crossover frequency and the distance between genes is used to construct **linkage maps,** which are diagrams that show the order of genes on chromosomes and the relative distances between them. The distance is represented using "map units" called centimorgans (cm), where 1 cm equals 1 percent recombination. The frequency of a crossover between any two linked genes is inferred from the proportion of offspring that are recombinant. Frequency of recombination is based on the percentage of meiotic divisions that break the linkage between two parental alleles. Genes at opposite ends of the same chromosome often cross over, generating a large recombinant class. Genes lying very close on the chromosome would only rarely be separated by a crossover. The probability that linked genes that are as far apart as possible on a chromosome will recombine due to their crossing over approaches the probability that, if on different chromosomes, they would independently assort—about 50 percent. **Figure 5.13** illustrates this distinction.

The situation with linked genes can be compared to a street lined with stores on both sides. There are more places to cross the street between stores at opposite ends on opposite sides than between two stores in the middle of the block. Similarly, more crossovers, or progeny with recombinant genotypes, are seen when two genes are farther apart on the same chromosome.

As the twentieth century progressed, geneticists in Columbia University's "fly room" mapped several genes on all four

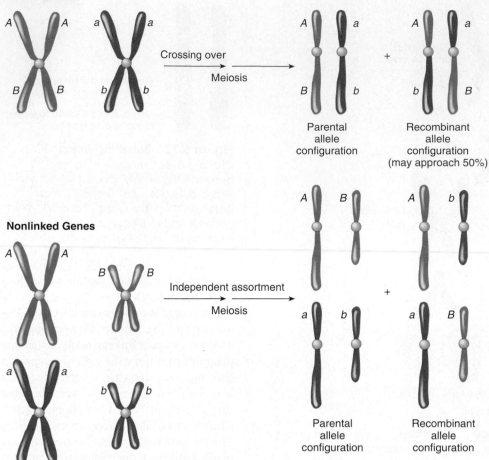

Linked Genes

Crossing over
Meiosis

Parental allele configuration

Recombinant allele configuration (may approach 50%)

Nonlinked Genes

Independent assortment
Meiosis

Parental allele configuration

Recombinant allele configuration

Figure 5.13 **Linkage versus non-linkage (independent assortment).** When two genes are widely separated on a chromosome, the likelihood of a crossover is so great that the recombinant class may approach 50 percent—which may appear to be the result of independent assortment.

chromosomes of the insect, and in other labs many genes were assigned to the human X chromosome. Localizing genes on the X chromosome was easier than doing so on the autosomes, because in human males, with their single X chromosome, recessive alleles on the X are expressed, a point we will return to in the next chapter.

By 1950, geneticists had begun to contemplate the daunting task of mapping genes on the 22 human autosomes. To start, a gene must be matched to its chromosome. This became possible, in a piecemeal fashion compared to today's genome sequencing, when people with a particular inherited condition or trait also had a specific chromosome abnormality. Matching phenotypes to chromosomal variants, a field called cytogenetics, is the subject of chapter 13.

In 1968, researchers assigned the first human gene to an autosome. R. P. Donohue was observing chromosomes in his own white blood cells when he noticed a dark area consistently located near the centromere of one member of his largest chromosome pair (chromosome 1). He then examined chromosomes from several family members for the dark area, noting also whether each family member had a blood type called Duffy. (Recall that blood types refer to the patterns of sugars on red blood cell surfaces.) Donohue found that the Duffy blood type was linked to the chromosome variant. He could predict a relative's Duffy blood type by whether or not the chromosome had the telltale dark area.

Finding a chromosomal variation and using it to detect linkage to another gene is a valuable but rare achievement. More often, researchers must rely on the sorts of experiments Sturtevant conducted on his flies—calculating percent recombination

(crossovers) between two genes whose locations on the chromosome are known. However, because humans do not have hundreds of offspring, as fruit flies do, nor do they produce a new generation every 10 days, obtaining sufficient data to establish linkage relationships requires observing the same traits in many families and pooling the information.

Solving a Problem: Linked Genes in Humans

As an example of determining the degree of linkage by percent recombination, consider the traits of Rh blood type and a form of anemia called elliptocytosis (OMIM 111700) (OMIM 130500). An Rh$^+$ phenotype corresponds to genotypes *RR* or *Rr*. (This is simplified.) The anemia corresponds to genotypes *EE* or *Ee*.

Suppose that in 100 one-child families, one parent is Rh negative with no anemia (*rree*), and the other parent is Rh positive with anemia (*RrEe*), and the *R* and *E* (or *r* and *e*) alleles are in cis. Of the 100 offspring, 96 have parental genotypes (*re/re* or *RE/re*) and four individuals are recombinants for these two genes (*Re/re* or *rE/re*). Percent recombination is therefore 4 percent, and the two linked genes are 4 cm (centimorgans) apart.

Consider another pair of linked genes in humans. Nail-patella syndrome (OMIM 161200) is a rare autosomal dominant trait that causes absent or underdeveloped fingernails and toenails, and painful arthritis, especially in the knee and elbow joints. The gene is 10 map units from the *I* gene that determines the ABO blood type, on chromosome 9. Geneticists determined the map distance by pooling information from many families. The information can be used to predict genotypes and phenotypes in offspring, as in the following example.

Greg and Susan each have nail-patella syndrome. Greg has type A blood, and Susan has type B blood. They want to know what the chance is that a child of theirs would inherit normal nails and knees and type O blood. Because information is available on Greg and Susan's parents, a genetic counselor can deduce their allele configurations (**figure 5.14**).

Greg's mother has nail-patella syndrome and type A blood. His father has normal

	Greg	Susan
Phenotype	nail-patella syndrome, type A blood	nail-patella syndrome, type B blood
Genotype	NnI^Ai	NnI^B__
Allele configuration	$\dfrac{N \quad I^A}{n \quad i}$	$\dfrac{N \quad i}{n \quad I^B}$

Gametes:	sperm	frequency	oocytes
Parental	$N\ I^A$	45%	$N\ i$
	$n\ i$	45%	$n\ I^B$
Recombinants	$N\ i$	5%	$N\ I^B$
	$n\ I^A$	5%	$n\ i$

N = nail-patella syndrome
n = normal

Figure 5.14 Inheritance of nail-patella syndrome. Greg inherited the N and I^A alleles from his mother; that is why the alleles are on the same chromosome. His n and i alleles must therefore be on the homolog. Susan inherited alleles N and i from her mother, and n and I^B from her father. Population-based probabilities are used to calculate the likelihood of phenotypes in the offspring of this couple. Note that in this figure, map distances are known and are used to predict outcomes.

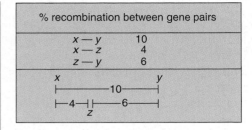

Figure 5.15 Recombination mapping. If we know the percent recombination between all possible pairs of three genes, we can determine their relative positions on the chromosome.

nails and type O blood. Therefore, Greg must have inherited the dominant nail-patella syndrome allele (N) and the I^A allele from his mother, on the same chromosome. We know this because Greg has type A blood and his father has type O blood—therefore, he couldn't have gotten the I^A allele from his father. Greg's other chromosome 9 must carry the alleles n and i. His alleles are therefore in cis.

Susan's mother has nail-patella syndrome and type O blood, and so Susan inherited N and i on the same chromosome. Because her father has normal nails and type B blood, her homolog bears alleles n and I^B. Her alleles are in trans.

Determining the probability that their child could have normal nails and knees and type O blood is the easiest question the couple could ask. The only way this genotype can arise from theirs is if an ni sperm (which occurs with a frequency of 45 percent, based on pooled data) fertilizes an ni oocyte (which occurs 5 percent of the time). The result—according to the product rule—is a 2.25 percent chance of producing a child with the $nnii$ genotype.

Calculating other genotypes for their offspring is more complicated, because more combinations of sperm and oocytes could account for them. For example, a child with nail-patella syndrome and type AB blood could arise from all combinations that include I^A and I^B as well as at least one N allele (assuming that the NN genotype has the same phenotype as the Nn genotype).

The Rh blood type and elliptocytosis, and nail-patella syndrome and ABO blood type, are examples of linked gene pairs. A linkage map begins to emerge when percent recombination is known between all possible pairs of three or more linked genes, just as a road map with more landmarks provides more information on distance and direction. Consider genes x, y, and z (**figure 5.15**). If the percent recombination between x and y is 10, between x and z is 4, and between z and y is 6, then the order of the genes on the chromosome is x-z-y (figure 5.15). This is the only order of the three genes that accounts for the percent recombination data.

Knowing the percent recombination between linked genes was useful in ordering them on genetic maps in a crude sense. Understanding the structure of DNA, and then sequencing the human genome, revealed an unexpected complexity in linkage mapping: crossing over is not equally likely to occur throughout the genome. Some DNA sequences are nearly always inherited together, more often than would be predicted from their frequency in the population. This nonrandom association between DNA sequences is called **linkage disequilibrium** (LD). The human genome consists of many "LD" blocks interspersed with areas where crossing over is prevalent.

The Evolution of Gene Mapping

Linkage mapping has had an interesting history. In the first half of the twentieth century, gene maps for nonhuman organisms, such as fruit flies, were constructed based on recombination frequencies between pairs of visible traits. In the 1950s, linkage data on traits in humans began to accumulate. At first, it was mostly a few visible or measurable traits linked to blood types or blood proteins.

In 1980 came a great stride in linkage mapping. Researchers began using DNA sequences near genes of interest as landmarks called genetic markers. These markers do not necessarily encode a protein that causes a phenotype—they might be differences that alter where a DNA cutting enzyme cuts, or differing numbers of short repeated sequences of DNA with no obvious function, or single nucleotide polymorphisms (SNPs) (see figures 7.9 and 14.4).

Computers tally how often genes and markers are inherited together. Gene mappers express the "tightness" of linkage between a marker and the gene of interest as a LOD score, which stands for "logarithm of the odds." A LOD score indicates the likelihood that particular crossover frequency data indicate linkage.

A LOD score of 3 or greater signifies linkage. It means that the observed data are 1,000 (10^3) times more likely to have occurred if the two DNA sequences (a disease-causing allele and its marker) are linked than if they reside on different chromosomes and just happen to often be inherited together by chance. It is somewhat like deciding whether two coins tossed together 1,000 times always come up both heads or both tails by chance, or because they are taped together side by side in that position, as linked genes are. If the coins land with the same side up in all 1,000 trials, it indicates they are very likely attached.

Before many disease-causing genes were discovered and the human genome sequence was known, genetic markers were used to predict which individuals in some families were most likely to have inherited a particular disorder, before symptoms began. Such tests are no longer necessary, because tests probe disease-causing genes directly.

Today, genetic markers are still used to distinguish parts of chromosomes. In pedigrees, marker designations are sometimes placed beneath the traditional symbols to further describe chromosomes. Such a panel of markers, called a **haplotype,** is a set of DNA sequences inherited together on the same chromosome due to linkage disequilibrium.

Haplotypes can look complicated, because markers are often given names that have meaning only to their discoverers. They read like license plates, bearing labels such as D9S1604. The haplotypes in the pedigree in **figure 5.16,** for a family with cystic fibrosis, are simplified. Each set of numbers beneath a symbol represents a

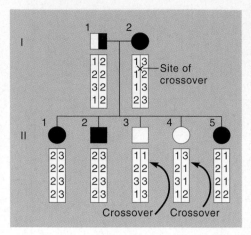

Figure 5.16 Haplotypes. The numbers in bars beneath pedigree symbols enable researchers to track specific chromosome segments with markers. Disruptions of a marker sequence indicate crossover sites.

"license plate" haplotype. Haplotypes make it possible to track which parent transmits which genes and chromosomes to offspring. In figure 5.16, knowing the haplotype of individual II–2 reveals which chromosome in parent I–1 contributes the mutant allele. Because Mr. II–2 received haplotype 3233 from his affected mother, his other haplotype, 2222, comes from his father. Since Mr. II–2 is affected and his father is not, the father must be a heterozygote, and 2222 must be the haplotype linked to the mutant CFTR allele.

It has been interesting and exciting to watch gene mapping evolve from the initial crude associations between blood types and chromosomal quirks, to today's maps

with their millions of signposts along the sequenced human genome. Throughout the 1990s, each October, *Science* magazine published a human genome map. The number of identified genes steadily grew as chromosome depictions became ever more packed with information. From that information, during your lifetime, will spring a revolution in understanding ourselves.

Key Concepts

1. Genes on the same chromosome are linked, and they are inherited in patterns that differ from the patterns of the unlinked genes Mendel studied.
2. Crosses involving linked genes produce a large parental class and a small recombinant class (caused by crossing over).
3. The farther apart two genes are on a chromosome, the more likely they are to recombine. Linkage maps are based on this relationship between crossover frequency and the distance between genes on the same chromosome.
4. Cytogenetic abnormalities revealed the first known linkage associations.
5. Linkage disequilibrium is a linkage combination that is stronger than that predicted by gene frequencies in a population.
6. Linkage maps reflect the percent recombination between linked genes. LOD scores describe the tightness of linkage and thereby the proximity of a gene to a marker. Haplotypes indicate linked DNA sequences.

Summary

5.1 When Gene Expression Appears to Alter Mendelian Ratios

1. Homozygosity for lethal recessive alleles stops development before birth, eliminating an offspring class.

2. A gene can have multiple alleles because its sequence can be altered in many ways. Different allele combinations produce different variations of the phenotype.

3. Heterozygotes of **incompletely dominant** alleles have phenotypes intermediate between those associated with the two

homozygotes. **Codominant** alleles are both expressed.

4. In **epistasis,** one gene affects the expression of another.

5. An incompletely **penetrant** genotype is not expressed in all individuals who inherit it. Phenotypes that vary in intensity among individuals are variable in **expressivity.**

6. **Pleiotropic** genes have several expressions.

7. A **phenocopy** is a characteristic that appears to be inherited but is environmentally caused.

8. In **genetic heterogeneity,** two or more genes specify the same phenotype.

9. The human genome sequence is explaining seeming exceptions to Mendel's laws.

5.2 Maternal Inheritance and Mitochondrial Genes

10. Only females transmit mitochondrial genes; males can inherit such a trait but cannot pass it on.

11. Mitochondrial genes do not cross over, and do not repair DNA.

12. The 37 mitochondrial genes encode tRNA, rRNA, or proteins involved in protein synthesis or energy reactions.

13. Many mitochondrial disorders are **heteroplasmic,** with mitochondria in a single cell harboring different alleles.

5.3 Linkage

14. Genes on the same chromosome are **linked** and, unlike genes that independently assort, produce many individuals with parental genotypes and a few with **recombinant** genotypes.

15. **Linkage maps** depict linked genes. Researchers can examine a group of known linked DNA sequences (a **haplotype**) to follow the inheritance of certain chromosomes.

16. Knowing whether linked alleles are in cis or trans, and using crossover frequencies from pooled data, one can predict the probabilities that certain genotypes will appear in progeny.

17. Genetic linkage maps assign distances to linked genes based on crossover frequencies.

Review Questions

1. Explain how each of the following phenomena can disrupt Mendelian phenotypic ratios.

   a. lethal alleles

   b. multiple alleles

   c. incomplete dominance

   d. codominance

   e. epistasis

   f. incomplete penetrance

   g. variable expressivity

   h. pleiotropy

   i. a phenocopy

   j. genetic heterogeneity

2. How does the relationship between dominant and recessive alleles of a gene differ from epistasis?

3. Why can transmission of an autosomal dominant trait with incomplete penetrance look like autosomal recessive inheritance?

4. How does inheritance of ABO blood type exhibit both complete dominance and codominance?

5. The lung condition emphysema may be caused by lack of an enzyme, or by smoking. Which cause is a phenocopy?

6. Describe why inheritance of mitochondrial DNA and linkage are exceptions to Mendel's laws.

7. How does a pedigree for a maternally inherited trait differ from one for an autosomal dominant trait?

8. What might be a confounding factor in attempting to correlate different genotypes with different expressions of a Mendelian illness?

9. If researchers could study pairs of human genes as easily as they can study pairs of genes in fruit flies, how many linkage groups would they detect?

Applied Questions

1. For each of the diseases described in situations *a* through *i*, indicate which of the following phenomena (A–H) is at work. A disorder may result from more than one of these causes.

   A. lethal alleles

   B. multiple alleles

   C. epistasis

   D. incomplete penetrance

   E. variable expressivity

   F. pleiotropy

   G. a phenocopy

   H. genetic heterogeneity

   a. A woman has severe neurofibromatosis type 1(OMIM 162200). She has brown spots on her skin and several large tumors beneath her skin. A gene test shows that her son has inherited the disease-causing autosomal dominant allele, but he has no symptoms.

   b. A man and woman have six children. They also had two stillbirths—fetuses that died shortly before birth.

   c. Most children with cystic fibrosis have frequent lung infections and digestive difficulties. Some people have mild cases, with onset of minor respiratory problems in adulthood. Some men have cystic fibrosis, but their only symptom is infertility.

   d. In Labrador retrievers, the *B* allele confers black coat color and the *b* allele brown coat color. The *E* gene controls expression of the *B* gene. If a dog inherits the *E* allele, the coat is golden no matter what the *B* genotype is. A dog of genotype *ee* expresses the *B* (black) phenotype.

   e. Two parents are heterozygous for genes that cause albinism, but each gene specifies a different enzyme in the biochemical pathway for skin pigment synthesis. Their children thus do not face a 25 percent risk of having albinism.

   f. Alagille syndrome (OMIM 118450), in its most severe form, prevents the formation of ducts in the gallbladder, causing liver damage. Affected children also usually have heart murmurs, unusual faces, a line in the eye, and butterfly-shaped vertebrae. Such children often have one otherwise healthy parent who has a heart murmur, unusual face, and butterfly vertebrae.

   g. Two young children in a family have terribly decayed teeth. Their parents think it is genetic, but the true cause is a babysitter who puts them to sleep with juice bottles in their mouths.

   h. A woman develops dark patches on her face. Her family physician suspects that she may have alkaptonuria. However, a dermatologist discovers that the woman has been using a facial cream containing hydroquinone, which causes dark skin patches in dark-skinned people.

   i. An apparently healthy 24-year-old basketball player dies suddenly during a game when her aorta, the largest artery, ruptures. A younger brother is nearsighted and has long and thin fingers, and an older sister is extremely tall, with long arms and legs. The older sister, too, has a weakened aorta. All of these siblings have Marfan syndrome (OMIM 154700).

2. If many family studies for a particular autosomal recessive condition reveal fewer affected individuals than Mendel's law predicts, the explanation may be either incomplete penetrance or lethal alleles. How might you use haplotypes to determine which of these two possibilities is the causative factor?

3. Nathan Milto, described in the opener to chapter 4, has late infantile Batten disease (OMIM 204500). His loss of motor skills, blindness, and seizures are caused by absence of a lysosomal enzyme that results in apoptosis of certain brain cells. In the juvenile form of Batten disease (OMIM 204200), a protein that normally binds a certain lipid to the plasma membrane is missing two amino acids. Because the lipid can't bind, the cell receives too many signals to divide. Loss of certain brain cells causes the same symptoms Nathan suffers. Describe how these two forms of this illness are pleiotropic and genetically heterogeneic.

4. A man who has type O blood has a child with a woman who has type A blood. The woman's mother has AB blood, and her father, type O. What is the probability that the child is of blood type

   a. O

   b. A

   c. B

   d. AB?

5. Two people who are heterozygous for familial hypercholesterolemia are concerned that a child might inherit the severe form of the illness. What is the probability that this will happen?

6. Enzymes are used in blood banks to remove the A and B antigens from blood types A and B. This makes the blood type O.

   a. Does this alter the phenotype or the genotype?

   b. Removing the A and B antigens from red blood cells is a phenocopy of what genetic phenomenon?

7. Ataxia-oculomotor apraxia syndrome (OMIM 208920), which impairs the ability to feel and move the limbs, usually begins in early adulthood. The molecular basis of the disease is impairment of ATP production in mitochondria, but the mutant gene is in the nucleus of the cells. Would this disorder be inherited in a Mendelian fashion? Explain your answer.

8. What is the chance that Greg and Susan, the couple with nail-patella syndrome, could have a child with normal nails and type AB blood?

9. A gene called secretor (OMIM 182100) is located 1 map unit from the H gene that confers the Bombay phenotype on chromosome 19. Secretor is dominant, and a person of either genotype $SeSe$ or $Sese$ secretes the ABO and H blood type antigens in saliva and other body fluids. This secretion, which the person is unaware of, is the phenotype. A man has the Bombay phenotype and is not a secretor. A woman does not have the Bombay phenotype and is a secretor. She is a dihybrid whose alleles are in cis. What is the chance that a child of theirs will have the same genotype as the father?

10. A Martian creature called a gazook has 17 chromosome pairs. On the largest chromosome are genes for three traits—round or square eyeballs (R or r); a hairy or smooth tail (H or h); and 9 or 11 toes (T or t). Round eyeballs, hairy tail, and 9 toes are dominant to square eyeballs, smooth tail, and 11 toes. A trihybrid male has offspring with a female who has square eyeballs, a smooth tail, and 11 toes on each of her three feet. She gives birth to 100 little gazooks, who have the following phenotypes:

   • 40 have round eyeballs, a hairy tail, and 9 toes

   • 40 have square eyeballs, a smooth tail, and 11 toes

   • 6 have round eyeballs, a hairy tail, and 11 toes

   • 6 have square eyeballs, a smooth tail, and 9 toes

   • 4 have round eyeballs, a smooth tail, and 11 toes

   • 4 have square eyeballs, a hairy tail, and 9 toes

   a. Draw the allele configurations of the parents.

   b. Identify the parental and recombinant progeny classes.

   c. What is the crossover frequency between the R and T genes?

Web Activities

Visit the Online Learning Center (OLC) at www.mhhe.com/lewisgenetics7. Select **Student Edition, chapter 5,** and **Web Activities** to find the website links needed to complete the following activities.

11. Go to the Family Village website. Family Village is a clearinghouse for disease information. Click on library. Explore the diseases, and identify one that exhibits pleiotropy.

12. Go to the United Mitochondrial Disease Foundation website and describe the phenotype of a mitochondrial disorder.

13. Browse the National Center for Biotechnology Information (NCBI) site, and list three sets of linked genes. Consult OMIM to describe the trait or disorder that each specifies.

14. Use OMIM to identify a genetically heterogeneic condition, and explain why this description applies.

Case Studies and Research Results

15. Connie Winslow is deaf. When she was old enough to attend school, she began having fainting spells, especially when she became excited. When she fainted while opening Christmas gifts, her parents took her to the emergency room, where doctors assured them, as they had in the past, that there wasn't a problem. The spells continued, and Connie became able to predict the attacks, telling her parents that her head hurt beforehand. Her parents took her to a neurologist, who checked Connie's heart and diagnosed long-QT syndrome with deafness, also known as Jervell and Lange-Nielsen syndrome (see OMIM 220400 and Reading 2.2). This is a severe form of inherited heartbeat irregularity that can be fatal. Seven different genes can cause long-QT syndrome. The doctor told them of a case described in a textbook from 1856: a young girl, called at school to face the headmaster for an infraction, became so agitated that she dropped dead. The

parents were not surprised; they had lost two other children to great excitement.

The Winslows visited a medical geneticist, who discovered that each had a mild heartbeat irregularity that did not produce symptoms. Connie's parents had normal hearing. Connie's younger brother Jim was also hearing-impaired and suffered night terrors, but had so far not fainted during the day. Like Connie, he had the full syndrome. Tina, still a baby, was also tested. She did not have either form of the family's illness; her heartbeat was normal.

Today, Connie and Jim are treated with beta blocker drugs, and each has a pacemaker to regulate heartbeat. Connie may receive an implantable defibrillator to automatically correct her heartbeat when it veers out of control. Diagnosing her may have saved her brother's life.

1. Which of the following applies to the condition in this family?

   a. genetic heterogeneity

   b. pleiotropy

   c. variable expressivity

   d. incomplete dominance

   e. a phenocopy

2. How is the inheritance pattern of Jervell and Lange-Nielsen syndrome similar to that of familial hypercholesterolemia?

3. How is it possible that Tina did not inherit either the serious or asymptomatic form of the illness?

4. Do the treatments for the condition affect the genotype or the phenotype?

Learn to apply the skills of a genetic counselor with these additional cases found in the *Case Workbook in Human Genetics:*

Cohen syndrome

Enamel hypoplasia

Hair and eye color

Thrombocytopenia and absent radius syndrome

CHAPTER

6

Matters of Sex

CHAPTER CONTENTS

BEING FEMALE—MORE THAN A "DEFAULT OPTION"

Ms. J. went to a hospital clinic because at 18 years old, she had not yet menstruated. She looked female. Her breasts and pubic hair had begun to develop at age 12, and her female hormone levels were normal, although her male hormone levels were slightly elevated. She also had acne, which is more common and severe in males. Medical imaging showed that her ovaries were positioned oddly and one kidney was underdeveloped. She lacked a uterus and vaginal canal.

The young woman had the chromosomes of a human female—two Xs. Those chromosomes appeared normal—they had not acquired the *SRY* gene from her father's Y chromosome that confers maleness, which might have explained her underdeveloped female organs and hints of masculinity. But the combination of abnormalities echoed a similar condition—in mice.

Female mice in which a gene called *Wnt4* is "knocked out" lack a uterus and vagina and make excess testosterone. Knockout mice of both sexes have underdeveloped single kidneys and abnormal adrenal glands, which normally produce sex hormones. Researchers inferred from the mice that *Wnt4* normally activates another gene that controls steroid hormone levels. Disable *Wnt4*, and hormone levels veer out of balance, disrupting sexual development.

Swiss researchers probed the *Wnt4* gene in Ms. J. and found a mutation also known in mice and other species, which means that the gene's function is vital. Her symptoms matched Mayer-Rokitansky-Kuster-Hauser syndrome (OMIM 277000), which, thanks to the mice, is now known to be caused by a mutation in the *Wnt4* gene. Perhaps more importantly, this case demonstrates that femaleness requires its own cascade of gene action—it is not merely the absence of activation of male-specific genes in the embryo.

A sex-reversed mouse. This mouse has the two X chromosomes that make it a genetic female, but it also has an *SRY* gene (courtesy of the researchers who bred her). The gene steered development toward maleness. Ms. J. has some signs of maleness due to a mutation in a different gene, *Wnt4*.

Whether we are male or female is enormously important in our lives, affecting our relationships, how we think and act, and how others perceive us. Gender is ultimately a genetic phenomenon, but is also layered with psychological and sociological components.

Maleness or femaleness is determined at conception, when he inherits an X and a Y chromosome, or she inherits two X chromosomes. Another level of sexual identity comes from the control that hormones exert over the development of reproductive structures. Finally, both biological factors and social cues influence sexual feelings, including the strong sense of whether we are male or female.

6.1 Sexual Development

Gender differences do not become apparent until the ninth week of prenatal development. During the fifth week, all embryos develop two unspecialized gonads, which are organs that will develop as either testes or ovaries. Each such "indifferent" gonad forms near two sets of ducts that present two developmental options. If one set of tubes, called the Müllerian ducts, continues to develop, they eventually form the sexual structures characteristic of a female. If the other set, the Wolffian ducts, persist, male sexual structures form. The choice of one of these developmental pathways occurs during the sixth week, depending upon the sex chromosome constitution. If a gene on the Y chromosome called *SRY* (for "sex-determining region of the Y") is activated, hormones steer development along a male route. In the absence of *SRY* activation, a female develops (**figure 6.1**). Femaleness was long considered a "default" option in human development, but sex determination is more accurately described as a fate imposed on ambiguous precursor structures. Several genes besides *SRY* guide early development of reproductive structures.

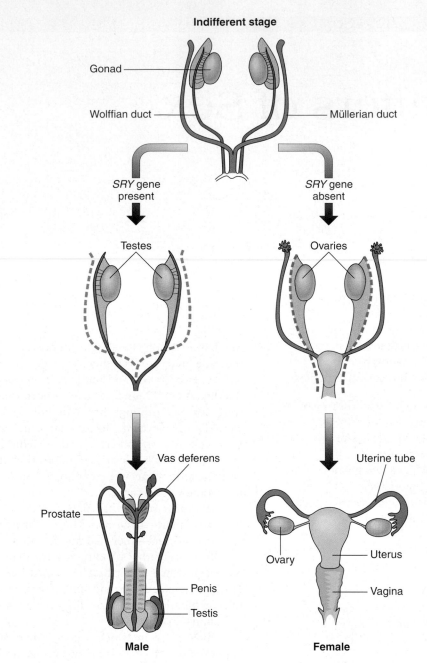

Figure 6.1 Male or female? The paired duct systems in the early human embryo may develop into male *or* female reproductive organs. The red tubes represent female structures and the blue tubes, male structures.

Sex Chromosomes

Human males and females have equal numbers of autosomes, but males have one X chromosome and one Y chromosome, and females have two X chromosomes (**figure 6.2**). The sex with two different sex chromosomes is called the **heterogametic sex,** and the other, with two of the same sex chromosomes, is the **homogametic sex.** In humans, males are heterogametic and females are homogametic. Some other species are different. In birds and snakes, for example, males are ZZ (homogametic) and females are ZW (heterogametic).

The X chromosome in humans contains more than 1,500 genes. The much smaller Y chromosome has 231 protein-encoding genes. In meiosis in a male, the X and Y chromosomes act as if they are a pair of

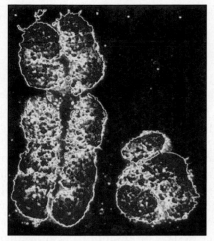

X chromosome   Y chromosome

Figure 6.2 The X and Y chromosomes.   In humans, females are the homogametic sex (XX) and males are the heterogametic sex (XY). (The chromosomes are in the replicated form because they were dividing when photographed.)

homologs. We will introduce the Y chromosome here, then consider the X in section 6.2.

Identifying genes on the human Y chromosome has been extremely difficult. Before the human genome sequence became available, researchers inferred the functions and locations of Y-linked genes by examining men who are missing parts of the chromosome and determining how they differ from normal. Creating linkage maps, which was possible for the other chromosomes, didn't work for the Y because it does not have a homolog with which to cross over (although its tips can cross over with the X chromosome).

Analysis of the genome sequence has revealed one source of the difficulty in mapping the Y chromosome: It has a very unusual organization. In the 95 percent of the chromosome that harbors male-specific genes, many DNA segments are palindromes, which in written languages are sequences of letters that read the same in both directions—"Madam, I'm Adam," for example. This symmetry of DNA sequence, described by researchers as "a hall of mirrors," destabilizes DNA replication. As a result, during meiosis, sections of a Y chromosome attract each other. This can loop out parts in between and may account for many cases of male infertility caused by missing parts of the

Y. Yet this organization may also provide a way for the chromosome to recombine with itself, essentially sustaining its structure. Two researchers—one an XX, one an XY—take a lighthearted look at the curious structure of the human Y chromosome in "In Their Own Words" on page 116.

The Y chromosome has a distinctive overall structure (**figure 6.3**) with a short arm and a long arm. At both tips of the Y chromosome are **pseudoautosomal regions,** termed PAR1 and PAR2. They comprise only 5 percent of the chromosome. The 63 pseudoautosomal genes are so-called because they have counterparts on the X chromosome and can cross over with them. These genes encode a variety of proteins that function in both sexes, participating in or controlling such activities as bone growth, cell division, immunity, signal transduction, the synthesis of hormones and receptors, fertility, and energy metabolism.

The bulk of the Y chromosome is termed the male-specific region, or MSY. (Until the Y chromosome was sequenced in 2003, this portion was called the nonrecombining region.) The MSY lies between the two pseudoautosomal regions, and it consists of three classes of DNA sequences. About 10 to 15 percent of the MSY consists of X-transposed sequences that are 99 percent identical to counterparts on the X chromosome. Protein-encoding genes are scarce here. Another 20 percent of the MSY consists of X-degenerate DNA sequences, which are somewhat similar to X chromosome sequences, and may be remnants of an ancient autosome that long ago gave rise to the X chromosome. The remainder of the MSY includes the palindrome-ridden regions, called amplicons. The genes in the MSY includes many repeats and specifies protein segments that combine in different ways—which is one reason why counting the number of protein-encoding genes on the Y chromosome has been difficult. Many of the genes in the MSY are essential to fertility, including *SRY*.

The Y chromosome was first seen under a light microscope in 1923, and researchers soon recognized its association with maleness. For many years, they sought to identify the gene or genes that determine sex. Important clues came from two very interesting types of people—men who have two X chromosomes (XX male syndrome), and

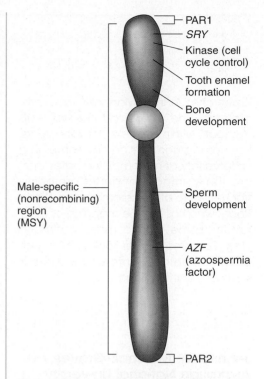

Figure 6.3 Anatomy of the Y chromosome.   The Y chromosome has two pseudoautosomal regions (PAR1 and PAR2) and a large central area (*MSY*) that comprises about 95 percent of the chromosome. A few genes are indicated here. *SRY* determines sex. *AZF* encodes a protein essential to producing sperm; mutations in it cause infertility.

women who have one X and one Y chromosome (XY female syndrome). A close look at their sex chromosomes revealed that the XX males actually had a small piece of a Y chromosome, and the XY females lacked a small part of the Y chromosome. The part of the Y chromosome present in the XX males was the same part that was missing in the XY females. This critical area accounted for half a percent of the Y chromosome. Finally, in 1990, two groups of researchers isolated and identified the *SRY* gene in this implicated area.

The Phenotype Forms

The *SRY* gene encodes a very important type of protein called a **transcription factor,** which controls the expression of other genes. The *SRY* transcription factor stimulates male development by sending signals to the indifferent gonads. In response, sustentacular

The Y Wars

Researcher Jennifer Marshall-Graves predicts that the Y chromosome will "self-destruct" within the next 10 million years. Her comparison of Y chromosomes in a wide variety of mammals indicates that, gradually, important genes are being transferred to other chromosomes. David Page, who has led the mapping of the Y chromosome, has a more optimistic view. Each researcher spoke out, in jest, at two scientific conferences. Here is some of what they had to say:

The Rise and Fall of the Human Y Chromosome

Jennifer A. Marshall-Graves, Australian National University

The Y chromosome is unique in the human genome. It is small, gene-poor, prone to deletion and loss, variable among species, and useless. You can lack a Y and not be dead, just female. It is impossible to understand why this chromosome is so weird without understanding where it came from. It is a sad decline, and I predict its imminent loss.

The X is a decent sort of chromosome. It accounts for 5 percent of the genome, with about 1,500 perfectly normal genes. The Y is a pathetic little chromosome that has few genes interspersed with lots of junk. And those genes are a weird lot. They are particularly concerned with male sexual development, so they are rather specialized. There are a number of important genes, but some are quite bizarre and many inactive. The Y

shares a lot of sequence with the X, and a lot of homology elsewhere, so the Y clearly diverged from the X.

There are several models of the Y (**figure 1**). The dominant Y model of a macho Y reflects the fact that the Y contains the male-determining *SRY* gene. The selfish Y model predicts that the Y kidnapped genes from elsewhere. The wimp Y model says that the Y is just a relic of the once glorious X chromosome. This model was first proposed by biologist Susumo Ohno in 1967 in the theory that the X and Y originated as a pair of autosomes. Then the Y acquired the male-determining locus, and other genes that are required for spermatogenesis gathered nearby. This led to suppressed recombination in this region of the Y,

Models of the Human Y

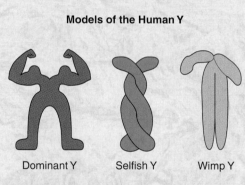

Dominant Y   Selfish Y   Wimp Y

Figure 1   Models of the human Y chromosome. Researcher Jennifer Marshall-Graves offers a tongue-in-cheek look at the Y chromosome, but her research findings are serious—the chromosome is shrinking.

which allowed all sorts of horrible genetic accidents to occur that could not be repaired. Mutations, deletions, and insertions accumulated until almost nothing was left, except bits at the top and bottom that still pair with the X. A few genes survived because they found a useful male-specific function, and many of these have made copies of themselves in a desperate race to stave off disappearing altogether.

The Y is degrading fast, losing genes at the rate of 5 per million years. I predict that it will be completely gone in 5 to 10 million years. Will we have males? The males in the audience can take comfort from the mole vole *Ellobius lutescens* (**figure 2**). It has no Y, but it does have males and females. It has no *SRY,* no Y chromosome at all. Both sexes are XO. How do they do it? We don't know. Clearly another gene takes over and new sex genes start evolving. Will there be new sex chromosome evolution in humans? Maybe it will happen in different ways in different populations, and we will split into two species.

Rethinking The Rotting Y Chromosome

David Page, Massachusetts Institute of Technology and Howard Hughes Medical Institute investigator

The Y chromosome has had a public relations problem for a long time. For most of the last half of the past century, people thought that the Y chromosome was a junk heap. The genomic junkyard view was the

cells in the developing testis secrete anti-Müllerian hormone, which destroys potential female structures (uterus, uterine tubes, and upper vagina). At the same time, interstitial cells in the testis secrete testosterone, which stimulates development of male structures (the epididymes, vasa deferentia, seminal vesicles, and ejaculatory ducts). Some testosterone is also converted to dihydrotestosterone (DHT), which directs the

development of the urethra, prostate gland, penis, and scrotum.

Because male prenatal sexual development is a multistep process, genetic abnormalities can intervene at several different points. The result may be an XY individual with a block in the gene- and hormone-controlled elaboration of male structures so that a chromosomal he is a phenotypic she. For example, in androgen insensitivity syn-

drome (OMIM 300068), caused by a mutation in a gene on the X chromosome, absence of receptors for testosterone stops cells in early reproductive structures from receiving the signal to develop as male. The person looks female, but is XY.

In a group of disorders called male pseudohermaphroditism (**figure 6.4**), testes are usually present (indicating that the *SRY* gene is functioning) and anti-Müllerian

Figure 2 Life without a Y? Males of all known mammals, with the exception of two species of mole voles, have Y chromosomes. Birds and reptiles do not. Evolutionary biologists think that the Y chromosome arose from an X chromosome about 310 million years ago, as the X lost many genes and gained a few that set their carriers on the road to maleness. This animal is a Y-less male mole vole—it reproduces just fine.

classic model for sex chromosome evolution. We can now update that model.

Back 300 million years ago, when we were reptiles, we had no sex chromosomes, only ordinary autosomes. Shortly after our ancestors parted company with the ancestors of birds, a mutation arose on one member of a pair of ordinary autosomes to give rise to *SRY*. The process of shutting down XY crossing over began, first in the vicinity of *SRY*, and then in an expanding region. Once a piece of the Y was no longer able to recombine with the X, its genes began to rot. The purpose of sex (recombination in meiosis) is not just to generate new gene combinations, but to allow genes to rid themselves of mildly deleterious mutations that accumulate. Y genes are not protected, because they have lots of areas of no crossing over. Genes decayed, except for *SRY* and the tips. It wasn't a very flattering model for the Y.

When Jennifer Marshall-Graves and John Aitken wrote their article in *Nature* on the future of sex, that the Y would self-destruct in 10 million years, it truly frightened the people in my lab. We decided we needed to pick up the pace. When the popular press

discovered the story of the impending death of the Y chromosome, they moved the date up to 5 million years from now.

Based on the sequencing of the Y, we've been able to rethink its evolution, and realized that the chromosome may have found a way around its seemingly inevitable problems. We looked closely at the male-specific region of the Y, reanalyzing sequences in a different way, chopped into smaller bits. And we found that each piece would find a match elsewhere on the Y. So segments on the Y are effectively functioning as alleles—30 percent have a perfect match elsewhere on the chromosome. These are not simple repeats, but highly complex sequences of tens to hundreds of kilobases. The region includes eight palindromes and one inverted repeat (**figure 3**). We propose that there is intense recombination within the palindromes. And so the Y has two forms of productive recombination: conventional routine recombination of crossing over with the X at pseudo-autosomal regions, and recombination within the Y. It's not that the Y doesn't recombine, it just does it its own way. The Y does copying that preserves its identity.

cen | u1 b1 t1 u2 t2 b2 u3 g1 r1 r2 b3 g2 r3 r4 g3 b4 | qter

1 Mb

Figure 3 The Y chromosome is highly repetitive. A section of the Y chromosome that David Page studies, called *AZFc* (for azoospermia factor c), consists of DNA sequences that read the same in either direction, an organization that can lead to instability as well as provide a mechanism to evolve new alleles. Other parts of the chromosome house similar repeats. Matching colors in this depiction represent identical sequences. Same-color arrows that point in opposite directions indicate inverted repeats.

hormone is produced, so the female set of tubes degenerates. However, a block in testosterone synthesis prevents the fetus from developing male structures. The child appears to be a girl. Then at puberty, the adrenal glands, which sit atop the kidneys, begin to produce testosterone (as they normally do in a male). This leads to masculinization: The voice deepens, and muscles build up into a masculine physique; breasts

do not develop, nor does menstruation occur. The clitoris may enlarge so greatly under the adrenal testosterone surge that it looks like a penis. Individuals with a form of this condition in the Dominican Republic are called *guevedoces,* for "penis at age 12." In the more common congenital adrenal hyperplasia due to 21-hydroxylase deficiency (OMIM 201910), an enzyme block leads to testosterone accumulation because testos-

terone cannot be converted to DHT. This causes overgrowth of the clitoris or penis, so that a girl may appear to be a boy.

A hermaphrodite is an individual with both male and female sexual structures. The word comes from the Greek god of war, Hermes, and the goddess of love, Aphrodite. "Pseudohermaphroditism" refers to the presence of both types of structures, but at different stages of life. Prenatal tests that

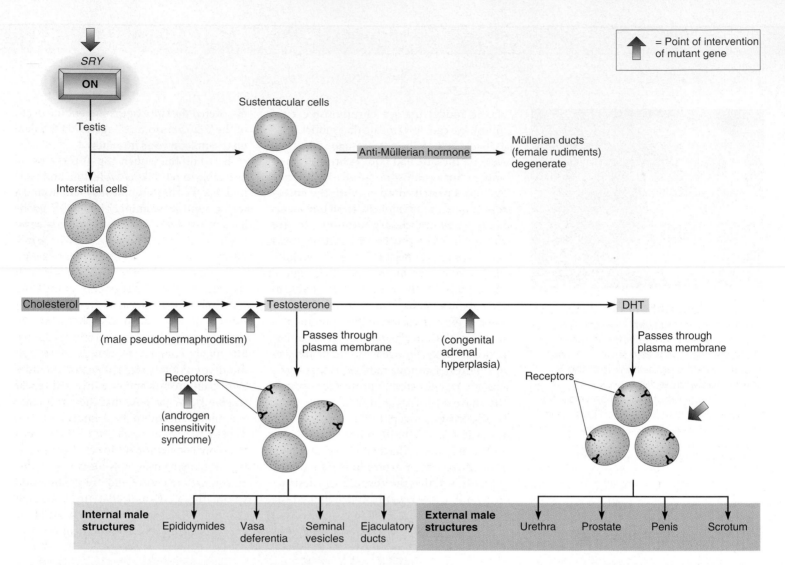

Figure 6.4 A chromosomal "he" develops as a phenotypic "she." In normal male prenatal development, activation of a set of genes beginning with *SRY* stimulates sustentacular cells to produce hormones that lead to destruction of female rudiments, and also stimulates interstitial cells to activate the biochemical pathway that produces testosterone, which yields DHT. Testosterone and DHT promote development of male structures. The green arrows indicate where mutations disrupt normal sexual development.

detect chromosomal sex have changed the way that pseudohermaphroditism is diagnosed. Before these tests were available, the condition was detected only after puberty, when masculinization occurred in a person who looked female. Today, pseudohermaphroditism is indicated when a prenatal chromosome check reveals an X and a Y chromosome, but the newborn is a phenotypic girl.

Transgender is a poorly understood condition related to sexual identity. A transgendered individual has the phenotype and sex chromosomes of one gender, but feels extremely strongly that he is a she, or vice versa. It is a much more profound condition than transvestitism, which refers to a male who prefers to wear women's clothing. The

genetic or physical basis of transgender is not known. Some affected individuals undergo surgery so that their physical selves match the gender that they feel certain they are.

Is Homosexuality Inherited?

No one really knows why we have feelings of belonging to one gender or the other, but these feelings are intense. Bioethics: Choices for the Future describes people whose gender identity persists even after surgery alters their phenotype in a way that contradicts their sex chromosome constitution.

In homosexuality, a person's phenotype and genotype are consistent, and physical

attraction is toward members of the same sex. Homosexuality is seen in all cultures and has been observed for thousands of years.

Evidence is accumulating that homosexuality is at least partially inherited. Earlier studies cite the feelings that homosexual individuals have as young children, well before they know of the existence or the meaning of the term. Studies with twins suggest a genetic influence. A 1991 study found that identical twins are more likely to both be homosexual than are both members of fraternal same-sex twin pairs. Specifically, in 52 percent of identical twin pairs in which one or both were homosexual, both brothers were homosexual, but this was true for only 22 percent of fraternal

Sex Reassignment: Making a Biological "He" into a Social "She"

Identical twins Bruce and Brian Reimer were born in 1965. At age eight months, most of Bruce's penis was accidentally burned off during a botched circumcision. On the advice of physicians and psychologists, the parents decided to "reassign" Bruce's gender as female. At 22 months of age, corrective surgery created Brenda from Bruce. The prominent psychologist in charge hailed the transformation a resounding success, and the case came to serve as a precedent for early surgical intervention for children born with "ambiguous genitalia" or structures characteristic of both sexes, a condition termed intersex. About 1 in 2,000 newborns are intersexes, with a few others the result of surgical accidents.

Gender Identity Can't Be Changed

Sex reassignment may not be the best treatment for some intersexual individuals. Reassessment of sex reassignment began when the Reimer case came to public attention. Reality for young Brenda was far different from the published descriptions.

Always uneasy in her dress-clad body, Brenda suffered ridicule and confusion, because it was always clear to her and others that she was more than a "tomboy"—she was a boy, despite her surgically altered appearance. Comparison to twin Brian worsened matters. When learning the truth at age 14, "Brenda" threatened suicide unless allowed to live as the correct gender. And so she became David Reimer. He eventually married, adopted stepchildren, and became a grandfather. He told his story to

Rolling Stone magazine. At about the same time, 1997, a groundbreaking paper by Keith Sigmundson, David's psychiatrist in his hometown of Winnipeg, and Milton Diamond of the University of Hawaii, supported David's contention that gender identity is due more to nature than nurture.

A study published in 2000 added more evidence. William Reiner, then at Johns Hopkins University, investigated fourteen children with a form of intersex called cloacal exstrophy. They were all XY, and had normal testicles and hormone levels, but no penis. Twelve were reassigned as female but behaved as boys throughout childhood. Six declared themselves male between the ages of 5 and 12 years. The two children who were not surgically converted into females are normal males who lack penises, something that surgery later in life may be able to correct. Other studies have confirmed Reiner's findings.

The Surgical Yardstick

In the past, physicians decided to remove a small or damaged penis and reassign sex as female using a literal yardstick: If a newborn's stretched organ exceeded an inch, he was deemed a he. If the protrusion was under three-eighths of an inch, she was deemed a she. Organs that fell in between were shortened into a clitoris during the first week of life, and girlhood officially began. Further plastic surgeries and hormone treatments during puberty completed the superficial transformation, with external female tissue sculpted from scrotal tissue. The reverse, creating a penis, is much

more difficult and was therefore usually delayed several months. These surgeries can destroy fertility and sexual sensation.

Easier to surgically treat are babies with congenital adrenal hyperplasia (CAH) due to 21-hydroxylase deficiency, the most common cause of intersex. The individual is XX, but overproduces masculinizing hormones (androgens). The result is a girl with a clitoris so large that it looks like a small penis. Thirty years ago, surgeons would cut away most of the extra tissue and create a vagina from skin flaps. Today, with the discovery that these females need a second surgery in adolescence anyway, treatment is postponed, giving these young women the chance to take part in decisions affecting their bodies.

Delaying surgery until a person can decide for him or herself may be the best approach for intersex individuals. Sex reassignment surgery is a bioethical issue that involves paternalism, confidentiality, the doctor-patient relationship, and the promise of physicians to "do no harm." Sums up Alice Dreger of Michigan State University, who has researched intersexuality extensively, "Gender identity is very complicated, and it looks like the various components interact and matter in different ways for different individuals. That's why unconsenting children and adults should never be subjected to cosmetic, medically unnecessary surgeries designed to alter their sexual tissue. We cannot predict what parts they may want later."

Perhaps the most compelling evidence that forcing someone to live in the wrong gender can do permanent damage is David Reimer's fate—he committed suicide in 2004.

---

twin pairs. Also, two brain areas are of different sizes in homosexual versus heterosexual men.

Research into the inheritance of homosexuality is controversial. In 1993, National Cancer Institute researcher Dean Hamer traced the inheritance of five genetic markers on the X chromosome in 40 pairs of

homosexual brothers. Although these DNA sequences are highly variable in the general population, they were identical in 33 of the sibling pairs. Hamer interpreted the finding to mean that genes causing or predisposing a male to homosexuality reside on the X chromosome. However, the work never identified a causative gene. One research

group confirmed and extended the work, finding that when two brothers are homosexual and have another brother who is heterosexual, the heterosexual brother does not share the X chromosome markers. This study also did not find the X chromosome markers between pairs of lesbian sisters. Several research groups have refuted

Hamer's findings. But a gene controlling homosexuality need not reside on a sex chromosome, where Hamer looked. Ongoing studies are searching among the autosomes for such genes.

In yet another approach to understanding the biological basis of homosexuality, researchers have genetically manipulated male fruit flies to display what looks like homosexual behavior. A mutant allele of an eye color gene called *white* causes the flies to have white eyes when expressed in cells of the eye only. Wild type eye color is red. Researchers altered male fly embryos so that the resulting adult insects expressed the *white* gene in every cell. The altered male flies displayed what appears to be mating behavior with each other (**figure 6.5**), presumably as a result of the altered gene expression.

The ability to genetically induce homosexual behavior suggests genetic control. The biochemical basis of the phenotype makes sense; the *white* gene's product, an enzyme that controls eye color, enables cells to use the amino acid tryptophan, which is required to manufacture the hormone serotonin. When all the fly's cells express the mutant *white* gene, instead of just eye cells,

Figure 6.5 Is homosexuality inherited? The ability to genetically alter male fruit flies, causing them to display mating behavior toward each other, adds to evidence that homosexuality is at least partially inherited.

Table 6.1

Sexual Identity

Level	Events	Timing
Chromosomal/ genetic	XY = male XX = female	Fertilization
Gonadal sex	Undifferentiated structure begins to develop as testis or ovary	6 weeks after fertilization
Phenotypic sex	Development of external and internal reproductive structures continues as male or female in response to hormones	8 weeks after fertilization, puberty
Gender identity	Strong feelings of being male or female develop	From childhood, possibly earlier
Sexual orientation	Attraction to same or opposite sex	From childhood

serotonin levels in the brain drop, and this may cause the unusual behavior. In other animals, lowered brain serotonin is associated with homosexual behavior.

Table 6.1 summarizes the several components of sexual identity.

Key Concepts

1. The human female is homogametic, with two X chromosomes, and the male is heterogametic, with one X and one Y.
2. The Y chromosome has 231 protein-encoding genes, and includes two small pseudoautosomal regions and the large male-specific region that does not recombine. Most of this region consists of palindromic sequences that lead to gene loss but also may maintain the chromosome's structure. Other DNA sequences here are similar to sequences on the X chromosome.
3. Activation of the *SRY* gene on the Y chromosome starts a cascade of gene action that causes the undifferentiated gonad to develop into a testis. Then, sustentacular cells in the testis secrete anti-Müllerian hormone, which stops development of female structures. Interstitial cells in the testis secrete testosterone, which stimulates development of male internal structures. Testosterone is also converted to DHT, which directs development of external structures.
4. Genes probably contribute to homosexuality.

6.2 Traits Inherited on Sex Chromosomes

Genes carried on the Y chromosome are said to be **Y-linked,** and those on the X chromosome are **X-linked.** Y-linked traits are rare, because the chromosome has few genes, and many have counterparts on the X chromosome. These traits are passed from male to male, because a female does not have a Y chromosome. No other Y-linked traits besides infertility (which obviously can't be passed on) are yet clearly defined, although certain gene products have been identified. Claims that "hairy ears" is a Y-linked trait did not hold up—it turned out that families hid their affected female members!

Genes on the X chromosome have different patterns of expression in females and males, because a female has two X chromosomes and a male just one. In females, X-linked traits are passed just like autosomal traits—that is, two copies are required for expression of a recessive allele, and one copy for a dominant allele. In males, however, a single copy of an X-linked allele causes expression of the trait or illness, because there is no copy of the gene on a second X chromosome to mask the other's effect. A man inherits an X-linked trait only from his mother, because he gets his Y chromosome from his father. The human male is considered **hemizygous** for X-linked traits, because he has only one set of X-linked genes.

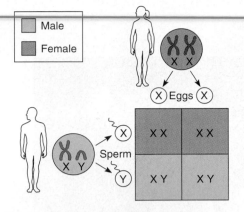

Male
Female

X Eggs X

Sperm

	X	X
X	X X	X X
Y	X Y	X Y

Figure 6.6 Sex determination in humans. An oocyte has a single X chromosome. A sperm cell has either an X or a Y chromosome. If a Y-bearing sperm cell with a functional *SRY* gene fertilizes an oocyte, the zygote is a male (XY). If an X-bearing sperm cell fertilizes an oocyte, then the zygote is a female (XX).

Understanding how sex chromosomes are inherited is important in predicting phenotypes and genotypes in offspring. A male inherits his Y chromosome from his father and his X chromosome from his mother (**figure 6.6**). A female inherits one X chromosome from each parent. If a mother is heterozygous for a particular X-linked gene, her son or daughter has a 50 percent chance of inheriting either allele from her. X-linked traits are always passed on the X chromosome from mother to son or from either parent to daughter, but there can be no direct male-to-male transmission of X-linked traits.

X-Linked Recessive Inheritance

An X-linked recessive trait is expressed in females if the causative allele is present in two copies. Many times, an X-linked trait passes from an unaffected heterozygous mother to an affected son. **Table 6.2** summarizes the transmission of an X-linked recessive trait.

If an X-linked condition is not lethal, a man may be healthy enough to transmit it to offspring. Consider the small family depicted in **figure 6.7,** an actual case. A middle-aged man who had rough, brown, scaly skin did not realize his condition was

---

Table 6.2

Criteria for an X-Linked Recessive Trait

1. Always expressed in the male.
2. Expressed in a female homozygote but very rarely in a heterozygote.
3. Passed from heterozygote or homozygote mother to affected son.
4. Affected female has an affected father and a mother who is affected or a heterozygote.

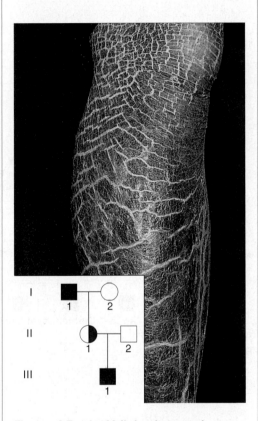

Figure 6.7 An X-linked recessive trait. Ichthyosis is transmitted as an X-linked recessive trait. A grandfather and grandson were affected in this family.

inherited until his daughter had a son. By a year of age, the boy's skin resembled his grandfather's. In the condition, called ichthyosis (OMIM 308100), an enzyme deficiency blocks removal of cholesterol from skin cells. The upper skin layer cannot peel off as it normally does, causing a brown, scaly appearance. A test of the

---

daughter's skin cells revealed that she produces half the normal amount of the enzyme, indicating that she is a carrier.

Colorblindness is another X-linked recessive trait that does not hamper the ability of a man to have children. About 8 percent of males of European ancestry are colorblind, as are 4 percent of males of African descent. Only 0.4 percent of females in both groups are colorblind. Reading 6.1, on page 124, takes a closer look at this interesting trait.

Figure 6.8 shows part of a very extensive pedigree for another X-linked recessive trait, the blood-clotting disorder hemophilia A (OMIM 306700). Note the combination of pedigree symbols and a Punnett square to trace transmission of the trait. Dominant and recessive alleles are indicated by superscripts to the X and Y chromosomes. In the royal families of England, Germany, Spain, and Russia, the mutant allele arose in one of Queen Victoria's X chromosomes; it was either a new mutation or she inherited it. In either case, she passed it on through carrier daughters and one mildly affected son. In Their Own Words in chapter 20 vividly describes one man's experience with hemophilia.

The transmission pattern of hemophilia A is consistent with the criteria for an X-linked recessive trait listed in table 6.2. A daughter can inherit an X-linked recessive disorder or trait if her father is affected and her mother is a carrier, because the daughter inherits one affected X chromosome from each parent. Without a biochemical test, though, an unaffected woman would not know she is a carrier for an X-linked recessive trait unless she has an affected son. A genetic counselor can estimate a potential carrier's risk using probabilities derived from Mendel's laws, combined with knowledge of X-linked inheritance patterns.

Consider a woman whose brother has hemophilia A. Both her parents are healthy, but her mother must be a carrier because her brother is affected. The woman's chance of being a carrier is 1/2 (or 50 percent), which is the chance that she has inherited the X chromosome bearing the hemophilia allele from her mother. The chance of the woman conceiving a son is 1/2, and of that son inheriting hemophilia is 1/2. Using the product rule, the risk that she will have a son with hemophilia, out of all the possible

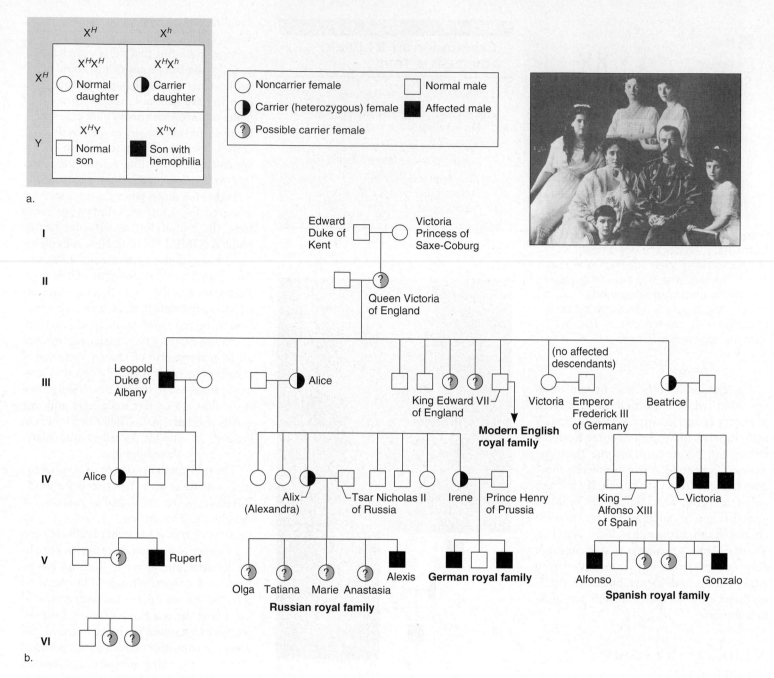

Figure 6.8 Hemophilia **(a)** This X-linked recessive disease usually passes from a heterozygous woman (designated X^HX^h, where X is the hemophilia-causing allele) to heterozygous daughters or hemizygous sons. The father is normal. **(b)** The disorder has appeared in the royal families of England, Germany, Spain, and Russia. The mutant allele apparently arose in Queen Victoria, who was either a carrier or produced oocytes in which the gene mutated. She passed the alleles to Alice and Beatrice, who were carriers, and to Leopold, who had a mild enough case that he fathered children. In the fourth generation, Alexandra was a carrier who married Nicholas II, Tsar of Russia. Alexandra's sister Irene married Prince Henry of Prussia, passing the allele to the German royal family, and Beatrice's descendants passed it to the Spanish royal family. This figure depicts only part of the extensive pedigree. The modern royal family in England does not carry hemophilia.

children she can conceive, is $1/2 \times 1/2 \times 1/2$, or 1/8.

Table 6.3 lists several X-linked disorders. Most genes on the X chromosome are not actually related to sex determination, and are necessary for normal development or physiology in both sexes.

X-Linked Dominant Inheritance

Dominant X-linked conditions and traits are rare. Again, gene expression differs between the sexes (**table 6.4**). A female who inherits a dominant X-linked allele has the associated trait or illness, but a male who inherits the allele is usually more severely affected because he has no other allele to offset it. The children of a normal man and a woman with a dominant, disease-causing gene on the X chromosome face the risks summarized in **figure 6.9**.

Table 6.3

Some Disease-Related Genes on the Human X Chromosome

Condition (r = recessive; D = dominant)	OMIM #	Symptoms
Agammaglobulinemia (r)	300300	Lack of certain antibodies
Alport syndrome (r)	301050	Deafness, kidney inflammation
Amelogenesis imperfecta (D)	301200	Abnormal tooth enamel
Anhidrotic ectodermal dysplasia (r)	305100	No teeth, hair, sweat glands
Chronic granulomatous disease (r)	306400	Skin and lung infections, enlarged liver and spleen
Diabetes insipidus (r)	304800	Copious urination
Duchenne muscular dystrophy (r)	310200	Progressive muscle weakness
Fabry disease (r)	301500	Abdominal pain, skin lesions, kidney failure
Hypophosphatemia (D and r)	307800	Vitamin-D-resistant rickets
Lesch-Nyhan syndrome (r)	300322	Mental retardation, self-mutilation, urinary stones, spastic cerebral palsy
Megalocornea (r)	249300	Enlarged cornea
Menkes disease (r)	309400	Kinky hair, brain degeneration, abnormal copper transport
Norrie disease (r)	310600	Eye degeneration
Ornithine transcarbamylase deficiency (rr)	311250	Mental deterioration, ammonia in blood
Retinitis pigmentosa (r and D)	312612	Constriction of visual field, nightblindness, clumps of pigment in eye
Rett syndrome (D)	312750	Mental retardation, neurodegeneration
Severe combined immune deficiency (r)	300400	Lack of T and B lymphocytes
Wiskott-Aldrich syndrome (r)	301000	Bloody diarrhea, infections, rash, bleeding

Table 6.4

Criteria for an X-Linked Dominant Trait

1. Expressed in female in one copy.
2. Much more severe effects in males.
3. High rates of miscarriage due to early lethality in males.
4. Passed from male to all daughters but to no sons.

An example of an X-linked dominant condition is incontinentia pigmenti (IP) (OMIM 308300). In affected females, swirls of skin pigment arise when melanin penetrates the deeper skin layers. A newborn girl with IP has yellow, pus-filled vesicles on her limbs that come and go over the first few weeks. Then the lesions become warty and eventually give way to brown splotches that may remain for life, although they fade with time. Males with the condition are so severely affected that they do not survive to be born. This is why women with the disorder have a miscarriage rate of about 25 percent.

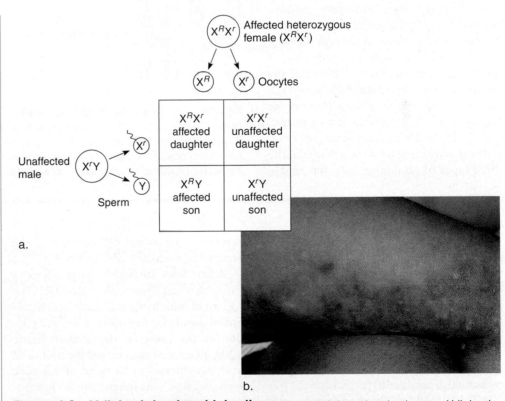

Figure 6.9   X-linked dominant inheritance. **(a)** A female who has an X-linked dominant trait has a 1 in 2 probability of passing it to her offspring, male or female. Males are generally more severely affected than females. **(b)** Note the characteristic patchy pigmentation on the leg of a girl who has incontinentia pigmenti.

Of Preserved Eyeballs and Duplicated Genes—Colorblindness

English chemist John Dalton saw things differently from most people. In a 1794 lecture, he described his visual world. Sealing wax that appeared red to other people was as green as a leaf to Dalton and his brother. Pink wildflowers were blue, and Dalton perceived the cranesbill plant as "sky blue" in daylight, but "very near yellow, but with a tincture of red," in candlelight. He concluded, "that part of the image which others call red, appears to me little more than a shade, or defect of light." The Dalton brothers had X-linked recessive colorblindness.

Curious about the cause of his colorblindness, Dalton asked his personal physician, Joseph Ransome, to dissect his eyes after he died. Ransome snipped off the back of one eye, removing the retina, where the cone cells that provide color vision are nestled among the more abundant rod cells that impart black-and-white vision. Because Ransome could see red and green normally when he peered through the back of his friend's eyeball, he concluded that it was not an abnormal filter in front of the eye that altered color vision. He stored the eyes in dry air, enabling researchers at the London Institute of Ophthalmology to analyze DNA in Dalton's eyeballs in 1994. Dalton's remaining retina lacked one of the three types of photopigments that enable cone cells to capture certain wavelengths of light.

Color Vision Basics

Cone cells are of three types, defined by the presence of any of three types of photopigments. An object appears colored because it reflects certain wavelengths of light, and each cone type captures a particular range of wavelengths with its photopigment. The brain then interprets the incoming information as a visual perception, much as an artist mixes the three primary colors to create many hues and shadings.

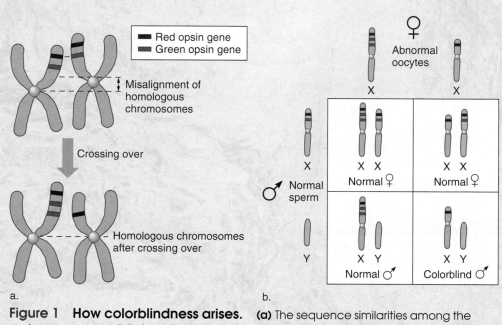

Figure 1   How colorblindness arises.   (a) The sequence similarities among the opsin genes responsible for color vision may cause chromosomes to misalign during meiosis in the female. Offspring may inherit too many, or too few, opsin genes. A son inheriting an X chromosome missing an opsin gene would be colorblind. A daughter, unless her father is colorblind, would be a carrier. **(b)** A missing gene causes X-linked colorblindness.

The gene that causes IP (called *NEMO*) encodes a transcription factor that activates genes that carry out the immune response and apoptosis in tissues that derive from ectoderm, such as skin, hair, nails, eyes, and the brain. Genetic tests can detect the deletion (missing DNA) that causes most cases.

Another X-linked dominant condition, congenital generalized hypertrichosis (CGH) (OMIM 307150), produces many extra hair follicles, and hence denser and more abundant upper body hair **(figure 6.10).** Hair growth is milder and patchier in females because of hormonal differences and the presence of a second X chromosome.

Researchers studied a large Mexican family that had 19 members with CGH. The pattern of inheritance was distinctive for X-linked dominant inheritance. In one portion of the pedigree, depicted in figure 6.10*b*, an affected man passed the trait to all four daughters, but to none of his nine sons. Because sons inherit the X chromosome from their mother, and only the Y from their father, they could not inherit CGH from their affected father.

The mutant gene that causes CGH controls a trait also present in ancestral species. At some time in our distant past, the functional form of the gene must have mutated in a way that enables humans to grow dense hair only on their heads and in areas dictated by sex hormones.

Solving a Problem: X-Linked Inheritance

Mendel's first law (segregation) applies to genes on the X chromosome. Therefore, the same logic is used to solve problems as to

Each photopigment has a vitamin A-derived portion called retinal and a protein portion called an opsin. The presence of retinal in photopigments explains why eating carrots, rich in vitamin A, promotes good vision. The presence of opsins—because they are controlled by genes—explains why colorblindness is inherited. The three types of opsins correspond to short, middle, and long wavelengths of light. Mutations in opsin genes cause three different types of colorblindness.

A gene on chromosome 7 encodes short-wave opsins, and mutations in it produce the rare autosomal "blue" form of colorblindness (OMIM 190900). Dalton had deuteranopia (green colorblindness), which means his eyes lacked the middle-wavelength opsin. In the third type, protanopia (red colorblindness), long-wavelength opsin is absent. Deuteranopia (OMIM 303800) and protanopia (OMIM 303900) are X-linked.

Molecular Analysis

Jeremy Nathans of Johns Hopkins University is another researcher who has personally contributed to our understanding of color vision. First, he used a cow version of a protein called rhodopsin that provides black-and-white vision to identify the human counterpart of the rhodopsin-related gene. Hypothesizing that the DNA sequence in the rhodopsin gene would be similar to that in the three opsin genes, and therefore able to bind to them, Nathans used the human rhodopsin gene as a "probe" to search his own DNA for genes with similar sequences. He found three. One was on chromosome 7, the other two on the X chromosome.

Although Nathans can see colors, his opsin genes are not entirely normal, which provided a big clue to how colorblindness arises and why it is so common. On his X chromosome, Nathans has one red opsin gene and two green genes, instead of the normal one of each. Because the red and green genes have similar sequences, Nathans reasoned, they can misalign during meiosis in the female (**figure 1**). The resulting oocytes would then have either two or none of one opsin gene type. An oocyte lacking either a red or a green opsin gene would, when fertilized by a Y-bearing sperm, give rise to a colorblind male.

People who are colorblind must get along in a multicolored world. To help them overcome the disadvantage of not seeing important color differences, computer algorithms can convert colored video pictures into shades they can see. **Figure 2** shows one of the tests typically used to determine whether someone is colorblind. Absence of one opsin type prevents affected individuals from seeing a different color in certain circles in the figure. These individuals cannot perceive a particular embedded pattern that other people can see.

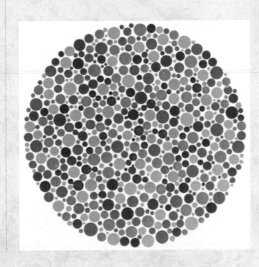

Figure 2   A test for colorblindness.
Males with red-green colorblindness cannot see the number 16 within this pattern of circles, as a person with normal color vision can.

Reproduced from *Ishihara's Tests for Colour Blindness*, published by Kanehara & Co. Ltd, Tokyo, Japan. Tests for colorblindness cannot be conducted with this material. For accurate testing, the original plates should be used.

trace traits transmitted on autosomes, with the added step of considering the X and Y chromosomes in Punnett squares. Follow these steps:

1. Look at the pattern of inheritance. Different frequencies of affected males and females in each generation may suggest X linkage.
   For an X-linked recessive trait:
   - An affected male has a carrier mother.
   - An unaffected female with an affected brother has a 50 percent (1 in 2) chance of being a carrier.
   - An affected female has a carrier or affected mother *and* an affected father.
   - A carrier (female) has a carrier mother *or* an affected father.

   For an X-linked dominant trait:
   - There may be no affected males, because they die early.
   - An affected female has an affected mother.

2. Draw the pedigree.
3. List all genotypes and phenotypes and their probabilities.
4. Assign genotypes and phenotypes to the parents in the problem. Look for clues in the phenotypes of relatives.
5. Determine how alleles might separate into gametes for the genes of interest on the X and Y chromosomes.
6. Unite the gametes in a Punnett square.
7. Determine the phenotypic and genotypic ratios for the F_1 generation.
8. To extend predictions to further generations, use the genotypes of the F_1 individuals and repeat steps 4 through 6.

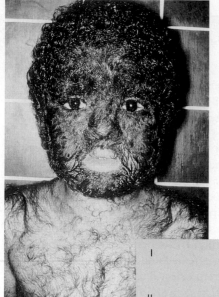

a.

Figure 6.10   An X-linked dominant condition. (a) This six-year-old child has congenital generalized hypertrichosis (CGH). (b) In this partial pedigree of a large Mexican family with CGH, the affected male in the second generation has passed the condition to all of his daughters and none of his sons. This is because he transmits his X chromosome only to females.

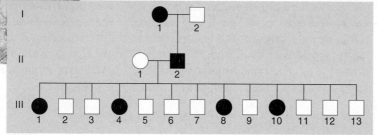

b.

6.3 X Inactivation Equalizes the Sexes

Females have two alleles for every gene on the X chromosome, whereas males have only one. In mammals, a mechanism called **X inactivation** balances this inequality. Early in the development of the female embryo, about 75 percent of the genes on one X chromosome in each cell are inactivated, and the remaining 25 percent are expressed to different degrees in different women. Which X chromosome is mostly turned off in each cell—the one inherited from the mother or the one from the father—is random. As a result, a female mammal mostly expresses the X chromosome genes inherited from her father in some cells and those from her mother in others (**figure 6.11**).

By studying rare human females who have lost a small part of one X chromosome, researchers identified a specific region, the X inactivation center, that shuts off much of the chromosome. Genes in the PARs and some others escape inactivation. A gene called *XIST* controls X inactivation. It encodes an RNA that binds to a specific site on the same (inactivated) X chromosome. From this point to the chromosome tip, the X chromosome is inactivated.

Once an X chromosome is inactivated in one cell, all its daughter cells have the same X chromosome inactivated. Because the inactivation occurs early in development, the adult female has patches of tissue that differ in their expression of X-linked genes. With each cell in her body having only one active X chromosome, she is chromosomally equivalent to the male.

X inactivation can alter the phenotype (gene expression), but not the genotype. It is not permanent, because the inactivation is reversed in germline cells destined to become oocytes. Therefore, a fertilized ovum does not have an inactivated X chromosome.

X inactivation is an example of epigenetics—a change that is passed from one cell generation to the next but that does not alter the DNA base sequence. We can observe X inactivation at the cellular level because the turned-off X chromosome absorbs a stain much faster than the active X. This differential staining occurs because inactivated DNA has chemical methyl groups that prevent it from being transcribed into RNA and also enable it to absorb stain.

As an example, consider Kallmann syndrome (OMIM 308700), which causes very poor or absent sense of smell and small gonads (testes or ovaries). It is X-linked recessive. Tanisha does not have Kallmann syndrome, but her brother Jamal and her maternal cousin Malcolm (her mother's sister's child) have it. Tanisha's and Malcolm's parents are unaffected, as is Tanisha's husband Sam. Tanisha and Sam wish to know the risk that a son of theirs would inherit the condition. Sam has no affected relatives.

Solution

1. Mode of inheritance: The trait is X-linked recessive because males are affected through carrier mothers.

2. K = wild type k = Kallmann syndrome

Genotypes	Phenotypes
$X^K X^K$, $X^K X^k$, $X^K Y$	normal
$X^k X^k$, $X^k Y$	affected

3.
Individual	Genotype	Phenotype	Probability
Tanisha	$X^K X^k$ or $X^K X^K$	normal (carrier)	50% each
Jamal	$X^k Y$	affected	100%
Malcolm	$X^k Y$	affected	100%
Sam	$X^K Y$	normal	100%

4. Tanisha's gametes if she is a carrier: X^K X^k
   Sam's gametes: X^K Y

5. Punnett Square

	X^K	X^k
X^K	$X^K X^K$	$X^K X^k$
Y	$X^K Y$	$X^k Y$

6. Interpretation: If Tanisha is a carrier, the probability that their son will have Kallmann syndrome is 50 percent, or 1 in 2. (Note that this is a conditional probability. The chance that any particular son will have the condition is actually 1 in 4, because Tanisha also has a 50 percent chance of being genotype $X^K X^K$ and therefore not a carrier.)

Key Concepts

1. Y-linked traits are passed on the Y chromosome, and X-linked traits on the X.
2. Because a male is hemizygous, he expresses all the genes on his X chromosome, whereas a female expresses recessive alleles on the X chromosome only if she is homozygous recessive.
3. X-linked recessive traits have a 50 percent probability of passing from carrier mothers to sons.
4. X-linked dominant conditions are expressed in both males and females but are more severe in males.
5. Mendel's first law can be used to solve problems involving X-linked genes.

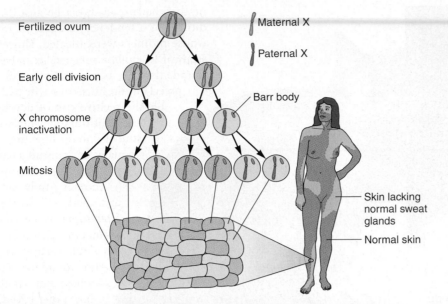

Figure 6.11 X inactivation. A female is a mosaic for expression of genes on the X chromosome because of the random inactivation of either the maternal or paternal X in each cell early in prenatal development. In anhidrotic ectodermal dysplasia, a woman has patches of skin that lack sweat glands and hair. (Colors distinguish cells with the inactivated X, not to depict skin color.)

The nucleus of a cell in a female, during interphase, has one dark-staining X chromosome called a **Barr body.** A cell from a male has no Barr body because his one X chromosome remains active (**figure 6.12**).

In 1961, English geneticist Mary Lyon proposed that the Barr body is the inactivated X chromosome and that it is turned off in early development. For homozygous X-linked genotypes, X inactivation would have no effect. No matter which X chromosome is turned off, the same allele is left to be expressed. For heterozygotes, however, X inactivation leads to expression of one allele or the other. Usually this doesn't affect health, because enough cells express the functional gene product. However, some traits reveal the X inactivation. The swirls of skin color in incontinentia pigmenti (IP) patients reflect patterns of X inactivation in cells in the skin layers. Where the normal allele for melanin pigment is shut off, pale swirls develop. Where pigment is produced, brown swirls result. The mosaic nature of the female due to X inactivation is also seen in anhidrotic ectodermal dysplasia (see figure 6.11).

A female who is heterozygous for an X-linked recessive gene can express the associated condition if the tissues that the illness affects have the normal allele inactivated. Consider a carrier of hemophilia A. If the X chromosome carrying the normal allele for the clotting factor is turned off in the liver, then the woman's blood will clot slowly, because the liver produces clotting factors—causing mild hemophilia. Luckily for her, slowed clotting time also greatly reduces her risk of cardiovascular disease caused by blood clots blocking circulation. A carrier of an X-linked trait who expresses the phenotype is called a **manifesting heterozygote.** In IP, X-inactivation is skewed in the opposite direction: The mutation-bearing X chromosome is preferentially silenced. Perhaps females in whom the normal X chromosome is silenced in most cells do not survive to be born.

A familiar example of X inactivation is the coat colors of tortoiseshell and calico cats. An X-linked gene confers brownish-black (dominant) or yellowish-orange (recessive) color. A female cat heterozygous for this gene has patches of each color, forming a tortoiseshell pattern that reflects different cells expressing either of the two alleles (**figure 6.13**). The earlier the X inactivation, the larger the patches, because more cell divisions can occur after the event, contributing more daughter cells. White patches may also occur due to epistasis by an autosomal gene that shuts off pigment synthesis. A cat with patches against such a white background is a calico. Tortoiseshell and calico cats are nearly always female. A male can have these coat patterns only if he inherits an extra X chromosome.

In humans, X inactivation can be used to detect carriers of some X-linked disorders. This is the case for Lesch-Nyhan syndrome, in which an affected boy has cerebral palsy, bites his fingers, shoulders, and lips to the

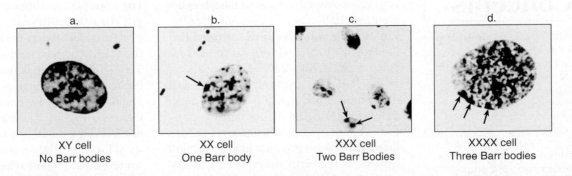

Figure 6.12 Barr bodies. A cell from a normal male has one X chromosome, which is active, and therefore no Barr body (**a**). A cell from a normal female has two X chromosomes and one Barr body (**b**). Rarely, a woman may have three X's and two Barr bodies (**c**), or even four X's and three Barr bodies (**d**).

a.

b.

Figure 6.13 Visualizing X inactivation.
X inactivation is obvious in tortoiseshell **(a)** and calico **(b)** cats. X inactivation is rarely observable in humans because most cells do not remain together during development, as a cat's skin cells do.

point of mutilation, is mentally retarded, and passes painful urinary stones. Mutation results in defective or absent HGPRT, an enzyme. A woman who carries Lesch-Nyhan syndrome can be detected when hairs from widely separated parts of her head are tested for HGPRT. (Hair is used for the test because it is accessible and produces the enzyme.) If some hairs contain HGPRT but others do not, she is a carrier. The hair cells that lack the enzyme have turned off the X chromosome that carries the normal allele; the hair cells that manufacture the normal enzyme have turned off the X chromosome that carries the disease-causing allele. The woman is healthy because her brain has enough HGPRT, but each son has a 50 percent chance of inheriting the disease.

Key Concepts

1. In female mammals, X inactivation compensates for differences between males and females in the numbers of gene copies on the X chromosome.
2. Early in development, one X chromosome in each cell of the female is turned off.
3. The effects of X inactivation can be noticeable when heterozygous alleles are expressed in certain tissues.

6.4 Sex-Limited and Sex-Influenced Traits

An X-linked recessive trait generally is more prevalent in males than females. Other situations, however, can affect gene expression in the sexes differently.

Sex-Limited Traits

A **sex-limited trait** affects a structure or function of the body that is present in only males or only females. Such a gene may be X-linked or autosomal.

Understanding sex-limited inheritance is important in animal breeding. For example, milk yield in cattle and horn development in sheep affect only one sex, but either parent can transmit the genes controlling these traits. In humans, beard growth and breast size are sex-limited traits. A woman does not grow a beard because she does not manufacture the hormones required for facial hair growth. She can, however, pass to her sons the genes specifying heavy beard growth.

An inherited medical condition that arises during pregnancy is obviously sex-limited, since males do not become pregnant. For example, preeclampsia is a sudden increase in blood pressure in the pregnant woman as the birth nears. It kills 50,000 women worldwide each year. Obstetricians have routinely asked their patients if their mothers had preeclampsia, because the condition has a tendency to occur in women whose mothers were affected. However, a study of 1.7 million pregnancies in Norway found that if a man's first wife had preeclampsia, his second wife had double the relative risk of developing the condition, too. The Norway study led to the hypothesis that a male can transmit a tendency to develop preeclampsia.

Another study on 298 men and 237 women in Utah supports the hypothesis that preeclampsia risk is sex-limited. This investigation found that women whose mothers-in-law had experienced preeclampsia when pregnant with the womens' husbands had approximately twice the relative risk of developing the condition themselves. One hypothesis to explain the results of the two studies is that a gene from the male affects the placenta in a way that elevates the pregnant woman's blood pressure. Genes that may confer the increased risk include those whose protein products participate in blood clotting, blood glucose control, and blood pressure. An alternative explanation might be that an infection transmitted by the male causes preeclampsia. Further investigations are warranted.

Sex-Influenced Traits

In a **sex-influenced trait,** an allele is dominant in one sex but recessive in the other. Such a gene may be X-linked or autosomal. The difference in expression can be caused by hormonal differences between the sexes. For example, an autosomal gene for hair growth pattern has two alleles, one that produces hair all over the head and another that causes pattern baldness **(figure 6.14)**. The baldness allele is dominant in males but recessive in females, which is why more men than women are bald. A heterozygous male is bald, but a heterozygous female is not. A bald woman is homozygous recessive. Even a bald woman tends to have some wisps of hair, whereas an affected male may be completely hairless on the top of his head.

a. b.

c. d.

Figure 6.14 Pattern baldness.
This sex-influenced trait was seen in the Adams family. John Adams (1735–1826) **(a)** was the second president of the United States and the father of John Quincy Adams (1767–1848) **(b)**, the sixth president. John Quincy was the father of Charles Francis Adams (1807–1886) **(c)**, a diplomat and the father of historian Henry Adams (1838–1918) **(d)**.

Key Concepts

1. A sex-limited trait affects body parts or functions present in only one gender.
2. A sex-influenced allele is dominant in one sex but recessive in the other.

6.5 Genomic Imprinting

In Mendel's pea experiments, it didn't matter whether a trait came from the male or female parent. For some genes in mammals, however, parental origin does influence the phenotype. These genes are said to be imprinted. In **genomic imprinting,** a molecule covers a gene or several linked genes and prevents them from being accessed to synthesize protein. The molecule, a methyl group, is a carbon atom bonded to three hydrogen atoms. For a particular imprinted gene, the copy inherited from either the father or the mother is always covered with methyls, even in different individuals. The result: a disease may be more severe, or different, depending upon which parent transmitted the gene. That is, a particular gene might function if it came from the father, but not if it came from the mother.

Silencing the Contribution From One Parent

Imprinting is an epigenetic alteration, in which a layer of meaning is stamped upon a gene without changing its DNA sequence. The imprinting pattern is passed from cell to cell in mitosis, but not from individual to individual through meiosis. When silenced DNA is replicated during mitosis, the pattern of blocked genes is exactly placed, or imprinted, on the new DNA, covering the same genes as in the parental DNA (**figure 6.15**). In this way, the "imprint" of inactivation is perpetuated, as if each such gene "remembers" which parent it came from. In meiosis, however, imprints are removed and reset. As oocyte and sperm form, the protective groups shielding their imprinted genes are stripped away, and new patterns are set down, depending upon whether the fertilized ovum is male or female. This is how women can have sons and men can have daughters without passing on their sex-specific parental imprints.

The function of genomic imprinting isn't well understood, but because many imprinted genes take part in early development, it may be a way to finely regulate the amounts of key proteins in the embryo. The fact that some genes lose their imprints after birth supports this idea of early importance. Also, imprinted genes occur in clusters, which are under the control of other regions of DNA called imprinting centers. Perhaps one gene in a cluster is essential for early development, and the others become imprinted simply because they are nearby—a bystander effect.

Genomic imprinting has implications for understanding early human development. It suggests that for mammals, it takes two opposite-sex parents to produce a healthy embryo and placenta. This was discovered in the early 1980s, through experiments on early mouse embryos and examination of certain rare pregnancy problems in humans. To investigate the roles of the genomes passed from the male and the female, researchers created fertilized mouse ova that contained two male pronuclei or two female pronuclei, instead of one from each. Results were strange. When the fertilized ovum contained two male genomes, a normal placenta developed, but the embryo was tiny and quickly stopped developing. A zygote with two female pronuclei, on the other hand, developed into an embryo, but the placenta was grossly abnormal. Therefore, it appeared that the male genome controls placenta development, and the female genome, embryo development.

The mouse results were consistent with abnormalities of human development. When two sperm fertilize an oocyte and the female pronucleus degenerates, an abnormal growth of placenta-like tissue called a hydatidiform mole forms (**figure 6.16a**). If a fertilized ovum contains only two female genomes but no male genome, a mass of random differentiated tissue, called a teratoma, grows. A teratoma, which means "monster cancer," may consist of a variety of tissues in a bizarre mix (figure 6.16b). With either a hydatidiform mole or a teratoma, no embryo results, although a pregnancy test may be positive, because the pregnancy hormone (hCG) may be produced.

These abnormalities indicate that, early in development, genes from a female parent direct different activities than genes from a male parent. The requirement that both male and female contribute to a zygote may explain why cloning mammals so rarely works, and when it does, the offspring are almost always unhealthy.

In a practical sense, manipulation of gamete nuclei may alter imprinting. Two imprinting disorders sometime arise in two types of assisted reproductive technologies: *in vitro* fertilization and injection of a sperm pronucleus into an oocyte. These technologies are associated with Angelman syndrome (OMIM 105830) and Beckwith-Wiedemann syndrome (OMIM 130650).

Genomic imprinting can explain incomplete penetrance, in which an individual is known to have inherited a genotype associated with a particular phenotype, but has no

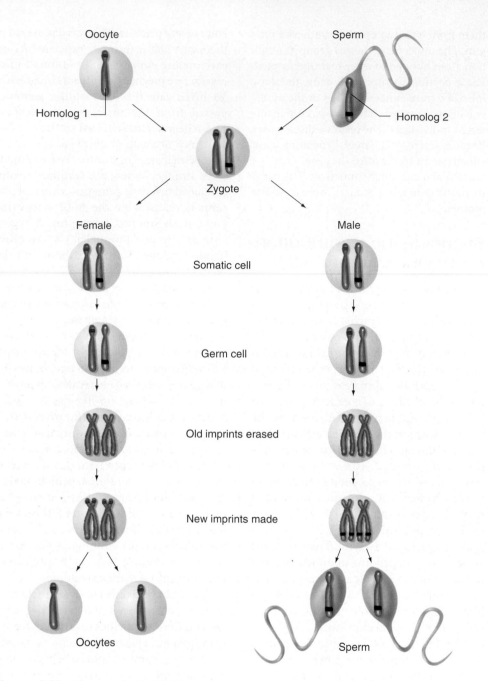

Figure 6.15 Genomic imprinting. Imprints are erased during meiosis, then reinstituted according to the sex of the new individual.

(a)

(b)

Figure 6.16 It takes pronuclei from a male and a female to start development. (a) When two male pronuclei form the first cell, an abnormal growth called a hydatidiform mole grows. It is similar to placental tissue. (b) When two female pronuclei form the first cell, an even more unusual structure forms, a teratoma ("monster cancer"). Teratomas may form in an ovary, and give rise to differentiated tissues, including hair, teeth, blood vessels, and nerve cells.

signs of the trait—such as a person with normal fingers whose parent and child have polydactyly. An imprinted gene that silences the dominant mutant allele could explain these cases: The predicted genotype is present, but the associated phenotype is not expressed.

Imprinting Disorders in Humans

At least 600 human genes are imprinted, and disruption of imprinting causes more than 30 known disorders. The effects of genomic imprinting are revealed only when an individual has one copy of a normally imprinted gene and the other copy is inactivated or deleted (absent). This chapter concludes with compelling examples of the effects of genomic imprinting gone awry.

In humans, a striking example of genomic imprinting involves two different syndromes that arise from small deletions in the same region of chromosome 15 (**figure 6.17**). A child with Prader-Willi syn-

drome (OMIM 176270) is obese, has small hands and feet, eats uncontrollably, and does not mature sexually. The other condition, Angelman syndrome, causes mental retardation, an extended tongue, large jaw, poor muscle coordination, and convulsions that make the arms flap. In many cases of Prader-Willi syndrome, only the mother's chromosome 15 region is expressed; the father's chromosome is deleted in that region. In Angelman syndrome, the reverse occurs: The father's gene (or genes) is expressed, and the mother's chromosome has the deletion.

Symptoms of Prader-Willi arise because several paternal genes that are not normally imprinted (that is, that are normally active)

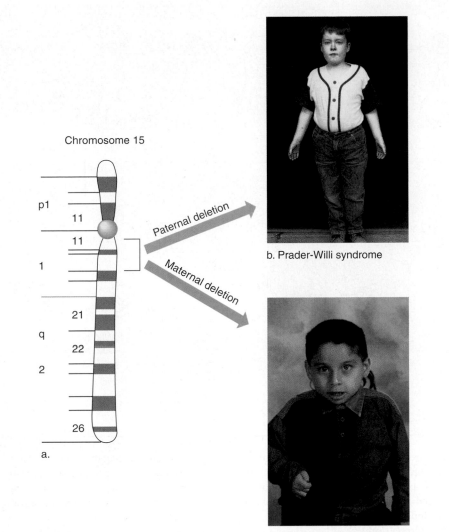

Chromosome 15

p1

11

11

1

21

q

22

2

26

a.

Paternal deletion

Maternal deletion

b. Prader-Willi syndrome

c. Angelman syndrome

Figure 6.17 Prader-Willi and Angelman syndromes. **(a)** Two distinct syndromes result from missing genetic material in the same region of chromosome 15. **(b)** Tyler has Prader-Willi syndrome, due to a deletion in the copy of the chromosome he inherited from his father. Note his small hands. **(c)** This child has Angelman syndrome, caused by a deletion in the chromosome 15 that he inherited from his mother. He is mentally retarded.

are missing. In Angelman syndrome, a normally active single maternal gene is deleted. This part of chromosome 15 is especially unstable because it includes highly repetitive DNA sequences, which bracket the genes that cause the symptoms.

Imprinting gone awry is associated with forms of diabetes mellitus, autism, Alzheimer disease, schizophrenia, and male homosexuality. Clues that indicate a condition is associated with genomic imprinting include increased severity depending on whether it is inherited from the father or mother and also a phenomenon called uniparental disomy. This term literally means

"two bodies from one parent," and refers to an offspring who inherits both copies of a gene from one parent. (Chapter 13 discusses uniparental disomy further.)

A Sheep With a Giant Rear End

Genomic imprinting is easier to observe in species that can be bred, so that traits can be followed over several generations. This is the case for Solid Gold, a ram with an overmuscled rear end (**figure 6.18a**), and his many offspring. Solid Gold was born in 1983 in Oklahoma, and by three weeks of age, his

hefty hindquarters were attracting attention. With visions of extra-meaty lamb chops, the breeder, instead of shipping Solid Gold to market, mated him to see if the trait was inherited. Sure enough, at three weeks of age, some of the lambs started to grow giant rears. Solid Gold became a favorite in the show ring. When a biology graduate student in search of a research project became interested in him, the ram's story continued in the genetics journals. Researchers named the mutation *callipyge*, which is Greek for "beautiful buttocks."

The trait was autosomal dominant, but people who tried to breed a stock from Solid Gold's lambs ran into a problem. The sheep seemed to be resisting Mendel's first law—the trait was passed only if it came from the father! In other words, a callipyge ewe did not yield a callipyge lamb (**figure 6.18b**). The situation turned out to be even more complex than straight genomic imprinting—not only must the ram be callipyge, but the ewe must be wild type to transmit the trait to all offspring.

Researchers eventually identified the callipyge gene, and also discovered that when it is overexpressed in the big-reared sheep, so are seven neighboring genes on sheep chromosome 18. (These other genes control wool quality and which muscles overgrow.) The involvement of several genes was another clue pointing to imprinting, which can affect a chromosome segment as well as an individual gene. Other researchers traced the entire suite of silenced genes to a single DNA base change, which disrupts imprinting.

Alas, the curious callipyge mutation did not turn out to be valuable to breeders—the meat was tough. However, callipyge genes were found in humans, prompting coverage of the ability to inherit a large rear end in several popular magazines.

Key Concepts

1. In genomic imprinting, the phenotype differs depending on whether a gene is inherited from the mother or the father.
2. Methyl groups may bind to DNA and temporarily suppress gene expression in a pattern determined by the individual's sex.
3. Imprinting may be a normal process in mammalian embryos.

♂ Callipyge × ♀ Wild type

♂ Wild type × ♀ Callipyge

Callipyge—
both sexes

Wild type—
both sexes

b.

Figure 6.18 Genomic imprinting in a sheep's rear. **(a)** Solid Gold astounded breeders with his overly muscled hindquarters. **(b)** Only callipyge males can pass on the trait, and only if the ewe is wild type.

Summary

6.1 Sexual Development

1. Sexual identity includes sex chromosome makeup; gonadal specialization; phenotype (reproductive structures); and gender identity.

2. The human male is the **heterogametic sex,** with an X and a Y chromosome. The female, with two X chromosomes, is the **homogametic sex.** The *SRY* gene on the Y chromosome and others determine maleness.

3. The human Y chromosome includes two **pseudoautosomal regions** and a large, male-specific region that does not recombine. Y-linked genes may correspond to X-linked genes, be similar to them, or be unique. Palindromic DNA sequences or inverted repeats can promote gene loss on the Y.

4. If the *SRY* gene is expressed, undifferentiated gonads develop as testes. If *SRY* is not expressed, the gonads develop as ovaries, under the direction of other genes.

5. Starting about eight weeks after fertilization, sustentacular cells in the testes secrete anti-Müllerian hormone, which prevents development of female structures, and interstitial cells produce testosterone, which triggers development of the epididymes, vasa differentia, seminal vesicles, and ejaculatory ducts.

6. Testosterone converted to DHT controls development of the urethra, prostate gland, penis, and scrotum. If *SRY* is not turned on, the Müllerian ducts continue to develop into female reproductive structures.

7. Evidence points to an inherited component to homosexuality.

6.2 Traits Inherited on Sex Chromosomes

8. Y-linked traits are rare and are passed from fathers to sons only.

9. Males are **hemizygous** for genes on the X chromosome and express phenotypes associated with these genes because they do not have another allele on a homolog. An X-linked trait passes from mother to son because he inherits his X chromosome from his mother and his Y chromosome from his father.

10. An X-linked allele may be dominant or recessive. X-linked dominant traits are more devastating to males than to females.

6.3 X Inactivation Equalizes the Sexes

11. **X inactivation** shuts off one X chromosome in each cell in female mammals, making them mosaics for heterozygous genes on the X chromosome. This phenomenon evens out the dosages of genes on the sex chromosomes between the sexes.

12. A female who expresses the phenotype corresponding to an X-linked gene she carries is a **manifesting heterozygote.**

6.4 Sex-Limited and Sex-Influenced Traits

13. **Sex-limited traits** may be autosomal or sex-linked, but they only affect one sex because of anatomical or hormonal gender differences.

14. A **sex-influenced gene** is dominant in one sex but recessive in the other.

6.5 Genomic Imprinting

15. In **genomic imprinting,** the phenotype corresponding to a particular genotype differs depending on whether the parent who passes the gene is female or male.

16. Imprints are erased during meiosis and reassigned based on the sex of a new individual.

17. Methyl groups that temporarily suppress gene expression are the physical basis of genomic imprinting.

Review Questions

1. How is sex expressed at the chromosomal, gonadal, phenotypic, and gender identity levels?

2. How do genes in the pseudoautosomal region of the Y chromosome differ from genes in the male-specific region (MSY)?

3. What are the phenotypes of the following individuals?

   a. a person with a mutation in the *SRY* gene, rendering it nonfunctional

   b. a normal XX individual

   c. an XY individual with a block in testosterone synthesis

4. List the events that must take place for a fetus to develop as a female.

5. Cite evidence that may point to a hereditary component to homosexuality.

6. Why is it unlikely one would see a woman who is homozygous for an X-linked dominant condition?

7. Why are male calico cats very rare?

8. How might X inactivation cause patchy hairiness on women who have congenital generalized hypertrichosis (CGH), even though the disease-causing allele is dominant?

9. How does X inactivation even out the "doses" of X-linked genes between the sexes?

10. Traits that appear more frequently in one sex than the other may be caused by genes that are inherited in an X-linked, sex-limited, or sex-influenced fashion. How might you distinguish among these possibilities in a given individual?

11. Cite evidence that genetic contributions from both parents are necessary for normal prenatal development.

12. Prader-Willi and Angelman syndromes are more common in children conceived with certain assisted reproductive technologies (*in vitro* fertilization and intracytoplasmic sperm injection) than among the general population. What process may these procedures disrupt?

Applied Questions

1. In Hunter syndrome (OMIM 309900), lack of the enzyme iduronate sulfate sulfatase leads to buildup of carbohydrates called mucopolysaccharides. In severe cases, the liver, spleen, and heart swell. In mild cases, deafness may be the only symptom. A child with this syndrome is deaf and has unusual facial features. Hunter syndrome is X-linked recessive. Intellect is usually unimpaired, and life span can be normal. Suppose a man who has mild Hunter syndrome has a child with a carrier.

   a. What is the probability that a male child would inherit Hunter syndrome?

   b. What is the chance that a female child would inherit Hunter syndrome?

   c. What is the chance that a girl would be a carrier?

   d. How might a carrier of this condition experience symptoms?

2. Coffin-Lowry syndrome (OMIM 303600) causes short, tapered fingers; abnormal finger and toe bones; puffy hands; soft, elastic skin; curved fingernails; facial anomalies; and sometimes hearing loss and heart problems. Evidence suggests that the syndrome is X-linked recessive, but girls are affected to a much lesser degree than boys. Suggest two explanations for why girls tend to have milder cases.

3. Amelogenesis imperfecta is an X-linked dominant condition that affects tooth enamel. Affected males have extremely thin enamel layers all over each tooth. Female carriers have grooved teeth from the uneven deposition of enamel. Explain the difference in phenotype between the sexes.

4. Huntington disease (see Bioethics: Choices for the Future, chapter 4) begins earlier and symptoms progress faster if the affected person inherits the disorder from his or her father. What phenomenon does this observation illustrate?

5. OMIM describes several families in which Mayer-Rokitansky-Kuster-Hauser syndrome (see the chapter opener) affects sisters, but not parents. In some families, the parents are blood relatives. The causative gene, *Wnt4*, is on chromosome 1.

   a. Suggest the most likely mode of inheritance for this syndrome.

   b. Is it sex-linked or sex-limited? How do you know?

   c. The condition affects 1 in 5,000 females, which is rather common for a Mendelian trait. If it renders a female incapable of normal reproduction, why is it so prevalent?

Web Activities

Visit the Online Learning Center (OLC) at www.mhhe.com/lewisgenetics7. Select **Student Edition, Chapter 6** and **Web Activities** to find the website links needed to complete the following activities.

6. Visit the National Center for Biotechnology Information (NCBI) website. Identify an X-linked disorder, then find it in OMIM and describe it.

7. At the Imprinted Gene Catalogue website, click on "search by species name" and then click on "complete list." Find two disorders that involve imprinting, one transmitted from the mother and one from the father, and use OMIM to describe them.

Case Studies and Research Results

8. Reginald has mild hemophilia A that he can control by taking a clotting factor. He marries Lydia, whom he met at the hospital where he and Lydia's brother, Marvin, receive their treatment. Lydia and Marvin's mother and father, Emma and Clyde, do not have hemophilia. What is the probability that Reginald and Lydia's son will inherit hemophilia A?

9. Harold works in a fish market, but the odor does not bother him because he has anosmia (OMIM 301700), an X-linked recessive lack of sense of smell. Harold's wife, Shirley, has a normal sense of smell. Harold's sister, Maude, also has a normal sense of smell, as does her husband, Phil, and daughter, Marsha, but their identical twin boys, Alvin and Simon, cannot detect

odors. Harold and Maude's parents, Edgar and Florence, can smell normally. Draw a pedigree for this family, indicating people who must be carriers of the anosmia gene.

10. Metacarpal 4–5 fusion is an X-linked recessive condition in which certain finger bones are fused. It occurs in many members of the Flabudgett family, depicted in the pedigree at the top of the page.

    a. Why are three females affected, considering that this is an X-linked condition?

    b. What is the risk that individual III-1 will have an affected son?

    c. What is the risk that individual III-5 will have an affected son?

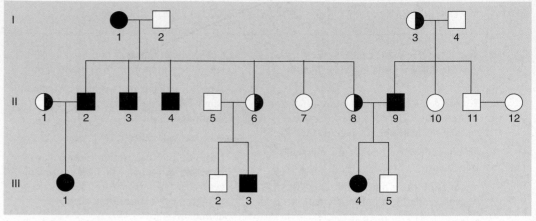

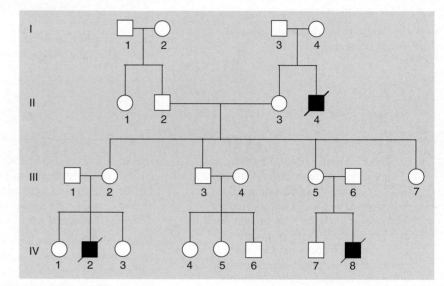

Metacarpal 4–5 fusion

11. Herbert is 58 years old and bald. His wife, Sheri, also has pattern baldness. What is the risk that their son, Frank, will lose his hair?

12. The Addams family knows of three relatives who had Menkes disease, an X-linked recessive disorder in which a child does not grow, has brain degeneration, and dies by age two. Affected children have peculiar white stubby hair.

    Wanda Addams is hesitant about having children because her two sisters have each had sons who died from Menkes disease. Her mother had a brother who died of the condition, too. The pedigree for the family is shown to the right.

    a. Fill in the symbols for the family members who must be carriers of Menkes disease.

    b. What is the chance that Wanda (III-7) is a carrier?

    c. If Wanda is a carrier, what is the chance that a son of hers would inherit the disease?

    d. Why don't any women in the family have Menkes disease?

Menkes disease

Learn to apply the skills of a genetic counselor with these additional cases found in the *Case Workbook in Human Genetics:*

Anhidrotic ectodermal dysplasia
Blue diaper syndrome
Chronic granulomatous disease
Congenital muscular dystrophies
Intersex

CHAPTER

7

Multifactorial Traits

CLEFT LIP AND PALATE

The young couple was shocked when they first saw their daughter. She had a cleft lip and palate—a hole between her nose and upper lip. The parents soon discovered that feeding Emily was difficult, because she could not maintain suction. Special nipples on her bottles helped.

Today, Emily is 14, and has a glorious smile. The defect occurred between weeks 4 and 12 of prenatal development, when the nose and jaw failed to meet and close. Emily endured many surgeries. Her first procedure, at 4 months, repaired her lip; the second, at a year, connected the edges of her palate (the roof of the mouth). The doctor also repositioned tissue at the back of her throat to correct her nasal speech. A speech and language therapist helped with Emily's early feeding problems and assisted her when frequent ear infections, due to openings at the back of her throat, caused hearing loss. At age seven Emily had orthodontia to make room for her permanent teeth, and at age 10, bone from her hip was used to bolster her palate so that it could support the teeth. At age 16, Emily can have surgery on her nose to build it up and straighten it.

Cleft lip, with or without cleft palate, is very variable in severity and has genetic and environmental components. Known causes include at least three genes; drugs used to treat seizures, anxiety, and high cholesterol; pesticide residues; and maternal smoking and infections. Emily's parents were nervous when expecting their second child, because population data indicated that the risk that he would be affected was about 4 percent. He wasn't. Since his birth, a test has become available that identifies a haplotype on chromosome 8 that is associated with a 12 percent risk of developing cleft lip and/or palate.

Cleft lip is more likely to occur in a person who has a relative with the condition.

A woman who is a prolific writer has a daughter who becomes a successful novelist. An overweight man and woman have obese children. A man whose father suffers from alcoholism has the same problem. Are these characteristics—writing talent, obesity, and alcoholism—inherited or learned? Or are they a combination of nature (genetics) and nurture (the environment)?

Most of the traits and medical conditions mentioned so far in this book are single-gene characteristics, inherited according to Mendel's laws, or linked on the same chromosome. Many single-gene disorders are very rare, each affecting one in hundreds or even thousands of individuals. Using Mendel's laws, geneticists can predict the probability that certain family members will inherit single-gene conditions. Most more common traits and diseases, though, seem to "run in families" with no pattern, or they occur sporadically, with just one case in a family.

Genes rarely act completely alone. Even single-gene disorders are modified by environmental factors and/or other genes. This chapter discusses non-Mendelian characteristics, and the tools used to study them. Chapter 8 focuses on the most difficult traits to assess for their inherited components—behaviors.

7.1 Genes and the Environment Mold Most Traits

On the first page of the first chapter of *On the Origin of Species*, Charles Darwin noted that two factors are responsible for biological variation—"the nature of the organism and the nature of the conditions." Darwin's thoughts were a nineteenth-century musing on heredity versus the environment. Though this phrase might seem to indicate that genes and the environment are adversaries, they are actually two forces that interact, and they do so in ways that mold many of our characteristics.

A trait can be described as either Mendelian or **polygenic.** A single gene causes a Mendelian trait. A polygenic trait, as its name implies, reflects the activities of more than one gene. Both Mendelian and polygenic traits can also be **multifactorial,** which means that they are influenced by the environment. Pure polygenic traits—those not influenced by the environment—are very rare.

Multifactorial traits, also called complex traits, include height, skin color, illnesses, and behavioral conditions and tendencies. Behavioral traits are not inherently different from other types of traits; they involve the functioning of the brain, rather than another organ.

In contrast to a single-gene disorder, a complex multifactorial condition may be caused by the additive contributions of several genes. Each gene confers a degree of susceptibility, but their contributions are not necessarily equal. For example, we know that multiple sclerosis (MS) has a genetic component, because siblings of an affected individual are 25 times as likely to develop MS as siblings of people who do not have MS. One model of MS origin suggests that five susceptibility genes have alleles, each of which increases the risk of developing the condition. Those risks add up, and, in the presence of an appropriate (and unknown) environmental trigger, the disease begins.

Polygenic Traits Are Continuously Varying

For a polygenic trait, the combined action of many genes often produces a "shades of grey" or "continuously varying" phenotype, also called a quantitative trait. DNA sequences that contribute to polygenic traits are called **quantitative trait loci,** or QTLs. A multifactorial trait is continuously varying if it is also polygenic. That is, it is the genetic component of the trait that contributes the continuing variation of the phenotype. The individual genes that confer a polygenic trait follow Mendel's laws, but together they do not produce Mendelian phenotypic ratios. They all contribute to the phenotype, without being dominant or recessive to each other. For example, the multiple genes that regulate height and skin color result in continuously varying phenotypes. Mendelian traits are instead discrete or qualitative, often providing a "black or white" phenotype such as "normal" versus "affected."

A polygenic trait varies in populations, as our many nuances of hair color, body weight, and cholesterol levels demonstrate.

Some genes contribute more to a polygenic trait than others. Within genes, alleles can have differing impacts depending upon exactly how they alter an encoded protein, as well as upon how common they are in a population. For example, a mutation in the gene that encodes the receptor that takes low-density lipoproteins (LDL cholesterol) into cells drastically raises a person's blood serum cholesterol level. But because fewer than 1 percent of the individuals in most populations have this mutation, it contributes very little to the variation in cholesterol level seen at the population level.

Although the expression of a polygenic trait is continuous, we can categorize individuals into classes and calculate the frequencies of the classes. When we do this and plot the frequency for each phenotype class, a bell-shaped curve results. Even when different numbers of genes affect the trait, the curve takes the same shape, as is evident in the following examples.

RELATING THE CONCEPTS

What evidence indicates that isolated cleft lip/palate involves quantitative trait loci?

Fingerprint Patterns

The skin on the fingertips folds into patterns of raised skin called dermal ridges that align to form loops, whorls, and arches. A technique called dermatoglyphics ("skin writing") compares the number of ridges that comprise these patterns to identify and distinguish individuals (**figure 7.1**). Dermatoglyphics is part of genetics because certain disorders (such as Down syndrome) include unusual ridge patterns. Dermatoglyphics is also part of forensics in fingerprint analysis.

The number of ridges in a fingerprint pattern is largely determined by genes, but also responds to the environment. Therefore it is a multifactorial trait. During weeks 6 through 13 of prenatal development, the ridge pattern can alter as the fetus touches the finger and toe pads to the wall of the amniotic sac. This early environmental effect explains why the fingerprints of identical twins, who share all genes, are in some cases not exactly alike.

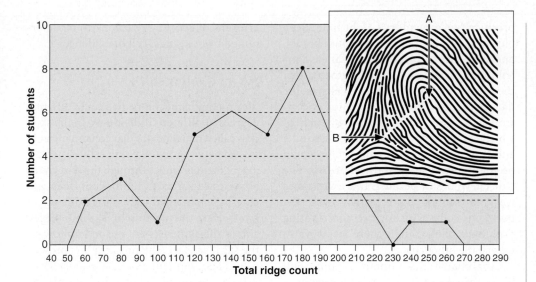

Figure 7.1 Anatomy of a fingerprint. Total ridge counts for a number of individuals, plotted on a bar graph, form an approximate bell-shaped curve. The number of ridges between landmark points A and B on this loop pattern is 12. Total ridge count includes the number of ridges on all fingers.

Data and print from Gordon Mendenhall, Thomas Mertens, and Jon Hendrix, "Fingerprint Ridge Count," in *The American Biology Teacher*, vol. 51, no. 4, April 1989, pp. 204–6.

Figure 7.2 The inheritance of height. The photograph in **(a)** illustrates the continuously varying nature of height. In the photo, taken around 1920, 175 cadets at the Connecticut Agricultural College lined up by height. In 1997, professor Linda Strausbaugh asked her genetics students at the school, today the University of Connecticut at Storrs, to recreate the scene **(b)**. They did, and confirmed the continuously varying nature of human height. But they also elegantly demonstrated how height increased during the twentieth century. The tallest people in the old photograph (a) are 5′9″ tall, whereas the tallest people in the more recent photograph (b) are 6′5″ tall.

We can quantify a fingerprint with a measurement called a total ridge count, which tallies the number of ridges comprising a whorl, loop, or arch part of the pattern for each finger. The average total ridge count in a male is 145, and in a female, 126. Plotting total ridge count reveals the bell curve characteristic of a continuously varying trait.

Height

The effect of the environment on height is obvious—people who do not eat enough do not reach their genetic potential for height. Students lined up according to height, but raised in two different decades and under different circumstances, vividly reveal the effects of genes and the environment on this continuously varying trait. Part *a* of **figure 7.2** depicts students from 1920, and part *b*, students from 1997. But also note that the tallest people in the old photograph are 5′9″, whereas the tallest people in the more recent photograph are 6′5″. The difference is attributed to such environmental factors as improved diet and better overall health.

We usually do not know exactly how many genes contribute to multifactorial traits that are also polygenic. However, geneticists can suggest models for a certain number of genes contributing to a trait based on the number of variants that can be discerned—although this is limited by what we can perceive.

Eye Color

Eye color is probably a pure polygenic trait—one with no environmental input (colored contact lenses don't count!). But eye color isn't only a matter of brown, blue, green, or hazel. Overlying the common tones are specks and flecks, streaks and rings, and regions of dark versus light. These modifications arise from the way pigment is laid down onto the distinctive peaks and valleys at the back of the iris, the colored portion of the eye.

Two genes specify greenish-blue pigments called lipochromes, and two or more other genes encode the brownish melanins. These genes interact in a hierarchy, with the brown genes masking the green/blues, and everything masking pure blue. Unlike the browns, which are caused by chemical pigments, pure blue is a "spectral color" that results from light scattering, much as the blue of the sky is an effect of the sun's rays penetrating the atmosphere. Hazel eyes have a mixture of lipochromes and melanins. Blue-eyed parents can have brown-eyed children if the parents do not have pure blue eyes—which few people do. Each parent contributes a slight ability to produce pigment, which adds in the child to color the irises a pale brown. Unlike pigment in the skin, melanin in the iris stays in the cell that produces it.

The topography at the back of the iris is as distinctive as fingerprints, and is an inherited trait. Thicker parts of this area darken the appearance of the pigments, rendering brown eyes nearly black in some parts, or blue eyes closer to purple. The bluest of blue eyes have thin irises with very little pigment. The effect of the iris surface on color is a little like the visual effect of a rough-textured canvas on paint.

For many years, eye color was thought to arise from two genes with two alleles each, as depicted in **figure 7.3**. Although this is a gross oversimplification, it does illustrate the bell curve that describes the phenotypes resulting from gene interaction. These alleles interact additively to produce five eye colors—light blue, deep blue or green, light brown, medium brown, and dark brown/black. If each allele contributes a certain amount of pigment, then the greater the number of such alleles, the darker the eye color. If eye color is controlled by two genes, *A* and *B*, each of which comes in two allelic forms—*A* and *a* and *B* and *b*—then the lightest color would be genotype *aabb;* the darkest, *AABB.* The bell curve arises because there are more ways to inherit light brown eyes, the midrange color, with any two contributing dominant alleles, than there are ways to inherit the other colors.

Analysis of the human genome will likely reveal additional eye color genes—the mouse has more than 60! Recognizing that there may be dozens of variations of eye color, with distinctions perhaps beyond our perception, one company has analyzed 300 sites within the two lipochrome and two melanin genes in hundreds of individuals. From this information, their researchers have developed a forensic tool called the "retinome," which is a database of eye colors that goes well beyond the standard four that appear on a driver's license. Using the retinome, a criminal suspect's eye color can be determined from a single cell of evidence.

Skin Color

Melanin pigment colors the skin to different degrees in different individuals. In the skin, cells called melanocytes that contain melanin extend between tile-like skin cells, distributing pigment granules through the skin layers. Some melanin exits the melanocytes and enters the hardened cells in the skin's upper layers. Here the melanin breaks into pieces, and as the skin cells are pushed up toward the skin's surface, the melanin bits provide color. The pigment protects against DNA damage from ultraviolet radiation, and exposure to the sun increases melanin synthesis. **Figure 7.4** shows a three-gene model for human skin color—an oversimplification of this highly variable trait.

Although people come in a wide variety of hues, we all have about the same number of melanocytes per unit area of skin. Differences in skin color arise from the number and distribution of melanin pieces in the skin cells in the uppermost layers. People with albinism cannot manufacture melanin (see figure 4.15).

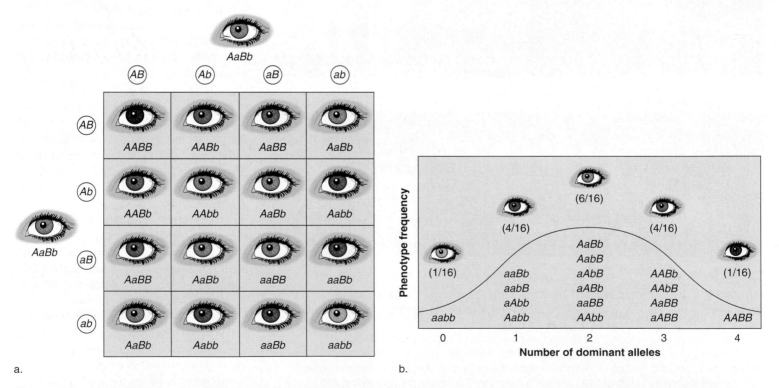

a.

b.

Figure 7.3 Variations in eye color. **(a)** A model of two genes, with two alleles each, can explain the existence of five eye colors in humans. **(b)** The frequency distribution of eye colors forms the characteristic bell-shaped curve for a polygenic trait.

Figure 7.4 Variations in skin color.
(a) A model of three genes, with two alleles each, can explain broad hues of human skin. In actuality, this trait likely involves many more than three genes. **(b)** Humans come in a great variety of skin colors.

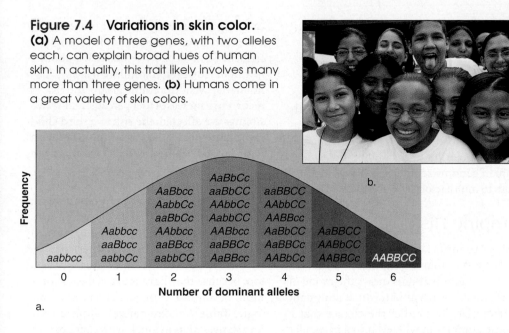

a.

b.

Skin color is one trait used to distinguish race. However, the definition of race based on skin color is more a social construct than a biological concept, for skin color is but one of thousands of variable traits. From a genetic perspective, when referring to other types of organisms, races are groups within species that are distinguished by different allele frequencies. Although we tend to classify people by skin color because it is an obvious visible way to distinguish individuals, skin color is not a reliable indicator of heritage.

When many genes are examined, two people with black skin may be less alike than either is to another person with white skin. On a population level, sub-Saharan Africans and Australian aborigines have dark skin, but are very dissimilar in other inherited characteristics. Their dark skins may reflect the same adaptation to life in a sunny, tropical climate. Overall, 93 percent of varying inherited traits are no more common in people of one skin color than any other.

In one telling investigation, 100 students in a sociology class in "Race and Ethnic Relations" at Pennsylvania State University demonstrated that skin color does not reflect ancestry. The students had their DNA tested for percent contribution from "European white", "black African", "Asian" and "Native American" gene variants. No student was pure anything, and many were quite surprised at what their DNA revealed about their ancestry. One student, a light-skinned black, learned that genetically he is 52 percent black African and 48 percent European white: approximately half black, half white. Another student who considered herself black was actually 58 percent white European. The U.S. census, in recognition of the complexity of classifying people into races, began to allow "mixed race" as a category in 2000. Many if not all of us fall into this category.

Although in a genetic sense the concept of race based on skin color has little meaning, in a practical sense, such groups do have different incidences of certain diseases. This reflects the tendency to choose partners within the group, which retains certain alleles, but it also may result from social inequalities, such as some groups' limited access to good nutrition or health care. Observations that members of particular races have a higher incidence of certain illnesses have influenced medical practice; some drug companies market hypertension and heart disease drugs to African Americans, who have a higher incidence of these conditions than do people in other populations. **Table 7.1** lists some drugs that seem to be more effective among either Americans of European descent or Americans of African descent. Very few drugs and population groups have been tested—these results are preliminary and will be refined as studies continue.

Offering medical treatments based on skin color may make sense on a population level, but on the individual level it may lead to errors. A white person might be denied a drug that would work, or a black person given one that doesn't, if the treatment decision is based on a superficial trait not

---

Table 7.1

Different Drug Responses Among European Americans and African Americans

Researchers have compared the effects of many drugs in different racial or ethnic groups. Most drugs tested either show no difference, or do show a difference but the physiological basis isn't known. For some drugs, one group will tend to respond to lower doses or in greater numbers than another. They are listed here. The difference in response is associated with inheriting particular gene variants or SNPs. (*EA* stands for European American and *AA* stands for African American.)

Drug Class/Name	Disorder	More Effective In: EA	More Effective In: AA
ACE inhibitor/Enalapril	hypertension	✓	
Antipsychotic/Clozapine	psychosis		✓
Antiviral/Alpha interferon	hepatitis	✓	
Beta blocker/Propranolol	hypertension	✓	
Calcium channel blocker/Diltiazem	hypertension		✓
Insulin	diabetes mellitus	✓	
Thiazide diuretic	hypertension		✓
Vasodilator combination/BiDil)	congestive heart failure		✓

directly related to how the body responds to a particular drug.

Genetics and genomics will address the problems inherent in race-based medicine by personalizing prescribing practices. The U.S. Food and Drug Administration published guidelines in 2005 for submission of data on correlations between genotypes and drug responses. Such data can reveal the bases of drug resistance and sensitivity. For example, in one study, researchers identified variants of a gene called *MDR* (for multidrug resistance) in four population groups. This gene encodes a protein that pumps poisons out of certain white blood cells and intestinal lining cells. When a variant results in a pump that works too well, the protein recognizes drugs used to treat cancer, AIDS, and other conditions as toxins, sending them out of the cell. Researchers have found this protein variant in 83 percent of West Africans, 61 percent of African Americans, 26 percent of Caucasians, and 34 percent of Japanese. MDR genotype could be used to prescribe certain drugs only for individuals whose cells would not pump the drugs out. Thus, MDR genotype is a more biologically meaningful basis for prescribing a drug than skin color.

In an even more compelling study, researchers cataloged 23 markers for genes that control drug metabolism in 354 people representing eight races: black (Bantu, Ethiopian, and Afro-Caribbean,), white (Norwegian, Armenian, and Ashkenazi Jews), and Asian (Chinese and New Guinean). The genetic markers fell into four very distinct groups that predict which of several blood thinners, chemotherapies, and painkillers will be effective—and these response groups did not at all match the traditional racial groups.

Key Concepts

1. Polygenic traits are determined by more than one gene and vary continuously in expression.
2. Multifactorial traits are determined by a combination of a gene or genes and the environment.
3. A bell curve describes the distribution of phenotypic classes of a polygenic trait.

7.2 Methods Used to Investigate Multifactorial Traits

It is much more challenging to predict recurrence risks for polygenic traits and disorders than for Mendelian traits. Geneticists evaluate the input of genes, using information from population and family studies. The human genome sequence is providing new clues to multifactorial traits and disorders.

Empiric Risk

Using Mendel's laws, it is possible to predict the risk that a single-gene trait will recur in a family if one knows the mode of inheritance—such as autosomal dominant or recessive. To predict the chance that a multifactorial trait will occur in a particular individual, geneticists use **empiric risks,** which are based on incidence in a specific population. Incidence is the rate at which a certain event occurs, such as the number of new cases of a particular disorder diagnosed per year in a population of known size.

Empiric risk is not a calculation, but a population statistic based on observation. The population might be broad, such as an ethnic group or community, or genetically more well-defined, such as families that have a particular disease. Empiric risk increases with the severity of the disorder, the number of affected family members, and how closely related a person is to affected individuals. As an example, consider using empiric risk to predict the likelihood of a neural tube defect (NTD). In the United States, the overall population risk of

carrying a fetus with an NTD is about 1 in 1,000 (0.1 percent). For people of English, Irish, or Scottish ancestry, the risk is about 3 in 1,000. However, if a sibling has an NTD, no matter what the ethnic group, the risk of recurrence increases to 3 percent, and if two siblings are affected, the risk to a third child is even greater. By determining whether a fetus has any siblings with NTDs, a genetic counselor can predict the risk to that fetus, using the known empiric risk.

If a trait has an inherited component, then it makes sense that the closer the relationship between two individuals, one of whom has the trait, the greater the probability that the second individual has the trait, too, because they have more genes in common. Studies of empiric risk support this logic. **Table 7.2** summarizes empiric risks for relatives of individuals with cleft lip.

Because empiric risk is based solely on observation, we can use it to derive risks for disorders with poorly understood transmission patterns. For example, certain multifactorial disorders affect one sex more often than the other. Pyloric stenosis is an overgrowth of muscle at the juncture between the stomach and the small intestine. It is five times more common among males than females. The condition must be corrected surgically shortly after birth, or the newborn will be unable to digest foods. Empiric data show that the risk of recurrence for the brother of an affected brother is 3.8 percent, but the risk for the brother of an affected sister is 9.2 percent. An empiric risk, then, is based on real-world observations—the mechanism of the illness or its cause need not be known.

Table 7.2

Empiric Risk of Recurrence for Cleft Lip

Relationship to Affected Person	Empiric Risk of Recurrence
Identical twin	40.0%
Sibling	4.1%
Child	3.5%
Niece/nephew	0.8%
First cousin	0.3%
General population risk (no affected relatives)	0.1%

Heritability

As Charles Darwin observed, some of the variation of a trait is due to inborn differences in populations, and some to differences in environmental influences. A measurement called **heritability,** designated H, estimates the proportion of the phenotypic variation for a particular trait that is due to genetic differences in a certain population at a certain time. The distinction between empiric risk and heritability is that empiric risk could result from nongenetic influences, whereas heritability focuses on the genetic component of a trait.

Figure 7.5 outlines the factors that contribute to observed variation in a trait. Heritability equals 1.0 for a trait whose variability is completely the result of gene action, such as in a population of laboratory mice. If there is no environmental variability, genetic differences determine expression of the trait in a population. Variability of most traits, however, reflects a combination of differences among genes and environmental components. **Table 7.3** lists some traits and their heritabilities.

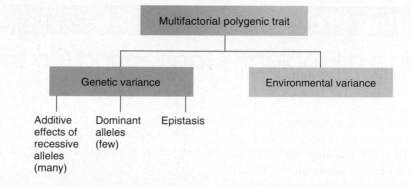

Figure 7.5   Heritability estimates the genetic contribution to a trait.
Observed variance in a polygenic, multifactorial trait or illness reflects genetic and environmental contributions. Genetic variants are mostly determined by the additive effects of recessive alleles of different genes, but they can also be influenced by the effects of a few dominant alleles and by epistasis (interactions between alleles of different genes).

Table 7.3
Heritabilities for Some Human Traits

Trait	Heritability
Clubfoot	0.8
Height	0.8
Blood pressure	0.6
Body mass index	0.5
Verbal aptitude	0.7
Mathematical aptitude	0.3
Spelling aptitude	0.5
Total fingerprint ridge count	0.9
Intelligence	0.5–0.8
Total serum cholesterol	0.6

Heritability changes as the environment changes. For example, the heritability of skin color would be higher in the winter months, when sun exposure is less likely to increase melanin synthesis. The same trait may be highly heritable in two populations, but certain variants much more common in one group due to environmental differences. Changing the environment can alter the differences between the two populations.

Researchers use several statistical methods to estimate heritability. One way is to compare the actual proportion of pairs of people related in a certain manner who share a particular trait, to the expected proportion of pairs that would share it if it were inherited in a Mendelian fashion. The expected proportion is derived by knowing the blood relationships of the individuals and using a measurement called the **coefficient of relatedness,** which is the proportion of genes that two people related in a certain way share (**table 7.4**).

Table 7.4
Coefficient of Relatedness for Pairs of Relatives

Relationship	Degree of Relationship	Percent Shared Genes (Coefficient of Relatedness)
Sibling to sibling	1°	50% (1/2)
Parent to child	1°	50% (1/2)
Uncle/aunt to niece/nephew	2°	25% (1/4)
Grandparent to grandchild	2°	25% (1/4)
First cousin to first cousin	3°	12 1/2% (1/8)

Reading 7.1

Solving a Problem: Connecting Cousins

With more genetic tests becoming available as the human genome sequence is analyzed, more people are learning that relatives beyond their immediate families have certain gene variants that might affect their health. Because the genetic closeness of the relationship impacts the risk of developing certain conditions, it may be helpful to be able to calculate the percentage of the genome that two relatives share.

The pedigree in **figure 1** displays an extended family, with "YOU" as the starting point. Calculate the percent of the genome shared for your first cousins once and twice removed (that is, removed from you by one or two generations, respectively)—in the figure, in generations III and II, while YOU are in generation IV. A second, third, or fourth cousin, by contrast, is in the same generation on a pedigree as the individual in question; see, for example, individual V-1 in figure 1.) **Table 1** summarizes the genetic relationships between cousins.

SOLUTION

The rules: Every step between parent and child, or sibling and sibling, has a value of 1/2, because these types of pairs share approximately 1/2 of their genes, according to Mendel's first law (chromosome segregation).

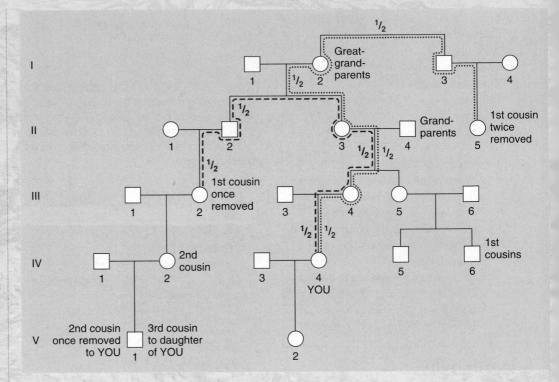

Figure 1 Pedigrees help determine the percentage of the genome two relatives share.

A parent and child, for example, share 50 percent of their genes, because of the mechanism of meiosis. Siblings share on average 50 percent of their genes, because they have a 50 percent chance of inheriting each allele for a gene from each parent. Genetic counselors use the designations of primary (1°), secondary (2°), and tertiary (3°) relatives when calculating risks (table 7.4 and **figure 7.6**). For extended or complicated pedigrees, the value of 1/2 or 50 percent between siblings and between parent-child pairs can be used to trace and calculate the percentage of genes shared between people related in other ways. **Reading 7.1** discusses how to calculate per-

centages of the genome shared for first cousins separated by generations, described as "removed" by one or more generations.

If the heritability of a trait is very high, then of a group of 100 sibling pairs, nearly 50 would be expected to have the same phenotype, because siblings share on average 50 percent of their genes. Height is a trait for which heritability reflects the environmental influence of nutrition. Of 100 sibling pairs in a population, for example, 40 might be the same number of inches tall. Heritability for height among this group of sibling pairs is .40/.50, or 80 percent, which is the observed phenotypic variation divid-

ed by the expected phenotypic variation if environment had no influence.

Genetic variance for a polygenic trait is mostly due to the additive effects of recessive alleles of different genes. For some traits, a few dominant alleles can greatly influence the phenotype, but because they are rare, they do not contribute greatly to heritability. This is the case for heart disease caused by a faulty LDL receptor, a rare dominant condition that is also influenced by many other genes. Epistasis (interaction between alleles of different genes) can also influence heritability. To account for the fact that different genes affect a phenotype

Table 1

Percent of Genome Shared by Cousins

Type of Cousin	Definition	Shared Genes
First	Share 2 grandparents	1/8
Second	Share great-grandparents but no grandparents	1/16
Third	Share great-great grandparents	1/32
Fourth	Share great-great-great grandparents	1/64
Once removed	One generation difference between relatives	1/16
Twice removed	Two generations difference between relatives	1/32

First cousin once removed (in green):

you to your mother	= 1/2
your mother to her mother (your grandmother)	= 1/2
your mother to her uncle (your great uncle)	= 1/2
your great uncle to his daughter (your first cousin)	= 1/2

$$1/2 \times 1/2 \times 1/2 \times 1/2 = 1/16$$

First cousin twice removed (in red):

you to your mother	= 1/2
your mother to her mother (your grandmother)	= 1/2
your grandmother to her mother (your greatgrandmother)	= 1/2
your greatgrandmother to her brother	= 1/2
your greatgrandmother's brother to his daughter	= 1/2

$$1/2 \times 1/2 \times 1/2 \times 1/2 \times 1/2 = 1/32$$

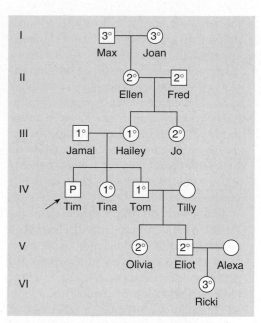

Figure 7.6 Tracing relatives. Tim has an inherited illness. A genetic counselor drew this pedigree to explain the approximate percentage of genes Tim shares with relatives. This information can be used to alert certain relatives to the risk that they or their offspring might be affected. ("P" is the proband, or affected individual who initiated the study. See table 7.4 for definitions of 1°, 2°, and 3° relationships. Individuals with no relationship indicated are not blood relatives of the proband.)

cists to tease apart the genetic and environmental components of multifactorial traits—adopted individuals and twins.

Key Concepts

1. Empiric risk applies population incidence data to predict risk of recurrence for a multifactorial trait or disorder.
2. Heritability measures the genetic contribution to a multifactorial trait; it is specific to a particular population at a particular time.
3. A coefficient of relatedness, the proportion of genes that individuals related in a certain way are expected to share, is used to calculate heritability.

to differing degrees, geneticists calculate a "narrow" heritability that considers only additive recessive effects, and a "broad" heritability that also considers the effects of rare dominant alleles and epistasis. For LDL cholesterol level, for example, the narrow heritability is 0.36, but the broad heritability is 0.96, reflecting the fact that a rare dominant allele has a large impact. The ability to taste bitter substances is another trait that is largely determined by one gene, on chromosome 7, but influenced by several others with much lesser, but additive, effects.

An understanding of multifactorial inheritance has many applications in agriculture. A breeder needs to know whether genetic or environmental variation mostly determines such traits as birth weight, milk yield, length of wool fiber, and egg hatchability. It is also valuable to know whether the genetic influences are additive or epistatic. The breeder can control the environmental input by adjusting the conditions under which animals are raised, and the inherited trait variant input by setting up matings between particular individuals.

Studying multifactorial traits in humans is difficult, because information must be obtained from many families. Two special types of people, however, can help geneti-

Adopted Individuals

A person adopted by people who are not blood relatives shares environmental influences, but typically not many genes, with the adoptive family. Conversely, adopted individuals share genes, but not the exact environment, with their biological parents. Therefore, biologists assume that similarities between adopted people and adoptive parents reflect mostly environmental influences, whereas similarities between adoptees and their biological parents reflect mostly genetic influences. Information on both sets of parents can reveal how heredity and the environment each contribute to a trait.

Many early adoption studies used the Danish Adoption Register, a database of all adopted Danish children and their families from 1924 to 1947. One study examined correlations between causes of death among biological and adoptive parents and adopted children. If a biological parent died of infection before age 50, the adopted child was five times more likely to die of infection at a young age than a similar person in the general population. This may be because inherited variants in immune system genes increase susceptibility to certain infections. In support of this hypothesis, the risk that an adopted individual would die young from infection did not correlate with adoptive parents' death from infection before age 50. Although researchers concluded that length of life is mostly determined by heredity, they did find evidence of environmental influences. For example, if adoptive parents died before age 50 of cardiovascular disease, their adopted children were three times as likely to die of heart and blood vessel disease as a person in the general population. What environmental factor might account for this correlation?

Twins

Studies that use twins to separate the genetic from the environmental contribution to a phenotype provide more meaningful information than studying adopted individuals (**figure 7.7**). Twin studies have largely replaced adoption methods.

Using twins to study genetic influence on traits dates to 1924, when German dermatologist Hermann Siemens compared school transcripts of identical versus fraternal twins. Noticing that grades and teachers' comments were much more alike for identical twins than for fraternal twins, he proposed that genes contribute to intelligence.

a.

b.

Figure 7.7 Twins. (a) Peyton and Riley Watnick are dizygotic (DZ; fraternal) twins. They share approximately half of their genes, as do any pair of siblings.
(b) These girls are monozygotic (MZ; identical) twins, with identical genomes. MZ twins can confound forensic techniques, because both individuals have the same DNA profiles. Forensic scientists can sometimes distinguish MZ twins by considering somatic mutations, which occur after the twins split early in development, but this is time-consuming and costly.

A trait that occurs more frequently in both members of identical (monozygotic or MZ) twin pairs than in both members of fraternal (dizygotic or DZ) twin pairs is at least partly controlled by heredity. Geneticists calculate the **concordance** of a trait as the percentage of pairs in which both twins express the trait.

In one study, 142 MZ twin pairs and 142 DZ twin pairs took a "distorted tunes test," in which 26 familiar songs were played, each with at least one note altered. A person was considered "tune deaf" if he or she failed to detect the mistakes in three or more tunes. Concordance for "tune deafness" was 67 percent for MZ twins, but only 44 percent for DZ twins, indicating a considerable inherited component in the ability to accurately perceive musical pitch. **Table 7.5** compares twin types for a variety of hard-to-measure traits. (Figure 3.16 shows how DZ and MZ twins arise.)

Diseases caused by single genes that are 100 percent penetrant, whether dominant or recessive, are 100 percent concordant in MZ twins. If one twin has the disease, so does the other. However, among DZ twins, concordance generally is 50 percent for a dominant trait and 25 percent for a recessive trait. These are the Mendelian values that apply to any two siblings. For a polygenic trait with little environmental input, concordance values for MZ twins are significantly greater than for DZ twins. A trait molded mostly by the environment exhibits similar concordance values for both types of twins.

Table 7.5

Concordance Values for Some Traits in Twins

| Trait | Concordance | |
	MZ (identical) twins	DZ (fraternal) twins
Acne	14%	14%
Alzheimer disease	78%	39%
Anorexia nervosa	55%	7%
Autism	90%	4.5%
Bipolar disorder	33–80%	0–8%
Cleft lip with or without cleft palate	40%	3–6%
Hypertension	62%	48%
Schizophrenia	40–50%	10%

An ongoing investigation called the Twins Early Development Study shows how concordance values indicate the degree to which heredity contributes to a trait. This project is following 7,756 pairs of twins born in England and Wales in 1994. One study looked at two-year-olds with language skills in the lowest 5 percent of children that age. The researchers recorded the number of words in the vocabularies of 1,044 pairs of MZ twins, 1,006 pairs of same-sex DZ twins, and 989 pairs of opposite-sex twins. The results clearly indicated a large genetic influence for children with poor language skills—concordance for MZ twins was 81 percent, but it was 42 percent for DZ twins. That is, if an MZ twin fell into the lowest 5 percent of two-year-olds for language acquisition, the chance that her twin would, too, was 81 percent. But if a DZ twin was in this category, the chance that her twin would also be was only 42 percent. When similar assessments were done on twin pairs of all levels of ability in language acquisition, the differences between the concordance values was much less. Therefore, the environment plays a larger role in most children's adeptness at learning vocabulary than it does for the 5 percent who struggle to do so.

Comparing twin types assumes that both types of twins share similar experiences. In fact, MZ twins are often closer emotionally than DZ twins. This discrepancy between the closeness of the two types of twins led to misleading results in studies conducted in the 1940s. One study concluded that tuberculosis is inherited because concordance among MZ twins was higher than among DZ twins. Actually, the infectious disease more readily passed between MZ twins because their parents kept them in close physical contact.

Even today, twin studies may be used to stretch a genetic connection. In 2005, political scientists used surveys on 8,000 sets of twins to demonstrate a genetic underpinning to whether twins agree or not on whether prayer should occur in school! They found concordance on the trait of "school prayer" of 66% for MZ twins and 46% for DZ.

For some traits for which the abnormality may begin before birth, the type of MZ twin may be important. That is, MZ twins with the same amnion may share more environmental factors than MZ twins who have separate amnions (see figure 3.16). Schizophrenia may begin prenatally and reflect environmental factors that trigger or aggravate it.

A more informative way to assess the genetic component of a multifactorial trait is to study MZ twins who were separated at birth, then raised in very different environments. Much of the work using this "twins reared apart" approach has taken place at the University of Minnesota. Here, since 1979, hundreds of sets of twins and triplets who were separated at birth have visited the laboratories of Thomas Bouchard. For a week or more, the twins and triplets undergo tests that measure physical and behavioral traits, including 24 different blood types, handedness, direction of hair growth, fingerprint pattern, height, weight, functioning of all organ systems, intelligence, allergies, and dental patterns. Researchers videotape facial expressions and body movements in different circumstances and probe participants' fears, vocational interests, and superstitions.

Twins and triplets separated at birth provide natural experiments for distinguishing nature from nurture. Many of their common traits can be attributed to genetics, especially if their environments have been very different. By contrast, their differences tend to reflect differences in upbringing, since their genes are identical (MZ twins and triplets) or similar (DZ twins and triplets).

The researchers have found that MZ twins and triplets separated at birth and reunited later are remarkably similar, even when they grow up in very different adoptive families (**figure 7.8**). Idiosyncrasies are particularly striking. For example, twins who met for the first time when they were in their thirties responded identically to questions; each paused for 30 seconds, rotated a gold necklace she was wearing three times, and then answered the question. Coincidence, or genetics?

The "twins reared apart" approach is not a perfectly controlled way to separate nature from nurture. MZ twins and other multiples share an environment in the uterus and possibly in early infancy that may affect later development. Siblings, whether adoptive or biological, do not always share identical home environments. Differences in sex, general health, school

Separated at birth, the Mallifert twins meet accidentally.

Figure 7.8

Originally published in the 4 May 1981 issue of *The New Yorker* Magazine, p. 43. © Tee and Charles Addams Foundation. Reprinted by permission.

and peer experiences, temperament, and personality affect each individual's perception of such environmental influences as parental affection and discipline.

Adoption studies, likewise, are not perfectly controlled experiments. In the past, adoption agencies tended to search for adoptive families with ethnic, socioeconomic, or religious backgrounds similar to those of the biological parents. Thus, even when different families adopted and raised separated twins, their environments were not as different as they might have been for two unrelated adoptees.

Association Studies

Empiric risk, heritability, and adoptee and twin studies are traditional ways of estimating the degree to which genes contribute to the variability of a trait or illness. With the availability of more types of genetic markers and human genome sequence data, researchers have more refined tools to identify DNA sequences that cause or confer susceptibility to disease.

Identification of the single genes underlying Mendelian traits and disorders has largely relied on linkage analysis, discussed in chapter 5. In a linkage study, researchers compiled data on families with more than one affected member, determining whether a nearby section of chromosome (a genetic marker) was inherited along with a disease-causing gene by inferring whether alleles of the two genes were in cis or trans (see figure 5.11). Linkage analysis in humans is difficult because the rarity of single-gene disorders makes it hard to find enough families and individuals to compare. Using linkage to identify the several genes that contribute to a polygenic trait is even harder. Fortunately, a newer method, SNP mapping, can detect DNA sequences that are inherited with polygenic traits. Because SNP mapping tracks large populations, it is easier to find subjects for study than for classic linkage studies based on extended families with rare condi-

tions. SNP studies are also better suited to track polygenic disorders.

Recall from chapter 1 that a SNP, or single nucleotide polymorphism, is a single DNA base at a specific site within the genome that varies in at least 1 percent of a population (**figure 7.9**). The human genome has a SNP at about every 600 to 1,200 bases among the nearly 3 billion bases. The number of SNPs is derived from an analysis of linkage disequilibrium (LD), which is the tendency for certain SNPs to be inherited together. LD generally occurs over areas that are 5,000 to 50,000 bases long. Several million SNPs, spread out over the genome, are enabling researchers to pair specific SNPs with specific gene variants that cause or contribute to a particular disease or trait, because the distance between the SNPs is less than the average length of a DNA sequence in linkage disequilibrium. Several SNPs that are transmitted together constitute a haplotype, which is short for "haploid genotype" (see figure 5.16).

In **association studies,** researchers compare SNP patterns between a group of individuals who have a particular disorder and a group who do not. An association study may use a case-control design, in which each individual in one group is matched to an individual in the other group who shares as many characteristics as possible, such as age, sex, activity level, and environmental exposures. SNP differences then correlate to presence or absence of the particular medical condition. For example, if 500 individuals with hypertension (high blood pressure) have particular DNA bases at six sites in the genome, and 500 matched individuals who do not have hypertension have different bases at only these six sites, then further investigation can probe these genome regions for genes whose protein products could control blood pressure. When many SNPs are considered, many susceptibility genes can be tracked, and patterns may emerge that can be used to predict the course of the illness.

An association study achieves greater power if it borrows from the older technique of looking at family members. In the "affected sibling pair" strategy, researchers scan genomes for markers that most siblings who have the same condition share, but that siblings who do not have the condition do not often share. Such genome regions may harbor genes that contribute to

the condition. The underlying logic is that because siblings share 50 percent of their genes, a trait or condition that many siblings share is likely to be inherited. Returning to the hypertension example, an affected sibling pair analysis would include 500 pairs of siblings who both have hypertension, and 500 pairs of siblings who do not, or in which only one does. Genome regions for which all (or most) of the 500 affected siblings have the same SNP, but few of the other group do, suggest where researchers might search for genes that contribute to the trait.

It will take a great deal of research to test whether SNP patterns are actually meaningful. For example, if a SNP pattern in a population is associated with breast cancer, then the next step might be for researchers to see if these tumors are different, at a cell and tissue level, from other cases of breast cancer. Software can compare SNP and histological data to see if the SNP pattern accurately reflects a physically distinct subtype of the disease. Then, eventually, a simple SNP scan might replace or augment microscopic analysis of the tumor.

SNP association studies have many applications. They provide a much faster way to identify DNA sequence differences among individuals than sequencing everyone's genome, a little like quickly noting distinctive items of clothing on two individuals instead of listing everything they are wearing. Correlations between SNP patterns and elevated disease risks may predict responses to particular drugs. Identifying the SNPs that travel with genes of interest will enable

```
A T G C T C G A G C C T A A T A
A T G C T C G A G C C T A A T A
A T G C T C G A G C C T A A T A
A T G C T C G A G C C T A A T A
A T G C T C G T G C C T A A T A
A T G C T C G A G C C T A A T A
A T G C T C G A G C C T A A T A
A T G C T C G A G C C T A A T A
```

Figure 7.9 SNPs are sites of variability in the genome. If this sequence represents eight individuals, one has a polymorphism at the eighth base, a T instead of an A. A SNP would affect the phenotype if it occurs in part of a protein-encoding gene that affects the protein's function.

researchers to search human genome data to identify nearby genes whose protein products could be absent or altered in a way that explains symptoms. This was the case, for example, for Crohn disease, in which the intestines become severely inflamed in response to the presence of certain bacteria.

Two different research groups identified a gene that increases the risk of developing Crohn disease. One group consulted older linkage data that indicated a causative gene on chromosome 16 in some families. These researchers considered all the genes identified so far in that region, and pursued one, called *NOD2*, that encodes a protein that takes part in the inflammatory response. Next they examined DNA from patients and controls and found that patients were much more likely to have mutations in this gene than were the healthy comparisons. The second research group looked at 11 SNPs in 235 families with affected members. They found that a certain pattern was overrepresented among the affected individuals, and that these people also had mutations in the *NOD2* gene.

The Crohn disease research reveals a limitation of association studies—they establish correlations, not causation. It is rare to find a DNA sequence that contributes to a polygenic trait exclusively in affected individuals. The *NOD2* allele that elevates risk of developing Crohn disease, for example, is found in 5 percent of the general population, but in 15 percent of individuals with the condition. Clearly, more than one gene contributes to the disease. Further studies are searching for associations of SNPs on other chromosomes to Crohn disease susceptibility.

The more complex a SNP association study, the more individuals are required to achieve statistical significance. Consider an investigation that examines 20 genes, each with 4 SNPs, that contribute to development of a particular polygenic disease. A screen would look for 160 data points per individual (20 genes × 4 SNPs/gene × 2 copies of each gene = 160 data points). With thousands of possible combinations (genotypes), it's clear that *many* thousands of individuals would have to be examined to note any correlations between the SNP pattern and disease.

Comparing SNP patterns might not be the only way to use genome sequence information to make disease/trait associations.

Individuals also differ in "large-scale copy variations," which are very long DNA base sequences present in some people but not in others. A 200,000 base sequence might be absent in one individual, present in one copy in another, and present in two copies in yet another. So far researchers have discovered 255 places in the human genome that harbor these mysterious sequences.

Figure 7.10 compares linkage analysis to association studies. **Table 7.6** reviews terms used in the study of multifactorial traits.

RELATING THE CONCEPTS

Researchers have identified a gene, *IRF6*, that causes isolated cleft lip/cleft palate in some families. An ongoing study is investigating 150 SNPs located among more than 140,000 DNA bases surrounding this gene. Explain how haplotypes based on these SNPs might be used to predict risk for the condition in the general population.

Linkage study

Follows allele configuration in large genome regions in families

Association Study

Follows SNP patterns over small genome regions in populations
(more powerful)

Figure 7.10 Differences between linkage and association studies. Geneticists have homed in on genes using linkage as a guide for more than half a century. Today association studies are more common. They are more powerful because they include many more people as well as many more data points.

Table 7.6

Terms Used in Evaluating Multifactorial Traits

Association study   Detecting correlation between SNP (or other marker) patterns and increased risk of developing a particular medical condition.

Empiric risk   The risk of recurrence of a trait or illness based on known incidence in a particular population.

Heritability   The percentage of phenotypic variation for a trait that is attributable to genetic differences. It equals the ratio of the observed phenotypic variation to the expected phenotypic variation for a population of individuals.

Coefficient of relatedness   The proportion of genes shared by two people related in a particular way. Used to calculate heritability.

Concordance   The percentage of twin pairs in which both twins express a trait.

Key Concepts

1. Researchers compare traits in adopted individuals to those in their adoptive and biological parents to assess the genetic contribution to a trait.
2. Concordance is the percentage of twin pairs in which both express a trait. For a trait largely determined by genes, concordance is higher for MZ than DZ twins.
3. Association studies seek correlations between SNP patterns and phenotypes in large groups of individuals.

7.3 Two Multifactorial Traits

Multifactorial traits include such common conditions as heart and blood vessel (cardiovascular) disease and obesity, as well as harder-to-define traits, including intelligence and aspects of personality, mood, and behavior.

Heart Health

Genes control cardiovascular functioning in several ways: how well the body handles lipids in the blood; how readily the blood clots; blood pressure; and how well cellular adhesion molecules enable white blood cells to stick to the walls of blood vessels (see figure 2.22). Lipids can only move in the circulation when bound to proteins to form large molecules called lipoproteins. Several genes encode the protein parts of lipoproteins, which are called apolipoproteins. Some types of lipoproteins carry lipids in the blood to tissues, where they are utilized, and other types of lipoproteins take lipids to the liver, where they are dismantled into biochemicals that the body can excrete more easily. One allele of a gene that encodes apolipoprotein E, called E4, increases the risk of a heart attack threefold in people who smoke. This is clear evidence that genes and environmental factors can interact in ways that cause illness.

Maintaining a healthy cardiovascular system requires a balance: Cells require sufficient lipid levels inside but cannot allow accumulation on the outside. Several dozen genes control lipid levels in the blood and tissues by specifying enzymes that process lipids, proteins that transport them, or receptor proteins that admit lipids into cells.

Much of what we know about genetic control of cardiovascular health comes from studying rare inherited conditions. Identifying and understanding a genetic cause of a rare disease can help the larger number of people who suffer from noninherited forms of the illness. For example, statins, the cholesterol-lowering drugs that millions of people take today, grew out of research on people with familial hypercholesterolemia, described in figure 5.2.

At least 50 genes regulate blood pressure. One gene encodes angiotensinogen, a protein that is elevated in the blood of people with hypertension. This protein controls blood vessel tone and fluid balance in the body. Certain alleles are found much more frequently among people with hypertension than chance would explain. Even though environmental factors, such as emotional stress, can raise blood pressure, knowing who is genetically susceptible to dangerously high blood pressure can alert doctors to monitor high-risk individuals.

An enzyme, lipoprotein lipase, is important in lipid metabolism. It lines the walls of the smallest blood vessels, where it breaks down fat packets released from the small intestine and liver. Lipoprotein lipase is activated by high-density lipoproteins (HDLs), and it breaks down low-density lipoproteins (LDLs). High HDL levels and low LDL levels are associated with a healthy cardiovascular system. In inborn errors of metabolism called type 1 hyperlipoproteinemias, too little lipoprotein lipase causes triglycerides (a type of fat) to reach dangerously high levels in the blood. Lipoprotein lipase also regulates fat cell size; fat cells contribute to obesity by enlarging, rather than dividing.

The fluidity of the blood is also critical to health. Overly active clotting factors or extra sticky white blood cells can induce formation of clots that block blood flow, usually in blood vessels in the heart or in the legs.

Genetic test panels detect multiple alleles in dozens of genes that cause or contribute to cardiovascular disease. DNA microarrays can monitor gene expression, assessing many contributing factors to heart and blood vessel health. For example, one gene expression microarray can indicate which cholesterol-lowering drugs are most likely to be effective and without side effects. The premise behind the utility of such information is that people have composite genetic risks based on the small contributions of several genes—the essence of polygenic inheritance. Computer analysis of the multigene tests accounts for environmental factors, such as those outlined in **table 7.7**. Some of these risk factors are controllable with lifestyle changes, such as exercising, not smoking, and maintaining a healthy weight.

Table 7.7

Risk Factors for Cardiovascular Disease

Uncontrollable	Controllable
Age	Fatty diet
Male sex	Hypertension
Genes	Smoking
Lipid metabolism	High serum cholesterol
Apolipoproteins	Low serum HDL
Lipoprotein lipase	High serum LDL
Blood clotting	Stress
Fibrinogen	Insufficient exercise
Clotting factors	Obesity
Homocysteine metabolism	Diabetes
Leukocyte adhesion	

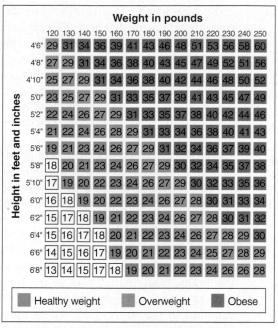

Weight in pounds

Height in feet and inches	120	130	140	150	160	170	180	190	200	210	220	230	240	250
4'6"	29	31	34	36	39	41	43	46	48	51	53	56	58	60
4'8"	27	29	31	34	36	38	40	43	45	47	49	52	51	56
4'10"	25	27	29	31	34	36	38	40	42	44	46	48	50	52
5'0"	23	25	27	29	31	33	35	37	39	41	43	45	47	49
5'2"	22	24	26	27	29	31	33	35	37	38	40	42	44	46
5'4"	21	22	24	26	28	29	31	33	34	36	38	40	41	43
5'6"	19	21	23	24	26	27	29	31	32	34	36	37	39	40
5'8"	18	20	21	23	24	26	27	29	30	32	34	35	37	38
5'10"	17	19	20	22	23	24	26	27	29	30	32	33	35	36
6'0"	16	18	19	20	22	23	24	26	27	28	30	31	33	34
6'2"	15	17	18	19	21	22	23	24	26	27	28	30	31	32
6'4"	15	16	17	18	20	21	22	23	24	26	27	28	29	30
6'6"	14	15	16	17	19	20	21	22	23	24	25	27	28	29
6'8"	13	14	15	17	18	19	20	21	22	23	24	26	26	28

■ Healthy weight   ■ Overweight   ■ Obese

Developed by the National Center for Health Statistics in collaboration with the National Center for Chronic Disease Prevention and Health Promotion

Figure 7.11 Body mass index (BMI). BMI equals weight/height^2, with weight measured in kilograms and height measured in meters. This chart provides a shortcut—the calculations have been done and converted to the English system of measurement.

Weight

Body weight reflects energy balance—the rate of food taken in versus the rate at which the body uses it for fuel. Excess food means, ultimately, excess weight. About 30 percent of all adults in the United States are obese, and another 35 percent are overweight. Being overweight or obese raises the risk of developing hypertension, diabetes, stroke, gallstones, sleep apnea, and some cancers.

Scientific studies of body weight use a measurement called body mass index (BMI), which is weight in proportion to height (**figure 7.11**). A BMI between 25 and 30 indicates overweight; a BMI over 30 indicates obesity. This measurement makes common sense—a person who weighs 170 pounds and is 6 feet tall is slim, whereas a person of the same weight who is 5 feet tall is obese. The tall person's BMI is 23; the short person's is 33.5.

Heritability for BMI is 0.55, which leaves room for environmental influences on our appetites and sizes. Yet the genetic picture is complex. Genomewide screens for SNPs associated with BMI point to at least 33 different regions likely to harbor genes that affect how much we eat, how we use calories, and how fat is distributed in the body. A picture is emerging of complex biochemical pathways and hormonal interactions, explaining some inherited disorders of body weight and energy metabolism, and suggesting points for drug intervention (**table 7.8**).

Leptin and Associated Proteins

Obesity research embraced genetics in 1994, when Jeffrey Friedman at Rockefeller University discovered a gene that encodes the protein hormone leptin in mice and in humans. Normally, eating stimulates fat cells (adipocytes) to secrete leptin. It travels in the bloodstream to a region of the brain's hypothalamus called the arcuate nucleus, which is a collection of nerve cells (neurons) and associated cells (**figure 7.12**). Here, leptin binds to receptors on neurons, signaling the cells to produce and release melanocyte stimulating hormone (MSH), which then leaves the arcuate nucleus and binds other receptors, called melanocortin-4 receptors, elsewhere in the hypothalamus. This binding sends signals that suppress appetite and increase metabolism to digest food already eaten. The melanocortin-4 receptor is an appetite "brake." Countering it is neuropeptide Y, an appetite "accelerator" that the arcuate nucleus releases when leptin levels are low.

When Friedman gave mice extra leptin, they ate less and lost weight. Headlines soon proclaimed the new magic weight loss elixir, a biotech company paid $20 million for rights to the hormone, and clinical trials ensued. The idea was to give obese individuals leptin, assuming that they suffered from a deficiency, and to therefore trick them into feeling full. But only about 15 percent of the people lost weight. The other 85 percent didn't lack leptin at all—their fat cells poured it out. Most of them instead had leptin resistance, a diminished ability to recognize the hormone due to defective leptin receptors. Although starvation lowers leptin levels, which increases appetite, the reverse doesn't happen—giving people leptin doesn't make them stop eating.

Despite disappointing clinical trials, the discovery of leptin fueled further research,

Table 7.8

Sites of Possible Genetic Control of Body Weight Related to Leptin

Protein	Function	OMIM	Effect on Appetite
Leptin	Stimulates cells in hypothalamus to decrease appetite and metabolize nutrients.	164160	↓
Leptin transporter	Enables leptin to cross from bloodstream into brain.	601694	↓
Leptin receptor	Binds leptin on hypothalamus cell surfaces, triggering hormone's effects.	601007	↓
Neuropeptide Y	Produced in hypothalamus when leptin levels are low and the individual loses weight.	162640	↑
Melanocortin-4 receptor	Activated when leptin levels are high and the individual gains weight.	155541	↓
Ghrelin	Signals hungers from stomach to brain in short term, stimulating neuropeptide Y.	605353	↑
PYY	Signals satiety from stomach to brain.	660781	↓
Stearoyl-CoA desaturase-1	Controls whether body stores or uses fat.	604031	↑

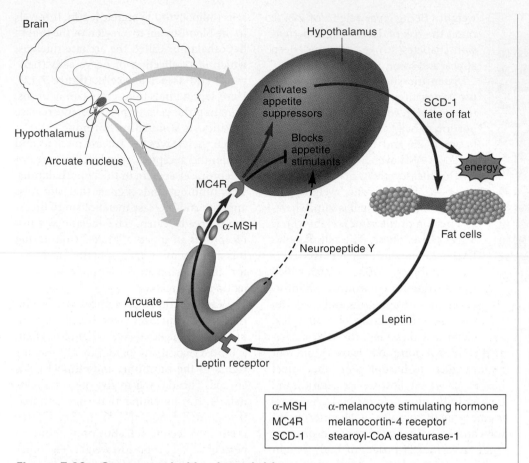

Figure 7.12 Genes control body weight. Fat cells (adipocytes) secrete leptin, which binds to receptors in the arcuate nucleus of the hypothalamus, triggering appetite suppressors and blocking appetite stimulants. The melanocortin-4 receptor stifles appetite, whereas neuropeptide Y stimulates it. Ghrelin and PYY are made in the stomach and provide short-term appetite control. Ghrelin activates neuropeptide Y, and PYY suppresses appetite.

α-MSH	α-melanocyte stimulating hormone
MC4R	melanocortin-4 receptor
SCD-1	stearoyl-CoA desaturase-1

Microarray experiments are useful for discovering weight-related genes. Researchers applied leptin to a DNA microarray containing 12,000 genes expressed in the liver. Leptin had the greatest effect on expression of the gene encoding an enzyme, stearoyl-CoA desaturase-1 (SCD-1), which determines whether the body stores fat or metabolizes it to release its energy.

The stomach is another source of obesity-related proteins. Ghrelin is a peptide (small protein) hormone produced in the stomach that responds to hunger, signaling the hypothalamus to produce more neuropeptide Y, which stimulates appetite. PYY is another peptide made in the stomach. Its action opposes ghrelin, signaling satiety to the brain. While leptin acts in the long term to maintain weight, ghrelin and PYY function in the short term. All of these hormonal signals impinge on the arcuate nucleus, where they are integrated to finely control appetite in a way that maintains weight.

Environmental Influences on Obesity

Many studies on adopted individuals and twins suggest that obesity has a heritability of 75 percent. Because the heritability for BMI is lower than this, the discrepancy suggests that genes play a larger role in those who tend to gain weight easily. This becomes obvious when populations that have a genetic tendency to obesity experience a drastic and sudden change in diet for the worse.

On the tiny island of Naura, in Western Samoa, the residents' lifestyles changed drastically when they found a market for the tons of bird droppings on their island as commercial fertilizer. The influx of money translated into inactivity and a high-calorie, high-fat diet, replacing an agricultural lifestyle and a diet of fish and vegetables. Within just a generation, two-thirds of the population had become obese, and a third suffered from diabetes.

The Pima Indians offer another example of environmental effects on body weight (**figure 7.13**). These people separated into two populations during the Middle Ages, one group settling in the Sierra Madre mountains of Mexico, the other in southern

and some people have benefited. A few severely obese children with leptin deficiency, for example, have attained normal weights with daily injections of leptin. One 9-year-old weighed more than 200 pounds and would eat in one meal what an adult would consume in half a day. Leptin injections dropped her weight to normal within four years. Children with leptin deficiency become intensely hungry. Some even eat food frozen because they cannot wait for it to defrost.

Also constantly ravenous, but with a markedly different phenotype, are individuals with lipodystrophy. In lipodystrophy, at puberty, the fat pads beneath the skin suddenly shrink away, leaving the person extremely skinny, yet constantly hungry. Adipocytes disappear, but fats are laid down beneath the skin

in painful lesions. Triglycerides enter the blood so swiftly that they must be cleansed from it often. Leptin injections have helped some affected individuals to gain weight and feel less hungry. At least two different genes are known to cause lipodystrophy, but their connection to leptin isn't yet understood.

Researchers are tracing some causes of obesity to proteins that interact with leptin. For example, most people who are heterozygotes (carriers) for a mutation in the gene that encodes the melanocortin-4 receptor are obese, and individuals lacking both functional alleles are morbidly obese. Melanocortin-4 receptor mutations account for about 6 percent of very obese children, and so far these mutations are the most common single-gene cause of obesity—much more common than leptin deficiency.

a.

b.

Figure 7.13 The environment influences gene expression. Comparison of average body weights among the Arizona population of Pima Indians **(a)** and the Mexican population **(b)** reveals the effects of the environment.

Arizona. By the 1970s, the Arizona Indians no longer farmed nor ate a low-calorie, low-fat diet, but instead consumed 40 percent of their calories from fat. With this extreme change in lifestyle, they developed the highest prevalence of obesity of any population on earth. (Prevalence is the total number of individuals with a certain condition in a particular population at a given time.) Half of the Arizona group had diabetes by age 35, weighing, on average, 57 pounds (26 kilograms) more than their southern relatives, who still eat a low-fat diet and are very active.

The Pima Indians demonstrate that future obesity is not sealed in the genes at conception, but instead is much more likely to occur if the environment provides too many calories and too much fat. They illustrate what geneticist James Neel termed the "thrifty gene hypothesis" in 1962. He suggested that long ago, the hunter-gatherers who survived famine had genes that enabled them to efficiently conserve fat. Today, with food plentiful, the genetic tendency to retain fat is no longer healthful, but harmful. Unfortunately, for many of us, our genomes hold an energy-conserving legacy that works too well—it is much easier to gain weight than to lose it, for sound evolutionary reasons.

Interactions and contributions of genes and the environment provide some of the greatest challenges in studying human genetics. Why does one smoker develop lung cancer, but another does not? Why can one person consistently overeat and never gain weight, while another gains easily? Because we exist in an environment, no gene functions in a vacuum. Subtle interactions of nature and nurture profoundly affect our lives and make us all—even identical twins—unique individuals.

Key Concepts

1. Genes that affect lipid metabolism, blood clotting, leukocyte adhesion, and blood pressure influence cardiovascular health.
2. Genes that encode leptin, the leptin receptor, and proteins that transmit or counter leptin's signals affect body weight.
3. Studies on adopted individuals and twins indicate a heritability of 75 percent for obesity.
4. Populations that suddenly become sedentary and switch to a fatty diet reflect environmental influences on body weight.

Summary

7.1 Genes and the Environment Mold Most Traits

1. **Multifactorial traits** reflect the environment and genes. A **polygenic trait** is determined by more than one gene and varies continuously in expression. The frequency distribution of phenotypes for a polygenic trait forms a bell curve.

7.2 Methods Used to Investigate Multifactorial Traits

2. **Empiric risk** measures the likelihood that a multifactorial trait will recur based on its prevalence in a population. The risk rises as genetic closeness to an affected individual increases, as the severity of the phenotype increases, and as the number of affected relatives rises.

3. **Heritability** estimates the proportion of variation in a multifactorial trait that is attributable to genetic variation. It describes a trait in a particular population at a particular time. Heritability compares the actual incidence of a shared trait among people related in a certain way to the expected incidence. Rare dominant alleles can contribute to heritability.

4. Characteristics shared by adopted people and their biological parents are mostly inherited, whereas similarities between adopted people and their adoptive parents reflect environmental influences.

5. Concordance measures the frequency of expression of a trait in both members of MZ or DZ twin pairs. The more influence genes exert over a trait, the higher the

differences in concordance between MZ and DZ twins.

6. Association studies correlate SNP patterns to increased risk of developing a disorder.

7.3 Two Multifactorial Traits

7. Genes that control lipid metabolism and blood clotting contribute to cardiovascular health.

8. Leptin, its receptor, its transporter, neuropeptide Y, and the melanocortin-4 receptor are proteins that affect body weight. Fat cells secrete leptin in response to starvation, and the protein acts in the hypothalamus. Populations that switch to a fatty, high-calorie diet and a less-active lifestyle reveal the effects of the environment on weight.

Review Questions

1. Consider the traits of eye color and body weight. Which is more likely to be inherited as a Mendelian trait, and which is multifactorial? Cite reasons for your answer.

2. Cite two examples from the chapter of a rare illness that helped researchers understand a process that could be applied to treat or help more people.

3. What is the difference between a Mendelian multifactorial trait and a polygenic multifactorial trait?

4. Which has a greater heritability—eye color or height? State a reason for your answer.

5. How can skin color have a different heritability at different times of the year?

6. In a large, diverse population, why are medium brown skin colors more common than very white or very black skin?

7. Using the information in figure 7.6, what percentage of genes does Tim share with

a. Joan?

b. Hailey?

c. Eliot?

d. Ricki?

8. Describe the type of information in a(n)

a. empiric calculation.

b. twin study.

c. adoption study.

d. association study.

9. Why does SNP mapping require extensive data?

10. Name three types of proteins that affect cardiovascular functioning and three that affect body weight.

11. Describe or sketch the circuitry for appetite control during starvation, based on figure 7.12.

Applied Questions

1. Rebecca breeds Maine coon cats. The partial pedigree to the right describes how her current cats are related—the umbrellalike lines indicate littermates, which are the equivalent of fraternal (DZ) twins in humans.

Cat lover Sam wishes to purchase a pair of Rebecca's cats to breed, but wants them to share as few genes as possible to minimize the risk that their kittens will inherit certain multifactorial disorders. Sam is quite taken with Farfel, but can't decide among Marbles, Juice, or Angie for Farfel's mate.

Calculate the percentage of genes that Farfel shares with each of these female relatives. With which partner would the likelihood of healthy kittens be greatest?

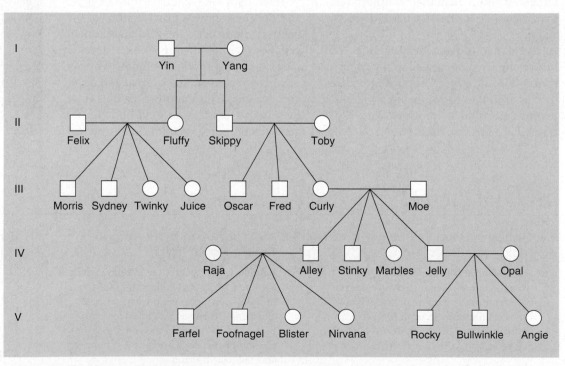

Maine coon cats

2. This chapter presented several types of data concerning the likelihood of isolated cleft lip/cleft palate recurring in a family due to inherited susceptibility. Consider the general population risk, empiric risk, and concordance and describe how these values suggest a considerable inherited component to this condition.

3. Attention deficit hyperactivity disorder (ADHD) affects 5 percent of children and adolescents and 3 percent of adults. Individuals with ADHD have difficulty learning in a classroom situation where they must remain still and controlled. Heritability ranges from .6 to .9 in different populations, and the relative risk to someone with an affected sibling ranges from .4 to .8. An adopted person is more likely to develop ADHD if a biological parent has the condition. An affected sibling pair association study using 270 pairs from the United States identified areas of chromosomes 16 and 17 that might harbor susceptibility genes. A study of 164 sib pairs in the Netherlands, however, pointed to sites on chromosomes 7 and 15.

    a. Explain how the data indicate that ADHD is either more likely caused by inherited factors or environmental factors.

    b. Why might the results differ for different populations?

    c. What should the next step be in understanding the biological basis of ADHD?

    d. Drug treatment is widely used for ADHD. What would be an advantage of knowing which genes predispose a person to the condition?

    e. Suggest a possible danger in developing a genetic test for ADHD.

4. Using figure 7.12, propose a drug treatment for obesity and explain how it would work.

5. The incidence of obesity in the United States has doubled over the past two decades. Is this due more to genetic or environmental factors? Cite a reason for your answer.

6. One way to calculate heritability is to double the difference between the concordance values for MZ versus DZ twins. For multiple sclerosis, concordance for MZ twins is 30 percent, and for DZ twins, 3 percent. What is the heritability? What does the heritability suggest about the relative contributions of genes and the environment in causing MS?

7. In chickens, weight gain is a multifactorial trait. Heritability accounts for several genes that contribute a small effect additively, as well as a few genes that exert a great effect. Is this an example of narrow or broad heritability?

Web Activities

Visit the Online Learning Center (OLC) at www.mhhe.com/lewisgenetics7. Select **Student Edition, chapter 7** and **Web Activities** to find the website links needed to complete the following activities.

8. Locate a website that deals with breeding show animals, farm animals, or crops to produce specific traits, such as litter size, degree of meat marbling, milk yield, or fruit ripening rate. Identify three traits with heritabilities that indicate a greater contribution from genes than the environment.

9. Visit the Centers for Disease Control and Prevention (CDC) website. From the leading causes of death, list three that have high heritabilities, and three that do not. Base your decisions on common sense or data, and explain your selections.

Case Studies and Research Results

10. Concordance for the eating disorder anorexia nervosa for MZ twins is 55 percent, and for DZ twins, 7 percent. Ashley and Maggie are DZ twins. Maggie has anorexia nervosa. Should Ashley be concerned about an inherited tendency to develop the condition? Explain your answer.

11. Lydia and Reggie grew up poor in New York City in the 1960s. Both took advantage of the City University of New York, then free, and went to medical school in Boston, where they met. Today, each has a thriving medical practice, and they are the parents of 18-year-old Jamal and 20-year-old Tanya.

    Jamal, taking a genetics class, wonders why he and Tanya do not resemble each other, or their parents, for some traits. The family is African American. Lydia and Reggie are short, 5'2" and 5'7" respectively, and each has medium brown eyes and skin, and dark brown hair. Tanya and Jamal are 5'8" and 6'1", respectively, and were often in the highest height percentiles since they were toddlers. Jamal has very dark skin, darker than his parents' skin, while Tanya's skin is noticeably lighter than that of either parent. Tanya's eyes are so dark that they appear nearly black.

    a. Give two explanations for why Tanya's eyes appear darker than those of her parents or brother.

    b. How can Jamal's skin be darker than that of his parents, and Tanya's be lighter?

    c. Which of the four traits considered—height, and eye, skin, and hair color—is most influenced by environmental factors?

    d. What is the evidence that Jamal and Tanya's height is due to environmental and genetic factors?

    e. Which of the four traits has the highest heritability?

Learn to apply the skills of a genetic counselor with these additional cases found in the *Case Workbook in Human Genetics*.

Complex traits among the Hutterites

Vitiligo

The Genetics of Behavior

CHAPTER

8

CHAPTER CONTENTS

ARE EATING DISORDERS HEREDITARY?

Meals were tense and regimented at the family dinner table. The children, although slim and athletic, wouldn't eat much, for fear that their mother would call them fat. Many a meal ended with the distinctive sounds, coming from the bathroom, of her vomiting, as her own mother had done.

The two daughters were gymnasts, constantly pressured to stay thin. All developed the eating disorder anorexia nervosa as teens. One of the two brothers had bouts of no appetite; the other was not affected. The daughters, now adults, still have eating issues, and they suffer permanent reminders of the family legacy, such as severe osteoporosis and dental damage from induced vomiting.

Some people with eating disorders may be of normal weight, because they counter gorging with exercising, vomiting, or using laxatives. Consider one woman's description of her behavior:

> For me a binge consists of a pound of cottage cheese; a head of lettuce; a steak; a loaf of Italian bread; a 10-ounce serving of broccoli, spinach, or a head of cabbage; a cake, an 18-ounce pie, with a quart or half gallon of ice cream. When my disease is at its worst, I eat raw oatmeal with butter, laden with mounds of sugar, or a loaf of white bread with butter and syrup poured over it.

Eating disorders are not just fads among people trying to copy stick-thin celebrities—they are psychiatric illnesses seen in people of many nations. Nor are these disorders new. Although they entered public awareness in the 1970s, very elderly people tell of anorexic behaviors in their parents.

Eating disorders are deadly psychiatric illnesses with genetic and environmental causes.

In 2001, a tennis promoter offered Steffi Graf and Andre Agassi $10 million if they would promise that their expected child would play a tennis match in the year 2017 against the offspring of another tennis great. *People* magazine called the child "the most DNA-advantaged prodigy in tennis" and quoted Agassi as saying, "I've got genetics on my side" when asked if Junior would win the match. Once Jaden was born, speculation about his future athletic abilities continued. By the time Jaden was three, a TV ad for a financial company featured him, portrayed by an actor, soundly beating tennis champ Taylor Dent. The ad ends, "The right genes make all the difference."

People who viewed the commercial objected because they wanted to see the *real* Jaden—but perhaps they were missing a larger point. That is, the idea that children of athletes will grow up to be athletes is flawed, and illustrates genetic determinism. Recall from chapter 1 that this is the idea that genes dictate every trait, even the ability to whack a ball over a net. Jaden Agassi may indeed have inherited fortuitous muscle anatomy, quick reflexes, athletic grace, and a competitive spirit from his parents. He will also experience powerful environmental cues. Still, whether or not he chooses to follow in his parents' footsteps—or how he will react if his athletic prowess does not match people's expectations—depends upon an unpredictable combination of many factors, both genetic and environmental.

This chapter explores how researchers are disentangling the genetic and environmental threads that contribute to several familiar behaviors and disorders.

8.1 Genes Contribute to Most Behavioral Traits

Behavioral traits include abilities, feelings, moods, personality, intelligence, and how a person communicates, copes with rage, and handles stress. Disorders with behavioral symptoms are wide-ranging and include phobias, anxiety, dementia, psychosis, addiction, and mood alteration. Very few medical conditions with behavioral components can be traced to a single gene. Most behavioral disorders fit the classic complex disease profile: They affect more than 1 in 1,000 individuals and are caused by several genes and the environment—that is, they are common, polygenic, and multifactorial.

Until recently, geneticists and social scientists studying behavior were limited to such tools as empiric risk estimates and adoptee and twin studies, discussed in chapter 7. These approaches clearly indicate that nearly all behaviors have inherited influences. Two powerful new approaches to understanding the biological basis of behavioral traits are:

1. Association studies that correlate genetic markers such as SNP (single nucleotide polymorphism) patterns with particular symptoms

2. Analysis of specific mutations that are present exclusively in individuals with the behavior

Behavioral genetics is, by definition, a study of nervous system variation and function, particularly of the brain. **Figure 8.1** depicts an inherited illness that affects brain structure in a very distinct way. Most inherited conditions that reflect abnormal brain structure or function are not nearly as obvious.

The human brain weighs about 3 pounds and resembles a giant gray walnut, but with the appearance and consistency of pudding. It consists of 100 billion nerve cells, or neurons, and at least a trillion other types of cells called neuroglia. Once thought to provide little more than scaffolding, neuroglial cells guide the development and movements of neurons in the embryo, and produce nerve growth factors that continue to nurture brain neurons throughout life. Neuroglial cells are essential, and, unlike neurons, can divide. Brain tumors can arise in neuroglial cells, and they are among the fastest-growing cancers.

The surfaces of extensions from each of the 100 billion neurons form close associations called synapses with some 1,000 to 10,000 other neurons. Neurons communicate across these tiny spaces using chemical signals called neurotransmitters. The neurons form networks and centers that oversee particular broad functions such as sensation and perception, memory, reasoning, and muscular movements.

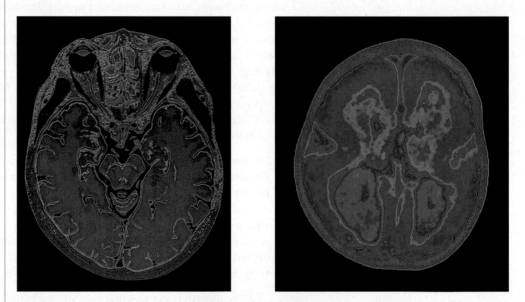

Figure 8.1   A very atypical brain.   In lissencephaly ("smooth brain" in Greek) (OMIM 607432), an affected newborn has difficulty feeding because of poor muscle coordination. Between 3 and 6 months, parents notice developmental delay. Eventually, severe mental retardation becomes apparent, and seizures begin. At some point in the diagnostic workup, a startling brain abnormality is revealed—lack of the characteristic convolutions on the cerebral cortex, the outer part of the brain. There are at least 20 ways to inherit this condition.

Genes control the synthesis, levels, and distribution of neurotransmitters. **Figure 8.2** indicates the points of gene control over sending and receiving nervous system information. Enzymes oversee the synthesis of neurotransmitters and their transport from the sending (presynaptic) neuron across the synapse to receptors on the plasma membrane of the receiving (postsynaptic) neuron. Genes also control the synthesis of myelin, a fatty substance that coats neuron extensions called axons. The myelin coats and insulates the neuron, which speeds neurotransmission. Signal transduction is also a key part of the function of the nervous system (see figure 2.21). Therefore, candidate genes for the inherited components of a variety of mood disorders and mental illnesses—as well as of normal variations in temperament and personality—affect neurotransmission and signal transduction.

Identifying the inherited and environmental contributors to a behavioral disorder is very challenging, because the same collections of symptoms may have different causes. Typically, traditional methods identify a large inherited component to a behavior, and further studies identify and describe candidate genes. Consider attention deficit hyperactivity disorder (ADHD). Applied Question 3 in chapter 7 presents evidence for an inherited component to ADHD. Linkage analysis on families with more than one affected member implicates the neurotransmitter dopamine. Two types of proteins control dopamine function: A "transporter" protein shuttles dopamine between neurons, and the dopamine D(4) receptor protein binds dopamine on the postsynaptic neuron. Development of drugs to treat ADHD focus on dopamine and the molecules that control its functioning.

Deciphering genetic components of most behavioral disorders is not as straightforward as ADHD analysis appears to be. Investigating the causes of autism illustrates the difficulty of reconciling empiric, adoptee, and twin data with molecular methods.

Autism is a spectrum of disorders of communication—the individual does not speak or interact with others and is comfortable only with restricted or repetitive behaviors. Asperger syndrome does not impair language ability and may be a mild form of autism. The general population incidence of autism is only 10 to 12 per 10,000 (<0.1 percent), but for a sibling of a person with the condition, risk of recurrence is 2 to 4 percent. Twin studies indicate high heritability. Yet the search for causative genes so far has yielded many candidates with weak linkage, rather than a few compelling candidate genes. Several whole genome scans, using many markers and hundreds of families that have more than one affected member, point to possible risk-raising genes on 14 different chromosomes! Autism is likely several different disorders that have similar symptoms.

Investigating the genetics of behavior is more challenging, for several reasons, than understanding a disorder in which an abnormal protein disrupts physiology in a clear way. Many behavioral disorders have symptoms in common, which can delay or obscure accurate diagnosis. However, many symptoms also fall within the range of normal behavior. Whether extreme anxiety is warranted depends upon the situation, and different individuals may react with different intensity to the same situation. Another complication of studying the genetics of behavior is that the reporting of symptoms may be highly subjective. A person can also unintentionally copy someone's unusual behavior, not realizing it is unusual. However, being too quick to assign a genetic cause to a behavior can be dangerous, as Bioethics: Choices for the Future, "Blaming Genes" discusses.

The examples that follow begin with traditional evaluation of siblings, adoptees, and twins, and conclude with a look at genes that underlie, contribute to, or influence behaviors. Identifying human behavioral genes may make it possible to subtype

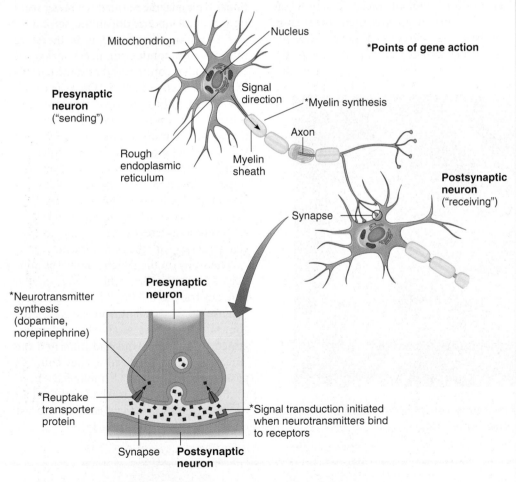

*Points of gene action

Mitochondrion

Nucleus

Presynaptic neuron ("sending")

Signal direction

*Myelin synthesis

Axon

Rough endoplasmic reticulum

Myelin sheath

Postsynaptic neuron ("receiving")

Synapse

Presynaptic neuron

*Neurotransmitter synthesis (dopamine, norepinephrine)

*Reuptake transporter protein

*Signal transduction initiated when neurotransmitters bind to receptors

Synapse **Postsynaptic neuron**

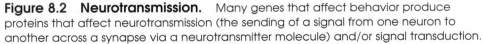

Figure 8.2 Neurotransmission. Many genes that affect behavior produce proteins that affect neurotransmission (the sending of a signal from one neuron to another across a synapse via a neurotransmitter molecule) and/or signal transduction.

Blaming Genes

It has become fashionable to blame genes for our shortcomings. A popular magazine's cover shouts "Infidelity: It May Be in Our Genes," advertising an article that actually has little to do with genetics. When researchers identify a gene that plays a role in fat metabolism, people binge on chocolate and forsake exercise, reasoning that if obesity is in their genes, they can't prevent it. Some behaviors have even been blamed on a gene for "thrill seeking" (**figure 1**).

Behavioral genetics has a checkered past. Early in the twentieth century, it was part of eugenics, the attempt to improve a population's collection of genes, or gene pool. The horrific experiments and exterminations the Nazis performed in the name of eugenics turned many geneticists away from studying the biology of behavior. Social scientists then dominated the field, attributing many behavioral disorders to environmental influences. For example, autism and schizophrenia were at one time blamed on "adverse parenting." By the 1960s, with a clearer concept of the gene, biologists reentered the debate. Today, researchers apply knowledge from biochemistry and neurobiology to identify genotypes that predispose a person to developing a clearly defined behavior.

Untangling the causes of human behavior remains highly controversial. A scientific conference to explore genetic aspects of violence was cancelled after a noted psychiatrist objected that "behavioral genetics is the same old stuff in new clothes. It's another way for a violent, racist society to say people's problems are their own fault, because they carry 'bad' genes." Genetic researchers on the trail of physical explanations for behaviors counter that their work can help uncover ways to alter or prevent dangerous behaviors.

Even in the rare instances when a behavior is associated with a particular DNA variant, environmental influences remain important. Consider a 1993 study of a Dutch family that had "a syndrome of borderline mental retardation and abnormal behavior." Family members had committed arson, attempted rape, and engaged in exhibitionism. Researchers found a mutation in a gene that made biological sense. Alteration of a single DNA base in the X-linked gene encoding an enzyme, monoamine oxidase A (MAOA), rendered the enzyme nonfunctional. This enzyme normally catalyzes reactions that metabolize dopamine, serotonin, and norepinephrine, and it is therefore important in conducting nerve messages. Other studies confirmed that some combinations of alleles of the MAOA gene correlate with highly aggressive behavior, and others with calmer temperaments. The direct effect of mutations in the MAOA gene still isn't known. Perhaps the inherited enzyme deficiency causes slight mental impairment, and this interferes with the person's ability to cope with certain frustrating situations, resulting in violence. Hence, the argument returns once again to how genes interact with the environment.

The study on the Dutch family was publicized and applied to other situations. An attorney tried to use the "MAOA deficiency defense" to free a client from execution for committing murder. A talk-show host suggested that people who had inherited the "mean gene" be sterilized so they couldn't pass on the tendency. This may have been meant as a joke, but it is frighteningly close to the eugenics practiced early in the last century, revisited in chapter 16.

Figure 1 A thrill-seeking gene? These air surfers were dropped from a helicopter over a mountain. Does a gene variant make them seek thrills?

Table 8.1

Prevalence of Behavioral Disorders in the U.S. Population

Condition	Prevalence (%)
Alzheimer disease	4.0
Anxiety	8.0
Phobias	2.5
Posttraumatic stress disorder	1.8
Generalized anxiety disorder	1.5
Obsessive compulsive disorder	1.2
Panic disorder	1.0
Attention deficit hyperactivity disorder	2.0
Autism	0.1
Drug addiction	4.0
Eating disorders	3.0
Mood disorders	7.0
Major depression	6.0
Bipolar disorder	1.0
Schizophrenia	1.3

Source: Psychiatric Genomics Inc., Gaithersburg, MD. The information was collated from the Surgeon General's 1999 Report on Mental Health.

Because eating disorders were once associated almost exclusively with females, most available risk estimates exclude males. Twin studies reveal a considerable genetic component, with heritability ranging from 0.5 to 0.8. Studies of eating disorders that recur in families without twins are more difficult to interpret. It's hard to determine whether a young girl is imitating her older sister by starving herself because genes predispose her to develop an eating disorder, or because she wants to be like her sister.

Genes that encode proteins that control appetite are candidate genes for eating disorders (see table 7.8). Genes that regulate the neurotransmitters dopamine and serotonin may also contribute to the risk of developing an eating disorder. It will be interesting to learn which genes affect body image, and how they do so.

In association studies, whole genome scans associate SNP patterns with eating disorders, as described in chapter 7. One biotechnology company is cataloging 60 SNP sites among the genomes of 2,000

mental disorders so that individualized treatment can begin early. **Table 8.1** lists some behavioral disorders.

Key Concepts

1. Most behavioral traits and disorders are fairly common, polygenic, and multifactorial.
2. Traditional methods to estimate the genetic contribution to a behavior include empiric risk estimates and adoptee and twin studies. Association studies and candidate gene analyses are now extending these data.
3. Behavioral disorders are difficult to study because symptoms overlap and behaviors can be imitated.

8.2 Eating Disorders

When 22-year-old gymnast Christy Henrich was buried on a Friday morning in July 1994, she weighed 61 pounds (**figure 8.3**). Three weeks earlier, she had weighed 47 pounds. Christy suffered from anorexia nervosa, a psychological disorder that is fairly common among professional athletes. The person perceives herself or himself as obese, even when obviously not, and intentionally starves. Christy's decline began when a judge at a gymnastics competition told her that at 90 pounds, she was too heavy to make the U.S. Olympic team. From then on, her life consisted of starving, exercising, and taking laxatives to hasten weight loss.

For economically advantaged females in the United States, the lifetime risk of developing anorexia nervosa is 0.5 percent. Anorexia has the highest risk of death of any psychiatric disorder—15 to 21 percent. The same population group has a lifetime risk of 2.5 percent of developing another eating disorder, bulimia. A person with bulimia eats huge amounts but exercises and vomits to maintain weight.

Five to 10 million people in the United States have eating disorders. About 10 percent of them are male. One survey of eight-year-old boys revealed that more than a third of them had attempted to lose weight. In an eating disorder called muscle dysmorphia, boys and young men take amino acid supplements to bulk up. Just as the person with anorexia looks in a mirror and sees herself as too large, a person with muscle dysmorphia sees himself as too small.

Figure 8.3   Eating disorders.
World-class gymnast Christy Henrich died of complications of anorexia nervosa in July 1994. In this photo, taken eleven months before her death, she weighed under 60 pounds. Concern over weight gain propelled her down the path of this deadly psychiatric illness.

individuals representing 600 families where more than one member has anorexia nervosa. The researchers are searching for a SNP pattern—if there is one—that appears disproportionately among individuals who have anorexia. Once the SNP maps highlight chromosome regions that seem to mark a predisposition to develop anorexia, researchers will look for genes in those regions whose protein products might affect appetite.

RELATING THE CONCEPTS

1. Construct a pedigree for the family described in the beginning of the chapter opening essay.
2. Does the family's pattern of eating disorders suggest single-gene inheritance? Cite a reason for your answer.

Key Concepts

1. Eating disorders are common.
2. Twin and heritability studies indicate a high genetic contribution to eating disorders.
3. Genes whose products control appetite or regulate the neurotransmitters dopamine and serotonin may cause or raise the risk of developing an eating disorder.

8.3 Sleep

Sleep has been called "a vital behavior of unknown function," and, indeed, without sleep, animals die. We spend a third of our lives in this mysterious state.

Genes influence sleep characteristics. When asked about sleep duration, schedule, quality, nap habits, and whether they are "night owls" or "morning people," MZ twins report significantly more in common than do DZ twins, even MZ twins separated at birth. Twin studies of brain wave patterns through four of the five stages of sleep confirm a hereditary influence. The fifth stage, REM sleep, is associated with dreaming and therefore may reflect the input of experience more than genes.

Narcolepsy

Researchers discovered the first gene related to sleep in 1999, for a condition called "narcolepsy with cataplexy" in dogs. Humans have the disorder (OMIM 161400), but it is rarely inherited as a single-gene trait—it is more often polygenic requiring an environmental trigger, or due to an autoimmune condition (the immune system attacking the body) that also involves a genetic susceptibility.

A person (or dog) with narcolepsy falls asleep suddenly several times a day. Extreme daytime sleepiness greatly disrupts ability to attend classes or work. People with narcolepsy have a tenfold higher rate of car accidents. Another symptom is sleep paralysis, the inability to move for a few minutes after awakening. The most dramatic manifestation of narcolepsy is cataplexy. During these short and sudden episodes of muscle weakness, the jaw sags, the head drops, knees buckle, and the person falls to the ground. This often occurs during a bout of laughter or excitement—which can be quite disturbing both for the affected individual and bystanders. People with narcolepsy and cataplexy cannot participate in even the most mundane of activities for fear of falling and injuring themselves. Narcolepsy with cataplexy affects only 0.02 to 0.06 percent of the general populations of North America and Europe, but the fact that it is much more common in certain families suggests a genetic component.

Dogs led the way to discovery of the narcolepsy gene. In 1999, Emmanuel Mignot and his team at Stanford University identified mutations in a gene that encodes a receptor for a neuropeptide called hypocretin (OMIM 602358). In Doberman pinschers and Labrador retrievers, the receptor does not arrive at the cell surfaces of certain brain cells. As a result, the cells cannot receive signals to promote a state of awakeness. Dachshunds have their own mutation—they make a misshapen, nonfunctional receptor. **Figure 8.4** shows a still frame of a film that Mignot made of narcoleptic dogs playing. Suddenly, they all collapse! A minute later, they get up and resume their antics. "You can't make dogs laugh, but you can make them so happy that they have attacks," says Mignot. To induce a narcoleptic episode in puppies, he lets them play with each other. He feeds older dogs meat, which excites them so much that they can take a while to finish a meal because they fall down in delight so often. Getting dogs to breed was difficult, too—sex proved even more exciting than play or food!

A year earlier, Masahi Yanagisawa, at the University of Texas Southwestern Medical Center in Dallas, discovered a protein called orexin, but thought it only sent signals to eat. Yanagisawa's orexin turned out to bind Mignot's hypocretin receptor.

Figure 8.4   Letting sleeping dogs lie. These Doberman pinschers have inherited narcolepsy. They suddenly fall into a short but deep sleep while playing. Research on dogs with narcolepsy led to discovery of a version of the gene in humans.

Yanagisawa bred mice that lacked the orexin gene, and then noticed something odd while watching the animals feed at night—the rodents suddenly fell down fast asleep! Researchers are now trying to figure out how one molecule can apparently control feeding as well as wakefulness. The hypocretin/orexin receptor gene, found on dog chromosome 12, is on human chromosome 6. The brains of humans with narcolepsy and cataplexy are remarkably deficient in hypocretin/orexin. Pharmaceutical researchers are trying to synthesize a drug that can mimic the missing molecule to treat narcolepsy.

Familial Advanced Sleep Phase Syndrome (OMIM 604348)

A multigenerational Utah family with many members who have an unusual sleep pattern led researchers to a "biological clock" gene that controls sleep. The subjects have familial advanced sleep phase syndrome (FASPS) and the effect is striking—they promptly fall asleep at 7:30 each night and awaken suddenly at 4:30 A.M. The family is a geneticist's dream—a distinctive behavioral phenotype, many affected individuals, and a clear mode of inheritance (autosomal dominant) (**figure 8.5**).

An analysis of the Utah family followed the standard approach to gene identification that chapter 7 described. A whole genome scan for short repeated DNA sequences revealed a variant area at the tip of the long arm of chromosome 2 found exclusively in the affected family members. Within that area is a gene, called *period*, that has a counterpart in golden hamsters and fruit flies that also disrupts the sleep-wake cycle. Humans with the condition have a single DNA base substitution in the gene. This mutation prevents the encoded protein from binding a phosphate chemical group, which it must do to pass on the signal that synchronizes the sleep-wake cycle with daily sunrise and sunset.

Despite the clear connection between sleep behavior and a specific gene in the Utah family, this gene is just one influence on this behavior. In others of the 50 known families with FASPS, linkage analysis did not point to this gene, meaning that the condition is genetically heterogeneic. Environmental influence is great, too. Daily rhythms such as the sleep-wake cycle are set by cells that form a "circadian pacemaker" in a part of the brain called the suprachiasmatic nuclei. Genes are expressed in these cells in response to light or dark in the environment.

Other environmental effects on sleeping and waking are more subtle. Knowing that the hour is late may trigger an "I should go to sleep" or "Yikes, I have to get up for class" response. Culture also affects the times that we retire and rise. Understanding how the *period* gene and others control the sleep-wake cycle may lead to new treatments for jet lag, insomnia, and the form of advanced sleep phase syndrome that is common among older individuals.

Key Concepts

1. Twin studies on sleep habits indicate a high heritability for sleep characteristics.
2. A single gene causes narcolepsy in dogs, and, more rarely, in humans.
3. A Utah family with a very unusual sleep-wake cycle led researchers to identify a "clock" gene in humans.

8.4 Intelligence

Intelligence is a vastly complex and variable trait that is subject to many genetic and environmental influences, and also to intense subjectivity. Sir Francis Galton, a half first cousin of Charles Darwin, investigated genius, which he defined as "a man endowed with superior faculties." He first identified successful and prominent people in Victorian-era English society, and then assessed success among their relatives. In his 1869 book, *Hereditary Genius*, Galton wrote that relatives of eminent people were more likely to also be successful than people in the general population. The closer the blood relationship, he concluded, the more likely the person was to succeed. This, he believed, established a hereditary basis for intelligence.

Definitions of intelligence vary. In general, intelligence refers to the ability to reason, learn, remember, connect ideas, deduce, and create. The first intelligence tests, developed in the late nineteenth century, assessed

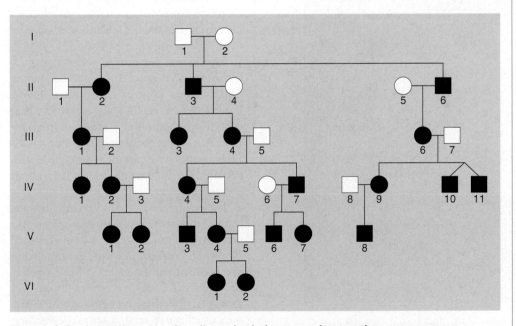

Figure 8.5 Inheritance of a disrupted sleep-wake cycle. This partial pedigree depicts a large family with familial advanced sleep phase syndrome. The condition is genetically heterogeneic—that is, different families have causative mutations in different genes. In this family from Utah, the condition is autosomal dominant.

sensory perception and reaction times to various stimuli. In 1904, Alfred Binet at the Sorbonne developed a test with verbal, numerical, and pictorial questions, used to predict the success of developmentally disabled youngsters in school. The test was subsequently modified at Stanford University to assess white, middle-class Americans. An average score on this "intelligence quotient," or IQ test, is 100, with two-thirds of all people scoring between 85 and 115 in a bell curve or normal distribution (**figure 8.6**). An IQ between 50 and 70 is considered mild mental retardation, and below 50, severe mental retardation.

Over the years, IQ has been a fairly accurate predictor of success in school and work. However, low IQ also correlates with many societal situations, such as poverty, a high divorce rate, failure to complete high school, incarceration (males), and having a child out of wedlock (females). In 1994, a book called *The Bell Curve* asserted that because certain minorities are overrepresented in these groups, they must be of genetically inferior intelligence and that is why they are prone to suffering social ills. It was a controversial thesis, to put it mildly.

The IQ test consists of short exams that measure verbal fluency, mathematical reasoning, memory, and spatial visualization skills. Because people tend to earn similar scores in all these areas, psychologists hypothesized that a general or global intelligence ability, called "g," must underlie the four basic skills that IQ encompasses. Statistical analysis indeed reveals one factor that accounts for general intelligence. In contrast, similar analysis of personality reveals five contributing factors. The g value is the part of IQ that accounts for differences between individuals based on a generalized intelligence, rather than on enhanced opportunities such as attending classes to boost test-taking skills.

Environment does not seem to play too great a role in IQ differences. Evidence includes the observation that IQ scores of adoptees, with time, become closer to those of their biological parents than to those of

Table 8.2	
Heritability of Intelligence Changes Over Time	
Age Group	**Heritability**
Preschoolers	0.4
Adolescents	0.6
Adults	0.8

their adoptive parents. Heritability studies also reveal a declining environmental impact with age (**table 8.2**). This makes sense. As a person ages, he or she has more control over the environment, so genetic contributions to intelligence become more prominent.

Researchers have long realized there must be a genetic explanation for intelligence differences because nearly all syndromes that result from abnormal chromosomes include some degree of mental retardation. Down syndrome and fragile X syndrome (see figure 13.1 and Reading 12.1) are two of the more common chromosomal causes of mental retardation. Down syndrome is usually caused by an extra chromosome, and fragile X syndrome by an expanding gene on the X chromosome. Mutations in genes located next to the tips of chromosomes account for many cases of mental retardation.

The search for single genes that contribute to intelligence differences focuses on proteins that control neurotransmission. For example, a certain SNP pattern in a gene encoding neural cellular adhesion molecule (N-CAM) correlates strongly with high IQ. Perhaps this gene variant eases certain neural connections that enhance an individual's learning ability.

Identifying the N-CAM variant illustrates the candidate gene approach—relating a gene with a known function to intelligence. In another approach, whole genome scans are locating other genes whose protein products affect neural connections. For example, a section of chromosome 4 harbors intelligence-related genes. Researchers identified three candidate genes here by comparing 147 markers in a group of children with very high IQ scores to children of average IQ.

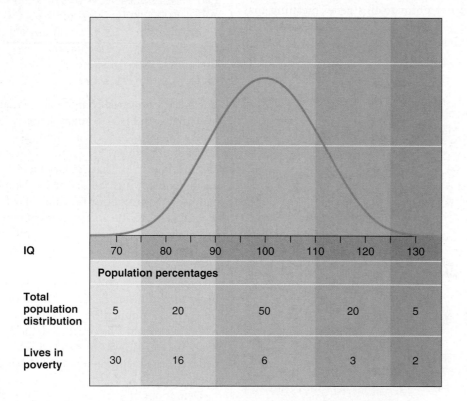

IQ	70	80	90	100	110	120	130
Population percentages							
Total population distribution	5		20	50	20		5
Lives in poverty	30		16	6	3		2

Figure 8.6 Success and IQ. IQ scores predict success in school and the workplace in U.S. society. The bell curve for IQ indicates that most people fall in the 85 to 115 range, shown in the total population distribution. However, when the population is stratified economically, those living in poverty tend to have lower IQs. Cause and effect are unclear.

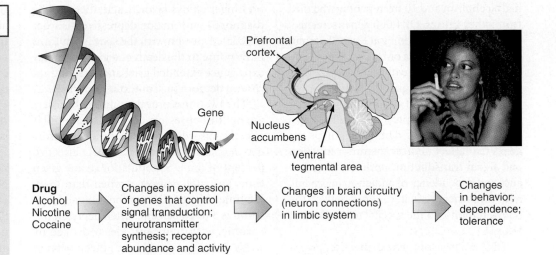

Figure 8.7   The events of addiction.   Addiction is manifest at several levels: at the molecular level, in neuron-neuron interactions in the brain, and in behavioral responses.

8.5 Drug Addiction

Drug addiction is compulsively seeking and taking a drug despite knowing its adverse effects. Drug addiction has two identifying characteristics: tolerance and dependence. Tolerance is the need to take more of the drug to achieve the same effects as time goes on. Dependence is the onset of withdrawal symptoms when a person ceases taking the drug. Both tolerance and dependence contribute to the biological and psychological components of craving the drug. The behavior associated with drug addiction can be extremely difficult to break.

Drug addiction produces long-lasting, rather than temporary, brain changes, because the craving and high risk of relapse remain even after a person has abstained for years. Heritability is 0.4 to 0.6, with a two- to threefold increase in risk among adopted individuals who have one affected biological parent. Twin studies also indicate an inherited component to drug addiction.

Brain imaging techniques have localized the "seat" of drug addiction in the brain by highlighting the cell surface receptors that bind neurotransmitters when a person craves the drug. The brain changes that contribute to addiction occur in a group of functionally related structures called the limbic system (**figure 8.7**). These structures are the nucleus accumbens, the prefrontal cortex, and the ventral tegmental area. The effects of cocaine seem to be largely con-fined to the nucleus accumbens, whereas alcohol affects the prefrontal cortex.

Although the specific genes and proteins that are implicated in addiction to different substances may vary, several general routes of interference in brain function are at play. Proteins involved in drug addiction are those that

- are part of the biosynthetic pathways of neurotransmitters, such as enzymes;

- form reuptake transporters, which remove excess neurotransmitter from the synapse;

- form receptors on the postsynaptic neuron that are activated or inactivated when specific neurotransmitters bind;

- are part of the signal transduction pathway in the postsynaptic neuron

Abused drugs are often plant-derived chemicals, such as cocaine, opium, and tetrahydrocannabinol (THC), the main active ingredient in marijuana. These substances bind to receptors on human neurons, which indicates that our bodies have versions of these substances. The human equivalents of the opiates are the endorphins and enkephalins, and the equivalent of THC is anandamide. The endorphins and enkephalins relieve pain. Anandamide modulates how brain cells respond to stimulation by binding to neurotransmitter receptors on presynaptic (sending) neu-rons. In contrast, neurotransmitters bind to receptors on postsynaptic neurons.

Amphetamines and LSDs produce their effects by binding to receptors on neurons that normally bind neurotransmitters called trace amines. Trace amines are found throughout the brain at low levels, compared to the more abundant neuro-transmitters such as dopamine and sero-tonin. LSD causes effects similar to the symptoms of schizophrenia (see section 8.7), suggesting that the trace amine recep-tors, which are proteins, may be implicated in the illness.

DNA microarray technology is probing the biology of addiction by revealing the expression of many genes at a time. In the past, research focused on individual genes whose encoded proteins fit a part of the picture—such as alcohol dehydrogenase, an enzyme in the pathway to metabolize ethanol. Another example of a gene whose encoded protein is implicated in predispo-sition to drug addiction is the dopamine D(2) receptor gene. People who are homo-zygous for the A1 allele are overrepresented among people with alcoholism and other addictions.

DNA microarray tests that detect expres-sion of thousands of genes at a time before and after exposure to a particular drug reveal that 1 to 5 percent of the genes are expressed. Consider a comparison of gene expression profiles in human brain cells, with 10 samples from deceased people who had alcoholism and 10 from people who died from other causes. Of 4,000 genes screened, 160 varied in expression by at least 40 percent between the two types of brains. In addition to identifying genes involved in signal transduction and neurotransmitter activity, the study also highlighted genes that function in the cell cycle and apoptosis, and in helping a cell survive oxidative damage. Perhaps the genes that affect neurotransmitter function and signal transduction underlie the tolerance and dependency of addiction in general, and the other activated genes reflect the body's response to the specific toxic effects of alcohol.

DNA microarray gene expression profiling also reveals a subset of genes whose actions are unique to brain cells, such as genes that control the synthesis of myelin, the fatty insulation on some neurons. Impairment of myelin synthesis with chronic alcohol use is consistent with imaging studies that show that the brain's white matter shrinks. Other microarray experiments compared expression of 6,000 genes in mouse brain neurons growing in culture, with or without exposure to alcohol. The expression of dozens of genes increased or decreased in the presence of alcohol. It will be interesting to determine if and how DNA expression profiles change with addiction to different drugs.

Key Concepts

1. Drug addiction is dependency on a drug despite knowing the activity is harmful.
2. Structures in the limbic system are directly involved in drug addiction.
3. A candidate gene for addiction encodes the dopamine D(2) receptor.
4. DNA microarray studies reveal many genes whose protein products affect neurotransmission, signal transduction, and myelin deposition on neurons.

8.6 Mood Disorders

Identifying genetic and environmental influences on mood disorders is challenging because the mood changes may appear to be extremes of normal behavior. For example, a person who has previously been happy but inexplicably becomes lethargic and sad, and no longer enjoys favorite activities may be diagnosed with major depressive disorder (MDD). A person with the same symptoms in response to the death of a loved one may experience extended grief and loss, but not clinical depression. Context is important.

The two most prevalent mood disorders are major depressive disorder and bipolar affective disorder (also called bipolar disorder or manic-depression). MDD affects 6 percent of the U.S. population at any given time, and affects more women than men. Lifetime risk of MDD for the general population is 5 to 10 percent. Often depression is chronic, with acute episodes provoked by stress. It is a serious illness. Fifteen percent of people hospitalized for severe, recurrent depression ultimately end their lives. About half of all people who experience a depressive episode will suffer others. Half of affected individuals do not seek medical help, and among those who do, a third do not respond to drug therapy; those who do may relapse when they discontinue taking an effective drug. Electroconvulsive (shock) therapy can fairly quickly help some patients who are drug-resistant. For many people, antidepressant treatment is very helpful if paired with psychotherapy.

Bipolar disorder is much rarer than MDD, affecting 1 percent of the population and with a general population lifetime risk of 0.5 to 1.0 percent. With this disorder, weeks or months of depression alternate with periods of mania, when the person is hyperactive and restless, and may experience a rush of ideas and excitement. Ideas may be fantastic, and behavior reckless. For example, a person who is normally quiet and frugal might, when manic, suddenly make large monetary donations and spend lavishly—very out-of-character behavior. In one subtype of bipolar disorder, the "up" times are termed hypomania, and they seem more a temporary reprieve from the doldrums than the starkly aberrant behavior of full mania. Bipolar disorder with hypomania may appear to be depression. This is an important distinction because different drugs are used to treat depression and bipolar disorder.

At the root of depression, and possibly of bipolar disorder too, is deficiency of the neurotransmitter serotonin, which affects mood, emotion, appetite, and sleep. Levels of norepinephrine, another type of neurotransmitter, are important as well. The abnormality occurs in transporter proteins that ferry neurotransmitters from the synapse to "reuptake pumps" in the presynaptic neuron. Overactive or overabundant transporters deplete neurotransmitter levels. Millions of people take drugs called selective serotonin reuptake inhibitors (SSRIs) to prevent presynaptic neurons from admitting serotonin from the synapse. This leaves more of the neurotransmitter available to stimulate the postsynaptic cell (**figure 8.8**), which apparently offsets the neurotransmitter deficit. Older antidepressants called tricyclics target norepinephrine, and newer dual-acting drugs affect both serotonin and norepinephrine levels.

The SSRIs and dual drugs may begin to produce effects after one week, often enabling a person with moderate or severe depression to return to some activities, but full response can take up to eight weeks. Other older drugs may take even longer to work. DNA microarray tests are being developed that can predict which drugs are most likely to help a particular patient with the fewest side effects.

The distribution of serotonin deficiency in the brain of a depressed person differs from normal. One study of the brains of 220 people who had died while clinically depressed revealed a generalized decrease in serotonin activity in many brain regions, and a concentrated area of poor activity in the prefrontal cortex of individuals who had been suicidal.

Assigning specific genes or chromosomal regions to bipolar disorder has lagged behind such efforts for depression. Over the past thirty years, linkage studies in large families, or association studies in very isolated populations such as the Amish, have indicated genome regions that may harbor genes that predispose to bipolar disorder, such as parts of chromosomes 4, 10, 18, 22, and mitochondrial DNA (because in some families only females pass on the trait). Evidence is more specific for depression, pointing strongly toward malfunction of the serotonin transporter coupled with an environmental trigger. For bipolar disorder, evidence is insufficient to support either a model of many genes, each with an additive effect (polygenic); roles for several genes, each with a large and independent effect (genetic heterogeneity); or different genes

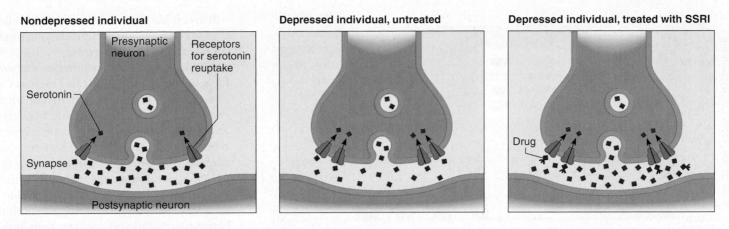

Figure 8.8 Anatomy of an antidepressant. Selective serotonin reuptake inhibitors (SSRIs) are antidepressant drugs that block the reuptake of serotonin, leaving more of the neurotransmitter in synapses. This corrects a neurotransmitter deficit that presumably causes the symptoms. Overactive or overabundant reuptake receptors can cause the deficit. The precise mechanism of SSRIs is not well understood, and the different drugs may work in slightly different ways.

interacting and controlling each other's expression (epistasis).

The National Institute of Mental Health ... nsortium of nine U.S. centers ...nome sequence ...mong

cent of the world's population, and 10 percent of affected individuals commit suicide.

Identifying genetic contributions to schizophrenia illustrates the difficulties in analyzing a behavioral condition. Some of the symptoms are also associated with other illnesses; many genes cause or contribute to ...al environmental factors may ...n.

...schizophrenia often affect ...hildhood or early adoles- ...ight suddenly have trouble ...in school, and learning may ...as memory falters and infor- ...sing skills lag. Symptoms of ...n between ages 17 and 27 for ...and 37 for females, including ...nd hallucinations—sometimes ...times seen. A person with schizo- ...y hear a voice giving instructions. ...ers perceive as irrational fears, such ...ollowed by monsters, are very real to ...on with schizophrenia. Meanwhile, ...e skills continue to decline. Speech ...s the garbled thought process; the per- ...ips from topic to topic with no obvious ...d of logic, or displays inappropriate ...tional responses, such as laughing at sad ...ws. Artwork by a person with schizophre- ...a can display the characteristic fragmenta- ...on of the mind (**figure 8.9**). (Schizophrenia ...eans "split mind," but it does not cause a split or multiple personality.)

The course of schizophrenia often plateaus (evens out) or becomes episodic. It is not a continuous decline, as is the case for

Figure 8.9 Schizophrenia alters thinking. People with schizophrenia communicate the disarray of their thoughts in characteristically disjointed drawings.

dementia. Schizophrenia is frequently misdiagnosed as depression or bipolar affective disorder. However, schizophrenia primarily affects thinking; these other conditions mostly affect mood. It is a very distinctive mental illness.

A heritability of 0.8 and empiric risk values indicate a strong role for genes in causing schizophrenia (**figure 8.10**). Because most of the symptoms are behavioral, however, it is possible to develop some of them—such as disordered thinking—from living with and imitating people who have schizophrenia. Although concordance is high, a person who has an identical twin with schizophrenia has a 52 percent chance of *not* developing schizophrenia. Therefore,

ability to o...
which leads to a wit...
Various forms of the condition a...

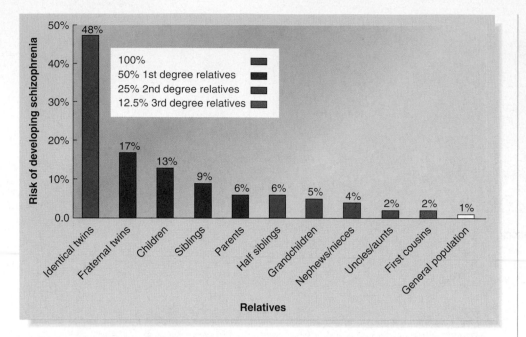

Figure 8.10   Schizophrenia has inherited and environmental components.

the condition has a significant environmental component, too.

Early investigations on the inheritance of schizophrenia focused on affected individuals who also had visible chromosome abnormalities, such as two Chinese brothers who had a duplication of part of chromosome 5. Assuming that the mental illness and the extra chromosomal material were related, researchers looked to this chromosome in other affected families, and identified five families in Iceland and two in England that had mutations in this part of the genome. But in other families, other chromosome regions or genes are associated with schizophrenia. For example, 30 percent of individuals missing the same small section of chromosome 22 develop schizophrenia, which is significantly greater than the 1 percent incidence in the general population. In some families, a gene on chromosome 6 that encodes a signal transduction protein called dystrobrevin binding protein 1 causes schizophrenia. In Iceland, mutation in a gene called neuregulin accounts for 30 percent of the people who have schizophrenia. This protein may normally control how neural connections form in response to environmental influences in certain parts of the brain.

The genes that encode neuregulin and dystrobrevin binding protein 1, and predis-posing DNA sequences on chromosomes 5 and 22, may be only a few of dozens of genes that can combine with an environmental trigger to cause schizophrenia. Genomewide screens of families with schizophrenia reveal at least twenty-four sites where affected siblings share alleles much more often than the 50 percent of the time that Mendel's first law predicts.

Identifying susceptibility genes for schizophrenia may be difficult because the environment may play an important causative role. **Table 8.3** lists some suggested environmental factors. One powerful candidate is infection during pregnancy. The woman's immune system bathes the brain of the embryo or fetus

---

Table 8.3

Environmental Risk Factors for Schizophrenia

Maternal malnutrition

Infection by Borna virus

Fetal oxygen deprivation

Obstetric or birth complication

Psychoactive drug use (phencyclidine)

Traumatic brain injury

Herpes infection at time of birth

---

with a combination of cytokines (molecules that function in signal transduction) that subtly alters brain development.

The idea that maternal infection can sow the seeds for schizophrenia first grew out of observations on the seasonality of birth dates. As far back as 1929, researchers noted an unusually high percentage of people with schizophrenia were born in the winter. One surge in cases dates back to the winter of the 1957 influenza pandemic. More recent studies that consult medical records or blood samples to confirm flu cases support the link between prenatal exposure to flu virus or the mother's immune response and schizophrenia—a few are discussed in an end-of-chapter question. The influenza virus can cross the placenta and alter brain cells.

Not all studies, however, link schizophrenia to winter births. A small association also occurs with births in June or July. To explain the exception, researchers suggest there are two subtypes of schizophrenia. The "deficit" subtype, characterized by "negative" symptoms such as lack of emotion, speech, facial expressions and socialization, correlates to a summer birth. The "nondeficit" subtype, with paranoia, hallucinations, delusions and disordered thinking, is more often associated with winter births. Ongoing DNA microarray gene expression studies may finally be able to sort out the causes and subtypes of this most puzzling disorder.

Behavioral traits have been much more difficult to describe, categorize, and attribute to genetic and/or environmental influences than other characteristics. **Table 8.4** lists the heritabilities and candidate genes for the behavioral traits and conditions discussed in this chapter.

Key Concepts

1. Schizophrenia affects thinking and causes delusions and hallucinations, usually beginning in young adulthood.
2. Studies have implicated several candidate genes and chromosomal regions as possible causes.
3. A possible environmental influence may be prenatal exposure to the maternal immune system's response to influenza.

Table 8.4

Review of Behavioral Traits and Disorders

Condition	Heritability	Candidate Genes
ADHD	0.80	Dopamine transporter
		Dopamine D(4) receptor (DRD4)
Eating disorders	0.50–0.80	Leptin
		Leptin transporter
		Leptin receptor
		Neuropeptide Y
		Melanocortin-4 receptor
Intelligence	0.80	Neural cellular adhesion molecule (N-CAM)
Addiction	0.40–0.60	Dopamine D(2) receptor (DRD2)
		Myelin synthesis
Depression	0.40–0.54	Serotonin synthesis, transporter, receptor
		Norepinephrine synthesis, transporter, receptor
Bipolar disorder	0.80	Serotonin transporter, receptor
		Monoamine oxidase A control
Schizophrenia	0.80	Dopamine synthesis, transporter, receptor
		Glutamate synthesis, transporter, receptor

Summary

8.1 Genes Contribute to Most Behavioral Traits

1. Most behavioral traits and conditions are multifactorial and are more common than most single-gene disorders.

2. Candidate genes for behavioral traits and disorders affect neurotransmission and signal transduction.

3. Analyzing behaviors is difficult because symptoms of different syndromes overlap, study participants can provide biased information, and behaviors can be imitated.

8.2 Eating Disorders

4. Eating disorders affect both sexes and are prevalent in the United States and other nations. Twin studies indicate high heritability.

5. Candidate genes for eating disorders include those whose protein products control appetite and the neurotransmitters dopamine and serotonin.

8.3 Sleep

6. Twin studies and single-gene disorders that affect the sleep-wake cycle reveal a large inherited component to sleep behavior.

7. A large family with familial advanced sleep phase syndrome enabled researchers to identify the first "clock" gene in humans. The *period* gene enables a person to respond to day and night environmental cues.

8.4 Intelligence

8. Intelligence is difficult to define and measure. The general intelligence (g) value measures the inherited portion of IQ that may underlie population variance in IQ test performance.

9. Heritability for intelligence increases with age, suggesting that environmental factors are more important early in life.

10. Many chromosomal disorders affect intelligence, suggesting high heritability. A gene that encodes N-CAM is a candidate gene for intelligence.

8.5 Drug Addiction

11. Defining characteristics of drug addiction are tolerance and dependence. Addiction produces stable brain changes, yet heritability is not as high as for some other behavioral conditions.

12. A candidate gene for drug addiction is the one that encodes the dopamine D(2) receptor. DNA microarray tests on gene expression in the brains of people with alcoholism help to identify genes involved in neurotransmission, signal transduction, cell cycle control, apoptosis, surviving oxidative damage, and myelination of neurons.

8.6 Mood Disorders

13. Major depressive disorder is relatively common and associated with deficits of serotonin and/or norepinephrine.

14. Bipolar affective disorder, which consists of depressive periods interspersed with times of mania or hypomania, is much rarer. Linkage and association studies implicate several chromosomal sites that may house genes that raise the risk of developing this disorder.

8.7 Schizophrenia

15. Schizophrenia greatly disrupts the ability to think and perceive the world. Onset is typically in early adulthood, and the course is episodic or steady but not degenerative.

16. Empiric risk estimates and heritability indicate a large genetic component, yet certain environmental associations exist, too.

17. Many candidate genes and genome regions are associated with schizophrenia.

Review Questions

1. In general, what types of gene products are responsible for variations in behaviors?

2. Why is the genetics of ADHD easier to analyze than that of autism?

3. Which behaviors are traced to altered activities in the following regions of the brain?

   a. suprachiasmatic nuclei

   b. nucleus accumbens

   c. prefrontal cortex

4. Why is identifying a candidate gene only a first step in understanding how behavior arises and varies among individuals?

5. Describe three factors that can complicate the investigation of a behavioral trait.

6. Why does the heritability of intelligence decline with age?

Applied Questions

1. Studies indicate that in the United States, the incidence of autism has dramatically increased since 1990.

   a. Does this finding better support a genetic cause or an environmental cause for autism?

   b. What is a nongenetic factor that might explain the increased incidence of autism?

2. Serotonin levels are implicated in eating disorders, major depressive disorder, and bipolar disorder.

   a. How can an abnormality in one type of neurotransmitter contribute to different disorders?

   b. What other neurotransmitter is implicated in more than one behavioral disorder?

3. How has DNA microarray technology changed the study of the genetics of behavior?

4. What might be the advantages and disadvantages of a SNP profile or other genotyping test done at birth that indicates whether a person is at high risk for developing a drug addiction?

5. The U.S. government prohibits recreational use of cocaine, marijuana, and opiates, which are physically addictive drugs, but not of alcohol and nicotine (cigarettes), also physically addictive. Do you think that the legal status of any of these drugs should be changed, and if so, how and why? What measures, if any, should the government use in deciding which drugs to outlaw?

6. Do you think that having a genotype known to predispose a person to aggressive or violent behavior should be a valid legal defense? Cite a reason for your answer.

7. In some association studies of depression and bipolar disorder, correlations to specific alleles are only evident when participants are considered in subgroups based on symptoms. What might be a biological basis for this finding?

8. Many older individuals experience advanced sleep phase syndrome. Even though this condition is probably a normal part of aging, how might research on the Utah family with an inherited form of the condition help researchers develop a drug to help the elderly sleep through the night and awaken later in the morning?

9. What alternate explanation besides genetic differences might account for the over-representation of minority groups among people with low IQ scores in the United States?

10. A study found that the risk of schizophrenia among spouses of people with schizophrenia who have no affected blood relatives is 2 percent. What might this indicate about the causes of schizophrenia?

11. Wolfram syndrome (OMIM 222300) is a rare autosomal recessive disorder that causes severe diabetes, impaired vision, and neurological problems. Examinations of hospital records and self-reports reveal that blood relatives of Wolfram syndrome patients have an eightfold risk over the general population of developing serious psychiatric disorders such as depression, violent behavior, and suicidal tendencies. Can you suggest further experiments and studies to test the hypothesis that these mental manifestations are a less severe expression of Wolfram syndrome?

12. A study of 2,685 twin pairs showed that female MZ twins are six times as likely as female DZ twins to both have alcoholism. Does this finding suggest a large genetic or environmental component to alcoholism?

Web Activities

Visit the Online Learning Center (OLC) at www.mhhe.com/lewisgenetics7. Select **Student Edition, chapter 8** and **Web Activities** to find the website links needed to complete the following activites.

13. Consult the Diagnostic and Statistical Manual of Mental Disorders (DSM-IV). Follow links and list three disorders for which candidate genes have been identified. Discuss how those genes might cause the phenotype.

Case Studies

14. Marjorie and Joyce are mothers of preteen girls and survivors of eating disorders. While growing up in the 1950s and 1960s, Marjorie and Joyce were careful about what they ate, because their mother, stick thin, was always dieting and constantly worrying about weight. Marjorie, like her mother, learned to control her weight by eating practically nothing. Joyce ate more, but exercised at least two hours a day. When either gave in to hunger and ate a large amount, they would induce vomiting or take laxatives. When the medical community and the media began to recognize eating disorders, the sisters learned that their lack of body fat could prevent them from becoming pregnant. They underwent psychotherapy and learned (slowly) how to eat normally.

    Now seeing familiar behavior patterns in their daughters terrifies Marjorie and Joyce. Marjorie's daughter Sherry seems to do nothing but exercise, living on smoothies and salads. It seems, to

Marjorie, a fine line between a healthy lifestyle and obsession with weight control. Joyce's daughter binges and purges, and was so secretive that it took Joyce many months to be certain there was a problem, even though she had once engaged in the same behavior.

The sisters learn of a company offering testing for gene variants that have been associated with increased risk of developing eating disorders. Marjorie encourages Sherry to take the test, feeling guilty that she passed on the causative genes. But Joyce feels that the testing will have no benefit, and attributes the development of an eating disorder to living in a society that pressures women to be skeletal.

Do you agree with Marjorie or Joyce? Cite a reason for your answer.

15. On the island of Fiji, women once valued having a full figure. Then, in 1995, television arrived, and with it, the show "Melrose Place," depicting skinny women as the ideal. Within three years, the incidence of eating disorders doubled, with a frightening percentage of the female population regularly vomiting on purpose so that they could continue to eat. Does this information argue more for a genetic or nongenetic cause of eating disorders? How could both influences contribute?

16. Elaine has obsessive compulsive disorder (OCD). Before she leaves the house for her job as a social worker each morning, she checks the stove exactly eight times, to be certain she hasn't left the burner on. She stops at the end of the driveway to be sure she hasn't run over a squirrel. At work, she washes her hands as often as once an hour, for she is constantly aware of dirt and germs. At her desk, Elaine is very organized. She keeps three pens on her desk and seven in her drawer, and they must be fine-tipped black pens, never blue. Her computer mouse must rest a set distance from the keyboard. She reads e-mail in the exact order that messages appear, filing them away meticulously. Elaine's thoughts seem as if they are constantly screaming out rules for her to follow—her repetitive behaviors are an attempt to satisfy the ever-present urges to organize and check things.

For years Elaine didn't really think that her behavior was abnormal enough to have a label, but after watching a television show, she realizes she may have OCD. She looks it up on the Internet, and finds the following facts:

- One to 3 percent of the population has OCD.
- In concordance studies, MZ twins both have OCD 80 to 87 percent of the time, while DZ twins both have it 47 to 50 percent of the time.
- Between 6.7 and 15 percent of first-degree relatives of young people with OCD also have symptoms.
- In OCD, levels of serotonin are low in some brain regions.

Elaine concludes that her condition is largely inherited, and is relieved that it is not due to something she has consciously done. She consults a physician, who prescribes an antidepressant. Elaine's symptoms improve somewhat.

a. Do you agree with Elaine's conclusion that OCD is mostly genetically influenced? How does the evidence support or refute this view?

b. What type of antidepressant drug might alleviate symptoms of OCD?

c. Suggest a type of study to investigate possible environmental factors that may contribute to the development of OCD.

17. A study from 2001 found a considerable increase in incidence of schizophrenia among the offspring of 60 women who had rubella while pregnant. In 2003, a study found that the offspring of 20,000 women who had influenza in early to mid-pregnancy during the 1950s and 1960s, confirmed from blood samples, had an incidence of schizophrenia 3 times that of matched controls whose mothers did not have the flu.

How can these prenatal infections be linked to increased risk of schizophrenia, since the viruses that cause rubella and flu are different?

18. In experiments with mice that have immune system genes that promote extra sensitivity to mercury, researchers exposed the mice, as well as mice with wild type genes, to four common vaccines given to children. The vaccines contained a mercury-based preservative, thimerosal, that many parents of children with autism claim caused the autism. The exposed sensitive mice displayed autistic-like behaviors and showed brain changes. As a result, the media reported a link between autism and vaccines.

What questions would you ask the researchers to determine whether they can extend their experimental results to people?

Learn to apply the skills of a genetic counselor with an additional case found in the *Case Workbook in Human Genetics.*

Alcoholism

CHAPTER

9

DNA Structure and Replication

CHAPTER CONTENTS

SICK BIRDS AND SNEEZING PIGS: FORESHADOWING A FLU PANDEMIC?

An unusual set of circumstances and RNA sequences have revealed possible beginnings of an influenza pandemic (global epidemic).

Influenza outbreaks usually begin in Asia. The virus incubates in birds, then infects pigs, sometimes exchanging genes with flu viruses in pig throats. When people are near sneezing pigs, the virus can "jump" to them, causing typical flu symptoms. Rarely, a flu virus jumps from birds to people, and the disease becomes deadly. If a bird flu virus mutates so that it can move from person to person, a pandemic can arise. This happened in 1918 and 1919, killing millions.

Conditions for a pandemic exist in China, Thailand, Vietnam, and Malaysia. Millions of wild and domesticated birds have already died, and people have contracted the virus from birds. Then in September 2004 in Thailand, a young mother died of bird flu after holding her dying daughter. Reconstructing events revealed that the mother acquired the virus from her daughter and comparing the RNA sequences of the virus from mother and daughter confirmed this conclusion—possibly signaling the long-feared human-to-human transmission.

During the 1918 flu pandemic, people wore masks to prevent spread of the disease.

A genetic material must carry out two jobs: duplicate itself and control the development of the rest of the cell in a specific way, wrote Francis Crick, codiscoverer with James Watson of the three-dimensional structure of DNA in 1953. Only DNA can do this (**figure 9.1**).

9.1 Experiments Identify and Describe the Genetic Material

DNA was first described in the mid-eighteenth century, when Swiss physician and biochemist Friedrich Miescher isolated nuclei from white blood cells in pus on soiled bandages. In the nuclei, he discovered an unusual acidic substance containing nitrogen and phosphorus. He and others found it in cells from a variety of sources. Because the material resided in cell nuclei, Miescher called it *nuclein* in an 1871 paper; subsequently, it was called a nucleic acid. But few people appreciated the importance of Miescher's discovery, because at the time, the study of heredity focused on the association between inherited disease and protein.

In 1902, English physician Archibald Garrod was the first to link human inheritance and protein. He noted that people who had certain inborn errors of metabolism lacked certain enzymes. One of the first inborn errors that he described was alkaptonuria, the subject of the chapter 5 opening essay. Other researchers added evidence of a link between heredity and enzymes from other species, such as fruit flies with unusual eye colors and bread molds with nutritional deficiencies. Both organisms had absent or abnormal specific enzymes. As researchers wondered what, precisely, was the connection between enzymes and heredity, they returned to Miescher's discovery of nucleic acids.

DNA Is the Hereditary Molecule

In 1928, English microbiologist Frederick Griffith took the first step in identifying DNA as the genetic material. Griffith noticed that mice with a certain variety of pneumonia harbored one of two types of *Diplococcus pneumoniae* bacteria. Type R

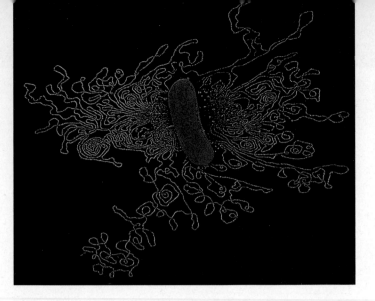

Figure 9.1 DNA is highly packaged. DNA bursts forth from this treated bacterial cell, illustrating how tightly DNA winds to fit into a single cell. The two copies of the human genome in a cell would each unravel to 1.8 meters, yet fit into a cell 6 millionths of a meter across.

bacteria are rough in texture. Type S bacteria are smooth because they are enclosed in a polysaccharide (a type of carbohydrate) capsule. Mice injected with type R bacteria did not develop pneumonia, but mice injected with type S did. The polysaccharide coat seemed to be necessary for infection.

When type S bacteria were heated—which killed them but left their DNA intact—they no longer could cause pneumonia in mice. However, when Griffith injected mice with a mixture of type R bacteria plus heat-killed type S bacteria—neither of which, alone, was deadly to the mice—the mice died of pneumonia (**figure 9.2**). Their bodies contained live type S bacteria, encased in polysaccharide. Griffith termed the apparent conversion of one bacterial type into another "transformation." How did it happen? What substance transformed type R to type S?

U.S. physicians Oswald Avery, Colin MacLeod, and Maclyn McCarty hypothesized that a nucleic acid might be the "transforming principle." They observed that treating broken-open type R bacteria with a protease—an enzyme that dismantles protein—did not prevent the transformation of a nonvirulent to a virulent strain, but treating it with deoxyribonuclease (or DNase), an enzyme that dismantles DNA only, did disrupt transformation. In 1944, they confirmed that DNA transformed the bacteria. They isolated DNA from heat-killed type S bacteria and injected it along with type R bacteria into mice (**figure 9.3**). The mice died, and their bodies contained active type S bacteria. The conclusion: DNA passed from type S bacteria into type R,

enabling the type R to manufacture the smooth coat necessary for infection.

Protein Is Not the Hereditary Molecule

Science seeks answers by eliminating explanations. For the search to identify the genetic material, researchers also had to show that protein does not transmit genetic information. To do this, in 1953, U.S. microbiologists Alfred Hershey and Martha Chase used *E. coli* bacteria infected with a virus that consisted of a protein "head" surrounding DNA. Viruses infect bacterial cells by injecting their DNA (or RNA) into them. Infected bacteria may then produce many more viruses. The viral protein coats remain outside the bacterial cells.

Researchers can analyze viruses by growing them on culture medium that contains a radioactive chemical that the viruses take up. The "labeled" nucleic acid emits radiation, which can be detected in several ways. When Hershey and Chase grew viruses with radioactive sulfur, the viral protein coats emitted radioactivity. When they repeated the experiment with radioactive phosphorus, the viral DNA emitted radioactivity. If protein is the genetic material, then the infected bacteria would have radioactive sulfur. But if DNA is the genetic material, then the bacteria would have radioactive phosphorus.

Hershey and Chase labeled two batches of virus, growing one in a medium containing radioactive sulfur (designated ^{35}S) and the other in a medium containing radioactive phosphorus (designated ^{32}P). The viruses grown on sulfur had their protein

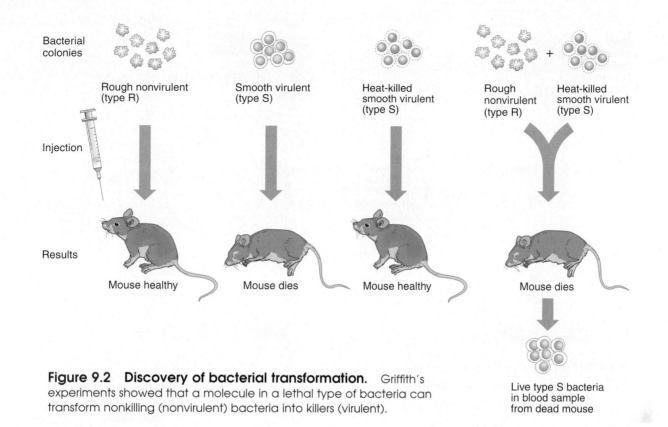

Figure 9.2   Discovery of bacterial transformation.   Griffith's experiments showed that a molecule in a lethal type of bacteria can transform nonkilling (nonvirulent) bacteria into killers (virulent).

marked, but not their DNA, because protein incorporates sulfur but DNA does not. Conversely, the viruses grown on labeled phosphorus had their DNA marked, but not their protein, because this element is found in DNA but not protein. (Miescher had noted phosphorus in DNA from soiled bandages.)

After allowing several minutes for the virus particles to bind to the bacteria and inject their DNA into them, Hershey and Chase agitated each mixture in a blender, shaking free the empty virus protein coats.

The contents of each blender were collected in test tubes, then centrifuged (spun at high speed). This settled the bacteria at the bottom of each tube because the much lighter virus coats drift down more slowly than bacteria.

At the end of the procedure, Hershey and Chase examined fractions containing the virus coats from the top of each test tube and the infected bacteria that had settled to the bottom (**figure 9.4**). In the tube containing viruses labeled with sulfur, the virus coats were radioactive, but the virus-

infected bacteria, containing viral DNA, were not. In the other tube, where the virus had incorporated radioactive phosphorus, the virus coats carried no radioactive label, but the infected bacteria were radioactive. Therefore, the part of the virus that could enter bacteria and direct them to mass produce more virus was the part that had incorporated phosphorus—the DNA. The genetic material was DNA, and not protein. (Some viruses use RNA as their genetic material, such as the influenza viruses described in the chapter opener. The RNA is copied into DNA in the infected cell.)

Discovering the Structure of DNA

In 1909, Russian-American biochemist Phoebus Levene identified the 5-carbon sugar **ribose** as part of some nucleic acids, and in 1929, he discovered a similar sugar—**deoxyribose**—in other nucleic acids. He had revealed a major chemical distinction between RNA and DNA: RNA contains ribose, and DNA contains deoxyribose.

Levene then discovered that the three parts of a nucleic acid—a sugar, a nitrogen-containing base, and a phosphorus-containing component—are present in

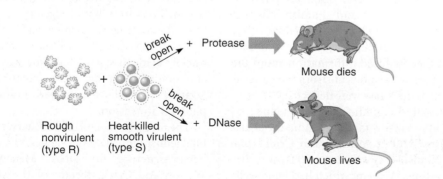

Figure 9.3   DNA is the "transforming principle."   Avery, MacLeod, and McCarty identified DNA as Griffith's transforming principle. By adding enzymes that either destroy proteins (protease) or DNA (deoxyribonuclease or DNase) to bacteria that were broken apart to release their contents, they demonstrated that DNA transforms bacteria—and that protein does not.

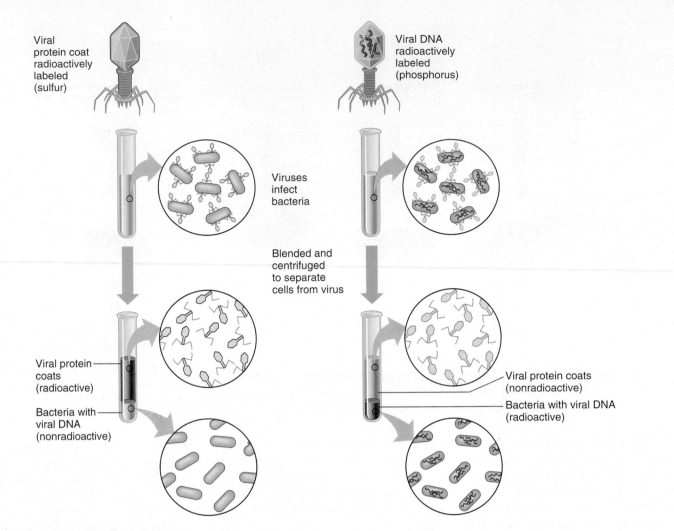

Viral protein coat radioactively labeled (sulfur)

Viral DNA radioactively labeled (phosphorus)

Viruses infect bacteria

Blended and centrifuged to separate cells from virus

Viral protein coats (radioactive)

Bacteria with viral DNA (nonradioactive)

Viral protein coats (nonradioactive)

Bacteria with viral DNA (radioactive)

Figure 9.4 **DNA is the hereditary material; protein is not.** Hershey and Chase used different radioactive molecules to distinguish the viral protein coat from the genetic material (DNA). These "blender experiments" showed that the virus transfers DNA, and not protein, to the bacterium. Therefore, DNA is the genetic material. The blender experiments used particular types of sulfur and phosphorus atoms that emit detectable radiation.

equal proportions. He deduced that a nucleic acid building block must contain one of each component. Furthermore, although the sugar and phosphate portions were always the same, the nitrogen-containing bases were of four types. Scientists at first thought that the bases were present in equal amounts, but if this were so, DNA could not encode as much information as it could if the number of each base type varied. Imagine how much less useful a written language would be if all the letters had to occur with equal frequency.

In the early 1950s, two lines of experimental evidence converged to provide the direct clues that finally revealed DNA's structure. Austrian-American biochemist Erwin Chargaff showed that DNA in several species contains equal amounts of the bases **adenine** (A) and **thymine** (T) and equal amounts of the bases **guanine** (G) and **cytosine** (C). Next, English physicist Maurice Wilkins and English chemist Rosalind Franklin bombarded DNA with X rays using a technique called X-ray diffraction, then deduced information about the structure of the molecule from the patterns in which the X rays were deflected.

Rosalind Franklin provided a clue that would prove pivotal in revealing the structure of DNA to Watson and Crick—she distinguished two forms of DNA, a dry, crystalline "A" form, which had been well-studied, and the wetter type seen in cells, the "B" form. It took her 100 hours to obtain "photo 51" of the B form in May 1952 (**figure 9.5**). Its remarkable symmetry told Franklin that the molecule was a sleek

helix, and revealed the position of the phosphates. She had long thought of DNA as a candidate for the genetic material. A lab notebook from her college days in 1939 bears the comment, "Geometrical basis for inheritance?" next to an illustration of a nucleic acid. By early 1953, she was very close to deducing the entire structure. On January 30, Wilkins showed Franklin's photo 51 to Watson.

The race was on. During February, famed biochemist Linus Pauling suggested a triple helix structure for DNA. Meanwhile, Watson and Crick, certain of the sugar-phosphate backbone largely from photo 51, turned their attention to the bases. Ironically, their eureka moment occurred not with sophisticated chemistry or crystallography, but while working with cardboard

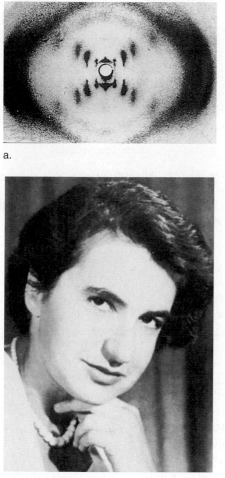

a.

b. Rosalind Franklin 1920–1958

Figure 9.5   Deciphering DNA structure. (a) Rosalind Franklin's "photo 51" of B DNA was critical to Watson and Crick's deduction of the three-dimensional structure of the molecule. The "X" in the center indicates a helix, and the darkened regions reveal symmetrically organized subunits. (b) Franklin died very young, of cancer. Willkins and Crick both died in 2004.

cutouts. On Saturday morning, February 28, Watson arrived early for a meeting with Crick. While he was waiting, he sat playing with cardboard cutouts of the four DNA bases, pairing A with A, then A with G. When he assembled A next to T, and G next to C, he noted the similar shapes, and suddenly all of the pieces fit. When Crick arrived forty minutes later, the two quickly realized they had solved the puzzle **(figure 9.6).** Watson, Crick, and Wilkins eventually received the Nobel prize. In 1958, Franklin died at the age of 37 from ovarian cancer, and the Nobel can only be awarded to a living person. In recent years, she has become a heroine for her role

Figure 9.6   Watson and Crick. Prints of this famed, if posed, photo fetched a high price when signed and sold at celebrations of DNA's fiftieth anniversary in 2003. Crick, told to point to the model, picked up a slide rule.

in deciphering the structure of DNA. **Table 9.1** summarizes some of the experiments that led to the discovery.

Key Concepts

1. DNA replicates, and contains information for protein synthesis.
2. Miescher first isolated DNA in 1869.
3. Garrod linked heredity to enzymes.
4. DNA was chemically characterized in the 1940s. Griffith identified a substance capable of transmitting infectiousness, which Avery, MacLeod, and McCarty showed was DNA.
5. Hershey and Chase confirmed that DNA, and not protein, is the genetic material.
6. Using Chargaff's discovery that the number of As equals the number of Ts, and the number of Gs equals the number of Cs, with Franklin's discovery that DNA is regular and symmetrical, Watson and Crick deciphered the structure of DNA.

9.2 DNA Structure

A **gene** is a section of a DNA molecule whose sequence of building blocks specifies the sequence of amino acids in a particular protein. The activity of the protein imparts the phenotype. The fact that different building blocks combine to form nucleic acids enables them to carry information, as the letters of an alphabet combine to form words. DNA may also encode RNA that does not specify a protein, but instead assists in protein synthesis or controls gene expression. These DNA sequences are discussed in chapters 10 and 11.

Table 9.1

The Road to the Double Helix

Investigator	Contribution	Timeline
Friedrich Miescher	Isolated nuclein in white blood cell nuclei	1869
Frederick Griffith	Transferred killing ability between types of bacteria	1928
Oswald Avery, Colin MacLeod, and Maclyn McCarty	Discovered that DNA transmits killing ability in bacteria	1940s
Alfred Hershey and Martha Chase	Determined that the part of a virus that infects and replicates is its nucleic acid and not its protein	1950
Phoebus Levene, Erwin Chargaff, Maurice Wilkins, and Rosalind Franklin	Discovered DNA components, proportions, and positions	1909–early 1950s
James Watson and Francis Crick	Elucidated DNA's three-dimensional structure	1953

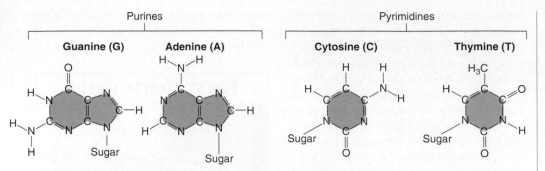

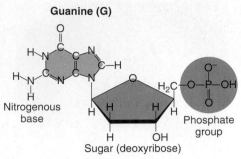

Figure 9.7 **DNA bases are the informational parts of nucleotides.** Adenine and guanine are purines, each composed of a six-membered organic ring plus a five-membered ring. Cytosine and thymine are pyrimidines, each with a single six-membered ring. (Within the molecules, C, H, N, O, and P are atoms of carbon, hydrogen, nitrogen, oxygen, and phosphorus, respectively.)

Figure 9.8 **Nucleotides.** A nucleotide of a nucleic acid consists of a 5-carbon sugar, a phosphate group, and an organic, nitrogenous base (G, A, C, or T).

Inherited traits are diverse because proteins have diverse functions. Biological proteins are extremely varied in form and function (see table 10.1). Enzymes are responsible for pea color, plant height, and the chemical reactions of metabolism. Collagen and elastin support connective tissues, and filaments of actin and myosin slide past each other to contract muscle. Hemoglobin transports oxygen, and antibodies protect against infection. Malfunctioning or inactive proteins, which reflect genetic defects, can devastate health. Most of the amino acids that assemble into proteins ultimately come from the diet; the body synthesizes the others.

The structure of DNA is easiest to understand if we begin with the smallest components. A single building block of DNA is a **nucleotide.** It consists of one deoxyribose sugar, one phosphate group (a phosphorus atom bonded to four oxygen atoms), and one nitrogenous base. **Figure 9.7** shows the chemical structures of the four types of bases, and **figure 9.8** shows one of them as part of a nucleotide. Adenine (A) and guanine (G) are **purines,** which have a two-ring structure. Cytosine (C) and thymine (T) are **pyrimidines,** which have a single-ring structure. Reading 9.1 explains how clues in DNA base sequences helped investigators solve a mystery of history.

Nucleotides join into long chains when chemical bonds form between the deoxyribose sugars and the phosphates. This creates a continuous **sugar-phosphate backbone (figure 9.9).** Two such chains of nucleotides align head-to-toe, as **figure 9.10a** depicts. M. C. Escher's drawing of hands in figure 9.10b resembles the spatial

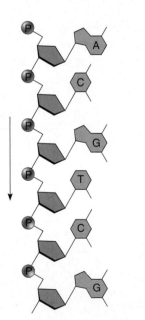

Figure 9.9 **A chain of nucleotides.**
A single DNA strand consists of a chain of nucleotides that forms when the deoxyribose sugars (green) and phosphates (yellow) bond to create a sugar-phosphate backbone. The bases A, C, G, and T are blue.

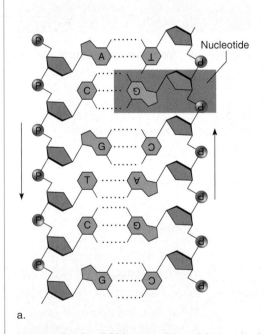

a.

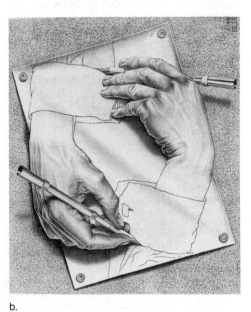

b.

Figure 9.10 **DNA consists of two chains of nucleotides.** **(a)** The nitrogenous bases of one strand are held to the nitrogenous bases of the second strand by hydrogen bonds (dotted lines). Note that the sugars point in opposite directions—that is, the strands are antiparallel. **(b)** Artist M. C. Escher captured the essence of antiparallelism in his depiction of hands.

DNA Makes History

One night in July 1918, Tsar Nicholas II of Russia and his family met gruesome deaths at the hands of Bolsheviks in a Ural mountain town called Ekaterinburg (**figure 1**). Captors led the tsar, tsarina, three of their daughters, the family physician, and three servants to a cellar and shot them, bayoneting those whose diamond jewelry deflected the bullets. The executioners then stripped the bodies and loaded them onto a truck, planning to hurl them down a mine shaft. But the truck broke down, and the killers instead placed the bodies in a shallow grave, then used sulfuric acid to damage the bodies and mask their identities.

In another July—many years later, in 1991—two Russian amateur historians found the grave. Because they knew that the royal family had spent its last night in Ekaterinburg, they alerted the government that they might have unearthed the long-sought bodies of the Romanov family. An official forensic examination soon determined that the skeletons represented nine individuals. The sizes of the skeletons indicated that three were children, and the porcelain, platinum, and gold in some of the teeth suggested royalty. Unfortunately, the acid had so destroyed the facial bones that some conventional forensic tests were not feasible. But one type of evidence survived—DNA. Thanks to DNA amplification technology, researchers obtained enough genetic material to solve the mystery.

British researchers examined DNA from cells in the skeletal remains. DNA sequences specific to the Y chromosome enabled the investigators to distinguish males from females. Then the genetic material of mitochondria, inherited from mothers only, established one woman as the mother of the children.

But a mother, her children, and companions were not necessarily a royal family. The researchers had to connect the skeletons to known relatives of Tsar Nicholas II. To do so, they again turned to DNA. However, an inherited quirk proved, at first, to be quite confusing.

Figure 1   DNA profiling sheds light on history.   DNA analysis identified the remains of the murdered Romanovs—and revealed an interesting genetic quirk.

The challenge in proving that the male remains with fancy dental work were once Tsar Nicholas II centered around nucleotide position 16169 of a mitochondrial gene whose sequence is highly variable among individuals. About 70 percent of the bone cells examined from the remains had cytosine (C) at this position, while the remainder had thymine (T). Skeptics at first suspected contamination or a laboratory error, but when the odd result was repeated, researchers realized that this historical case had revealed a genetic phenomenon called heteroplasmy. The bone cells of this man apparently harbored two populations of mitochondria, one type with C at this position, the other with T.

The DNA of a living blood relative of the tsar, Countess Xenia Cheremeteff-Sfiri, had only T at nucleotide site 16169. Xenia is the great-granddaughter of Tsar Nicholas II's sister. However, mitochondrial DNA from Xenia and the murdered man matched at every other site. DNA of another living relative, the Duke of Fife, the great-grandson of Nicholas's maternal aunt, matched Xenia at

the famed 16169 site. A closer relative, Nicholas's nephew Tikhon Kulikovsky, refused to lend his DNA, citing anger at the British for not assisting the tsar's family during the Bolshevik revolution.

But the story wasn't over. In yet another July, in 1994, researchers would finally solve the mystery.

Attention turned to Nicholas's brother, Grand Duke of Russia Georgij Romanov. In 1899, Georgij had died at age 28 of tuberculosis. His body was exhumed in July 1994, and researchers sequenced the troublesome mitochondrial gene in bone cells from his leg. They found a match! Georgij's mitochondrial DNA had the same variable site as the man murdered in Siberia, who was, therefore, Tsar Nicholas II. The researchers calculated the probability that the remains are truly those of the tsar, rather than resembling Georgij's unusual DNA sequence by chance, as 130 million to 1. At least part of the murdered Russian royal family can finally rest in peace, thanks to DNA analysis—the bodies of the two youngest children, Alexis and Anastasia, were not found.

relationship of the two strands of the DNA double helix.

The opposing orientation of the two nucleotide chains in a DNA molecule is called **antiparallelism.** It derives from the structure of the sugar-phosphate backbone. Antiparallelism becomes evident when the carbons of the sugars are assigned numbers to indicate their positions in the molecule (**figure 9.11**). The carbons are numbered from 1 to 5, starting with the first carbon moving clockwise from the oxygen in each sugar in **figure 9.12.** One chain runs from the 5 carbon (top of the figure) to the 3 carbon, but the chain aligned with it runs from the 3 to the 5 carbon. These ends are called 5′ ("5 prime") and 3′ ("3 prime").

The symmetrical DNA double helix forms when nucleotides containing A pair with those containing T, and nucleotides containing G pair with those carrying C. Because purines have two rings and pyrimidines one, the consistent pairing of a purine with a pyrimidine ensures that the double helix has the same width throughout, as Watson discovered using cardboard cut-outs. These specific purine-pyrimidine couples are called **complementary base pairs.** Chemical attractions called hydrogen bonds hold the base pairs together. Two hydrogen bonds join A and T, and three hydrogen bonds join G and C, as **figure 9.13** shows. Finally, DNA forms a double helix when the antiparallel, base-paired strands twist about one another in a regular fashion. The double-stranded, helical structure of DNA gives it great strength—50 times the strength of single-stranded DNA, which would not form a helix. The many negative charges of the phosphate groups on the outside of the molecule attract positively charged DNA binding proteins, whose interactions are critical to using genetic information.

DNA molecules are incredibly long. The DNA of the smallest human chromosome, if stretched out, would be 14 millimeters long. But it is packed into a chromosome that, during cell division, is only 2 micrometers long. This means that the DNA molecule must fold so tightly that its compacted length shrinks by a factor of 7000:

$$\left(\frac{14 \times 10^{-3} \text{ meters}}{2 \times 10^{-6} \text{ meters}} \right)$$

Various types of proteins compress the DNA without damaging or tangling it.

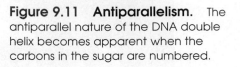

Figure 9.11 Antiparallelism. The antiparallel nature of the DNA double helix becomes apparent when the carbons in the sugar are numbered.

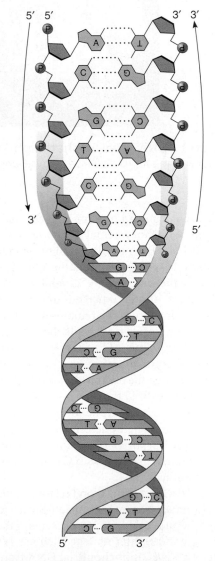

Figure 9.12 DNA is directional. Antiparallelism in a DNA molecule arises from the orientation of the deoxyribose sugars. One half of the double helix runs in a 5′ to 3′ direction, and the other half runs in a 3′ to 5′ direction.

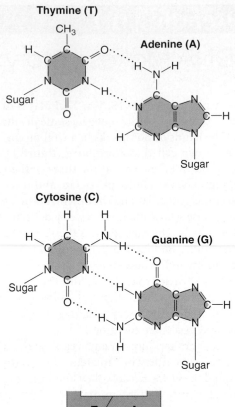

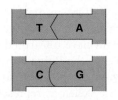

Figure 9.13 DNA base pairs. The key to the constant width of the DNA double helix is the pairing of purines with pyrimidines. Two hydrogen bonds join adenine and thymine; three hydrogen bonds link cytosine and guanine.

Scaffold proteins form frameworks that guide DNA strands. Then, the DNA coils around proteins called **histones,** forming a beads-on-a-string-like structure. The bead part is called a **nucleosome.** It is a little like wrapping a very long, thin piece of thread around your fingers, to keep it from unraveling and tangling. DNA wraps at several levels, until it is compacted into a chromosome (**figure 9.14**). Specifically, a nucleosome forms around packets of eight histone proteins (a pair of each of four types). A fifth type anchors nucleosomes to short "linker" regions of DNA, which then tighten the nucleosomes into fibers 30 nanometers (nm) in diameter. As a result, at any given time, only small portions of the DNA double helix are exposed. Chemical modifi-

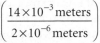

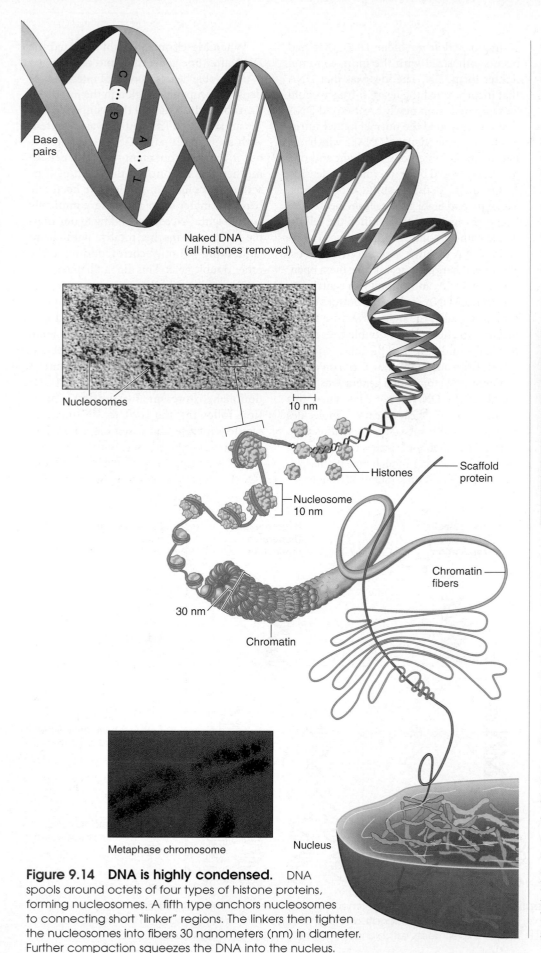

cation of the histones controls when particular DNA sequences are accessible. (This is discussed further in chapter 11.) DNA also unwinds locally when it replicates.

Altogether, the chromosome substance is called **chromatin,** which means "colored material." Chromatin is not just DNA; it is about 30 percent histone proteins, 30 percent DNA binding proteins, 30 percent DNA, and 10 percent RNA. Points along the chromatin attach it, in great loops, to nuclear matrix material on the inner face of the nuclear membrane. Without the proteins that are part of chromatin, it is unlikely that DNA would be biologically useful.

Key Concepts

1. The DNA double helix's backbone is alternating deoxyribose and phosphate held together by complementary pairs of A-T and G-C bases. A and G are purines; T and C are pyrimidines.
2. The DNA double helix is antiparallel, its strands running in an opposite head-to-toe manner.
3. DNA winds tightly about histone proteins, forming nucleosomes, which in turn wind tighter to form chromatin.

9.3 DNA Replication—Maintaining Genetic Information

As soon as Watson and Crick deciphered the structure of DNA, its mechanism for replication became obvious. They ended their report on the structure of DNA with the statement, *It has not escaped our notice that the specific pairing we have postulated immediately suggests a possible copying mechanism for the genetic material.*

Replication Is Semiconservative

Watson and Crick envisioned the two halves of the DNA double helix unwinding and separating, exposing unpaired bases that would attract their complements. Two double helices would thus form from one. This route to replication is called **semiconservative,** because each new DNA double helix conserves half of the original. But separating the

Base pairs

Naked DNA (all histones removed)

Nucleosomes

10 nm

Histones

Scaffold protein

Nucleosome 10 nm

30 nm

Chromatin fibers

Chromatin

Metaphase chromosome

Nucleus

Figure 9.14   DNA is highly condensed. DNA spools around octets of four types of histone proteins, forming nucleosomes. A fifth type anchors nucleosomes to connecting short "linker" regions. The linkers then tighten the nucleosomes into fibers 30 nanometers (nm) in diameter. Further compaction squeezes the DNA into the nucleus.

long strands posed a huge physical challenge, a little like having to keep two pieces of thread the length of a football field from tangling!

Some researchers suggested that DNA might replicate in any of three possible ways: semiconservative; conservative, with one double helix specifying creation of a second double helix; or dispersive, with a double helix shattering into pieces that would join with newly synthesized DNA pieces to form two molecules (**figure 9.15**).

An experimental approach to reveal how DNA replicates was first suggested in 1941, when English geneticist J. B. S. Haldane wrote, *How can one distinguish between model and copy? Perhaps you could use heavy nitrogen atoms in the food supplied to your cell, hoping that the "copy" genes would contain it while the models did not.*

In 1957, two young researchers, Matthew Meselson and Franklin Stahl, tried Haldane's experiment using bacteria. Their experiments beautifully illustrated scientific inquiry, because their evidence not only supported one hypothesis, but disproved the other two.

Meselson and Stahl labeled DNA newly synthesized by bacteria with heavy nitrogen (^{15}N) in the media. The DNA could then be distinguished from older DNA that had been synthesized with the more common lighter form, ^{14}N. The idea was that DNA that incorporated the heavy nitrogen could be separated from newly synthesized DNA that incorporated the normal lighter nitrogen by its greater density. DNA in which one half of the double helix was light and one half heavy would be of intermediate density.

In their density shift experiments, Meselson and Stahl grew cells on media with heavy nitrogen and then shifted the cells to media with light nitrogen. They traced replicating DNA through several cell divisions. The researchers grew cells, broke them open, extracted DNA, and spun it in a centrifuge. The heavier DNA sank to the bottom of the centrifuge tube, the light DNA rose to the top, and the heavy-light double helices settled in the middle area of the tube.

Meselson and Stahl grew *E. coli* on media containing ^{15}N for several generations, making all of the DNA heavy. They knew this because only "heavy-heavy" molecules appeared in the tube after centrifugation. They then shifted the bacteria to media containing ^{14}N, allowing enough time for the bacteria to divide only once (about 30 minutes).

When Meselson and Stahl collected the DNA after one generation and centrifuged it, the double helices were all of intermediate density, occupying a region in the middle of the tube, indicating that they contained half ^{14}N and half ^{15}N. This pattern was consistent with semiconservative DNA replication— but it was also consistent with a dispersive mechanism. In contrast, the result of conservative replication would have been one band of material in the tube completely labeled with ^{15}N, corresponding to one double helix, and another totally "light" band containing ^{14}N only, corresponding to the other double helix. This did not happen.

To definitively distinguish among the three routes to DNA replication, supporting the semiconservative mode and disproving the others, Meselson and Stahl extended the experiment one more generation. If the semiconservative mechanism held up, each hybrid (half ^{14}N and half ^{15}N) double helix present after the first generation following the shift to ^{14}N medium would separate and assemble a new half from bases labeled only with ^{14}N. This would produce two double helices with one ^{15}N (heavy) and one ^{14}N (light) chain, plus

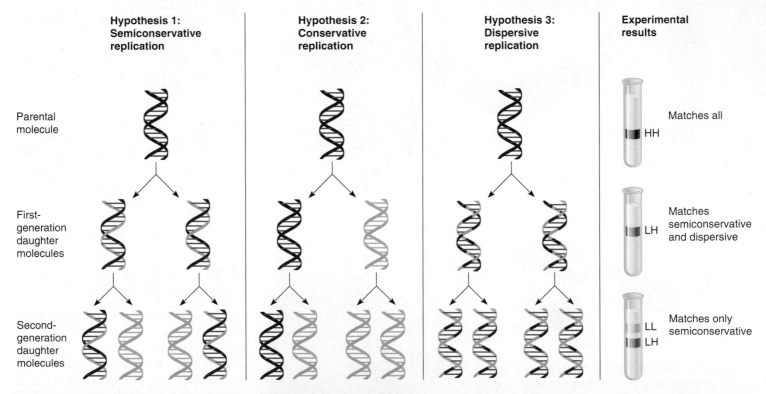

Figure 9.15 Three models for DNA replication. Density shift experiments distinguished the three hypothesized mechanisms of DNA replication. DNA molecules containing light nitrogen are designated "LL" and those with heavy nitrogen, "HH." Molecules containing both isotopes are designated "LH." These experiments established that DNA replication is semiconservative.

two double helices containing only ^{14}N. The tube would have one heavy-light band and one light-light band. This is indeed what Meselson and Stahl saw.

The conservative mechanism would have yielded two bands in the tube in the third generation, indicating three completely light double helices for every completely heavy one. The third generation for the dispersive model would have been a single large band, somewhat higher than the second-generation band because additional ^{14}N would have been randomly incorporated.

Steps of DNA Replication

After experiments demonstrated the semiconservative nature of DNA replication, the next challenge was to decipher the steps of the process.

When DNA replicates, it unwinds, breaks, builds a new nucleotide chain, and mends (**figure 9.16**). Enzymes called helicases unwind and hold apart replicating DNA so that other enzymes can guide the assembly of a new DNA strand. Helicases can also repair errors in replicated DNA.

Human DNA replicates about 50 bases per second. To get the job done, a human chromosome replicates simultaneously at hundreds of points along its length, and the pieces join. A site where DNA is locally opened, resembling a fork, is called a **replication fork.**

DNA replication begins when a helicase breaks the hydrogen bonds that connect a base pair (**figure 9.17**). Binding proteins hold the two strands apart. Another enzyme, primase, then attracts complementary RNA nucleotides to build a short piece of RNA, called an RNA primer, at the start of each segment of DNA to be replicated. The RNA primer is required because the major replication enzyme, **DNA polymerase** (DNAP), can only add bases to an existing nucleic acid strand. (A polymerase is an enzyme that builds a polymer, which is a chain of chemical building blocks.) Next, the RNA primer attracts DNAP, which brings in DNA nucleotides complementary to the exposed bases on the parental strand; this strand serves as a mold, or template. New bases are added one at a time, starting at the RNA primer. The new DNA strand grows as hydrogen bonds form between the complementary bases.

DNAP works directionally, adding new nucleotides to the exposed 3′ end of the sugar in the growing strand. Overall, replication proceeds in a 5′ to 3′ direction, because this is the only chemical configuration in which DNAP can add bases. How can the growing fork proceed in one direction, when both parental strands must be replicated? The answer is that on at least one strand, replication is discontinuous. It is accomplished in small pieces from the inner part of the fork outward, in a pattern similar to backstitching. **Ligase** is an enzyme that then seals the sugar-phosphate backbones of the pieces, building the new strand. These pieces, up to 150 nucleotides long, are called Okazaki fragments, after their discoverer (see figure 9.17).

DNA polymerase also "proofreads" as it goes, excising mismatched bases and inserting correct ones, and removes the RNA primer and replaces it with the correct DNA bases. Finally, ligases seal the sugar-phosphate backbone. Ligase comes from a Latin word meaning "to tie."

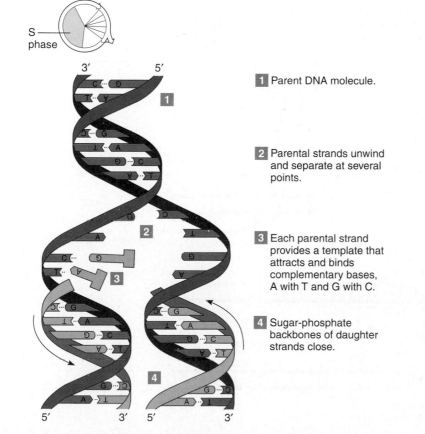

S phase

1 Parent DNA molecule.

2 Parental strands unwind and separate at several points.

3 Each parental strand provides a template that attracts and binds complementary bases, A with T and G with C.

4 Sugar-phosphate backbones of daughter strands close.

Figure 9.16 Overview of DNA replication.

RELATING THE CONCEPTS

Explain how nucleic acid replication is part of the sequence of events that occurs as avian (bird) flu causes epidemics in humans.

Key Concepts

1. Experiments that followed the distribution of labeled DNA showed that DNA replication is semiconservative, not conservative or dispersive.
2. Enzymes orchestrate DNA replication.
3. DNA replication occurs simultaneously at several points on each chromosome, and the pieces join.
4. At each initiation site, primase directs synthesis of a short RNA primer, which DNA eventually replaces. DNA polymerase adds complementary bases to the RNA primer. Finally, ligase joins the sugar-phosphate backbone.
5. DNA is synthesized in a 5′ to 3′ direction, discontinuously on one strand.

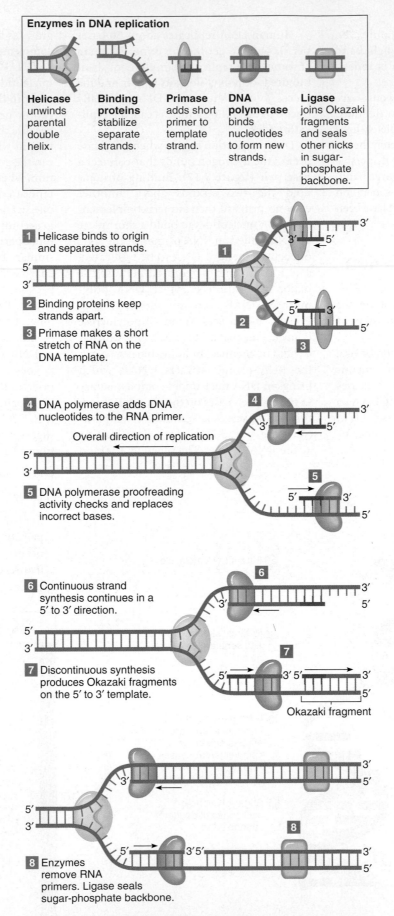

Enzymes in DNA replication

Helicase unwinds parental double helix.

Binding proteins stabilize separate strands.

Primase adds short primer to template strand.

DNA polymerase binds nucleotides to form new strands.

Ligase joins Okazaki fragments and seals other nicks in sugar-phosphate backbone.

1 Helicase binds to origin and separates strands.

2 Binding proteins keep strands apart.

3 Primase makes a short stretch of RNA on the DNA template.

4 DNA polymerase adds DNA nucleotides to the RNA primer.

Overall direction of replication

5 DNA polymerase proofreading activity checks and replaces incorrect bases.

6 Continuous strand synthesis continues in a 5′ to 3′ direction.

7 Discontinuous synthesis produces Okazaki fragments on the 5′ to 3′ template.

Okazaki fragment

8 Enzymes remove RNA primers. Ligase seals sugar-phosphate backbone.

Figure 9.17   DNA replication takes many steps.

Summary

9.1 Experiments Identify and Describe the Genetic Material

1. DNA encodes information that the cell uses to synthesize protein. DNA can also replicate, passing on its information.

2. Many experimenters described DNA as the hereditary material. Miescher identified DNA in white blood cell nuclei. Garrod connected heredity to enzyme abnormalities. Griffith identified a "transforming principle" that transmitted infectiousness in pneumonia-causing bacteria; Avery, MacLeod, and McCarty discovered that the transforming principle is DNA; and Hershey and Chase confirmed that the genetic material is DNA and not protein.

3. Levene described the three components of a DNA building block and found that they appear in DNA in equal amounts. Chargaff discovered that the amount of **adenine** (A) equals the amount of **thymine** (T), and the amount of **guanine** (G) equals that of **cytosine** (C). A and G are **pyrimidines**; C and T are **purines.** Rosalind Franklin showed that the molecule is a certain type of helix. Watson and Crick deduced DNA's structure.

9.2 DNA Structure

4. A **nucleotide** is a DNA building block. It consists of a **deoxyribose,** a phosphate, and a nitrogenous base.

5. The rungs of the DNA double helix consist of hydrogen-bonded **complementary base pairs** (A with T, and C with G). The rails are chains of alternating sugars and phosphates that run **antiparallel** to each other. DNA is highly coiled, and complexed with protein to form **chromatin.**

9.3 DNA Replication—Maintaining Genetic Information

6. Meselson and Stahl demonstrated the **semiconservative** nature of DNA replication with density shift experiments.

7. During replication, the DNA unwinds locally at several sites. **Replication forks** form as hydrogen bonds break between base pairs. **Primase** builds short RNA primers, which DNA sequences eventually replace. Next, **DNA polymerase** fills in DNA bases, and **ligase** seals the sugar-phosphate backbone.

8. Replication proceeds in a 5′ to 3′ direction, so the process must be discontinuous in short stretches on one strand.

Review Questions

1. DNA specifies and regulates the cell's synthesis of protein. If a cell contains all the genetic material required to carry out protein synthesis, why must its DNA be replicated?

2. List the components of a nucleotide.

3. How does a purine differ from a pyrimidine?

4. Why would a DNA structure in which each base type could form hydrogen bonds with any of the other three base types not produce a molecule that is easily replicated?

5. What part of the DNA molecule encodes information?

6. Explain how DNA is a directional molecule in a chemical sense.

7. Match the experiment described in the left column to a concept it illustrates in the right column (more than one answer may be possible).

   1. Density shift experiments

   2. Discovery of an acidic substance that includes nitrogen and phosphorus on dirty bandages

   3. "Blender experiments" that showed that the part of a virus that infects bacteria contains phosphorus, but not sulfur

   4. Determination that DNA contains equal amounts of guanine and cytosine, and of adenine and thymine

   5. Discovery that bacteria can transfer a "factor" that transforms a harmless strain into a lethal one

   a. DNA is the hereditary material

   b. Complementary base pairing is part of DNA structure and maintains a symmetrical double helix

   c. Identification of nuclein

   d. DNA, not protein, is the hereditary material

   e. DNA replication is semiconservative, not conservative or dispersive

8. Place the following enzymes in the order in which they begin to function in DNA replication.

   ligase   primase

   exonuclease   helicase

   DNA polymerase

9. How can incredibly long DNA molecules fit into a cell's nucleus?

10. Place in increasing size order:

    nucleosome

    histone protein

    chromatin

11. How are very long strands of DNA replicated without twisting into a huge tangle?

12. List the steps in DNA replication.

13. Why must DNA be replicated continuously as well as discontinuously?

14. How does RNA participate in DNA replication?

15. Describe two experiments that supported one hypothesis while also disproving another.

Applied Questions

1. In Bloom syndrome, ligase malfunctions. As a result, replication forks move too slowly. Why?

2. DNA contains the information that a cell uses to synthesize a particular protein. How do proteins assist in DNA replication?

3. A person with deficient or abnormal ligase may have an increased cancer risk and chromosome breaks that cannot heal. The person is, nevertheless, alive. Why are there no people who lack DNA polymerase?

4. Write the sequence of a strand of DNA replicated from each of the following base sequences:

   a. T C G A G A A T C T C G A T T
   b. C C G T A T A G C C G G T A C
   c. A T C G G A T C G C T A C T G

5. Which do you think was the more far-reaching accomplishment, determining the structure of DNA, or sequencing the human genome? State a reason for your answer.

6. Several people have suggested storing DNA from all of Earth's species in a repository on the moon, to serve as, in the words of one researcher, "a mini Noah's Ark for repopulating the Earth after a catastrophe."

   a. Suggest a way that DNA might be stored in the moon repository.

   b. What techniques would be needed to grow organisms from stored DNA? (Hint: see chapter 3.)

   c. Would storing computers with genome sequences be a viable alternative to storing DNA? Cite a reason for your answer.

Web Activities

Visit the Online Learning Center (OLC) at www.mhhe.com/lewisgenetics7 to complete the following web activities. After you have logged onto the Student Edition of the OLC, select **chapter 9** and **Web Activities.** There you will find links to the websites needed to complete the activities.

7. Visit the Cystic Fibrosis Mutation Database website. Select twenty contiguous bases of the sequence for the cystic fibrosis gene and write the complementary sequence.

Case Studies and Research Results

8. A very expensive diamond necklace was ripped off the neck of a celebrity who had borrowed it to wear for the opening of her new film. She noticed right away, but the robber had already escaped. Unfortunately for him, he had a cold, and had dribbled a few drops of nasal secretion onto the back of the celebrity's neck—which a nearby police officer had the presence of mind to sample. In it were cells from the perpetrator's nose lining. Analysis of mitochondrial DNA, however, revealed two different DNA sequences. Detective Stabler concluded that there were two robbers, but the celebrity insisted that she had seen only one man run away. What is another explanation for finding two types of mitochondrial DNA?

Learn to apply the skills of a genetic counselor with this additional case found in the *Case Workbook in Human Genetics:*

DNA replication

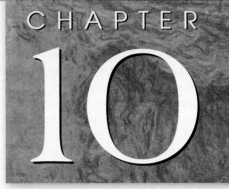

CHAPTER 10

Gene Action: From DNA to Protein

CHAPTER CONTENTS

THE IMPORTANCE OF PROPERLY FOLDED PROTEIN

For his first year of life, Julien F. seemed like any child—chatty, curious, and constantly moving and exploring. But by seventeen months, Julien had stopped talking and his inability to stand unassisted alarmed his pediatrician, who also detected poor muscle tone.

Julien continued to sleep more and speak less. At Children's National Medical Center in Washington, D.C., he was finally diagnosed with GM1 gangliosidosis (OMIM 230500).

Julien had a lysosomal storage disease. His cells lacked an enzyme, beta-galactosidase, that normally breaks down a lipid that is part of plasma membranes, particularly the neurons of the brain and cells in the liver, spleen, and kidneys. Without the enzyme, the lipid builds up. It pours out of the lysosome and accumulates in the endoplasmic reticulum, impairing ability to correct misfolded proteins. In addition to the buildup of lipid on brain cells, abnormally folded proteins trigger an "unfolded protein response" that leads to cell death.

Julien had the juvenile form of the condition. By age five, he was blind and could not communicate or move. He had several seizures a day, and his body had stiffened, even though his parents moved him often. Julien's only movement was to smile during dreams. He passed away in November 2000.

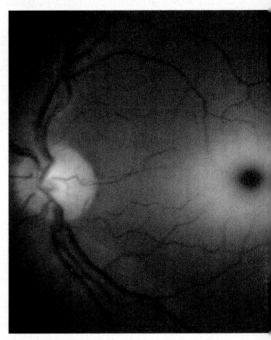

A "cherry red spot" in the eye indicates the biochemical buildup that signals a lysosomal storage disease.

DNA replication preserves genetic information by giving each new cell a complete set of operating instructions. A cell uses some of the information to manufacture proteins, which have a great variety of functions (**table 10.1**). To use the genetic information in the nucleus, the process of **transcription** first copies a gene into an RNA molecule that is complementary to one strand of the DNA double helix. The copy is taken out of the nucleus and into the cytoplasm. There, the process of **translation** uses the information in three types of RNA to manufacture a protein by aligning and joining specified amino acids. Finally, the protein must fold into a specific three-dimensional form in order to function.

Cells replicate their DNA only during S phase of the cell cycle. In contrast, transcription and translation occur continuously, except during M phase. These processes supply the proteins essential for life, as well as those that give a cell its specialized characteristics. This chapter considers the steps of transcription and translation, and the next chapter discusses control of these processes.

10.1 Transcription

Watson and Crick, shortly after publishing their structure of DNA in 1953, described the relationship between nucleic acids and proteins as a directional flow of information called the "central dogma" (**figure 10.1**). As Francis Crick explained in 1957, *"The specificity of a piece of nucleic acid is expressed solely by the sequence of its bases, and this sequence is a code for the amino acid sequence of a particular protein."* This statement inspired more than a decade of intense research to identify the participants in protein synthesis and discover how they interact. At center stage: RNA.

RNA is the bridge between gene and protein. RNA and DNA share an intimate relationship, as **figure 10.2** depicts. RNA is synthesized against (is complementary to) one strand of the double helix, called the **template strand**, with the assistance of an enzyme, **RNA polymerase.** The other strand of the DNA double helix is the **coding strand.**

Table 10.1

Protein Diversity in the Human Body

Protein	Function
Actin, myosin, dystrophin	Muscle contraction
Antibodies, antigens, cytokines	Immunity
Carbohydrases, lipases, proteases, nucleases	Digestion (digestive enzymes)
Casein	Milk protein
Collagen, elastin, fibrillin	Connective tissue
Colony stimulating factors, erythropoietin	Blood cell formation
DNA and RNA polymerase	DNA replication, gene expression
Ferritin	Iron transport in blood
Fibrin, thrombin	Blood clotting
Growth factors, kinases, cyclins	Cell division
Hemoglobin, myoglobin	Oxygen transport
Insulin, glucagon	Control of blood glucose level
Keratin	Hair structure
Tubulin, actin	Cell movements
Tumor suppressors	Cancer prevention

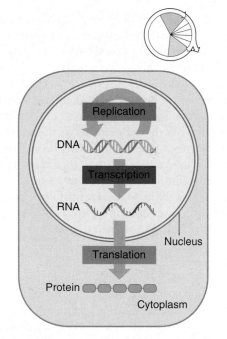

Figure 10.1 DNA to RNA to protein.
The central dogma of molecular biology states that information stored in DNA is copied to RNA (transcription), which is used to assemble proteins (translation). DNA replication perpetuates genetic information. This figure repeats within the chapter, with the part under discussion highlighted.

RNA Structure and Types

RNA and DNA have similarities and differences (**figure 10.3** and **table 10.2**). Both are nucleic acids, consisting of sequences of nitrogen-containing bases joined by sugar-phosphate backbones. However, RNA is usually single-stranded, whereas DNA is double-stranded. Also, RNA has the pyrimidine base **uracil** in place of DNA's thymine. As their names imply, RNA nucleotides include the sugar ribose, rather than DNA's deoxyribose. Functionally, DNA stores genetic information, whereas RNA controls how that information is used.

As RNA is synthesized along DNA, it folds into a three-dimensional shape, or **conformation,** that is determined by complementary base pairing within the same RNA molecule. For example, a sequence of AAU-UUCC might hydrogen bond to a sequence of UUAAAGG elsewhere in the molecule. These shapes are very important for RNA's functioning. The three major types of RNA are messenger RNA, ribosomal RNA, and transfer RNA (**table 10.3**). Table 11.2 describes other types of RNA molecules.

Messenger RNA (mRNA) carries the information that specifies a particular

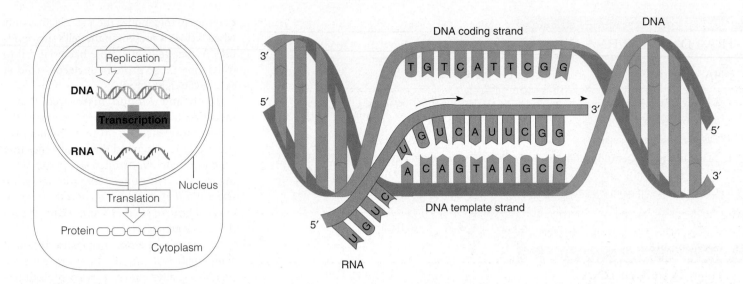

Figure 10.2 The relationship among RNA, the DNA template strand, and the DNA coding strand.   The RNA sequence is complementary to that of the DNA template strand and so is the same sequence as the DNA coding strand, with uracil (U) in place of thymine (T).

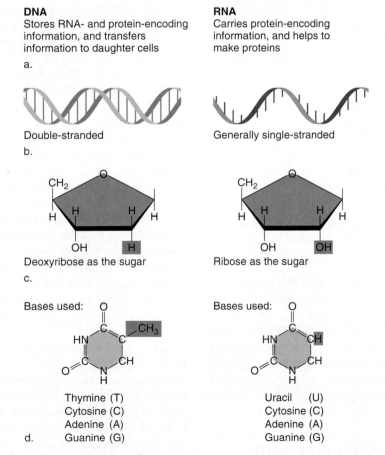

DNA
Stores RNA- and protein-encoding information, and transfers information to daughter cells
a.

Double-stranded
b.

Deoxyribose as the sugar
c.

Bases used:

Thymine (T)
Cytosine (C)
Adenine (A)
d.   Guanine (G)

RNA
Carries protein-encoding information, and helps to make proteins

Generally single-stranded

Ribose as the sugar

Bases used:

Uracil   (U)
Cytosine (C)
Adenine (A)
Guanine (G)

Figure 10.3   DNA and RNA differences.   **(a)** DNA is double-stranded; RNA is usually single-stranded **(b)**. DNA nucleotides include deoxyribose, whereas RNA nucleotides have ribose **(c)**. Finally, DNA nucleotides include the pyrimidine thymine, whereas RNA has uracil **(d)**.

protein. Each three mRNA bases in a row form a genetic code word, or **codon,** that specifies a certain amino acid. Because genes vary in length, so do mature mRNA molecules. Most such mRNAs are 500 to 4,500 bases long. Specialized cells can carry out particular functions because they "express" certain subsets of genes—that is, they produce certain mRNA molecules, or transcripts. The information in the transcripts is then used to manufacture the encoded proteins. A muscle cell, for example, has many mRNAs that specify the contractile proteins actin and myosin, whereas a skin cell contains many mRNAs that specify the scaly protein keratin.

To use the information in an mRNA sequence, a cell requires two other major classes of RNA. **Ribosomal RNA** (rRNA) molecules range from 100 to nearly 3,000 nucleotides long. Ribosomal RNAs associate with certain proteins to form a ribosome. Recall from chapter 2 that a ribosome is a structural support for protein synthesis (**figure 10.4**).

A ribosome has two subunits that are separate in the cytoplasm but join at the site of initiation of protein synthesis. The larger ribosomal subunit has three types of rRNA molecules, and the small subunit has one. Ribosomal RNA, however, is more than a structural support. Certain rRNAs catalyze

Table 10.2

How DNA and RNA Differ

DNA	RNA
1. Usually double-stranded	1. Usually single-stranded
2. Thymine as a base	2. Uracil as a base
3. Deoxyribose as the sugar	3. Ribose as the sugar
4. Maintains protein-encoding information	4. Carries protein-encoding information and controls how information is used
5. Cannot function as an enzyme	5. Can function as an enzyme

Table 10.3

Major Types of RNA

Type of RNA	Size (number of nucleotides)	Function
mRNA	500–4,500+	Encodes amino acid sequence
rRNA	100–3,000	Associates with proteins to form ribosomes, which structurally support and catalyze protein synthesis
tRNA	75–80	Transports specific amino acids to the ribosome for protein synthesis.

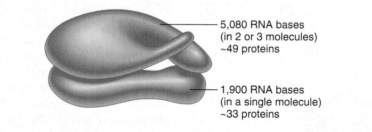

5,080 RNA bases
(in 2 or 3 molecules)
~49 proteins

1,900 RNA bases
(in a single molecule)
~33 proteins

Figure 10.4 The ribosome. A ribosome from a eukaryotic cell, shown here, has two subunits; together, they consist of 82 proteins and four rRNA molecules.

the formation of the peptide bonds between amino acids. Such an RNA with enzymatic function is called a ribozyme. Other rRNAs help to align the ribosome and mRNA.

The third major type of RNA molecule, **transfer RNA** (tRNA), binds an mRNA codon at one end and a specific amino acid at the other. A tRNA molecule is only 75 to 80 nucleotides long. Some of its bases weakly bond with each other, folding the tRNA into loops in a characteristic cloverleaf shape (**figure 10.5**). One loop of the tRNA has three bases in a row that form the **anticodon,** which is complementary to an mRNA codon. The end of the tRNA opposite the anticodon strongly bonds to a spe-

cific amino acid. A tRNA with a particular anticodon sequence always carries the same amino acid. (Organisms have 20 types of amino acids.) For example, a tRNA with the anticodon sequence GAA always picks up the amino acid phenylalanine. Special enzymes attach amino acids to tRNAs that bear the appropriate anticodons (**figure 10.6**).

Transcription Factors

Study of the control of gene expression began in 1961, when French biologists François Jacob and Jacques Monod described the remarkable ability of *E. coli* to produce the

enzymes to metabolize the sugar lactose only when lactose is present in the cell's surroundings. What "tells" a simple bacterial cell to transcribe those proteins it needs—and at exactly the right time?

Jacob and Monod discovered that a modified form of lactose "turned on" the genes whose encoded proteins break down the sugar. They named the set of genes that are coordinately controlled an operon, writing in 1961, *The genome contains not only a series of blueprints, but a coordinated program of protein synthesis and means of controlling its execution.*

In bacteria, operons turn transcription of a few genes on or off. In more complex organisms, different cell types express different subsets of genes. To manage this, groups of proteins called **transcription factors** come together, forming an apparatus that binds DNA at certain sequences and initiates transcription at specific sites on chromosomes. (Bacterial regulatory proteins are also called transcription factors.) The transcription factors, activated by signals from outside the cell such as hormones and growth factors, set the stage for transcription by forming a pocket for RNA polymerase—the enzyme that builds an RNA chain.

Several types of transcription factors interact to transcribe a gene. Because transcription factors are proteins, they themselves are gene-encoded. The DNA sequences that transcription factors bind may be located near the genes they control, or as far as 40,000 bases away. DNA may form loops so that the genes encoding proteins that interact come near each other for transcription. Other proteins in the nucleus may help bring certain genes and their associated transcription factors in close proximity, much as books on a specialized topic might be grouped together in a library for easier access.

Many transcription factors have regions in common, called motifs, that fold into similar conformations. These motifs enable the transcription factor to bind DNA. They have very colorful names, such as "helix-turn-helix," "zinc fingers," and "leucine zippers," that reflect their distinctive shapes.

The human genome encodes at least 2,000 transcription factors. Overall, they control gene expression and link the genome to the environment. For example, lack of oxygen, such as from choking or

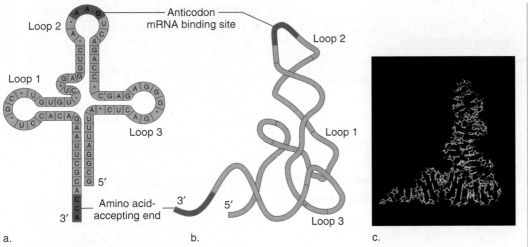

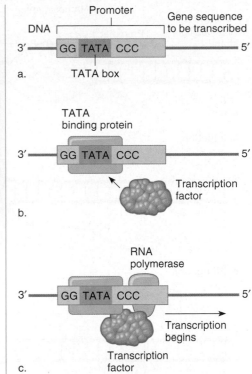

a.　　　　　　　　　　b.　　　　　　　　c.

Figure 10.5　Transfer RNA.　(a) Certain nucleotide bases within a tRNA hydrogen bond with each other to give the molecule a "cloverleaf" conformation that can be represented in two dimensions. The darker bases at the top form the anticodon, the sequence that binds a complementary mRNA codon. Each tRNA terminates with the sequence CCA, where a particular amino acid covalently bonds. Three-dimensional representations of a tRNA **(b)** and **(c)** depict the loops that interact with the ribosome.

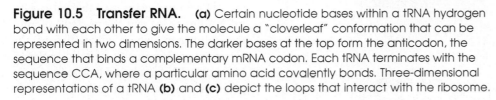

Figure 10.7　Setting the stage for transcription to begin.　(a) Proteins that initiate transcription recognize specific sequences in the promoter region of a gene. **(b)** A binding protein recognizes the TATA region and binds to the DNA. This allows other transcription factors to bind. **(c)** The bound transcription factors form a pocket that allows RNA polymerase to bind and begin making RNA.

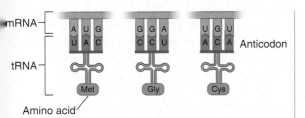

Figure 10.6　A tRNA with a particular anticodon sequence always binds the same type of amino acid.

smoking, sends signals that activate transcription factors to turn on dozens of genes that enable cells to handle the stress of low-oxygen conditions.

Mutations in transcription factor genes can have wide-ranging effects, because the factors control many genes. If the mutation is in all cells, it may affect several organ systems.

Steps of Transcription

How do transcription factors and RNA polymerase (RNAP) "know" where to bind to DNA to begin transcribing a specific gene? Transcription factors and RNA polymerase are attracted to a **promoter,** a special sequence that signals the start of the gene. Signals from outside the cell alter the chromatin structure in a way that exposes the promoter of a gene whose transcription is required under the particular conditions. **Figure 10.7** shows a simplified view of transcription factor binding, which sets up a site called a preinitiation complex to receive RNA polymerase. The first transcription factor to bind, called a TATA binding protein, is attracted to a DNA sequence called a TATA box—the base sequence TATA surrounded by long stretches of G and C. Once the first transcription factor binds, it attracts others in groups, and finally RNA polymerase joins the complex, binding just in front of the start of the gene sequence. The coming together of these components is transcription initiation.

Complementary base pairing underlies transcription. In the next stage, transcription elongation, enzymes unwind the DNA double helix locally, and RNA nucleotides bond with exposed complementary bases on the DNA template strand (see figure 10.2). RNA polymerase adds the RNA nucleotides in the sequence the DNA specifies, moving along the DNA strand in a 3' to 5' direction, synthesizing the RNA molecule in a 5' to 3' direction. A terminator sequence in the DNA indicates where the gene's RNA-encoding region ends. When this spot is reached, transcription termina-

tion occurs (**figure 10.8**). A typical rate of transcription in humans is 20 bases per second.

RNA is transcribed using only the gene's template strand. The coding strand is so-called because its sequence is identical to that of the RNA, except with thymine (T) in place of uracil (U). Several RNAs may be transcribed from the same DNA template strand simultaneously (**figure 10.9**). Since mRNA is relatively short-lived, with about half of it degraded every 10 minutes, a cell must constantly transcribe certain genes to maintain supplies of essential proteins. However, different genes on the same chromosome may be transcribed from different strands of the double helix.

To determine the sequence of RNA bases transcribed from a gene, write the RNA bases that are complementary to the template DNA strand, using uracil opposite

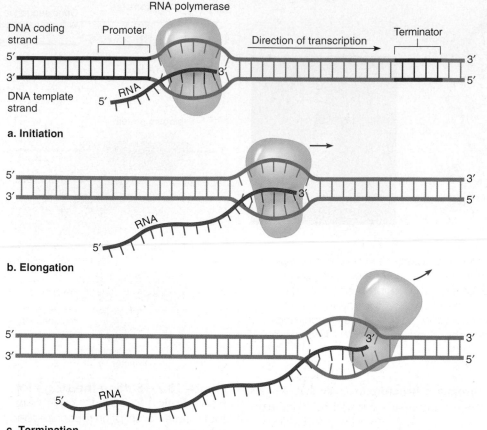

Figure 10.8 Transcription of RNA from DNA. Transcription occurs in three stages: initiation, elongation, and termination. Initiation is the control point that determines which genes are transcribed. RNA nucleotides are added during elongation. A terminator sequence in the gene signals the end of transcription.

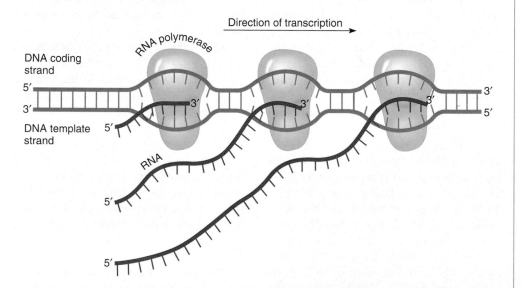

Figure 10.9 Many identical copies of RNA are simultaneously transcribed.
Usually 100 or more DNA bases lie between RNA polymerases.

adenine. For example, a DNA template strand that has the sequence

CCTAGCTAC

is transcribed into RNA with the sequence

GGAUCGAUG

and the coding DNA sequence is

GGATCGATG.

RNA Processing

In bacteria, RNA is translated into protein as soon as it is transcribed from DNA because a nucleus does not physically separate the two processes. In eukaryotic cells, mRNA must first exit the nucleus to enter the cytoplasm, where organelles of the secretory pathway assist protein synthesis. Messenger RNA is altered before it participates in protein synthesis in these more complex cells. (Some protein synthesis may occur in the nucleus, although the evidence is controversial.)

First, after mRNA is transcribed, a short sequence of modified nucleotides, called a cap, is added to the 5′ end of the molecule. The cap consists of a backwardly inserted guanine (G), which attracts an enzyme that adds methyl groups (CH_3) to the G and one or two adjacent nucleotides. This methylated cap is a recognition site for protein synthesis. At the 3′ end, a special polymerase adds about 200 adenines, forming a "poly A tail." The poly A tail is necessary for protein synthesis to begin, and may also stabilize the mRNA so that it stays intact.

Further changes occur to the capped, poly A tailed mRNA before it is translated into protein. Parts of mRNAs called **introns** (short for "intervening sequences") that were transcribed are removed. The ends of the remaining molecule are spliced together before the mRNA is translated. The parts of mRNA that are translated are called **exons** (**figure 10.10**).

Once introns are spliced out, enzymes check the remaining mRNA in a process called proofreading. Messenger RNAs that are too short or too long may be held in the nucleus. Proofreading also monitors tRNAs, ensuring that they assume the correct conformation.

Prior to intron removal, the mRNA is called pre-mRNA. Introns may function as

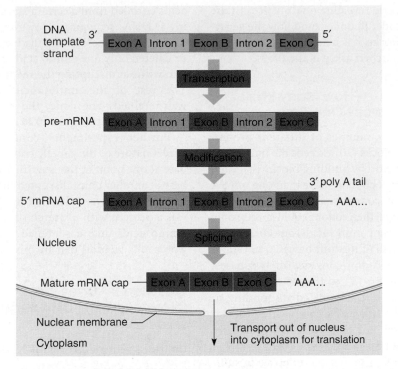

Figure 10.10 Messenger RNA processing—the maturing of the message.
Several steps process pre-mRNA into mature mRNA. First, a large region of DNA containing the gene is transcribed. Then a modified nucleotide cap and poly A tail are added, and introns are spliced out. Finally, the intact, mature mRNA is sent out of the nucleus.

ble, may be one way that our genome maximized its informational content over evolutionary time. Introns may have enabled exons to combine in different ways. The fact that some disease-causing mutations disrupt intron/exon splice sites suggests that this cutting and pasting of gene parts is essential to health, a subject discussed in chapter 12.

For some genes, mRNA is cut to different sizes in different tissues. This may explain how cell types use the same protein in slightly different ways in different tissues. Consider apolipoprotein B (apo B), which transports fats. In the small intestine, the mRNA that encodes apo B is short, and the protein binds and carries dietary fat. In the liver, however, the mRNA is not shortened, and the longer encoded protein transports fats manufactured in the liver, which do not come from food.

ribozymes, associating with proteins to form small nuclear ribonucleoproteins (snRNPs), or "snurps." Four snurps form a structure called a spliceosome that cuts introns out and attaches exons to form the mature mRNA that exits the nucleus.

Introns range in size from 65 to 10,000 or more bases; the average intron is 3,365 bases. The average exon, in contrast, is 145 bases long. The number, size, and organization of introns vary from gene to gene. The coding portion of the average human gene is 1,340 bases, whereas the average total size of a gene is 27,000 bases. The dystrophin gene is 2,500,000 bases, but its corresponding mRNA sequence is only 14,000 bases! The gene contains 80 introns.

The discovery of introns surprised geneticists, who likened gene structure to a sentence in which all of the information contributes to the meaning. In the 1980s, some geneticists called introns "junk DNA." The prevalence of introns, however, suggests that they must have a function, or they would not have persisted through evolution. Said one speaker at a genomics confer-

ence, "Anyone who still thinks that introns have no function, please volunteer to have them removed, so we can see what they do." He had no takers.

The human genome harbors more and larger introns than the genomes of our closest relatives, and introns have complicated analysis of our genome sequence. Computer programs must probe raw DNA sequences for telltale signs of a protein-encoding gene, and then distinguish exons from introns. For example, a specific short sequence that indicates the start of a protein-encoding gene is called an open reading frame. Another clue is that dinucleotide repeats, such as CGCGCG, flank some introns, forming splice sites that signal the spliceosome where to cut and paste the RNA.

We do not know why some genes have introns and some do not. Introns may be ancient genes that have lost their original function, or they may be remnants of the DNA of viruses that once infected the cell. Combining genes in discrete pieces, which an intron/exon organization makes possi-

Key Concepts

1. RNA is single-stranded, contains uracil instead of thymine and ribose instead of deoxyribose, and has different functions than DNA.
2. Messenger RNA transmits information to build proteins. Each three mRNA bases in a row forms a codon that specifies a particular amino acid.
3. Ribosomal RNA and proteins form ribosomes, which physically support protein synthesis and help catalyze bonding between amino acids.
4. Transfer RNAs connect mRNA codons to amino acids.
5. Bacterial operons are simple gene control systems. In more complex organisms, cascades of transcription factors control gene expression.
6. RNA polymerase inserts complementary RNA bases opposite the DNA template strand.
7. Messenger RNA (mRNA) gains a modified nucleotide cap and a poly A tail.
8. Introns are transcribed and cut out, and exons are reattached. Introns are common and large in human genes.
9. Certain genes are transcribed into different-sized RNAs in different cell types.

10.2 Translation of a Protein

Transcription copies the information in DNA into the complementary language of RNA. The next step is translating mRNA into the precise sequence of amino acids that forms a protein. This is possible because particular mRNA codons correspond to particular amino acids (**figure 10.11**). This correspondence between the chemical languages of mRNA and protein is the **genetic code.**

Francis Crick hypothesized that an "adaptor" molecule would enable the RNA message to attract and link amino acids into proteins. He envisioned *20 different kinds of adaptor molecule, one for each amino acid, and 20 different enzymes to join the amino acids to their adaptors.* In the 1960s, researchers deciphered the genetic code, determining which mRNA codons correspond to which amino acids.

The news media often mentions the recent deciphering of the "human genetic code." This term is incorrect. The genetic code is not unique to humans, and it was cracked decades ago. The code is the correspondence between nucleic acid triplet and amino acid, not the sequence itself.

Deciphering the Genetic Code

The researchers who deciphered the genetic code used logic and experiments. More recently, annotation of the human genome sequence has confirmed and extended the earlier work, revealing new nuances in the genetic code. To understand how the genetic code works, it is helpful to ask the questions researchers asked in the 1960s.

Question 1—How Many RNA Bases Specify One Amino Acid?

Because the number of different protein building blocks (20) exceeds the number of different mRNA building blocks (4), each codon must contain more than one mRNA base. If a codon consisted of only one mRNA base, then codons could specify only four different amino acids, one corresponding to each of the four bases: A, C, G, and U. If each codon consisted of two bases, then only 16 (4^2) different amino acids could be specified, one corresponding to each of the 16 possible combinations of two RNA bases. If a codon consisted of three bases, then the genetic code could specify as many as 64 (4^3) different amino acids, sufficient to encode the 20 different amino acids that make up proteins. Therefore, the minimum number of bases in a codon is three.

Francis Crick and his coworkers conducted experiments on a type of virus called T4 that confirmed the triplet nature of the genetic code. They exposed the virus to chemicals that add or remove one, two, or three bases, and examined a viral gene with a sequence and protein product they knew. Altering the sequence by one or two bases produced a different amino acid sequence, because it disrupted the **reading frame,** which is the sequence of amino acids encoded from a certain starting point in a DNA sequence. However, adding or deleting three contiguous bases added or deleted only one amino acid in the protein without disrupting the reading frame. The rest of the amino acid sequence was retained. The code, the researchers deduced, is triplet (**figure 10.12**).

Further experiments confirmed the triplet nature of the genetic code. Adding a base at one point in the gene and deleting a base at another point disrupted the reading frame only between these sites. The result was a protein with a stretch of the wrong amino acids, like a sentence with a few words in the middle that are misspelled.

Question 2—Does the Information in a DNA Sequence Overlap?

Consider the hypothetical mRNA sequence:

AUGCCCAAG

If the genetic code is triplet and a DNA sequence is "read" in a nonoverlapping manner, then this sequence has only three codons and specifies three amino acids:

AUGCCCAAG
AUG (methionine)
    CCC (proline)
        AAG (lysine)

If the DNA sequence is overlapping, however, the sequence specifies seven codons:

AUGCCCAAG
AUG (methionine)
 UGC (cysteine)
  GCC (alanine)
   CCC (proline)
    CCA (proline)
     CAA (glutamine)
      AAG (lysine)

An overlapping DNA sequence seems to pack maximal information into a limited number of bases, but this would constrain protein structure because certain amino acids must always follow certain others. For example, AUG would always be followed by an amino acid whose codon begins with UG. This does not happen. Therefore, the protein-encoding DNA sequence is not overlapping.

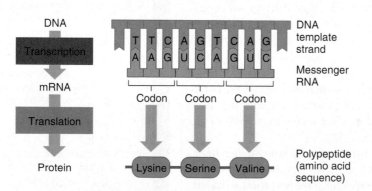

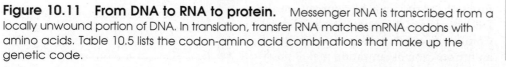

Figure 10.11 From DNA to RNA to protein. Messenger RNA is transcribed from a locally unwound portion of DNA. In translation, transfer RNA matches mRNA codons with amino acids. Table 10.5 lists the codon-amino acid combinations that make up the genetic code.

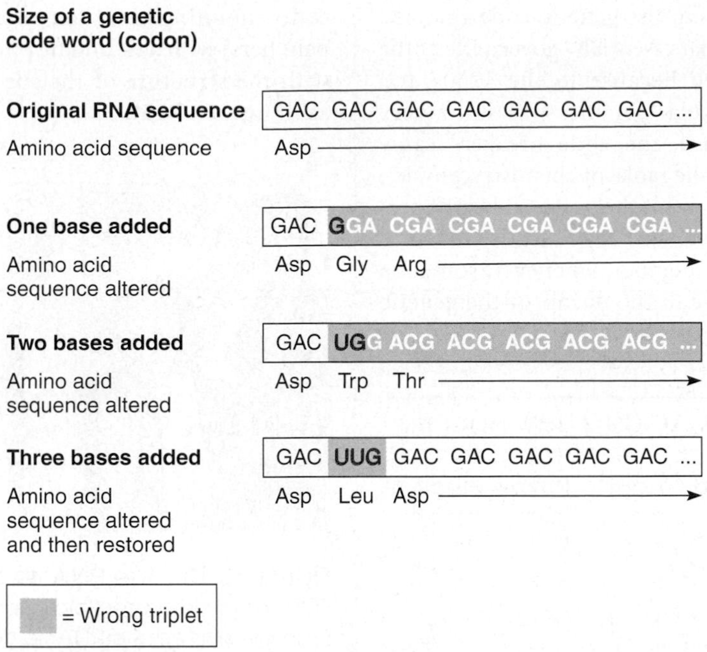

Size of a genetic code word (codon)

Original RNA sequence

GAC GAC GAC GAC GAC GAC GAC ...

Amino acid sequence   Asp ⟶

One base added

GAC **G**GA CGA CGA CGA CGA CGA ...

Amino acid sequence altered   Asp Gly Arg ⟶

Two bases added

GAC **UG**G ACG ACG ACG ACG ACG ...

Amino acid sequence altered   Asp Trp Thr ⟶

Three bases added

GAC **UUG** GAC GAC GAC GAC GAC ...

Amino acid sequence altered and then restored   Asp Leu Asp ⟶

▓ = Wrong triplet

Figure 10.12 Three at a time. Adding or deleting one or two nucleotides in a DNA sequence results in a frame shift that disrupts the encoded amino acid sequence. However, adding or deleting three bases does not disrupt the reading frame. Therefore, the code is triplet. This is a simplified representation of the Crick experiment. First a G is added, then a U, then another U so that the altered codons are GGA, UGG, and UUG, corresponding to the amino acids glycine, tryptophan, and leucine. (Table 10.5 includes the full names of the amino acids abbreviated here.)

Question 3—Can mRNA Codons Specify Anything Other Than Amino Acids?

Chemical analysis eventually showed that the genetic code includes directions for starting and stopping translation. The codon AUG signals "start," and the codons UGA, UAA, and UAG signify "stop." Another form of "punctuation," a short sequence of bases at the start of each mRNA, enables the mRNA to hydrogen bond with rRNA in a ribosome. It is called a leader sequence.

Question 4—Do All Species Use the Same Genetic Code?

All species use the same mRNA codons to specify the same amino acids. This universality of the genetic code is evidence that all life evolved from a common ancestor. No other mechanism as efficient at directing cellular activities has emerged and persisted.

The only known exceptions to the "universality" of the genetic code are a few codons in mitochondria and in certain single-celled eukaryotes (ciliated protozoa). These deviations may be tolerated because they do not affect the major repositories of DNA. The mitochondrial genome is small, and the affected ciliated protozoa have a second, smaller nucleus that houses some genes with one or two alternate codon-amino acid associations. In both cases, the major DNA sites adhere to the universal genetic code. Some types of single-celled organisms translate a stop codon into a twenty-first type of amino acid. Overall, however, the genetic code is considered universal—which has implications for genetic technologies.

The ability of mRNA from one species to be translated in the cell of another made recombinant DNA technology possible, in which bacteria manufacture proteins normally made in the human body. Chapter 19 explains the role of biotechnology in developing protein-based drugs.

Question 5—Which Codons Specify Which Amino Acids?

In 1961, Marshall Nirenberg and Heinrich Matthaei at the National Institute of Health began deciphering which codons specify which amino acids, using a precise and logical series of experiments. First they synthesized mRNA molecules in the laboratory. Then they added them to test tubes that contained all the chemicals and structures needed for translation, extracted from *E. coli* cells. Which amino acid would each synthetic RNA specify?

The first synthetic mRNA they made had the sequence UUUUUU. . . . In the test tube, this was translated into a peptide consisting entirely of one amino acid type: phenylalanine. This was the first entry in the genetic code dictionary: The codon UUU specifies the amino acid phenylalanine. The next experiments revealed that AAA codes for the amino acid lysine and CCC for proline. (GGG was unstable.)

Other researchers synthesized chains of alternating bases. Synthetic mRNA of sequence AUAUAU . . . introduced codons AUA and UAU. When translated, the mRNA yielded an amino acid sequence of alternating isoleucines and tyrosines. But was AUA the code for isoleucine and UAU for tyrosine, or vice versa? Another experiment with a more complex sequence answered the question.

The mRNA UUUAUAUUUAUA, when translated from the first U of a UUU, encoded alternating phenylalanine and isoleucine. Because the first experiment had showed that UUU codes for phenylalanine, the researchers deduced that AUA must code for isoleucine. If AUA codes for isoleucine, then UAU must code for tyrosine (**table 10.4**).

By the end of the 1960s, researchers had deciphered the entire genetic code (**table 10.5**). Sixty of the possible 64 codons specify particular amino acids, three indicate "stop," and one encodes both the amino acid methionine and "start." This means that some amino acids are specified by more than one codon. For example, both UUU and UUC encode phenylalanine. Different codons that specify the same amino acid are termed synonymous, just as synonyms are words with the same meaning. The genetic code is said to be degenerate because most amino acids are not uniquely specified.

Synonymous codons often differ from one another by the base in the third position. The corresponding base of a tRNA's anticodon is called the "wobble" position because it can bind to more than one type of base in synonymous codons. The degeneracy of the genetic code protects against mutation, because changes in the DNA that substitute a synonymous codon do not alter the protein's amino acid sequence.

Deciphering the genetic code revealed the "rules" that essentially govern life at the cellular level. Because in the 1950s and 1960s molecular genetics was still a very young science, the code breakers came largely from the ranks of chemistry, physics, and math. Some of the more exuberant personalities organized an "RNA tie club" and inducted a member whenever someone added a piece to the puzzle of the genetic code, anointing him (there were no prominent hers) with a tie and tie pin emblazoned with the structure of the specified amino acid (figure 10.13).

Figure 10.13 The RNA tie club. In 1953, physicist-turned-biologist George Gamow started the RNA tie club, to "solve the riddle of RNA structure and to understand the way it builds proteins." The club had 20 members, and each received a tie and tie pin bearing the name of the particular amino acid he had worked on. Francis Crick (upper left) was tyrosine; James Watson (lower right) was proline.

Table 10.4

Deciphering RNA Codons and the Amino Acids They Specify

Synthetic RNA	Encoded Amino Acid Chain	Puzzle Piece
UUUUUUUUUUUUUUUUUU	Phe-Phe-Phe-Phe-Phe-Phe	UUU = Phe
AAAAAAAAAAAAAAAAAA	Lys-Lys-Lys-Lys-Lys-Lys	AAA = Lys
GGGGGGGGGGGGGGGGGG	Gly-Gly-Gly-Gly-Gly-Gly	GGG = Gly
CCCCCCCCCCCCCCCCCC	Pro-Pro-Pro-Pro-Pro-Pro	CCC = Pro
AUAUAUAUAUAUAUAUAU	Ile-Tyr-Ile-Tyr-Ile-Tyr	AUA = Ile or Tyr
		UAU = Ile or Tyr
UUUAUAUUUAUAUUUAUA	Phe-Ile-Phe-Ile-Phe-Ile	AUA = Ile
		UAU = Tyr

Table 10.5

The Genetic Code

		Second Letter							
		U		**C**		**A**		**G**	
U	UUU	Phenylalanine (Phe)	UCU	Serine (Ser)	UAU	Tyrosine (Tyr)	UGU	Cysteine (Cys)	U
	UUC		UCC		UAC		UGC		C
	UUA	Leucine (Leu)	UCA		UAA	"stop"	UGA	"stop"	A
	UUG		UCG		UAG	"stop"	UGG	Tryptophan (Trp)	G
C	CUU	Leucine (Leu)	CCU	Proline (Pro)	CAU	Histidine (His)	CGU	Arginine (Arg)	U
	CUC		CCC		CAC		CGC		C
	CUA		CCA		CAA	Glutamine (Gln)	CGA		A
	CUG		CCG		CAG		CGG		G
A	AUU	Isoleucine (Ile)	ACU	Threonine (Thr)	AAU	Asparagine (Asn)	AGU	Serine (Ser)	U
	AUC		ACC		AAC		AGC		C
	AUA		ACA		AAA	Lysine (Lys)	AGA	Arginine (Arg)	A
	AUG	Methionine (Met) and "start"	ACG		AAG		AGG		G
G	GUU	Valine (Val)	GCU	Alanine (Ala)	GAU	Aspartic acid (Asp)	GGU	Glycine (Gly)	U
	GUC		GCC		GAC		GGC		C
	GUA		GCA		GAA	Glutamic acid (Glu)	GGA		A
	GUG		GCG		GAG		GGG		G

First Letter (left column) / Third Letter (right column)

The human genome project picked up where the genetic code experiments of the 1960s left off by identifying the DNA sequences that are transcribed into tRNAs. That is, 61 different tRNAs could theoretically exist, one for each codon that specifies an amino acid (the 64 triplets minus 3 stop codons). However, only 49 different genes encode tRNAs. This is because the same type of tRNA can detect synonymous codons that differ only in whether the wobble (third) position is U or C. The same type of tRNA, for example, binds to both UUU and UUC codons, which specify the amino acid phenylalanine. Synonymous codons ending in A or G use different tRNAs. Sequencing of other genomes reveals that some types of organisms preferentially use particular codons for amino acids specified by more than one type of codon. Researchers do not yet understand the significance, if any, of such codon usage bias.

Building a Protein

Protein synthesis requires mRNA, tRNA molecules carrying amino acids, ribosomes, energy-storing molecules such as adenosine triphosphate (ATP) and guanosine triphos-phate (GTP), and various protein factors. These pieces meet in a stage called translation initiation (**figure 10.14**).

First, the mRNA leader sequence hydrogen bonds with a short sequence of rRNA in a small ribosomal subunit. The first mRNA codon to specify an amino acid is always AUG, which attracts an initiator tRNA that carries the amino acid methionine (abbreviated *met*). This methionine signifies the start of a polypeptide. The small ribosomal subunit, the mRNA bonded to it, and the initiator tRNA with its attached methionine form the initiation complex.

To start the next stage, elongation, a large ribosomal subunit attaches to the initiation complex. The codon adjacent to the initiation codon (AUG), which is GGA in **figure 10.15,** then bonds to its complementary anticodon, which is part of a free tRNA that carries the amino acid glycine. The two amino acids (*met* and *gly* in the example), still attached to their tRNAs, align.

The part of the ribosome that holds the mRNA and tRNAs together can be described as having two sites. The positions of the sites on the ribosome remain the same with respect to each other as transla-tion proceeds, but they cover different parts of the mRNA as the ribosome moves. The P site holds the growing amino acid chain, and the A site right next to it holds the next amino acid to be added to the chain. In fig-ure 10.15, when the forming protein consists of only the first two amino acids, *met* occupies the P site and *gly* the A site.

With the help of rRNA that functions as a ribozyme, the amino acids link by forming peptide bonds. Then the first tRNA is released. It will pick up another amino acid of the same type and be used again. The ribosome and its attached mRNA are now bound to a single tRNA, with two amino acids extending from it at the P site. This is the start of a polypeptide.

Next, the ribosome moves down the mRNA by one codon. The region of the mRNA that was at the A site is thus now at the P site. A third tRNA enters, carrying its amino acid (*cys* in figure 10.15b). This third amino acid aligns with the other two and forms a peptide bond to the second amino acid in the growing chain, now extending from the P site. The tRNA attached to the second amino acid is released and recycled. The polypeptide continues to build, one

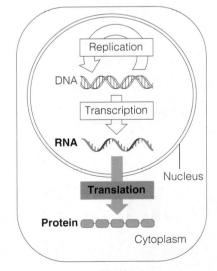

Figure 10.14 Translation begins as the initiation complex forms.
Initiation of translation brings together a small ribosomal subunit, mRNA, and an initiator tRNA, and aligns them in the proper orientation to begin translation.

Translation initiation

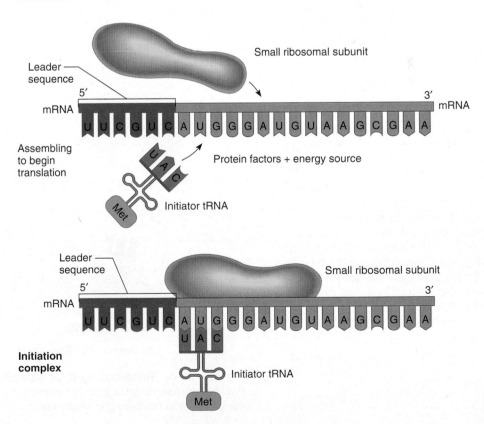

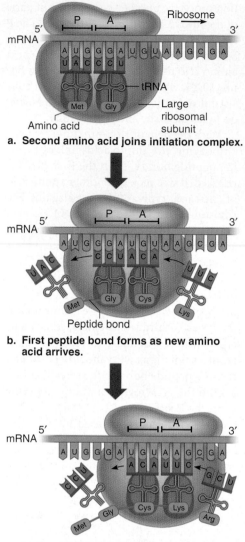

a. Second amino acid joins initiation complex.

b. First peptide bond forms as new amino acid arrives.

c. Amino acid chain extends.

Figure 10.15   Building a polypeptide.   (a) A large ribosomal subunit binds to the initiation complex, and a tRNA bearing a second amino acid (glycine, in this example) forms hydrogen bonds between its anticodon and the mRNA's second codon at the A site. The first amino acid, methionine, occupies the P site. **(b)** The methionine brought in by the first tRNA forms a peptide bond with the amino acid brought in by the second tRNA, and a third tRNA arrives, in this example carrying the amino acid cysteine. **(c)** A fourth amino acid is linked to the growing polypeptide chain, and the process continues until a termination codon is reached.

amino acid at a time. Each piece is brought in by a tRNA whose anticodon corresponds to a consecutive mRNA codon as the ribosome moves down the mRNA (figure 10.15*c*).

Elongation halts when the A site of the ribosome contains a "stop" codon (UGA, UAG, or UAA), because no tRNA molecules correspond to these codons. The last tRNA leaves the ribosome, the ribosomal subunits separate from each other and are recycled, and the new polypeptide is released (**figure 10.16**). The polypeptide is made up of a long chain of amino acids joined by peptide bonds (**figure 10.17**).

Ribosomes are critical to protein synthesis. Antibiotic drugs whose names end in "mycin" work by destroying bacterial ribosomes, which are smaller than those in eukaryotic cells.

Protein synthesis is economical. A cell can produce large amounts of a particular protein from just one or two copies of a gene. A plasma cell in the immune system, for example, manufactures 2,000 identical antibody molecules per second. To mass produce proteins at this rate, RNA, ribosomes, enzymes, and other proteins must be continually recycled. In addition, transcription always produces multiple copies of a particular mRNA, and each mRNA may be bound to dozens of ribosomes, as **figure 10.18** shows. As soon as one ribosome has moved far enough along the mRNA, another ribosome attaches. In this way, many copies of the encoded protein are made from the same mRNA.

As complex as protein synthesis is, stringing together a chain of amino acids is only a first step. The chain must fold for the protein to assume its three-dimensional form. To do this, **chaperone proteins** stabilize partially folded regions of the amino acid chain, as is discussed in the next section. Finally, certain proteins must be altered before they can function. Insulin, which is 51 amino acids long, for example, is initially translated as the polypeptide proinsulin, which is 80 amino acids long. Enzymes cut it to 51. Some proteins must have sugars attached for them to become functional, or must aggregate.

Solving A Problem

A good way to review how DNA replicates and uses genetic information is to follow molecular sequences from DNA to RNA to protein. **Figure 10.19** shows the "rules" for a sample sequence.

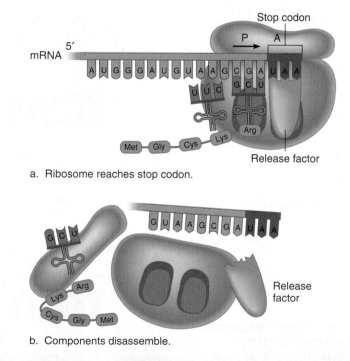

a. Ribosome reaches stop codon.

b. Components disassemble.

Figure 10.16   Terminating a polypeptide.   (a) A protein release factor binds to the stop codon, releasing the completed protein from the tRNA and **(b)** freeing all of the components of the translation complex.

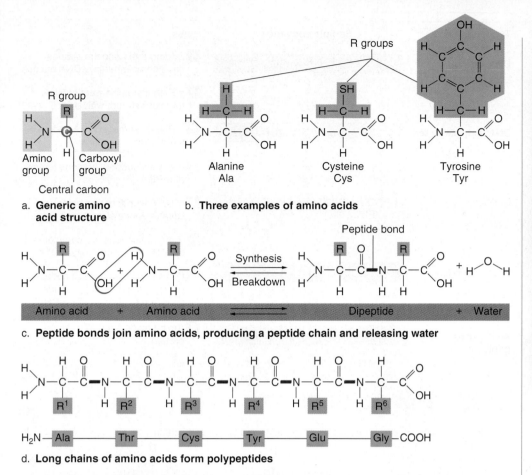

a. **Generic amino acid structure**

b. **Three examples of amino acids**

Alanine
Ala

Cysteine
Cys

Tyrosine
Tyr

R group

Amino group

Central carbon

Carboxyl group

R groups

c. **Peptide bonds join amino acids, producing a peptide chain and releasing water**

Peptide bond

Synthesis

Breakdown

Amino acid + Amino acid ⇌ Dipeptide + Water

d. **Long chains of amino acids form polypeptides**

H_2N—Ala—Thr—Cys—Tyr—Glu—Gly—COOH

Figure 10.17 Amino acids join by peptide bonds to form polypeptides.

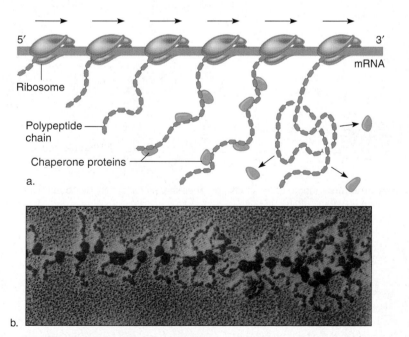

5′

3′

mRNA

Ribosome

Polypeptide chain

Chaperone proteins

a.

b.

Figure 10.18 Making multiple copies of a protein. Several ribosomes can simultaneously translate a protein from a single mRNA. **(a)** These ribosomes hold different-sized polypeptides—the closer to the end of a gene, the longer the polypeptide. Chaperone proteins help fold the polypeptide. **(b)** In the micrograph, the ribosomes on the left have just begun translation and the polypeptides are short. Further along the polypeptides are longer.

10.3 Protein Folding

Proteins must fold into one (or more) specific three-dimensional shape(s), or conformation(s), to function. This folding occurs because of attractions and repulsions between atoms. In addition, thousands of water molecules surround a growing chain of amino acids, and, because some amino acids are attracted to water and some are repelled by it, the water contorts the protein's shape. Sulfur atoms also affect conformation by bridging the two types of amino acids that contain them. Proteins fold in the endoplasmic reticulum.

The conformation of a protein is described at several levels (**figure 10.20**). The amino acid sequence of a polypeptide chain determines its **primary (1°) structure.** Chemical attractions between amino acids that are close together in the 1° structure fold the polypeptide chain into its **secondary (2°) structure,** which may form loops, coils, barrels, helices, sheets, or other distinctive shapes. (The two most common secondary structures are an alpha helix and a beta-pleated sheet.) Secondary structures wind into larger **tertiary (3°) structures** as more widely separated amino acids attract or repel in response to water molecules. Finally, proteins consisting of more than one polypeptide form a **quaternary (4°) structure.** Hemoglobin, the blood protein

Figure 10.19 Solving a Problem: from DNA to RNA to protein.
DNA specifies RNAs, which work together to align amino acids to form proteins.

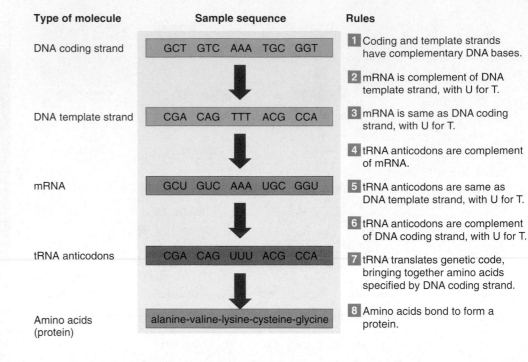

Type of molecule	Sample sequence	Rules
DNA coding strand	GCT GTC AAA TGC GGT	**1** Coding and template strands have complementary DNA bases.
DNA template strand	CGA CAG TTT ACG CCA	**2** mRNA is complement of DNA template strand, with U for T.
		3 mRNA is same as DNA coding strand, with U for T.
mRNA	GCU GUC AAA UGC GGU	**4** tRNA anticodons are complement of mRNA.
		5 tRNA anticodons are same as DNA template strand, with U for T.
		6 tRNA anticodons are complement of DNA coding strand, with U for T.
tRNA anticodons	CGA CAG UUU ACG CCA	**7** tRNA translates genetic code, bringing together amino acids specified by DNA coding strand.
Amino acids (protein)	alanine-valine-lysine-cysteine-glycine	**8** Amino acids bond to form a protein.

Figure 10.20 Four levels of protein structure.
(a) The amino acid sequence of a polypeptide forms the primary structure, while **(b)** hydrogen bonds between non-R groups create secondary structures such as helices and sheets. The tertiary structure **(c)** is formed when R groups interact, folding the polypeptide in three dimensions and forming a unique shape. **(d)** If different polypeptide units must interact to be functional, the protein forms a quaternary structure.

H_2N — Ala — Thr — Cys — Tyr — Glu — Gly — COOH

a. **Primary structure**—the sequence of amino acids in a polypeptide chain

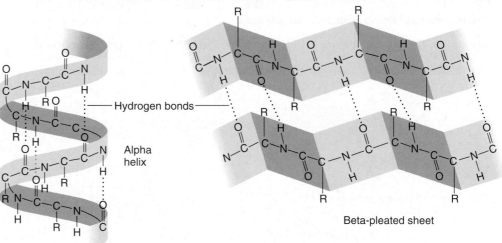

b. **Secondary structure**—loops, coils, sheets, or other shapes formed by hydrogen bonds between neighboring carboxyl and amino groups

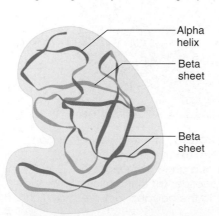

c. **Tertiary structure**—three-dimensional forms shaped by bonds between R groups, interactions between R groups and water

d. **Quaternary structure**—protein complexes formed by bonds between separate polypeptides

that carries oxygen, has four polypeptide chains (see figure 11.1). The liver protein ferritin has 20 identical polypeptides of 200 amino acids each. In contrast, the muscle protein myoglobin is a single polypeptide chain.

Protein folding begins within a minute after the amino acid chain winds away from the ribosome. A small protein might contort into its final, functional form in one quick step, taking only a few microseconds. Larger proteins may fold into a series of short-lived intermediates.

Various proteins assist in this precise folding. Chaperone proteins stabilize partially folded regions in their correct form, and prevent a protein from getting "stuck" in an intermediate form—which would be disastrous for the cell. Proteins called folding catalysts help new chemical bonds to form as the final shape arises, and folding sensor proteins monitor the accuracy of folding. Should a protein misfold, the folding sensors activate a cellular reaction to stress called the unfolded protein response. (In Julien F.'s cells, the unfolded protein response could not save the affected cells; they died by apoptosis.) The folding sensors start the response by adding phosphates to a type of translation initiation factor, which slows or even stops protein synthesis. Meanwhile, transcription speeds up for genes that encode chaperone proteins and folding catalysts, to restore proper folding.

As the cell readies to begin folding accurately, misfolded proteins are sent out of the endoplasmic reticulum. Back in the cytoplasm, the misfolded proteins are "tagged" with yet another protein, called ubiquitin. A misfolded protein bearing just one ubiquitin tag may straighten and refold correctly, but a protein with more than one tag is taken to a cellular garbage disposal called a **proteasome** (**figure 10.21**). A proteasome is a tunnel-like multi-protein structure. Through it, the protein is stretched out, chopped up, and its peptide pieces degraded into amino acids.

Misfolded proteins that are not destroyed can cause disease. Some mutations that cause cystic fibrosis, for example, prevent the encoded protein from assuming its final form and anchoring in the plasma membrane, where it normally controls the flow of chloride ions. Instead, the protein builds up in the cell.

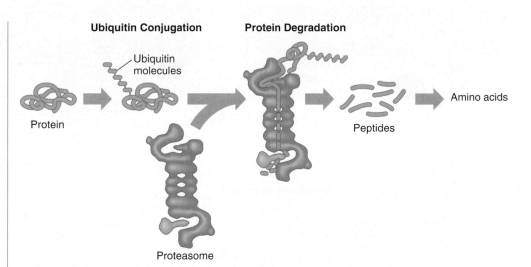

Ubiquitin Conjugation   **Protein Degradation**

Figure 10.21   Protein folding quality control. Ubiquitin binds to a misfolded protein and escorts it to a proteasome. The proteasome, which is composed of several proteins, encases the misfolded protein, straightening and dismantling it.

In several disorders that affect the brain, misfolded proteins—different proteins in different conditions—aggregate, forming masses that clog the proteasomes and block them from processing any malformed proteins. In Huntington disease, for example, extra glutamines in the protein huntingtin cause it to obstruct proteasomes, which eventually kills the cell. Misfolded proteins that clog proteasomes also occur in Alzheimer disease, amyotrophic lateral sclerosis (Lou Gehrig's disease), Parkinson disease, and Lewy body dementia. **Table 10.6** lists proteins that misfold and aggregate in these disorders. Some are discussed further in chapter 12.

Another type of protein folding disorder arises in glycoproteins called prions (pronounced *pree-ons*). Diseases called transmissible spongiform encephalopathies result when one conformation of the prion glycoprotein (PrP) is infectious, which means that it causes others to misfold like it (**figure 10.22**). Unlike other misfolded proteins that cause disease, the variant forms of prion protein have the same primary structure, but they can fold into at least eight conformations (**figure 10.23**). Transmissible spongiform encephalopathies are known in 85 types of mammals, including humans. The affected brain becomes riddled with holes,

Table 10.6

Disorders of Protein Folding and Aggregation

Disease	Misfolded Protein	OMIM (protein)
Alzheimer disease	amyloid beta precursor protein	104760
Familial amyotrophic lateral sclerosis	superoxide dismutase	147450
Huntington disease	huntingtin	143100
Parkinson disease	alpha synuclein	163890
Lewy body dementia		
Prion disorders	prion protein	176640

(All but Huntington disease are genetically heterogeneic; that is, abnormalities in different proteins cause similar syndromes.)

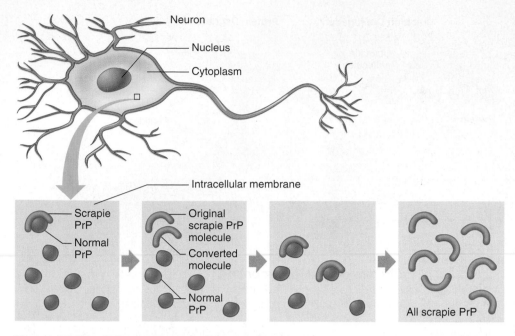

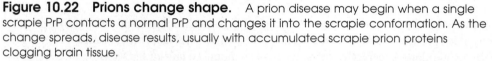

Figure 10.22 Prions change shape. A prion disease may begin when a single scrapie PrP contacts a normal PrP and changes it into the scrapie conformation. As the change spreads, disease results, usually with accumulated scrapie prion proteins clogging brain tissue.

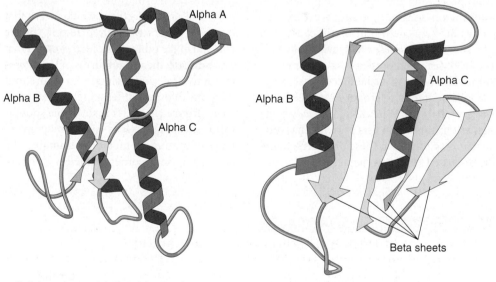

a. Cellular prion protein (noninfectious)   b. Scrapie prion protein (infectious)

Figure 10.23 One protein, multiple conformations. A cellular form of prion protein does not cause disease **(a).** The scrapie form is infectious—it converts the cellular form to produce more of itself **(b).** Infectious prions cause scrapie in sheep, bovine spongiform encephalopathy in cows, and variant Creutzfeldt-Jakob disease in humans.

resembling a sponge. Nerve cells die, and star-shaped supportive cells overgrow. Reading 10.1 describes prion disorders in humans.

The "rules" by which DNA sequences specify protein shapes are still not well understood, even as we routinely decipher the sequences of entire genomes. The straightforward linear relationship between gene and protein that emerged from the experiments of the 1960s was merely an opening chapter to the story of how a cell builds its proteins.

RELATING THE CONCEPTS

1. Describe the point in the protein folding process that was impaired in Julien F.'s cells.
2. How can a disease such as GM1 gangliosidosis be pleiotropic (cause several symptoms)?

Key Concepts

1. Protein folding begins as translation proceeds, with enzymes and chaperone proteins assisting.
2. Misfolded proteins are tagged with ubiquitin and sent through a proteasome for dismantling.
3. A protein can fold in more than one way.
4. Infectious conformations of prion proteins cause disease.

Reading 10.1

Considering Kuru

Prion diseases cause extreme weight loss and poor coordination, with other symptoms, such as dementia or relentless insomnia, reflecting the part of the brain that degenerates. These diseases are typically fatal within 18 months of symptom onset.

About 10 percent of people who suffer from prion diseases have mutations in the gene that encodes prion protein, called *PrP*. Most cases, however, are acquired. A person is exposed to prions that are in the infectious conformation, triggering conversion of the person's own normal prions. These rare illnesses were first discovered in sheep, which develop a disease called scrapie when they eat prion-infected brains from other sheep.

A prion disease affecting humans was kuru, which struck the Foré people in a remote mountainous area of New Guinea (**figure 1**). In the Foré language, *kuru* means to tremble. The disease began with wobbling legs, quickly followed by trembling hands and fingers. Gradually, the entire body became wracked with shaking. A peculiar symptom was uncontrollable laughter, leading to the nickname "laughing disease." Speech slurred and faded, thinking slowed, and after several months, the person could no longer walk or eat. Death typically came within a year.

The fact that only women and young children developed kuru at first suggested that the disease might be inherited, but D. Carleton Gajdusek, a physician who has spent much of his lifetime studying the Foré, learned that the preparation of human brain for a cannibalism ritual probably passed on the infectious prions. After the people abandoned the ritual in the 1970s, the disease gradually disappeared. Gajdusek vividly described the Foré preparation of human brains at a time when he thought the cause was viral:

> Children participated in both the butchery and the handling of cooked meat, rubbing their soiled hands in their armpits or hair, and elsewhere on their bodies. They rarely or never washed. Infection with the kuru virus was most probably through the cuts and abrasions of the skin or from nose picking, eye rubbing, or mucosal injury.

Although kuru vanished, other prion diseases surfaced. In the 1970s and 1980s, several people acquired Creutzfeldt-Jakob disease (CJD). This time, the route of transmission was either through corneal transplants, in which infectious prions entered the brain through the optic nerve, or from human growth hormone taken from cadavers and used to treat short stature in children. The most familiar prion disease is "mad cow disease" and the variant CJD it has caused in more than 120 people in the United Kingdom since 1995. People likely acquired the infectious prions by eating infected beef.

Researchers have studied the *PrP* gene in great detail, and have discovered that several specific polymorphisms (variants) that affect different sites in the protein interact in ways that make some people resistant to prion diseases, yet others highly susceptible. These mutations are discussed further in chapter 12. The persistence of the protective gene variants, some researchers say, is evidence

Figure 1 Kuru. Kuru is a prion disease that affected the Foré people of New Guinea until they gave up a cannibalism ritual that spread an infectious form of prion protein.

that cannibalism may have been common in some of our prehistoric ancestors—protected individuals survived. This hypothesis is consistent with anthropological evidence of cannibalism, such as human bite marks on human bones. The function of normal prion protein isn't known, but it resides in the plasma membranes of brain neurons.

Summary

10.1 Transcription

1. Some DNA is **transcribed** into RNA, which is then **translated** into protein.

2. RNA is transcribed from the **template strand** of DNA. The other DNA strand is called the **coding strand.**

3. RNA is a single-stranded nucleic acid similar to DNA but containing uracil and ribose rather than thymine and deoxyribose.

4. Several types of RNA participate in protein synthesis. **Messenger RNA** (mRNA) carries a protein-encoding gene's information. **Ribosomal RNA** (rRNA) associates with certain proteins to form ribosomes, which physically support protein synthesis. **Transfer RNA** (tRNA) is cloverleaf-shaped, with a three-base **anticodon** that is complementary to mRNA on one end and bonds to a particular amino acid on the other end.

5. Operons control gene expression in bacteria. In more complex organisms, **transcription factors** regulate which genes are transcribed in a particular cell type.

6. Transcription begins when transcription factors help **RNA polymerase** (RNAP) bind to a gene's **promoter.** RNAP then adds RNA nucleotides to a growing chain, in a sequence complementary to the DNA template strand.

7. After a gene is transcribed, the mRNA receives a "cap" of modified nucleotides at the 5′ end and a poly A tail at the 3′ end.

8. Many genes do not encode information in a continuous manner. After transcription, segments called **exons** are translated into protein, but segments called **introns** are removed. Introns may outnumber and outsize exons. Alternate splicing can increase protein diversity.

10.2 Translation of a Protein

9. Each three consecutive mRNA bases form a **codon** that specifies a particular amino acid. The **genetic code** is the correspondence between each codon and the amino acid it specifies. Of the 64 different possible codons, 60 specify amino acids, one specifies the amino acid methionine and "start," and three signal "stop." Because 61 codons specify the 20 amino acids, more than one type of codon may encode a single amino acid. The genetic code is nonoverlapping, triplet, universal, and degenerate.

10. In the 1960s, researchers used logic and clever experiments using synthetic RNAs to decipher the genetic code.

11. Translation requires tRNA, ribosomes, energy-storage molecules, enzymes, and protein factors. An initiation complex forms when mRNA, a small ribosomal subunit, and a tRNA carrying methionine join. The amino acid chain elongates when a large ribosomal subunit joins the small one. Next, a second tRNA binds by its anticodon to the next mRNA codon, and its amino acid bonds with the first amino acid. Transfer RNAs add more amino acids, forming a polypeptide. The ribosome moves down the mRNA as the chain grows. The P site bears the amino acid chain, and the A site holds the newest tRNA. When the ribosome reaches a "stop" codon, it falls apart into its two subunits and is released. The new polypeptide breaks free.

12. After translation, some polypeptides are cleaved, have sugars added, or aggregate. The cell uses or secretes the protein.

10.3 Protein Folding

13. A protein must fold into a particular conformation to be active and functional.

14. A protein's **primary structure** is its amino acid sequence. Its **secondary structure** forms as amino acids close in the primary structure attract one another. **Tertiary structure** appears as more widely separated amino acids attract or repel in response to water molecules. **Quaternary structure** forms when a protein consists of more than one polypeptide.

15. **Chaperone proteins** help conformation. Folding catalysts help new bonds form and folding sensor proteins oversee folding accuracy.

16. Ubiquitin attaches to misfolded proteins, and escorts them to **proteasomes** for dismantling. Protein misfolding causes disease.

17. Some proteins can fold into several conformations, some of which can cause disease.

18. At least one conformation of prion protein is infectious, causing transmissible spongiform encephalopathies.

Review Questions

1. Explain how complementary base pairing is responsible for
   a. the structure of the DNA double helix.
   b. DNA replication.
   c. transcription of RNA from DNA.
   d. the attachment of mRNA to a ribosome.
   e. codon/anticodon pairing.
   f. tRNA conformation.

2. A retrovirus has RNA as its genetic material. When it infects a cell, it uses enzymes to copy its RNA into DNA, which then integrates into the host cell's chromosome. Is this flow of genetic information consistent with the central dogma? Why or why not?

3. Genomics is highly dependent upon computer algorithms that search DNA sequences for indications of specialized functions. Explain the significance of detecting the following sequences:
   a. a promoter
   b. a sequence of 75 to 80 bases that folds into a cloverleaf shape
   c. RNAs with poly A tails

4. Many antibiotic drugs work by interfering with protein synthesis in the bacteria that cause infections. Explain how each of the

following antibiotic mechanisms disrupts genetic function in bacteria.

  a. Transfer RNAs misread mRNA codons, binding with the incorrect codon and bringing in the wrong amino acid.

  b. The first amino acid is released from the initiation complex before translation can begin.

  c. Transfer RNA cannot bind to the ribosome.

  d. Ribosomes cannot move.

  e. A tRNA picks up the wrong amino acid.

5. How is the bacterial lactose operon similar to the transcription factor response to low-oxygen conditions?

6. List the differences between RNA and DNA.

7. Where in a cell do DNA replication, transcription, and translation occur?

8. How does transcription control cell specialization?

9. How can the same mRNA codon be at an A site on a ribosome at one time, but at a P site at another time?

10. Describe the events of transcription initiation.

11. List the three major types of RNA and their functions.

12. Describe three ways RNA is altered after it is transcribed.

13. What are the components of a ribosome?

14. Why would an overlapping genetic code be restrictive?

15. How are the processes of transcription and translation economical?

16. How does the shortening of proinsulin to insulin differ from the shortening of apolipoprotein B?

17. Explain how protein misfolding conditions and illnesses that result from abnormal transcription factors might each produce many different symptoms.

18. What factors determine how a protein folds into its characteristic conformation?

19. Why would two-nucleotide codons be insufficient to encode the number of amino acids in biological proteins?

20. Cite two ways RNA helps in its own synthesis, and two ways proteins help in their own synthesis.

21. How do a protein's primary, secondary, and tertiary structures affect conformation? Which is the most important determinant of conformation?

22. How is timing important in protein folding?

Applied Questions

1. The *BRCA1* gene, when missing several bases, causes a form of breast cancer. The gene has 24 exons and 23 introns.

  a. How many splice sites does the gene contain? (A splice site is the junction of an exon and an intron.)

  b. In a woman with *BRCA1* breast cancer, an entire exon is missing, or "skipped." How many splice sites does her affected copy of the gene have?

2. List the RNA sequences that would be transcribed from the following DNA template sequences.

  a. TTACACTTGCTTGAGAGTC

  b. ACTTGGGCTATGCTCATTA

  c. GGCTGCAATAGCCGTAGAT

  d. GGAATACGTCTAGCTAGCA

3. Given the following partial mRNA sequences, reconstruct the corresponding DNA template sequences.

  a. GCUAUCUGUCAUAAAAGAGGA

  b. GUGGCGUAUUCUUUUCCGGGUAGG

  c. GAGGGAAUUCUUUCUCAACGAAGU

  d. AGGAAAACCCCUCUUAUUAUAGAU

4. List three different mRNA sequences that could encode the following amino acid sequence:

  histidine-alanine-arginine-serine-leucine-valine-cysteine

5. Write a DNA sequence that would encode the following amino acid sequence:

  valine-tryptophan-lysine-proline-phenylalanine-threonine

6. In the film *Jurassic Park,* which is about cloned dinosaurs, a cartoon character named Mr. DNA talks about the billions of genetic codes in DNA. Why is this statement incorrect?

7. When researchers investigating the genetic code examined synthetic RNA of sequence ACACACACACACACA, they found that it encoded the amino acid sequence *thr-his-thr-his-thr-his.* How did the researchers determine the codon assignments for ACA and CAC?

8. Titin is a muscle protein named for its gargantuan size—its gene has the largest known coding sequence of 80,781 DNA bases. How many amino acids long is it?

9. An extraterrestrial life form has a triplet genetic code with five different bases. How many different amino acids can this code specify, assuming no degeneracy?

10. In malignant hyperthermia, a person develops a life-threateningly high fever after taking certain types of anesthetic drugs. In one family, the mutation deletes three contiguous bases in exon 44. How many amino acids are missing from the protein?

11. A mutation in a gene that encodes RPGR-interacting protein causes visual loss. The protein is 1,259 amino acids long. What is the minimal size of this gene?

12. Some biotechnology companies are developing chaperone proteins as a treatment for certain disorders. Name a disease that a chaperone protein might be able to treat, and suggest how it would do so.

Web Activities

Visit the Online Learning Center (OLC) at www.mhhe.com/lewisgenetics7 to complete the following web activities. After you have logged onto the Student Edition of the OLC, select **chapter 10** and **Web Activities.** There you will find links to the websites needed to complete the activities.

13. Go to the Harvard University website. Scroll down to the lists of "noncanonical" codes in organisms other than humans. (*Noncanonical* means it differs from the universal genetic code.) Find three examples of deviations from the universal code, and list what the codon-amino acid assignment is in most organisms. (Replace the T's on the website with the U's to correspond to the genetic code chart in the textbook.)

14. Use the Web to find out how the ubiquitin-proteasome system is overtaxed or disabled in a neurodegenerative disease such as

Alzheimer disease, Parkinson disease, Huntington disease, amyotrophic lateral sclerosis, or Lewy body dementia. (Find websites for these disorders and discuss how the mechanism involves proteasomes.)

Case Studies and Research Results

15. Five patients meet at a clinic for families in which several members have early-onset Parkinson disease. This condition causes rigidity, tremors, and other motor symptoms. Only 2 percent of cases of Parkinson disease are inherited. The five patients all have mutations in a gene that encodes the protein parkin, which has 12 exons. For each patient, indicate whether the mutation shortens, lengthens, or does not change the size of the protein.

   a. Manny Filipo's parkin gene is missing exon 3.

   b. Frank Myer's parkin gene has a duplication in intron 4.

   c. Theresa Ruzi's parkin gene lacks six contiguous nucleotides in exon 1.

   d. Elyse Fitzsimmon's parkin gene has an altered splice site between exon 8 and intron 8.

   e. Scott Shapiro's parkin gene is deleted.

Learn to apply the skills of genetic counselor with this additional case found in the *Case Workbook in Human Genetics:*

   Alpha-antitrypsin deficiency

CHAPTER

11

Control of Gene Expression

DNA MICROARRAYS UNCLOAK A CANCER

The symptoms of leukemia were hard to spot. Timothy was often bruised, although he was usually too tired to play, or had a fever. He had frequent nosebleeds. A blood test revealed the abnormal, abundant white blood cells of leukemia. The cancer cells were crowding other blood components in the bone marrow. Timothy had acute lymphoblastic leukemia (ALL), but fortunately was among the 90 percent of children who respond to chemotherapy. At the clinic, he and his parents met a family that wasn't so lucky. Their baby daughter Emily worsened after each treatment, and she died.

Researchers at the Dana Farber Cancer Institute were concerned about the hundred or so babies in the United States each year who, like Emily, develop ALL but do not respond to drugs that work well in others. So the investigators analyzed what the cancer cells *do*, rather than how they appear. Gene expression profiling using DNA microarrays compared 12,000 genes, and showed that in sicker infants, 1,000 genes were underexpressed and 200 overexpressed compared to the vast majority of children with ALL. These children had an entirely different disease, named "mixed lineage leukemia," or MLL.

MLL is "primitive," affecting bone marrow cells that are very early progenitors. Their high rate of division may explain how they resist standard treatments.

Considering MLL in the diagnosis may alert physicians to which patients are unlikely to benefit from ALL chemotherapy. Understanding the functions of the genes that are under- or overexpressed may reveal new molecular targets for treatment.

Understanding gene expression has improved diagnosis of leukemia, a group of blood cancers.

Rarely do all of the instruments in an orchestra sound at once. Instead, the musical composition dictates precisely how the instruments interact, with only some sounding at any one time, their intensity building and diminishing in a controlled manner. So it is with genomes. Once a new genome is launched shortly after conception, a program of gene expression unfolds that oversees the sculpting of the body. Yet the program is flexible enough that signals from the environment can also influence the process of building the organism.

The discoveries of the 1950s and 1960s on DNA structure and function answered some questions about this process while raising many more. How does a bone cell "know" to transcribe the genes that control the synthesis of collagen and not to transcribe genes that specify muscle proteins? How does the balance of blood cell types shift when a person has leukemia?

Watson and Crick's depiction of DNA inspired the idea that a single gene specifies a single protein. But the one-gene-one-protein picture is a great oversimplification. Sequencing of the human genome revealed that protein types in the human body outnumber the genes that encode them. However, much of the genome does not encode protein at all. This raises a compelling question: How can a genome specify many more proteins than there are genes, yet use only a tiny portion of its information to do this?

Although knowing the human genome sequence enables researchers to ask and answer new types of questions, the challenge to explain how cells differentiate persists. The basic question: How does a multicellular organism control when and where particular genes are transcribed, and how the resulting mRNAs are translated into protein?

11.1 Gene Expression Through Time and Tissue

The globin proteins that transport oxygen in the blood vividly illustrate the exquisite control of gene expression. The molecule's changing composition through development was discovered half a century ago.

Globin Chain Switching

A hemoglobin molecule in an adult has four polypeptide chains, each wound into a globular conformation (**figure 11.1**). Two of the chains are 146 amino acids long and are called "beta" (β). The other two chains are 141 amino acids long and are termed "alpha" (α). The genes for beta subunits are clustered on chromosome 11, and the alpha genes are grouped on chromosome 16.

The subunits of the hemoglobin molecule change with changes in oxygen concentration, which in turn depend upon whether oxygen arrives through the placenta or the newborn's lungs. The chemical basis for this "globin chain switching" is that different polypeptide subunits attract oxygen molecules to different degrees. Parts of the globin gene clusters, called locus control regions, oversee the changes in the molecule's composition and assembly.

The subunit makeup of the hemoglobin molecule differs in the embryo, fetus, and adult (**figure 11.2**). In the embryo, as the placenta forms, hemoglobin consists first

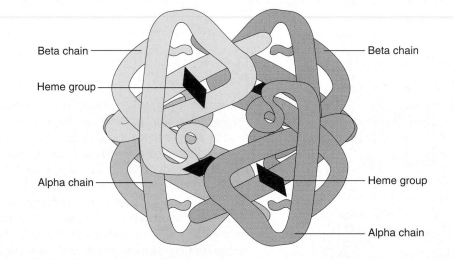

Figure 11.1 **The structure of hemoglobin.** A hemoglobin molecule is made up of two globular protein chains from the beta (β) globin group and two from the alpha (α) globin group. Each globin surrounds an iron-containing chemical group called a heme.

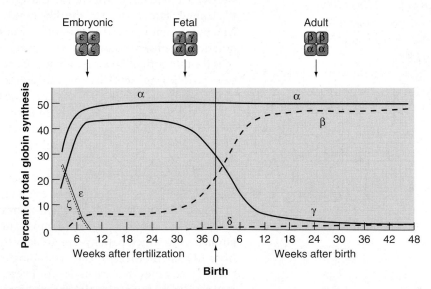

Figure 11.2 **Globin chain switching.** The subunit composition of human hemoglobin changes as the concentration of oxygen in the environment changes. With the switch from the placenta to the newborn's lungs to obtain oxygen, beta (β) globin begins to replace gamma (γ) globin.

of two epsilon (ε) chains, which are in the beta globin group, and two zeta (ζ) chains, which are in the alpha globin group. About 4 percent of the hemoglobin in the embryo includes beta chains. This percentage gradually increases.

As the embryo develops into a fetus, the epsilon and zeta chains decrease in number, as gamma (γ) and alpha chains accumulate. Hemoglobin consisting of two gamma and two alpha chains is called fetal hemoglobin. The gamma globin subunits bind very strongly to oxygen released from maternal red blood cells into the placenta, so that fetal blood carries 20 to 30 percent more oxygen than an adult's blood. As the fetus matures, beta chains gradually replace the gamma chains. At birth, however, the hemoglobin is not fully of the adult type—fetal hemoglobin (two gamma and two alpha chains) comprises from 50 to 85 percent of the blood. By four months of age, the proportion drops to 10 to 15 percent, and by four years, it is less than 1 percent.

In a condition called hereditary persistence of fetal hemoglobin, gamma globin continues to be made into adulthood. Because this condition does not affect health, drugs that activate gamma globin genes can treat disorders that result from certain mutations in the beta globin gene. These are discussed further in chapter 12.

RELATING THE CONCEPTS

Different leukemias affect cells in the bone marrow in different stages of specialization, from stem and progenitor cells to cells that are just a few divisions away from being mature white blood cells. How is this phenomenon similar to the normal control of the composition of the hemoglobin molecule?

Building Tissues and Organs

The globin chains affect one type of molecule, hemoglobin. Changing gene expression and the resulting production of proteins can also be observed on a larger scale. For example, blood plasma contains about 40,000 different types of proteins. (Plasma is the liquid portion of blood that the red blood cells packed with hemoglobin travel through, along with white blood cells and platelets.) Ten types of proteins account for 90 percent of all the plasma protein molecules, and nearly half of those are one type, albumin. This means that many thousands of types of proteins are present in vanishingly small amounts, which is why only 300 or so have been described. But change the conditions—the person develops an infection or allergic reaction—and the protein profile of the plasma can change dramatically. Behind it all is differential gene expression.

Blood is a structurally simple tissue that is easy to obtain and study. A solid gland or organ, constructed from specialized cells and tissues, is much more complex. Its unique organization must be maintained throughout a lifetime of growth, repair, and changing external conditions. Stem cell biology is beginning to shed light on how genes are turned on and off during the development of an organ or gland. Researchers isolate individual stem cells and then see which combinations of growth factors, hormones, and other biochemicals must be added to steer development towards a particular cell type. These manipulations may mimic natural development.

Consider the pancreas. It is a dual gland, with two types of cell clusters that have exocrine and endocrine functions (**figure 11.3**). An exocrine gland secretes into ducts; the exocrine portion of the pancreas does this with digestive enzymes. In contrast, an endocrine gland secretes directly into the bloodstream. The other portion of the pancreas does this for polypeptide hormones that control nutrient utilization

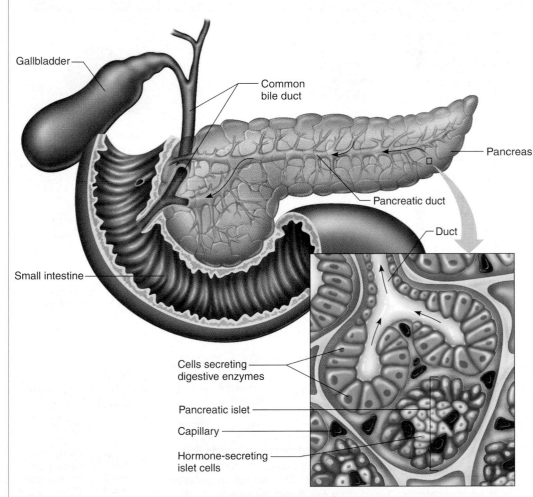

Gallbladder
Common bile duct
Pancreas
Pancreatic duct
Duct
Small intestine
Cells secreting digestive enzymes
Pancreatic islet
Capillary
Hormone-secreting islet cells

Figure 11.3 The pancreas is both an exocrine and an endocrine gland.
The expanded drawing shows the hormone-secreting pancreatic islets next to enzyme-secreting exocrine cells.

(table 11.1). The endocrine cell clusters are called pancreatic islets (once called islets of Langerhans).

As the pancreas develops in the embryo, ducts form first. Within their walls reside rare stem cells and progenitor cells (see figures 2.23 and 2.24). When a transcription factor called pdx-1 is activated, some of the progenitor cells divide. (Recall from chapter 10 that a transcription factor controls which genes are transcribed under certain conditions.) The progenitor cells give rise to daughter cells that follow an exocrine pathway; they are destined to produce digestive enzymes (**figure 11.4**). Other progenitor cells respond to different signals and divide to yield daughters that follow the endocrine pathway. The most familiar pancreatic hormone is insulin—its absence (or the inability of cells to recognize it) causes diabetes mellitus.

Researchers can observe the specialization of pancreas cells by taking individual progenitor cells from human pancreas ducts and supplying specific growth factors at particular times. This treatment stimulates certain progenitor cells to give rise to clusters that look and function like pancreatic islets. When exposed to glucose, the cells secrete insulin! Researchers project that within ten years, physicians will be able to coax the body of a person with diabetes to develop its own new and functional pancreatic beta cells.

Proteomics

A more complete portrait of gene expression emerges through **proteomics,** the consideration of all proteins made in a cell, tissue, gland, organ, or entire body. **Figure 11.5** depicts a global way of tracking the proteome—comparing the relative representations of fourteen categories of proteins from conception through old age to their expression before birth. This is accomplished using either gene expression microarrays of cells from different stages of prenatal development or postnatal life, or using a chemical technique called mass spectrometry to identify proteins directly.

The differences in gene expression at different times make sense. For example, transcription factors are more highly expressed in the embryo and fetus, presumably because of the extensive cell differentiation that is a hallmark of this period. During the prenatal period, enzymes are less emphasized, perhaps because the fetus receives some enzymes through the placenta. Such proteomic profiles shift with time in different tissues, and during periods of health and disease.

Another way to look at the complete proteome is by specific functions, which has led to the creation of various "ome" words. Genes whose encoded proteins control lipid synthesis constitute the "lipidome," and those that monitor carbohydrate production and use form the "glycome." Researchers make fun of all the "omics," but the designations are helpful in sorting out the thousands of proteins a human cell can manufacture. However, identifying proteins is only a first step. The next hurdle is to determine how proteins with related functions interact—forming "interactomes."

Table 11.1

Pancreatic Hormones

Hormone	Function	Cell Type
Glucagon	Stimulates production of glucose	Alpha
Insulin	Stimulates cells to take up glucose	Beta
Somatostatin	Controls rate of carbohydrate absorption in blood	Delta
Pancreatic polypeptide	Controls secretion of digestive enzymes	F

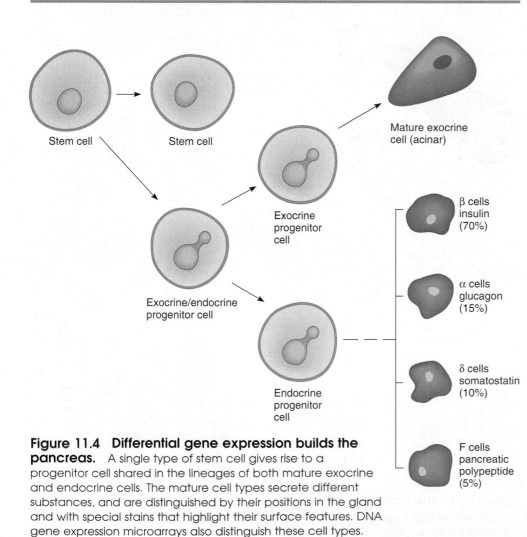

Figure 11.4 Differential gene expression builds the pancreas.   A single type of stem cell gives rise to a progenitor cell shared in the lineages of both mature exocrine and endocrine cells. The mature cell types secrete different substances, and are distinguished by their positions in the gland and with special stains that highlight their surface features. DNA gene expression microarrays also distinguish these cell types.

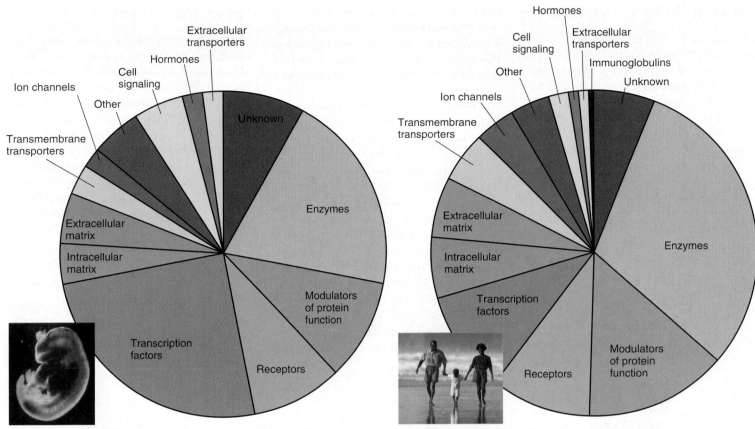

a. Distribution of health-related proteins from conception to birth

b. Distribution of health-related proteins from conception through old age

Figure 11.5 Proteomics meets medicine. One way to analyze the effects of genes is to categorize them by the functions of their protein products, and then to chart the relative abundance of each class at different stages of development, in sickness and in health. The pie chart in **(a)** considers 14 categories of proteins that when abnormal or missing cause disease, and their relative abundance from conception to birth. The pie chart in **(b)** displays the same protein categories from conception to old age. These depictions represent just one of the many new ways of looking at differential gene expression.

Researchers are accumulating gene expression profiles for many cell types under many conditions. Comparing gene expression profiles over time is particularly useful clinically, to chart disease progression and response to therapy. For example, 55 genes are overexpressed and 480 underexpressed in cells of a prostate cancer that has a very high likelihood of spreading. Cells from a prostate cancer that will not spread have a different gene expression profile.

RELATING THE CONCEPTS

What are two advantages of using gene expression profiling as a diagnostic test for cancer?

Key Concepts

1. Gene expression patterns change over time and in different cell types.
2. The subunit composition of hemoglobin changes in the embryo, fetus, and after birth.
3. As a pancreas forms, progenitor cells diverge from shared stem cells and their daughters specialize.
4. Proteomics tracks all of the proteins in a cell, tissue, organ, or organism under specific conditions.

11.2 Mechanisms of Gene Expression

We have already seen in a general sense how gene expression is controlled: signals instruct cells to activate combinations of transcription factors, which control the genes that are transcribed. The transcription factors interact, positioning DNA to ease its interactions with yet other proteins.

Individuals may vary greatly in the degree to which a particular gene is expressed. This variability arises from differences in DNA sequences that control when, where, and how much a particular gene is transcribed. Also, for several hundred genes, people have different numbers of copies, which affects expression. We now take a closer look at two mechanisms that control gene expression: changing chromatin structure that regulates DNA accessibility, and small "interfering" RNA molecules that seek and destroy selected mRNA transcripts.

Chromatin Remodeling

For many years, biologists thought that histone proteins were simple scaffolds that wind long DNA molecules into nucleosomes, little more than tiny spools (see figure 9.14). However, histones do much more: They play a major role in exposing DNA when it is to be transcribed, and shielding it when it is to be silenced. To do this, enzymes add or delete small organic chemical groups to histones in a process called **chromatin remodeling.** The resulting patterns of added chemical groups control the effect of histones on their associated protein-encoding genes.

The three major types of small molecules that bind to histones are acetyl groups, methyl groups, and phosphate groups (**figure 11.6**). For example, when acetyl groups bind to the tails of specific histones, the DNA wound around the histone packet shifts slightly, which exposes the area where RNA polymerase binds to start transcription. Different enzymes add acetyl groups to turn on or shut off gene expression. Phosphate groups also activate transcription. Methyl groups turn off transcription when they bind histones.

The modified state of the chromatin can be passed on when DNA replicates. This is an example of an epigenetic change, or change "outside conventional genetics." That is, these changes are heritable, but they do not directly affect the DNA sequence.

Enzymes that add or delete acetyl, methyl, and phosphate groups must be in a balance that controls which genes are expressed and which are silenced. One limitation to altering chromatin remodeling to treat inherited disease is that this action could affect the expression of many genes—not just the one implicated in the disease. **Table 11.2** lists disorders that result from abnormal chromatin remodeling.

RNA Interference

Genetics continues to hold surprises. Although usually only one strand of DNA is transcribed, up to 8 percent of genes may actually be transcribed from both strands, leading to a phenomenon called **RNA interference** (or RNAi) that destroys specific mRNA molecules. Transcribing both DNA strands of certain genes leads to formation of single-stranded RNAs that bind in places within themselves, generating double-stranded, hairpin-shaped structures (**figure 11.7**). Through a series of interactions with proteins, these RNAs are shortened to form "small interfering RNAs," known as siRNAs. These RNAs are opened up and then find and bind their mRNA complements, tagging them for dismantling by enzymes.

The siRNAs affect gene expression in the nucleus and in the cytoplasm. In the nucleus, siRNAs help add methyls to histones, shutting off transcription at its start.

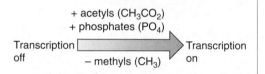

Figure 11.6   Chromatin remodeling. Chromatin remodeling adds or removes certain organic chemical groups to or from histones. The pattern of binding controls whether the DNA wrapped around the histones is transcribed or not.

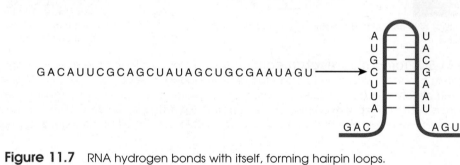

Figure 11.7   RNA hydrogen bonds with itself, forming hairpin loops.

Table 11.2

Disorders of Chromatin Remodeling

Disease	OMIM	Protein	Symptoms	Defect
α-thalassemia mental retardation syndrome	301040	ATRX	Anemia, mental retardation	Undermethylation of heterochromatin
ICF syndrome	242860	DNMT3B	Immunodeficiency, unstable centromeres, facial anomalies	Undermethylation of repeats
Rett syndrome	312750	MECP2	Repetitive movements, irregular breathing, seizures, loss of motor control: profound neurodegeneration starting at 6 months	Failure to remove acetyls from histones on gene *DLX5* expressed in brain
Rubinstein-Taybi syndrome	180849	CBP	Mental retardation, short stature, facial anomalies	Adds acetyl groups to certain histones, causing inappropriate transcription of some genes

Slightly smaller siRNAs in the cytoplasm bind mature mRNAs, acting after transcription ends. RNA interference is not the normal dismantling of a used mRNA, but a distinct mechanism that suppresses expression of certain genes.

As in other activities in the cell, several proteins and protein complexes orchestrate the steps of RNA interference (**figure 11.8**). An enzyme called dicer first cuts long double-stranded RNAs into 21- to 24-base-long pieces. Then a large protein complex called RISC (for RNA-induced silencing complex) binds the pieces and unwinds them, exposing single strands. The antisense RNA strands—so-called because they are complementary to the targeted RNA—then attract the mRNAs, which are chewed up by enzymes (nucleases) that are part of RISC.

Introducing siRNAs into cells is a useful research tool. An siRNA is said to "knock down" a specific mRNA, compared to other techniques, discussed in chapter 19, that "knock out" or "knock in" a function. "Knocking down" depletes the mRNA molecules representing a particular gene, whereas "knocking out" completely silences a gene and "knocking in" adds a gene.

The discovery of RNAi has spawned a new biotechnology, in which researchers synthesize short RNAs to intentionally destroy particular mRNAs. This approach can be used to discover the functions of specific genes, or to silence dangerous ones. For example, experimental vaccines using RNAi "knock-down" (deplete) expression of key genes in the viruses that cause SARS, AIDS, polio, and hepatitis C. To treat cancer, siRNAs might knock down genes whose protein products destabilize the cell cycle. SiRNAs can also knock down an enzyme required for caffeine synthesis in coffee plants, creating a better-tasting decaf.

Key Concepts

1. Acetyl, phosphate, and methyl groups bind to histone proteins, controlling transcription.
2. In RNA interference, short double-stranded RNAs separate, and the antisense strands bind specific mRNAs, marking them for destruction.

11.3 Proteins Outnumber Genes

Our 25,000 or so genes encode 200,000 or more proteins. But this apparent paradox wasn't entirely unexpected.

The discovery of introns in 1977 first raised the possibility that some genes could specify more than one protein by mixing and matching exons. It might be advantageous to do this in different tissues or under different circumstances. Combining the exons of one gene in different ways is called **alternate splicing (figure 11.9)**. For exam-

ple, an antibody-secreting cell of the immune system at first produces a shortened version of the antibody that is presented on the cell's surface, which alerts other cells. As the infection progresses, the cell transcribes a different exon that adds a portion to the antibody that enables it to be secreted into the bloodstream, where it attacks the pathogen.

Alternate splicing explains how a long sequence of DNA can specify more mRNAs than genes. On a part of chromosome 22, for example, 245 genes yield 642 mRNA transcripts. About half of all human genes are alternately spliced.

Introns present another way to maximize the protein-encoding information content of DNA. Introns may seem wasteful, little more than vast stretches of DNA bases that outnumber and outsize exons. But a DNA sequence that is an intron in one context may encode protein in another. Consider prostate specific antigen (PSA), a protein found on certain cell surfaces that is overproduced in some prostate cancers. The gene for PSA has five exons and four introns, but it also is alternately spliced to encode a different protein called PSA-linked molecule (PSA-LM), that includes only the first exon, but also the fourth intron! The proteins seem to work against each other. When the level of one is high, the other is low. New blood tests to detect elevated risk of prostate cancer may consider levels of both proteins.

In another situation where introns may account for the overabundance of proteins compared to genes, a DNA sequence that is an intron in one gene's template strand may encode protein on the coding strand. This is the case for the gene for neurofibromin,

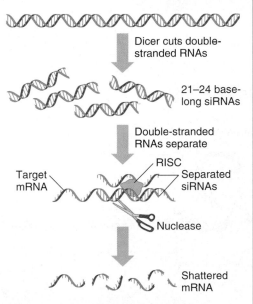

Figure 11.8 RNA Interference.
Dicer cuts double-stranded portions of RNA molecules, which then associate with RNA-induced silencing complexes (RISCs). The RNAs open, revealing single strands that locate and bind specific mRNAs. Nucleases then break down the targeted mRNAs.

Figure 11.8 labels: Dicer cuts double-stranded RNAs; 21–24 base-long siRNAs; Double-stranded RNAs separate; RISC; Target mRNA; Separated siRNAs; Nuclease; Shattered mRNA

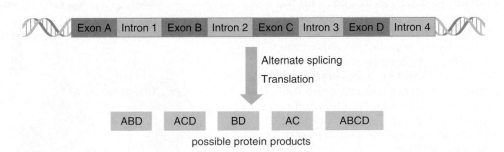

Figure 11.9 labels: Exon A | Intron 1 | Exon B | Intron 2 | Exon C | Intron 3 | Exon D | Intron 4; Alternate splicing; Translation; ABD | ACD | BD | AC | ABCD; possible protein products

Figure 11.9 Exons provide flexibility in gene structure that expands gene number.   Alternate splicing enables a cell to manufacture different versions of a protein by adding or deleting parts.

which when mutant causes neurofibromatosis type 1 (OMIM 162200), an autosomal dominant condition that causes benign tumors beneath the skin and "café au lait" spots on the skin surface. Within an intron of the neurofibromin gene, but on the coding strand, are instructions for three other genes.

Still another way a gene can maximize its informational content is for its encoded protein to be cut to yield two products. This happens in dentinogenesis imperfecta (OMIM 125490), which causes discolored, misshapen teeth with peeling enamel (**figure 11.10**). The dentin, which is the bone-like substance beneath the enamel that forms the bulk of the tooth, is abnormal. Dentin is a complex mixture of extracellular matrix proteins. Dentin protein is 90 percent collagen, and this and most of the rest of the proteins are also found in bone. However, two proteins are unique to dentin: dentin phosphoprotein (DPP) and dentin sialoprotein (DSP). A single gene encodes these two proteins.

Researchers had associated abnormal DPP with dentinogenesis imperfecta. However, DPP, because it is much more abundant than DSP, may have overshadowed the rarer protein. Both DPP and DSP are translated from a single mRNA molecule as the precursor protein dentin sialophosphoprotein (DSPP). DPP may be much more abundant because it is longer-lived than DSP; that is, DSP is degraded faster.

Table 11.3 summarizes mechanisms that maximize genomic information.

Table 11.3

Maximizing the Informational Content of Genes

Mechanism	Example
Alternate splicing	antibody-producing cell
Use of introns	PSA and PSA-LM; NF1
Two proteins split from precursor	Dentinogenesis imperfecta

11.4 Most of the Human Genome Does *Not* Encode Protein

When the first generation of molecular geneticists worked out the details of transcription and translation in the 1960s, they never imagined that only 1.5 percent of the human genome encodes protein. What does the "other" 98.5 percent do? In general, this noncoding DNA falls into four categories: (1) RNAs other than mRNA (called noncod-

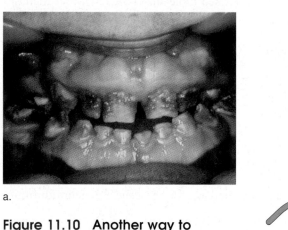

a.

Figure 11.10 Another way to encode two genes in one. **(a)** The misshapen, discolored, and enamel-stripped teeth of a person with dentinogenesis imperfecta were at first associated with deficiency of the protein DPP. Then researchers discovered that DSP is deficient, too, but is usually present in such small amounts that its role wasn't recognized. **(b)** Both DPP and DSP are cut from the same larger protein, but DSP is degraded faster.

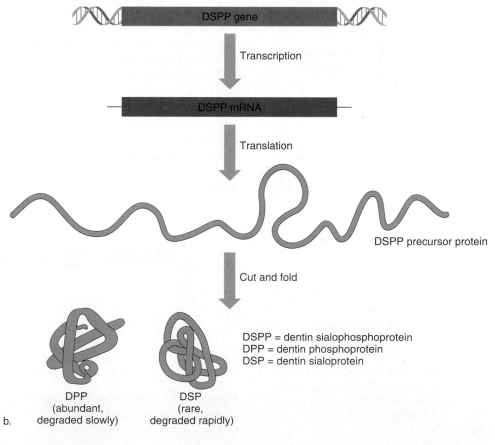

DSPP = dentin sialophosphoprotein
DPP = dentin phosphoprotein
DSP = dentin sialoprotein

Table 11.4

The Nonprotein Encoding Parts of the Human Genome

	Function or Characteristic
Noncoding RNA genes	
tRNA genes	Connect mRNA codon to amino acid
rRNA genes	Parts of ribosomes
Pseudogenes	DNA sequences very similar to known gene sequences that may be transcribed but are not translated
Small nucleolar RNAs	Process rRNA in nucleolus
Small nuclear RNAs	Parts of spliceosomes
Telomerase RNA	Part of ribonucleoprotein that adds bases to chromosome tips
Xist RNA	Inactivates one X chromosome in cells of females
Vault RNA	Part of "vault," a large ribonucleoprotein complex of unknown function
Introns	Parts of genes that are transcribed but cut out before translation
Promoters and other control sequences	Guide enzymes that carry out DNA replication, transcription, or translation
Small interfering RNAs	Control transcription
Micro RNAs	Control transcription of more than one-third of genes
Repeats	
Transposons	Repeats that move around the genome
Telomeres	Chromosome tips whose lengths control the cell cycle
Centromeres	Provide backdrop for proteins that form attachments for spindle fibers
Duplications of 10 to 300 kilobases	Unknown
Simple short repeats	Unknown

ing or ncRNAs), (2) introns, (3) promoters and other control sequences, including siRNAs, and (4) repeated sequences (table 11.4).

Noncoding (nc) RNAs

About a third of the human genome is transcribed into RNA types other than mRNA, such as tRNA and rRNA. The rate of transcription of a cell's tRNA genes is attuned to cell specialization. The collection of proteins characteristic of a skeletal muscle cell, for example, would require different amounts of certain amino acids than the proteins of a white blood cell, and therefore different amounts of the corresponding tRNAs too.

Human tRNA genes are dispersed among the chromosomes in clusters—25 percent of them are on the sixth largest chromosome, for example. Altogether, tRNAs account for 0.1 percent of the genome. Our 500 or so types of tRNA genes may seem like a lot, but frogs have thousands! This may reflect the fact that frog eggs are huge and contain many types of proteins.

The 243 types of rRNA genes are grouped on six chromosomes, each cluster harboring 150 to 200 copies of a 44,000-base repeat sequence. Once transcribed from these clustered genes, the rRNAs go to the nucleolus, where yet another type of ncRNA called small nucleolar RNA (snoRNA) cuts them into their final forms.

Hundreds of thousands of ncRNAs are neither tRNA nor rRNA, nor snoRNAs, nor the other less abundant types described in table 11.4. Instead, they are transcribed from DNA sequences called **pseudogenes.** A pseudogene is very similar in sequence to a particular protein-encoding gene, and it may be transcribed into RNA, but it is not translated into protein. Presumably it is altered in sequence from the original gene in a way that impairs its translation—perhaps the encoded amino acids cannot fold into a functional protein. Pseudogenes may be remnants of genes past, once-functional variants that diverged from the normal sequence too greatly to encode a working protein. Pseudogenes are incredibly common in the human genome. For example, at least 324 pseudogenes shadow our tRNA genes.

Repeats

The human genome is riddled with highly repetitive sequences that may be a different type of information than a protein's amino acid sequence. Perhaps they are a language in which meaning lies in repeat size or number. Or, perhaps some types of repeats help to hold a chromosome together.

The most abundant type of repeat is a sequence of DNA that can jump about the genome, called a transposable element, or **transposon** for short. Barbara McClintock originally identified transposons in corn in the 1940s, and then they were discovered in bacteria in the 1960s. Transposons comprise about 45 percent of the human genome sequence, typically repeated in many copies. Some transposons include parts that encode enzymes that cut them out of one chromosomal site and integrate them into another.

Transposons are classified by size, whether they are transcribed into RNA, which enzymes they use to move, and whether they resemble bacterial transposons. For example, a class of transposons called long interspersed elements (LINEs) are 6,000 bases long and are transcribed and then trimmed to 900 bases before they are "reverse transcribed" back into DNA (by an enzyme called reverse transcriptase) and reinserted into a chromosome. In contrast, short interspersed elements (SINEs) are 100 to 500 bases long and use enzymes that are encoded in LINEs to reinsert.

A major class of SINEs are called Alu repeats. Each Alu repeat is about 300 bases long, and a human genome may contain 300,000 to 500,000 of them. Alu repeats comprise 2 to 3 percent of the genome, and they have been increasing in number over time because they can copy themselves. Alu repeats may serve as attachment points for proteins called cohesins that bind newly replicated DNA to parental strands before anaphase, when replicated chromosomes pull apart in mitosis.

Other rarer classes of repeats include those that comprise telomeres, centromeres, and rRNA gene clusters; duplications of 10,000 to 300,000 bases (10 to 300 kilobases); copies of pseudogenes; and simple repeats of one, two, or three bases. Many repeats arise from RNAs that are reverse transcribed into DNA and are then inserted into chromosomes. In fact, the entire human genome may have duplicated once or even twice, as is discussed further in chapter 16.

Our understanding of the functions of repeats lags far behind our knowledge of the roles of the various noncoding RNA genes. Repeats may make sense in light of evolution, past and future. Pseudogenes are likely vestiges of genes that functioned in our nonhuman ancestors. Perhaps the repeats that seem to have no obvious function today will serve as raw material from which future genes may arise.

Key Concepts

1. Most of the genome encodes many types of RNA as well as introns, promoters, and other control sequences and repeats.
2. We do not know the functions of some repeats.

Summary

11.1 Gene Expression Through Time and Tissue

1. Changes in gene expression occur over time at the molecular level (globin switching), at the tissue level (blood plasma), and at the organ/gland level (pancreas development).

2. **Proteomics** uses analytical chemistry techniques and gene expression DNA microarrays to catalog the types of proteins in particular cells, tissues, organs, or entire organisms under specified conditions.

11.2 Mechanisms of Gene Expression

3. The pattern of chemical groups on histones forms an epigenetic code that spreads, can be transmitted when the cell divides, and controls gene expression.

4. Acetylation, phosphorylation, and methylation control gene expression. They carry out **chromatin remodeling.**

5. **RNA interference** silences genes in the nucleus and removes certain mRNAs in the cytoplasm.

11.3 Proteins Outnumber Genes

6. Only 1.5 percent of the human genome encodes protein, yet those 25,000 or so genes specify up to 200,000 proteins.

7. Mechanisms to explain the mismatch between gene and protein diversity include **alternate splicing,** use of introns, and cutting proteins translated from a single gene.

11.4 Most of the Human Genome Does *Not* Encode Protein

8. The rest of the genome includes noncoding RNAs, introns, promoters and other controls, and repeats.

Review Questions

1. Why is control of gene expression necessary?

2. Describe three types of cells and how they differ in gene expression.

3. What questions about DNA were raised after the genetic code was worked out in the 1960s, and then after sequencing the human genome?

4. What is the environmental signal that stimulates globin switching?

5. How does development of the pancreas illustrate differential gene expression?

6. Distinguish between a genetic and an epigenetic change. What do they have in common? How do they differ?

7. How do histones control gene expression, yet genes also control histones?

8. Name two types of chemical reactions that silence transcription.

9. What controls whether histones allow DNA wrapped around them to be transcribed?

10. What information is needed to use RNAi to treat a viral infection?

11. How does alternate splicing generate more than one type of protein from the information in a gene?

12. The media often call DNA sequences that do not encode protein "junk." Give three reasons why this DNA should not be considered junk.

13. In the 1960s, a gene was defined as a continuous sequence of DNA, located

permanently at one place on a chromosome, that specifies a sequence of amino acids from one strand. List three ways this definition has changed.

14. Give three examples of discoveries mentioned in the chapter that changed the way we think about the genome.

15. How can one of the two dental proteins implicated in dentinogenesis imperfecta be much more abundant than the other if they are both transcribed and translated from the same gene?

16. State four roles of DNA other than encoding protein.

Applied Questions

1. Invent a new "omics" to investigate genes that are functionally related in a particular way.

2. Drug companies are synthesizing compounds that inhibit the enzymes that either put acetyl groups on histones or take them off. Would you use a drug that adds or removes acetyl groups to combat a cancer caused by too little expression of a gene that normally suppresses cell division?

3. Chromosome 7 has 863 protein-encoding genes, but many more proteins. The average gene is 69,877 bases, but the average mRNA is 2,639 bases. Explain both of these observations.

4. CHARGE syndrome (OMIM 21400) causes heart defects, visual problems, facial palsy, blocked nostrils, and difficulty swallowing. A mutation in a gene called *Chd1* causes the condition. The product of this gene recognizes and binds methyl groups on certain histones. Explain how this mutation leads to pleiotropy (multiple symptoms).

5. In chronic myelogenous leukemia, an exchange between two nonhomologous chromosomes fuses two genes. Expression of the fused gene increases the rate of synthesis of tyrosine kinase, which lifts control of the cell cycle. How might RNA interference be used to treat this cancer?

6. Which is a more targeted approach to treating cancer, removing methyl groups to reactivate genes that normally suppress cancer, or using RNAi?

7. How many different proteins encompassing two exons can be produced from a gene that has three exons?

8. When researchers compared the number of mRNA transcripts that correspond to a part of chromosome 19 to the number of protein-encoding genes in the region, they found 1,859 transcripts and 544 genes. Account for the discrepancy.

9. Figure 11.5 shows the distribution of types of proteins that, when abnormal or absent from a certain cell type, cause disease. Such charts have been constructed for different stages of development—prenatal, under a year, childhood, puberty to age 50, and over age 50. Explain the observation that transcription factors account for:

- 9 percent of proteins overall (throughout development and life)

- 25 percent of proteins before birth

- 7 percent of proteins from birth to one year of age

- 6 percent of proteins from childhood to age 50 years

- 5 percent of proteins for those over 50 years of age

Web Activities

Visit the Online Learning Center (OLC) at www.mhhe.com/lewisgenetics7. Select **Student Edition, Chapter 11,** and **Web Activities** to find the website links needed to complete the following activities.

10. Many companies are offering products based on RNA interference to use in research to "knock down" gene expression. Go to one of the following websites for the following companies, or find others. Research a particular RNAi product, and suggest how it might be used. (The companies listed have all existed for many years. There are many newer ones.)

Ambion

Invitrogen

New England Biolabs

Novagen

Qiagen

Stratagene

Case Studies

11. Jerrold is 38 years old. His body produces too much of the hormone estrogen and as a result, he has gynecomastia—well-developed breasts. He had a growth spurt and development of pubic hair by age 5, and then his growth dramatically slowed so that his adult height is well below normal. He had his breasts removed, but has a very high-pitched voice and no facial hair, which are lingering signs of his excess estrogen production. Jerrold's son, Timmy, is 8 years old and has the same symptoms—breast enlargement and early, rapid growth.

Jerrold and Timmy have an overactive gene for aromatase, an enzyme required to synthesize estrogen. Five promoters control expression of the gene in different tissues, and each promoter is activated by a different combination of hormonal signals. The five promoters lead to estrogen production in skin, fat, brain, gonads (ovaries and testes) and placenta. In premenopausal women, the ovary-specific promoter is highly active, and estrogen is abundant. In men and postmenopausal women, however, only small amounts of estrogen are normally produced, in skin and fat. The father and son have a wild type aromatase gene, but high levels of estrogen in several tissues, particularly fat, skin, and blood. They do, however, have a mutation that turns around an adjacent gene so that the aromatase gene falls under the control of a different promoter. Suggest how this phenotype arises.

12. Margaret is 102 years old, and she still walks at least half a mile a day, albeit slowly. She is a trim vegetarian who has rarely been ill her entire life. Morris is an obese, balding 62-year-old man who has high blood pressure and colon cancer. How might their proteome portraits, such as the one in figure 11.5, differ? (Hint: Reread Reading 3.1, The Centenarian Genome.)

Learn to apply the skills of genetic counselor with this additional case found in the *Case Workbook in Human Genetics:*

Hypoxia-inducible factor 1

Rett syndrome

SAGE

Gene Mutation

TWO MUTATIONS STRIKE ONE GENE—AND ONE LITTLE GIRL

Newborn screening for sickle cell disease indicated that Juanita was a carrier, like her father. Unlike most carriers, though, Juanita was sick. She was hospitalized a few times for an enlarged spleen and anemia, and some of her red blood cells were sickled.

Juanita ran into trouble at 19 months of age, on a plane. Midflight, she suddenly became ill, her spleen swelling enormously and her blood pressure plummeting. When Juanita turned blue, a doctor on the flight administered oxygen. The child safely reached a hospital, where her spleen was removed. She had severe hemolytic anemia—her cells were sickling *and* bursting. This double danger hinted at the cause of Juanita's problem: she had two mutations in a beta globin gene.

The mutation Juanita inherited from her father changed the sixth amino acid in the beta globin gene from glutamic acid to valine, but the gene had a second mutation that affected another part of the encoded protein. Since her father's gene did not have a second mutation, it must have occurred in the sperm that was to join an ovum to become, eventually, Juanita. The second mutation changed the 68th amino acid in beta globin from leucine to phenylalanine. This change exposed part of the molecule that is normally shielded, destabilizing the entire globin molecule. Although Juanita's mother contributed a normal allele, the overall effect in the double-carrier child was a drastic lowering of the attraction of oxygen to hemoglobin. However, this happens only in the presence of a particular metabolic by-product made in the spleen only at high altitudes!

Normal red blood cells (*top*) and sickled red blood cells (*bottom*). Several mutations can cause red blood cells to sickle. The abnormal cells block circulation where and when oxygen levels fall, causing great pain.

A **mutation** is a change in a gene's nucleotide base sequence. It can occur at the DNA level, substituting one DNA base for another or adding or deleting a few bases, or at the chromosome level, the subject of chapter 13. Chromosomes can exchange parts, and genetic material can even jump from one chromosome to another. These events can cause mutation. This chapter discusses mutations at the DNA level. They can occur in the part of a gene that encodes a protein, in a sequence that controls transcription, in an intron, or at a site critical to intron removal and exon splicing.

The effects of mutation vary. A mutation can stop or slow production of a protein, overproduce it, or impair the protein's function—such as altering its secretion, location, or interaction with another protein. Not all mutations are harmful. For example, a mutation protects against HIV infection. About 1 percent of the general population is homozygous for a recessive allele that encodes a cell surface protein called CCR5 (see figure 17.13). To infect an immune system cell, HIV must bind CCR5 and another protein. Because the mutation prevents CCR5 from moving from the cytoplasm to the cell surface, HIV cannot bind. Heterozygotes are partially protected against HIV infection.

The term *mutation* refers to genotype— that is, a change at the DNA or chromosome level. The familiar term **mutant** refers to an unusual phenotype. The nature of a mutant phenotype depends upon how the mutation affects the gene's product or activity, and usually connotes an abnormal or unusual characteristic. However, a mutant phenotype may also be an unusual variant that is nevertheless "normal," such as red hair. Detecting mutations forms the basis of the several types of genetic tests described in table 20.2.

In an evolutionary sense, mutation has been essential to life, because it produces individuals with variant phenotypes who are better able to survive specific environmental challenges, including illnesses. Disease-resistant gene variants that arise by mutation tend to become more common in populations over time when they are protective, because they give people with the mutation a survival advantage. Chapter 15 further discusses the role of mutations in populations.

A mutation may be present in all the cells of an individual or just in some cells. In a **germline mutation,** the change occurs during the DNA replication that precedes *meiosis.* The resulting gamete and all the cells that descend from it after fertilization have the mutation. In contrast, in a **somatic mutation,** the change happens during DNA replication before a *mitotic* cell division. All the cells that descend from the original changed cell are altered, but they might only comprise a small part of the body. Somatic mutations are responsible for certain cancers (see Reading 18.1 and figure 18.5).

12.1 Mutations Can Alter Proteins—Three Examples

Identifying how a mutation causes symptoms has clinical applications, and also reveals the workings of biology. Following are three examples of mutations that cause disease.

The Beta Globin Gene

The first genetic illness to be understood at the molecular level was sickle cell disease (**figure 12.1**). In 1904, young medical intern Ernest Irons noted "many pear-shaped and elongated forms" in a blood sample from a dental student in Chicago who had anemia. Irons sketched this first view of sickle cell disease at the cellular level, and reported his findings to his supervisor, physician James Herrick. Alas, Herrick published the work but did not mention Irons. Herrick has been credited with the discovery ever since.

In 1949, Linus Pauling discovered that hemoglobin from healthy people and from people with the anemia, when placed in a solution in an electrically charged field (a technique called electrophoresis), moved to different positions. Hemoglobin from the parents of people with the anemia, who were carriers, moved to both positions.

The difference between the two types of hemoglobin lay in beta globin. Recall from figure 11.1 that adult hemoglobin consists of two alpha polypeptide subunits and two beta subunits. Protein chemist V. M. Ingram took a shortcut to localize the mutation in the 146-amino-acid-long protein. He cut normal and sickle hemoglobin with a protein-digesting enzyme, separated the pieces, stained them, and displayed them on filter paper. The patterns of fragments— known as peptide fingerprints—were different for the two types of hemoglobin. This meant, Ingram deduced, that the two molecules differ in amino acid sequence. Then he homed in on the difference. One piece of the molecule in the fingerprint, fragment four, occupied a different position in each of the two types of hemoglobin. Because this peptide was only 8 amino acids long, Ingram needed to decipher only that short sequence to find the site of the mutation. It was a little like knowing which sentence on a page contains a typographical error.

Ingram identified the tiny mutation responsible for sickle cell disease: a substitution of the amino acid valine for the glutamic acid that is normally the sixth amino acid

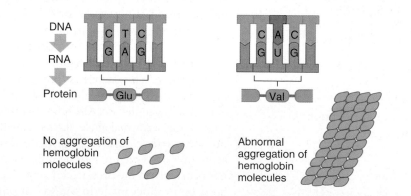

Figure 12.1 Sickle cell disease results from a single DNA base change that substitutes one amino acid in the protein (valine replaces glutamic acid). The result is a change in the surfaces of the molecules that causes aggregation into long, curved rods that deform the red blood cell.

in the beta globin polypeptide chain. At the DNA level, the change was even smaller—a CTC to a CAC, corresponding to RNA codons GAG and GUG. This was learned after researchers deciphered the genetic code. The valine at this position changes the surfaces of hemoglobin molecules so that in low-oxygen conditions they attach at many more points than they would if the wild type glutamic acid were at the site. The aggregated hemoglobin molecules form ropelike cables that bend red blood cells into rigid, fragile, sickle-shaped structures. The misshapen cells lodge in narrow blood vessels, cutting off local blood supplies. Once a blockage occurs, sickling speeds up and spreads, as the oxygen level falls. The result is great pain in the blocked body parts, particularly the hands, feet, and intestines. The bones ache, and depletion of normal red blood cells causes the great fatigue of anemia.

The sickle cell mutation arose in three places in Africa and in the Arab-India region (see figure 15.12). The mutant allele was spread around the globe by the slave trade.

Sickle cell disease was the first inherited illness linked to a molecular abnormality, but it wasn't the first known condition that results from a mutation in the beta globin genes. In 1925, Thomas Cooley and Pearl Lee described severe anemia in Italian children,

and in the decade following, others described a milder version of "Cooley's anemia," also in Italian children. The disease was named thalassemia, from the Greek for "sea," in light of its high prevalence in the Mediterranean area. The two disorders turned out to be the same. The severe form, sometimes called thalassemia major, results from a homozygous mutation in the beta globin gene. The milder form, called thalassemia minor, affects some individuals who are heterozygous for the mutation.

Once researchers had worked out the structure of the hemoglobin molecule, and learned that different globins function in the embryo and fetus, the molecular basis of thalassemia became clear. The disorder that is common in the Mediterranean is more accurately called beta thalassemia, because the symptoms result from too few beta globin chains. Without them, not enough hemoglobin molecules are assembled to effectively deliver oxygen to tissues. Fatigue and bone pain arise during the first year of life as the child depletes fetal hemoglobin, and the "adult" beta globin genes are not transcribed and translated on schedule.

As severe beta thalassemia progresses, red blood cells die because the excess of alpha globin chains prevents formation of hemoglobin molecules. Liberated iron slowly destroys the heart, liver, and endocrine

glands. Periodic blood transfusions can control the anemia, but they hasten iron buildup and organ damage. Drugs called chelators that entrap the iron can extend life past early adulthood, but they are very costly and not available in developing nations.

RELATING THE CONCEPTS

How would an electrophoresis test on Juanita's father's blood differ from test results from her mother's blood?

Disorders of Orderly Collagen

Much of the human body consists of the protein collagen, a major component of connective tissue. It accounts for more than 60 percent of the protein in bone and cartilage and provides 50 to 90 percent of the dry weight of skin, ligaments, tendons, and the dentin of teeth. Collagen is in parts of the eyes and the blood vessel linings, and it separates cell types in tissues.

Mutations in the genes that encode collagen, not surprisingly, lead to a variety of medical problems (**table 12.1**). These disorders are particularly devastating because collagen has an extremely precise conformation that is easily disrupted, even by

Table 12.1

Collagen Disorders

Disorder	OMIM	Genetic Defect (Genotype)	Signs and Symptoms (Phenotype)
Alport syndrome	203780	Mutation in type IV collagen interferes with tissue boundaries	Deafness and inflamed kidneys
Aortic aneurysm	100070	Missense mutation substitutes *arg* for *gly* in α1 gene	Aorta bursts
Chondrodysplasia	302950	Deletion, insertion, or missense mutation replaces *gly* with bulky amino acids	Stunted growth, deformed joints
Dystrophic epidermolysis bullosa	226600	Collagen fibrils that attach epidermis to dermis break down	Skin blisters on any touch
Ehlers-Danlos syndrome	130050	Missense mutations replace *gly* with bulky amino acids; deletions or missense mutations disrupt intron/exon splicing	Stretchy, easily scarred skin, lax joints
Osteoarthritis	165720	Missense mutation substitutes *cys* for *arg* in α1 gene	Painful joints
Osteogenesis imperfecta type I	166200	Inactivation of α allele reduces collagen triple helices by 50%	Easily broken bones; blue eye whites; deafness
Stickler syndrome	108300	Nonsense mutation in procollagen	Joint pain, degeneration of vitreous gel and retina

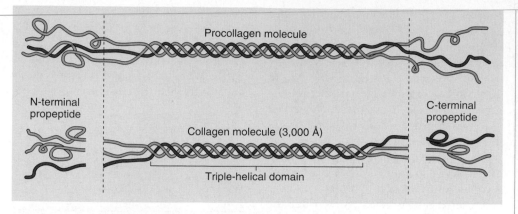

Figure 12.2 Collagen has a very precise conformation. The α1 collagen gene encodes the two blue polypeptide chains, and the α2 procollagen gene encodes the third (red) chain. The procollagen triple helix is shortened before it becomes functional, forming the fibrils and networks that comprise much of the human body.

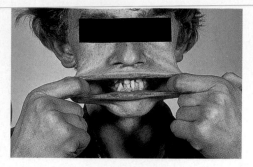

Figure 12.3 A disorder of connective tissue produces stretchy skin. A mutation that blocks trimming of procollagen chains to produce collagen causes the stretchy skin of Ehlers-Danlos syndrome type I.

slight alterations that might have little effect in proteins with other shapes (**figure 12.2**).

Collagen is sculpted from a longer precursor molecule called procollagen, which consists of many repeats of the amino acid sequence glycine-proline-modified proline. Three procollagen chains entwine. Two of the chains are identical and are encoded by one gene, and the other is encoded by a second gene. The electrical charges and interactions of these amino acids with water coil the procollagen chains into a very regular triple helix, with space in the middle only for tiny glycine. The ragged ends of the polypeptides are snipped off by enzymes to form mature collagen. The collagen fibrils continue to associate with each other outside the cell, building the fibrils and networks that hold the body together.

The boy in **figure 12.3** has a form of Ehlers-Danlos syndrome. A mutation prevents his procollagen chains from being cut, and collagen molecules cannot assemble. They form ribbonlike fibrils that lack the tensile strength to keep the skin from becoming too stretchy. Other collagen mutations cause missing procollagen chains, kinks in the triple helix, and defects in aggregation outside the cell.

Aortic aneurysm is a serious connective tissue disorder. Detection of the causative mutation before symptoms arise can be lifesaving. An early sign is a weakened aorta (the largest blood vessel in the body, which emerges from the heart), which can suddenly burst. A person who knows that he or she has inherited the mutant collagen gene can have frequent ultrasound exams to detect aortic weakening early enough to treat it surgically.

Early-Onset Alzheimer Disease

The story of the discovery of a mutation that causes an early-onset, autosomal dominant form of Alzheimer disease began in the 1880s, when a woman named Hannah, born in Latvia, developed progressive dementia. Hannah's condition was highly unusual; she was only in her early forties when the classic forgetfulness that heralds the disease's onset began. Apparently this form of the illness originated, in this family, in Hannah. Many of her descendants also experienced dementia, some as early as in their thirties.

In 1974, Hannah's grandson and greatgrandson, both physicians, constructed an extensive pedigree tracing Alzheimer disease in their family. They circulated the pedigree among geneticists, hoping to elicit interest in identifying the family's mutation, offering their own and relatives' DNA for testing. Research teams in Mexico, the United States, and Canada began the search in 1983. By 1992, they narrowed the investigation to a portion of chromosome 14, and three years later, they pinpointed the gene. It encodes a protein called presenilin 1 that acts as a receptor anchored in the membrane of a Golgi apparatus or a vesicle (**figure 12.4**). Normally, the protein monitors the cell's storage or use of beta amyloid, the

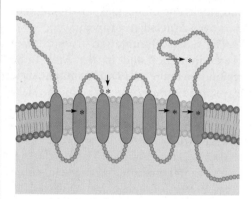

Figure 12.4 One cause of Alzheimer disease. When geneticists searched the DNA of people with very early-onset inherited Alzheimer disease, they identified a gene on chromosome 14 whose protein product, shown here, is a receptor anchored into a membrane at seven points. This protein resides in vesicles derived from the Golgi apparatus. When abnormal, it cuts amyloid precursor proteins into abnormal-sized pieces that fuse and accumulate outside cells. Asterisks indicate sites where mutations in the gene disrupt the protein, causing symptoms.

substance that accumulates in the brains of people with Alzheimer disease. Members of families that have early-onset Alzheimer disease due to mutation in this gene have elevated levels of presenilin 1 in their bloodstreams before symptoms begin. Somehow the abnormality in presenilin disrupts amyloid production, folding, or function.

Many mutations substitute one amino acid for another in the gene for presenilin 1, impairing its function sufficiently to cause the beta amyloid buildup that eventually causes the symptoms of Alzheimer disease. Mutations in at least four other genes can cause or increase the risk of developing Alzheimer disease. **Table 12.2** offers other examples of mutations that impair health.

Multiple Mutations Cause Confusion

The usually precise language of science falls somewhat short when describing the consequences of mutations. Mutations in one gene may cause differing degrees of the same syndrome, whereas for other genes, different mutations may be associated with distinct disorders.

Cystic fibrosis (CF) is associated with a gene with many mutations, but one diagnosis. A man whose only symptom is infertility and occasional respiratory infections could have the same CF diagnosis as an extremely ill teenager with near-constant lung infections and severe malnutrition.

Mutations in the beta globin gene, in contrast, cause clinically distinct illnesses, such as sickle cell disease or beta thalassemia. These two disorders affect the same tissue, blood. More confusing are mutations in a gene called *lamin A,* which affect different parts of the body. The lamin A proteins form a network beneath the inner nuclear membrane that interacts with other proteins. Mutations in the *lamin A* gene cause the rapid-aging disorder Hutchinson-Gilford progeria syndrome (see figure 3.22 and table 3.4) and at least six other conditions, including muscular dystrophies and a heart condition.

Another source of confusion in assigning mutations to specific medical conditions is genetic heterogeneity—the same symptoms caused by different mutant genes. This situation arises when a gene is matched to a syndrome, and then exceptions are noted: That is, people have the symptoms, but not the mutation. This is the case for combined deficiencies of factors V and VIII (OMIM 227300), which impairs blood clotting. The disorder was initially associated with a protein called LMAN1, which is essential for secretion of both clotting factors. But large studies revealed that 30 percent of the affected individuals have a wild type *LMAN1* gene and a mutation in a different gene, *MCFD2.*

The proteins the two genes encode—LMAN1 and MCFD2—act together, and are thus known as co-transporters. Both are necessary to move the clotting factors from the endoplasmic reticulum to the Golgi apparatus. Impair either, and blood does not clot normally.

Table 12.2

How Mutations Cause Disease

Disorder	OMIM	Protein	Genetic Defect (Genotype)	Signs and Symptoms (Phenotype)
Cystic fibrosis	602421	Cystic fibrosis transmembrane regulator (CFTR)	Missing amino acid or other defect alters conformation of chloride channels in certain epithelial cell plasma membranes. Water enters cells, drying out secretions.	Frequent lung infection, pancreatic insufficiency
Duchenne muscular dystrophy	310200	Dystrophin	Deletion eliminates dystrophin, which normally binds to inner face of muscle cell plasma membranes, maintaining cellular integrity. Cells and muscles weaken.	Gradual loss of muscle function
Familial hypercholesterolemia	143890	LDL receptor	Deficient LDL receptors cause cholesterol to accumulate in blood.	High blood cholesterol, early heart disease
Hemophilia A	306700	Factor VIII	Absent or deficient clotting factor causes hard-to-control bleeding.	Slow or absent blood clotting
Huntington disease	143100	Huntingtin	Extra bases in the gene add amino acids to the protein product, which impairs certain transcription factors and proteasomes.	Uncontrollable movements, personality changes
Marfan syndrome	154700	Fibrillin	Deficient connective tissue protein in lens and aorta.	Long limbs, weakened aorta, spindly fingers, sunken chest, lens dislocation
Neurofibromatosis type 1	162200	Neurofibromin	Defect in protein that normally suppresses activity of a gene that causes cell division.	Benign tumors of nervous tissue beneath skin

12.2 Causes of Mutation

A mutation can occur spontaneously or be induced by exposure to a chemical or radiation. An agent that causes mutation is called a **mutagen.**

Spontaneous Mutation

A spontaneous mutation can be a surprise. For example, two healthy people of normal height may have a child with achondroplasia, an autosomal dominant form of dwarfism. How could this happen when no other family members are affected? If the mutation is dominant, why are the parents of normal height? The child has a genetic condition, but he did not inherit it. Instead, he originated it. His siblings have no higher risk of inheriting the condition than anyone in the general population, but each of his children will face a 50 percent chance of inheriting it. The boy's achondroplasia arose from a *de novo,* or new, mutation in his mother's oocyte or father's sperm cell. This is a spontaneous mutation—that is, it is not caused by a mutagen. A spontaneous mutation usually originates as an error in DNA replication.

One cause of spontaneous mutation stems from the chemical tendency of free nitrogenous bases to exist in two slightly different structures, called tautomers. For extremely short times, each base is in an unstable tautomeric form. If, by chance, such an unstable base is inserted into newly forming DNA, an error will be generated and perpetuated when that strand replicates. **Figure 12.5** shows how this can happen.

RELATING THE CONCEPTS

Which of Juanita's beta globin gene variants is the result of spontaneous mutation, and how do you know this?

Spontaneous Mutation Rate

The spontaneous mutation rate varies for different genes. The gene that, when mutant, causes neurofibromatosis type 1 (NF1), for example, has a very high mutation rate, arising in 40 to 100 of every million gametes

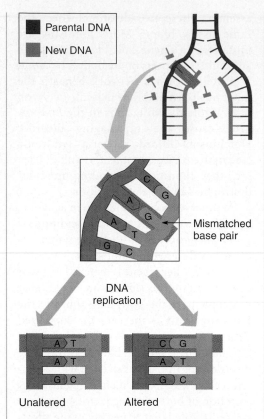

Figure 12.5 Spontaneous mutation. DNA bases are very slightly chemically unstable, and for brief moments they exist in alternate forms. If a replication fork encounters a base in its unstable form, a mismatched base pair can result. After another round of replication, one of the daughter cells has a different base pair than the one in the corresponding position in the original DNA. (This figure depicts two rounds of DNA replication.)

(**table 12.3**). NF1 affects 1 in 3,000 births, about half in families with no prior cases. The gene's large size may contribute to its high mutability—there are more ways for its sequence to change, just as there are more opportunities for a misspelling to occur in a long sentence than in a short one.

Based on the prevalence of certain disease-causing genes, geneticists estimate that each human gene has about a 1 in 100,000 chance of mutating. Each of us probably carries a few new spontaneously mutated genes. Mitochondrial genes mutate at a higher rate than nuclear genes because they cannot repair DNA (see section 12.6.)

Estimates of the spontaneous mutation rate for a particular gene are usually derived from observations of new, domi-

nant conditions, such as achondroplasia in the boy. This is possible because a new dominant mutation is detectable simply by observing the phenotype. In contrast, a new recessive mutation would not be obvious until two heterozygotes produced a homozygous recessive offspring with a noticeable phenotype.

The spontaneous mutation rate for autosomal genes can be estimated using the formula: number of *de novo* cases/2X, where X is the number of individuals examined. The denominator has a factor of 2 to account for the nonmutated homologous chromosome.

Spontaneous mutation rates in human genes are difficult to assess because our generation time is long—usually 20 to 30 years. In bacteria, a new generation arises every half hour or so, and mutation is therefore much more frequent. The genetic material of viruses also spontaneously mutates rapidly.

Mutational Hot Spots

In some genes mutations are more likely to occur in regions called hot spots, where sequences are repetitive. It is as if the molecules that guide and carry out replication become "confused" by short repeated sequences, much as an editor scanning a manuscript might miss the spelling errors in the words "happpiness" and "bananana" (**figure 12.6**). For example, more than one-third of the many mutations that cause alkaptonuria occur at or near one or more CCC repeats, even though these repeats account for only 9 percent of the gene (see the opening essay for chapter 5).

The increased incidence of mutations in repeats has a physical basis. Within a gene, when DNA strands locally unwind to replicate in symmetrical or repeated sequences, bases located on the same strand may pair. For example, a stretch of ATATAT pairs with TATATA elsewhere on the same strand. This pairing interferes with replication and repair enzymes, increasing the chance of an error. For example, mutations in the gene for clotting factor IX, which causes hemophilia B, occur 10 to 100 times as often at any of 11 sites in the gene that have extensive direct repeats of CG.

Small additions and deletions of DNA bases are more likely to occur near sequences called palindromes (see figure 12.6). These

Table 12.3

Mutation Rates of Some Genes That Cause Inherited Disease

Disorder	OMIM	Mutations per Million Gametes	Signs and Symptoms (Phenotype)
X-linked			
Duchenne muscular dystrophy	310200	40–105	Muscle atrophy
Hemophilia A	306700	30–60	Severe impairment of blood clotting
Hemophilia B	306900	0.5–10	Mild impairment of blood clotting
Autosomal Dominant			
Achondroplasia	100800	10	Very short stature
Aniridia	106200	2.6	Absence of iris
Huntington disease	143100	<1	Uncontrollable movements, personality changes
Marfan syndrome	154700	4–6	Long limbs, weakened blood vessels
Neurofibromatosis type 1	162200	40–100	Brown skin spots, benign tumors under skin
Osteogenesis imperfecta	166200	10	Easily broken bones
Polycystic kidney disease	600666	60–120	Benign growths in kidneys
Retinoblastoma	180200	5–12	Malignant tumor of retina

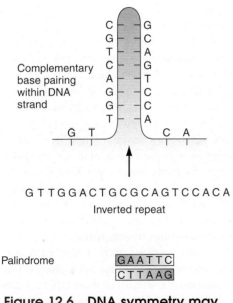

Figure 12.6 DNA symmetry may increase the likelihood of mutation. These examples show repetitive and symmetrical DNA sequences that may "confuse" replication enzymes, causing errors.

sequences read the same, in a 5′ to 3′ direction, on complementary strands. Put another way, the sequence on one strand is the reverse of the sequence on the complementary strand. Palindromes probably increase the spontaneous mutation rate by disturbing replication.

The blood disorder alpha thalassemia (OMIM 141800) illustrates the confusing effect of direct (as opposed to inverted) repeats of an entire gene. A person who does not have the disorder has four genes that specify alpha globin chains, two next to each other on each chromosome 16. Homologs with repeated genes can misalign during meiosis when the first sequence on one chromosome lies opposite the second sequence on the homolog. If crossing over occurs, a sperm or oocyte can form that has one or three of the alpha globin genes instead of the normal two (**figure 12.7**). Fertilization with a normal gamete then results in a zygote with one extra or one missing alpha globin gene.

A person with only three alpha globin genes produces enough hemoglobin, and is a healthy carrier. Rarely, individuals arise with only two copies of the gene, and they are mildly anemic and tire easily. A person with a single alpha globin gene is severely anemic, and a fetus lacking alpha globin genes does not survive.

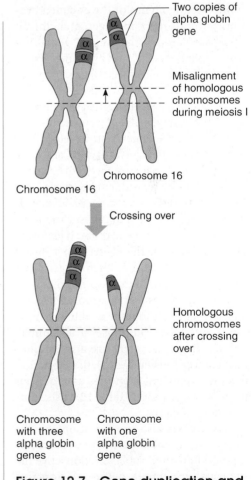

Figure 12.7 Gene duplication and deletion. The repeated nature of the alpha globin genes makes them prone to mutation by mispairing during meiosis.

Induced Mutation

Researchers can sometimes infer a gene's normal function by observing what happens when mutation alters it. Because the spontaneous mutation rate is far too low to be a practical source of genetic variants for experiments, researchers make mutants. Geneticists have used many mutagens on experimental organisms to infer normal gene functions, yielding many collections.

Intentional Use of Mutagens

Chemicals or radiation induce mutation. Alkylating agents, for example, are chemicals that remove a DNA base, which is replaced with any of the four bases—three of which are a mismatch against the complementary strand. Dyes called acridines add or remove a single DNA base. Because the DNA sequence is read three bases in a row, adding or deleting a single base can destroy a gene's information, altering the amino acid sequence of the encoded protein. Several other mutagenic chemicals alter base pairs, so that an A-T replaces a G-C, or vice versa. X rays and other forms of radiation delete a few bases or break chromosomes.

Researchers have developed several ways to test the mutagenicity of a substance. The best known, the Ames test, developed by Bruce Ames of the University of California, assesses how likely a substance is to harm the DNA of rapidly reproducing bacteria. One version of the test uses a strain of *Salmonella* that cannot grow when the amino acid histidine is absent from its medium. If exposure to a substance enables bacteria to grow on the deficient medium, then a gene has mutated that allows it to do so. Another variation of the Ames test incorporates mammalian liver tissue into the medium to make the results more like those of an animal. Because many mutagens are also carcinogens (cancer-causing agents), the substances that the Ames test identifies as mutagens may also cause cancer. **Table 12.4** lists some common mutagens.

Researchers also use variations on the Ames test. In one experiment, researchers exposed human fibroblasts (connective tissue cells) growing in culture to liquefied cigarette smoke. The chemicals from the smoke cut chromosomes through both DNA strands. This is an especially damaging insult because broken chromosomes can join with each other in different ways that can activate cancer-causing genes. Hence, the experiment may have modeled one way that cigarettes cause cancer.

A limitation of using a mutagen is that it cannot cause a specific mutation. In contrast, a technique called site-directed mutagenesis changes a gene in a desired way. A gene is mass-produced, but it includes an intentionally substituted base, just as an error in a manuscript is printed in every copy of a book. Site-directed mutagenesis is faster and more precise than waiting for nature or a mutagen to produce a useful variant. It also makes it possible to study lethal mutations that can theoretically exist, but never do because they are so drastic. Researchers can study such a lethal mutation in cell culture, or in model organisms before they cease developing.

Accidental Exposures to Mutagens

Some mutagen exposure is unintentional. This occurs from workplace contact before the danger is known; from industrial accidents; from medical treatments such as chemotherapy and radiation; and from exposure to weapons that emit radiation.

An environmental disaster that released mutagenic radiation was a steam explosion at a nuclear reactor in the former Soviet Union on April 25, 1986. Between 1:23 and 1:24 A.M., Reactor 4 at the Chernobyl Nuclear Power Station in Ukraine exploded, sending a great plume of radioactive isotopes into the air that spread for thousands of miles. The reactor had been undergoing a test, its safety systems temporarily disabled, when it overloaded and rapidly flared out of control. Twenty-eight people died of acute radiation exposure in the days following the explosion.

Acute radiation poisoning is not a genetic phenomenon. Evidence of a mutagenic effect has come from the increased rate of thyroid cancer among children who were living in nearby Belarus. Rates have multiplied tenfold. The thyroid glands of young people soak up iodine, which in a radioactive form bathed the area in the days after the explosion. Cancer rates have also risen among workers who cleaned up the disaster. Analysis of radiation exposure in their teeth is being used to assess whether cancer risk rises with degree of exposure.

Another way researchers tracked mutation rates after the Chernobyl explosion was to compare the lengths of short DNA repeats called minisatellite sequences in children born in 1994 and in their parents, who lived in the Mogilev district of Belarus at the time of the accident and have remained there. Minisatellites are the same length within all cells of an individual. A minisatellite size in a child that does not match the size of either parent indicates that a mutation occurred in a parent's gamete. Such a mutation was twice

Table 12.4

Commonly Encountered Mutagens

Mutagen	Source
Aflatoxin B	Fungi growing on peanuts and other foods
2-amino 5-nitrophenol	Hair dye components
2,4-diaminoanisole	"
2,5-diaminoanisole	"
2,4-diaminotoluene	"
p-phenylenediamine	"
Furylfuramide	Food additive
Nitrosamines	Pesticides, herbicides, cigarette smoke
Proflavine	Antiseptic in veterinary medicine
Sodium nitrite	Smoked meats
Tris (2,3-dibromopropyl phosphate)	Flame retardant in children's sleepwear

as likely to occur in exposed families than in control families living elsewhere. Mutation rates of nonrepeated DNA sequences are too low to provide useful information on the effects of radiation exposure, so investigators track minisatellites as a sensitive test of change.

Researchers learned of a new type of mutation from a young man conceived within a week of the Chernobyl accident, near the disaster site. He has extra digits, an abnormal epiglottis, and a benign growth on the hypothalamus, a group of symptoms called Pallister-Hall syndrome (OMIM 146510). On his way to a camp for "children of Chernobyl" in the summer of 2002, he stopped at the National Institutes of Health to provide a DNA sample. Researchers indeed found a mutation in the gene on chromosome 7 known to cause the syndrome—a 72-base insertion that causes a "stop" codon to form, shortening the encoded protein. Oddly, the insertion matched mitochondrial DNA sequences. Apparently, the radiation damaged mitochondria in the sperm or oocyte, sending some mitochondrial DNA into the nucleus, where it inserted into the Pallister-Hall gene. Another clue to the unusual origin of this young man's condition is that it is autosomal dominant, but neither of his parents have it. Since his case, researchers have discovered 27 such "nuclear DNA sequences of mitochondrial origin."

Natural Exposure to Mutagens

The simple condition of being alive exposes us to radiation that can cause mutation. Natural environmental sources of radiation include cosmic rays, sunlight, and radioactive minerals in the earth's crust, such as radon. Contributions from medical X rays and occupational radiation hazards are comparatively minor (**table 12.5**). Job sites with increased radiation exposure include weapons facilities, research laboratories, health care facilities, nuclear power plants, and certain manufacturing plants. Radiation exposure is measured in units called millirems; the average annual exposure in the northern hemisphere is 360 millirems.

Most of the potentially mutagenic radiation we are exposed to is of the ionizing type, which means that it has sufficient energy to remove electrons from atoms.

Table 12.5	
Sources of Radiation Exposure	
Source	**Percentage of Total**
Natural (cosmic rays, sunlight, earth's crust)	81%
Medical X rays	11%
Nuclear medicine procedures	4%
Consumer products	3%
Other (nuclear fallout, occupational)	<1%

Unstable atoms that emit ionizing radiation both exist naturally and are made by humans. Ionizing radiation breaks the DNA sugar-phosphate backbone.

Ionizing radiation is of three major types. Alpha radiation is the least energetic and most short-lived, and the skin absorbs most of it. Uranium and radium emit alpha radiation. Beta radiation can penetrate the body farther, and emitters include tritium (a form of hydrogen), carbon-14, and strontium-70. Both alpha and beta rays tend not to harm health, although they can do damage if inhaled or eaten. In contrast is the third type of ionizing radiation, gamma rays. These can penetrate all the way through the body, damaging tissues. Plutonium and cesium isotopes used in weapons emit gamma rays, and this form of radiation is used to kill cancer cells.

X rays are the major source of exposure to human-made radiation, and they are not a form of ionizing radiation. They have less energy and do not penetrate the body to the extent that gamma rays do.

The effects of radiation damage to DNA depend upon the functions of the mutated genes. Mutations in oncogenes or tumor suppressor genes, discussed in chapter 18, can cause cancer. Radiation damage can be widespread, too. Exposing cells to radiation and then culturing them causes a genome-wide destabilization, so that mutations may occur even after the cell has divided a few times. Cell culture studies have also identified a "bystander effect," when radiation seems to harm even cells not directly exposed. Researchers have noted which cells in an experiment received radiation, then detected chromosome breakage in those cells and in cells located nearby. This effect is not well understood.

Chemical mutagens exist in the environment, too. Evaluating the risk that a specific chemical exposure will cause a mutation is very difficult, largely because people vary greatly in inherited susceptibilities, and are exposed to many chemicals. The risk that exposure to a certain chemical will cause a mutation is often less than the natural variability in susceptibility within a population, making it nearly impossible to track the true source and mechanism of any mutational event. Human genome sequence information can be used to determine specific inherited risks for specific employees who might encounter a mutagen in the workplace. However, such testing raises ethical concerns. The Bioethics box discusses testing for sensitivity to the element beryllium.

Key Concepts

1. Genes have different mutation rates.
2. Spontaneous mutations result when rare base tautomers are incorporated during replication.
3. Spontaneous mutations are more frequent in microorganisms and viruses because they reproduce often and lack DNA repair.
4. Mutations are more likely to happen when the nearby DNA is repetitive or symmetrical.
5. Mutagens are chemicals or radiation that increase risk of mutation. Researchers use mutagens to more quickly obtain mutants, which reveal normal gene function. Site-directed mutagenesis creates and amplifies specific mutations.
6. Mutagen exposure can be accidental.
7. Some radiation sources are natural.

Beryllium Screening

Screening workers for a genetic variant that predisposes them to develop a possibly fatal reaction to a substance they may contact on the job may seem like a good idea, but some workers say the risks outweigh uncertain benefits. The case in point: screening for chronic beryllium disease (CBD), also called berylliosis. This condition causes the person to react to the metal beryllium, which is used in nuclear power plants, in electronics, and in manufacturing fluorescent powders. Exposed workers include those who mine beryllium, nuclear power plant employees, and anyone who inhales beryllium dust.

A few people exposed to beryllium dust or vapor develop an immune response that damages the lungs, producing cough, shortness of breath, fatigue, loss of appetite, and weight loss. Fevers and night sweats indicate the immune system is responding to the exposure. The steroid drug prednisone can control symptoms, but it isn't used until symptoms begin, which can be anywhere from a few months to forty years after first exposure.

The Department of Energy and some private companies have screened more than ten thousand workers exposed to beryllium on the job using a test based on immune system response. In people who have symptoms, certain white blood cells divide in the presence of beryllium. About 45 percent of people without symptoms who test positive

develop the condition. A more precise predictive genetic test will eventually replace the immune system test, but even this test is imperfect. It detects homozygosity for a rare genetic variant that is part of the human leukocyte antigen complex, a group of genes that control immune system function. A person who tests positive on this genetic test has an 85 percent chance of developing CBD if exposed to beryllium (**figure 1**).

Genetic screening for beryllium sensitivity will be controversial. So far, workers at the Department of Energy who test positive on the immune system test are not allowed to work near beryllium. Some of them prefer to decide for themselves where they work. In addition, both the immune system test and genetic test are inconclusive—they detect susceptibility, so some people who test positive will not develop the condition. Finally, people fear that test results might have negative effects on health insurance coverage.

The goal of CBD screening at the Department of Energy and the companies is to protect workers. Not only are sensitive individuals kept away from beryllium, but efforts have been underway for several years to minimize the dust and to make sure all beryllium workers use protective clothing and devices. Still, some people object to identifying susceptible individuals. This is

Figure 1 Beryllium screening.
Screening for beryllium sensitivity aims to protect workers, but some people see it as an invasion of their genetic privacy.

yet another area where "political correctness" clashes with the reality of genetics—individuals *do* vary in their responses to many stimuli, including environmental lung irritants. To deny that is to expose certain people to potentially harmful surroundings. Genetic screening for beryllium sensitivity may set a precedent for other types of susceptibility testing being developed using human genome information.

12.3 Types of Mutations

Mutations are classified by exactly how they alter DNA. **Table 12.6** summarizes the types of genetic changes described in this section using an analogy to an English sentence.

Point Mutations

A **point mutation** is a change in a single DNA base. It is a **transition** if a purine replaces a purine (A to G or G to A) or a pyrimidine replaces a pyrimidine (C to T or

T to C). It is a **transversion** if a purine replaces a pyrimidine or vice versa (A or G to T or C). A point mutation can have any of several consequences—or it may have no obvious effect at all on the phenotype, acting as a silent mutation.

Missense and Nonsense Mutations

A point mutation that changes a codon that normally specifies a particular amino acid into one that codes for a different amino acid is called a **missense mutation.** If the

substituted amino acid alters the protein's conformation sufficiently or occurs at a site critical to its function, signs or symptoms of disease or an observable variant of a trait may result.

The point mutation that causes sickle cell disease (see figure 12.1) is a missense mutation. The DNA sequence CTC encodes the mRNA codon GAG, which specifies glutamic acid. In sickle cell disease, the mutation changes the DNA sequence to CAC, which encodes GUG in the mRNA, which specifies valine. This mutation changes the protein's shape, which alters its function.

Table 12.6

Types of Mutations

A sentence comprised of three-letter words can provide an analogy to the effect of mutations on a gene's DNA sequence:

Normal	THE ONE BIG FLY HAD ONE RED EYE
Missense	TH**Q** ONE BIG FLY HAD ONE RED EYE
Nonsense	THE ONE BIG
Frameshift	THE ONE **QBI GFL YHA DON ERE DEY**
Deletion	THE ONE BIG ▢ HAD ONE RED EYE
Insertion	THE ONE BIG **WET** FLY HAD ONE RED EYE
Duplication	THE ONE BIG FLY **FLY** HAD ONE RED EYE
Expanding mutation	
generation 1	THE ONE BIG FLY HAD ONE RED EYE
generation 2	THE ONE BIG FLY **FLY FLY** HAD ONE RED EYE
generation 3	THE ONE BIG FLY **FLY FLY FLY FLY** HAD ONE RED EYE

A point mutation that changes a codon specifying an amino acid into a "stop" codon—UAA, UAG, or UGA in mRNA—is a **nonsense mutation.** A premature stop codon shortens the protein product, which can profoundly influence the phenotype. Nonsense mutations are predictable by considering which codons can mutate to a "stop" codon.

The most common cause of factor XI deficiency (OMIM 264900), a blood clotting disorder, is a nonsense mutation that changes one GAA codon specifying glutamic acid to UAA, signifying "stop." The shortened clotting factor cannot halt the profuse bleeding that occurs during surgery or from injury. In the opposite situation, when a normal stop codon mutates into a codon that specifies an amino acid, the resulting protein is longer than normal, because translation continues through what is normally a stop codon.

Point mutations may exert profound effects by controlling how transcription proceeds. For example, in 15 percent of people who have Becker muscular dystrophy (OMIM 310200)—a milder adult form of the condition—the muscle protein dystrophin is normal, but its levels are reduced. The mutation causing the protein shortage is in the promoter for the dystrophin gene, which slows the transcription rate. Since cells then produce fewer mRNAs that encode dystrophin, the protein is scarce. Muscle function suffers. In contrast, the other 85 percent of individuals who have Becker muscular dystrophy have shortened proteins, not a deficiency of normal-length proteins.

Another way that point mutations can affect protein production is to disrupt the trimming of long precursor molecules. Such a mutation causes the type of Ehlers-Danlos syndrome that affects the boy in figure 12.3.

RELATING THE CONCEPTS

Is each of Juanita's mutations a missense or a nonsense mutation?

Splice Site Mutations

A point mutation can greatly affect a gene's product if it alters a site where introns are normally removed from the mRNA. This is called a splice site mutation. It can affect the phenotype if an intron is translated into amino acids, or if an exon is skipped instead of being translated, shortening the protein.

Retaining an intron adds bases to the protein coding portion of an mRNA. For example, in one family with severe cystic fibrosis, a missense mutation alters an intron site so that it is not removed. The encoded protein is too bulky to move to its normal position in the plasma membrane, where it should enable salt to exit the cell. As a result, chloride (a component of salt) accumulates in cells and water moves in, drying and thickening the mucus outside the cells.

A missense mutation need not alter the amino acid sequence to cause harm if it disrupts intron/exon splicing. For example, a missense mutation in the BRCA1 gene that causes a form of breast cancer (OMIM 113705) went undetected for a long time because it does not alter the amino acid sequence. Instead, the protein is missing several amino acids. The missense mutation creates an intron splicing site where there should not be one, and an entire exon is "skipped" when the mRNA is translated into protein, as if it were an intron. This mutation, therefore, is a deletion (missing material), but is caused by a missense mutation.

A disorder called familial dysautonomia (OMIM 223900)(FD) usually results from a splice site mutation that causes exon skipping. An exon in the gene encoding an enzyme (I-kappa beta-kinase-associated protein) is not translated because a point mutation in one of its splice sites signals the spliceosome not to translate that segment. Symptoms reflect loss of certain neurons that control sensation and involuntary responses. The In Their Own Words box in this chapter describes life for a child with FD.

A peculiarity of some disorders caused by exon-skipping mutations is that some cells seem to ignore the problem, manufacturing a normal protein from the affected gene—after all, the amino acid sequence information is still there. Depending upon which cells function, the phenotype may be less severe than in individuals with the same disorder but with a different type of mutation in the coding portion of the gene.

Studies on various cell types from individuals with FD or who have died from the disease reveal that the cells where the exon is skipped are the cells that contribute to symptoms. That is, many cells from the brain and spinal cord are missing the exon, but cells from muscle, lung, liver, white

Familial Dysautonomia: Rebekah's Story

Our daughter Rebekah has familial dysautonomia. This is a rare genetic disorder that affects the functioning of the autonomic and peripheral nervous systems. Rebekah was born in 1992, but she was not diagnosed until she was almost three. Her diagnosis came as both a shock and a relief, putting her bizarre and terrifying symptoms into perspective.

Rebekah, who appeared to be healthy at birth, developed medical problems that became more serious with each passing month. Always a discontented, gassy baby, Rebekah began to decline rapidly by nine months, and within a year our lifestyle included twice monthly visits to the hospital. She suffered from frequent pneumonia, vomiting and retching, extremely high fevers, chills, rapid heartbeat, and seizures. At times, she would become covered with hot, red blotches. Other times, her hands and feet got very cold and appeared puffy and blue. Episodes of crying would precipitate breathholding, when she would turn blue and lose consciousness. As she lost ground on the growth and development charts, medical testing failed to reveal a cause for these symptoms. As we watched our baby suffer and become more ill, we wondered if we would identify the problem before she died. Our physicians, in their frustration, sometimes hinted that perhaps we were the cause.

After more than twelve local hospitalizations and a variety of tests, we traveled to a major children's teaching hospital, hoping that a fresh team of doctors would identify Rebekah's condition. To our surprise, one doctor knew immediately that she had FD. He recognized the pattern of "dysautonomic crises"—a hallmark of FD. Two more symptoms, which we hadn't even noticed, were diagnostic indicators. Individuals with FD do not cry tears, and they lack papillae (bumps) on the tip of the tongue. Our Eastern European, Jewish heritage was also a clue, because FD is one of a number of diseases primarily affecting this population.

To a varying degree, FD reduces sensation of pain, heat, and cold. There are problems with balance and coordination, including motor difficulties that affect feeding, swallowing, and breathing. Most people with FD have a feeding tube, and must limit what they eat or drink by mouth due to danger of aspiration. FD causes fluctuations in blood pressure, digestive problems, and learning disabilities. Most individuals develop scoliosis, usually requiring corrective spine surgery before growth is complete. In short, FD affects every organ and system in the body.

Confronting this diagnosis was a shock, but the alternative of not knowing was even worse. Having a diagnosis has allowed us to finetune Rebekah's therapies and activities to maximize her health and well-being. With improved nutrition, excellent therapies, and wonderful teachers, Rebekah has made tremendous progress, but we are always poised for a hospital stay. Even a minor illness can set off a crisis. A team of pediatric specialists monitors her lungs, heart, eyes, back, and growth and development.

Rebekah is a happy, good-natured child who makes friends easily and is sensitive to the needs of others. She works hard in school, and is able to keep up with her classmates when her health is good. She has learned to overcome her learning challenges, using assistive technology in school to help her with writing and organizing. When Rebekah got a back brace to try to slow the scoliosis, and I tried to steer her to choose clothes that would deemphasize the bulk of the brace, she told me, "Mom, just relax. They're going to see it sooner or later!"

Figure 1 Rebekah with her dog, Tracy.

We don't know what the future holds. FD is a progressive, degenerative disease with life-threatening complications and a shortened lifespan. Any major stress, including developmental changes, surgery, a serious illness, and increased emotional stress, can exacerbate the severity. Yet, we feel hopeful for our daughter's future. With the discovery of the genetic mutations that cause FD in 2001, we know scientists are working on developing effective treatments and a cure. We are most encouraged by Rebekah herself. Her positive outlook on life, her willingness to find the good in any situation, and her ability to overcome challenges with spunk and humor inspire everyone around her. We can't predict the future, but we can say with confidence that Rebekah will be able to experience a life filled with joy and achievement. Isn't this what we all want for our children?

Lynn Lieberman

blood cells, and various glands produce normal protein. This means there may be a way to coax nervous system cells in affected children to also produce the protein. Current clinical trials are examining the ability of several natural compounds to restore normal processing of the FD gene's information.

Deletions and Insertions Can Shift the Reading Frame

In genes, the number three is very important, because triplets of DNA bases specify amino acids. Adding or deleting a number of bases that is not a multiple of three devastates a gene's function because it disrupts the gene's reading frame, which refers to the nucleotide position where the DNA begins to encode protein. Such a change that alters the reading frame is called a **frameshift mutation.** The mutation that disables the CCR5 HIV receptor is a frameshift. It is described toward the beginning of the chapter.

A **deletion mutation** removes genetic material. A deletion that removes three or a multiple of three bases will not cause a frameshift, but can still alter the phenotype. Deletions range from a single DNA nucleotide to thousands of bases to larger pieces of chromosomes. The next chapter considers large deletions. Many common inherited disorders result from deletions. About two-thirds of people with Duchenne muscular dystrophy, for example, are missing large sections of the huge gene that encodes dystrophin. Many cases of male infertility are caused by tiny deletions in the Y chromosome.

An **insertion mutation** adds DNA and it, too, can offset a gene's reading frame. In one form of Gaucher disease (OMIM 230800), for example, an inserted single DNA base prevents production of an enzyme that normally breaks down glycolipids in lysosomes. The resulting buildup of glycolipid enlarges the liver and spleen and causes easily fractured bones and neurological impairment. Gaucher disease is common among Jewish people of eastern European descent. Although most cases arise from a missense mutation, some families have the insertion mutation. Gaucher disease illustrates how different types of mutations in the same gene cause the same or a similar phenotype.

Another type of insertion mutation repeats part of a gene's sequence. The insertion is usually adjacent or close to the original sequence, like a typographical error repeating a word word. Two copies of a gene next to each other is called a **tandem duplication.** A form of Charcot-Marie-Tooth disease (OMIM 118200), which causes numb hands and feet, results from a one-and-a-half-million-base-long tandem duplication.

Figure 12.8 compares the effects on protein sequence of missense, nonsense, and frameshift mutations in the gene that encodes the LDL receptor, causing familial hypercholesterolemia (OMIM 144010)(see figure 5.2). These three mutations exert very different effects on the protein. A missense mutation replaces one amino acid with another, bending the protein in a way that impairs its function. A nonsense mutation is much more drastic, removing part of the protein. A frameshift mutation introduces a section of amino acids not normally part of the protein.

Pseudogenes and Transposons Revisited

Recall from chapter 11 that a pseudogene is a DNA sequence that is very similar to that of a protein-encoding gene. A pseudogene is not translated into protein, although it may be transcribed. The pseudogene may have descended from the original gene sequence, which was duplicated when DNA strands misaligned during meiosis, similar to the situation depicted in figure 12.7 for the alpha globin gene. When this happens, a gene and its pseudogene end up right next to each other on the chromosome. The original gene or the copy then mutates to such an extent that it is no longer functional and becomes a pseudogene. Its duplicate lives on as the functional gene.

Although a pseudogene is not translated, its presence can interfere with the expression of the functional gene and cause a mutation. For example, some cases of Gaucher disease can result from a crossover between the working gene and its pseudogene, which has 96 percent of the same

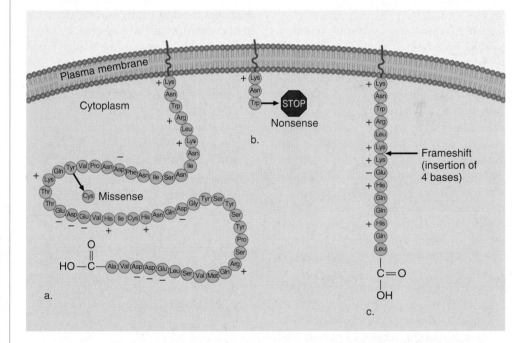

Figure 12.8 Different mutations in a gene can cause the same disorder.
In familial hypercholesterolemia, several types of mutations may disrupt the portion of the LDL receptor normally anchored in the cytoplasm. LDL receptor **(a)** bears a missense mutation—a substitution of a cysteine for a tyrosine. The receptor is bent enough to impair its function, causing disease. The short LDL receptor in **(b)** results from a nonsense mutation, in which a stop codon replaces a tryptophan codon. In **(c)**, a 4-base insertion alters the reading frame, so that a sequence of amino acids not normally in this protein forms until a stop codon occurs.

sequence located 16,000 bases away. The result is a fusion gene, which is a sequence containing part of the functional gene and part of the pseudogene. The fusion gene does not retain enough of the normal gene sequence to enable the cell to synthesize the enzyme. Gaucher disease results. As previously mentioned, this is a lysosomal storage disease whose symptoms include fatigue, bruising, anemia, and weak bones. For many patients, supplying the enzyme eliminates symptoms.

Chapter 11 also considered transposons, or "jumping genes." Transposons can disrupt the site they jump from, shut off transcription of the gene they jump into, or alter the reading frame of their destination if they are not a multiple of three bases. For example, a boy with X-linked hemophilia A had a transposon in his factor VIII gene—a sequence that was also in his carrier mother's genome, but on her chromosome 22. Apparently, in the oocyte, the transposon jumped into the factor VIII gene on the X chromosome, causing the boy's hemophilia.

Expanding Repeats

Until 1992, myotonic dystrophy was very puzzling because it worsened and began at an earlier age as it passed from one generation to the next. This phenomenon is called "anticipation," and for many years it was thought to be psychological. A grandfather might experience only mild weakness in his forearms, and cataracts. His daughter might have more noticeable arm and leg weakness, and a flat facial expression. Her children who inherit the genes might experience severe muscle impairment.

With the ability to sequence genes, researchers found that myotonic dystrophy indeed worsens with each generation because the gene expands! The gene, on chromosome 19, has an area rich in repeats of the DNA triplet CTG. A person who does not have myotonic dystrophy usually has from 5 to 37 copies of the repeat, whereas a person with the disorder has from 50 to thousands of copies (**figure 12.9**). Myotonic dystrophy is an example of an expanding triplet repeat disorder.

So far, expanding triplet repeats have been discovered in more than fifteen human inherited disorders. Usually, a repeat num-

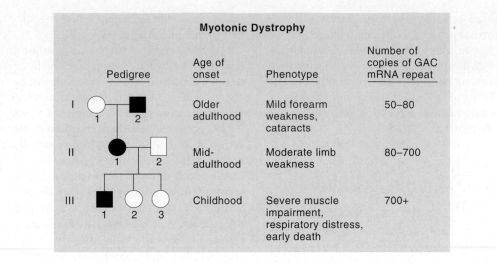

Figure 12.9 **Expanding genes explain anticipation.** In some disorders, symptoms that worsen from one generation to the next—a phenomenon termed *anticipation*—have a physical basis: The gene is expanding as the number of repeats grows.

ber of fewer than 40 copies is stably transmitted to the next generation and doesn't produce symptoms. Larger repeats are unstable, increasing in number with each generation and causing symptoms that are more severe and begin sooner. Reading 12.1 describes the first triplet repeat disorder to be discovered, fragile X syndrome.

The mechanism behind triplet repeat disorders lies in the DNA sequence. The bases of the repeated triplets implicated in the expansion diseases, unlike others, bond to each other in ways that bend the DNA strand into shapes, such as hairpins. These shapes then interfere with replication, which causes the expansion. Once these repeats are translated, the extra-long proteins shut down cells in various ways:

- binding to parts of transcription factors that have stretches of amino acid repeats similar to or matching the expanded repeat

- blocking proteasomes and thereby enabling misfolded proteins to persist

- directly triggering apoptosis.

Triplet repeat proteins may also enter the nucleus when their wild type versions function only in the cytoplasm, or vice versa.

The triplet repeat disorders are said to cause a "dominant toxic gain of function." This means that they cause something novel to happen, rather than removing a function, such as is often associated with recessive enzyme deficiencies. The idea of a gain of function arose from the observation that deletions of these genes do not cause symptoms. **Table 12.7** describes several triplet repeat disorders. Particularly common among them are the "polyglutamine diseases" that have repeats of the mRNA codon CAG, which encodes the amino acid glutamine.

For some triplet repeat disorders, the mutation thwarts gene expression before a protein is even manufactured. In myotonic dystrophy type 1—the gene variant on chromosome 19 in which triplet repeats were discovered—the expansion occurs in the initial untranslated region of the gene, resulting in a huge mRNA. When genetic testing became available for myotonic dystrophy, researchers discovered a second form of the illness in people who had symptoms, but wild type alleles for the myotonic dystrophy gene on chromosome 19. They have myotonic dystrophy type 2, which is caused by an expanding *quadruple* repeat of CCTG in a gene on chromosome 3. Affected individuals have more than 100 copies of the repeat, compared to the normal fewer than 10 copies.

When researchers realized that this second repeat mutation for myotonic dystrophy was also in a non-protein-encoding part of the gene—an intron—a mechanism of disease became apparent: The mRNA is

Fragile X Syndrome—The First of the Triplet Repeat Disorders

In the 1940s, geneticists hypothesized that a gene on the X chromosome confers mental retardation, because more affected individuals are male. It wasn't until 1969, though, that a clue emerged to the genetic basis of X-linked mental retardation. Two retarded brothers and their mother had an unusual X chromosome. The tips at one chromosome end dangled, separated from the rest of each chromatid by a thin thread (**figure 1**). When grown under specific culture conditions (lacking folic acid), this part of the X chromosome was very prone to breaking—hence, the name fragile X syndrome. Although fragile X syndrome was first detected at the chromosomal level, the cause is a mutation at the DNA level—but of a type never seen before.

Fragile X syndrome is second only to Down syndrome in genetic or chromosomal causes of mental retardation. Worldwide, it affects 1 in 2,000 males, accounting for 4 to 8 percent of all males with mental retardation. One in 4,000 females is affected. They usually have milder cases because of the presence of a second, normal X chromosome.

Youngsters with fragile X syndrome do not appear atypical, but by young adulthood, certain similarities emerge. The fragile X patient has a very long, narrow face. The ears protrude, the jaw is long, and the testicles are very large. Mental impairment and behavioral problems vary, and relate to difficulty in handling environmental stimuli. They include mental retardation, learning disabilities, repetitive speech, hyperactivity, shyness, social anxiety, a short attention span, language delays, and temper outbursts.

Fragile X syndrome is inherited in an unusual pattern. Because the fragile chromosome is the X, the associated syndrome should be transmitted as any X-linked trait is, from carrier mother to affected son. However, penetrance is incomplete. One-fifth of males who inherit the chromosomal abnormality have no symptoms. However, because they pass on the affected chromosome to all their daughters—half of whom have some degree of mental impairment—they are called "transmitting males." A transmitting male's grandchildren may inherit the condition.

Researchers in the 1980s were on the right track when they proposed two states of the X chromosome region responsible for fragile X signs and symptoms—a premutation form that does not cause symptoms but does transmit the condition, and a full mutation form that usually causes mental impairment. Still, it took a molecular-level look in 1991 to begin to clarify the inheritance pattern of fragile X syndrome.

In unaffected individuals, the fragile X area contains 6 to 50 repeats of the DNA sequence CGG, as part of a gene called the fragile X mental retardation gene (*FMR1*). In people who have the fragile chromosome and show its effects, this region is greatly expanded to 200 to 2,000 CGG repeats. Transmitting males, as well as females with mild symptoms, often have a premutation consisting of an intermediate number of repeats—50 to 200 copies.

The *FMR1* gene encodes fragile X mental retardation protein (FMRP). This protein, when abnormal, binds to and disables several different mRNA molecules whose encoded proteins are crucial for brain neuron function. The fact that a mutation in *FMR1* ultimately affects several proteins explains the several signs and symptoms of fragile X syndrome.

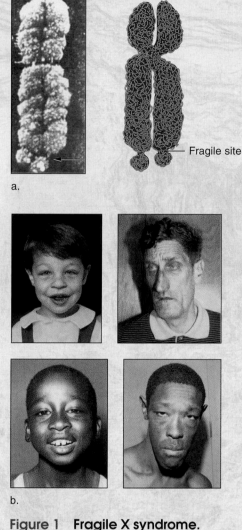

a.

b.

Figure 1 Fragile X syndrome.
A fragile site on the tip of the long arm of the X chromosome **(a)** is associated with mental retardation and a characteristic long face that becomes pronounced with age **(b)**.

too big to leave the nucleus. In myotonic dystrophy type 1, the excess material is added to the start of the gene; in type 2, it appears in an intron that is not excised. The bulky mRNAs bind to a protein that, in turn, alters intron splicing in several other genes. Deficiency of the proteins encoded by these final affected genes causes the symptoms.

A lesson learned from the expanding repeat disorders is that a DNA sequence is more than just one language that can be translated into another. Whether a sequence is random—CGT CGT ATG CAT CAG, for example—or highly repetitive—such as CAG CAG CAG CAG on and on—can affect transcription, translation, or the ways that proteins interact.

Table 12.7

Triplet Repeat Disorders

Disorder	OMIM	mRNA Repeat	Normal Number of Copies	Disease Number of Copies	Signs and Symptoms (Phenotype)
Fragile X syndrome	309550	CGG or CCG	6–50	200–2,000	Mental retardation, large testicles, long face
Friedreich ataxia	229300	GAA	6–29	200–900	Loss of coordination and certain reflexes, spine curvature, knee and ankle jerks
Haw River syndrome	140340	CAG	7–25	49–75	Loss of coordination, uncontrollable movements, dementia
Huntington disease	143100	CAG	10–34	40–121	Personality changes, uncontrollable movements, dementia
Jacobsen syndrome	147791	CGG	11	100–1,000	Poor growth, abnormal face, slow movement
Myotonic dystrophy type I	160900	CTG	5–37	80–1,000	Progressive muscle weakness; heart, brain, and hormone abnormalities
Myotonic dystrophy type II	602668	CCTG	<10	>100	Progressive muscle weakness; heart, brain, and hormone abnormalities
Spinal and bulbar muscular atrophy	313200	CAG	14–32	40–55	Muscle weakness and wasting in adulthood
Spinocerebellar ataxia (5 types)	271245	CAG	4–44	40–130	Loss of coordination

Key Concepts

1. A point mutation alters a single DNA base and can occur in any part of a gene.
2. In a transversion, a purine replaces a pyrimidine, or vice versa; in a transition, a purine replaces a purine or a pyrimidine replaces a pyrimidine.
3. A missense mutation replaces one amino acid with another.
4. A nonsense mutation changes an amino-acid-coding codon into a "stop" codon, shortening the protein. A stop codon that is changed to an amino-acid-coding codon lengthens the protein.
5. Mutations in intron/exon splice sites, promoters, or other control regions affect gene function.
6. Inserting or deleting bases can cause a frameshift mutation.
7. Tandem duplications repeat a section of a gene.
8. Pseudogenes are nonfunctional sequences very similar to nearby functional genes.
9. Transposons can move, insert into genes, and cause illness.
10. Expanded repeats exert effects that arise from protein misfolding.
11. The degree of repetition in a DNA sequence may affect its function.

12.4 The Importance of Position

The degree to which a mutation alters the phenotype depends upon where in the gene the change occurs, and how the mutation affects the conformation, activity, or expression of an encoded protein. A mutation that replaces an amino acid with a very similar one would probably not affect the phenotype greatly, because it wouldn't substantially change the conformation of the protein. Even substituting a very different amino acid would not have much effect if the change is in part of the protein not crucial to its function. In contrast, Juanita's mutations in the beta globin gene critically changed the protein, under certain conditions.

The effects of specific mutations are well-studied in hemoglobin. They are less understood, but still fascinating, in the gene that encodes prion protein.

Globin Variants

Because the globin gene mutations were the first to be analyzed in humans, and because some variants are easily detected using electrophoresis, hundreds of globin gene mutations have been known for years. Mutations in these genes can cause anemia with or without sickling, or cause cyanosis (a blue pallor due to poor oxygen binding). Rarely, a mutation boosts the molecule's affinity for oxygen. Some globin gene variants exert no effect and are thus considered "clinically silent" (**Table 12.8**).

Oddly, hemoglobin S and hemoglobin C are variants that result from mutations that change the sixth amino acid in the beta globin polypeptide, but in different ways. Homozygotes for hemoglobin S have sickle cell disease, yet homozygotes for hemoglobin C are healthy. Both types of homozygotes are resistant to malaria because the unusual hemoglobin alters the shapes and surfaces of red blood cells in ways that keep out the parasite that causes the illness, discussed in chapter 15.

An interesting consequence of certain mutations in either the alpha or beta globin chains is hemoglobin M. Normally, the iron in hemoglobin is in the ferrous form, which means that it has two positive charges. In hemoglobin M, the mutation stabilizes the ferric form, which has three positive charges and cannot bind oxygen. Fortunately, an enzyme converts the abnormal ferric iron to the normal ferrous form, so that the only symptom is usually cyanosis. The condition

Table 12.8

Globin Mutations

Associated Phenotype	Name	Mutation
Clinically silent	Hb Wayne	Single-base deletion in alpha gene causes frameshift, changing amino acids 139–141 and adding amino acids
	Hb Grady	Nine extra bases add three amino acids between amino acids 118 and 119 of alpha chain
Oxygen binding	Hb Chesapeake	Change from arginine to leucine at amino acid 92 of beta chain
	Hb McKees Rock	Change from tyrosine to STOP codon at amino acid 145 in beta chain
Anemia	Hb Constant Spring	Change from STOP codon to glutamine elongates alpha chain
	Hb S	Change from glutamic acid to valine at amino acid 6 in beta chain causes sickling
	Hb Leiden	Amino acid 6 deleted from beta chain
Protection against malaria	Hb C	Change from glutamic acid to lysine at amino acid 6 in beta chain causes sickling

has been known for more than two hundred years in a small town in Japan. Many people there have "blackmouth" because of the cyanosis caused by the faulty hemoglobin. It is autosomal dominant.

Even more noticeable than people with blackmouth are the "blue people of Troublesome Creek." Seven generations ago, a French orphan who settled in this area of Kentucky brought in a recessive gene that causes a form of methemoglobinemia. He was missing an enzyme (cytochrome b5 reductase) that normally converts a type of hemoglobin with poor oxygen affinity, called methemoglobin, back into normal hemoglobin by adding an electron. This man chose a wife who was a carrier for the same disease. After extensive inbreeding in the isolated community, a large pedigree of "blue people" of both sexes arose.

In "blue person disease," the excess oxygen-poor hemoglobin causes a dark blue complexion. Carriers may have frighteningly bluish lips and fingernails at birth, which usually improve. This form of methemoglobinemia also affects the Navajo and Eskimos. A second form of methemoglobinemia results from absence of a protein that must be present for cytochrome b5 reductase to function. This is another example of genetic heterogeneity, because

the same phenotype—blueness—results from mutations in different genes. Treatment is simple: A tablet of methylene blue, a commonly used dye, adds the electron back to methemoglobin, converting it to normal hemoglobin.

Susceptibility to Prion Disorders

For the prion protein gene, as with the globin genes, certain mutations exert drastic effects, while others don't. Recall from chapter 10 that a prion is a protein that assumes both stable and infectious conformations. A prion disease can be inherited, such as fatal familial insomnia, or acquired, such as developing variant Creutzfeldt-Jakob disease from eating beef from a cow that had bovine spongiform encephalopathy. The prion protein has at least eight distinct conformations. The normal form of the protein has a central core made up of helices. In a disease-causing form, the helices open into a sheet (see figure 10.23). Precise genetic changes control the plasticity of the prion protein—and the person's health.

The nature of the 129th amino acid in the prion protein is key to developing the disease. In people who inherit these disor-

ders, amino acid 129 is either valine in all copies of the protein (genotype VV) or methionine in all copies (genotype MM). These people are homozygous for this small part of the gene. Most people, however, are heterozygous, with valine in some prion proteins and methionine in others (genotype VM). Perhaps having two different amino acids at this position enables the proteins to assemble and to carry out their normal functions without damaging the brain.

A mutation at a different site raises the risk of brain disease even higher. Normally prion protein folds so that amino acid 129 is near amino acid 178, which is aspartic acid. People who inherit prion diseases are homozygous for the gene at position 129, and have another mutation that changes amino acid 178 to asparagine. Interestingly, people with two valines at position 129 develop fatal familial insomnia, whereas those with two methionines develop a form of Creutzfeldt-Jakob syndrome. Other genes affect susceptibility to prion disorders, too.

Key Concepts

1. Whether a mutation alters the phenotype, and how it does so, depends upon where in the protein the change occurs.
2. Mutations in globin genes are well-studied and diverse; they may cause anemia or cyanosis, or they may be silent. Hemoglobin M affects the ability of the iron in the molecule to bind oxygen.
3. Mutations in two parts of the prion protein gene predispose an individual to developing a prion disorder.

12.5 Factors That Lessen the Effects of Mutation

Mutation is a natural consequence of DNA's ability to change. This flexibility is essential for evolution because it generates new variants, some of which may resist environmental change and enable a population or even a species to survive. However, many factors minimize the deleterious effects of mutations on phenotypes.

The genetic code imparts built-in protection against mutation. Synonymous codons render many alterations in the third codon position "silent." For example, a change from RNA codon CAA to CAG does not alter the designated amino acid, glutamine, so a protein whose gene contains the change would not be altered. Other genetic code nuances prevent synthesis of drastically altered proteins. For example, mutations in the second codon position sometimes replace one amino acid with another that has a similar conformation. Often, this does not disrupt the protein's form too much. GCC mutated to GGC, for instance, replaces alanine with equally small glycine.

A **conditional mutation** affects the phenotype only under certain conditions. This can be protective if an individual avoids the exposures that trigger symptoms. Consider a common variant of the X-linked gene that encodes glucose 6-phosphate dehydrogenase (G6PD), an enzyme that immature red blood cells use to extract energy from glucose. One hundred million people worldwide have G6PD deficiency (OMIM 305900), which can cause life-threatening hemolytic anemia, but only under rather unusual conditions—eating fava beans, inhaling pollen in Baghdad, or taking a certain antimalarial drug.

In the fifth century B.C., the Greek mathematician Pythagoras wouldn't allow his followers to consume fava beans—he had discovered that it would make some of them ill. During the second World War, several soldiers taking the antimalarial drug primaquine developed hemolytic anemia. A study began shortly after the war to investigate the effects of the drug on volunteers at the Stateville Penitentiary in Joliet, Illinois. Researchers soon identified abnormal G6PD in people who developed anemia when they took the drug.

What do fava beans, antimalarial drugs, and dozens of other triggering substances have in common? They "stress" red blood cells by exposing them to oxidants, chemicals that strip electrons from other compounds. Without the enzyme, the stress bursts the red blood cells.

Another protection against mutation occurs in stem cells. When a stem cell divides to yield another stem cell and a progenitor or differentiated cell, the oldest DNA strands segregate with the stem cell, and the most recently replicated DNA strands go to the more specialized daughter cells. This makes sense in organs where stem cells very actively yield specialized daughter cells, such as the skin and small intestine. Because mutations occur when DNA replicates, this skewed distribution of chromosomes sends the DNA most likely to harbor mutations into cells that will soon be shed (from a towel rubbed on skin or in a bowel movement) while keeping mutations away from the stem cells that must continually regenerate the tissues.

Key Concepts

1. Genetic code degeneracy ensures that some third-codon-position mutations do not alter the specified amino acid. Changes in the second codon position often substitute a structurally similar amino acid.
2. Conditional mutations are expressed only in certain environments.
3. Preferential segregation of the oldest DNA strands to stem cells rather than daughter cells protects against mutation.

12.6 DNA Repair

Any manufacturing facility tests a product in several ways to see whether it has been assembled correctly. Mistakes in production are rectified before the item goes on the market—at least, most of the time. The same is true for a cell's manufacture of DNA.

DNA replication is incredibly accurate—only about 1 in 100 million bases is incorrectly incorporated. DNA polymerase as well as repair enzymes oversee the fidelity of replication.

All eukaryotes can repair their nuclear DNA, although some species do so more efficiently than others. Mitochondrial DNA cannot repair itself, which accounts for its higher mutation rate. The master at DNA repair is a large, reddish microbe. *Deinococcus radiodurans* was discovered in a can of spoiled ground meat at the Oregon Agricultural Experiment Station in Corvallis in 1956, where it had withstood radiation used to sterilize the food. It tolerates 1,000 times the radiation level that a person can, and it can even live amidst the intense radiation of a nuclear reactor. The bacterium realigns its radiation-shattered pieces of DNA. Then enzymes bring in new nucleotides and assemble the pieces.

The discovery of DNA repair systems began with observations in the late 1940s that when fungi were exposed to ultraviolet radiation, those cultures later placed nearest a window grew best. The researchers who noted these effects were not investigating DNA repair, but using UV light in other experiments. Therefore, DNA repair was inadvertently discovered before the structure of DNA was. The DNA-damaging effect of ultraviolet radiation, and the ability of light to correct it, was soon observed in a variety of organisms.

Types of DNA Repair

Since its beginning, the Earth has been periodically bathed in ultraviolet radiation. Volcanoes, comets, meteorites, and supernovas all depleted ozone in the atmosphere, which allowed ultraviolet wavelengths of light to reach organisms. The shorter wavelengths—UVA—are not dangerous, but the longer UVB wavelengths damage DNA by forming an extra covalent bond between adjacent (same-strand) pyrimidines, particularly thymines (**figure 12.10**). The linked thymines are called thymine dimers. Their extra bonds kink the double helix sufficiently to disrupt replication and permit insertion of a noncomplementary base. For example, an A might be inserted opposite a G or C, instead of opposite a T. Thymine dimers also disrupt transcription.

Early in the evolution of life, organisms that could handle UV damage had a survival advantage. Enzymes enabled them to do this, and because enzymes, as proteins, are gene-encoded, DNA repair came to persist. In many modern species, three types of DNA repair peruse the genetic material for mismatched base pairs. In the first type of DNA repair, enzymes called photolyases absorb energy from visible light and use it to detect and bind to pyrimidine dimers, then break the extra bonds. This type of repair, called photoreactivation, is what enables ultraviolet-damaged fungi to recover when exposed to sunlight. Humans do not have this type of DNA repair.

In the early 1960s, researchers discovered a second type of DNA self-mending, called

---

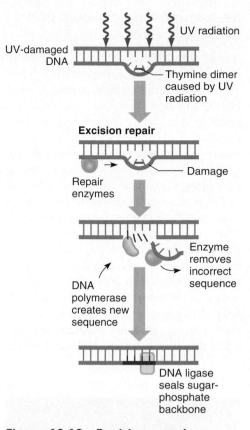

Figure 12.10 Excision repair.
Human DNA damaged by UV light is repaired by excision repair, which removes and replaces the pyrimidine dimer and a few surrounding bases.

excision repair, in mutant *E. coli* that were unable to repair ultraviolet-induced DNA damage. The enzymes that carry out excision repair cut the bond between the DNA sugar and base and snip out—or excise—the pyrimidine dimer and surrounding bases (see figure 12.10). Then, a DNA polymerase fills in the correct nucleotides, using the exposed template as a guide. DNA polymerase also detects and corrects mismatched bases in newly replicated DNA.

Humans have two types of excision repair. **Nucleotide excision repair** replaces up to 30 nucleotides and removes errors that result from several types of insults, including exposure to chemical carcinogens, UVB in sunlight, and oxidative damage. Thirty different proteins carry out nucleotide excision repair, functioning together as a structure called a repairosome. The second type of excision repair, **base excision repair,** replaces one to five

nucleotides at a time, but specifically corrects errors that result from oxidative damage. Oxygen free radicals are highly reactive forms of oxygen that arise during chemical reactions such as those of metabolism and transcription. Free radicals damage DNA. Genes that are very actively transcribed face greater oxidative damage from free radicals; base excision repair targets this type of damage.

A third mechanism of DNA sequence correction is **mismatch repair.** Enzymes "proofread" newly replicated DNA for small loops that emerge from the double helix. The enzymes excise the mismatched base so that it can be replaced (**figure 12.11**). These loops emerge from where the two strands do not precisely align, as they do if complementary base pairing occurs at every point. Such slippage and mismatching can occur in chromosome regions where very short DNA sequences repeat. These sequences, called microsatellites, are scattered throughout the genome. Like minisatellites, microsatellite lengths can vary from person to person, but within an individual, they are usually the same length. Excision and mismatch repair differ in the cause of the error—ultraviolet-induced pyrimidine dimers versus replication errors—and in the types of enzymes involved.

The three forms of DNA repair in human cells relieve the strain on thymine dimers or replace incorrectly inserted bases. Another form of repair can heal a broken sugar-phosphate backbone in both strands, which can result from exposure to ionizing radiation or oxidative damage. This insult breaks a chromosome, which can cause cancer. At least two types of multiprotein complexes reseal the backbone, either by rejoining the broken ends or recombining with DNA on the unaffected homolog.

In another type of DNA repair called damage tolerance, a "wrong" DNA base is left in place, but replication and transcription proceed. "Sloppy" DNA polymerases, with looser adherence to the base-pairing rules, read past the error, randomly inserting any other base, although it could affect the encoded protein. It is a little like retaining a misspelled wrod in a sentence—usually the meaning remains clear.

Figure 12.12 summarizes DNA repair mechanisms.

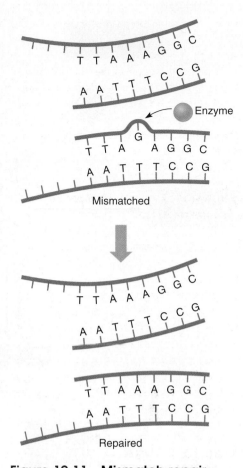

Figure 12.11 Mismatch repair.
In this form of DNA repair, enzymes detect loops and bulges in newly replicated DNA that indicate mispairing. The enzymes correct the error. Highly repeated sequences are more prone to this type of error.

DNA Repair Disorders

The ability to repair DNA is crucial to health. Mutations in any of the genes whose protein products take part in DNA repair can cause problems. A particular repair disorder may be genetically heterogeneic because it can be caused by mutations in any of several genes that participate in the same repair mechanism. That is, different single-gene defects can cause the same symptoms.

A protein called p53 controls whether DNA is repaired and the cell salvaged, or the cell dies by apoptosis. Signal transduction activates the p53 protein, stabilizing it and causing it to aggregate into complexes consisting of four proteins. These quartets bind to the DNA by recognizing four palindromic repeats that indicate genes that slow

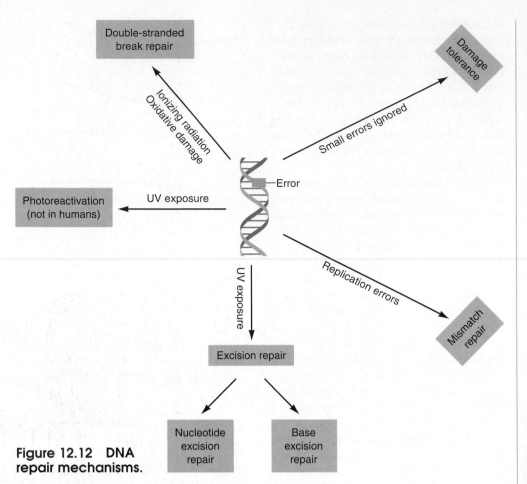

Figure 12.12 DNA repair mechanisms.

Labels in figure:
- Double-stranded break repair
- Ionizing radiation / Oxidative damage
- Damage tolerance
- Small errors ignored
- Error
- Photoreactivation (not in humans)
- UV exposure
- UV exposure
- Replication errors
- Mismatch repair
- Excision repair
- Nucleotide excision repair
- Base excision repair

Xeroderma Pigmentosum (XP) (OMIM 278700)

A child with XP lives, intentionally, indoors in artificial light, because even the briefest exposure to sunlight causes painful blisters. Failing to cover up and use sunblock can result in skin cancer (**figure 12.13**). More than half of all children with XP develop the cancer before they reach their teens. People with XP have a 10,000-fold increased risk of developing skin cancer compared to others.

XP is autosomal recessive, and results from mutations in any of seven genes. It can reflect malfunction of nucleotide excision repair or deficient "sloppy" DNA polymerase, both of which allow thymine dimers to stay and block replication. Only about 250 people in the world are known to have it. A family living in upstate New York runs a special summer camp for children with XP, where they turn night into day. Activities take

the cell cycle. The cycle must slow for repair to take place. If the damage is too severe, the p53 protein quartets instead increase the rate of transcription of genes that promote apoptosis.

In DNA repair disorders, chromosome breakage caused by factors such as radiation cannot be repaired. Mutations in repair genes therefore greatly increase susceptibility to certain types of cancer following exposure to ionizing radiation or chemicals that affect cell division. These conditions develop because errors in the DNA sequence accumulate and are perpetuated to a much greater extent than they are in people with functioning repair systems. We conclude this chapter with a closer look at repair disorders.

Trichothiodystrophy (OMIM 601675)

At least five genes can cause trichothiodystrophy. At its worst, this condition causes dwarfism, mental retardation, and failure to develop, in addition to the scaly hair with low sulfur content that gives the illness its name. Although the child may appear to be normal for a year or two, growth soon slows dramatically, signs of premature aging begin, and life ends early. Hearing and vision may fail. Interestingly, the condition does not increase the risk of cancer. Symptoms reflect accumulating oxidative damage. Individuals have faulty nucleotide excision repair, base excision repair, or both.

Inherited Colon Cancer

Hereditary nonpolyposis colon cancer (HNPCC) (OMIM 120435) was linked to a DNA repair defect when researchers discovered different-length microsatellites within an individual. Because mismatch repair normally keeps a person's microsatellites all the same length, people with this type of colon cancer might have a breakdown in this form of DNA repair. The causative gene is located on chromosome 2 and is remarkably similar to a corresponding mismatch repair gene in *E. coli*. HNPCC is common, affecting 1 in 200 people.

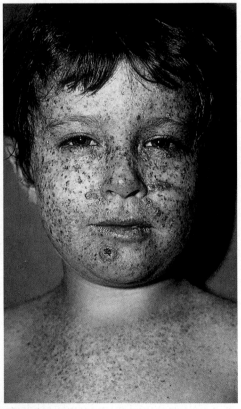

Figure 12.13 A DNA repair disorder. The marks on this child's face result from sun exposure. He is highly sensitive because he has inherited xeroderma pigmentosum (XP), an impairment of excision repair. The large lesion on his chin is a skin cancer.

place at night, or in special areas where the windows are covered and light comes from low-ultraviolet incandescent lightbulbs.

Ataxia Telangiectasis (AT) (OMIM 208900)

This multisymptom disorder is the result of a defect in a kinase that functions as a cell cycle checkpoint (see figure 2.18). Cells proceed through the cell cycle without pausing just after replication to inspect the new DNA and to repair any mispaired bases. Some cells die through apoptosis if the damage is too great to repair. Because of the malfunctioning cell cycle, individuals who have this autosomal recessive disorder have extremely high rates of cancer, particularly of the blood. Additional symptoms include poor balance and coordination (ataxia), red marks on the face (telangiectasia), delayed sexual maturation, and high risk of contracting lung infections and developing diabetes mellitus. These symptoms probably arise from disruption of other functions of the kinase.

AT is rare, but heterozygotes are not. They make up from 0.5 to 1.4 percent of various populations. Carriers may have mild radiation sensitivity, which causes a two- to sixfold increase in cancer risk over the general population. Some physicians advise people who know they are AT carriers to avoid or limit medical X rays, because for them even low exposure may cause cancer.

DNA's changeability, so vital for evolution of a species, comes at the cost of occasional harm to individuals. The DNA repair systems are, like many genetic functions, more complex than researchers originally envisioned. The human genome sequence will probably reveal several dozen DNA repair genes, which will perhaps be studied as a "repairome."

We continue looking at mutation in chapter 13, at the chromosomal level.

Summary

12.1 Mutations Can Alter Proteins— Three Examples

1. A **mutation** is a change in a gene's nucleotide base sequence that is rare and can cause a **mutant** phenotype.

2. A **germline mutation** originates in meiosis and affects all cells of an individual. A **somatic mutation** originates in mitosis and affects a subset of cells.

3. A mutation causes illness by disrupting the function or amount of a protein. In sickle cell disease, beta globin is misshapen; in beta thalassemia, it is absent or reduced. Mutations readily disrupt the highly symmetrical gene encoding collagen. One form of Alzheimer disease is caused by mutation in a receptor protein. Whether different mutations in a gene cause the same or distinct illnesses varies.

12.2 Causes of Mutation

4. A spontaneous mutation arises due to chemical phenomena or to an error in DNA replication. Spontaneous mutation rate is characteristic of a gene and is more likely to occur in repeated or symmetrical DNA sequences.

5. **Mutagens** are chemicals or forms of radiation that can induce mutation by deleting, substituting, or adding bases. An organism may be exposed to a mutagen intentionally, accidentally, or naturally.

12.3 Types of Mutations

6. A **point mutation** alters a single DNA base. It may be a **transition** (purine to purine or pyrimidine to pyrimidine) or a **transversion** (purine to pyrimidine or vice versa). A **missense mutation** substitutes one amino acid for another, while a **nonsense mutation** substitutes a "stop" codon for a codon that specifies an amino acid, shortening the protein product. Point mutations in splice sites can lead to many extra or missing amino acids.

7. Adding or deleting genetic material may upset the reading frame or otherwise alter protein function.

8. A pseudogene results when a duplicate of a gene mutates. It may disrupt chromosome pairing, causing mutation.

9. Transposons may disrupt the functions of genes they jump into.

10. Expanding triplet repeat mutations add stretches of the same amino acid to a protein, usually one that functions in the brain. They expand because they attract each other, which affects replication. This type of mutation may add a function, often leading to a neurodegenerative disease when the number of repeats exceeds a threshold level.

12.4 The Importance of Position

11. Several types of mutations can affect a gene.

12. Mutations in the globin genes may affect the ability of the blood to transport oxygen, or have no effect.

13. Susceptibility to prion disorders requires one to inherit two mutations that affect different parts of the protein that interact as the amino acid chain folds.

12.5 Factors That Lessen the Effects of Mutation

14. Synonymous codons limit the effects of mutation. Changes in the second codon position often substitute a similarly shaped amino acid, so the protein's function may not be impaired.

15. **Conditional mutations** are expressed only in response to certain environmental triggers.

16. Sending the most recently replicated DNA into cells headed for differentiation, while sending older strands into stem cells, protects against mutation.

12.6 DNA Repair

17. DNA polymerase proofreads DNA, but repair enzymes correct errors in other ways.

18. Photoreactivation repair uses light energy to split pyrimidine dimers that kink the DNA.

19. In **excision repair,** pyrimidine dimers are removed and the area is filled in correctly. **Nucleotide excision repair** replaces up to 30 nucleotides from various sources of mutation. **Base excision repair** fixes up to five bases that paired incorrectly due to oxidative damage.

20. **Mismatch repair** proofreads newly replicated DNA for loops that indicate noncomplementary base pairing.

21. DNA repair also fixes the sugar-phosphate backbone. Damage tolerance enables replication to continue beyond a mismatch.

22. Mutations in repair genes lead to chromosome breakage and increased cancer risk.

Review Questions

1. Distinguish between a germline and a somatic mutation. Which is likely to be more severe? Which can be transmitted to offspring?

2. Why is the collagen gene prone to mutation?

3. How can DNA spontaneously mutate?

4. What is the physical basis of a mutational hot spot?

5. What are three different types of mutations that cause Gaucher disease?

6. Cite three ways in which the genetic code protects against the effects of mutation.

7. List four ways that DNA can mutate without affecting the phenotype.

8. What is a conditional mutation?

9. List two types of mutations that can alter the reading frame.

10. Why can a mutation that retains an intron's sequence and a triplet repeat mutation have a similar effect on a gene's encoded protein?

11. Cite two ways a jumping gene can disrupt gene function.

12. List two reasons it takes many years to detect induction of recessive mutations in a human population.

13. What is a physical, molecular explanation for anticipation, the worsening of an inherited illness over successive generations?

14. Compare and contrast how short repeats within a gene, long triplet repeats within a gene, and repeated genes can cause disease.

15. What criteria should be used to determine whether mutations in a gene are likely to cause different disorders or differing degrees of the same disorder?

16. How do excision and mismatch repair differ?

17. In trichothiodystrophy, brittle hair and nails and scaly skin arise in some patients only during periods of fever that persist long enough for hair, nail, and skin changes to become noticeable. What type of mutation causes this disorder?

18. Explain how semiconservative DNA replication makes it possible for stem cells to receive the DNA least likely to bear mutations.

19. Consult the genetic code (table 10.5).

    a. Describe a point mutation (a change of one codon into another) that would not affect a protein's primary structure.

    b. Look up amino acid structures and identify a point mutation that would drastically alter a protein.

Applied Questions

1. The average life span for people in the United States with sickle cell disease is 42 for men and 48 for women, but in Africa it is younger than 30 for both genders. Do you think that the distinction between nations is due more to heredity or the environment? Cite a reason for your answer.

2. Retinitis pigmentosa causes night blindness and loss of peripheral vision before age 20. A form of X-linked retinitis pigmentosa is caused by a frameshift mutation that deletes 199 amino acids. How can a simple mutation have such a drastic effect?

3. One form of Ehlers-Danlos syndrome (not the "stretchy skin" type described in the chapter) can be caused by a mutation that changes a C to a T. This change results in the formation of a "stop" codon and premature termination of procollagen. Consult the genetic code (table 10.5) and suggest how this can happen.

4. Townes-Brocks syndrome causes several unrelated problems, including extra thumbs, a closed anus, hearing loss, and malformed ears. The causative mutation occurs in a transcription factor. How can a mutation in one gene cause such varied symptoms?

5. Susceptibility to developing prion diseases entails a mutation from aspartic acid (*asp*) to asparagine (*asn*). Which nucleotide base changes make this happen?

6. Two teenage boys meet at a clinic to treat muscular dystrophy. The boy who is more severely affected has a two-base insertion at the start of his dystrophin gene. The other boy has the same two-base insertion but also has a third base inserted a few bases away. Explain why the second boy's illness is milder.

7. About 10 percent of cases of amyotrophic lateral sclerosis (also known as ALS and

Lou Gehrig disease) are inherited. This disorder causes loss of neurological function over a five-year period. Two missense mutations cause ALS. One alters the amino acid asparagine (*asn*) to lysine (*lys*). The other changes an isoleucine (*ile*) to a threonine (*thr*). List the codons involved and describe how single-base mutations alter the amino acids they specify.

8. In one family, Tay-Sachs disease stems from a four-base insertion, which changes an amino-acid-encoding codon into a "stop" codon. What type of mutation is this?

9. Epidermolytic hyperkeratosis is an autosomal dominant condition that produces scaly skin. It can be caused by a missense mutation that substitutes a histidine (*his*) amino acid for an arginine (*arg*). Write the mRNA codons that could account for this change.

10. Fanconi anemia is an autosomal recessive condition that causes bone marrow abnormalities and an increased risk of certain cancers. It is caused by a transversion mutation that substitutes a valine (*val*) for an aspartic acid (*asp*) in the amino acid sequence. Which mRNA codons are involved?

11. Aniridia is an autosomal dominant eye condition in which the iris is absent. In one family, an 11-base insertion in the gene causes a very short protein to form. What kind of mutation must the insertion cause?

12. A biotechnology company has encapsulated DNA repair enzymes in fatty bubbles called liposomes. Why would this be a valuable addition to a suntanning lotion?

Web Activities

Visit the Online Learning Center (OLC) at www.mhhe.com/lewisgenetics7. Select **Student Edition, chapter 12,** and **Web Activities** to find the website links needed to complete the following activities.

13. Children with Hutchinson-Gilford progeria syndrome age extremely rapidly. In 2003, researchers identified the gene that encodes lamin A as the cause of the disorder. In 18 of 20 children whose DNA was sequenced, a single base change alters a C to a T, but this mutation removes 50 amino acids from the encoded protein. In all 20 children, the parents do not have the mutation.

    a. Is the mutation in the 18 children *de novo* or induced? What is the evidence for this distinction?

    b. How can a change in a single base remove 50 amino acids?

    c. Using OMIM, list and describe six other disorders caused by mutation in the lamin A gene.

Case Studies

14. Jan and Marcia meet at a clinic for college students who have cystic fibrosis. They are both studying genetics, and they become interested in learning about the particular mutations in their families. Jan's mutation results in exon skipping. Marcia's mutation is a nonsense mutation. Which young woman probably has more severe symptoms? Cite a reason for your answer.

15. Marshall and Angela have skin cancer resulting from xeroderma pigmentosum. They meet at an event for teenagers with cancer. However, their mutations affect different genes. They decide to marry but not to have children because they believe that each child would have a 25 percent chance of inheriting XP because it is autosomal recessive. Are they correct? Why or why not?

16. Life is dangerous for a kindergartner who can't feel pain. Ashlyn Blocker, of Patterson, Georgia, has congenital insensitivity to pain with anhidrosis (CIPA) (OMIM 256800). Ashlyn must have hot food served on ice, or she'll eat it when it is too hot and burn her mouth. If she falls in the schoolyard and scrapes her knee, she'll get up and keep running, unaware of the injury. If it's too hot in the schoolyard, she may pass out because she cannot sweat.

    Ashlyn's parents and doctor discovered her condition when she had a painful eye test at age 8 months and didn't cry. When teething started, she bit through her lips and tongue.

    The genetic cause of CIPA is a mutation that replaces a glycine (*gly*) with an arginine (*arg*). List every type of mutation that might cause this change.

Learn to apply the skills of a genetic counselor with additional cases found in the *Case Workbook in Human Genetics:*

    Bloom syndrome

    DNA repair

    Gyrate atrophy

    Open-angle glaucoma

    Otospondylomegaepiphyseal dysplasia

    Tay-Sachs disease

    von Willebrand disease

VISIT YOUR ONLINE LEARNING CENTER

Visit your online learning center for additional resources and tools to help you master this chapter. See us at

www.mhhe.com/lewisgenetics7.

Chromosomes

CHAPTER

13

LIVING WITH DOWN SYNDROME

When David G. was born in 1986, doctors told his 19-year-old mother, Toni, to put him into an institution. "They told me he wouldn't walk, talk, or do anything. Today, I want to bring him back and say look, he walks and talks and runs track and is graduating high school," recalls Toni.

Like other teens, David has held part-time jobs, gone to dances, and uses a laptop computer to do homework. But he is unlike other teens in that some of his cells have an extra chromosome 21, which limits his intellectual abilities. "Maybe he's not book smart, but when you look around at what he can do, he's smart," Toni says. David has indeed come far, thriving at his local high school. His speech is difficult to understand, and he has some facial features characteristic of Down syndrome, but he has a winning personality and close friends.

Toni fought hard to ease David's school experience. He went to a special preschool, but Toni assumed he'd be placed in a regular kindergarten class. Instead, he was put into a "self-contained" classroom, where he picked up negative behaviors from classmates with more severe problems. Toni had to eventually sue the school district to get him into a regular classroom. David needs one-on-one help for math and reading, but for other subjects he enjoys being in class.

Sometimes David gets into unusual situations because he takes things literally. He once dialed 911 when he stubbed his toe, because he'd been told to do just that when he was hurt. Another time he wandered off and walked into a strange house, asking the elderly residents for a drink of water. Toni had told him to wait 5 minutes while she finished a phone call, and at the sixth minute, David left. But for the most part, he's done so well that he will probably move into a group home and find a job after high school.

Toni is very glad that she didn't for a moment consider her original doctor's advice. "David teaches me something every day," she says.

Many people who have Down syndrome go to school or have jobs. This young woman is learning how to be a pastry chef.

Genetic health is largely a matter of balance—inheriting the "correct" number of genes, usually on the "correct" number of chromosomes (46, for humans). Too much or too little genetic material, particularly among the autosomes, can cause syndromes (groups of signs and symptoms). The cells of most people who have Down syndrome, for example, have an extra chromosome 21 and, therefore, extra copies of all the genes on that chromosome (**figure 13.1**). The extra genes cause mental retardation and medical problems, but, as the chapter opener illustrates, people with Down syndrome can lead full and productive lives.

Abnormal numbers of genes or chromosomes are a form of mutation. Mutations range from the single-base changes described in chapter 12, to missing or extra pieces of chromosomes or entire chromosomes, to entire extra sets of chromosomes. A mutation is considered a chromosomal aberration if it is large enough to see with a light microscope using stains and/or fluorescent tags to highlight missing, extra, or moved material. The mutations described in chapter 12 and this chapter represent a continuum—they differ in scale and in our ability to detect them.

In general, excess genetic material has milder effects on health than a deficit. Still,

most chromosomal abnormalities are so harmful that prenatal development ceases in the embryo. As a result, only a few—0.65 percent—of all newborns have chromosomal abnormalities that produce symptoms. An additional 0.20 percent have chromosomal rearrangements; their chromosome parts have flipped or been swapped, but they do not produce symptoms unless they disrupt genes that are crucial to health.

Cytogenetics is the subdiscipline within genetics that links chromosome variations to specific traits, including illnesses. Human genome sequence information is adding to our cytogenetics knowledge by identifying which genes contribute which symptoms to chromosome-related syndromes, and by comparing the gene contents of the chromosomes. For example, for decades geneticists did not understand why the most frequently seen extra autosomes in newborns are chromosomes 13, 18, and 21. The human genome sequence revealed that these chromosomes have the lowest gene densities—that is, they carry considerably fewer protein-encoding genes than the other autosomes, compared to their total amount of DNA. Therefore, extra copies of these chromosomes are tolerated well enough for some fetuses with them to survive to be born.

This chapter explores several ways that chromosome structure can deviate from normal and the consequences of these variations.

13.1 Portrait of a Chromosome

A chromosome consists primarily of DNA and proteins, and is duplicated and transmitted—via mitosis or meiosis—to the next cell generation. Cytogeneticists have long described and distinguished chromosome types by size and shape, using stains and dyes to contrast dark **heterochromatin,** which is mostly repetitive DNA sequences, with lighter **euchromatin,** which harbors more protein-encoding genes (**figure 13.2**).

Telomeres and Centromeres Are Essential

A chromosome must include structures that enable it to replicate and remain intact—everything else is essentially informational cargo (protein-encoding genes and their controls) and DNA sequences that impart stability to the overall structure. The essential parts of a chromosome, in terms of navigating cell division, are:

- telomeres

- origin of replication sites, where replication forks begin to form

- the centromere

Recall from figure 2.19 that **telomeres** are chromosome tips. In humans, each consists of many repeats of the sequence TTAGGG. In most cells types, telomeres are whittled down with each mitotic cell division.

The **centromere** is the largest constriction of a chromosome. It is the place where spindle fibers attach. A chromosome without a centromere is no longer a chromosome. It vanishes from the cell as soon as division begins, because there is no way for it to attach to the spindle.

In humans, many of the hundreds of thousands of DNA bases that form the centromere are repeats of a 171-base DNA sequence called an alpha satellite. (In this usage, *satellite* refers to the fate of these

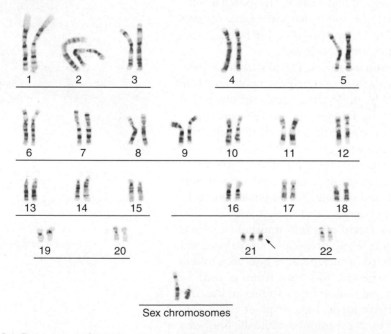

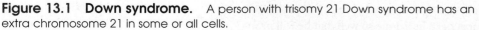
Sex chromosomes

Figure 13.1 **Down syndrome.** A person with trisomy 21 Down syndrome has an extra chromosome 21 in some or all cells.

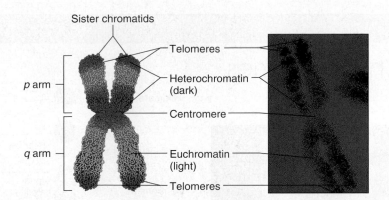

Sister chromatids

Telomeres

p arm

Heterochromatin (dark)

Centromere

q arm

Euchromatin (light)

Telomeres

Figure 13.2 Portrait of a chromosome. Tightly wound, highly repetitive heterochromatin forms the centromere (the largest constriction) and the telomeres (the tips) of chromosomes. Elsewhere, lighter-staining euchromatin includes protein-encoding genes. The centromere divides this chromosome into a short arm (*p*) and a long arm (*q*).

sequences when chromosomal DNA is shattered and different pieces settle in a density gradient. Because this part of the chromosome settles out at a density separate from the rest of the chromosome, it is called a satellite, just as the moon is a satellite derived from the Earth.) The size and number of repeats in alpha satellites are similar in many species, although the sequence differs. This suggests that satellites have a structural role in maintaining chromosomes, rather than an informational role such as encoding protein.

Centromeres also include centromere-associated proteins. Some of these, synthesized only when mitosis is imminent, form a structure called a kinetochore that contacts the spindle fibers. The kinetochore appears at prophase and apparently vanishes during telophase.

Centromeres are replicated toward the end of S phase. A protein that may control their duplication is called centromere protein A, or CENP-A. Molecules of CENP-A stay with centromeres as chromosomes are replicated, covering about half a million DNA base pairs. When the replicated (sister) chromatids separate at anaphase, each member of the pair retains some CENP-A. The protein therefore passes to the next cell generation, but it is *not* DNA. The amino acid sequence of CENP-A is nearly identical in diverse species, indicating that it has persisted through evolution and is thus important. CENP-A and other centromere-associated proteins are likely the critical parts of centromeres, rather than the alpha satellite DNA sequences. Evidence for the importance of

CENP-A comes from similar DNA sequences that function as "neocentromeres." They are found throughout the genome in noncentromeric regions, and they can function as centromeres if moved, even if they lack alpha satellites.

Centromeres lie within vast stretches of heterochromatin. The arms of the chromosome lie outward from the centromere. Gradually, the DNA includes more protein-encoding sequences as distance from the centromere increases. Gene density varies greatly among chromosomes. Chromosome 21 is a gene "desert," harboring a million-base stretch with no protein-encoding genes at all. Chromosome 22, in contrast, is a gene "jungle." These two tiniest chromosomes are remarkably similar in size, but chromosome 22 contains 545 genes to chromosome 21's 225! **Table 13.1** compares some basic characteristics of the first five autosomes sequenced. Many websites, including www.ornl.gov/hgmis/posters/chromosome, offer detailed chromosome maps that depict the locations of genes associated with particular traits or illnesses. **Reading 13.1** describes artificial chromosomes.

The chromosome parts that lie between protein-rich areas and the telomeres are

Table 13.1

Five Autosomes

Chromosome	Size in Megabases (millions of bases)	Percentage of Genome	Genes of Interest
5	194.00	6	Acute myelogenous leukemia Basal cell carcinoma Colorectal cancer Dwarfism Salt resistant hypertension
16	98.00	3	Adult polycystic kidney disease Breast cancer Crohn disease Prostate cancer
19	60.00	2	Atherosclerosis Type I diabetes mellitus DNA repair
21	33.55	1	Alzheimer disease Amyotrophic lateral sclerosis Bipolar disorder susceptibility Homocystinuria Usher syndrome
22	33.46	1	Cat eye syndrome Chronic myelogenous leukemia DiGeorge syndrome Schizophrenia susceptibility

HACs—Human Artificial Chromosomes

What are the minimal building blocks necessary to form a chromosome? A chromosome consists of three basic parts:

1. Telomeres

2. Origins of replication occurring every 50 to 350 kilobases (thousands of bases) in human chromosomes

3. Centromeres

These elements enable the entire unit to replicate during cell division and the original and replicated DNA double helices to be distributed into two cells from one.

The 24 types of human chromosomes range from 50 to 250 megabases (millions of bases) long. Researchers considered two ways to construct a chromosome—pare down an existing chromosome to see how small it can get and still hold together, or build a new chromosome from DNA pieces.

To shorten an existing chromosome, researchers exchanged in a piece of DNA that included telomere sequences. New telomeres formed at the insertion site, like periods added to the middle of a sentence, prematurely ending it. This technique formed chromosomes as small as 3.5 megabases—but researchers couldn't get them out of cells for further study.

Huntington Willard (now at Duke University) and his colleagues at Case Western Reserve University tried the building-up approach. They sent separately into cultured cells telomere DNA alpha satellites and random pieces of DNA from the human

Figure 1   Human artificial chromosomes.   The arrow indicates a human artificial chromosome. Note how small it is compared to the natural chromosomes around it.

(John Harrington, Huntington Willard, et al. 1997. *Nature Genetics* 4:345–55.)

genome containing origin-of-replication sites (**figures 1 and 2**). In the cells, some of the pieces assembled in a correct orientation to form structures 6 to 10 megabases long. These "human artificial chromosomes"—HACS—withstand repeated rounds of cell division. They have the integrity of a natural chromosome.

Other researchers have shortened combinations of telomeres, neocentromeres, and other sequences to 0.7 to 1.8 megabases.

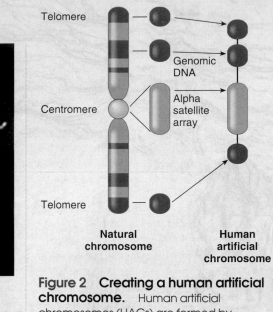

Figure 2   Creating a human artificial chromosome.   Human artificial chromosomes (HACs) are formed by combining isolated telomeres, centromeric DNA from alpha satellite arrays, and genomic DNA derived from natural chromosomes.

(Modified from Willard, 1998. *Curr Opin Genet Dev* 8:219–25.)

Neocentromeres are centromere DNA sequences minus the alpha satellite repeats. The human genome contains several dozen of them, scattered among the chromosomes.

Constructing ever-smaller artificial chromosomes is revealing what a chromosome is—an autonomous nucleic acid/protein partnership that can replicate. More practically, artificial chromosomes may one day ferry healing genes to cells where gene activity is missing or abnormal.

---

termed subtelomeres (**figure 13.3**). These areas extend from 8,000 to 300,000 bases inward toward the centromere from the telomeres. Subtelomeres include some protein-encoding genes and therefore bridge the gene-rich regions and the telomere repeats. The transition is gradual. Areas of 50 to 250 bases, right next to the telomeres, consist of 6-base repeats, many of them very similar to the TTAGGG of the

telomeres. Then, moving inward from the 6-base zone are many shorter repeats, each present in a few copies. Their function isn't known. Finally the sequence diversifies and protein-encoding genes appear.

At least 500 protein-encoding genes lie in the subtelomere regions. About half are members of multigene families (groups of genes of very similar sequence next to each other) that include pseudogenes. These

multigene families may reflect recent evolution: Apes and chimps have only one or two genes for many of the large gene families in humans. Such gene organization is one explanation for why our genome sequence is so very similar to that of our primate cousins—but we are clearly different animals. Our genomes differ more in gene copy number and chromosomal organization than in DNA base sequence.

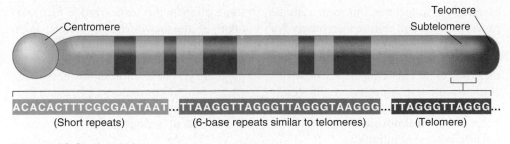

ACACACTTTCGCGAATAAT...TTAAGGTTAGGGTTAGGGTAAGGG...TTAGGGTTAGGG...
(Short repeats)   (6-base repeats similar to telomeres)   (Telomere)

Figure 13.3 Subtelomeres.   The repetitive sequence of a telomere gradually diversifies toward the centromere. A subtelomere consists of from 8,000 to 300,000 bases from the telomere inward on a chromosome arm. In this and other figures, the centromere is depicted as a buttonlike structure to more easily distinguish it. However, the centromere is composed of DNA, just like the rest of the chromosome.

Karyotypes Are Chromosome Charts

Even in this age of genomics, the standard chromosome chart, or **karyotype,** remains a major clinical tool. A karyotype displays chromosomes by size and by physical landmarks that appear during mitotic metaphase, when DNA coils tightly.

The 24 human chromosome types are numbered from largest to smallest—1 to 22—although chromosome 21 is actually the smallest. The other two chromosomes are the X and the Y. Early attempts to size-order chromosomes resulted in generalized groupings because many of the chromosomes are of similar size.

Centromere position is one distinguishing feature of chromosomes. A chromosome is **metacentric** if the centromere divides it into two arms of approximately equal length. It is **submetacentric** if the centromere establishes one long arm and one short arm, and **acrocentric** if it pinches

off only a small amount of material toward one end (**figure 13.4**). Some species have telocentric chromosomes that have only one arm, but humans do not. The long arm of a chromosome is designated *q,* and the short arm *p* (*p* stands for "petite").

Five human chromosomes (13, 14, 15, 21, and 22) are distinguished further by bloblike ends, called satellites, that extend from a thinner, stalklike bridge from the rest of the chromosome. (This use of the word *satellite* differs from the usage of the term in centromeric repeats.) The stalklike regions do not bind stains well. The stalks carry many copies of genes encoding ribosomal RNA and ribosomal proteins. These areas are called nucleolar organizing regions. They coalesce to form the nucleolus, a structure in the nucleus where ribosomal building blocks are produced and assembled.

Karyotypes are useful at several levels. When a baby is born with the distinctive facial features of Down syndrome, a kary-

otype confirms the clinical diagnosis. Within families, karyotypes are used to identify relatives with a particular chromosomal aberration that can affect health. For example, in one family, several adult members died from a rare form of kidney cancer. Researchers karyotyped the affected individuals and found that they all had an exchange, called a **translocation,** between chromosomes 3 and 8. When karyotypes showed that two young family members had the translocation, physicians examined and monitored their kidneys, detecting cancer very early and treating it successfully.

Karyotypes of individuals from different populations can reveal the effects of environmental toxins, if abnormalities appear only in a group exposed to a particular contaminant. Because chemicals and radiation that can cause cancer and birth defects often break chromosomes into fragments or rings, detecting this genetic damage can alert physicians to the possibility that certain cancers may appear in the population.

Karyotypes compared among species can clarify evolutionary relationships. The more recent the divergence of two species from a common ancestor, the more closely related we presume they are, and the more alike their chromosome banding patterns should be. Our closest relative, according to karyotypes, is the pygmy chimpanzee (bonobo). The human karyotype is also remarkably similar to that of the domestic cat, and somewhat less similar to those of mice, pigs, and cows. Among mammals, it is least like the karyotype of the aardvark, indicating that this is a primitive placental mammal.

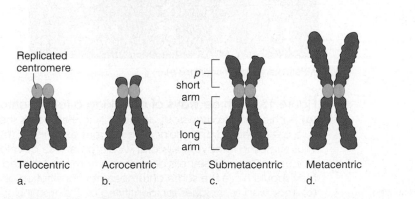

Figure 13.4 Centromere position distinguishes chromosomes.   **(a)** A telocentric chromosome has the centromere at one end. Humans do not have any telocentric chromosomes. **(b)** An acrocentric chromosome has the centromere near an end. **(c)** A submetacentric chromosome's centromere creates a long arm (*q*) and a short arm (*p*). **(d)** A metacentric chromosome's centromere establishes equal-sized arms.

Key Concepts

1. A chromosome minimally includes telomeres, origins of replication, and centromeres.
2. A centromere consists of alpha satellite repeats and associated proteins, some of which form the kinetochore, where spindle fibers attach. Centromere protein A enables the centromere to replicate.
3. Subtelomeres contain telomerelike repeats and protein-encoding multigene families.
4. Chromosomes differ by size, centromere location, satellites, and staining. Karyotypes are size-order chromosome charts.

13.2 Visualizing Chromosomes

Extra or missing chromosomes are easily detected by counting a number other than 46. Identifying chromosome rearrangements, such as an inverted sequence or an exchange of parts between two chromosomes, requires a way to distinguish among the chromosomes. A combination of stains and DNA probes applied to chromosomes allows this. A **DNA probe** is a labeled piece of DNA that binds to its complementary sequence on a particular chromosome.

Obtaining Cells for Chromosome Study

Any cell other than a mature red blood cell (which lacks a nucleus) can be used to examine chromosomes, but some cells are easier to obtain and culture than others. For adults, white blood cells separated from a blood sample or skinlike cells collected from the inside of the cheek are usually used for a chromosome test. A person might require such a test if he or she has a family history of a chromosomal abnormality or seeks medical help because of infertility.

To identify blood-borne cancers (leukemias and lymphomas), cytogeneticists examine chromosomes from bone marrow cells, which give rise to blood cells. DNA microarray tests are replacing karyotypes in matching cancers to the most effective chemotherapies, as the opener to chapter 11 described.

Chromosome tests are commonly performed on cells from fetuses. Couples who receive a prenatal diagnosis of a chromosome abnormality can arrange for treatment of the newborn, if possible; learn more about the condition and contact support groups and plan care; or terminate the pregnancy. These choices are best made after a genetic counselor or physician provides information on the medical condition and treatment options.

Amniocentesis

The first fetal karyotype was constructed in 1966 by a technique called **amniocentesis.** A doctor removes a small sample of fetal cells and fluids from the uterus with a needle passed through the woman's abdominal wall (**figure 13.5a**). The cells are cultured for a week to 10 days, and typically 20 cells are karyotyped. DNA probes can detect chromosomes in a day or two. The sampled

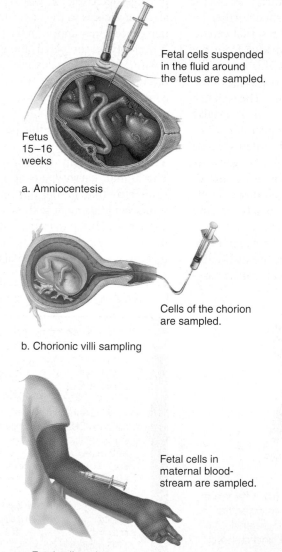

Fetal cells suspended in the fluid around the fetus are sampled.

Fetus 15–16 weeks

a. Amniocentesis

Cells of the chorion are sampled.

b. Chorionic villi sampling

Fetal cells in maternal bloodstream are sampled.

c. Fetal cell sorting

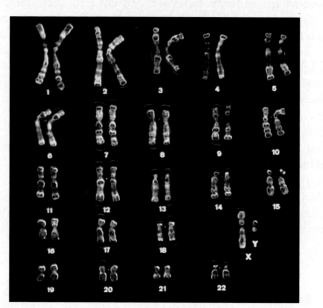

d. Fetal karyotype (normal male)

Figure 13.5 Three ways of checking a fetus's chromosomes. (a) Amniocentesis draws out amniotic fluid. Fetal cells shed into the fluid are collected and their chromosomes examined. **(b)** Chorionic villus sampling removes cells that would otherwise develop into the placenta. Since these cells descended from the fertilized ovum, they should have the same chromosomal constitution as the fetus. **(c)** Improved techniques for identifying and extracting specific cells allow researchers to detect fetal cells in a sample of blood from the woman. **(d)** For all three techniques, the harvested cells are allowed to reach metaphase, when chromosomes are most visible, and are then broken open on a slide. The chromosomes are stained or their DNA probed, then arranged into a karyotype.

Figure 13.6 Ultrasound. In an ultrasound exam, sound waves are bounced off the embryo or fetus, and the pattern of deflected sound waves is converted into a three-dimensional-appearing image. "4D ultrasound" provides a video of an embryo or fetus.

amniotic fluid is also examined for deficient, excess, or abnormal biochemicals that could indicate an inborn error of metabolism. Ultrasound is used to follow the needle's movement and to visualize fetal parts, such as the profile in **figure 13.6.**

Amniocentesis can detect approximately 800 of the more than 5,000 known chromosomal and biochemical problems. Additional tests for single-gene disorders must be requested. The most common chromosomal abnormality detected is one extra chromosome, called a **trisomy.** Amniocentesis is usually performed between 14 and 16 weeks gestation, when the fetus isn't yet very large but amniotic fluid is plentiful. Amniocentesis can be carried out anytime after this point.

Doctors recommend amniocentesis if the risk that the fetus has a detectable condition exceeds the risk that the procedure will cause a miscarriage, which is about 1 in 350 (**table 13.2**). The most common candidate for the test is a pregnant woman over age 35. This "advanced maternal age" is statistically associated with increased risk that the fetus will have an extra or missing chromosome. Amniocentesis is also warranted if a couple has had several spontaneous abortions or children with birth defects or a known chromosome abnormality.

Another reason to seek amniocentesis is if a blood test on the pregnant woman reveals low levels of a fetal liver protein called alpha fetoprotein (AFP) and high levels of human chorionic gonadotropin (hCG). These signs may indicate a fetus with a small liver, which may reflect a trisomy. Such maternal serum marker tests may assess a third or fourth biochemical, too (estriol or inhibin). Yet another maternal serum marker, pregnancy-associated plasma protein A (PAPP), is detectable only during the first trimester.

Maternal serum marker tests are useful for pregnant women younger than 35 who would not routinely undergo age-related amniocentesis. Doctors use maternal serum marker tests to screen their patients to identify those who may require genetic counseling and perhaps further, more invasive testing. For example, one four-marker test measures AFP, hCG, estriol, and inhibin during the second trimester. Considering maternal age along with the four markers reveals risk of trisomy 18 and trisomy 21. A risk greater than 1 in 270 usually indicates the need for a more definitive test, such as amniocentesis. About 7 percent of pregnant women tested are above this risk level, but only one in six of them actually carries a fetus with trisomy 21 Down syndrome. The AFP, hCG, and estriol values can also reflect elevated risk of trisomy 18. About 0.3 percent of all women tested are found to carry a fetus at high risk of having trisomy 18, and one in 15 of these fetuses is actually affected. Both trisomies are discussed later in the chapter.

In addition to the high number of false positives (high risk on the screen but normal fetus), maternal serum markers also have false negatives (low risk on the screen but an affected fetus). The four-marker test detects only 80 percent of trisomy 18 cases. This lack of precision is why maternal serum marker tests are considered screens rather than diagnostic tests.

Chorionic Villus Sampling

During the 10th through 12th week of pregnancy, **chorionic villus sampling** (CVS) obtains cells from the chorionic villi, the structures that develop into the placenta (figure 13.5*b*). A karyotype is prepared directly from the collected cells, rather than first culturing them, as in amniocentesis. Results are ready in days.

Because chorionic villus cells descend from the fertilized ovum, their chromosomes should be identical to those of the embryo and fetus. Occasionally, a chromosomal aberration occurs only in a cell of the embryo, or only in a chorionic villus cell. This results in chromosomal mosaicism—the karyotype of a villus cell differs from that of an embryo cell. Chromosomal mosaicism has great clinical consequences. If CVS indicates an aberration in villus cells that is not also in the fetus, then a couple may elect to terminate the pregnancy based on misinformation—the fetus is actually chromosomally normal, although CVS indicates otherwise. In the opposite situation, the results of the CVS may be normal, but the fetus has abnormal chromosomes.

CVS is slightly less accurate than amniocentesis, and in about 1 in 1,000 to 3,000 procedures, it halts development of the feet and/or hands, a condition termed transverse limb defects. Also, CVS does not sample amniotic fluid, so tests for inborn errors of metabolism are not possible.

Couples expecting a child are sometimes asked to choose between amniocentesis and CVS. The advantage of CVS is earlier results, but the disadvantage is a greater risk

Table 13.2				
Amniocentesis or Chorionic Villus Sampling (CVS)?				
Procedure	Gestation Time (weeks)	Cell Source	Route	Added Risk of Miscarriage
CVS	10–12	Chorionic villi	Vagina	0.8%
Amniocentesis	14–16	Skin, bladder, digestive system cells in amniotic fluid	Needle in abdomen	0.3%

of spontaneous abortion. Although CVS is slightly more invasive and dangerous to the fetus, its greater risk reflects the fact that CVS is done earlier in pregnancy. Since most spontaneous abortions occur early in pregnancy, more will follow CVS than amniocentesis. The spontaneous abortion rate after the 12th week of pregnancy is about 5 percent, and the additional risk that CVS poses is 0.8 percent. In contrast, the spontaneous abortion rate after the 14th week is 3.2 percent, and amniocentesis adds 0.3 percent to the risk.

Fetal Cell Sorting

Fetal cell sorting, a newer technique that separates fetal cells from the woman's bloodstream, is safer than amniocentesis and CVS but is still experimental (figure 13.5c) in the United States. The technique traces its roots to 1957, when a pregnant woman died when cells from a very early embryo lodged in a major blood vessel in her lung, blocking blood flow. The fetal cells were detectable because they were from a male, and contained the telltale Y chromosome. This meant that fetal cells could enter a woman's circulation.

By studying the blood of other pregnant women, researchers found that fetal cells enter the maternal circulation in up to 70 percent of pregnancies. Cells from female embryos, however, cannot be distinguished from the cells of the pregnant woman on the basis of sex chromosome analysis. But fetal cells from either sex can be distinguished from maternal cells using a device called a fluorescence-activated cell sorter. It separates fetal cells from maternal blood by identifying surface characteristics that differ from those on the woman's cells. The fetal cells are then karyotyped (figure 13.5d) and fetal DNA extracted and amplified for specific gene tests. Free fetal DNA can also be isolated from maternal blood.

Rarely, other techniques are used to sample fetal blood, skin, liver, or muscle. These biopsy procedures are usually done by using ultrasound to guide a hollow needle through the woman's abdominal wall to the fetus. Such an invasive test is performed if the family has a disease affecting the particular tissue, and a DNA-based test is not available.

Instead of examining chromosomes, ultrasound can identify physical features that are part of chromosomal syndromes. For example, ultrasound scans can enable a physician to detect increased fluid at the back of the neck (called nuchal translucency) and absent or underdeveloped nasal bones, both characteristic of Down syndrome and part of its initial description in 1866. In one study, 75 percent of fetuses with Down syndrome had these characteristics, compared to 0.5 percent of fetuses who did not have Down syndrome. Yet in another study, 70 to 90 percent of fetuses with increased nuchal translucency were normal. Ultrasound scanning is not precise, but it is safer than obtaining chromosomes with amniocentesis or CVS.

Ultrasound scans and maternal serum marker tests are sometimes combined. Particularly valuable, because it can be done in the first trimester, is nuchal translucency plus detection of two serum markers. This combination can identify 90 percent of trisomy 21 Down syndrome cases.

Preparing Cells for Chromosome Observation

Cytogeneticists have tried to describe and display human chromosomes since the late nineteenth century (**figure 13.7**). Then, the prevailing view held that humans had an XO sex determination system, with females having an extra chromosome (XX). Estimates of the human chromosome number ranged from 30 to 80. In 1923, Theophilus Painter published sketches of human chromosomes from three patients at a Texas state mental hospital. The patients had been castrated in an attempt to control their abusive behavior, and Painter was able to examine the tissue. He could not at first tell whether the cells had 46 or 48 chromosomes, but finally decided that he saw 48. Painter later showed that both sexes have the same chromosome number.

The difficulty in distinguishing between 46 or 48 chromosomes was physical—it is challenging to prepare a cell in which chromosomes do not overlap. To easily count the chromosomes, scientists had to find a way to capture them when they are most condensed—during cell division—and also spread them apart. Since the 1950s, cytogeneticists have used colchicine, an extract of the chrysanthemum plant, to arrest cells during division.

Swelling, Squashing, and Untangling

How to untangle the spaghettilike mass of chromosomes was solved by accident in 1951. A technician mistakenly washed white blood cells being prepared for chromosome analysis in a salt solution that was less con-

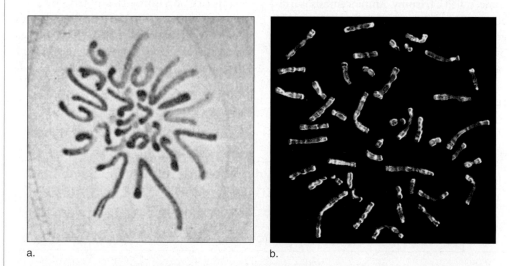

a.   b.

Figure 13.7 Viewing chromosomes, then and now. (a) The earliest drawings of chromosomes, by German biologist Walter Flemming, date from 1882. His depiction captures the random distribution of chromosomes as they splash down on a slide. **(b)** A micrograph of actual human chromosomes.

centrated than the interiors of the cells. Water rushed into the cells, swelling them and separating the chromosomes.

Two years later, cell biologists Albert Levan and Joe-Hin Tjio found that when they drew cell-rich fluid into a pipette and dropped it onto a microscope slide prepared with stain, the cells burst open and freed the mass of chromosomes. Adding a glass coverslip spread the chromosomes enough that they could be counted. By 1956, after many studies of chromosomes from cultured cells, researchers finally agreed that the number of chromosomes in a diploid human cell is 46. Shortly after, the number 23 was seen in gametes. The Technology Timeline lists these and other events in the history of cytogenetics.

Until recently, a karyotype was constructed using a microscope to locate a cell where the chromosomes were not touching, photographing the cell, developing a print, cutting out the individual chromosomes, and arranging them into a size-ordered chart. Today, a computer scans ruptured cells in a drop of stain and selects one in which the chromosomes are the most visible and well-spread. Then image analysis software recognizes the band patterns of each stained chromosome pair, sorts the structures into a size-ordered chart, and prints the karyotype. If the software recognizes an abnormal band pattern, a database pulls out identical or similar karyotypes from records of other patients.

Staining

In the earliest karyotypes, dyes stained chromosomes a uniform color. Chromosomes were grouped into size classes, designated A through G, in decreasing size order. In 1959, scientists described the first chromosomal abnormalities—Down syndrome (an extra chromosome 21), Turner syndrome (also called XO syndrome, a female with only one X chromosome), and Klinefelter syndrome (also called XXY syndrome, a male with an extra X chromosome). Before this, women with Turner syndrome were thought to be genetic males because they lack Barr bodies (see figure 6.12), while men with Klinefelter syndrome were thought to be genetic females because their cells have Barr bodies. Visualizing and distinguishing the sex chromosomes revealed the causes of these conditions.

The first chromosome stains highlighted large deletions and duplications, but usually researchers only vaguely understood the nature of a chromosomal syndrome. In 1967, a mentally retarded child with material missing from chromosome 4 would have been diagnosed as having a "B-group chromosome" disorder. Today, geneticists can identify the exact genes that are missing.

Describing smaller-scale chromosomal aberrations required better ways to distinguish chromosomes. In the 1970s, Swedish scientists developed stains that create banding patterns unique to each chromosome. These stains are specific for AT-rich or GC-rich stretches of DNA, or for heterochromatin, which stains darkly at the centromere and telomeres.

The ability to detect missing, extra, inverted, or misplaced bands allowed researchers to link many more syndromes with specific chromosome aberrations. In the late 1970s, researchers found that synchronizing the cell cycle of cultured cells revealed even more bands per chromosome. Today, another improvement, **fluorescence *in situ* hybridization,** or FISH, highlights individual genes.

FISHing

One drawback of conventional chromosome stains is that they are not specific to particular chromosomes, but instead generate different banding patterns among the 24 human chromosome types. In contrast, FISH uses DNA probes complementary to specific DNA sequences, and if those sequences are unique to a particular chromosome, the technique can identify it. FISH probes are attached to molecules that fluoresce when illuminated, producing a flash of color precisely where the probe binds to a chromosome in a patient's sample.

FISH is based on a technique, developed in 1970, called *in situ* hybridization, which originally used radioactive rather than fluorescent labels. *In situ* hybridization took weeks to work, because it relied on exposing photographic film to reveal bound DNA probes. The danger of working with radioactivity, and the crudeness of the results, prompted researchers to seek alternative ways to detect DNA probes.

FISH can "paint" entire karyotypes. Each chromosome is probed with several different fluorescent molecules. A computer integrates the images and creates a unique false color for each chromosome (see figure 13.5d). Many laboratories that perform amniocentesis or chorionic villus sampling use FISH probes specific to chromosomes 13, 18, 21, and the sex chromosomes to quickly identify the most common chromosome abnormalities. In

Technology Timeline

1923	Theophilus Painter's chromosome sketches are published; human chromosome number thought to be 48
1951	Method to detangle chromosomes discovered by accident
1953	Albert Levan and Joe-Hin Tjio develop "squash and stain" technique for chromosome preparation
1956	Using tissue culture cells, Levan, Tjio, and Biesele determine chromosome number to be 46
1956	J. L. Hamerton and C. E. Ford identify 23 chromosomes in human gametes
1959	First chromosome abnormalities identified
1960	Kidney bean extract called phytohemagglutinin added to chromosome preparation protocol to separate and stimulate division in white blood cells
1970s	Several chromosome stains implemented to improve resolution of karyotypes
1970s	FISH developed
1990s	Spectral karyotyping combines FISH probes to distinguish each chromosome

Figure 13.8 FISHing for genes and chromosomes. FISH shows three fluorescent dots that correspond to three copies of chromosome 21.

figure 13.8, FISH reveals the extra chromosome 21 in cells from a fetus with trisomy 21 Down syndrome.

A new type of prenatal chromosome analysis amplifies certain repeated sequences on chromosomes 13, 18, 21, X, and Y. The technique distinguishes paternally derived from maternally derived repeats on each homolog for these five chromosomes. An abnormal ratio of maternal to paternal repeats indicates a numerical problem, such as two copies of one parent's chromosome 21. Combined with the one chromosome 21

from the other parent, this situation would produce a fertilized ovum with three copies of chromosome 21, which causes Down syndrome.

Chromosomal Shorthand

Geneticists abbreviate the pertinent information in a karyotype by listing chromosome number, then sex chromosome constitution, then abnormal autosomes. Symbols describe the type of aberration, such as a deletion or translocation; numbers correspond to specific bands. A normal male is 46,XY; a normal female is 46,XX. Geneticists use this notation to describe gene locations. For example, the β-globin subunit of hemoglobin is located at 11p15.5. **Table 13.3** gives some examples of chromosomal shorthand.

Chromosome information is displayed in an ideogram, which is a graphical representation of a karyotype (**figure 13.9**). Bands appear as stripes, and they are divided into numbered regions and subregions. Specific gene loci known from mapping data are listed on the righthand side with information from the human genome sequence. Ideograms are becoming so crowded with notations indicating specific genes that they may soon become obsolete.

Table 13.3

Chromosomal Shorthand

Abbreviation	What It Means
46,XY	Normal male
46,XX	Normal female
45,X	Turner syndrome (female)
47,XXY	Klinefelter syndrome (male)
47,XYY	Jacobs syndrome (male)
46,XY del (7q)	A male missing part of the long arm of chromosome 7
47,XX,+21	A female with trisomy 21 Down syndrome
46,XY t (7;9)(p21.1; q34.1)	A male with a translocation between the short arm of chromosome 7 at band 21.1 and the long arm of chromosome 9 at band 34.1

Figure 13.9 Ideogram.
An ideogram is a schematic chromosome map. It indicates chromosome arm (*p* or *q*), major regions delineated by banding patterns, and the loci of selected genes. This is a partial map of human chromosome 3.

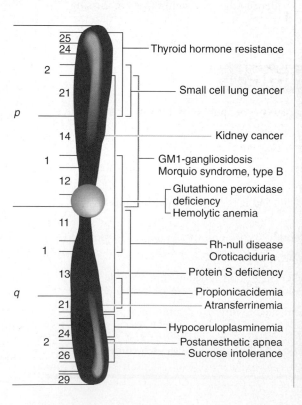

Thyroid hormone resistance

Small cell lung cancer

Kidney cancer

GM1-gangliosidosis
Morquio syndrome, type B

Glutathione peroxidase deficiency
Hemolytic anemia

Rh-null disease
Oroticaciduria

Protein S deficiency

Propionicacidemia
Atransferrinemia

Hypoceruloplasminemia
Postanesthetic apnea
Sucrose intolerance

Key Concepts

1. Karyotypes display chromosomes in size order.
2. Chromosomes can be visualized in any cell that has a nucleus and can be cultured.
3. Fetal karyotypes are made from cells obtained by amniocentesis, chorionic villus sampling, or fetal cell sorting from maternal blood. Ultrasound can detect physical problems associated with chromosome abnormalities.
4. Cytogeneticists obtain cells; display, stain, and probe chromosomes with fluorescent molecules; and then arrange them in a karyotype.
5. Chromosomal shorthand summarizes the number of chromosomes, sex chromosome constitution, and type of aberration. Ideograms display features of individual chromosomes.

13.3 Abnormal Chromosome Number

A human karyotype is abnormal if the number of chromosomes is not 46, or if individual chromosomes have extra, missing, or rearranged genetic material. **Table 13.4** summarizes the types of chromosome abnormalities in the order in which they are discussed.

Abnormal chromosomes account for at least 50 percent of spontaneous abortions. Yet only 0.5 to 0.7 percent of newborns have abnormal chromosomes. Therefore, most embryos and fetuses with abnormal chromosomes stop developing before birth.

Polyploidy

The most drastic upset in chromosome number is an entire extra set. A cell with extra sets of chromosomes is **polyploid.** An individual whose cells have three copies of each chromosome is a triploid (designated 3N, for three sets of chromosomes). Two-thirds of all triploids result from fertilization of an oocyte by two sperm. The other cases arise from formation of a diploid gamete, such as when a normal haploid sperm fertilizes a diploid oocyte. Triploids account for 17 percent of spontaneous abortions (**figure 13.10**). Very rarely, an infant survives as long as a few days, with defects in nearly all organs.

Polyploids are very common among flowering plants, including roses, cotton, barley, and wheat, and in some insects. Certain human cells may be polyploid. The liver, for example, has some tetraploid (4N) and even octaploid (8N) cells.

Aneuploidy

Cells missing a single chromosome or having an extra one are **aneuploid,** which means "not good set." Rarely, aneuploids can have more than one missing or extra chromosome, indicating defective meiosis in a parent. A normal chromosome number is **euploid,** which means "good set."

Most autosomal aneuploids (with a missing or extra non-sex chromosome) are spontaneously aborted. Those that survive have specific syndromes, with symptoms depending upon which chromosomes are missing or extra. Mental retardation is

Table 13.4

Chromosome Abnormalities

Type of Abnormality	Definition
Polyploidy	Extra chromosome sets
Aneuploidy	An extra or missing chromosome
Monosomy	One chromosome absent
Trisomy	One chromosome extra
Deletion	Part of a chromosome missing
Duplication	Part of a chromosome present twice
Translocation	Two chromosomes join long arms or exchange parts
Inversion	Segment of chromosome reversed
Isochromosome	A chromosome with identical arms
Ring chromosome	A chromosome that forms a ring due to deletions in telomeres, which cause ends to adhere

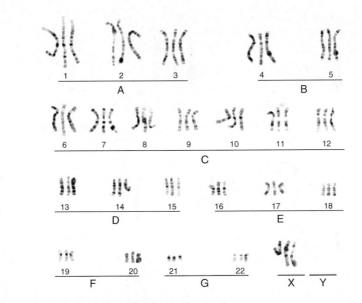

Figure 13.10 Polyploids in humans are lethal. Individuals with three copies of each chromosome (triploids) in every cell account for 17 percent of all spontaneous abortions and 3 percent of stillbirths and newborn deaths.

common in an individual who survives with aneuploidy, because development of the brain is so complex and of such long duration that nearly any chromosome-scale disruption involves genes whose protein products affect the brain. Sex chromosome aneuploidy usually produces milder symptoms.

Most children born with the wrong number of chromosomes have an extra chromosome (a trisomy) rather than a missing chromosome (a monosomy). Most monosomies are so severe that an affected

embryo ceases developing. Trisomies and monosomies are named according to the chromosome involved, and the associated syndrome has traditionally been named for the investigator who first described it. Today, cytogenetic terminology is used because it is more precise. For example, Down syndrome can result from an extra chromosome 21 (a trisomy) or a translocation. The distinction is important in genetic counseling. Translocation Down syndrome, although accounting for only 4 percent of cases, has a much higher recurrence risk within a family than the trisomy 21 form, a point we will return to later in the chapter.

The meiotic error that causes aneuploidy is called **nondisjunction.** Recall that in normal meiosis, homologs separate, and each of the resulting gametes receives only one member of each chromosome pair. In nondisjunction, a chromosome pair fails to separate at anaphase of either the first or second meiotic division. This produces a sperm or oocyte that has two copies of a particular chromosome, or none, rather than the normal one copy (**figure 13.11**). When such a gamete fuses with its partner at fertilization, the zygote has either 45 or 47 chromosomes, instead of the normal 46. Different trisomies tend to be caused by nondisjunction in the male or female, at meiosis I or II.

A cell can have a missing or extra chromosome in 49 ways—an extra or missing copy of each of the 22 autosomes, plus the five abnormal types of sex chromosome combinations—Y, X, XXX, XXY, and XYY. (Sometimes individuals have four or even five sex chromosomes.) However, only nine types of aneuploids are recognized in newborns. Others are seen in spontaneous abortions or fertilized ova intended for *in vitro* fertilization.

Most of the 50 percent of spontaneous abortions that result from extra or missing

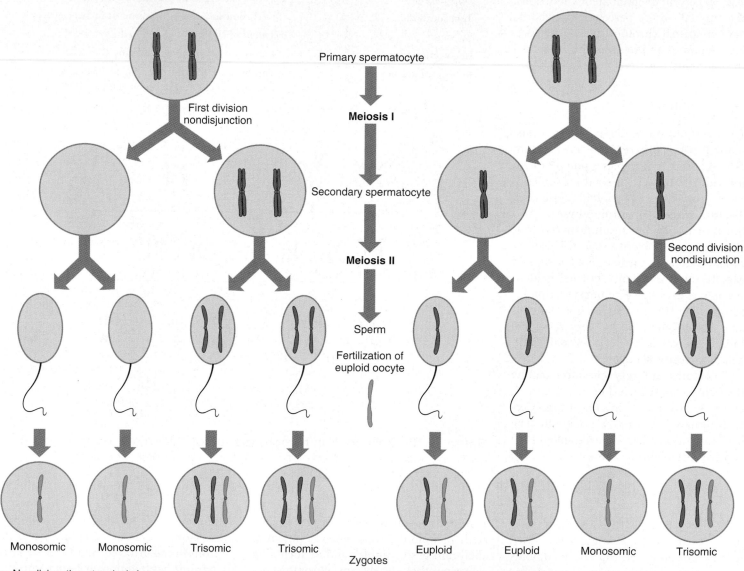

a. Nondisjunction at meiosis I

b. Nondisjunction at meiosis II

Figure 13.11   Extra and missing chromosomes—aneuploidy.   Unequal division of chromosome pairs can occur at either the first or second meiotic division. **(a)** A single pair of chromosomes is unevenly partitioned into the two cells arising from meiosis I in a male. The result: two sperm cells have two copies of the chromosome, and two sperm cells have no copies. When a sperm cell with two copies of the chromosome fertilizes a normal oocyte, the zygote is trisomic; when a sperm cell lacking the chromosome fertilizes a normal oocyte, the zygote is monosomic. **(b)** This nondisjunction occurs at meiosis II. Because the two products of the first division are unaffected, two of the mature sperm are normal and two are aneuploid. Oocytes can undergo nondisjunction as well, leading to zygotes with extra or missing chromosomes when normal sperm cells fertilize them.

chromosomes are 45,X individuals (missing an X chromosome), triploids, or trisomy 16. About 9 percent of spontaneous abortions are trisomy 13, 18, or 21. More than 95 percent of newborns with abnormal chromosome numbers have an extra 13, 18, or 21, or an extra or missing X or Y chromosome. These conditions are all rare at birth—together they affect only 0.1 percent of all children. But nondisjunction occurs in 5 percent of recognized pregnancies.

Types of chromosome abnormalities seem to differ between the sexes. Abnormal oocytes mostly have extra or missing chromosomes, whereas abnormal sperm more often have structural variants, such as inversions or translocations, discussed later in the chapter.

Aneuploidy and polyploidy also arise during mitosis, producing groups of somatic cells with the extra or missing chromosome. An individual with two chromosomally distinct cell populations is a mosaic. If only a few cells are altered, health may not be affected. However, a mitotic abnormality that occurs early in development, so that many cells descend from the unusual one, can affect health. A chromosomal mosaic for a trisomy may have a mild version of the associated condition. This is usually the case for the 1 to 2 percent of people with Down syndrome who are mosaic. The phenotype depends upon which cells have the extra chromosome. Unfortunately, prenatal testing cannot reveal which cells are affected.

Autosomal Aneuploids

Most autosomal aneuploids cease developing long before birth. Following are descriptions of the most common autosomal aneuploids among liveborns. The information is summarized in **table 13.5.**

Trisomy 21 Down Syndrome The most common autosomal aneuploid among liveborns is trisomy 21. The characteristic extra folds in the eyelids, called epicanthal folds, and flat face of a person with trisomy 21 prompted Sir John Langdon Haydon Down to term the condition *mongoloid* when he described it in 1866. As the medical superintendent of a facility for the profoundly mentally retarded, Down noted that about 10 percent of his patients resembled people of Mongolian heritage. The resemblance is superficial and meaningless. Males and females of all ethnic groups can have Down syndrome.

Down syndrome may have been recognized as early as 1515 in the Flemish painting "The Adoration of the Christ Child." An angel next to Mary has the characteristic facial features and short fingers of the condition.

Researchers suspected a link between Down syndrome and an abnormal chromosome number as long ago as 1932. In 1958, improved chromosome visualization techniques revealed 47 chromosomes in cells of a person with Down syndrome. By 1959, researchers had implicated chromosome 21. In 1960, they discovered Down syndrome caused by a translocation between chromosome 21 and another chromosome, and in 1961, researchers identified mosaic Down syndrome. The affected girl had physical signs of the condition, but normal intelligence.

A person with Down syndrome is usually short and has straight, sparse hair and a tongue protruding through thick lips. The hands have an abnormal pattern of creases, the joints are loose, and poor reflexes and muscle tone give a "floppy" appearance. Developmental milestones (such as sitting, standing, and walking) come slowly, and toilet training may take several years. Intelligence varies greatly. Parents of a child with Down syndrome can help their child reach maximal potential by providing a stimulating environment.

Many people with Down syndrome have physical problems, including heart and kidney defects and hearing and vision loss. A suppressed immune system can make influenza deadly. Digestive system blockages are common and may require surgical correction. A child with Down syndrome is 15 times more likely to develop leukemia than a child who does not have the syndrome, but this is still only a 1 percent risk. Many of the medical problems associated with Down syndrome are treatable, so that more than 70 percent of affected individuals live beyond age 30, and many live much longer. In 1910, life expectancy was only nine years.

Some people with Down syndrome who pass age 40 develop the black fibers and tangles of amyloid protein in their brains characteristic of Alzheimer disease, although they usually do not become severely demented. The chance of a person with trisomy 21 developing Alzheimer disease is 25 percent, compared to 6 percent for the general population. A gene on chromosome 21 causes one inherited form of Alzheimer disease. Perhaps the extra copy of the gene in trisomy 21 has a similar effect to a mutation in the gene that causes Alzheimer disease. In a person with Down syndrome, Alzheimer disease seems to accelerate the forgetfulness that can accompany aging.

Before the human genome sequence became available, researchers studied people who have a third copy of only part of chromosome 21 to identify specific genes that could cause symptoms. That region has been narrowed down to about 2.5 million bases, but the responsible area may not be just one stretch of genes—that is, the genes that exert effects when present in an extra copy may be distributed over this area. Experiments in mice that have third copies of part of the chromosome that corresponds to human chromosome 21 show that the syndrome is more complex than a

Table 13.5		
Comparing and Contrasting Trisomies 13, 18, and 21		
Type of Trisomy	Incidence at Birth	Percent of Conceptions That Survive 1 Year After Birth
13 (Patau)	1/12,500–1/21,700	<5%
18 (Edward)	1/6,000–1/10,000	<5%
21 (Down)	1/800–1/826	85%

Table 13.6

Genes Associated with Trisomy 21 Down Syndrome

Gene Product	OMIM	Signs and Symptoms (Phenotype)
Amyloid precursor protein (APP)	104760	Protein deposits in brain
Chromatin assembly factor I (CAF1A)	601245	Impaired DNA synthesis
Collagen type VI (COL6A1)	120220	Heart defects
Crystallin (CRYA1)	123580	Cataracts
Cystathione beta synthase (CBS)	236200	Impaired metabolism and DNA repair
Interferon receptor 1 (IFNAR)	107450	Impaired immunity
Kinase 1 (DYRK1A)	600855	Mental retardation
Oncoprotein ETS2 (ETS2)	164740	Skeletal abnormalities, cancer
Phosphoribosylglycinamide formyltransferase (GART)	138440	Impaired DNA synthesis and repair
Superoxide dismutase (SOD1)	147450	Premature aging

single region conferring symptoms. **Table 13.6** lists some genes known to contribute to trisomy 21 Down syndrome symptoms.

The likelihood of giving birth to a child with Down syndrome increases dramatically with the age of the mother (**figure 3.12**). However, 80 percent of children with trisomy 21 are born to women under age 35, because younger women are more likely to become pregnant and less likely to have amniocentesis. About 90 percent of trisomy 21 conceptions are due to nondisjunction during meiosis I in the female. The 10 percent of cases due to the male result from nondisjunction during meiosis I or II. The chance that trisomy 21 will recur in a fami-

ly, based on empirical data (how often it actually does recur in families), is 1 percent.

The age factor in trisomy 21 Down syndrome and other trisomies may reflect the fact that meiosis in the female ends after conception. The older a woman is, the longer her oocytes have been arrested on the brink of completing meiosis, a time period of 15 to 45 years! This may be why most trisomies originate in maternal meiosis I. The cause or causes of nondisjunction in the female aren't known. One hypothesis is that during their long existence in the ovary, oocytes may have been exposed to toxins, viruses, and radiation. A variation on this idea suggests that females have a pool of aneuploid oocytes resulting from nondisjunction, which for an unknown reason do not mature. As a woman ages, selectively releasing normal oocytes each month, the abnormal ones remain, much as black jellybeans accumulate as people preferentially eat the colored ones. Yet a third possible explanation is that trisomies result from gametes in which a homolog pair do not extensively cross over during meiosis I. This has been observed in both oocytes and spermatocytes. For reasons unknown, such chromosomes may migrate to the same pole, resulting in a gamete with an extra chromosome.

The association between maternal age and Down syndrome has been recognized since the nineteenth century, when physicians noticed that affected babies were often the youngest children in large families. The syndrome was thought to be caused by

syphilis, tuberculosis, thyroid malfunction, alcoholism, or emotional trauma. In 1909, a study of 350 affected infants revealed an overrepresentation of older mothers, prompting some researchers to attribute the link to "maternal reproductive exhaustion." In 1930, another study found that the increased risk of Down syndrome correlated to maternal age, and not to the number of children in the family.

RELATING THE CONCEPTS

Why is David's Down syndrome less severe than many other people who have trisomy 21?

Trisomy 18—Edward Syndrome

Trisomies 18 and 13 were described in the same research report in 1960 (**figure 13.13**).

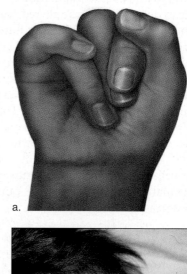

a.

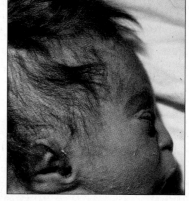

b. Trisomy 13

Figure 13.13   Trisomies 18 and 13.
(a) An infant with trisomy 18 clenches its fist in a characteristic manner, with fingers overlapping. In trisomy 13 **(b)**, the face is deformed.

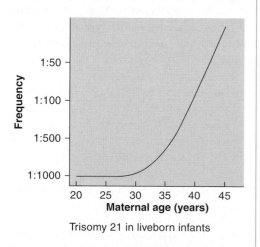

Trisomy 21 in liveborn infants

Figure 13.12   The risk of conceiving an offspring with trisomy 21 rises dramatically with maternal age.

Only 1 in 6,000 to 10,000 newborns has trisomy 18, but as table 13.5 indicates, most affected individuals do not survive to be born. The severe symptoms of trisomy 18 explain why few affected fetuses survive and also make the syndrome relatively easy to diagnose prenatally using ultrasound—yet the symptoms are presumably milder than those associated with most aneuploids, which are spontaneously aborted. Affected children have great physical and mental disabilities, with developmental skills stalled at the six-month level. Major abnormalities include heart defects, a displaced liver, growth retardation, and oddly clenched fists. Milder signs include overlapping placement of fingers, a narrow and flat skull, abnormally shaped and low-set ears, a small mouth and face, unusual or absent fingerprints, short large toes with fused second and third toes, and "rocker-bottom" feet. Most cases of trisomy 18 are traced to nondisjunction in meiosis II of the oocyte.

Trisomy 13—Patau Syndrome

Trisomy 13 is very rare, but, as is the case with trisomy 18, the number of newborns with the anomaly reflects only a small percentage of affected conceptions. Trisomy 13 has a different set of signs and symptoms than trisomy 18. Most striking, although rare, is a fusion of the developing eyes, so that a fetus has one large eyelike structure in the center of the face. More common is a small or absent eye. Major abnormalities affect the heart, kidneys, brain, face, and limbs. The nose is often malformed, and cleft lip and/or palate is present in a small head. Extra fingers and toes may occur. Appearance of a facial cleft and extra digits on an ultrasound exam are sufficient evidence to pursue chromosome analysis of the fetus.

Ultrasound examination of an affected newborn often reveals more extensive anomalies, such as an extra spleen, abnormal liver, rotated intestines, and an abnormal pancreas. A few individuals have survived until adulthood, but they do not progress developmentally beyond the six-month level.

Sex Chromosome Aneuploids

People with sex chromosome aneuploidy have extra or missing sex chromosomes. **Table 13.7** indicates how these aneuploids

Table 13.7

How Nondisjunction Leads to Sex Chromosome Aneuploids

Situation	Oocyte	Sperm	Consequence
Normal	X	Y	46,XY normal male
	X	X	46,XX normal female
Female nondisjunction	XX	Y	47,XXY Klinefelter syndrome
	XX	X	47,XXX triplo-X
		Y	45,Y nonviable
		X	45,X Turner syndrome
Male nondisjunction (meiosis I)	X		45,X Turner syndrome
	X	XY	47,XXY Klinefelter syndrome
Male nondisjunction (meiosis II)	X	XX	47,XXX triplo-X
	X	YY	47,XYY Jacobs syndrome
	X		45,X Turner syndrome

can arise. Some conditions can result from nondisjunction in meiosis in the male *or* female.

Turner Syndrome (45,X)   In 1938, at a medical conference, a U.S. endocrinologist named Henry Turner described seven young women, aged 15 to 23, who were sexually undeveloped, short, had folds of skin on the back of their necks, and had malformed elbows. (Eight years earlier, an English physician named Ullrich had described the syndrome in young girls, so it is called Ullrich syndrome in the U.K.) Alerted to what would become known as Turner syndrome in the United States, other physicians soon began identifying such patients in their practices. Physicians assumed that a hormonal insufficiency caused the symptoms. They were right, but there was more to the story—a chromosomal imbalance caused the hormone deficit.

In 1954, at a London hospital, a physician discovered that cells from Turner patients lacked a Barr body, the dark spot that indicates a second X chromosome (see figure 6.12). Might lack of a sex chromosome cause the symptoms, particularly failure to mature sexually? By 1959, karyotyping confirmed the absence of an X chromosome in cells of Turner syndrome patients. Later, researchers learned that only 50 percent of people with Turner syndrome are XO. The rest have partial deletions or are mosaics, with only some cells affected.

Like the autosomal aneuploids, Turner syndrome is found more frequently among spontaneously aborted fetuses than among newborns—99 percent of affected fetuses are not born. The syndrome affects 1 in 2,000 female births. However, people with the condition usually do not know they have a chromosome abnormality until they lag in sexual development. Two X chromosomes are necessary for normal sexual development in females.

In childhood, signs of Turner syndrome include wide-set nipples, slight webbing at the back of the neck, short stature, coarse facial features, and a low hairline at the back of the head. About half of people with Turner syndrome have impaired hearing and frequent ear infections due to a small defect in the shape of the coiled part of the inner ear. They cannot hear certain frequencies of sound. At sexual maturity, sparse body hair develops, but the girls do not ovulate or menstruate, and their breasts do not develop. The uterus is very small, but the vagina and cervix are normal. In the ovaries, oocytes speed through development, depleting the supply during infancy. Intelligence is normal, and life can be fairly normal if the women receive hormone supplements. Using growth hormone increases height.

Although women with Turner syndrome are infertile, individuals who are mosaics may have children, but they are at high risk of conceiving offspring that have abnormal

A Personal Look at Klinefelter Syndrome

I was diagnosed with Klinefelter syndrome (KS) at age 25, in February 1996. Being diagnosed has been . . . a big sigh of relief after a life of frustrations. Throughout my early childhood, teens, and even somewhat now, I was very shy, reserved, and had trouble making friends. I would fly into rages for no apparent reason. My parents knew when I was very young that there was something about me that wasn't right.

I saw many psychologists, psychiatrists, therapists, and doctors, and their only diagnosis was "learning disabilities." In the seventh grade, I was told by a psychologist that I was stupid and lazy, and I would never amount to anything. After barely graduating high school, I started out at a local community college. I received an associate degree in business administration, and never once sought special help. I transferred to a small liberal arts college to finish up my bachelor of science degree, and spent an extra year to complete a second degree. Then I started a job as a software engineer for an Internet-based company. I have been using comput-

ers for 20 years and have learned everything I needed to know on my own.

To find out my KS diagnosis, I had gone to my general physician for a physical. He noticed that my testes were smaller than they should be and sent me for blood work. The karyotype showed Klinefelter syndrome, 47,XXY. After seeing the symptoms of KS and what effects they might have, I found it described me perfectly. But, after getting over the initial shock and dealing with the denial, depression, and anger, I decided that there could be things much worse in life. I decided to take a positive approach.

There are several types of treatments for KS. I give myself a testosterone injection in the thigh once every two weeks. My learning and thought processes have become stronger, and I am much more outgoing and have become more of a leader. Granted, not all of this is due to the increased testosterone level, some of it is from a new confidence level and from maturing.

I feel that parents who are finding out prior to the birth of their son [that he will

Stefan Schwarz

have Klinefelter syndrome] or parents of affected infants or young children are very lucky. There is so much they can do to help their child have a great life. I have had most all of the symptoms at some time in my life, and I've gotten through and done well.

Stefan Schwarz

(Stefan Schwarz runs a Boston-area support group for KS.)

numbers of chromosomes. Turner syndrome is the only aneuploid condition that is unrelated to the age of the mother.

For many years, it was thought that Turner syndrome had no effects in adulthood, but this was largely because most studies of common adult disorders did not consider chromosome status. Researchers in Edinburgh have been tracking the health of 156 women with Turner syndrome for more than 25 years, and have found that having Turner syndrome shortens lifespan. For example, 68 percent of the 156 participants reached age 60, compared to 88 percent of the general British population. Adults with Turner syndrome are more likely to develop certain disorders than the general population, including osteoporosis, types 1 and 2 diabetes, and colon cancer.

The many signs and symptoms of Turner syndrome result from the loss of specific genes. For example, loss of a gonadal dys-

genesis gene accounts for the ovarian failure, whereas absence of a homeobox gene, which is a transcription factor, causes short stature. Another gene causes the unusual hearing defect.

Extra X Chromosomes   About 1 in every 1,000 females has an extra X chromosome in each of her cells, a condition called triplo-X. The only symptoms are tall stature and menstrual irregularities. Although triplo-X females are rarely mentally retarded, they tend to be less intelligent than their siblings. The lack of symptoms reflects the protective effect of X inactivation—all but one of the X chromosomes is inactivated.

About 1 in 1,000 males has an extra X chromosome, which causes Klinefelter syndrome (XXY). Severely affected men are underdeveloped sexually, with rudimentary testes and prostate glands and sparse pubic and facial hair. They have very long arms

and legs, large hands and feet, and may develop breast tissue. Klinefelter syndrome is the most common genetic or chromosomal cause of male infertility, accounting for 4 to 6 percent of cases.

Testosterone injections during adolescence can limit limb lengthening and prompt development of secondary sexual characteristics. Boys and men with Klinefelter syndrome may be slow to learn, but they are usually not mentally retarded unless they have more than two X chromosomes, which is rare.

Many men with Klinefelter syndrome discover it only when they have an infertility problem and their chromosomes are checked. Some affected men probably never learn that they have Klinefelter syndrome. Affected individuals can look like anyone else (see In Their Own Words.)

Some men with Klinefelter syndrome have fathered children. Doctors select sperm that contain only one sex chromosome and use them to fertilize oocytes. However, sperm from men with Klinefelter syndrome are more likely to have extra chromosomes—usually X or Y, but also autosomes—than sperm from men who do not have Klinefelter syndrome.

XYY Syndrome

One male in 1,000 has an extra Y chromosome. Awareness of this condition arose in 1961, when a tall, healthy, middle-aged man, known for his boisterous behavior, underwent a routine chromosome check after fathering a child with Down syndrome. The man had an extra Y chromosome. A few other cases were detected over the next several years.

In 1965, researcher Patricia Jacobs published results of a survey among 197 inmates at Carstairs, a high-security prison in Scotland. Of 12 men with unusual chromosomes, seven had an extra Y. Might their violent or aggressive behavior be linked to their extra Y chromosome? Jacobs's findings were repeated in studies in English and Swedish mental institutions. Soon after, *Newsweek* magazine ran a cover story on "congenital criminals." In 1968, defense attorneys in France and Australia pleaded their violent clients' cases on the basis of an inherited flaw, the extra Y. The condition became known as Jacobs syndrome. Meanwhile, the National Institute of Mental Health, in Bethesda, Maryland, held a conference on Jacobs syndrome, lending legitimacy to the hypothesis that an extra Y predisposes to violent behavior.

In the early 1970s, newborn screens began in hospital nurseries in England, Canada, Denmark, and Boston. Social workers and psychologists visited XYY children and offered "anticipatory guidance" to the parents on how to deal with their toddling future criminals. By 1974, geneticists and others halted the program, pointing out that singling out these boys on the basis of a few statistical studies was inviting self-fulfilling prophecy.

Today, we know that 96 percent of XYY males are apparently normal. The only symptoms attributable to the extra chromosome may be great height, acne, and perhaps speech and reading problems. An explanation for the continued prevalence of XYY among mental-penal institution populations may be more psychological than biological. Large body size may lead teachers, employers, parents, and others to expect more of these people, and a few of them may deal with this stress aggressively.

Jacobs syndrome can arise from nondisjunction in the male, producing a sperm with two Y chromosomes that fertilizes an X-bearing oocyte. Geneticists have never observed a sex chromosome constitution of one Y and no X. Since the Y chromosome carries little genetic material, and the gene-packed X chromosome would not be present, the absence of so many genes makes development beyond a few cell divisions in a YO embryo impossible.

Key Concepts

1. Polyploids have extra sets of chromosomes and do not survive for long.
2. Aneuploids have extra or missing chromosomes. Nondisjunction during meiosis causes aneuploidy.
3. Trisomies are less severe than monosomies, and sex chromosome aneuploidy is less severe than autosomal aneuploidy.
4. Mitotic nondisjunction produces chromosomal mosaics.
5. Down syndrome (trisomy 21) is the most common autosomal aneuploid, followed by trisomies 18 and 13.
6. Sex chromosome aneuploid conditions include Turner syndrome (XO), triplo-X, Klinefelter syndrome (XXY), and Jacobs syndrome (XYY).

13.4 Abnormal Chromosome Structure

Structural chromosomal defects include missing, extra, or inverted genetic material within a chromosome or combined or exchanged parts of nonhomologs (translocations) (**figure 13.14**).

Deletions and Duplications

A **deletion** is missing genetic material. Deletions range greatly in size, from a few bases to large expanses of chromosomes. The larger deletions tend to have greater effects because they remove more genes. Consider cri-du-chat syndrome (French for

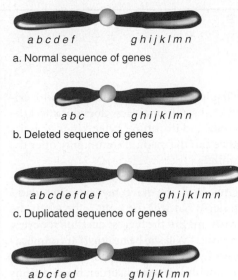

a. Normal sequence of genes

a b c   *g h i j k l m n*

b. Deleted sequence of genes

a b c d e f d e f   *g h i j k l m n*

c. Duplicated sequence of genes

a b c f e d   *g h i j k l m n*

d. Inverted sequence of genes

Figure 13.14 Chromosome abnormalities. If a hypothetical normal gene sequence appears as shown in **(a)**, then **(b)** represents a deletion, **(c)** a duplication, and **(d)** an inversion.

"cat's cry"), caused by deletion of part of the short arm of chromosome 5 (also called 5p⁻ syndrome). Affected children have a high-pitched cry similar to the mewing of a cat, have pinched facial features, and are mentally retarded and developmentally delayed. The chromosome region responsible for the catlike cry is distinct from the region that causes mental retardation and developmental delay. The cri-du-chat deletion also removes the gene for telomerase reverse transcriptase, which normally keeps telomeres long in cells that divide often. The gene's absence may contribute to the shortened lifespan of affected individuals.

A cytogeneticist can determine by examining a detailed karyotype whether a child will have only the catlike cry and perhaps poor weight gain, or will have all of the signs and symptoms, which include low birth weight, poor muscle tone, a small head, and impaired language skills. In Their Own Words on page 258 describes a child who had 5p⁻ syndrome.

A **duplication** is a region of a chromosome where genes are repeated. Duplications, like deletions, are more likely to cause symptoms if they are extensive. For example, duplications of chromosome 15 do not produce a phenotype unless they repeat several

Ashley's Message of Hope

What is it like to have a child born with cri-du-chat syndrome? How does this affect the family and its future? What kinds of assistance can the medical community offer the family?

The birth of any child raises many questions. Will she have my eyes, her dad's smile? What will she want to be when she grows up? But the biggest question for every parent is "Will she be healthy?" If complications occur during birth or if the child is born with a genetic disorder, the questions become more profound and immediate. "How did this happen?" "Where do we go from here?" "Will this happen again?"

Our daughter, Ashley Elizabeth Naylor (**figure 1**), was born August 12, 1988. We had a lot of mixed emotions the day of her birth, but mainly we felt fear and despair. The doctors suspected complications, which led to a cesarean section, but the exact problem was not known. Two weeks after her birth, chromosome analysis revealed cri-du-chat (cat cry) syndrome, also known as 5p⁻ syndrome because part of the short arm of one copy of chromosome 5 is missing. The prognosis was uncertain. This is a rare disorder, we were told, and little could be offered to help our daughter. The doctors used the words "profoundly retarded," which cut like a knife through our hearts and our hopes. It

wasn't until a few years later that we realized how little the medical community actually knew about cri-du-chat syndrome and especially about our little girl!

Ashley defied all the standard medical labels, as well as her doctors' expectations. Her spirit and determination enabled her to walk with the aid of a walker and express herself using sign language and a communication device. With early intervention and education at United Services for the Handicapped, Ashley found the resources and additional encouragement she needed to succeed. In return, Ashley freely offered one of her best-loved and sought-after gifts— her hugs. Her bright eyes and glowing smile captured the hearts of everyone she met.

In May of 1994, Ashley's small body could no longer support the spirit that inspired so many. She passed away after a long battle with pneumonia. Her physical presence is gone, but her message remains: hope.

If you are a parent faced with similar profound questions after the birth of your child, do not assume one doctor has all the answers. Search for doctors who respect your child enough to talk to her, not just about her. Above all, find an agency or a school that can help you give your child a chance to succeed. Early education for your child and support for yourself are crucial.

Figure 1 Ashley Naylor brought great joy to her family and community during her short life.
Courtesy of Kathy Naylor.

If you are a student in a health field, become as knowledgeable as possible and stay current with the latest research, but most importantly, be sensitive to those who seek your help. Each word you speak is taken to heart. Information is important, but hope can make all the difference in a family's future.

Kathy Naylor

---

genes. **Figure 13.15** shows three duplicated chromosome 15s, with increasing amounts of material repeated. Many people have the first two types of duplications and have no symptoms. However, several unrelated individuals with the third, larger duplication have seizures and mental retardation.

Duplications have been very important in evolution. The human genome includes many duplications that are single genes in other organisms, even our closest primate relatives. Sometimes the duplicated gene mutates and takes on a new function. Chapter 16 considers chromosome evolution.

FISH can detect tiny deletions and duplications that are smaller than the bands of conventional chromosome staining.

Small duplications are generally not dangerous, but some "microdeletions" are associated with a number of syndromes. Certain microdeletions in the Y chromosome, for example, cause male infertility.

Deletions and duplications can arise from chromosome rearrangements. These include translocations, inversions, and ring chromosomes.

Translocations

In a translocation, different (nonhomologous) chromosomes exchange or combine parts. Translocations can be inherited because they can be present in carriers, who have the normal amount of genetic material, but

rearranged. Deletions and duplications affect the amount of genetic material, so they are not usually inherited because they typically adversely affect a person's health or fertility. Deletions and duplications therefore cannot usually be "carried," but instead arise spontaneously. Exposure to certain viruses, drugs, and radiation can cause translocations, but often their cause isn't known.

There are two major types of translocations. In a **Robertsonian translocation,** the short arms of two different acrocentric chromosomes break, leaving sticky ends that cause the two long arms to join, forming a single, large chromosome with two long arms. The tiny short arms are lost, but their DNA sequences are repeated else-

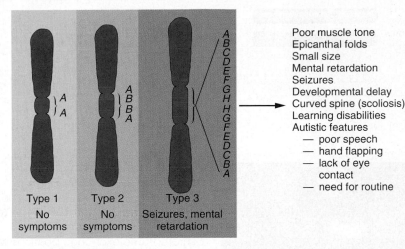

Poor muscle tone
Epicanthal folds
Small size
Mental retardation
Seizures
Developmental delay
Curved spine (scoliosis)
Learning disabilities
Autistic features
— poor speech
— hand flapping
— lack of eye
contact
— need for routine

ABCDEFGHHGFEDCBA

Type 1
No
symptoms

Type 2
No
symptoms

Type 3
Seizures, mental
retardation

Figure 13.15 A duplication. A study of duplications of parts of chromosome 15 revealed that small duplications do not affect the phenotype, but larger ones may. The letters indicate specific DNA sequences, which serve as markers to compare chromosome regions. Note that the duplication is also inverted.

where in the genome, so their absence does not cause symptoms. The person with the large, translocated chromosome, called a **translocation carrier,** has 45 chromosomes, but may not have symptoms if no crucial genes have been deleted or damaged. Even so, he or she may produce unbalanced gametes—sperm or oocytes with too many or too few genes. This can lead to spontaneous abortion or birth defects.

In 1 in 20 cases of Down syndrome, a parent has a Robertsonian translocation between chromosome 21 and another, usually chromosome 14. The individual with the translocation produces some gametes that lack either of the involved chromosomes and some gametes that have extra material from one of the translocated chromosomes (**figure 13.16**). In such a case,

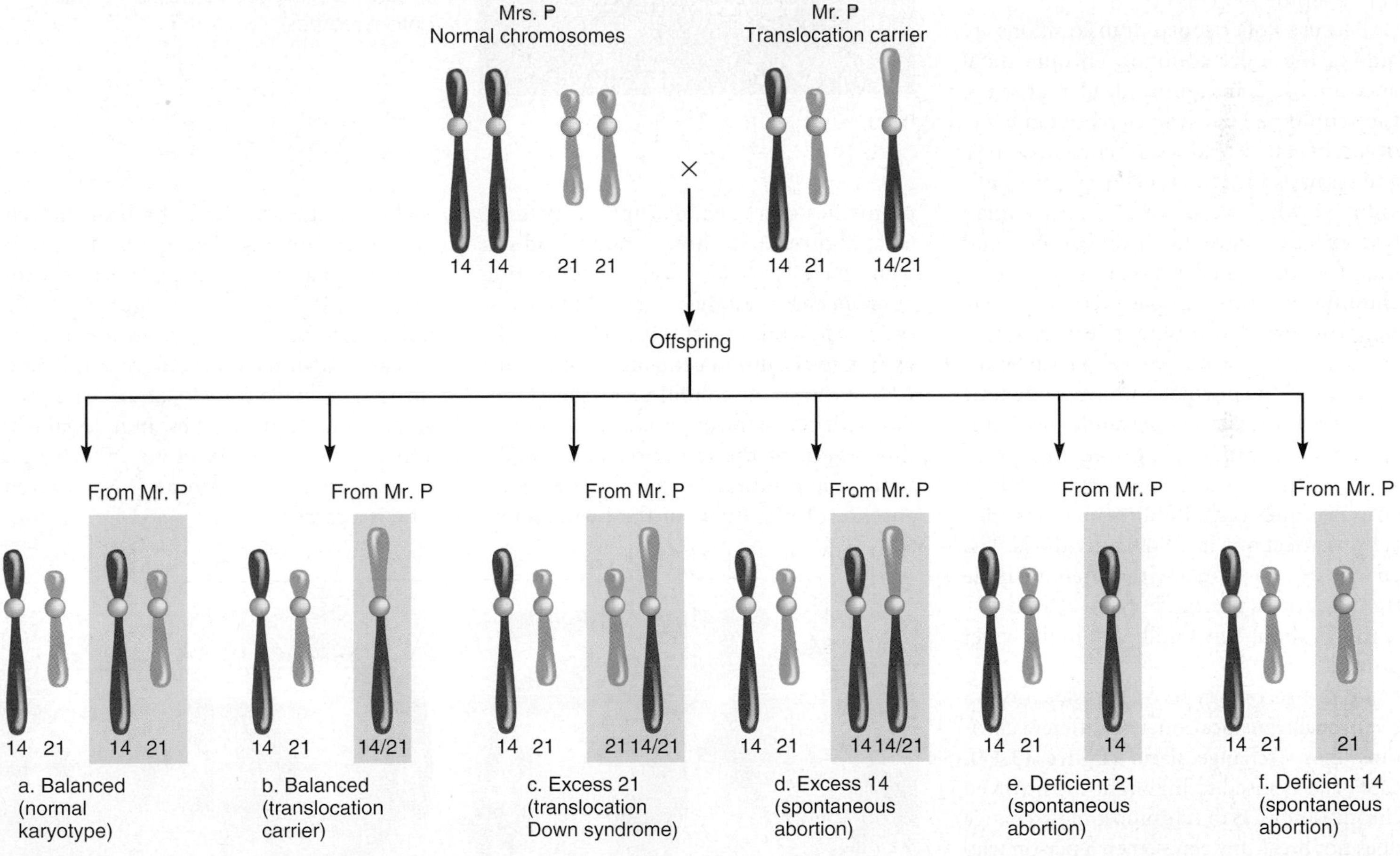

Mrs. P
Normal chromosomes

Mr. P
Translocation carrier

×

14 14   21 21

14   21   14/21

Offspring

From Mr. P

From Mr. P

From Mr. P

From Mr. P

From Mr. P

From Mr. P

14 21   14 21

14 21   14/21

14 21   21 14/21

14 21   14 14/21

14 21   14

14 21   21

a. Balanced
(normal
karyotype)

b. Balanced
(translocation
carrier)

c. Excess 21
(translocation
Down syndrome)

d. Excess 14
(spontaneous
abortion)

e. Deficient 21
(spontaneous
abortion)

f. Deficient 14
(spontaneous
abortion)

Figure 13.16 A Robertsonian translocation. Mr. P. has only 45 chromosomes because the long arm of one chromosome 14 has joined the long arm of one chromosome 21. He has no symptoms. Mr. P. makes six types of sperm cells, and they determine the fates of offspring. **(a)** A sperm with one normal chromosome 14 and one normal 21 yields a normal child. **(b)** A sperm carrying the translocated chromosome produces a child who is a translocation carrier, like Mr. P. **(c)** If a sperm contains Mr. P.'s normal 21 and his translocated chromosome, the child receives too much chromosome 21 material and has Down syndrome. **(d)** A sperm containing the translocated chromosome and a normal 14 leads to excess chromosomal 14 material, which is lethal in the embryo or fetus. If a sperm lacks either chromosome 21 **(e)** or 14 **(f),** it leads to monosomies, which are lethal prenatally. (Chromosome arm lengths are not precisely accurate.)

each fertilized ovum has a 1 in 2 chance of ending in spontaneous abortion, and a 1 in 6 chance of developing into an individual with Down syndrome. The risk of giving birth to a child with Down syndrome is theoretically 1 in 3, because the spontaneous abortions are not births. However, because some Down syndrome fetuses spontaneously abort, the actual risk of a couple in this situation having a child with Down syndrome is about 15 percent. The other two outcomes—a fetus with normal chromosomes or a translocation carrier like the parent—have normal phenotypes. Either a male or a female can be a translocation carrier, and the condition is not related to age. The second most common type of Robertsonian translocation occurs between chromosomes 13 and 14, causing symptoms of Patau syndrome because of an excess of chromosome 13 material.

Because Robertsonian translocations are among the more common chromosomal aberrations, an intriguing idea has arisen—they could one day lead to a human karyotype of 44 instead of 46 chromosomes, and perhaps even two types of people! Individuals who have one Robertsonian translocation have 45 chromosomes, and therefore may make gametes missing a chromosome, which impairs fertility. A person who has two different Robertsonian translocations would have 44 chromosomes, but the normal amount of genetic material. Two such people could have children together, the male producing sperm and the female producing oocytes with 22 chromosomes each. Robertsonian translocations affect one in 1,000 individuals. The chance of two people with different single translocations passing both to shared offspring is about 1 in 4 million—unlikely, yet possible.

In the second type of translocation, a **reciprocal translocation,** two different chromosomes exchange parts (**figure 13.17**). FISH can be used to highlight the involved chromosomes. If the chromosome exchange does not break any genes, then a person who has both translocated chromosomes is healthy and a translocation carrier. He or she has the normal amount of genetic material, but it is rearranged.

A reciprocal translocation carrier can have symptoms if one of the two break-

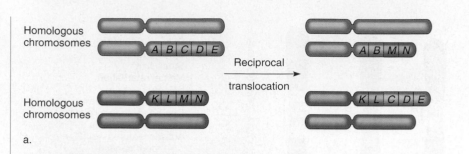

a.

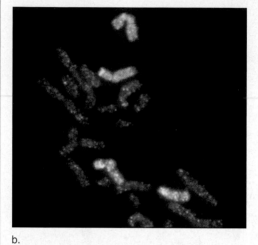

b.

Figure 13.17 A reciprocal translocation.   In a reciprocal translocation, two nonhomologous chromosomes exchange parts. In **(a)**, genes C, D, and E on the blue chromosome exchange positions with genes M and N on the red chromosome. Part **(b)** highlights a reciprocal translocation using FISH. The pink chromosome with the dab of blue, and the blue chromosome with a small section of pink, are the translocated chromosomes.

points lies in a gene, disrupting its function. **Figure 13.18** shows a father and son who have a reciprocal translocation between chromosomes 2 and 20 that causes Alagille syndrome (OMIM 118450). The exchange disrupts a gene on chromosome 20 that causes the condition, because families with the syndrome have deletions in this region of the chromosome. Alagille syndrome produces a characteristic face, absence of bile ducts in the liver, abnor-

malities of the eyes and ribs, heart defects, and severe itching. The symptoms are so variable that some people do not know they have it. The father in figure 13.18 did not realize he had the syndrome until he had a child with a more severe case. Sometimes, a *de novo* translocation arises in a gamete that leads to a new individual with a disorder, as opposed to inheriting a translocated chromosome from a parent who is a carrier.

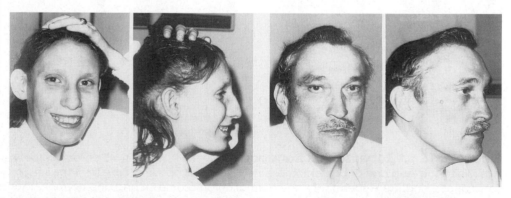

Figure 13.18 A translocation syndrome.   In one family with Alagille syndrome, a reciprocal translocation occurs between chromosomes 2 and 20. Distinctive facial features are part of the condition.

A translocation carrier produces some unbalanced gametes—sperm or oocytes that have deletions or duplications of some of the genes in the translocated chromosomes. The resulting phenotype depends upon the particular genes that the chromosomal rearrangement disrupts and whether they are extra or missing.

Information from the human genome sequence and from other chromosomal abnormalities can explain how some translocations affect health. For example, a child was born with a reciprocal translocation between chromosomes 12 and 22. The distinctive symptoms of language delay, mild mental retardation, loose joints, minor facial anomalies, and a narrow, long head matched those of another chromosome problem, called 22q13.3 deletion syndrome (OMIM 606232). That condition is caused by absence of a gene (called *ProSAP2*) that forms scaffolds for neurons in the cerebral cortex and cerebellum. The translocation cuts this gene, abolishing its function just as a deletion does. As a result, these parts of the brain malfunction.

A genetic counselor suspects a translocation if a family has had multiple birth defects and spontaneous abortions. Studies to identify disease-causing genes often began with people whose translocations pointed the way toward a gene of interest. The availability of the human genome sequence is enabling researchers to identify the DNA sequences that are disrupted in translocations. This is more useful information, in terms of explaining associated symptoms, than chromosome banding.

Inversions

An inverted sequence of chromosome bands indicates that part of the chromosome has flipped. Five to 10 percent of inversions affect health, probably because they disrupt important genes. When a fetal chromosome test reveals an inversion, it may not be known whether symptoms will be associated with the problem. Usually, the parents then have their chromosomes checked, and if one has the inversion and is healthy, it is assumed the child will most likely not have symptoms. If neither parent has the inversion, then the anomaly arose in a gamete, and effects may depend on which genes are involved. The human genome sequence can be consulted to identify genes that might be implicated.

Like a translocation carrier, an adult heterozygous for an inversion can be healthy, but have reproductive problems. One woman had an inversion in the long arm of chromosome 15 and had two spontaneous abortions, two stillbirths, and two children with multiple problems who died within days of birth. She did eventually give birth to a healthy child. How did the inversion cause these problems?

Inversions with such devastating effects can be traced to meiosis, when a crossover occurs between the inverted chromosome segment and the noninverted homolog. To allow the genes to align, the inverted chromosome forms a loop. When crossovers occur within the loop, some areas are duplicated and some deleted in the resulting recombinant chromosomes. In inversions, the abnormal chromosomes result from the chromatids that crossed over.

Two types of inversions are distinguished by the position of the centromere relative to the inverted section. A **paracentric inversion** does not include the centromere (**figure 13.19**). A single crossover within the inverted segment gives rise to two very abnormal chromosomes. The other two chromosomes are normal. One abnormal chromosome retains both centromeres and is termed dicentric. When the cell divides, the two centromeres are pulled to opposite sides of the cell, and the chromosome breaks, leaving pieces with extra or missing segments. The second type of abnormal chromosome resulting from a crossover within an inversion loop is a small piece that lacks a centromere, called an acentric fragment. When the cell divides, the fragment is lost because a centromere is required for cell division.

A **pericentric inversion** includes the centromere within the loop. A crossover in it produces two chromosomes that have

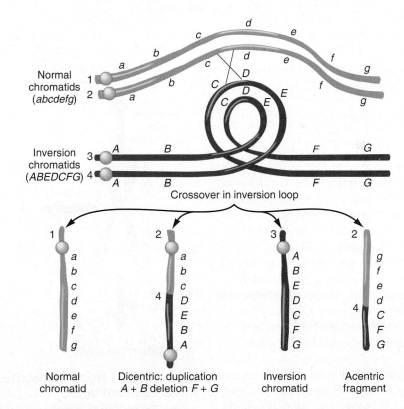

Figure 13.19 Paracentric inversion. A crossover between a chromosome with a paracentric inversion and its normal homolog, when in the region of the inversion, produces one normal chromatid, one inverted chromatid, one with two centromeres (dicentric), and one with no centromere (an acentric fragment). The letters *a* through *g* denote genes.

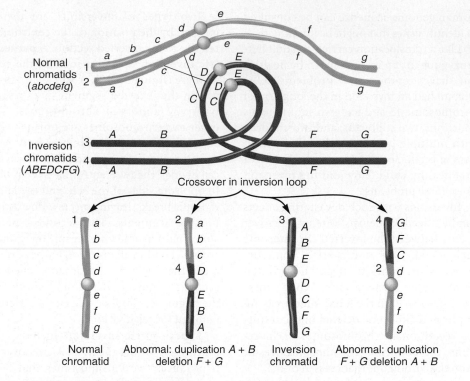

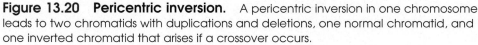

Figure 13.20 Pericentric inversion. A pericentric inversion in one chromosome leads to two chromatids with duplications and deletions, one normal chromatid, and one inverted chromatid that arises if a crossover occurs.

duplications and deletions, but one centromere each (**figure 13.20**).

Isochromosomes and Ring Chromosomes

Another meiotic error that leads to unbalanced genetic material is the formation of an isochromosome, which is a chromosome that has identical arms. This occurs when, during division, the centromeres part in the wrong plane (**figure 13.21**). Isochromosomes are known for chromosomes 12 and 21 and for the long arms of the X and the Y. Some women with Turner syndrome are not the more common XO, but have an isochromosome with the long arm of the X chromosome duplicated but the short arm absent.

Chromosomes shaped like rings form in 1 out of 25,000 conceptions. Ring chromosomes may arise when telomeres are lost, leaving sticky ends that adhere. Exposure to radiation can also form rings. They can involve any chromosome, and may occur in addition to a full diploid

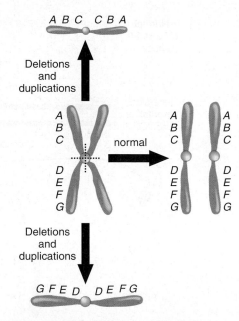

Figure 13.21 Isochromosomes have identical arms. They form when chromatids divide along the wrong plane (in this depiction, horizontally rather than vertically).

chromosome set, or account for one of the 46 chromosomes.

Ring chromosomes can produce symptoms when they add genetic material. For example, a small ring chromosome of DNA from chromosome 22 causes cat eye syndrome (OMIM 607576). Affected children have vertical pupils, are mentally retarded, have heart and urinary tract anomalies, and have skin growing over the anus. They have 47 chromosomes—the normal two chromosome 22s and a ring.

Ring chromosomes detected on routine amniocentesis present a challenging problem in genetic counseling, because rings usually do not affect health. Usually they consist of highly repeated DNA sequences that do not encode proteins.

Table 13.8 summarizes the causes of different types of chromosomal aberrations.

Key Concepts

1. Chromosome rearrangements can cause deletions and duplications.
2. In a Robertsonian translocation, the long arms of two different acrocentric chromosomes join.
3. In a reciprocal translocation, chromosomes exchange parts.
4. If a translocation leads to a deletion or duplication, or disrupts a gene, symptoms may result.
5. Gene duplications and deletions can occur in isochromosomes and ring chromosomes, and when crossovers involve inversions.
6. An isochromosome has two identical arms, introducing duplications and deletions.
7. Ring chromosomes can add genetic material.

13.5 Uniparental Disomy—Two Genetic Contributions from One Parent

If nondisjunction occurs in both of the gametes that join to become a fertilized ovum, a pair of homologs (or parts of them) can come solely from one parent, rather than one from each parent, as Mendel's law of segregation predicts. For example, if a sperm lacking a chromosome

Table 13.8

Causes of Chromosomal Aberrations

Abnormalities	Causes
Numerical Abnormalities	
Polyploidy	Error in cell division (meiosis or mitosis) in which not all chromatid pairs separate in anaphase Multiple fertilization
Aneuploidy	Nondisjunction (in meiosis or mitosis) leading to lost or extra chromosomes
Structural Abnormalities	
Deletions and duplications	Translocation Crossover between a chromosome that has a pericentric inversion and its noninverted homolog
Translocation	Exchange between nonhomologous chromosomes
Inversion	Breakage and reunion of fragment in same chromosome, but with wrong orientation
Dicentric and acentric	Crossover between a chromosome with a paracentric inversion and its noninverted homolog
Ring chromosome	A chromosome loses telomeres and the ends fuse, forming a circle

Figure 13.22   Uniparental disomy.
Uniparental disomy doubles part of one parent's genetic contribution. In this family, the woman with CF inherited two copies of her mother's chromosome 7, and neither of her father's. Unfortunately, it was the chromosome with the disease-causing allele that she inherited in a double dose.

14 fertilizes an ovum with two copies of that chromosome, an individual with the normal 46 chromosomes results, but one homologous pair comes only from the female. Inheriting two chromosomes or chromosome segments from one parent is called **uniparental disomy** (UPD) ("two bodies from one parent"). The alternative state—normal segregation of chromosome pairs—is termed biparental inheritance. UPD can also arise from a trisomic embryo in which some cells lose an extra chromosome, restoring diploidy, but leaving two homologs from one parent.

Because UPD requires the simultaneous occurrence of two very rare events—either nondisjunction of the same chromosome in sperm and oocyte, or trisomy followed by chromosome loss—it is very rare. In addition, many cases are probably never seen, because bringing together identical homologs inherited from one parent could give the fertilized ovum a homozygous set of lethal alleles. Development would halt. Other cases of UPD may go undetected if they cause known recessive conditions and both parents are assumed to be carriers,

when actually only one parent contributed to the offspring's illness. This was how UPD was discovered.

In 1988, Arthur Beaudet of the Baylor College of Medicine saw a very unusual patient with cystic fibrosis. Beaudet was comparing CF alleles in the patient to those in her parents, and he found that only the mother was a carrier—the father had two normal alleles. Didn't both parents have to be carriers for a child to inherit this autosomal recessive disorder? Beaudet did further testing and constructed haplotypes for each parent's chromosome 7, which includes the CF gene. He found that the daughter had two copies from her mother, and none from her father (**figure 13.22**). How did this happen?

Apparently, in the patient's mother, nondisjunction of chromosome 7 in meiosis II led to formation of an oocyte bearing two identical copies of the chromosome, instead of the usual one. A sperm that had also undergone nondisjunction and lacked a chromosome 7 then fertilized the abnormal oocyte. The mother's extra genetic material compensated for the father's

deficit, and an offspring developed. Unfortunately, she inherited a double dose of the mother's chromosome that carried the mutant CF allele. In effect, inheriting two of the same chromosome from one parent shatters the protection that combining genetic material from two individuals offers, a protection that is the defining characteristic of sexual reproduction.

UPD may also cause disease if it removes the contribution of the important parent for an imprinted gene. Recall from chapter 6 that an imprinted gene is expressed if it comes from one parent, but silenced if it comes from the other (see figure 6.15). If UPD removes the parental genetic material that must be present for a critical gene to be expressed, a mutant phenotype results. The classic example is the 20 to 30 percent of Prader-Willi syndrome and Angelman syndrome cases caused by UPD (see figure 6.17). These disorders arise from mutations in different genes that are closely linked in a region of the long arm of chromosome 15, where imprinting occurs. They both cause mental retardation and a variety of other symptoms, but are quite distinct.

In 1989, researchers found that some children with Prader-Willi syndrome have two parts of the long arm of chromosome 15 from their mothers. The disease results because the father's Prader-Willi gene must be expressed for the child to avoid the associated illness. For Angelman syndrome, the situation is reversed. Children have a double dose of their father's DNA in the same chromosomal region implicated in Prader-Willi syndrome, with no maternal contribution. The mother's gene must be present for health.

People usually learn their chromosomal makeup only when something goes wrong—when they have a family history of reproductive problems, exposure to a toxin, cancer, or symptoms of a known chromosomal disorder. While researchers analyze the human genome sequence, chromosome studies will continue to be part of medical care—beginning before birth.

Key Concepts

1. Uniparental disomy (UPD) results when two chromosomes or chromosome parts are inherited from the same parent.
2. It can arise from a trisomy and subsequent chromosome loss, or from two nondisjunction events.
3. Uniparental disomy can cause disease if it creates a homozygous recessive condition, or if it disrupts imprinting.

Summary

13.1 Portrait of a Chromosome

1. Mutation can occur at the chromosomal level. **Cytogenetics** is the study of chromosome aberrations and their effects on phenotypes.

2. **Heterochromatin** stains darkly and harbors many DNA repeats. **Euchromatin** is light staining and contains many protein-encoding genes.

3. A chromosome consists of DNA and proteins. Essential parts are the **telomeres, centromeres,** and origin of replication sites.

4. Centromeres include alpha satellites and centromere-associated proteins, some of which form kinetochores that contact spindle fibers. CENP-A is a protein that may control centromere duplication.

5. Subtelomeres have telomerelike repeats that gradually disappear, as some protein-encoding genes occur. Chromosomes vary in gene density.

6. Chromosomes are distinguishable by size, centromere position, satellites, DNA probes to specific sequences, and staining patterns.

7. A **karyotype** is a size-ordered chromosome chart. A **metacentric** chromosome has two fairly equal arms. A **submetacentric** chromosome has a large arm and a short arm. An **acrocentric** chromosome's centromere is near a tip, so that it has one long arm and one very short arm.

13.2 Visualizing Chromosomes

8. Chromosomes can be obtained from any cell that has a nucleus. Prenatal diagnostic techniques that obtain fetal chromosomes include **amniocentesis, chorionic villus sampling,** and fetal cell sorting.

9. Hand-cut karyotypes and stains to view chromosomes are giving way to computerized karyotyping and chromosome-specific **fluorescence** *in situ* **hybridization.** Ideograms are diagrams that display chromosome bands, FISH data, and gene loci.

10. Chromosomal shorthand indicates chromosome number, sex chromosome constitution, and the nature of the specific chromosomal abnormality.

13.3 Abnormal Chromosome Number

11. A **euploid** somatic human cell has 22 pairs of autosomes and one pair of sex chromosomes.

12. **Polyploid** cells have extra chromosome sets.

13. **Aneuploids** have extra or missing chromosomes. **Trisomies** (an extra chromosome) are less harmful than **monosomies** (lack of a chromosome), and sex chromosome aneuploidy is less severe than autosomal aneuploidy. **Nondisjunction** is uneven distribution of chromosomes in meiosis. It causes aneuploidy. Most autosomal aneuploids cease developing as embryos. The most common at birth are trisomies 21, 13, and 18, because these chromosomes are gene-poor. Sex chromosome anomalies (XXY; 45,X; XXX; XYY) are less severe.

13.4 Abnormal Chromosome Structure

14. **Deletions** and/or **duplications** can result from crossing over after pairing errors occur in synapsis. Crossing over in an inversion heterozygote can also generate deletions and duplications.

15. In a **Robertsonian translocation,** the short arms of two acrocentric chromosomes break, leaving sticky ends on the long arms that join to form an unusual, large chromosome.

16. In a **reciprocal translocation,** two nonhomologous chromosomes exchange parts. In both types of translocation, a **translocation carrier** may have an associated phenotype if the translocation disrupts a vital gene. A translocation carrier also produces a predictable percentage of unbalanced gametes, which can lead to birth defects and spontaneous abortions.

17. A heterozygote for an inversion may have reproductive problems if a crossover occurs between the inverted region and the noninverted homolog, generating deletions and duplications. A **paracentric inversion** does not include the centromere; a **pericentric inversion** does.

18. Isochromosomes repeat one chromosome arm but delete the other. They form when the centromere divides in the wrong plane during meiosis. Ring chromosomes form when telomeres are removed, leaving sticky ends that adhere.

13.5 Uniparental Disomy—Two Genetic Contributions from One Parent

19. In **uniparental disomy,** a chromosome, or a part of one, doubly represents one parent. It can result from nondisjunction in both gametes, or from a trisomic cell that loses a chromosome.

20. Uniparental disomy causes symptoms if it creates a homozygous recessive state associated with an illness, or if it affects an imprinted gene.

Review Questions

1. What are the essential components of a chromosome? of a centromere?

2. How does the DNA sequence change with distance from the telomere?

3. How are centromeres and telomeres alike?

4. What happens during meiosis to produce each of the following?

   a. an aneuploid

   b. a polyploid

   c. the increased risk of trisomy 21 Down syndrome in the offspring of a woman over age 40 at the time of conception

   d. recurrent spontaneous abortions to a couple in which the man has a pericentric inversion

   e. several children with Down syndrome in a family where one parent is a translocation carrier

5. A human liver has patches of cells that are octaploid—that is, they have eight sets of chromosomes. Explain how this might arise.

6. Describe an individual with each of the following chromosome constitutions. Mention the person's sex and possible phenotype.

   a. 47,XXX

   b. 45,X

   c. 47,XX, trisomy 21

7. Which chromosomal anomaly might you expect to find more frequently among the members of the National Basketball Association than in the general population? Cite a reason for your answer.

8. List three examples illustrating the idea that the amount of genetic material involved in a chromosomal aberration affects the severity of the associated phenotype.

9. List three types of chromosomal aberrations that can cause duplications and/or deletions, and explain how they do so.

10. Why would having the same inversion on both members of a homologous chromosome pair *not* lead to unbalanced gametes, as having the inversion on only one chromosome would?

11. Define or describe the following technologies:

    a. FISH

    b. amniocentesis

    c. chorionic villus sampling

    d. fetal cell sorting

12. Why are trisomies 13 and 18 more common at birth than trisomies 5 or 16?

13. How many chromosomes would a person have who has Klinefelter syndrome and also trisomy 21?

14. Explain why a female cannot have Klinefelter syndrome and a male cannot have Turner syndrome.

15. List three causes of Turner syndrome.

Applied Questions

1. The following is part of a chart used to provide genetic counseling on maternal age effect on fetal chromosomes. Answer questions a-e based on this chart.

Maternal Age	Trisomy 21 Risk	Risk for Any Aneuploid
20	1/1,667	1/526
24	1/1,250	1/476
28	1/1,053	1/435
30	1/952	1/385
32	1/769	1/322
35	1/378	1/192
36	1/289	1/156
37	1/224	1/127
38	1/173	1/102
40	1/106	1/66
45	1/30	1/21
48	1/14	1/10

   a. If the risk that amniocentesis will cause spontaneous abortion is 1 in 250 at a particular obstetrical practice, at what age should patients in this practice undergo the test?

   b. The Willoughbys have a son who has trisomy 21 Down syndrome. The mother, Suzanne, is 24 years old and pregnant. The Martinis do not have any relatives who have Down syndrome or any other chromosomal condition. Karen Martini is pregnant, and is 32 years old. Who has the lower risk of having a child with Down syndrome, Suzanne Willoughby or Karen Martini?

   c. Why are the risks in the righthand column higher than those in the middle column?

   d. Sam and Alice Dekalb receive genetic counseling because of "advanced maternal age"—Alice is 40 years old. When amniocentesis reveals trisomy 13, the couple is shocked, explaining that they thought the risk of a chromosomal problem was less than 1 percent. How have they misinterpreted the statistics?

   e. A 40-year-old woman wants to have children after she is 45. How much will her risk of conceiving a child with trisomy 21 increase in that time?

2. Amniocentesis indicates that a fetus has the chromosomal constitution 46, XX,del(5)(p15). What does this mean? What might the child's phenotype be?

3. What type of test could determine whether a triploid infant resulted from a diploid oocyte fertilized by a haploid sperm, or from two sperm fertilizing one oocyte?

4. For an exercise in a college genetics laboratory course, a healthy student constructs a karyotype from a cell in a drop of her blood. She finds only one chromosome 3 and one chromosome 21, plus two unusual chromosomes that do not seem to have matching partners.

   a. What type of chromosomal abnormality does she have?

   b. Why doesn't she have any symptoms?

   c. Would you expect any of her relatives to have any particular medical problems? If so, which medical conditions?

5. A fetus ceases developing in the uterus. Several of its cells are karyotyped. Approximately 75 percent of the cells are diploid, and 25 percent are tetraploid (four copies of each chromosome). What do you think happened? When in development did it probably occur?

6. Distinguish among Down syndrome caused by aneuploidy, mosaicism, and translocation.

7. A couple has a son diagnosed with Klinefelter syndrome. Explain how the son's chromosome constitution could have arisen from either parent.

8. DiGeorge syndrome (OMIM 188400) causes abnormal parathyroid glands, disrupting blood calcium levels; heart defects; and an underdeveloped thymus gland, impairing development of the immune system. About 85 percent of patients have a microdeletion of a particular area of chromosome 22. In one family, a girl, her mother, and a maternal aunt have very mild cases of DiGeorge syndrome, and they also all have a reciprocal translocation involving chromosomes 22 and 2.

   a. How can a microdeletion and a translocation cause the same symptoms?

   b. Why were the people with the translocation less severely affected than the people with the microdeletion?

   c. What other problems might arise in the family with the translocation?

9. Refer to this list of all of the human chromosomes, the number of protein-encoding genes on each, and the number of bases to answer these questions:

   a. Which two chromosomes are out of order, and why?

   b. Which chromosome has the greatest proportion of its DNA sequence that encodes protein?

   c. Which chromosome has the greatest proportion of noncoding sequence?

   d. How much larger is the largest chromosome compared to the smallest?

Chromosome Number	Number of Genes	Number of Bases (in millions)
1	2,968	279
2	2,288	251
3	2,032	221
4	1,297	197
5	1,643	198
6	1,963	176
7	1,443	163
8	1,127	148
9	1,299	140
10	1,440	143
11	2,093	148
12	1,652	142
13	748	118
14	1,098	107
15	1,122	100
16	1,098	104
17	1,576	88
18	766	86
19	1,454	72
20	927	66
21	303	45
22	288	48
X	1,184	163
Y	231	51

Web Activities

Visit the Online Learning Center (OLC) at www.mhhe.com/lewisgenetics7. Select **Student Edition, chapter 13,** and **Web Activities** to find the website links needed to complete the following activities.

10. Go to the website for the Genetic Science Learning Center at the Eccles Institute of Human Genetics at the University of Utah. Follow the instructions to create a karyotype.

11. Visit the website for the Human Genome Landmarks poster. Select a chromosome, and use Online Mendelian Inheritance in Man (OMIM) to describe four traits or disorders associated with it. Or, consult the website for the Human Chromosome Launchpad for information on four genes carried on a specific chromosome.

Case Studies

12. The medical literature includes 18 cases of children with a syndrome consisting of poor growth before birth, developmental delay, premature puberty, loose joints, a large head, short stature, and small hands. In a different syndrome, children have a small chest, ears, and facial features as well as rib and finger defects. Children with the first condition have both copies of the entire long arm of chromosome 14 from their mothers, whereas children with the second condition inherit the same chromosome part from their fathers.

   a. What type of chromosomal aberration is responsible for these two disorders?

   b. Describe how each of the conditions might arise.

   c. Describe how these conditions might result from a deletion mutation.

13. Two sets of parents who have children with Down syndrome meet at a clinic. The Phelps know that their son has trisomy 21. The Watkins have two affected children, and Mrs. Watkins has had two spontaneous abortions. Why should the Watkins be more concerned about future reproductive problems than the Phelps? How are the offspring of the two families different, even though they have the same symptoms?

Learn to apply the skills of a genetic counselor with additional cases found in the *Case Workbook in Human Genetics.*

   DiGeorge syndrome

   Down syndrome

   Tetrasomy 12*p*

   Turner syndrome

   Williams syndrome

VISIT YOUR ONLINE LEARNING CENTER

Visit your online learning center for additional resources and tools to help you master this chapter. See us at

www.mhhe.com/lewisgenetics7.

CHAPTER

14

When Allele Frequencies Stay Constant

CHAPTER CONTENTS

IDENTIFYING THE "VOODOO CHILD"

On September 21, 2001, a boy's torso was found floating in the River Thames in east London. Detectives at Scotland Yard named him Adam. He had probably died from beheading. Contents of the stomach—clay, quartz crystals, gold, and ground bone—suggested that Adam had drunk a special brew about two days earlier. This evidence, as well as candles that washed ashore a few days later, suggested to detectives that Adam had been the victim of a ritualistic killing. Media reports dubbed him the "voodoo child."

Pollen in Adam's stomach came from Africa, but chemicals in hair and skin revealed he'd been in the U.K. for about a month. The detectives suspected that Adam might have been a victim of kidnapping followed by forced labor—a slave. To investigate whether Adam was a local child, forensic investigators sampled DNA from the nuclei of his cells and obtained a DNA profile, or fingerprint, which looks at a few chosen sites in the genome. Then they compared the profile to others from parents living in the area whose children were missing. Nothing matched.

Finally a key clue came from DNA profiling, but not of nuclear DNA. To analyze remains from the September 11 terrorist attacks in the United States, vast improvements had been made in DNA profiling, particularly working with mitochondrial DNA (mtDNA), which is often the only genetic material available following a violent death. By comparing the boy's sequences against a geographic database of mtDNA, investigators found the general area of his home—southwestern Nigeria. Analysis of the element strontium in his bone mineral further localized the area to a 50- to 100-mile stretch along a major road, near Benin City. Scotland Yard detectives scoured the town, putting up posters and talking to officials and parents of missing children, to no avail. However, they have arrested several gang members who may be responsible for many kidnappings, including Adam.

Columns of bonds reveal identities, courtesy of DNA profiling techniques.

So far, we've considered the gene as a "character" that transmits traits, and as a biochemical blueprint for a specific protein. Genes can also be considered at the population level.

A **population** is any group of members of the same species in a given geographical area who are potentially capable of mating and producing fertile offspring. Human populations might include the students in a class, a stadium full of people, or the residents of a community, state, or nation. **Population genetics** is a branch of genetics that considers all the alleles in a population, which constitute the **gene pool.** The "pool" in gene pool refers to a collection of gametes, and an offspring represents two gametes from the pool. Alleles can move between populations when individuals migrate and mate. This movement, termed gene flow, underlies evolution.

At the population level, genetics reflects history, anthropology, human behavior, and sociology, enabling us to trace our beginnings and to understand our diversity. This chapter introduces the major principle of population genetics, and the next two chapters explore the impact of population genetics on evolution.

14.1 The Importance of Knowing Allele Frequencies

Thinking about genes at the population level begins by considering frequencies—that is, how often a particular gene variant occurs in a particular population. Such frequencies can be calculated for alleles, genotypes, or phenotypes. For example, an allele frequency for the cystic fibrosis (CF) gene might be the number of $\Delta F508$ alleles among the residents of San Francisco. $\Delta F508$ is the most common allele that, when homozygous, causes the disorder. The allele frequency derives from the two $\Delta F508$ alleles in each person with CF, plus those carried in heterozygotes, as a proportion of all alleles for that gene in the gene pool of San Francisco. The genotype frequencies are the proportions of heterozygotes and the two types of homozygotes in the population. Finally, a phenotypic frequency is simply the percentage of people in the population who have CF (or who do not). With

multiple alleles for a single gene, the situation becomes more complex.

Phenotypic frequencies are determined empirically—that is, by observing how common a condition or trait is in a population. These figures have value in genetic counseling in estimating the risk that a particular inherited disorder will occur in an individual when there is no family history of the illness. **Table 14.1** shows disease incidence for phenylketonuria (PKU), an inborn error of metabolism that causes mental retardation unless the person follows a special low-protein diet from birth. Note how the frequency differs in different populations.

On a broader level, shifting allele frequencies in populations reflect the small steps of genetic change, called **microevolution,** that collectively constitute evolution. Genotype frequencies can change when any of the following conditions are met:

1. Individuals of one genotype are more likely to produce offspring with each other than with those of other genotypes (*nonrandom mating*).

2. Individuals *migrate* between populations.

3. Reproductively isolated small groups form within or separate from a larger population (*genetic drift*).

4. *Mutation* introduces new alleles into a population.

5. People with a particular genotype are more likely to produce viable, fertile offspring under a specific environmental condition than individuals with other genotypes (*natural selection*).

Table 14.1

Frequency of PKU in Various Populations

Population	Frequency of PKU
Chinese	1/16,000
Irish, Scottish, Yemenite Jews	1/5,000
Japanese	1/119,000
Swedes	1/30,000
Turks	1/2,600
United States Caucasians	1/10,000

Because these conditions are operating more often than not, genetic equilibrium—when allele frequencies are *not* changing—is rare. Thus, microevolution is not only possible, but also nearly unavoidable. Chapter 15 considers these factors in depth.

When enough microevolutionary changes accumulate to keep two fertile organisms of opposite sex from successfully producing fertile offspring together, **macroevolution,** or the formation of a new species, has occurred. In contrast to evolution, this chapter discusses the interesting, but unusual, situation in which allele frequencies stay constant, a condition called **Hardy-Weinberg equilibrium.**

Key Concepts

1. Population genetics is the study of allele frequencies in groups of organisms of the same species in the same geographic area.
2. The genes in a population comprise its gene pool.
3. Microevolution reflects changes in allele frequencies in populations. It is not occurring if allele frequencies stay constant from generation to generation (called Hardy-Weinberg equilibrium).
4. Hardy-Weinberg equilibrium happens when mating is random and the population is large, with no migration, genetic drift, mutation, or natural selection.

14.2 When Allele Frequencies Stay Constant

Population genetics looks at phenotypes and genotypes among large numbers of individuals. Allele frequencies reveal the underlying rules.

Hardy-Weinberg Equilibrium

In 1908, a Cambridge University mathematician named Godfrey Harold Hardy (1877–1947) and Wilhelm Weinberg (1862–1937), a German physician interested in genetics, independently used algebra to explain how allele frequencies can be used to predict phenotypic and genotypic frequencies in populations of diploid, sexually reproducing organisms.

Hardy unintentionally cofounded the field of population genetics with a simple letter published in the journal *Science*—he did not consider his idea to be worthy of the more prestigious British journal *Nature*. The letter began with a curious mix of modesty and condescension:

I am reluctant to intrude in a discussion concerning matters of which I have no expert knowledge, and I should have expected the very simple point which I wish to make to have been familiar to biologists.

Hardy continued to explain how mathematically inept biologists had deduced from Mendel's work that dominant traits would increase in populations, while recessive traits would become rarer. This seems logical, but it is untrue, because recessive alleles are introduced by mutation or migration, maintained in heterozygotes, and increase in frequency when they confer a reproductive advantage (natural selection). Hardy and Weinberg disproved the assumption that dominant traits increase while recessive traits decrease using the language of algebra.

The expression of population genetics in algebraic terms begins with the simple equation

$$p + q = 1.0$$

where p represents all dominant alleles for a gene, and q represents all recessive alleles. The expression "$p + q = 1.0$" simply means that all the dominant alleles and all the recessive alleles comprise all the alleles for that gene in a population.

Next, Hardy and Weinberg described the possible genotypes for a gene with two alleles using the binomial expansion

$$p^2 + 2pq + q^2 = 1.0$$

In this equation, p^2 represents the percentage of homozygous dominant individuals, q^2 represents the percentage of homozygous recessive individuals, and $2pq$ represents the percentage of heterozygotes (**table 14.2**). The letter p designates the frequency of a dominant allele, and q is the frequency of a recessive allele. **Figure 14.1** shows how the binomial expansion is derived from allele frequencies. Note that the derivation is conceptually the same as tracing alleles in a monohybrid cross.

Table 14.2

The Hardy-Weinberg Equation

Algebraic Expression	What It Means
$p + q = 1.0$ (allele frequencies)	All dominant alleles plus all recessive alleles add up to all alleles for a particular gene in a population.
$p^2 + 2pq + q^2 = 1.0$ (genotype frequencies)	For a particular gene with 2 alleles, all homozygous dominant individuals (p^2) plus all heterozygotes ($2pq$) plus all homozygous recessives (q^2) add up to all of the individuals in the population.

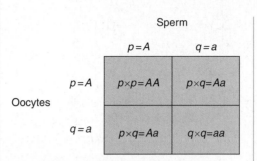

Figure 14.1 Source of the Hardy-Weinberg equation. A variation on a Punnett square reveals how random mating in a population in which gene A has two alleles—A and a—generates genotypes aa, AA, and Aa, in the relationship $p^2 + 2pq + q^2$.

The binomial expansion used to describe genes in populations became known as the Hardy-Weinberg equation. It can reveal the allelic frequency changes that underlie evolution. If the proportion of genotypes remains the same from generation to generation, as the equation indicates, then that gene is not evolving (changing). This situation, Hardy-Weinberg equilibrium, is an idealized state. It is possible only if the population is large, if its members mate at random, and if no migration, genetic drift, mutation, or natural selection takes place.

Hardy-Weinberg equilibrium is rare for protein-encoding genes that affect the phenotype, because an organism's appearance and health affect its ability to reproduce. That is, genes that affect the phenotype are subject to natural selection—harmful allelic combinations are weeded out of the population. However, Hardy-Weinberg equilibrium is seen in DNA repeats and other sequences that do not affect the phenotype, and therefore are not subject to natural selection. DNA profiling actually analyzes repeats.

Solving a Problem: The Hardy-Weinberg Equation

We can follow the frequency of two alleles of a particular gene from one generation to the next to understand Hardy-Weinberg equilibrium. Mendel's laws underlie such population-based calculations.

Consider an autosomal recessive trait: a middle finger shorter than the second and fourth fingers. If we know the frequencies of the dominant and recessive alleles, then we can calculate the frequencies of the genotypes and phenotypes and trace the trait through the next generation. The dominant allele D confers normal-length fingers; the recessive allele d confers a short middle finger (**figure 14.2**). We can figure out the frequencies of the dominant and recessive alleles by observing the frequency of homozygous recessives, because this phenotype reflects only one genotype. If 9 out of 100 individuals in a population have short fingers—genotype dd—the frequency is 9/100 or 0.09. Since dd equals q^2, then q equals 0.3. Since $p + q = 1.0$, knowing that q is 0.3 tells us that p is 0.7.

Next, we can calculate the proportions of the three genotypes that arise when gametes combine at random:

Homozygous dominant = DD
= $0.7 \times 0.7 = 0.49$
= 49 percent of individuals in generation 1

Homozygous recessive = dd
= $0.3 \times 0.3 = 0.09$
= 9 percent of individuals in generation 1

Heterozygous = $Dd + dD$
= $2pq = (0.7)(0.3) + (0.3)(0.7) = 0.42$
= 42 percent of individuals in generation 1

The proportion of homozygous individuals is calculated simply by multiplying the allele frequency for the recessive or dominant allele by itself. The heterozygous calculation is $2pq$ because there are two ways of combining a D with a d gamete—a D sperm with a d egg, and a d sperm with a D egg.

In this population, 9 percent of the individuals have a short middle finger. Now jump ahead a few generations, and assume that people choose mates irrespective of finger length. This means that each genotype of a female (DD, Dd, or dd) is equally likely to mate with each of the three types of

males (DD, Dd, or dd), and vice versa. **Table 14.3** multiplies the genotype frequencies for each possible mating, which leads to offspring in the familiar proportions of 49 percent DD, 42 percent Dd, and 9 percent dd. This gene, therefore, is in Hardy-Weinberg equilibrium—the allele and genotype frequencies do not change from one generation to the next.

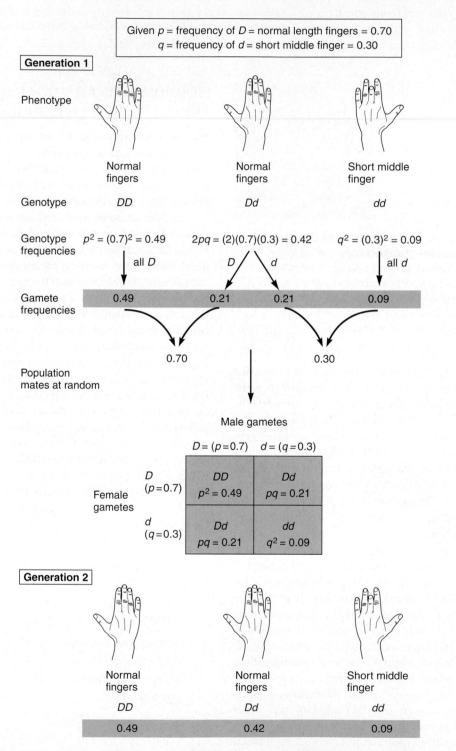

Figure 14.2 Hardy-Weinberg equilibrium. In Hardy-Weinberg equilibrium, allele frequencies remain constant from one generation to the next.

14.3 Applying Hardy-Weinberg Equilibrium

The Hardy-Weinberg equation is applied to population statistics on genetic disease incidence to derive carrier risks. To determine allele frequencies for autosomal recessively inherited characteristics, we need to know the frequency of one genotype in the population. This is typically the homozygous recessive class, because its phenotype indicates its genotype.

The incidence of an autosomal recessive disorder can be used to help calculate the risk that a particular person is a heterozygote. Returning to the example of CF, the incidence of the disease, and therefore also of carriers, may vary greatly in different populations (**table 14.4**). **Figure 14.3** provides another example of an illness common in one population, but exceedingly rare elsewhere.

Cystic fibrosis affects 1 in 2,000 Caucasian newborns. Therefore, the homozygous reces-

Table 14.3

Hardy-Weinberg Equilibrium—When Allele Frequencies Stay Constant

Possible Matings			Frequency of Offspring Genotypes		
Male	**Female**	**Proportion in Population**	**DD**	**Dd**	**dd**
0.49 *DD*	0.49 *DD*	0.2401 (*DD* × *DD*)	0.2401		
0.49 *DD*	0.42 *Dd*	0.2058 (*DD* × *Dd*)	0.1029	0.1029	
0.49 *DD*	0.09 *dd*	0.0441 (*DD* × *dd*)		0.0441	
0.42 *Dd*	0.49 *DD*	0.2058 (*Dd* × *DD*)	0.1029	0.1029	
0.42 *Dd*	0.42 *Dd*	0.1764 (*Dd* × *Dd*)	0.0441	0.0882	0.0441
0.42 *Dd*	0.09 *dd*	0.0378 (*Dd* × *dd*)		0.0189	0.0189
0.09 *dd*	0.49 *DD*	0.0441 (*dd* × *DD*)		0.0441	
0.09 *dd*	0.42 *Dd*	0.0378 (*dd* × *Dd*)		0.0189	0.0189
0.09 *dd*	0.09 *dd*	0.0081 (*dd* × *dd*)			0.0081
		Resulting offspring frequencies:	0.49	0.42	0.09
			DD	*Dd*	*dd*

sive frequency—*cc* if *c* represents the disease-causing allele—is 1/2,000, or 0.0005 in this population. This equals q^2. The square root of q^2 is about 0.022, which equals the frequency of the *c* allele. If q equals 0.022, then p, or $1 - q$, equals 0.978. Carrier frequency is equal to $2pq$, which equals (2)(0.978)(0.022), or 0.043—about 1 in 23. **Figure 14.4** summarizes these calculations.

If there is no CF in a family, a person's risk of having an affected child, based on population statistics, is low. For example, a Cau-casian couple with no family history of CF asks a genetic counselor to calculate the risk that they could conceive a child with this illness. The genetic counselor tells them that the chance of *each* potential parent being a carrier is about 4.3 percent, or 1 in 23. But this is only part of the picture. The chance that *both* of these people are carriers is 1/23 multiplied by 1/23—or 1 in 529—because the probability that two independent events will occur equals the product of the probability that each event will happen alone. However, if they *are* both carriers, each of their children would face a 1 in 4 chance of inheriting the illness, based on Mendel's first law of gene segregation. Therefore, the risk that two unrelated Caucasian individuals with no family history of CF will have an affected child is 1/4 × 1/23 × 1/23, or 1 in 2,116.

For X-linked traits, different predictions of allele frequencies apply to males and females. For a female, who can be homozygous recessive, homozygous dominant, or a heterozygote, the standard Hardy-Weinberg equation of $p^2 + 2pq + q^2$ applies. However, in males, the allele frequency is the phenotypic frequency, because a male who inherits an X-linked recessive allele exhibits it in his phenotype.

Figure 14.3 Disease incidence varies in populations.
Many of the 60 known cases of Crigler-Najjar syndrome are in the Mennonite and Amish communities of Lancaster County, Pennsylvania. This enzyme deficiency causes the buildup of a substance called bilirubin, producing severe jaundice (yellowing of the skin and eyes). If untreated, the syndrome causes fatal brain damage. Fortunately, exposure to ultraviolet light breaks down excess bilirubin, but children must sleep unclothed under the lights every night. The Mennonite and Amish populations have many other autosomal recessive illnesses that are extremely rare elsewhere, because they descended from a few founding families and marry among themselves.

Table 14.4

Carrier Frequency for Cystic Fibrosis

Population Group	Carrier Frequency
African Americans	1 in 66
Asian Americans	1 in 150
Caucasians of European descent	1 in 23
Hispanic Americans	1 in 46

Cystic Fibrosis

incidence (autosomal recessive class) = 1/2000 = 0.0005

$$\therefore q^2 = 0.0005$$

$$\therefore q = \sqrt{0.0005} = 0.022$$

$$\therefore p = 1 - q = 1 - 0.022 = 0.978$$

$$\therefore \text{carrier frequency} = 2pq = (2)\,(0.978)\,(0.022) = 0.043 = 1/23$$

Figure 14.4 Calculating the carrier frequency given population incidence: Autosomal recessive.

The incidence of X-linked hemophilia (OMIM 306700) (X^hY), for example, is 1 in 10,000 male births. Therefore, q (the frequency of the h allele) equals 0.0001. Using the formula $p + q = 1$, the frequency of the wild type allele is 0.9999. The incidence of carriers (X^HX^h), who are all female, equals $2pq$, or $(2)(0.0001)(0.9999)$, which equals 0.00019; this is 0.0002, or 0.02 percent, which equals about 1 in 5,000. The incidence of a female having hemophilia (X^hX^h) is q^2, or $(0.0001)^2$, or about 1 in 100 million. **Figure 14.5** summarizes these calculations.

Neat allele frequencies such as 0.6 and 0.4, or 0.7 and 0.3, are unusual. Mendelian disorders are usually very rare, and the q component of the Hardy-Weinberg equation contributes little. Because this means that the value of p approaches 1, the carrier frequency, $2pq$, is very close to $2q$. Thus, the carrier frequency is approximately twice the frequency of the rare, disease-causing allele.

Consider Tay-Sachs disease, which occurs in 1 in 3,600 Ashkenazim (Jewish people of eastern European descent). This means that q^2 equals 1/3,600, or about 0.0003. The square root, q, equals 0.017. The frequency of the dominant allele is then $1 - 0.017$, or 0.983. What is the likelihood that an Ashkenazi carries Tay-Sachs disease? It is $2pq$, or $(2)(0.983)(0.017)$, or 0.033. Note that this is very close to double the frequency of the mutant allele, 0.017. The Hardy-Weinberg equation can be modified to analyze genes that have more than two alleles.

Hemophilia A

incidence = 1/10,000 male births = 0.0001

$\therefore q = 0.0001$

$\therefore p = 1 - q \quad 1 - 0.0001 = 0.9999$

$\therefore$ carrier frequency (females) = $2pq$ = (2) (0.9999) (0.0001) = 0.00019 = about 1/5000

$\therefore$ affected females = q^2 = (0.0001) (0.0001) = 1/100 million

Figure 14.5 Calculating the carrier frequency given population incidence: X-linked recessive.

14.4 DNA Profiling Is Based on Hardy-Weinberg Assumptions

Hardy-Weinberg equilibrium also applies to parts of the genome that do not affect the phenotype, and are therefore not subject to natural selection. Short repeated sequences that are not part of a protein-encoding gene fall into this category. Repeats are scattered throughout the genome. A certain repeated sequence resides at the same chromosomal site in everyone, but people vary in the numbers of copies of that repeated sequence. Repeat copy number can be followed as a trait of sorts, called a marker, to identify an individual. That is, repeat number is another DNA variation among individuals, and is therefore a population polymorphism.

A person can have the same number of copies of a particular repeat on each homolog, or different numbers of copies, such as four copies of GCATC on one chromosome four, and six copies on its homolog, as shown in **figure 14.6.** This is analogous to being a homozygote or heterozygote for a visible inherited trait.

DNA profiling (originally called DNA fingerprinting) is a technology based on detecting differences in repeat copy number. In general, DNA profiling calculates the probability that certain combinations of repeat numbers will occur in two DNA sources by chance. For example, if a DNA profile of skin cells taken from under the fingernails of an assault victim matches the profile from a suspect's hair, and the chances are very low that those two samples would match by chance, that is strong evidence of guilt rather than a coincidental similarity. DNA evidence is more often valuable in excluding a suspect, and should be considered along with other types of evidence.

Although obtaining a DNA profile is a molecular technique, interpreting it requires statistical analysis of population data. Two types of repeats are used in forensics (legal applications) and in identifying

Key Concepts

1. Allele frequencies in populations can be inferred from the frequency of homozygous recessive individuals (q^2). The values of q and p can then be deduced and the Hardy-Weinberg equation applied to predict the frequency of carriers.
2. Genotype frequencies for X-linked traits are calculated differently in males and females. The frequency of the recessive phenotype in males is q, and in females q^2, as in autosomal recessive inheritance.
3. For very rare inherited disorders, p approaches 1, so the carrier frequency is approximately twice the frequency of the disease-causing allele ($2q$).

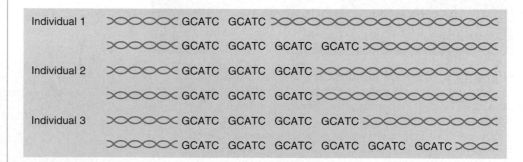

Figure 14.6 DNA profiling detects differing numbers of repeats at specific chromosomal loci. Individuals 1 and 3 are heterozygotes for the number of copies of a 5-base sequence at a particular chromosomal locus. Individual 2 is a homozygote, with the same number of repeats on the two copies of the chromosome. (Repeat number is considered an allele.)

Table 14.5

Characteristics of Repeats Used in DNA Profiling

Type	Repeat Length	Distribution	Example	Fragment Sizes
VNTRs (minisatellites)	10–80 bases	not uniform	TTCGGGTTG	50–1500 bases
STRs (microsatellites)	2–10 bases	more uniform	ACTT	50–500 bases

victims of disasters: **variable number of tandem repeats** (VNTRs), and **short tandem repeats** (STRs). **Table 14.5** compares them.

DNA Profiling Began with Forensics

Sir Alec Jeffreys at Leicester University in the United Kingdom invented DNA profiling in the 1980s. He detected differences in numbers of VNTRs among individuals by cutting these regions of DNA with restriction enzymes. These enzymes naturally protect bacteria by cutting foreign DNA, such as from viruses, at specific short sequences. (see chapter 19). Jeffreys measured DNA fragments using a technique called agarose gel electrophoresis, described in Reading 14.1. The different-sized fragments that result from "digesting" DNA are called restriction fragment length polymorphisms (RFLPs).

DNA pieces migrate through a jellylike material (agarose or the more discriminat-ing polyacrylamide) when an electrical field is applied. A positive electrode is placed at one end of the gel strip, and a negative electrode at the other. The DNA pieces, carrying negative charges because of their phosphate groups, move toward the positive pole. The pieces migrate according to size, with the shorter pieces moving faster and thus traveling farther in a given time. The pattern that forms when the different-sized fragments stop moving, with the shorter fragments closer to the positive pole and the longer ones farther away, creates a distinctive DNA pattern, or profile, that looks like a strip of black smears. An individual who is heterozygous for a repeat copy number will have two bands for that locus, as shown in **figure 14.7** for Individuals 1 and 3 from figure 14.6. A locus for which an individual is homozygous has only one corresponding band (Individual 2).

Jeffreys' first cases proved that a boy was the son of a British citizen so that he could enter the country, and freed a man jailed for raping two schoolgirls. Then in 1988, Jeffreys' approach matched DNA profiles from suspect Tommie Lee Andrews's blood cells to sperm cells left on his victim in a notorious rape case. More recently, Jeffreys used DNA profiling to demonstrate that Dolly, the Scottish sheep, was truly a clone of the six-year-old ewe that donated her nucleus (**figure 14.8**).

Variations on DNA profiling based on sequences other than VNTRs are used when sample DNA is scarce. Short tandem repeats (STRs) are used when DNA is fragmented, such as in the evidence from the terrorist

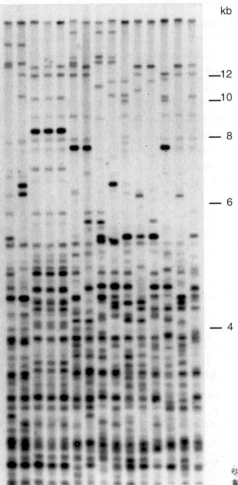

Figure 14.8 DNA profiles are most informative when compared. This group of DNA profiles compares the DNA of Dolly the cloned sheep (lane D) to that of the donor udder tissue (U), to that of cultured udder tissue (C). The other twelve lanes represent a dozen other sheep. The match among Dolly and the two versions of her nucleus donor is obvious.

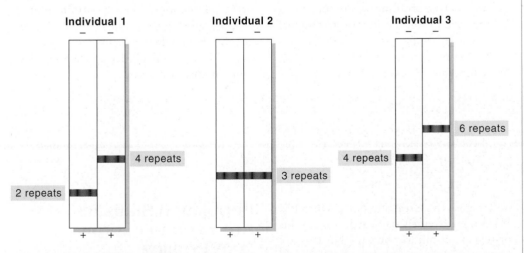

Figure 14.7 DNA profiles. DNA fragments that include differing numbers of copies of the same repeat migrate at different speeds and stop moving at different points on a polyacrylamide gel. These gels correspond to the individuals represented in figure 14.6. Actual DNA profiles typically scan at least 10 to 15 repeats on different chromosomes.

DNA Profiling: Molecular Genetics Meets Population Genetics

DNA profiling has rapidly become a standard and powerful tool in forensic investigations, agriculture, paternity testing, and historical investigations. Until 1986, DNA profiling was unheard of outside of scientific circles. A dramatic rape case changed that.

Tommie Lee Andrews was the first person in the United States to be convicted of a crime on the basis of DNA evidence. Andrews picked his victims months before he attacked and watched them so that he knew when they would be home alone. On a balmy Sunday night in May 1986, Andrews awaited Nancy Hodge, a young computer operator at Disney World in Orlando, Florida. The burly man surprised her when she was in her bathroom removing her contact lenses. He covered her face, then raped and brutalized her repeatedly.

Andrews was very careful not to leave fingerprints, threads, hairs, or any other indication that he had ever been in Hodge's home. But he had not counted on the new technology of DNA profiling, then called DNA fingerprinting. Thanks to a clear-thinking crime victim and scientifically informed lawyers, Andrews was soon at the center of a trial—not only his trial, but one that would judge the technology that helped to convict him.

After the attack, Hodge went to the hospital, where she provided a vaginal secretion sample containing the rapist's sperm cells. Two district attorneys who had read about DNA testing sent some of the sperm to a biotechnology company that extracted DNA and cut it with restriction enzymes. The sperm's DNA pieces were then mixed with labeled DNA probes that bound to complementary sequences.

The same procedure of extracting, cutting, and probing DNA was done on white blood cells from Hodge and Andrews, who had been apprehended and held as a suspect in several assaults. When the radioactive DNA pieces from each sample, which were the sequences where the probes had bound, were separated and displayed by size, the resulting pattern of bands—the DNA profile—matched exactly for the sperm sample and Andrews's blood, differing from Hodge's DNA (**figure 1**).

Tommie Lee Andrews's allele frequencies were compared to those for a representative African American population. At his first trial in November 1987, the judge, perhaps fearful that too much technical information would overwhelm the jury, did not allow the prosecution to cite population-based statistics. Without the appropriate allele frequencies, DNA profiling was reduced to a comparison of smeary lines on test papers to see whether the patterns of DNA pieces in the forensic sperm sample looked like those for Andrews's white blood cells. The probabilities determined from population-based statistics indicated that the possibility that Tommie Lee Andrews's DNA would match the evidence by chance was 1 in 10 billion. But the prosecution could not mention this.

After a mistrial was declared, the prosecution cited the precedent of using population statistics to derive databases on standard blood types. So when Andrews stood trial just three months later for raping a different woman, the judge permitted population analysis. Andrews was convicted. Today, in jail, he keeps a copy of the *Discover* magazine article (written by this author) that describes his role in the first case tried using DNA profiling.

The sizes of the DNA pieces in the type of DNA profile used in the Andrews case vary from person to person because of differences in DNA sequence in the regions surrounding the probed genes. The discriminating power of the technology stems from the fact that there are many more ways for the 3 billion bases of the human genome to vary than there are people. However, this theoretical variation is tempered by the fact that certain gene combinations are more prevalent in some populations. That is, within ethnic groups, some people may be more alike genetically, and distinguishing among them might be more difficult. Today, DNA profiles typically analyze repeats.

attacks on September 11, 2001, from the Asian tsunami of December 26, 2004, and the London train and bus bombings of July 7, 2005. The smaller size of these repeats makes them more likely to persist in degraded DNA, even in as little as a few hundred DNA bases. STRs are amplified using the polymerase chain reaction (see chapter 19). If DNA is extremely damaged, two regions of repeats in mitochondrial DNA that are highly variable in populations may yield information. MtDNA evidence led Scotland Yard to trace their headless corpse to Africa, as described in the chapter opener.

Population Statistics Are Used to Interpret DNA Profiles

The power of DNA profiling is greatly expanded by tracking repeats on several chromosomes—as in forensics in general,

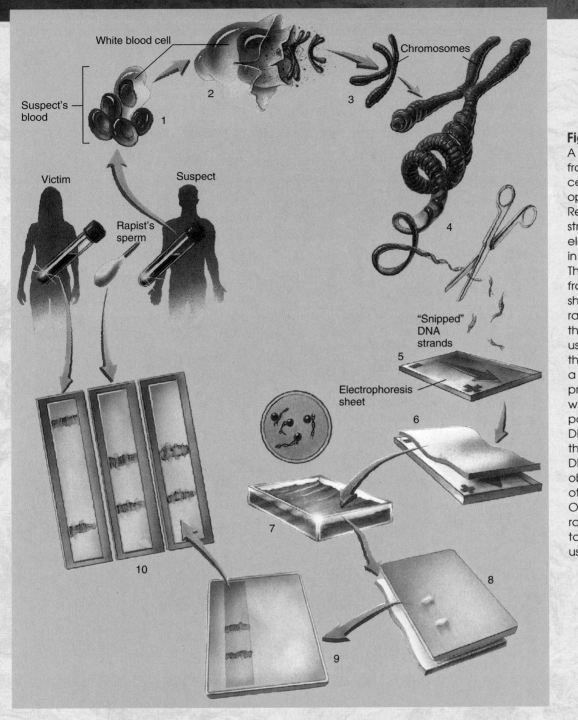

Figure 1   DNA profiling.
A blood sample (1) is collected from the suspect. White blood cells are separated and burst open (2), releasing DNA (3). Restriction enzymes snip the strands into fragments (4), and electrophoresis aligns them by size in a groove on a sheet of gel (5). The resulting pattern of DNA fragments is transferred to a nylon sheet (6). It is then exposed to radioactively tagged probes (7) that home in on the DNA areas used to establish identity. When the nylon sheet is placed against a piece of X-ray film (8) and processed, black bands appear where the probes stuck (9). This pattern of bands constitutes a DNA profile (10). This profile may then be compared to the victim's DNA pattern, the rapist's DNA obtained from sperm cells, and other biological evidence. Originally, DNA profiling used radioactively tagged probes, but today fluorescent labels are often used.

the more clues, the better. The numbers of copies of a repeat are assigned probabilities based on their observed frequencies in a particular population. Considering repeats on different chromosomes makes it possible to use the product rule to calculate the probabilities of particular combinations of repeat numbers occurring in a population, based on Mendel's law of independent assortment.

The Hardy-Weinberg equation and the product rule are used to derive the statistics that back up a DNA profile. First, the pattern of fragments indicates whether an individual is a homozygote or a heterozygote for each repeat, because a homozygote only has one band representing that locus. Genotype frequencies are then calculated using parts of the Hardy-Weinberg equation. That is, p^2 and q^2 denote the two homozygotes for a two-allele repeat, and $2pq$ represents the heterozygote. Then the

Table 14.6

Multiplied Frequencies of Different Repeat Numbers

The Case: A famous painting has been stolen from a gallery. The thief planned the crime carefully, but as she was removing the painting from its display, she sneezed. She averted her face, but a few tiny droplets hit the wall. Detectives obtained a DNA profile using 6 repeat alleles, from different chromosomes, for DNA in nose lining cells in the droplets. Then they compared the profile to those compiled for eight people in the vicinity, all women, who had been identified by hidden camera. (Assume the suspects are in the same ethnic group.) Most of the samples matched at 2 to 4 sites, but one matched at all 6. She was the crook. Notice how the probability of guilt increases with the number of matches. Matching for the very rare allele #3 is particularly telling.

Allele	Repeat	Frequency	Cumulative Multiplied Frequencies
1	ACT on chromosome 4	1/60	
2	GGC on chromosome 17	1/24	$1/60 \times 1/24 = 1/1,440$
3	AAGCTA on chromosome 14	1/1,200	$1/1,440 \times 1/1,200 = 1/1,728,000$
4	GGTCTA on chromosome 6	1/11	$1/1,728,000 \times 1/11 = 1/19,008,000$
5	ATACGAGG on chromosome 9	1/40	$1/19,008,000 \times 1/40 = 1/760,320,000$
6	GTA	1/310	$1/760,320,000 \times 1/310 = 1/235,699,200,000$

frequencies are multiplied. **Table 14.6** shows an example of multiplying frequencies of different repeat numbers. The result is the probability that this particular combination of repeat sizes would occur in a particular population. Logic then enters the equation. If the combination is very rare in the population the suspect comes from, and if it is found both in the suspect's DNA and in crime scene evidence, such as a rape victim's body or the stolen property in table 14.6, the suspect's guilt appears highly likely. **Figure 14.9** summarizes the procedure.

For DNA profiling, Hardy-Weinberg equilibrium is assumed, and when it isn't operating, problems can arise. For example, the requirement of nonrandom mating wouldn't be met in a community with a few very large families where distant relatives might inadvertently marry each other—a situation that happens in small towns. A particular DNA profile for one person might be shared by his or her cousins. In one case, a young man was convicted of rape based on a DNA profile—which he shared with his father, the actual rapist. Considering a larger number of repeat sites can guard against such complications. If more repeat sites had been considered in the rape case, chances are that they would have revealed a polymorphism that the son had inherited from his mother, but that the father lacked. This would have indicated that the son was not guilty, but a close male relative might be.

The accuracy and meaning of a DNA profile depend upon the population that is the source for the allele frequencies. If populations are too broadly defined, then allele frequencies are typically low, leading to very large estimates of the likelihood that a suspect matches evidence based on chance. In one oft-quoted trial, the prosecutor concluded, *The chance of the DNA fingerprint of the cells in the evidence matching blood of the defendant by chance is 1 in 738 trillion.* The numbers were not at fault, but some population geneticists questioned the validity of the databases. Did they really reflect the gene pool compositions of actual populations? By 1991, several judges had rejected DNA evidence because population geneticists had testified that the databases greatly oversimplify human population structure. Therefore, the odds that crime scene DNA matched suspect DNA were not as reliable as originally suggested.

The first DNA profiling databases neatly shoehorned many different groups into just three—Caucasian, black, or Hispanic—designations not necessarily biologically meaningful. People from Poland, Greece, or Sweden would all be considered white, and a dark-skinned person from Jamaica and one from Somalia would be lumped together as blacks. Perhaps the most incongruous of all were the Hispanics. Cubans and Puerto Ricans are part African, whereas people from Mexico and Guatemala have mostly

Native American gene variants. Spanish and Argentinians have neither black African nor Native American genetic backgrounds. Yet these diverse peoples were considered a single population! Other groups were left out, such as Native Americans and Asians. Ultimately, analysis of these three databases revealed significantly more homozygous recessives for certain polymorphic genes than the Hardy-Weinberg equation would predict, confirming what many had suspected—allele frequencies were not in equilibrium.

Giving meaning to the allele frequencies necessary to interpret DNA profiles requires more restrictive ethnic databases. A frequency of 1 in 1,000 for a particular allele in all whites may actually be much higher or lower in, for example, only Italians, because they (and many others) tend to marry among themselves. On the other hand, narrowly defined ethnic databases may be insufficient to interpret DNA profiles from people of mixed heritages, such as someone whose mother was Scottish/French and whose father was Greek/German.

We may need to develop mathematical models to account for real population structures. Perhaps the first step will be to understand the forces that generate genetic substructures within more broadly defined populations, which means taking into account history and human nature. Chapter 15 explores these factors.

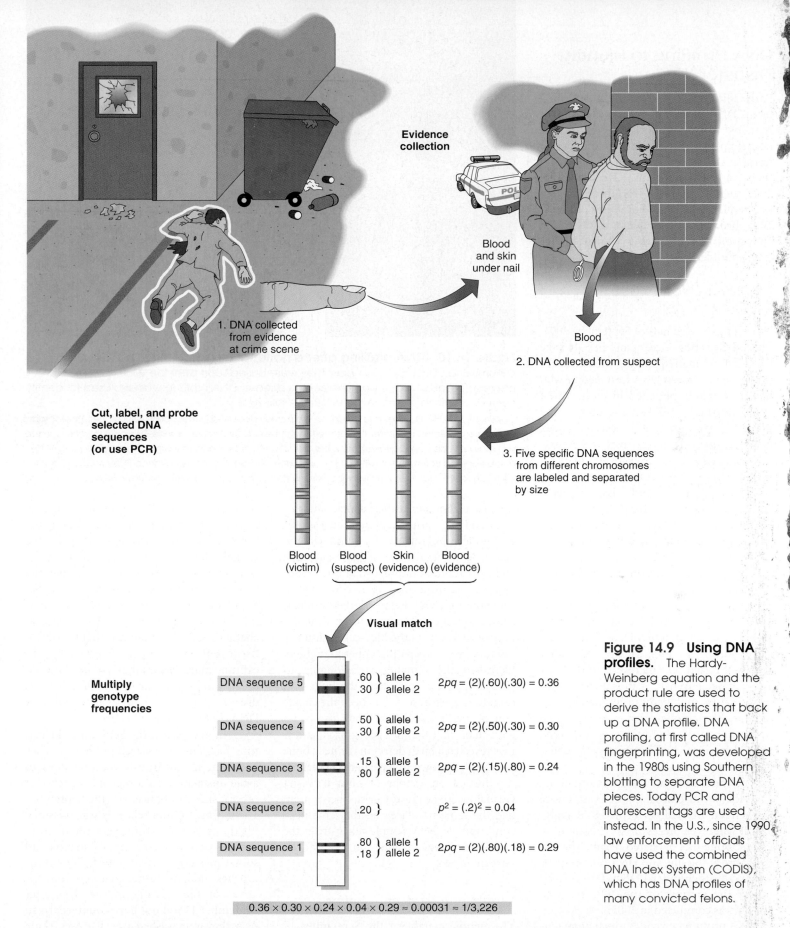

Evidence collection

Blood and skin under nail

Blood

1. DNA collected from evidence at crime scene

2. DNA collected from suspect

Cut, label, and probe selected DNA sequences (or use PCR)

3. Five specific DNA sequences from different chromosomes are labeled and separated by size

Blood (victim) Blood (suspect) Skin (evidence) Blood (evidence)

Visual match

Multiply genotype frequencies

DNA sequence 5   .60 } allele 1 $2pq = (2)(.60)(.30) = 0.36$
                .30 } allele 2

DNA sequence 4   .50 } allele 1 $2pq = (2)(.50)(.30) = 0.30$
                .30 } allele 2

DNA sequence 3   .15 } allele 1 $2pq = (2)(.15)(.80) = 0.24$
                .80 } allele 2

DNA sequence 2   .20 } $p^2 = (.2)^2 = 0.04$

DNA sequence 1   .80 } allele 1 $2pq = (2)(.80)(.18) = 0.29$
                .18 } allele 2

$0.36 \times 0.30 \times 0.24 \times 0.04 \times 0.29 \approx 0.00031 \approx 1/3,226$

Conclusion: The probability that another person in the suspect's population group has the same pattern of these alleles is approximately 1 in 3,226.

Figure 14.9 Using DNA profiles. The Hardy-Weinberg equation and the product rule are used to derive the statistics that back up a DNA profile. DNA profiling, at first called DNA fingerprinting, was developed in the 1980s using Southern blotting to separate DNA pieces. Today PCR and fluorescent tags are used instead. In the U.S., since 1990, law enforcement officials have used the combined DNA Index System (CODIS), which has DNA profiles of many convicted felons.

DNA Profiling to Identify Disaster Victims

Early editions of this textbook described using DNA profiling to identify human remains from plane crashes—then the largest application of the technology. The terrorist attacks of September 11, 2001, and then the Asian tsunami of December 26, 2004, took the scope of DNA profiling to a new level. The London bombinbs of July 7, 2005, brought the challenge of obtaining and profiling DNA from burnt bodies trapped underground.

Identifying World Trade Center Victims

During the second half of September, 2001, Myriad Genetics, a company in Salt Lake City that normally provides breast cancer tests, received frozen DNA from people who had presumably perished in the terrorist attack on the World Trade Center on the 11th. The laboratory also received cheek brush scrapings from relatives of the missing, amassed at DNA collection centers set up throughout New York City in the days after the disaster, and tissue labeled "reference samples" from the victims' toothbrushes, razors, and hairbrushes. Technologists determined the numbers of copies of STRs at 13 loci, as well as the sex chromosome constitution. The probability that any two individuals would have the same 13 markers by chance is 1 in 250 trillion. Therefore, if the STR pattern of crime scene evidence matched DNA from a victim's toothbrush, identification was fairly certain. Myriad sent its results to the New York State Forensic Laboratory, where investigators matched family members to victims. Meticulous records were kept so that when an individual was identified more than once, the bad news wasn't delivered to the grieving family again.

The most difficult part was obtaining samples, because most of the bodies were incinerated. Myriad performed STR analysis on whatever pieces of soft tissue were found. Bone bits persisted despite the ongoing fire at the site, and DNA profiling was done on mitochondrial DNA. DNA profiling of the victims of the 2001 terrorist attacks was completed in 2005.

DNA profiling provides much more reliable information on identity than traditional

Figure 14.10   DNA profiling after a national disaster. These bodies were exhumed to obtain DNA samples. They were buried soon after the 2004 tsunami to prevent spread of water-borne infectious disease. DNA profiling was also used to identify hurricane Katrina victims in the U.S. in the fall, 2005.

forensic techniques such as dental patterns, scars, and fingerprints, and clues such as jewelry, wallets, and rolls of film found with the victim. Consider the case of Jose Guadalupe and Christopher Santora, two of fifteen firefighters lost from one engine company on September 11, 2001. Rescue workers brought a body found beside a fire truck next to the destroyed towers to the Medical Examiner's office on September 13th. Other firefighters identified the remains as belonging to Guadalupe based on where the body had been found, because he had been the chauffeur of the fire truck. The body also had a gold chain that the men recognized, and X rays revealed a birth defect in the neck bone that he was known to have had. Guadalupe was buried on October 1—but it wasn't Guadalupe in the grave. Christopher Santora had the same necklace and the same neck condition! A DNA sample taken from the buried man's remains and from Santora's relatives matched.

Identifying Tsunami Victims

The people caught in the September 11 tragedy died by fire; those caught in the tsunami of 2004 died by water. Whereas New York workers had to search among the rubble of skyscrapers to find body parts, the roughly 200,000 bodies strewn about by the tsunami were everywhere. Even though the technology to identify bodies in the two disasters was the same, the scope of the tsunami disaster introduced new technical challenges. Rather than hunting for bits of tissue out in the open, workers had to exhume many bodies that had been buried quickly to prevent the spread of infectious disease in order to obtain tissue samples for DNA profiling (**figure 14. 10**).

DNA analyses of the September 11 victims had shown researchers that matches had to be sought from several relatives, to avoid mistakes when two people matched at several sites by chance. For the September 11 victims, finding relatives was possible; for the tsunami victims, very often entire families were washed away, leaving few and sometimes no relatives to test. The toothbrushes and hairbrushes collected from relatives of the missing in the days after September 11 did not have counterparts in Asia; they were washed away like everything else. To compensate for these deficiencies,

Sir Alec Jeffreys advised assessing 15 to 20 repeat sites, rather than the usual 10 to 15; other forensic scientists suggested comparing up to 50 to derive the DNA profiles. Hundreds of teams of forensic specialists rushed to the scene to collect evidence, or did so in other countries from relatives of tourists who had been caught in the disaster while traveling in the affected nations. Fortunately, DNA profiling was extremely useful because so many people began the process so soon after the disaster, and techniques had improved so greatly since 2001.

RELATING THE CONCEPTS

How was DNA profiling used to help identify Adam, the "voodoo child?" What information did it provide, what were its limitations, and how did other types of evidence help the investigation?

Key Concepts

1. DNA profiles are based on SNPs or differences in the numbers of DNA repeats among individuals, based on the idea that the human genome sequence can vary in more ways than there are people.
2. Population statistics are applied to determine the probability that the same pattern would occur by chance in two individuals.
3. A limitation of the method is that databases may not adequately represent real populations. Developing narrower ethnic databases and considering historical and social factors may make population statistics more realistic.
4. DNA profiling of nuclear and mitochondrial DNA was performed on evidence from the September 11 terrorist attacks and the 2004 tsunami.

14.5 Genetic Privacy

Before the information age, population genetics was an academic discipline that was more theoretical than practical. Today, with the combination of information technology, whole genome sequencing, and shortcuts to identify people by SNP or repeat number patterns, population genetics represents a powerful way to identify individuals.

The human genome is made up of 3.2 billion units of information, each of which can be one of four possibilities—that's a huge capacity for diversity. The human genome can vary many more ways than there are people—about 10 billion worldwide. Given these daunting numbers, one only need consider 30 to 80 genome sites to uniquely describe each person. This is why forensic tests typically only compare 10 to 15 or so loci (sites) to rule out or establish identity.

The ease of assigning highly individualized genetic nametags may be helpful in forensics, but it poses privacy issues. Consider a "DNA dragnet," a forensic approach of taking DNA profiles of all residents of a town where a violent crime is unsolved. Sir Alec Jeffreys in the U.K. conducted some of the first DNA dragnets in the late 1980s. The largest to date occurred in 1998 in Germany, where more than 16,000 men had their DNA profiled in a search for the man who raped and murdered an 11-year-old. The dragnet indeed caught the killer.

A more recent case involving a DNA dragnet happened in the small town of Truro, near the tip of Cape Cod, Massachusetts. Writer Christa Worthington was brutally murdered in January 2002, a knife driven completely through her heart into the floorboards, and her toddler daughter found at her side, trying to mop up the blood. Worthington was murdered in her home in the seaside summer village where only 790 men lived in the winter. DNA from semen in her body did not match samples in any criminal databases. Three years later, on the advice of the Federal Bureau of Investigation, police began asking men at Truro's few gathering places—a pizza restaurant, the garbage dump—to provide cheek swabs for DNA testing. There was no requirement to do so, but a record was being kept of all who refused, and everyone knew it. Several citizens filed complaints with the American Civil Liberties Union, but most of Truro's male residents complied—including the trash collector who was arrested and charged with the crime in April, 2005.

Genetic privacy is also an issue in health care. In some facilities, patients can meet with a geneticist and have tests, but have the results kept in a "shadow file" that is not part of the official medical record. In this way, the results remain confidential.

Legislation concerning genetic privacy in the United States has not caught up with bioinformatics (the use and analysis of biological information). In fact, the U.S. lags behind other nations in establishing biobanks that would store DNA profiles (see Reading 1.1). Some protection comes from the Health Insurance Portability and Accountability Act (HIPAA) of 1996, which requires a patient's consent for sharing health-related information or test results. This would not apply to DNA profiles obtained in forensic investigations. In 2003, the Privacy Rules reinforced HIPAA, but still did not address use and prevention of abuse of genetic data.

Key Concepts

1. Each person has a unique genetic signature (except multiples).
2. DNA profiling introduces privacy issues.

Summary

1. A **population** is a group of interbreeding members of the same species in a particular area. Their genes constitute the **gene pool.**

14.1 The Importance of Knowing Allele Frequencies

2. Population genetics considers allele, genotype, and phenotype frequencies to reveal whether microevolution is

occurring. Phenotypic frequencies can be determined empirically, then used in algebraic expressions to derive other frequencies.

3. Genotype frequencies change if migration, nonrandom mating, genetic drift, mutations, or natural selection operate. In **Hardy-Weinberg equilibrium,** frequencies are not changing.

14.2 When Allele Frequencies Stay Constant

4. Hardy and Weinberg proposed an algebraic equation to explain the constancy of allele frequencies. This would show why dominant traits do not increase and recessive traits do not decrease in populations. The Hardy-Weinberg equation is a binomial expansion used to represent genotypes in a population.

5. Hardy-Weinberg equilibrium is demonstrated by following gamete frequencies as they recombine in the next generation. In equilibrium, these genotypes remain constant if evolution is not occurring. When the equation

$p^2 + 2pq + q^2$ represents a gene with one dominant and one recessive allele, p^2 corresponds to the frequency of homozygous dominant individuals; $2pq$ stands for heterozygotes; and q^2 represents the frequency of the homozygous recessive class. The frequency of the dominant allele is p, and of the recessive allele, q.

14.3 Applying Hardy-Weinberg Equilibrium

6. If we know either p or q, we can calculate genotype frequencies, such as carrier risks. Often such information comes from knowing the q^2 class, which corresponds to the frequency of homozygous recessive individuals in a population.

7. For X-linked recessive traits, the mutant allele frequency for males equals the trait frequency. For very rare disorders or traits, the value of p approaches 1, so the carrier frequency ($2pq$) is approximately twice the frequency of the rare trait (q).

14.4. DNA Profiling Is Based on Hardy-Weinberg Assumptions

8. Repeats (**VNTRs** and **STRs**) that do not encode protein are presumably in Hardy-Weinberg equilibrium and can be compared to establish individual DNA profiles.

9. To obtain a **DNA profile,** determine repeat numbers (using RFLPs or PCR) and multiply population-based allele frequencies to derive the probability that profiles from two sources match by chance.

14.5 Genetic Privacy

10. People vary genetically in more ways than there are people.

11. Individuals can be distinguished genetically, introducing privacy concerns about identifying people using genetic information.

Review Questions

1. "We like him, he seems to have a terrific gene pool," say the parents upon meeting their daughter's boyfriend. Why doesn't their statement make sense?

2. What is *not* happening in a population in Hardy-Weinberg equilibrium?

3. Why is knowing the incidence of a homozygous recessive condition in a population important in deriving allele frequencies?

4. Two couples want to know their risk of conceiving a child with cystic fibrosis. In one couple, neither partner has a family history of the disease; in the other, one partner knows he is a carrier. How do their risks differ?

5. State which of the following choices is more informative in creating a DNA profile, and why.

   a. short DNA repeats versus protein-encoding genes

   b. VNTRs versus STRs

   c. a rare repeat versus a common one

6. Why are specific population databases necessary to interpret DNA profiles?

7. How is the Hardy-Weinberg equation used to predict the recurrence of X-linked recessive traits?

Applied Questions

1. Glutaric aciduria type I (OMIM 231680) causes progressive paralysis and brain damage. It is very common in the Amish of Lancaster County, Pennsylvania—0.25 percent of newborns have the disorder. Calculate the percentage of newborns that are carriers for this condition, and the percentage that do not have the disease-causing allele.

2. Torsion dystonia (OMIM 128100) is a movement disorder that affects 1 in 1,000 Jewish people of eastern European descent (Ashkenazim). What is the carrier frequency in this population?

3. Factor IX deficiency (hemophilia B, OMIM 306900) is a clotting disorder affecting 1 in 190 Ashkenazim living in Israel. It affects 1 in 1,000,000 Japanese, Korean, Chinese, German, Italian, African American, English, Indian, and Arab people.

   a. What is the frequency of the mutant allele in the Israeli population?

   b. What is the frequency of the normal allele in this population?

   c. Calculate the proportion of carriers in the Israeli population.

   d. Why might the disease incidence be very high in the Israeli population but very low in others?

4. The Finnish population has a 1 percent carrier frequency for a seizure disorder called myoclonus epilepsy (OMIM 607876). Two people who have no relatives with the illness ask a genetic counselor to calculate the risk that they will conceive an affected child, based on their belonging to this population group. What is the risk?

5. Maple syrup urine disease (MSUD) (OMIM 248600) is an autosomal recessive inborn error of metabolism that causes mental and physical retardation, difficulty feeding, and a sweet odor to urine. In Costa Rica, 1 in 8,000 newborns inherits the condition. Calculate the carrier frequency of MSUD in this population.

6. The amyloidoses are a group of inborn errors of metabolism in which sticky protein builds up in certain organs. Amyloidosis caused by a mutation in the gene encoding a blood protein called transthyretin (OMIM 176300) affects the heart and/or nervous system. It is autosomal recessive. In a population of 177 healthy African Americans, four proved, by blood testing, to have one mutant allele of the transthyretin gene. What is the carrier frequency in this population?

7. Ability to taste phenylthiocarbamide (PTC) (OMIM 607751) is mostly determined by the *T* gene. *TT* individuals taste a strong, bitter taste; *Tt* people experience a slightly bitter taste; *tt* individuals taste nothing.

   A fifth-grade class of 20 students tastes PTC that has been applied to small pieces of paper, rating the experience as "very yucky" (*TT*), "I can taste it" (*Tt*), and "I can't taste it" (*tt*). For homework, the students test their parents, with these results:

   Of 6 *TT* students, 4 have 2 *TT* parents; and two have one parent who is *TT* and one parent who is *Tt*.

   Of 4 students who are *Tt*, 2 have 2 parents who are *Tt*, and 2 have one parent who is *TT* and one parent who is *tt*.

   Of the 10 students who can't taste PTC, 4 have 2 parents who also are *tt*, but 4 students have one parent who is *Tt* and one who is *tt*. The remaining 2 students have 2 *Tt* parents.

   Calculate the frequencies of the *T* and *t* alleles in the two generations. Is Hardy-Weinberg equilibrium maintained, or is this gene evolving?

8. DNA dragnets have been so successful in catching criminals in several countries that some people in law enforcement have suggested simply storing DNA samples of everyone at birth, so that a DNA profile could be obtained from anyone at any time. Do you think that this is a good idea or not? Cite reasons for your answer.

Web Activities

Visit the Online Learning Center (OLC) at www.mhhe.com/lewisgenetics7. Select **Student Edition, chapter 14,** and **Web Activities** to find the website links needed to complete the following activity.

9. On December 5, 1984, Theresa Fusco was raped and strangled near a roller-skating rink on Long Island, New York. Two similar crimes had occurred in previous months. Three young men were charged with the crime and then convicted, all the while proclaiming their innocence, maintaining that their confessions had been coerced and witnesses had lied. At their trial in 1990, defense lawyers requested DNA profiling, but the judge ruled that the technology was too unproven to use. In 2003, the case was reopened. Stored semen was taken from the "rape kit" and subjected to DNA testing, and the three men were exonerated. They had not killed Theresa Fusco after all—but they had spent more than a decade in prison.

   a. Why might the judge have refused to consider DNA testing in 1990?

   b. List the types of cells that could have been used to settle this case.

   c. What information on the three suspects would be needed to interpret DNA patterns?

   d. Do you think it is fair to decide whether or not a science-based forensic test or tool be used based on how well a judge, jury, lawyers, or the public—who may have little or no training in genetics—understands how it works?

   e. In 1992, lawyers Barry Scheck and Peter Neufeld, of the Cardozo School of Law in New York City, founded the nonprofit Innocence Project, a legal clinic that reopens cases where DNA profiling could have made a difference in the verdict. They have vindicated nearly 200 individuals. Consult the Innocence Project website, click on "Case Profiles," and select a case, describing how the DNA evidence exonerated a prisoner.

Case Studies

10. An extra row of eyelashes is an autosomal recessive trait that occurs in 900 of the 10,000 residents of an island in the south Pacific. Greta knows that she is a heterozygote for this gene, because her eyelashes are normal, but she has an affected parent. She wants to have children with a homozygous dominant man, so that the trait will not affect her offspring. What is the probability that a person with normal eyelashes in this population is a homozygote for this gene?

11. In a true crime that took place in Israel, a man knocked a woman unconscious with a cement block and then raped her. He was careful not to leave any hairs at the crime scene. But he left behind eyeglasses with unusual frames, and an optician helped police locate him. The man also left a half-eaten lollipop at the scene. DNA from blood taken from the suspect matched DNA from cheek-lining cells collected from the base of the telltale lollipop at four repeat loci on different chromosomes. Allele frequencies from the man's ethnic group in Israel are listed beside the profile pattern below:

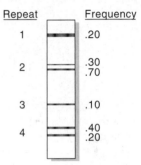

Repeat		Frequency
1		.20
2		.30 .70
3		.10
4		.40 .20

   a. For which of the tested repeats is the person a homozygote? How do you know this?

   b. What is the probability that the suspect's DNA matches that of the lollipop rapist by chance? (Do the calculation.)

   c. The man's population group is highly inbred—many people have children with relatives. How does this information affect the accuracy or reliability of the DNA profile?

   (P.S.—He was so frightened by the DNA analysis that he confessed!)

Learn to apply the skills of a genetic counselor with this additional case found in the *Case Workbook in Human Genetics:*

The Ice Maiden

CHAPTER

15

Changing Allele Frequencies

CHAPTER CONTENTS

THE EVOLUTION OF LACTOSE TOLERANCE

For millions of people who have lactose intolerance, dairy food causes cramps, bloating, gas, and diarrhea. Production of lactase, an enzyme secreted in the small intestine, declines from early childhood. It breaks down the milk sugar lactose.

Gene sequencing suggests why lactose intolerance is very common in some populations but not in others. For four linked genes on chromosome 7 that regulate digestion of the milk components lactose and calcium, variants are distinctly different in populations where most people *can* digest dairy. According to our genes, inability to digest lactose is the more ancient, wild type condition.

The alleles enabling adults to digest milk likely arose by chance, perhaps in Turkey or the Ural mountains of western Russia, according to clues in DNA sequences. (The more variable the milk-digesting gene variants on chromosome 7, the more time has elapsed, and the more ancient the people.) These people migrated to Europe between 3,500 and 6,600 years ago. When they introduced herding, individuals with milk-tolerant alleles were healthier, leaving more children—many of whom could digest milk. Over time, populations that drank milk accumulated the lactose-tolerant alleles, whereas those that did not, such as in Asia and Africa, continued to have many individuals who could not digest milk. Their genotype was not a handicap because milk was not a dietary staple.

Lactose intolerance is variable in ethnic populations in the United States today. Asian-Americans are 90 percent lactose intolerant; African-Americans, 75 percent; Native Americans 75 percent; and European-Americans, just 10 percent.

The ability to digest lactose (milk sugar) became more prevalent in populations after agriculture introduced dairy foods—thanks to evolution.

Historically, we seem to have gone out of our way to ensure that the very specific conditions necessary for Hardy-Weinberg equilibrium—unchanging allele frequencies from generation to generation—do not occur, at least for some genes. Wars and persecution kill certain populations. Economic and political systems enable some groups to have more children. Religious restrictions and personal preferences guide our choices of mates. We travel, shuttling genes in and out of populations. Natural disasters and new diseases reduce populations to a few individuals, who then rebuild their numbers, at the expense of genetic diversity. These factors, plus mutation and a reshuffling of genes at each generation, make a gene pool very fluid. The ever-present and interacting forces of nonrandom or selective mating, migration, genetic drift, mutation, and natural selection work to differing degrees to shape populations. The action is at the allele level. As we saw in Chapter 14, changing allele frequencies can change genotype frequencies—which in turn can change phenotype frequencies. In a series of illustrations, colored shapes represent individuals who have specific genotypes. Figure 15.14 then combines the illustrations to summarize the chapter.

15.1 Nonrandom Mating

In the theoretical state of Hardy-Weinberg equilibrium, individuals of all genotypes are presumed equally likely to mate and to choose partners at random. In reality, we give great thought to selecting mates. We choose partners based on physical appearance, ethnic background, intelligence, and shared interests. We marry people similar to ourselves about 80 percent of the time. Worldwide, about one-third of all marriages occur between people who were born fewer than ten miles apart! Nonrandom mating is a major factor in changing allele frequencies in human populations.

Another form of nonrandom mating occurs when certain individuals contribute disproportionately to the next generation (**figure 15.1**). This is common in agriculture when an animal or plant with valuable characteristics is bred extensively. Semen from one prize bull may be used to artificially

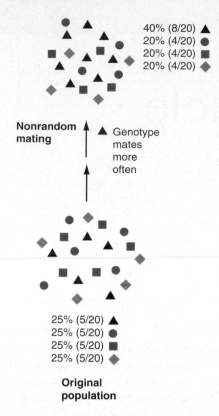

40% (8/20) ▲
20% (4/20) ●
20% (4/20) ■
20% (4/20) ◆

Nonrandom mating → ▲ Genotype mates more often

25% (5/20) ▲
25% (5/20) ●
25% (5/20) ■
25% (5/20) ◆

Original population

Figure 15.1 Nonrandom mating alters allele frequencies. The different-colored shapes represent individuals with distinctive genotypes. If Hardy-Weinberg equilibrium exists for these genes in this population, then the percentages would remain the same through the generations. However, the blue triangle genotype is more reproductively successful, skewing the allele frequencies in the next generation.

inseminate thousands of cows. Such an extreme situation can arise in a human population when a man fathers many children. A striking mutation can reveal such behavior. In the Cape population of South Africa, for example, a Chinese immigrant known as Arnold had a very rare dominant mutation that causes one's teeth to fall out before age 20. Arnold had seven wives. Of his 356 living descendants, 70 have the dental disorder. The frequency of this allele in the Cape population is exceptionally high, thanks to Arnold.

The high frequency of autosomal recessive albinism among Arizona's Hopi Indians also reflects nonrandom mating. Albinism is uncommon in the general U.S. population, but it affects 1 in 200 Hopi Indians. The reason for the trait's prevalence is cultural—men with albinism often stay back

and help the women, rather than risk severe sunburn in the fields with the other men. They contribute more children to the population because they have more contact with the women.

The events of history reflect nonrandom mating patterns. When a group of people is subservient to another, genes tend to "flow" from one group to the other as the males of the ruling class have children with females of the underclass—often forcibly. Historical records and chromosome DNA sequences show this directional gene flow phenomenon. For example, Y chromosome analysis reveals that Genghis Khan, a Mongolian warrior who lived from 1162 to 1227, was so attentive to his many wives that today, 1 in every 200 males living between Afghanistan and northeast China shares his Y—that's 16 million men (**figure 15.2**)! Today, gene pools are still changed intentionally by mass rape. *In Their Own Words* on page 285 lets some of the victims speak.

Despite our partner preferences, many traits do mix randomly in the next generation. This may be because we are unaware

Figure 15.2 A prevalent Y.
Genghis Khan left his mark, in the form of his Y chromosome, on many men. Rape on a sweeping scale spread the chromosome in certain Asian populations.

Genocide by Rape in Sudan

Throughout history, male warriors have left their genetic legacy by raping the women they have conquered. This is sometimes a momentary act of violence; at other times it is a calculated attempt to alter the conquered peoples' gene pool. One of the places where this is happening today is Darfur, in western Sudan (see figure 15.12). Since 2003, Arab militia called the Janjaweed ("a man with a horse and a gun") have systematically attacked black Africans, killing men and children, and repeatedly raping women. Women permitted to live after rape have their thighs slashed, so others will see their taint. Those who conceive are ostracized, for the people believe that pregnancy cannot result from rape, and the women have been promiscuous with the enemy.

Here, in their own words to Amnesty International, are views from women in Darfur:

> I was with another woman, Aziza, aged 18, who had her stomach slit on the night we were abducted. She was pregnant and was killed as they said: "It is the child of the enemy."
>
> —S., from Disa

After six days some of the girls were released. But the others, as young as eight years old, were kept there. Five to six men would rape us in rounds, one after the other for hours during six days, every night. My husband could not forgive me after this, he disowned me.

—S., from Silaya, who was five months pregnant when she was raped

Soldiers arrived by car, camels and horses. Some fifteen women and girls who had not fled quickly enough were raped in different huts in the village. The Janjaweed broke the limbs of some women and girls to prevent them from escaping.

—N., from Um Baru

© Amnesty International Publications

The situation in Darfur is not unique. Similar genocide by rape has been ongoing in the eastern Democratic Republic of the Congo since 1996. Husbands flee; children are killed for defending their mothers; girls and women are serially raped, then injured or killed. Both in Darfur and the Congo, the conquerors have claimed that their intent is

Figure 1 This Sudanese woman is lucky— she has escaped the violence of her homeland for a refugee camp in eastern Chad.

to diminish the genetic contributions of their victims and spread their own genes (**figure 1**).

To learn more and to help, see www. womenforwomen.org.

---

of these characteristics or because we do not consider them in choosing a partner. In populations where AIDS is extremely rare or nonexistent, for example, the two mutations that render a person resistant to HIV infection are in Hardy-Weinberg equilibrium. This would change, over time, if HIV arrives, because the people with these mutations would become more likely to survive to produce offspring—some of whom would perpetuate the protective mutation. Natural selection would intervene, ultimately altering allele frequencies.

Many blood types are in Hardy-Weinberg equilibrium because we do not select life partners on the basis of blood type. Yet sometimes the opposite occurs. People with mutations in the same gene meet when their families participate in programs for people with the associated

inherited condition, such as summer camps for children with cystic fibrosis. More than two-thirds of the relatives visiting such a camp are likely to be carriers for CF, compared to the 1 in 23 or fewer in large population groups. In the reverse situation, in certain religious Jewish communities, young people take tests for the dozen or so genetic disorders that are much more common among Jewish people of eastern European descent (Ashkenazim). Results are stored in a database with a numerical identifier, and when two people contemplate marriage, they can find out if they are carriers for the same disorder. If so, they may elect not to marry or not to have children. More than 100,000 young people have been tested, and the program, called Dor Yeshorim, is partly responsible for the near-nonexistence of Tay-Sachs disease among Ashkenazi Jews—

today most cases are in other population groups that have not been tested.

A population that practices consanguinity has very nonrandom mating. Recall from chapter 4 that a consanguineous relationship is one in which "blood" relatives have children together. On the family level, this practice increases the likelihood that harmful recessive alleles from shared ancestors will be combined and passed to offspring, causing disease. The birth defect rate in offspring is 2.5 times the normal rate of about 3 percent. On a population level, consanguinity decreases genetic diversity. The proportion of homozygotes rises as that of heterozygotes falls.

Some populations encourage marriage between cousins, resulting in an increase in the incidence of certain recessive disorders. In certain parts of the middle east, Africa, and India, from 20 to 50 percent of all marriages are between cousins, or uncles and nieces. The tools of molecular genetics can reveal these relationships. Researchers traced DNA sequences on the Y chromosome and in mitochondria among residents of an ancient, geographically isolated "micropopulation" on the island of Sardinia, near Italy. They consulted archival records dating from the village's founding by 200 settlers around 1000 A.D. to determine familial relationships. Between 1640 and 1870, the population doubled, reaching 1,200 by 1990. Fifty percent of the present population descends from just two paternal and four maternal lines, and 86 percent have the same X chromosome. Researchers are analyzing medical conditions that are especially prevalent in this population, which include hypertension and a kidney disorder.

Worldwide, about 960 million married couples are related, and know of their relationship. Also contributing to nonrandom mating is endogamy, which is marriage within a community. In this case, spouses may be distantly related and unaware of the connection. Endogamy helped Scotland Yard detectives identify the area of Nigeria where "Adam," the murdered boy in the opener to chapter 14, came from. The nonrandom mating there caused some gene variants to be much more prevalent than they are elsewhere.

Key Concepts

1. People choose mates for many reasons, and they do not contribute the same numbers of children to the next generation. This changes allele frequencies in populations.
2. Traits lacking obvious phenotypes may be in Hardy-Weinberg equilibrium.
3. Consanguinity and endogamy in populations increase the proportion of homozygotes at the expense of heterozygotes.

15.2 Migration

Large cities, with their pockets of ethnicity, defy Hardy-Weinberg equilibrium by their very existence. Waves of immigrants formed the population of New York City, for example. The original Dutch settlers of the 1600s had different alleles than those in today's metropolis, which include alleles from the English, Irish, Slavics, Africans, Hispanics, Italians, Asians, and others. **Figure 15.3** depicts the effect on allele and genotype frequencies when individuals join a migrating population.

Historical Clues

We can trace the genetic effects of migration by correlating allele frequencies in present-day populations to events in history, or by tracking how allele frequencies change from one geographical region to another, and then inferring in which directions ancient peoples traveled. **Figure 15.4** depicts the great changes in frequency of the allele that causes galactokinase deficiency in several European populations (OMIM 230200). This autosomal recessive disorder causes cataracts (clouding of the lens) in infants. It is very common among a population of 800,000 gypsies, called the Vlax Roma, who live in Bulgaria. It affects 1 in 1,600 to 2,500 people among them, and 5 percent of the people are carriers. But among all gypsies in Bulgaria as a whole, the incidence drops to 1 in 52,000. As the map in figure 15.4 shows, the disease becomes rarer to the west. This pattern may have arisen when people with the allele settled in Bulgaria, with only a few individuals or families moving westward.

Allele frequencies reflect who rules whom. The frequency of ABO blood types in certain parts of the world today mirrors past Arab rule. The distribution of ABO blood types is very similar in northern Africa, the Near East, and southern Spain. These are precisely the regions where Arabs ruled until 1492. The uneven distribution of allele frequencies can also reveal when and where nomadic peoples stopped. For example, in the eighteenth century, European caucasians called trekboers migrated to the Cape area of South Africa. The men stayed and had children with the native women of the Nama tribe. The mixed society remained fairly isolated, leading to the distinctive allele frequencies found in the present-day people of color of the area.

Genetic analyses can corroborate migration patterns in the historical records, as is the case for Creutzfeldt-Jakob disease (CJD). Recall from chapters 10 and 11 that this disorder can be caused by a mutation in the prion protein gene on chromosome 20 (OMIM 176640). It is rare, but more than 70 percent of affected families worldwide share the same mutation, suggesting a common origin. Researchers examined the implicated section of the chromosome in 62 affected families from 11 populations, looking at a haplotype that included repeated DNA sequences and a SNP. They identified the same haplotype in families from certain groups in Libya, Tunisia, Italy, Chile, and Spain—the exact populations that were expelled from Spain in the Middle Ages. These groups apparently took the CJD gene with them, where it persists today, causing the rare inherited form of this disease.

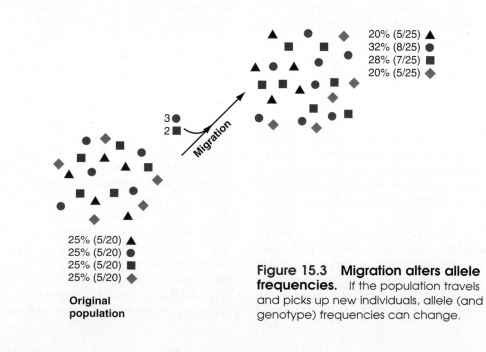

Figure 15.3 Migration alters allele frequencies. If the population travels and picks up new individuals, allele (and genotype) frequencies can change.

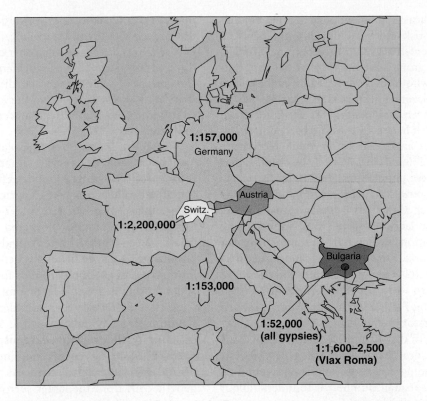

Figure 15.4 Galactokinase deficiency in Europe. This autosomal recessive disorder that causes blindness varies in prevalence across Europe. It is most common among the Vlax Roma gypsies in Bulgaria. The condition becomes much rarer to the west, as indicated by the shading from dark to light green.

Geographical and Linguistic Clues

Sometimes allele frequencies change from one neighboring population to another in a gradient termed a **cline.** Changing allele frequencies usually reflect migration patterns, as immigrants introduced alleles and emigrants removed them. Clines may be gradual, reflecting unencumbered migration paths, but barriers often cause more abrupt changes in allele frequencies. Geographical formations such as mountains and bodies of water may block migration, maintaining population differences in allele frequencies. Language differences may isolate alleles, as people who cannot communicate verbally tend not to have children together.

Allele frequencies up and down the lush strip of fertile land that hugs the Nile River illustrate the concept of clines. In one study, researchers analyzed specific DNA sequences in the mitochondrial DNA of 224 people who live on either side of the Nile, an area settled for fifteen thousand years. The researchers found a gradual change in mitochondrial DNA sequences. The farther apart two individuals live along the Nile, the less alike their mitochondrial DNA. This is consistent with evidence from mummies and historical records that the area once consisted of a series of kingdoms separated by wars and language differences. If the area had been one large interacting settlement, then the DNA sequences would have been more mixed. Instead, the researchers suggest, the Nile may have served as a "genetic corridor" between Egypt and sub-Saharan Africa.

Another pattern of changing allele frequencies reflects the human dependence on communication. In one study, population geneticists correlated twenty blood types to geographically defined regions of Italy and to areas where a single dialect is spoken. They chose Italy because it is rich in family history records and linguistic variants. Six of the blood types varied more consistently with linguistically defined subregions of the country than with geographical regions. Perhaps differences in language prevent people from socializing, sequestering alleles within groups that speak the same dialect because these people marry each other.

Key Concepts

1. Migration alters genotype frequencies by adding and removing alleles from populations.
2. Clines are gradual changes in allele frequencies between neighboring populations.
3. Geographical barriers and language differences often create more abrupt divisions.

15.3 Genetic Drift

When a small group of individuals separate from a larger population, or reproduce only among themselves within a larger population, allele frequencies may change as a result of chance sampling from the whole (**figure 15.5**). This change in allele frequency

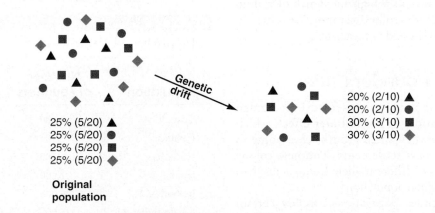

Figure 15.5 Genetic drift alters allele frequencies. If members of a population leave or do not reproduce, allele frequencies can change. When half of this population does not contribute to the next generation, two genotypes increase in frequency and two decrease.

that occurs when a small group separates from the larger whole is termed **genetic drift.** It can be compared to reaching into a bag of jellybeans and, by chance, grabbing only green and yellow ones. The allele frequency changes that occur with genetic drift are random and therefore unpredictable, just as reaching into the jellybean bag a second time might yield mostly black and orange candies.

Genetic drift occurs when the population size plummets, due either to migration, to a natural disaster that isolates small pockets of people, or to the consequences of human behavior. In a common scenario, members of a small community choose to reproduce only among themselves, which keeps genetic variants within their ethnic group. Pittsburgh, Pennsylvania, for example, is made up of many distinct neighborhoods whose residents are more similar to each other genetically than they are to others in the city. New York City, too, is more a hodgepodge of groups with distinct ethnic flavors, rather than a "melting pot" of mixed heritage.

Some groups of people become isolated in several ways—geographically, linguistically, and by choice of partners. Such populations often have a high incidence of several otherwise rare inherited conditions. The native residents of the Basque country in the western part of the Pyrenees Mountains between France and Spain, for example, still speak remnants of Euskera, a language the first European settlers brought in more than 10,000 years ago. The Basques have unusual frequencies of certain ABO and Rh blood types, rare mitochondrial DNA sequences and cell surface antigen patterns, and a high incidence of a mild form of muscular dystrophy. We return to them at the end of the chapter.

The Founder Effect

A common type of genetic drift in human populations is the **founder effect,** which occurs when small groups leave home to found new settlements. The new colony may have different allele frequencies than the original population.

Founder populations amplify certain alleles while maintaining great stretches of uniformity in other DNA sequences. This shows up in increased disease frequencies. Among a population of 18,000 who live in northeastern Finland, for example, the lifetime risk of developing schizophrenia is 3.2 percent, nearly triple the national average. This group traces its ancestry to forty families who settled in the region at the end of the seventeenth century. The population has been easy to study because the Finnish church has records of births, deaths, marriages, and moves, and hospital records are available. **Table 15.1** lists some other founder populations.

A powerful founder effect appears in the French Canadian population of Quebec. They lack diversity in disease-causing mutations, which reflects a long history of isolation. Consider breast cancer caused by the *BRCA1* gene. More than 500 alleles are known worldwide, yet only four are seen among French Canadians. Several inborn errors of metabolism are also more common in this group. The French Canadians have ideal characteristics for gene discovery, including many generations since founding (14), a small number of founders (about 2,500), a high rate of population expansion (74 percent increase per generation), a large present-day population (about 6 million), and most marriages occurring within the group.

The French Canadian population exemplifies genetic drift because the people have kept mostly to themselves within a larger population. The French founded Quebec City in 1608. Until 1660, the population grew as immigrants arrived from France, and then began to increase from births. More than 10,000 French had arrived by the time the British took over in 1759, but many of them had headed westward, taking their genes with them. Meanwhile, in Quebec, religious, language, and other cultural differences separated the French and English gene pools. The French Canadian population of Quebec grew from the 2,500 or so founding genotypes to about 6 million individuals today.

The cultural and physical isolation in Canada created an unusual situation—a founder effect within a founder effect. In the nineteenth century, when agricultural lands opened up about 150 miles north of Quebec, some families migrated north. Their descendants, who remained in the remote area, form an incredibly genetically homogeneous subpopulation of founders split off from the original set of founders.

A classic example of a founder effect within a larger population is the Dunker community of Germantown, Pennsylvania. Excellent historical records combined with distinctive traits enabled geneticists to track genetic drift from the larger surrounding population. The Dunkers came from Germany between 1719 and 1729, but they have lived among others since that time. Still, the frequencies of some genotypes are different among the Dunkers than among their non-Dunker neighbors, and they are also different from the frequencies seen among people living in their native German village. The Dunkers have a different distribution of blood types (**table 15.2**) and much higher incidence of attached earlobes, hyperextensible thumbs, hairs in the middle of their fingers, and left-handedness compared to the other two groups.

Table 15.1

Founder Populations

Population	Number of Founders	Number of Generations	Population Size Today
Costa Rica	4,000	12	2,500,000
Finland	500	80–100	5,000,000
Hutterites	80	14	36,000
Japan	1,000	80–100	120,000,000
Iceland	25,000	40	300,000
Newfoundland	25,000	16	500,000
Quebec	2,500	12–16	6,000,000
Sardinia	500	400	1,660,000

Table 15.2

Genetic Drift and the Dunkers

Blood Type	Population		
	U.S.	Dunker	European
ABO System			
A	40%	60.0%	45%
B, AB	15%	5.0%	15%
Rh^-	15%	11.0%	15%
MN System			
M	30%	44.5%	30%
MN	50%	42.0%	50%
N	20%	13.5%	20%

Table 15.3

Inherited Conditions Common Among the Amish and Mennonites of Lancaster County, Pennsylvania

Disorder	OMIM	Signs and Symptoms (Phenotype)
Ataxia telangiectasia	208900	Increased sensitivity to radiation, loss of balance and coordination, red marks on face, delayed sexual maturation, lung infections, diabetes, high risk of cancer
Bipolar affective disorder	Several	Mood swings (manic depression)
Cartilage-hair hypoplasia (metaphyseal chondrodysplasia, McKusick type)	250250	Dwarfism, sparse hair, anemia, poor immunity
Crigler-Najjar syndrome	218800	Bilirubin buildup, jaundice, brain damage
Ellis-van Creveld syndrome	225500	Dwarfism, short fingers, underdeveloped nails, polydactyly, hair "blaze" pattern, heart disease, fused bones, teeth at birth
Glutaric aciduria type I	231670	Paralysis, brain damage
Homocystinuria	236200	Damaged blood vessels, stroke, heart attack
Limb-girdle muscular dystrophy	253600	Progressive muscle weakness in limbs
Maple syrup urine disease	248600	Sweet-smelling urine, sleepiness, vomiting, mental retardation
Metachromatic leukodystrophy	250100	Rigid muscles, convulsions, mental deterioration
Morquio syndrome	252300	Clouded corneas, abnormal skeleton and aortic valve
Sudden infant death syndrome with dysgenesis of testes	608800	Sudden cessation of heartbeat and breathing; underdeveloped testes

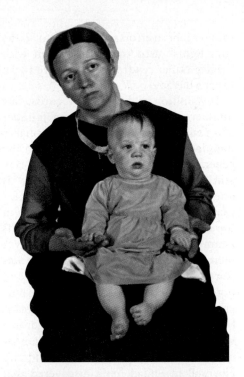

Figure 15.6 Ellis-van Creveld syndrome. This Amish child has inherited Ellis-van Creveld syndrome. He has short-limbed dwarfism, extra fingers, heart disease, fused wrist bones, and had teeth at birth. The condition is autosomal recessive, and the mutant allele occurs in 7 percent of the people of this community. Affected individuals have severe dwarfism, but heterozygotes have the milder condition Weyers acrodental dysostosis. These were thought to be different disorders until the gene was discovered in early 2000.

Founder effects can be studied at the phenotypic and genotypic levels. Phenotypically, a founder effect is indicated when a community of people, known from local history to have descended from a few founders, have a collection of inherited traits and illnesses that are rare elsewhere in the world. This is striking among the Old Order Amish and Mennonites of Lancaster County, Pennsylvania. Often, worried parents would bring their ill children to medical facilities in Philadelphia. Over the years, researchers realized that these people are subject to an array of extremely rare conditions (**table 15.3** and **figure 15.6**). For example, Victor McKusick, founder of *Online Mendelian Inheritance in Man,* discovered and described cartilage-hair hypoplasia. In 1965, six Amish children died at a Philadelphia hospital from chickenpox. Part of their inherited syndrome was impaired immunity, and the children succumbed to this usually mild illness. Until McKusick made the connection, the other symptoms—including dwarfism, sparse hair, and anemia—were not recognized as part of a syndrome. Today, as many geneticists study inherited diseases common among the Amish and Mennonites, treatments are becoming available, from special diets to counter inborn errors of metabolism, to the "bili lights" used to treat children with Crigler-Najjar syndrome (see figure 14.3), to gene therapy.

In addition to historical records, raw numbers provide evidence of a founder effect when allele frequencies in the smaller population are compared to those in the general population. The incidence of certain diseases in Lancaster County is astounding. Maple syrup urine disease, for example, affects 1 in 225,000 newborns in the United States, but 1 in 400 newborns among the Lancaster families! Similarly, a member of a Lancaster research team was startled when he took a walk at night and noted house after house where an eerie blue glow emanated from a window. Inside each house, a child with Crigler-Najjar syndrome was being treated with bili lights. In another example, a research fellow at Children's Hospital in Philadelphia discovered that several young children from Lancaster County with cerebral palsy presumably caused by oxygen deprivation at birth actually had an inborn error of metabolism called glutaric aciduria type I. He went from farm to farm, tracking cases against genealogical records, and found that *every family* that could trace its roots back to the founders had members who had the disease! Today, 0.5 percent of newborns in this population have the condition.

A mutation that is the same in all affected individuals in a population is strong evidence of a founder effect due to descent from shared ancestors. The Bulgarian gypsies who have galactokinase deficiency, for example, all have a mutation that is extremely rare elsewhere.

Very often a disease-associated allele is identical in DNA sequence among people in the same population, and so is the DNA surrounding the gene. This pattern indicates that a portion of a chromosome has been passed among the members of the population from its founders. For this reason, many studies that trace founder effects examine haplotypes that include tightly linked genes.

When historical or genealogical records are particularly well kept, founder effects can sometimes be traced to the very beginning. This is the case for the Afrikaner population of South Africa. The 2.5 million Afrikaners descended from a small group of Dutch, French, and German immigrants who had huge families, often with as many as ten children. In the nineteenth century, some Afrikaners migrated northeast to the Transvaal Province, where they lived in isolation until the Boer War in 1902 introduced better transportation.

Today, 30,000 Afrikaners have porphyria variegata (OMIM 176200), an autosomal dominant deficiency of one of the enzymes required to manufacture heme, the iron-containing part of hemoglobin. Symptoms include nervous attacks, abdominal pain, very fragile and sun-sensitive skin, and a severe reaction to a particular barbiturate anesthetic. All affected people descended from one couple who came from the Netherlands in 1688! Today's allele frequency in South Africa is far higher than that in the Netherlands because the founding couple had many children—who, in turn, had large families, passing on and amplifying the dominant gene.

In a similar extreme example of a founder effect, all of the cases of polydactyly (extra digits) among the Old Order Amish in Lancaster stem from one founder (see figure 15.6). Today, thanks to large families and restricted marriages, the Amish have more cases than the rest of the world combined!

Founder effects are also evident in more common illnesses, where different populations have different mutations in the same gene. This is the case for *BRCA1* breast cancer. The disease is most prevalent among Ashkenazi Jewish people. Nearly all affected individuals in this population have the same 3-base deletion. In contrast, *BRCA1* breast cancer is quite rare in blacks, but it has affected families from the Ivory Coast in Africa, the Bahamas, and the southeastern United States. These families all share a 10-base deletion in the *BRCA1* gene, probably inherited from West Africans who were ancestors of all three modern groups. Slaves brought the disease to the United States and the Bahamas between 1619 and 1808, but some of their relatives who stayed in Africa have perpetuated the gene variant there.

Population Bottlenecks

A **population bottleneck** occurs when many members of a group die, and only a few are left to replenish the numbers. A bottleneck is genetically significant because the new population has only those alleles in the small group that survived the catastrophe. An allele in the remnant population might become more common in the replenished population than it was in the original larger group. Therefore, the new population has a much more restricted gene pool than the larger ancestral population, with some variants amplified, others diminished.

Population bottlenecks sometimes occur when people (or other animals) colonize islands. An extreme example is seen among the Pingelapese people of the eastern Caroline islands in Micronesia. Between 4 and 10 percent are born with "Pingelapese blindness," an autosomal recessive combination of colorblindness, nearsightedness, and cataracts. It is also called achromatopsia (OMIM 603096). Nearly 30 percent of the Pingelapese are carriers. Elsewhere, only 1 in 20,000 to 50,000 people inherits the condition. The prevalence of the blindness among the Pingelapese stems from a typhoon in 1780 that killed all but 9 males and 10 females who founded the present-day population. This severe population bottleneck, combined with geographic and cultural isolation, increased the frequency of the blindness gene as the population resurged.

Figure 15.7 illustrates schematically the dwindling genetic diversity that results from a population bottleneck, shown against a backdrop of a cheetah. Today's cheetahs live in just two isolated populations of a few thousand animals in South and East Africa. Their numbers once exceeded 10,000. The South African cheetahs are so alike genetically that even unrelated animals can accept skin grafts from each other. Researchers attribute the cheetahs' genetic uniformity to two bottlenecks—one that occurred at the end of the most recent ice age, when habitats changed, and another following mass slaughter by humans in the nineteenth century. However, the good health of the animals today indicates that the genes that have survived enable the cheetahs to thrive in their environment.

Human-wrought disasters that kill many people can also cause population bottlenecks—perhaps even more severely, because aggression is typically directed at particular groups, while a typhoon indiscriminately kills anyone in its path. The

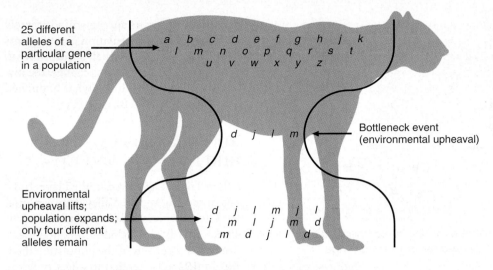

25 different alleles of a particular gene in a population

a b c d e f g h i k
l m n o p q r s t
u v w x y z

Bottleneck event (environmental upheaval)

d j l m

Environmental upheaval lifts; population expands; only four different alleles remain

d j l m j l
j m l j m d
m d j l d

Figure 15.7 Population bottlenecks. A population bottleneck occurs when the size of a genetically diverse population drastically falls, remains at this level for a time, and then expands again. The new population loses some genetic diversity if alleles are lost. Cheetahs are difficult to breed in zoos because sperm quality is poor and many newborns die—both due to lack of genetic diversity.

Chmielnicki massacre was one of many attacks against the Ashkenazi Jewish people. Overall, these acts have left a legacy of several inherited diseases that are at least ten times more common among Jewish people than in other populations (**table 15.4**).

The Chmielnicki massacre began in 1648, when a Ukrainian named Bogdan Chmielnicki led a massacre against the Polish people, including peasants, nobility, and the Jewish people, in retaliation for a Polish nobleman's seizure of his possessions. By 1654, Russians, Tartars, Swedes, and others joined the Ukrainians in wave after wave of violence against the Polish people. Thousands perished, with only a few thousand Jewish people remaining.

The Jewish people have survived many massacres, and therefore many population bottlenecks; after the Chmielnicki massacre, like the others, their numbers grew again. From 1800 to 1939, the Jewish population in Eastern Europe swelled to several million. Yet massacres continued. Jewish people also tended to have children only with each other. Both of these factors—nonrandom mating and population bottlenecks—changed allele frequencies and contributed to the high incidence of certain inherited diseases seen among the Ashkenazim today. Several genetic testing companies offer "Jewish genetic disease" panels that are not meant to discriminate or stereotype, but are based on a genetic fact of life—some illnesses are more common in certain populations.

Key Concepts

1. Genetic drift occurs when a subset of a population has different allele frequencies than the larger population.
2. The founder effect occurs when a few individuals leave a community to start a new settlement. The resulting population may, by chance, either lack some alleles from the original population or have high frequencies of others.
3. In a population bottleneck, many members die, and only a few contribute to the next generation.

15.4 Mutation

A major and continual source of genetic variation is mutation—when one allele changes into another. (**figure 15.8**). Genetic variability also arises from crossing over and independent assortment during meiosis, but these events recombine existing traits rather than introduce new ones. If a DNA base change occurs in a part of a gene that encodes part of a protein necessary for its function, then an altered trait may result. If the mutation occurs in a gamete, then the change can pass to future generations and affect an allele's frequency in the population.

Table 15.4

Autosomal Recessive Genetic Diseases Prevalent Among Ashkenazi Jewish Populations

Disorder	OMIM	Signs and Symptoms (Phenotype)	Carrier Frequency
Bloom syndrome	210900	Sun sensitivity, short stature, poor immunity, impaired fertility, increased cancer risk	1/110
Breast cancer	113705, 600185	Malignant breast tumor caused by mutant *BRCA1* or *BRCA2* genes	3/100
Canavan disease	271900	Brain degeneration, seizures, developmental delay, death by 18 months of age	1/40
Familial dysautonomia	223900	No tears, cold hands and feet, skin blotching, drooling, difficulty swallowing, excess sweating	1/32
Gaucher disease	231000	Enlarged liver and spleen, bone degeneration, nervous system impairment	1/12
Niemann-Pick disease type A	257200	Lipid accumulation in cells, particularly in the brain; mental and physical retardation, death by age three	1/90
Tay-Sachs disease	272800	Brain degeneration causing developmental retardation, paralysis, blindness, death by age four	1/26
Fanconi anemia type C	227650	Deficiencies of all blood cell types, poor growth, increased cancer risk	1/89

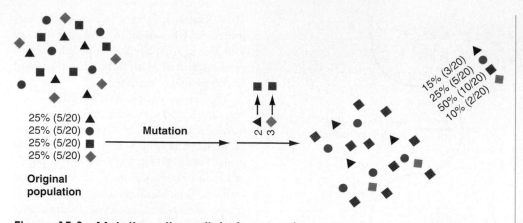

Figure 15.8 Mutation alters allele frequencies. If one allele changes into another from one generation to the next, genotype frequencies can change.

Natural selection eliminates deleterious alleles that are expressed in the phenotype and affect an individual's ability to reproduce. Yet these alleles are maintained in heterozygotes, where they do not exert a noticeable effect, and are reintroduced by new mutation. Therefore, all populations have some alleles that would be harmful if homozygous. The collection of such deleterious alleles in a population is called its **genetic load.**

Overall, the contribution that mutation makes to counter Hardy-Weinberg equilibrium is quite small compared to the influence of migration and nonrandom mating. Natural selection has the greatest influence of all. The spontaneous mutation rate is only about 30 bases per haploid genome in each gamete. Each of us probably has only 5 to 10 recessive lethal alleles. Fortunately, most mutations are "silent"—that is, they do not alter the phenotype due to the degeneracy of the genetic code and changes that do not alter protein function.

Key Concepts

1. Mutation alters genotype frequencies by introducing new alleles.
2. Heterozygotes and mutations maintain the frequencies of deleterious alleles in populations.

15.5 Natural Selection

Environmental change can alter allele frequencies when individuals with certain phenotypes are more likely to survive and reproduce than others (**figure 15.9**). This differential survival based on phenotype, and therefore genotype, is **natural selection.** It may be negative—removal of alleles—or positive—retention of alleles. For example, gene variants that confer the ability to taste bitter substances tend to be more prevalent where the environment includes toxin-containing plants that people might eat. Since toxins often taste bitter, people with more sensitive palates would be more likely to spit out poisons—and live to transmit their protective genes.

The appearance or reemergence of infectious diseases can reveal the effect of natural selection on allele frequencies. If infection kills before reproductive age or impairs fertility, its spread will ultimately remove from the population gene variants that make an individual susceptible to infection. But if conditions change, the disease may resurge. This is what happened with tuberculosis (TB).

Tuberculosis Ups and Downs—and Ups

When TB first appeared in the Plains Indians of the Qu'Appelle Valley Reservation in Saskatchewan, Canada in the mid-1880s, it struck swiftly and lethally, infecting many organs. Ten percent of the population died. But by 1921, TB tended to affect only the lungs, and only 7 percent of the population died annually from it. By 1950, mortality was down to 0.2 percent.

Outbreaks of TB ran similar courses in other human populations. The disease appeared in crowded settlements where the bacteria easily spread in exhaled droplets. In the 1700s, TB raged through the cities of Europe. Immigrants brought it to the United States in the early 1800s, where it also swept the cities. Many people thought TB was hereditary until German bacteriologist Robert Koch identified the causative bacterium in 1882.

As in the Plains Indians, TB incidence and virulence fell dramatically in the cities of the industrialized world in the first half of the twentieth century—before anti-

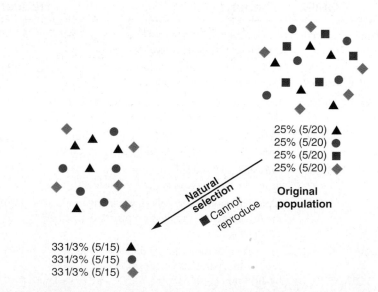

Figure 15.9 Natural selection alters allele frequencies. If health conditions impair the ability of individuals of a certain genotype to reproduce, allele frequencies can change.

biotic drugs were discovered. What tamed tuberculosis?

Natural selection, operating on both the bacterial and human populations, lessened the virulence of the infection. Some people inherited resistance and passed this beneficial trait on. At the same time, the most virulent bacteria killed their hosts so quickly that the victims had no time to spread the infection. As the deadliest bacteria were selected out of the population (negative selection), and as people who inherited resistance mutations contributed disproportionately to the next generation (positive selection), TB gradually evolved from a severe, acute, systemic infection to a rare chronic lung infection. This was true until the late 1980s.

Recent events created conditions just right for the resurgence of TB, but in a form resistant to many of the eleven drugs used to treat it. Some health officials trace the return of tuberculosis to complacency; researchers turned to other projects when funding became scarce for this seemingly controlled disease. Patients became complacent, too. When antibiotics eased symptoms in two to three months, patients felt cured and stopped taking the drugs, even though they unknowingly continued to spread live bacteria for up to 18 months. The *Mycobacterium tuberculosis* bacteria had time to mutate, and mutant strains to flourish, eventually evolving the drug resistances that make the newest cases so difficult to treat (**Reading 15.1**). Tuberculosis treatment in the 1950s was actually more effective; many patients were isolated for a year or longer in rest homes called sanitaria and were not released until the bacteria were gone (**figure 15.10**).

Today, 1 in 7 new tuberculosis cases is resistant to several drugs, and 5 percent of these patients die. People living in crowded, unsanitary conditions with poor health care are especially susceptible to drug-resistant TB. Another reservoir of new TB infection is persons infected with HIV. Tuberculosis develops so quickly in people with suppressed immunity that it can kill before physicians have determined which drugs to use—physicians have even died of the infection while treating patients. The resurgence of TB should remind us never to underestimate the evolution that operates in all organisms—and does so unpredictably.

Evolving HIV

Because the RNA or DNA of viruses replicates often and errors are not repaired, viral mutations accumulate rapidly. Like bacteria, the viruses in a human body form a population, including naturally occurring genetic variants. In HIV infection, natural selection controls the diversity of HIV genetic variants within a human body as the disease progresses. The human immune system and drugs to slow the infection become the environmental factors that select (favor) resistant viral variants.

HIV infection can be divided into three stages, both from the human and the viral perspective (**figure 15.11**). A person infected with HIV may experience an initial acute phase, with symptoms of fever, night sweats, rash, and swollen glands. In a second period, lasting from 2 to 15 years, health usually returns. In a third stage, immunity collapses,

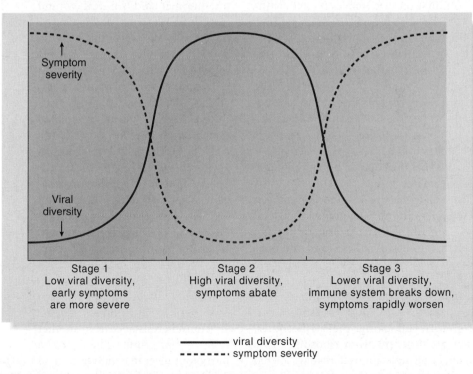

Figure 15.10 Infectious diseases involve allele frequency changes in the pathogen and the host. Early in the twentieth century, tuberculosis was controlled by isolating infected people in sanitaria. TB has returned, due to complacency and ease of infection among immunosuppressed individuals.

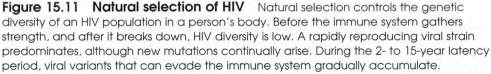

— viral diversity
------ symptom severity

Figure 15.11 Natural selection of HIV Natural selection controls the genetic diversity of an HIV population in a person's body. Before the immune system gathers strength, and after it breaks down, HIV diversity is low. A rapidly reproducing viral strain predominates, although new mutations continually arise. During the 2- to 15-year latency period, viral variants that can evade the immune system gradually accumulate.

Antibiotic Resistance: Genomics to the Rescue?

We take antibiotics for granted—we remember the bottles of thick, pink medicine that cured our ear infections as kids, and the packets of pills that we take today to vanquish a strep throat or help heal an infected wound. But many antibiotics are no longer effective. The reason is the interplay between mutation and natural selection.

Bacteria in a human body constitute a population of organisms that represent different genetic variants. Some bacterial strains inherit the ability to survive in the presence of a particular antibiotic drug. When the bacteria infect a person, the immune system responds, causing symptoms such as inflammation and fever. The person goes to the doctor, and a course of antibiotics seems to help; but a month later, symptoms return. What has happened?

The drug probably killed most of the bacteria, but a few survived because they have a mutation that enables them to withstand the antibiotic assault. Certain antibiotics actually induce mutation in bacteria, and may contribute to resistance in this way. As sensitive bacteria die, resistant mutants reproduce, taking over the niche the antibiotic-sensitive bacteria vacated. Soon, the person has enough antibiotic-resistant bacteria to feel ill again. The next step is to try a drug that works differently— and hope that the bacteria haven't mutated around that one, too.

Bacteria with drug-resistance mutations circumvent antibiotic actions in several ways. Penicillin kills bacteria by tearing apart their cell walls. Resistant microbes produce enzyme variants that dismantle penicillin, or have altered cell walls that the drug cannot bind. Erythromycin, streptomycin, tetracycline, and gentamicin kill bacteria by attacking their ribosomes, which are different from ribosomes in a human. Drug-resistant bacteria have altered ribosomes that the drugs cannot bind.

Bacteria become resistant in two ways. In vertical transmission, bacteria become resistant by mutation, then pass the resistance from one bacterial generation to the next by cell division. In horizontal (also called lateral) transmission, groups of resistance genes are passed on transposons, described in chapters 11 and 12. Transposons move from cell to cell as part of DNA circles called plasmids. Bacteria usually pass transposons to close relatives, but in a hospital environment, genes may be transmitted horizontally to even distantly-related types of bacteria. With horizontal gene transfer, resistance passes quickly. This is what has happened with infection by the bacterium *Staphylococcus aureus*.

S. aureus is normally present in low numbers in the nose and on the skin. But in high numbers, it causes pimples and boils, food poisoning, toxic shock syndrome, pneumonia, and surgical wound infections. It is particularly dangerous in hospitals, where the infection spreads rapidly among people unable to fight it. This common bacterium became resistant to penicillin soon after the drug was introduced in the 1940s. A related penicillin, methicillin, worked for a time, but resistant bacterial strains appeared suddenly in 2000, at such an alarming rate that the microorganism has earned its own acronym: MRSA, for methicillin-resistant *Staph aureus*. Doctors have turned to another antibiotic, vancomycin, to treat MRSA, but the effort may be too late.

DNA sequencing revealed that in one hospital, *S. aureus* picked up its vancomycin resistance from another type of bacterium. The scene: a nasty foot ulcer in a dialysis patient in Detroit. The wound harbored vancomycin-resistant *Enterococcus faecalis* as well as two types of *S. aureus*, one resistant to the antibiotic and one sensitive. By sequencing the plasmids that included the resistance gene, investigators deduced that *S. aureus* picked up an *E. faecalis* plasmid bearing a resistance gene called *vanA*. Then the *vanA* gene jumped to an *S. aureus* plasmid. Further research showed that *S. aureus* strains resistant to both methicillin and vancomycin have arisen independently in many countries.

Clues for developing effective new antibiotics lie in the genomes of the microorganisms that make us sick. For example, British researchers sequenced the genomes of two strains of *S. aureus*—one resistant to methicillin and one not. Now they are looking among the several genes only in the resistant strain to deduce how the bacterium evades the drug (**figure 1**).

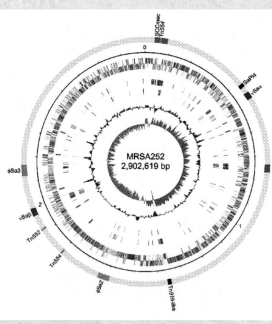

Figure 1   A genome map of methicillin-resistant *S. aureus*.

the virus replicates explosively, and opportunistic infections and cancer eventually cause death.

The HIV population changes and expands throughout the course of infection, even when the patient seems to stay the same for a long time. New mutants continuously arise, and they alter such traits as speed of replication and the patterns of molecules on the viral surface.

In the first stage of HIV infection, as the person battles acute symptoms, viral variants that replicate swiftly predominate. In the second stage, the immune system starts to fight back and symptoms abate, as viral replication slows and many viruses are destroyed. Now natural selection acts—those viral variants that persist reproduce and mutate, giving rise to a diverse viral population. Ironically, drugs used to treat AIDS may further select against the weakest HIV variants. Gradually, the HIV population overtakes the immune system cells, but so huge is the army of cells that make up the immune system that years may pass before immunity begins to noticeably decline.

The third stage, full-blown AIDS, occurs when the virus overwhelms the immune system. Now, with the selective pressure off, viral diversity again diminishes, and the fastest-replicating variants predominate. HIV wins. The entire scenario of HIV infection reflects the value of genetic diversity—to enable the survival of a population or species in the face of an environmental threat. When that threat—an immune system attack or drugs—wipes out sensitive variants, one genotype may prevail.

Knowing that HIV diversifies early in the course of infection has yielded clinical benefits. Patients now take combinations of drugs as soon as they are diagnosed. These drugs act in different ways to squelch several viral variants simultaneously, slowing the course of the infection.

Balanced Polymorphism

If natural selection eliminates individuals with detrimental phenotypes from a population, then how do harmful mutant alleles remain in a gene pool? Harmful recessive alleles are replaced in two ways: by new mutation, and by persistence in heterozygotes.

Sometimes, a recessive condition remains particularly prevalent because the heterozygote enjoys some unrelated health advantage, such as being resistant to an infectious disease or able to survive an environmental threat. This "heterozygous advantage" that maintains a recessive, disease-causing allele in a population is called **balanced polymorphism**. Recall that *polymorphism* means variant; the effect is *balanced* because the protective effect of the noninherited condition counters the negative effect of the deleterious allele, maintaining its frequency in the population. (Balanced polymorphism is a type of balancing selection, which more generally refers to maintaining heterozygotes in a population.)

Sickle Cell Disease and Malaria

Sickle cell disease is an autosomal recessive disorder that causes anemia, joint pain, a swollen spleen, and frequent, severe infections. It is the classic example of balanced polymorphism: carriers are resistant to malaria, or develop very mild cases. Malaria is an infection by the parasite *Plasmodium falciparum* that causes debilitating cycles of chills and fever. The parasite spends the first stage of its life cycle in the salivary glands of the mosquito *Anopheles gambiae*. When an infected female mosquito draws blood from a human, malaria parasites enter the human's red blood cells, which transport the parasites to the liver. The red blood cells burst, releasing parasites throughout the body.

In people with sickle cell disease, many of the red blood cells burst prematurely, which expels the parasites before they can cause rampant infection. The blood of a person with sickle cell disease is also thicker than normal, which may hamper the parasite's ability to infect. A sickle cell disease carrier's blood is abnormal enough to be inhospitable to the malaria parasite—but usually not abnormal enough to cause the blocked circulation of sickle cell disease.

A clue to the protective effect of being a carrier for sickle cell disease came from striking differences in the incidence of the two diseases in different parts of the world (**figure 15.12**).

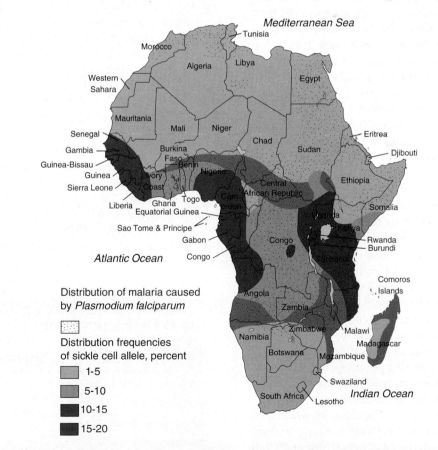

Figure 15.12 Balanced polymorphism. Comparing the distribution of people with malaria and people with sickle cell disease in Africa reveals balanced polymorphism. Carriers for sickle cell disease are resistant to malaria because changes in the blood caused by the sickle cell allele are not severe enough to impair health, but do inhibit the malaria parasite.

In the United States, 8 percent of African Americans are sickle cell carriers, whereas in parts of Africa, up to 45 to 50 percent are carriers. Although Africans had known about a painful disease that shortened life, the sickled cells weren't recognized until 1910 (see chapter 12). In 1949, British geneticist Anthony Allison found that the frequency of sickle cell carriers in tropical Africa was higher in regions where malaria rages all year long. Blood tests from children hospitalized with malaria showed that nearly all were homozygous for the wild type sickle cell allele. The few sickle cell carriers among them had the mildest malaria. Was malaria enabling the sickle cell allele to persist by felling people who did not inherit it? The fact that sickle cell disease is rarer where malaria is rare supports the idea that sickle cell heterozygosity protects against the infection.

Further evidence of a sickle cell carrier's advantage in a malaria-ridden environment is the fact that the rise of sickle cell disease parallels the cultivation of crops that provide breeding grounds for *Anopheles* mosquitoes. About 1000 B.C., Malayo-Polynesian sailors from southeast Asia traveled in canoes to East Africa, bringing new crops of bananas, yams, taros, and coconuts. When the jungle was cleared to grow these crops, the open space provided breeding grounds for the mosquitoes. The insects, in turn, offered a habitat for part of the life cycle of the malaria parasite.

The sickle cell allele may have been brought to Africa by people migrating from Southern Arabia and India, or it may have arisen directly by mutation in East Africa. However it happened, people who inherited one copy of the sickle cell allele survived or never contracted malaria. These carriers had more children and passed the protective allele to approximately half of them. Gradually, the frequency of the sickle cell allele in East Africa rose from 0.1 percent to 45 percent in 35 generations. Carriers paid the price for this genetic protection, however, whenever two of them produced a child with sickle cell disease.

A cycle set in. Settlements with large numbers of sickle cell carriers escaped debilitating malaria. They were strong enough to clear even more land to grow food—and support the disease-bearing mosquitoes.

Glucose-6-Phosphate Dehydrogenase Deficiency and Malaria

Recall from chapter 12 that G6PD deficiency is an X-linked recessive enzyme deficiency that causes life-threatening hemolytic anemia under specific conditions, such as eating fava beans or taking certain drugs. Among African children with severe malaria, heterozygous females ($X^G X^g$) and affected (hemizygous) males ($X^g Y$) for G6PD deficiency are underrepresented. This suggests that carrying or inheriting G6PD deficiency protects against malaria. The parasite enters the red blood cells of carriers or affected males but cannot reproduce sufficiently to cause infection.

The fact that G6PD deficiency is X-linked introduces a possibility not seen with sickle cell disease, which is autosomal recessive. Because in G6PD deficiency heterozygotes and hemizygotes (males with the disease) have an advantage, the mutant allele should eventually predominate in a malaria-exposed population as homozygotes and hemizygotes for the normal allele die of malaria. However, this doesn't happen—there are still males hemizygous and females homozygous for the normal allele. The reason again relates to natural selection.

Table 15.5 shows how natural selection acts in two directions on the two types of hemizygous males—selecting for the mutant allele because it protects against malarial infection, yet also selecting for the normal allele because it protects against an enzyme deficiency. This is the "balance" of balanced polymorphism.

Studies of different G6PD mutations in different populations confirm that the beginning of a mutation's prevalence coincides with the onset of agriculture. These protective mutations are only seen where agriculture and malaria coexist.

PKU and Fungal Infection

Being a heterozygote for PKU (see section 5.1) may protect a fetus from fungal infection. Carriers have elevated phenylalanine levels in their blood that are not sufficient to cause symptoms of PKU, but are high enough to inactivate a fungal poison, ochratoxin A, that harms fetuses.

Historical evidence links PKU heterozygosity to protection against the fungal toxin. PKU is most common in Ireland and western Scotland, and many affected families living elsewhere trace their roots to this part of the world. PKU spread eastward in Europe when the Vikings brought wives and slaves back from the Celtic lands. In the moist environment of Ireland and Scotland, the fungi that produce ochratoxin A—*Aspergillis* and *Penicillium*—grow on grains. During the famines that have plagued these nations, starving people ate moldy grain. If PKU carriers were more likely to have children than noncarriers because of the protective effects of the PKU gene, over time, the disease-causing allele would have increased in this population.

Prion Disease and Cannibalism

Being a heterozygote for the prion protein gene may protect against the disorders of protein folding called transmissible spongiform encephalopathies (see figures 10.22 and 10.23, Reading 10.1, and section

---

Table 15.5

G6PD Deficiency Protects Against Malaria

Genotypic Class	Enzyme Deficiency	Malaria Susceptibility
Normal male $X^G Y$	no	yes
G6PD male $X^g Y$	yes	no
Heterozygous female $X^G X^g$	no	no
Homozygous female $X^G X^G$	no	yes
Homozygous female $X^g X^g$	yes	no

12.4). The best studied such illness is kuru, which caused brain degeneration among the Foré people in Papua New Guinea until the Australian government halted the practice of ritual cannibalism in the mid-1950s. A recent investigation of the prion protein gene among 30 elderly Foré women who had participated in many of the brain-eating feasts revealed more heterozygotes than would exist if the gene were in Hardy-Weinberg equilibrium. That is, the excess suggests natural selection at work. Specifically, of the 30 women, 23 were heterozygotes; 15 would have been expected if selection was not acting on this gene, based on Hardy-Weinberg equilibrium observed among 140 Foré who had not participated in the ritual. In the heterozygotes, some of the normal prion proteins have a valine at amino acid position 129, and some a methionine. The presence of two amino acids, encoded one on each homolog in a heterozygote, apparently prevents the infectious misfolding that occurs to the prion protein when a person encounters an abnormal prion protein—as happens in cannibalism. (All of the people in the United Kingdom who have developed variant CJD, the human form of "mad cow disease," have only methionine at position 129.)

The overrepresentation of heterozygotes among the Foré survivors led to the hypothesis that balancing selection has favored this genotype in the population, and that cannibalism may have been the driving force. That is, homozygotes who were cannibals died of a prion disorder before reproducing, leaving the resistant heterozygotes to slowly accumulate in the population.

This new genetic view of cannibalism supports anthropological evidence that eating human flesh has occurred in many times and places, from Neanderthal caves in France and Croatia to the American Southwest. Evidence of cannibalism includes human bones damaged in ways similar to the bones of animals prepared for consumption, such as scratch marks to remove muscle and signs of breaking and crushing to obtain marrow. Biochemical evidence for past cannibalism includes human myoglobin, found only in human muscle, in fossilized human excrement.

Cystic Fibrosis and Diarrheal Disease

Balanced polymorphism may explain why CF is so common—its cellular defect protects against diarrheal illnesses such as cholera and typhus. Diarrheal disease epidemics have left their mark on many human populations, and continue to be a major killer in the developing world.

Severe diarrhea rapidly dehydrates the body and leads to shock, kidney and heart failure, and death in days. In cholera, bacteria produce a toxin that opens chloride channels in cells of the small intestine. As salt (NaCl) leaves the intestinal cells, water rushes out, producing diarrhea. Cholera opens chloride channels, releasing chloride and water. The CFTR protein does just the opposite, closing chloride channels and trapping salt and water in cells, which dries out mucus and other secretions. A person with CF is very unlikely to contract cholera, because the toxin cannot open the chloride channels in the small intestine cells.

CF carriers enjoy the mixed blessing of balanced polymorphism. They do not have enough abnormal chloride channels to cause the labored breathing and clogged pancreas of cystic fibrosis, but they do have enough of a defect to prevent the cholera toxin from taking hold. During the devastating cholera epidemics that have occurred throughout history, individuals carrying mutant CF alleles had a selective advantage, and they disproportionately transmitted those alleles to future generations.

However, because CF arose in western Europe and cholera originated in Africa, an initial increase in CF heterozygosity may have been a response to a different diarrheal infection—typhoid fever. The causative bacterium, *Salmonella typhi,* rather than producing a toxin, enters cells lining the small intestine—but only if CFTR channels are present. The cells of people with severe CF manufacture CFTR proteins that never reach the cell surface, and therefore no bacteria get in. Cells of CF carriers admit some bacteria.

Diabetes Mellitus and Surviving Famine

Type 2 diabetes mellitus is a gradual failure of cells to respond to insulin and take up glucose from the bloodstream. It is a complex trait—a first-degree relative has a tenfold increased risk of developing the condition. A person can inherit a susceptibility to the disease. Symptoms begin if the diet is unhealthy or lifestyle sedentary, such as among Arizona's Pima Indians (see figure 7.13).

The "thrifty genotype" hypothesis suggests that type 2 diabetes is common today because susceptibility genes might once have been beneficial, preventing the breakdown of fat and altering the body's ability to store glucose. In times past, extra fat stores and altered glucose metabolism enabled people to survive famine. Today, these abilities cause weight gain and diabetes.

P-Glycoprotein and Resistance to AIDS Drugs

Selection is a response to environmental change. When an environment changes, a trait that has been selected for may no longer be advantageous. This is the case for a gene, *MDR1,* that encodes the protein portion of a glycoprotein (called P-glycoprotein) that dots the surfaces of intestinal lining cells and T lymphocytes. An allele that causes overexpression of the gene became prevalent in some populations because the encoded protein enables cells to pump out poisons, such as naturally occurring toxins in plants. But when someone intentionally takes a toxin, such as a drug to combat cancer, these pumps lose their value. Variants in this gene may explain why drugs to fight HIV infection are less effective in Africans than in other groups.

Researchers identified a polymorphism in an exon of the *MDR1* gene that correlates with overexpression, creating an allele called *C.* Another allele that leads to very few P-glycoproteins is called *T.* The resulting genotypes and phenotypes are therefore:

TT   (very few P-glycoproteins)

TC   (intermediate number of P-glycoproteins)

CC   (many P-glycoproteins)

Researchers genotyped 172 individuals from West Africa, 41 African Americans, 537 Caucasians, and 50 Japanese, all healthy.

The results were startling (**figure 15.13**). The *CC* genotype is clearly overrepresented among West Africans (83%) and African Americans (61%), compared to Caucasians (26%) and Japanese (34%). The finding makes sense in terms of natural selection—the *CC* genotype enables people to evade infection by bacteria and viruses that cause gastroenteritis, a major killer in Africa. But the same genotype resists drugs used to treat cancer and AIDS, along with antirejection transplant drugs. Using this information, physicians will be able to identify people with the *CC* genotype and perhaps give them higher doses of certain drugs.

The forces of nonrandom mating, migration, genetic drift, mutation, and natural selection interact in complex ways. **Figure 15.14** reviews and summarizes the

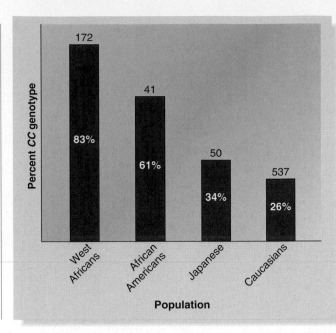

**Figure 15.13
Resistance genotypes.**
An advantageous trait can turn disadvantageous when the environment changes. The *CC* genotype of the *MDR1* gene enables a person's cells to pump out poisons. The genotype became prevalent in populations with African ancestry, possibly because it protected against viral and bacterial infections of the gastrointestinal tract in the African environment. Today, the same genotype may hamper the efficacy of AIDS drugs.

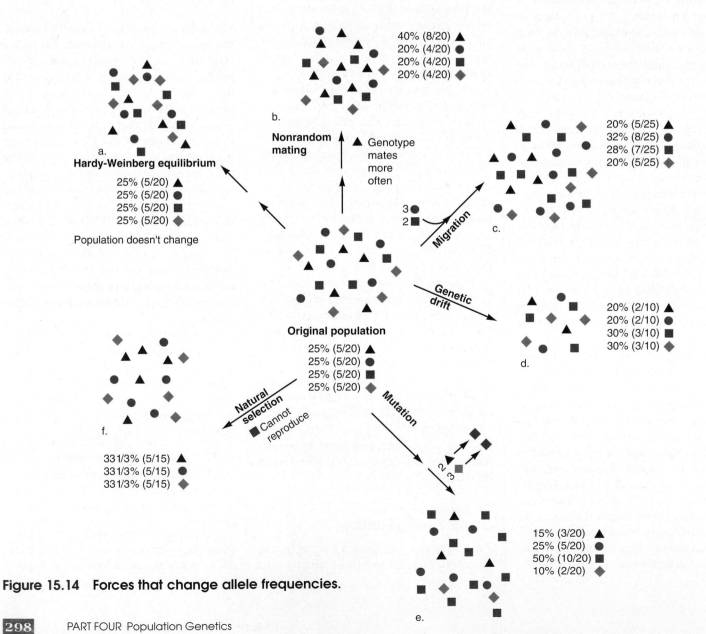

Figure 15.14   Forces that change allele frequencies.

Dogs and Cats: Products of Artificial Selection

The pampered poodle may win in the show ring, but it is a poor specimen in terms of genetics and evolution. Human preferences in pets can lead to breeds that might never have evolved naturally. Behind carefully bred traits lurk small gene pools and extensive inbreeding—all of which can harm the health of highly prized and highly priced show animals. Purebred dogs suffer from more than 300 inherited disorders!

The sad eyes of the basset hound make him a favorite in advertisements, but his runny eyes can hurt. Short legs make him prone to arthritis, his long abdomen causes back injuries, and his floppy ears can hide infection. The eyeballs of the Pekingese protrude so much that a mild bump can pop them out of their sockets. The tiny jaws and massive teeth of bulldogs cause dental and breathing problems, sinusitis, and "dog breath."

The modern dog closest to what canines were like before people controlled their breeding is the dingo, the wild dog of Australia. It was introduced from southeast Asia in a tiny founder population, about 5,000 years ago. Today, breeders are using

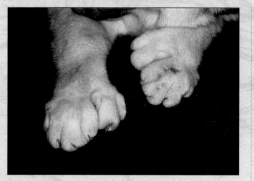

Figure 1   Multitoed cats are common in New England but rare elsewhere.

genomics to identify disease-causing genes before animals are mated. The U.K. Kennel Club is collecting cheek cell samples from dogs and identifying the DNA sequences associated with diseases, using the extensive pedigrees that have been kept for many years.

We artificially select natural oddities in cats, too. One of every 10 New England cats has extra toes, thanks to a multitoed ancestor in colonial Boston (**figure 1**). Elsewhere, these cats are rare. The sizes of the blotched

Figure 2   An American curl cat.

tabby populations in New England, Canada, Australia, and New Zealand correlate to the time since cat-loving Britons colonized each region. The Vikings brought the orange tabby to the islands off the coast of Scotland, rural Iceland, and the Isle of Man.

The American curl cat's origin traces back to a stray female who wandered into a home in Lakewood, California, in 1981 and passed her curled-up ears to kittens (**figure 2**). A dominant mutation forms extra cartilage in the outer ear.

---

forces that alter allele frequencies and therefore impact evolution. Reading 15.2 looks at intentional shrinking of gene pools—the "artificial selection" used to breed dogs and cats.

RELATING THE CONCEPTS

1. How did natural selection mold the differing abilities of people to digest milk in different populations?
2. Ability to digest milk arose from positive selection. Cite an example of negative selection. (You can invent one.)
3. How can lactose intolerance be the wild type phenotype in a population?

Key Concepts

1. Because of natural selection, different alleles are more likely to confer a survival advantage in different environments. Cycles of infectious disease prevalence and virulence often reflect natural selection.
2. Balanced polymorphism is a type of natural selection in which a particular disease-causing allele is maintained because heterozygotes resist a certain infectious illness or environmental condition.

15.6 Gene Genealogy

Identifying different mutations in a gene is useful in charting the evolution and spread of genetic variants. An assumption is that the more prevalent an allele is, the more ancient it is, because it has had more time to spread and accumulate in a population. Correlating allele frequencies with historical, archeological, and linguistic evidence provides fascinating peeks at the evolution of modern peoples.

PKU Revisited

The diversity of PKU mutations suggests that the disease has arisen more than once. Mutations common to many groups of people probably represent more ancient mutations, which perhaps occurred before many groups spread into disparate populations. This was the case for CJD among groups that left Spain in the Middle Ages, mentioned earlier in the chapter. In contrast, mutations found only in a small geographical region, or perhaps in a single family, are more likely to be of recent origin. They have had less time to spread. For example, Turks, Norwegians, French Canadians, and Yemeni Jews have their own PKU alleles. Analysis of the frequencies of PKU mutations in different populations, plus logic, can reveal the roles that genetic drift, mutation, and balanced polymorphism have played in maintaining the mutation.

A high mutation rate cannot be the sole reason for the continued prevalence of PKU because some countries, such as Denmark, continue to have only one or two mutations. If the gene were unstable, so that it mutated frequently, all populations would have several different types of PKU mutations. This is not the case.

In some isolated populations, such as French Canadians and Yemeni Jews, migration and the founder effect have maintained certain PKU alleles. Consider the history of PKU among Yemeni Jews. In most populations, point mutations in the phenylalanine hydroxylase (*PAH*) gene cause PKU. Virtually all of the Yemeni Jews in Israel who have PKU instead have a 6,700-base deletion in the third exon of the *PAH* gene. Researchers traced the spread of this mutation from North Africa to Israel.

The researchers tested for the telltale deletion in the grandparents of the 22 modern Yemeni Jewish families with PKU in Israel. By asking questions and consulting court and religious records, which this close-knit community kept meticulously, the team found that all clues pointed to San'a, the capital of Yemen. The earliest records identify two families with PKU in San'a, and indicate that the mutation originated in one person before 1800. By 1809, religious persecution and hard economic times led nine families carrying the mutation to migrate north and settle in three towns (**figure 15.15**). Four of the families then moved farther northward, into four more towns. Twenty more families spread from San'a to inhabit 17 other towns. All of this migration took place from 1762 through the mid-1900s, and eventually led to Israel.

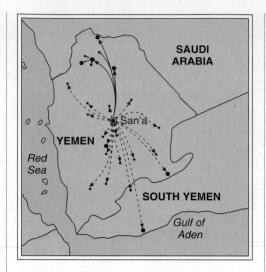

Figure 15.15 The origin of PKU.
The exon 3 deletion in Israeli Yemeni Jews probably arose in San'a, Yemen, in the mid–18th century. The allele spread northward as families moved from San'a in 1809 (solid arrows) and subsequently spread to other regions (broken arrows).
Source: Data from Smadar Avigad, et al., A single origin of phenylketonuria in Yemenite Jews, *Nature* 344:170, March 8, 1990.

A more recent example of the effect history and politics can have on gene frequency is the influx of families with PKU into northwest Germany after the second world war, when Germans from the east moved westward. Future shifts in allele frequencies may parallel the breakdown of the former Soviet Union.

CF Revisited

Tracing allele distributions in modern populations known to have very ancient roots offers clues to how early genetic disorders plagued humankind. For example, the CF allele ΔF508 is very prevalent among northern Europeans, but not as common in the south. This distribution might mean that early farmers migrating from the Middle East to Europe in the Neolithic period, up until about 10,000 years ago, brought the allele to Europe. At this time, people were just beginning to give up a hunter-gatherer lifestyle for semi-permanent settlements, exhibiting the first activities of agriculture.

The origin, or at least the existence, of ΔF508 may go farther back, to the Paleolithic age more than 10,000 years ago. People then were hunter-gatherers who occasionally lived in caves and tents. They used tools of chipped stone, followed a lunar calendar, and created magnificent cave art.

Geneticists were led back to the Paleolithic by an intriguing group of people, the Basques. Researchers studied 45 families from the Basque country (in the mountains between France and Spain) with cystic fibrosis and identified affected children with four pure Basque grandparents by their distinctive double surnames. For 87 percent of the pure Basque CF patients, ΔF508 was the causative mutation—a frequency higher than for most European populations, indicating a more ancient origin. Families of "mixed Basque" background, whose Basque ancestors interbred with French or Spanish neighbors, showed ΔF508 only 58.3 percent of the time. Among the nearby Spanish population, the frequency of ΔF508 is 50 percent. Today's remaining Basque people, then, may carry a pocket of cystic fibrosis mutations that arose long ago.

RELATING THE CONCEPTS

Explain how geography played a role in the evolution of genes that enable people to digest cow's milk.

Key Concepts

1. PKU originated more than once. Genetic drift, balanced polymorphism, and perhaps mutation have affected its prevalence.
2. Studies of allele prevalence place the origin of CF farther back than previously thought.

Summary

15.1 Nonrandom Mating

1. Hardy-Weinberg equilibrium assumes all individuals mate with the same frequency and choose mates without regard to phenotype. This rarely, if ever, happens in real human populations. We choose mates based on certain characteristics, and some individuals have many more children than others.

2. DNA sequences that do not cause a phenotype important in mate selection or reproduction may be in Hardy-Weinberg equilibrium.

3. Consanguinity increases the proportion of homozygotes in a population, which may lead to increased incidence of recessive illnesses or traits.

15.2 Migration

4. **Clines** are changes in allele frequencies from one area to another.

5. Clines may reflect geographical barriers or linguistic differences and may be either abrupt or gradual.

6. Human migration patterns through history explain many cline boundaries.

Forces behind migration include escape from persecution and a nomadic lifestyle.

15.3 Genetic Drift

7. **Genetic drift** occurs when a small population separates from a larger one, and its members breed only among themselves, perpetuating allele frequencies not characteristic of the ancestral population.

8. Genetic drift is random and may occur within a larger group or apart from it.

9. A **founder effect** is genetic drift that occurs when a few individuals found a settlement and their alleles form a new gene pool, amplifying their alleles and eliminating others.

10. A **population bottleneck** is a narrowing of genetic diversity that occurs after many members of a population die and the few survivors rebuild the gene pool.

15.4 Mutation

11. Mutation continually introduces new alleles into populations. It occurs as a consequence of DNA replication errors.

12. Mutation does not have as great an influence on disrupting Hardy-Weinberg equilibrium as the other factors.

13. The **genetic load** is the collection of deleterious alleles in a population.

15.5 Natural Selection

14. Environmental conditions influence allele frequencies via **natural selection,** as the rise and fall of infectious disease indicates. Alleles that do not enable an individual to reproduce in a particular environment are selected against and diminish in the population, unless conditions change. Beneficial alleles are retained.

15. In **balanced polymorphism,** the frequencies of some deleterious alleles are maintained when heterozygotes have a reproductive advantage under certain conditions.

15.6 Gene Genealogy

16. Frequencies of different mutations in different populations provide information on the natural history of alleles and on the relative importance of nonrandom mating, genetic drift, and natural selection in deviations from Hardy-Weinberg equilibrium.

Review Questions

1. Give examples of how each of the following can alter gene frequencies from Hardy-Weinberg equilibrium:
   a. nonrandom mating
   b. migration
   c. a population bottleneck
   d. mutation

2. Explain the influence of natural selection on
   a. the virulence of tuberculosis.
   b. bacterial resistance to antibiotics.
   c. the changing degree of genetic diversity in an HIV population during infection.
   d. the prevalence of cystic fibrosis.

3. Why can increasing homozygosity in a population be detrimental?

4. How can genomics be used to develop new antibiotics?

5. Why might a mutant allele that causes an inherited illness in homozygotes persist in a population?

6. Give an example of an inherited disease allele that protects against an infectious illness.

7. Porphyria variegata, which resulted from a founder mutation in the Afrikaner population, is inherited as an autosomal dominant with incomplete penetrance. How can this mode of inheritance complicate analysis of the condition?

8. Provide two examples of how molecular evidence confirms the presence of genetic uniformity.

9. Explain how table 15.2 indicates that genetic drift has occurred among the Dunkers.

10. Would a carrier test to detect the common cystic fibrosis allele $\Delta F508$ be more accurate in France or Finland? Cite a reason for your answer.

11. What type of molecular evidence indicates a founder effect?

12. How does a founder effect differ from a population bottleneck?

13. Describe two scenarios in human populations, one of which accounts for a gradual cline, and one for an abrupt cline.

14. How do genetic drift, nonrandom mating, and natural selection interact?

15. Define:
   a. founder effect
   b. balanced polymorphism
   c. genetic load

16. How does a knowledge of history, sociology, and anthropology help geneticists to interpret allele frequency data?

Applied Questions

1. About half of the Melanesian people of Papua New Guinea are resistant to malaria and have shortened glycophorin C proteins, which are on the surfaces of red blood cells. These people are homozygous recessive for a deletion in part of the gene. One way the malaria parasite enters red blood cells is through glycophorin C. Normally, the protein anchors the plasma membrane to the cytoskeleton. However, because other proteins do this, too, no symptoms arise from being homozygous recessive for the glycophorin C deletion mutation. Is this an example of balanced polymorphism? Give a reason for your answer.

2. The ability to taste bitter substances is advantageous in avoiding poisons, but might keep people from eating bitter vegetables that contain chemicals that protect against cancer. Devise an experiment, perhaps based on population data, to test either hypothesis—that the ability to taste bitter substances is either protective or harmful.

3. Many people think that evolution is the transformation of one species into another, such as chimpanzees to humans, and is "just a theory." State the genetic definition of microevolution, and give three examples, either from the chapter or from the news, that show evolution going on right now.

4. The high prevalence of Tay-Sachs disease among the Ashkenazim was once attributed to balanced polymorphism in which being a carrier protects against respiratory infections. This hypothesis arose from the observation that survivors of the Warsaw ghetto, where Jews were massacred during World War II, did not succumb to tuberculosis and other respiratory illnesses as frequently as other people did. A recent study, however, concluded that the high incidence of Tay-Sachs disease is due to genetic drift, not balanced polymorphism. The evidence is that a dozen other genetic diseases are about equally prevalent in the Ashkenazim. How does this evidence argue against balanced polymorphism?

5. A mutation that removes the receptor for HIV on human cells also blocks infection by the bacterium that causes plague. Seven centuries ago, in Europe, the "Black Death" plague epidemic increased the protective allele in the population. Today it makes 3 million people in the United States and the United Kingdom resitant to HIV infection. Is the increase in incidence of this allele due to nonrandom mating or natural selection?

6. Use the information in chapters 14 and 15 to explain why

   a. porphyria variegata is more prevalent among Afrikaners than other South African populations.

   b. many people among the Cape population in South Africa lose their teeth before age 20.

   c. cystic fibrosis and sickle cell disease remain common Mendelian illnesses.

   d. the Pima Indians have an extremely high incidence of type 2 diabetes.

   e. the Amish in Lancaster County and certain Pakistani groups have a high incidence of genetic diseases that are very rare elsewhere.

   f. the frequency of the allele that causes galactokinase deficiency varies across Europe.

   g. a haplotype associated with Creutzfeldt-Jakob disease is the same in populations from Spain, Chile, Libya, Italy, and Tunisia.

   h. mitochondrial DNA sequences vary gradually in populations along the Nile River valley.

   i. disease-causing *BRCA1* alleles are different in Jewish people of eastern European descent and African Americans.

7. Which principles discussed in this chapter do the following science fiction plots illustrate?

   a. In *When Worlds Collide,* the Earth is about to be destroyed. One hundred people are selected to colonize a new planet.

   b. In *The Time Machine,* set in the distant future on Earth, one group of people is forced to live on the planet's surface while another group is forced to live in caves. Over many years, they come to look and behave differently. The Morlocks that live below ground have dark skin, dark hair, and are very aggressive, whereas the Eloi that live aboveground are blond, fair-skinned, and meek.

   c. In *Children of the Damned,* all of the women in a small town are suddenly made pregnant by genetically identical beings from another planet.

   d. In *The War of the Worlds,* Martians cannot survive on Earth because they are vulnerable to infection by terrestrial microbes.

   e. In Dean Koontz's novel *The Taking,* giant mutant killer fungi kill nearly everyone on Earth, sparing only young children and the few adults who protect them. The human race must re-establish itself from the survivors.

8. Treatment for PKU has been so successful that, over the past 30 years, many people who would otherwise have been profoundly mentally retarded have led normal lives and become parents. How has this treatment altered Hardy-Weinberg equilibrium for the mutant alleles that cause PKU?

9. Ashkenazim, French Canadians, and people who live in southwestern Louisiana all have a much higher incidence of Tay-Sachs disease than other populations.

   a. Each of these groups has a different mutation. How is this possible?

   b. A controversial hypothesis proposes that the high incidence of Tay-Sachs disease and other genetic disorders that harm brain cells among the Ashkenazim reflects balanced polymorphism. Because brain cells are affected, carriers are, for reasons unknown, more intelligent and therefore had a survival advantage during periods of persecution because they could better use their wits to escape violence. What evidence might indicate whether the hypothesis has validity?

10. Syndrome X consists of obesity, type 2 diabetes, hypertension, and heart disease. Researchers surveyed and sampled blood from nearly all of the 2,188 residents of the Pacific Island of Kosrae, and found that 1,709 of them are part of the same pedigree. The incidence of all of the symptoms of syndrome X is much higher in this population than for other populations. Suggest a reason for this finding, and indicate why it would be difficult to study these particular traits, even in an isolated population.

11. People with familial Mediterranean fever (OMIM 249100) have an unusually low

incidence of asthma. What force may help maintain this disorder in populations?

12. By which mechanisms discussed in this chapter do the following situations alter Hardy-Weinberg equilibrium?

a. Ovalocytosis (OMIM 166910) is a rare genetic abnormality that is not only symptomless, but seems to be beneficial. A protein that anchors the red blood cell plasma membrane to the cytoplasm is abnormal, making the membrane unusually rigid. As a result, the parasites that cause malaria cannot enter the red blood cells of individuals with ovalocytosis.

b. In the mid-1700s, a multitoed male cat from England crossed the sea and settled in Boston, where he left behind quite a legacy of kittens—about half of whom also had six, seven, eight, or even nine digits on their paws. Today, in Boston and nearby regions, multitoed cats are far more common than in other parts of the United States.

c. Many slaves in the United States arrived in groups from Nigeria, which is an area in Africa with many ethnic subgroups. They landed at a few sites and settled on widely dispersed plantations. Once emancipated, former slaves in the South were free to travel and disperse.

Web Activities

Visit the Online Learning Center (OLC) at www.mhhe.com/lewisgenetics7. Select **Student Edition, chapter 15,** and **Web Activities** to find the website link needed to complete the following activity.

13. Go to the Centers for Disease Control and Prevention website, and the journal *Emerging Infectious Diseases.* Using this resource, describe an infectious disease that is evolving, and cite the evidence for this.

Case Studies

14. The human population of India is divided into many castes, and the people follow strict rules governing who can marry whom. Researchers from the University of Utah compared several genes among 265 Indians of different castes and 750 people from Africa, Europe, and Asia. The study found that the genes of higher Indian castes most closely resembled those of Europeans, and that the genes of the lowest castes most closely resembled those of Asians. In addition, the study found that maternally inherited genes (mitochondrial DNA) more closely resembled Asian versions of those genes, but paternally inherited genes (on the Y chromosome) more closely resembled European DNA sequences. Construct an historical scenario to account for these observations.

Learn to apply the skills of a genetic counselor with these additional cases found in the *Case Workbook in Human Genetics:*

3-methyl glutaconic aciduria type III

Jewish genius?

CHAPTER

16

Human Origins and Evolution

CHAPTER CONTENTS

LONELY HUMANITY

It's odd, in the animal world, to be the only ones of our kind. That wasn't always so, and perhaps this is why the theme of a dual humanity persists in science fiction. H. G. Wells's classic, *The Time Machine* looked far ahead at two battling breeds of people. In the novel *Darwin's Children*, a virus scrambles the genomes of a cohort of newborns, sowing the seeds of a new species. Various books describe holdouts from times past—a Neanderthal in modern-day Tajikastan, or a caveman encountered by an intrepid paleontologist in a Kenyan jungle.

Fossils indicate that from 2 to 6 million years ago, humans and pre-humans overlapped, in time if not in space. The discovery of preserved bones of several ancient humans on the island of Flores in Indonesia in 2004, however, may indicate a much more recent coexistence. Researchers discovered a female skeleton about 17 feet beneath a cave floor, with pieces of others nearby. *Homo floresiensis* was about half as tall as a modern human, with a brain about a third of the size. The small brain, if this means lower intelligence, is inconsistent with the sophisticated tools found with the skeletons. The "little lady of Flores" lived about 18,000 years ago—and she remains mysterious.

Was she a descendant of *Homo erectus,* who lived in the area 1.6 million years ago? Or was she a direct descendant of our own species, as evidenced by her fancy tools? A long-ago founder effect might have stranded a group of very short people on the island, but small body sizes happen, over time, to other animals isolated on islands. With limited resources, gene variants are gradually selected that sculpt a body that needs less food. Understanding who the little people of Flores were will require comparison of their genomes to those of other modern populations.

Comparing skulls among modern humans, our modern primate cousins, and fossilized hominids can reveal much about our forebears and our evolution.

Imagine being asked to build a story from the following elements:

1. A pumpkin that turns into a coach

2. A prince who hosts a ball

3. A poor but beautiful young woman with dainty feet who has two mean and ugly stepsisters with large feet

Chances are that unless you're familiar with the fairytale "Cinderella," you wouldn't come up with that exact story. Ten people given the same pieces of information might construct ten very different tales.

So it is with the sparse evidence we have of our own beginnings—pieces of a puzzle in time, some out of sequence, many missing. Traditionally, paleontologists (scientists who study evidence of ancient life) have consulted the record in the earth's rocks—fossils—to glimpse the ancestors of *Homo sapiens*, our own species. Researchers assign approximate ages to fossils by observing which rock layers fossils are in, and by extrapolating the passage of time from the ratios of certain radioactive chemicals in surrounding rock.

Fossils aren't the only way to peek into species' origins and relationships. Modern organisms also provide intriguing clues to the past in their DNA. The premise for such studies is that closeness of relationship is reflected in greater similarity of DNA sequence, because similar sequences are more likely to have arisen from individuals or species sharing ancestors than from the exact same set of spontaneous mutations. By analogy, it is more likely that two young women wearing the same combination of clothes and accessories purchased them at the same store, than that each happened to assemble the same collection of items from different sources.

Treelike diagrams are used to depict evolutionary relationships, based on fossil evidence and/or inferred from DNA sequence similarities. Branchpoints on the diagrams represent divergence from shared ancestors. Overall, evolution can be depicted as a series of branches as species diverged, driven by allele frequencies changing in response to the forces discussed in chapter 15: nonrandom mating, genetic drift, migration, mutation, and natural selection. Evolution is *not* a linear morphing of one type of organism into another—a common misunderstanding.

In this chapter, we explore human origins, genetic and genomic evidence for evolution, and how we attempt to alter the evolution of our own species and others.

16.1 Human Origins

A species includes organisms that can successfully produce healthy offspring. *Homo sapiens* ("the wise human") probably first appeared during the Pleistocene epoch, about 200,000 years ago. Our ancestry reaches farther back, to about 60 million years ago when rodentlike insect eaters flourished. These first primates diverged to give rise to many new species. Their ability to grasp and to perceive depth provided the flexibility and coordination necessary to dominate the treetops.

Hominoids and Hominids

About 30 to 40 million years ago, a monkey-like animal the size of a cat, *Aegyptopithecus*, lived in the lush tropical forests of Africa. Although the animal probably spent most of its time in the trees, fossilized remains of limb bones indicate it could run on the ground, too. Fossils of different individuals found together indicate that they were social animals. *Aegyptopithecus* had fangs it might have used for defense. The large canine teeth seen only in males suggest that males may have provided food for their smaller female mates. *Propliopithecus* was a monkeylike contemporary of *Aegyptopithecus*. Both animals are possible ancestors of gibbons, apes, and humans.

From 22 to 32 million years ago, Africa was home to the first **hominoids,** animals ancestral to apes and humans only. One such resident of southwestern and central Europe was called *Dryopithecus,* meaning "oak ape," because its fossilized bones were found with oak leaves (**figure 16.1**). The way the bones fit together suggests that this animal lived in the trees but could swing and walk farther than *Aegyptopithecus.*

More abundant fossils represent the middle-Miocene apes of 11 to 16 million years ago. These apes were about the size of a human seven-year-old and had small brains and pointy snouts. (*Miocene* refers to the geologic time period).

Apelike animals similar to *Dryopithecus* and the mid-Miocene apes flourished in Europe, Asia, and the Middle East during the same period. Because of the large primate population in the forest, selective pressure to venture onto the grasslands in search of food and habitat space must have been intense. Many primate species probably vanished as the protective forests shrank. Of all of the abundant middle-Miocene apes, one survived to give rise to humans and African apes. Eventually, animals ancestral to humans only, called **hominids,** arose and eventually thrived. (Some researchers use *hominim* instead of *hominid*.)

Hominoid and hominid fossils from 4 to 19 million years ago are scarce, and are often just fragments of tooth and jaw. About 6 million years ago, the hominid lineage split from the apes. There are at least three candidates for this first primate one step closer to humanity from the chimp: *Ardipithecus kadabba* from Ethiopia, *Sahelanthropus tchadensis* from Chad, and *Orrorin tugenensis* from Kenya. Note the position of these hominids at the base of the evolutionary tree diagram in **figure 16.2,** which depicts probable relationships among some of our relatives, past and present. This evolutionary tree is based on fossil evidence and DNA sequence comparisons for the modern species.

Australopithecus

Four million years ago, human forebears diversified as bipedalism—the ability to walk upright—opened up vast new habitats on the plains. Several species of a hominid called *Australopithecus* lived at this time, from 2 to 4 million years ago, probably following a hunter-gatherer lifestyle. Figure 16.2 depicts the probable relationships among some known australopithecines, members of the genus *Homo*, and our closest modern primate relatives.

Australopithecines had flat skull bases, as do all modern primates except humans. They stood about 1.2 to 1.5 meters (four to five feet) tall and had brains about the size of a gorilla's, with humanlike teeth. The angle of preserved pelvic bones, plus the discovery of *Australopithecus* fossils with

a. *Dryopithecus*   b. *Australopithecus*   c. *Homo erectus*

Figure 16.1 Human forerunners.

(a) The "oak ape" *Dryopithecus,* who lived from 22 to 32 million years ago, was more dextrous than his predecessors. **(b)** Several species of *Australopithecus* lived from 2 to more than 4 million years ago, and walked upright. **(c)** *Homo erectus* made tools out of bone and stone, used fire, and dwelled communally in caves from 1.6 million years ago to about 35,000 years ago.

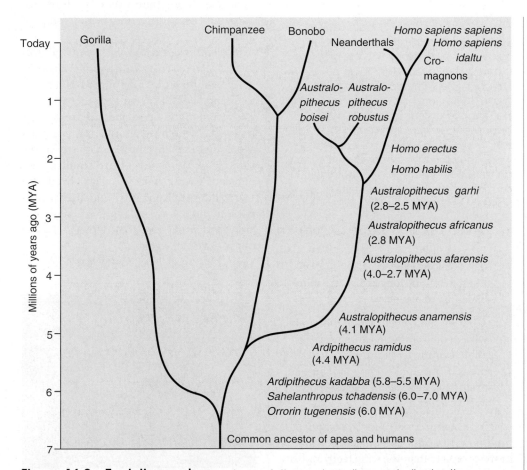

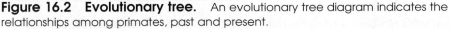

Figure 16.2 Evolutionary tree.
An evolutionary tree diagram indicates the relationships among primates, past and present.

those of grazing animals, indicate that this ape-human had left the forest.

The most ancient species of australopithecine known, *Australopithecus anamensis,* lived about 4.1 million years ago. It led to *A. afarensis,* represented by a famous fossilized partial skeleton named "Lucy" discovered in 1974. She lived about 3.6 million years ago in the grasses along a lake in the Afar region of Ethiopia **(figure 16.3).** This is the only place known on Earth where fossil evidence of our ancestors covers 6 million years. Lucy's skull was shaped more like a human's than an ape's, with a less prominent face and larger brain than her predecessors had. The condition of her skeleton indicated that she died, with arthritis, at about the age of 20.

Much of what we know about the australopithecines comes from two parallel paths of footprints, preserved in volcanic ash in the Laetoli area of Tanzania. Archaeologist Mary Leakey and her team discovered them by accident in 1976. The 89-foot-long trail of footprints, left about 3.6 million years ago, was probably made by a large and small individual walking close together, and a third following in the steps of the larger animal in front. The shape of the prints indicates that their feet and gait were remarkably like ours.

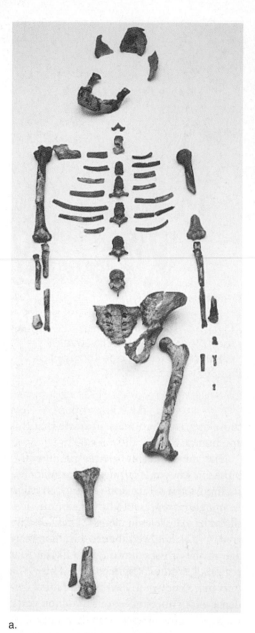

a.

b.

Figure 16.3 Lucy. About 3.6 million years ago, a small-brained human ancestor walked upright in the grasses along a lake in the Afar region of Ethiopia. She skimmed the shores for crabs, turtles, and crocodile eggs to eat. Her discoverers named her "Lucy" because they were listening to the Beatles song "Lucy in the Sky with Diamonds" when they found her bones **(a)**. **(b)** shows an artist's interpretation of what this animal on the road to humanity may have looked like.

Australopithecus garhi may have overlapped with the earliest members of *Homo.* Its fossils from the Afar region date from about 2.5 million years ago. Trying to understand how this human forebear lived is a little like piecing together the Cinderella story. Among the *A. garhi* fossils, researchers found:

1. Remains of a butchered antelope. The ends of the long bones had been cleanly cut with tools, the marrow removed, meat stripped, and the tongue cleanly sliced off.

2. Limb bones from an individual who stood about 1.4 meters (about 4.5 feet)

tall. The long legs were like those of a human, but the long arms were more like those of an ape.

3. A partial skull of a different individual, with a small cranium (holding a 450-cubic-centimeter brain, compared to the 1,350- to 1,400-cc modern human brain) and large teeth, suggesting an apelike lower face.

The *A. garhi* fossil finds are important for many reasons. The evidence is much more complete than the stray teeth and jaw bits of other australopithecines. More importantly, this hominid lived in the right time and place to be, and had characteristics

consistent with, the long-sought bridge between *Australopithecus* and *Homo.* The fossils were found in a desert in eastern Ethiopia, near later *Homo* fossils. The limb lengths and cranial capacity suggest a transitional form. In addition, *A. garhi* hunted.

Homo

Like our Cinderella story, the tale of how *Australopithecus* was replaced by *Homo* is built on sparse clues. Some australopithecines were "dead ends," dying off or leading to the Neanderthals, a branch of the *Homo* family tree that eventually vanished. Clues suggest that by 2.3 million years ago, *Australopithecus* coexisted with *Homo habilis*—a more humanlike cave dweller who cared intensively for its young. *Habilis* means handy, and this primate was the first to use tools for tasks more challenging than stripping meat from bones. *H. habilis* may have descended from a group of australopithecines who ate a greater variety of foods than other ape-humans, allowing them to live in a wider range of habitats.

H. habilis coexisted with and was followed by *Homo erectus* during the Paleolithic Age (**table 16.1**). One famed *H. erectus* fossil, named "Daka" for the place where it was found in the Afar region, represents an individual who lived about a million years ago. It had a shallow forehead, massive brow ridges, a brain about a third smaller than ours, and strong, thick legs. Daka lived on a grassland, with elephants, wildebeests, hippos, antelopes, many types of pigs, and giant hyenas. **Figure 16.4** depicts what he might have looked like.

Figure 16.4 *Homo erectus.* This artist's rendition is based on many fossils.

Table 16.1

Cultural Ages

Age	Time (years ago)	Defining Skills
Paleolithic	750,000 to 15,000	Earliest chipped tools
Mesolithic	15,000 to 10,000	Cutting tools, bows and arrows
Neolithic	10,000 to present	Complex tools, agriculture

H. erectus left fossil evidence of cooperation, social organization, tools, and use of fire. Fossilized teeth and jaws suggest that these hominids ate meat. The distribution of fossils indicates that they lived in families of male-female pairs (most primates have harems). The male hunted, and the female nurtured the young. They were the first to have an angled skull base that permitted them to produce a greater range of sounds, making speech possible. *H. erectus* also has the honor of being our oldest ancestor for whom we have evidence of genetic disease, amelogenesis imperfecta (OMIM 104500), in the teeth of a fossil from Ethiopia. *H. erectus* fossils are very widespread. They have been found in China, Java, Africa, Europe, and southeast Asia, indicating that these animals could migrate farther than earlier primates.

We have a glimpse of our ancestors who lived about 156,000 years ago from skulls discovered near the town of Herto in the Afar region of Ethiopia (**figure 16.5**). Driving by Herto in 1997, after a season of punishing rains, paleoanthropologist Tim White of the University of California, Berkeley, spotted a skull emerging from the sand near the Awash River. The skull came from a hippo, and it bore evidence of butchering. When White returned a few days later with helpers, the team found three human skulls in pieces. These and other fossils had survived because the El Niño rains had driven the modern-day residents and their cattle from Herto; otherwise they would have trampled the evidence.

One adult skull was from a young man; the other was damaged beyond recognition. The third skull was from a child about seven years of age. All were smooth, with decorative cutmarks, and found alone, with no other body parts. These clues suggest a ritualistic burial. The team also found hundreds of stone blades, axes, and flaking tools, and

Figure 16.5 *Homo sapiens idaltu.*
Discovery of three skulls made possible this artist's depiction of what this early member of our species might have looked like—not very much different from us.

other human remains. The researchers envisioned a band of early humans, with some features much like our own, that lived near a shallow lake that formed when the river overflowed. The lake was home to hippos and fish, and cattle lived near lush vegetation nearby. It isn't clear whether our forebears hunted hippos or simply ate carcasses they found.

The people, named *H. sapiens idaltu*, which means "elder" in the local Afar language, are our oldest known ancestors, their faces much more like our own than like the mysterious Neanderthals. The *H. sapiens idaltu* fossils provide evidence that the Neanderthals split off from the line leading

to us at least 300,000 years ago. The fossils are also remarkably consistent in time with other evidence of humans having originated in Africa, a point we return to soon.

By 70,000 years ago, humans used more intricately carved tools made of bones, and red rock that bore highly symmetrical hatchmarks, which may indicate early counting. However, given that groups of hominids may have been very isolated on the vast continent, and some of them were already leaving Africa, it's possible that even as *H. sapiens idaltu* and perhaps others yet to be discovered were far along the road to modern humanity, pockets of *H. erectus* may have persisted, perhaps as recently as 35,000 years ago.

The ill-fated Neanderthals were also contemporaries of *H. erectus* and members of genus *Homo*. They appeared in Europe about 150,000 years ago, but may have lived elsewhere long before that. Neanderthals and modern people may have coexisted in what is now Israel about 90,000 years ago. By 70,000 years ago, the Neanderthals had spread to western Asia. The most recent Neanderthals that we know of lived about 30,000 years ago. They had slightly larger brains than we do, prominent brow ridges, gaps between certain teeth, very muscular jaws, and large, barrel-shaped chests.

The Neanderthals take their name from Neander Valley, Germany, where quarry workers blasting in a limestone cave on a summer day in 1856 discovered the first preserved bones of this hominid. A Neanderthal discovered in France fifty years later, the "Old Man" of La Chapelle-aux-Saints, led to their common depiction as primitive and slow-witted, stooped perhaps due to arthritis. When paleoanthropologists at the American Museum of Natural History in New York recently reconstructed a complete Neanderthal skeleton from fossils collected all over the world and compared it to that of a modern human, the differences were stark (**figure 16.6**). A Neanderthal had a wider pelvis, shoulders, and ribcage, and shorter forearms and shins; prominent brow ridges, a forward pointing face, and a sloping forehead. The characteristic heavy brow bones might have resulted from genetic drift acting over time on populations isolated in cave systems, not breeding with others. The stocky skeletons might reflect natural selection in a cold climate. Anthropologists still

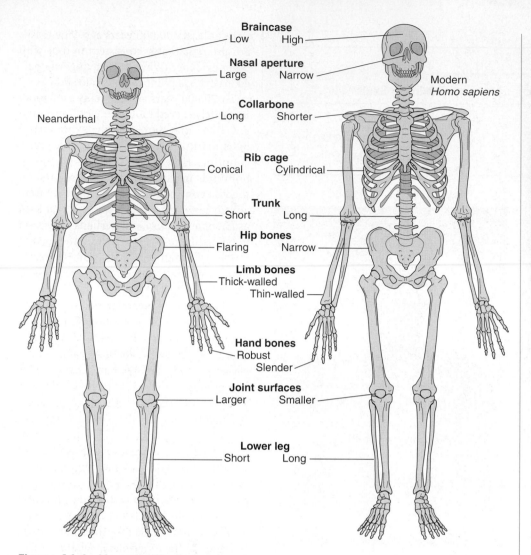

Figure 16.6 **Neanderthals.** Neanderthals probably looked a lot like us, but there were many subtle skeletal differences.

Braincase
Low   High

Nasal aperture
Large   Narrow

Neanderthal

Modern
Homo sapiens

Collarbone
Long   Shorter

Rib cage
Conical   Cylindrical

Trunk
Short   Long

Hip bones
Flaring   Narrow

Limb bones
Thick-walled
Thin-walled

Hand bones
Robust
Slender

Joint surfaces
Larger   Smaller

Lower leg
Short   Long

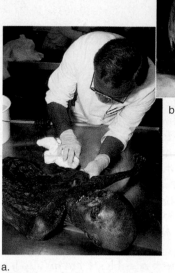

a.

b.

Figure 16.7   A 5,300-year-old man.
(a) Hikers discovered Ötzi, the Ice Man, in the Austrian/Italian Alps in 1991. He lived 5,300 years ago. **(b)** Ötzi wore well-made clothing, including a hat; used intricate arrows that demonstrate familiarity with ballistics and engineering; and carried mushrooms with antibiotic properties. He had tattoos, indentations in his ears that suggest he wore earrings, and evidence of a haircut. This depiction is derived from the evidence found on and near Ötzi's preserved body.

debate whether the Neanderthals were truly different from us, or had a few exaggerated human features. A fossilized, deformed skeleton buried with flowers in Shanidar Cave, Iraq, reveals that the Neanderthals may have been religious hunter-gatherers that were either clever enough or lucky enough to have survived a brutal ice age.

Fossil evidence indicates that from 30,000 to 40,000 years ago, the Neanderthals coexisted with the lighter-weight, finer-boned Cro-Magnons. The newcomers had high foreheads and well-developed frontal brain regions, and signs of culture that we do not see for Neanderthals. The first Cro-Magnon fossils were found in a French cave. Five adults and a baby were arranged in what appeared to be a communal grave. Nearby were pierced seashells that

may have been used as jewelry. Intricate art decorated the cave walls. In contrast, the few Neanderthal graves show no evidence of ritual, just quick burial.

We don't know how it happened, but by about 28,000 years ago, the Neanderthals were gone, and the Cro-Magnons presumably continued on the path to humanity. Fossil evidence suggests that it wasn't the weather that felled the Neanderthals, so perhaps the Cro-Magnons outcompeted them for resources. We return to Neanderthals later in the chapter to examine DNA evidence of their relationship to us.

Modern Humans

Cave art from about 14,000 years ago indicates that by that time, our ancestors had

developed fine hand coordination and could use symbols—milestones in cultural evolution. By 10,000 years ago, people had expanded from the Middle East across Europe, bringing agricultural practices.

In 1991, hikers in the Ötztaler Alps of northern Italy discovered an ancient man frozen in the ice (**figure 16.7**). After amateurs hacked away at him, causing much damage, the Ice Man, named Ötzi, ended up in the hands of several research groups. The Ice Man was on a mountain more than 10,000 feet high 5,300 years ago when he perished. He was dressed for the weather, with a bearskin cap and cloak, and shoes. Berries found with him place the season as late summer or early fall.

Ötzi may have died following a fight. When found he had a knife in one hand, cuts and bruises, and an arrowhead embedded in his left shoulder from the rear. The wound bore blood from two other individuals, according to DNA profiling. His bearskin cloak had the blood of a third person on it. One interpretation is that he killed or wounded two people, and perhaps carried a wounded comrade on his back. Suffering from blood loss or an infected wound, he probably fell into a ditch, where he froze to death and was soon covered by snow. After this safe burial, which preserved his body intact, a glacier sealed the natural tomb. DNA analyses on Ötzi's tissues suggest that he belonged to the same gene pool as modern people living in the area, which is near the Italian-Austrian border.

Another way that anthropologists try to glimpse what humans were like a few thousand years ago is by studying vanishing indigenous peoples, such as the San (bushmen) and pygmies of Africa, the Basques of Spain, the Etas of Japan, the Hill People of New Guinea, the Yanomami of Brazil, and another Brazilian tribe, the Arawete, who number only 130 individuals. Studying DNA sequences within these populations provides information on their origins, as we'll see later in the chapter.

Yet another way to look back in time is to compare genetic diversity in modern human populations. In one landmark study, researchers determined the pattern of alleles at 377 short tandem repeats (STRs) on the autosomes of 1,056 people representing 52 populations defined by geography, language, or culture. Recall that STRs are 2-to-10-base repeats that are not subject to natural selection because they do not affect the phenotype. An "allele" is simply the number of repeats. The total number of such alleles for the 377 STRs is 4,682. Computer algorithms searched for similarities in allele patterns, without considering other characteristics, and revealed that the people fell into five clusters—which corresponded exactly with what is known about ancient human migration patterns out of Africa.

RELATING THE CONCEPTS

Give two examples of hominids that overlapped in time.

Key Concepts

1. Monkeylike *Aegyptopithecus* lived about 30 to 40 million years ago and was ancestral to gibbons, apes, and humans. The first hominoid (ape and human ancestor), *Dryopithecus*, lived 22 to 32 million years ago and may have walked onto grasslands.
2. Hominids (human ancestors) appeared about 19 million years ago.
3. About 4 million years ago, bipedalism opened up new habitats for *Australopithecus*. *A. garhi* may have been a direct forebear of *Homo*.
4. By 2 million years ago, *Australopithecus* coexisted with the more humanlike *Homo habilis*. Later, *H. habilis* coexisted with *H. erectus*, who used tools in more complex societies. *H. erectus* then coexisted with *H. sapiens*. *H. sapiens idaltu* lived 156,000 years ago.
5. The Neanderthals were a side branch from modern humans, who disappeared about 30,000 years ago.
6. A preserved man from 5,300 years ago is genetically like us.

16.2 Molecular Evolution

Fossils paint an incomplete picture of the past because only certain parts of certain organisms were preserved, and very few have been discovered. Additional information on the past comes from within the cell, where the informational molecules of life, DNA and amino acid sequences, change over time as mutations occur. The more alike a gene or protein sequence is in two species, the more closely related the two are presumed to be—that is, the more recently they shared an ancestor. The assumption is that it is highly unlikely that two unrelated species would evolve precisely the same sequence of DNA nucleotides or amino acids by chance. Determining and comparing genome, DNA or protein sequences, and chromosome banding patterns is the field of **molecular evolution. Figure 16.8** shows how tree diagrams depict genetic change.

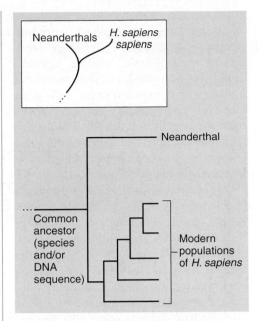

Figure 16.8 Evolutionary tree diagrams reflect differences in DNA sequence. The branchpoints represent mutational events, which result in two DNA sequences—the original, and the changed sequence. This evolutionary tree is based on comparisons of key mtDNA sequences that indicate Neanderthals diverged from our lineage before modern human populations became established. Times can be assigned if the mutation rate is known for the DNA sequences under comparison. The inset shows the corresponding information on the stylized version of a tree in figure 16.2.

Comparing Genes and Genomes

We can assess similarities in DNA sequences between two species for a piece of DNA, a single gene, a chromosome segment, or even an entire genome. Most such efforts address evolutionary questions, but they can have practical applications, too. Animal models of human disease provide an example of the utility of knowing our closest genetic relatives.

Animal Models

Identifying corresponding genes in different species is very important in medical research; animal models of human disease are used to test experimental treatments. It is important that an animal model of a human disease have the same signs and symptoms. Because of differences in physiology, development, and lifespan, another mammal might not have the same phenotype as a human with a mutation in the same gene. However, corresponding genes can reveal basic abnormalities at the cell or molecular level. For example, two-thirds of the genes known to cause Mendelian disorders in humans have counterparts in fruit flies!

For some genes, a close correspondence in phenotype can be seen among species. People with Waardenburg syndrome (OMIM 148820), for example, have a characteristic white forelock of hair; wide-spaced, light-colored eyes; and hearing impairment (**figure 16.9**). The gene responsible is very similar in sequence to one in cats who have white coats and blue eyes and who are deaf. Horses, mice, and minks also have this combination of traits, which is thought to stem from abnormal movements of pigment cells in the embryo's outermost layer.

Targeted Comparative Sequencing

Comparing completely sequenced genomes among different species could, of course, reveal much about evolutionary relationships, but is very time-consuming for organisms with large genomes. A shortcut is a technique called targeted comparative sequencing, which aligns representative sections from the genomes of different species—a little like reading excerpts rather than whole books to get an idea of how similar two books are.

In general, DNA sequences that encode protein tend to be very similar among closely related species, because natural selection weeds out nonfunctional proteins that have vital functions. Such similar sequences are termed "highly conserved." Sequences that are similar in different, closely related species but that do not encode protein often control transcription or translation, and so are also vital and therefore subject to natural selection. In contrast, genome regions that vary widely among species typically do not affect the phenotype, and are therefore not subject to natural selection. Within a protein-encoding gene, the exons tend to be highly conserved, but the introns, which are removed, are not.

The type of information gained from targeted comparative sequencing depends upon the species probed. Eric Green, a researcher at the National Institutes of Health, had been working on the cystic fibrosis region of human chromosome 7 for many years, when he decided to investigate corresponding sequences in 17 other vertebrate species. These species included chimps, baboons, cats, dogs, cows, mice, rats, chickens, and pufferfish. The DNA sequence similarities paralleled phenotypic similarities and presumably evolutionary relationships. For example, for a 20,000-base portion of the *CFTR* gene, humans had 94 percent of the DNA sequence in common with baboons, and increasingly less of the sequence in common with cows, mice, and pufferfish.

Ironically, targeted comparative sequencing is not very informative when restricted to humans and chimps, because our genomes are so similar. The two genomes are like two introductory biology textbooks, presenting basically the same information overall, perhaps in a slightly different order. To highlight the differences between us, Edward Rubin, at the U.S. Department of Energy Joint Genome Institute, compared four genome regions among a different set of eighteen vertebrates, but included only primates and hominoids, our closest relatives. The reasoning: Those sequences that all primates, but not other vertebrates, share reveal the functions necessary to be a primate. The DNA sequences that are unique to humans help to define us at the genetic level. Green's work is like comparing excerpts from a dictionary, a science fiction novel, a cookbook, and a children's story; Rubin's approach is like comparing only mystery novels. **Figure 16.10** depicts one way to display similarities and differences among the genomes of different species.

Comparing Chimps and Humans

We have more in common with chimpanzees than with any other animals, but just how similar we are at the genome level depends upon the types of DNA sequences considered. The commonly repeated estimate of 98.7 percent similarity at the

a.

b.

c.

Figure 16.9   The same mutation can cause similar effects in different species.   A mutation in mice **(a)**, cats **(b)**, humans **(c)**, and other types of mammals causes light eye color, hearing or other neurological impairment, and a fair forelock.

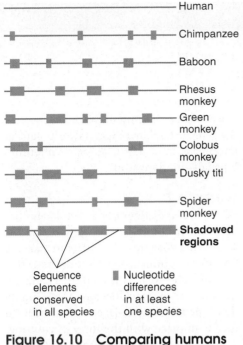

Figure 16.10 **Comparing humans to their closest relatives.** For this section of a chromosome, the thin lines represent regions where the DNA sequences correspond among species.

Sequence elements conserved in all species

Nucleotide differences in at least one species

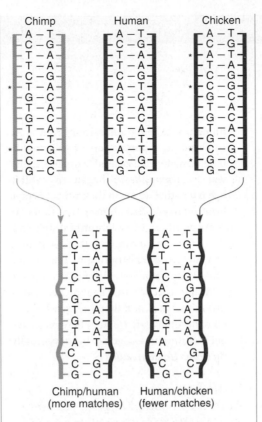

Figure 16.11 **The rate of DNA hybridization reflects the degree of evolutionary relatedness.** This highly schematic diagram shows why DNA from a human hybridizes more rapidly with chimpanzee DNA than with chicken DNA. Each * refers to a site where chimp or chicken DNA differs from human DNA.

genome level between human and chimp originated from studies conducted in the 1970s using a technique called DNA hybridization. DNA from two species is cut and mixed. Complementary pieces bind, and some hybrid molecules form, with one strand of the double helix from one species, the other from the other species. The higher the temperature required to separate hybrid double helices, the more of the sequence they share, because they bind at more complementary base pairs (**figure 16.11**). These studies used DNA segments present in single copies in the genome, which indicates that they likely encode protein. Protein comparisons support the 98.7 percent sequence identity.

Most comparisons of the human and chimp genomes at first considered only protein-encoding genes. But another way that the genomes differ is in the number of copies of certain DNA sequences. Roy Britten, a researcher at the California Institute of Technology who coinvented DNA hybridization technology, took a different view—he included "indels," for "insertions and deletions," in the calculations of genome similarity. If small insertions and deletions that distinguish the human and chimp versions of the same gene are counted, then our degree of genome similarity diminishes to about 96.6 percent. A simple calculation on a short, hypothetical DNA sequence demonstrates how this happens.

Consider an ancestral sequence of 15 bases:

G A T A C G A G C T C T A A C

"Ancestral" means that the most recent common ancestor of humans and chimps had this sequence. If a single-base substitution occurred after the divergence from the shared ancestor, then the correspondence would be less than 100 percent:

Chimp   G A T A C G A G C T C T A A C
Human   G A T A C G A G C T A T A A C

After the C-to-A point mutation occurs, humans and chimps share 14 of these 15 bases, for a correspondence of 93.3 percent, rather than 100 percent identity. But what happens when three bases at a time (so as not to offset the reading frame) are added or deleted in one of the evolutionary lines? Imagine that three bases insert into the human lineage as follows:

Chimp   G A T A C G A G C T C T A A C
Human   G A T **G C A** A C G A G C T C T A A C

The correspondence is now 15 out of 18 bases, or 83.3 percent.

If the human lineage lost three bases, the correspondence would also diminish:

Chimp   G A T A C G A G C T C T A A C
Human   G A T A G C T C T A A C

The sequence now shares 12 out of 15 bases, or 80 percent.

The similarity between the human and chimp genome decreases even more if noncoding regions, such as introns and repeats, are considered. However, differences in indels and repeats do not explain how we differ from chimps on a whole organism level. To assess phenotypic distinctions, it is more helpful to look at individual traits, sometimes determined by single genes (**Reading 16.1**).

Uniquely human traits include spoken language, abstract reasoning ability, highly opposable thumbs, and larger frontal lobes of the brain. Individual genes can actually have great effects on appearance, physiology, and development. For example, one stark difference between chimp and human that could stem from a single gene is hairiness. Chimpanzees and gorillas express a keratin gene whose counterpart in humans has been silenced into pseudogene status by a nonsense mutation. The protein isn't made. When our ancestors left the forests, natural selection might have favored loss of body hair to provide more efficient cooling or as a way to shed skin parasites such as lice. Speech may also be due to a single gene difference between humans and chimps. A family in London whose members have unintelligible speech led to discovery of a single gene that controls speaking ability—and is present, but different, in chimps.

Another single gene that accounts for great differences among primates controls the switch from embryonic to fetal hemoglobin (see figure 11.2). More primitive primates lack or have very little fetal hemoglobin. In more recently evolved and

What Makes Us Human?

The sequencing of the chimpanzee genome enabled researchers to compare humans with our closest modern relatives on a gene-by-gene basis, adding to fossil evidence about how we came to be. Differences might provide peeks into our evolution. Below are just a few examples of ways that we differ from chimps—some controversial.

Tool Use

A paleontologist enjoys a different view of the origin of humanity than that of a geneticist. University of California, Berkeley, paleontologist Tim White heads a team that, for months at a time, calls the Afar region of Ethiopia home. Here, scattered and at different levels, lie remains of our ancestors stretching back some 6 million years, teasing at the time when our forebears split from an ancestor shared with the chimpanzee.

White led the teams that discovered *Ardipithecus, Australopithecus garhi*, "Daka", and *H. sapiens idaltu*. He sums up what distinguishes our species in one word: culture. His imagination takes him back in time, filling in what he has seen with what he envisions:

It all started about 2.5 million years ago, when guys started banging rocks together. That's what allowed the niche to expand in the beginning, the start of culture. Tool making, utilizing stone, probably began in *Australopithecus,* such as Lucy. They were very adaptable and very widespread, all over Africa. These bipeds were small-brained, and they weren't busy becoming human, but being australopithecines. At some point, a population of that highly intelligent, bipedal generalist organism that was *Australopithecus* began to exhibit behaviors that we see in the chimp. Chimps hunt monkeys. But chimps lack tools. At some point, an early hominid didn't lack those tools anymore, and that particular sect formed the beginning of the lineage that would ultimately diverge from other australopithecines that kept on being australopithecines, and that lineage would go on to become early *Homo.* There may have been different varieties, but eventually there was *Homo erectus.*

Walking

Diseases of modern humanity can provide glimpses into traits of evolutionary import. Consider Joubert syndrome (OMIM 608629). In this disorder, nerve cell fibers in the brain cannot cross from where they originate on one side of the brain to the other, so a person cannot control whether she moves the right or left arm in response to a command—both arms move. The part of the brain that controls posture, balance, and coordination is compromised. The person is also mentally retarded and very fatigued. The gene that causes Joubert syndrome, called *AHI1,* is identical in all modern human groups examined, but has different sequence variants in chimps, gorillas, and orangutans. In the lineage leading to humanity, the gene came to control movement. Researchers hypothesize that the *AHI1* gene is necessary to be able to place one foot in front of the other. To study the gene further, investigators are looking at another animal that is at least sometimes bipedal—the kangaroo!

Running

Homo erectus distinguished itself in another key way: It could run for long distances, something that no other primate can do. Specific anatomical adaptations made this possible. The nuchal ligament that connects the skull to the neck became more highly developed in *H. erectus,* enabling the head to stay in place with the force of running. The leg muscles were also more highly developed than those of chimps or australopithecines, acting as springs. *H. erectus* also originated a large buttocks, whose muscles contract during running. All three of these structures are not merely the result of being able to walk, but enabled early *Homo,* and us, to run. This skill would have enabled our ancestors to escape predators, find food, and locate new homes faster than if they could not run.

A Big Brain

The difference between a big-brained human and a small-brained chimp may be a single gene. About 2.4 million years ago, a gene called *MYH16* underwent a nonsense mutation, which prevented production of a type of muscle protein called a myosin. The mutation is seen in all modern human populations, but not in other primates. Without this particular type of myosin, jaw development is not as great. With a diminished jaw, more complex primates, fetal hemoglobin lengthened the fetal period, which maximized brain growth—and with larger brains came greater skills. Single genes can also explain the longer childhood and adolescence in humans compared to chimpanzees.

Single genes that distinguish humans from chimps appear to be few, but they tend to be implicated in Mendelian disorders. Perhaps this reflects the fact that the genes that distinguish us have recently taken on their new functions, and the genome has not yet had time for redundancies to have evolved.

In 1975, Mary-Claire King and the late Allan Wilson, at the University of California, Berkeley, developed the "regulatory hypothesis" to explain why humans and chimps are genetically so similar, but look and behave so differently. They suggested that the key underlying difference is in gene expression, not in genome sequence. Today, DNA gene expression microarrays are providing data that backs up their hypothesis. One study contrasted gene expression in the liver and brain in the two species. The differences in the brain were far greater than in

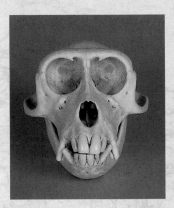
Macaque

Gorilla

Human

Figure 1   Did a shrinking jaw expand the human brain?   A nonsense mutation that disables a gene that encodes a type of myosin (a muscle protein) is present in all modern human populations. The gene is expressed in all other primates. Lack of this particular protein may result in a smaller jaw, which may have altered the structure of the skull in a way that permitted the bony plates of the head to move apart at their sutures. A more pliable skull may have enabled the brain within to expand. With larger brains, presumably, came greater intelligence.

the bony plates of the skull could expand, allowing greater brain growth (**figure 1**). Researchers nicknamed the mutation RFT, for "room for thought." Fossil evidence indicates that the switch from "big jaw, small brain" to "small jaw, big brain" happened when *Homo* gradually replaced *Australopithecus*, about 2 million years ago. The genetic analysis may be new, but the idea wasn't. Charles Darwin wrote in 1871 that different-sized jaw muscles were at the root of the distinction between apes and humans.

Cognition

At the genetic level, humans and chimps may differ more in the numbers of copies of particular genes than in the nature of the genes. Researchers used a microarray-based technique to identify genes with an increased "copy number" in the human genome compared to the genome of the great apes, identifying 134. A large proportion of these genes are involved in brain structure or function. Some of the genes promote the signal transduction that underlies long-term memory; others, when mutant, cause mental retardation or impair language skills.

Sense of Smell

Natural selection both retains useful traits and weeds out harmful or useless ones. Both of these forces may have operated to fine-tune our sense of smell.

The sense of smell derives from a one-inch-square patch of tissue high in the nose that consists of 12 million cells that bear odorant receptor (OR) proteins. (In contrast, a bloodhound has 4 *billion* such cells!) Molecules given off by something smelly bind to distinct combinations of these receptors, which then signal the brain in a way that creates the perception of an associated odor.

Our odorant receptor genes number 906, comprise about 1 percent of the genome, and occur in clusters. About 60 percent of them are pseudogenes—their sequences are similar to those of functional "smell" genes, but are riddled with mutations that prevent translation of complete proteins. Perhaps they are remnants of a distant past when we depended more upon our sense of olfaction. Natural selection may have, over time, eliminated OR genes no longer essential to survival. Yet other genetic evidence indicates that natural selection also has acted positively to retain the OR genes that continue to function. While the pseudogenes harbor many diverse SNPs (sites where more than one base is common), the functional OR genes are remarkably like one another. In addition, the nucleotide differences that persist among the retained genes actually alter the encoded amino acid. Geneticists interpret this unusual finding to suggest that natural selection favored these particular sequences.

As researchers continue to painstakingly compare the human and chimp genomes, many other distinctions between the two are certain to arise.

the liver, which is consistent with the cognitive and behavioral distinctions between human and chimpanzee—presumably, our livers are more alike.

Comparisons of the human genome sequence to those of other species are interesting too. Our close relationship to the other vertebrates is revealed by comparing the human genome sequence to that of the pufferfish *Tetraodon nigroviridis*. Its genome is like ours, minus many of the repeats and introns. So similar are the two genomes that researchers used the pufferfish genome to round down the original estimate of the number of human genes. On the other hand, it is odd to think that the protein-encoding portion of our genome is nearly the same as that of a fish!

Overall, the human genome has a more complex organization of the same basic parts as the fruit fly and roundworm genomes. For example, the human genome harbors thirty copies of the gene that encodes fibroblast growth factor, compared to two copies in the fly and worm genomes. This growth factor is important for the development of highly complex organs.

Genome studies indicate that over deep evolutionary time, genes and gene pieces provided vertebrates, including humans, with defining characteristics of complex neural networks, blood clotting pathways, and acquired immunity. In addition, the genomes of vertebrates make possible refined apoptosis, greater control of transcription, complex development, and more intricate signaling both within and between cells.

Did the Human Genome Duplicate?

Comparing the human genome to itself provides clues to evolution, too. The many duplicated genes and chromosome segments in the human genome suggest that it doubled, at least once, since diverging from a vertebrate ancestor about 500 million years ago. Researchers infer what might have happened from the number and organization of duplicated sequences. The duplications that riddle the genome are consistent with either a double doubling, followed by the loss of some genes, or, more likely, a single doubling followed by additional duplication of certain DNA sequences, a scenario one investigator calls "the big bang" followed by "the slow shuffle." Sequence information from gene families, which are clusters of genes with similar sequences and functions, supports the idea of a lone complete doubling at the dawn of vertebrate life, followed by a continual turnover of about 5 to 10 percent of the genome beginning 30 to 50 million years ago. (Section 16.3 explains how approximate dates are assigned to genetic changes.)

The extensive duplication within the human genome distinguishes us from other primates. Some of the doublings are vast. Half of chromosome 20 repeats, rearranged, on chromosome 18. Much of chromosome 2's short arm reappears as almost three-quarters of chromosome 14, and a block on its long arm is echoed on chromosome 12. The gene-packed yet tiny chromosome 22 includes eight huge duplications and several gene families. The human genome is riddled with redundancy—but it may not be wasted.

Duplications in a genome provide raw material and flexibility for future evolution. A copy of a DNA sequence can mutate, allowing a cell to "try out" a new function while the old one carries on. More often, though, the twin mutates into a silenced pseudogene, leaving a ghost of the gene behind as a similar but untranslated DNA sequence.

A duplication can be located near the original DNA sequence it was copied from, or away from it. A sequence repeated right next to itself is called a tandem duplication, and it usually results from mispairing during DNA replication. A copy of a gene on a different chromosome may arise when messenger RNA is copied (reverse transcribed) into DNA, which then inserts elsewhere among the chromosomes.

Duplication of an entire genome results in polyploidy, discussed in chapter 13. It is common in plants and some insects, but not vertebrates. (Polyploidy versus duplications can be compared to burning an entire CD versus copying only certain songs.) If a polyploid event was followed by the loss of some genes, then peppered with additional gene duplications, the result would look much like the modern human genome (**figure 16.12**). The remnants of such an ancient whole-genome duplication would have become further muddled with time, as inversions and translocations altered the ancestral DNA sequence.

Ancient DNA

When comparing DNA of modern species, a researcher can easily repeat an experiment—ample samples of DNA are available directly from the sources. This isn't so for ancient DNA, such as genetic material from insects preserved in amber, which is hardened resin from pine trees. The mix of

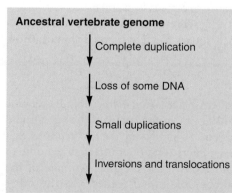

Ancestral vertebrate genome

Complete duplication

Loss of some DNA

Small duplications

Inversions and translocations

Figure 16.12 The evolution of the modern human genome.

chemicals in amber entombed whatever fell into it when it was the consistency of maple syrup. Alcohols and sugars in the resin dried out the specimen, and other organic molecules acted as fixatives, keeping cellular contents in place. The resin sealed out oxygen and bacteria, preventing decomposition. Finally, other organic molecules hardened the resin over 4 to 5 million years. Today, the DNA is extracted and amplified.

Probing ancient, preserved DNA for clues to past life is romanticized by the media. The novel and film *Jurassic Park,* for example, described cloning dinosaurs from blood in mosquitoes trapped in amber—not a very likely scenario. In reality, researchers are exploring how to bring back a mammoth from preserved DNA in mammoths that were flash-frozen at high altitudes. These elephant ancestors roamed the grasslands of Siberia from 1.8 million years ago until about 11,000 years ago. Starvation following the last ice age drove their extinction, although a few isolated populations survived until about 3,800 years ago.

Biologists are searching frozen mammoth remains for cell nuclei that could yield DNA. If they find some, researchers may attempt to recreate a mammoth by using a mammoth somatic cell nucleus to direct development in an enucleated elephant oocyte. An alternative strategy is to use mammoth sperm bearing X chromosomes to artificially inseminate an elephant, which may give birth to a female that is half elephant, half mammoth. Some 13 or so years later, she can then be inseminated by more mammoth sperm, producing a baby that is three-quarters mammoth, and so on.

The first successful extraction of bits of ancient DNA occurred in 1990, from a 17-million-year-old magnolia leaf entombed in amber. The quest to probe ancient DNA has sent researchers into the back rooms of museums, dusting off specimens of pressed leaves, pinned insects, and old bones and pelts, in search of nucleic acid clues to past life.

Comparing Chromosomes

Before gene and genome sequencing, researchers recognized that similarities in chromosome banding patterns reflect evolutionary relatedness. Human chromosome

Table 16.2

Percent of Common Chromosome Bands Between Humans and Other Species

Chimpanzees	99$^+$%
Gorillas	99$^+$%
Orangutans	99$^+$%
African green monkeys	95%
Domestic cats	35%
Mice	7%

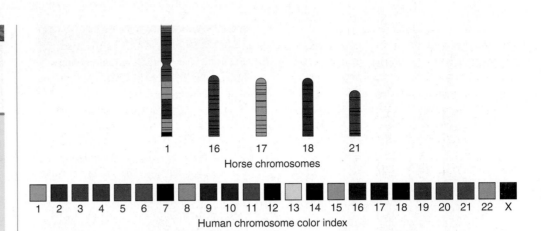

Figure 16.13 Comparing horse and human chromosomes. This color-coded display of the horse genome indicates, at a glance, regions that are highly conserved. Many horse chromosomes are very similar to human chromosomes, but note that horse chromosome 1 is made up of several segments corresponding to different human chromosomes.

banding patterns most closely match those of chimpanzees, then gorillas, and then orangutans (**table 16.2**). The karyotypes of humans, chimpanzees, and apes differ from each other mostly by inversions, which are changes that occur within chromosomes.

If both copies of human chromosome 2 were broken in half, we would have 48 chromosomes, as the three species of apes do, instead of 46. The banding pattern of chromosome 1 in humans, chimps, gorillas, and orangutans matches that of two small chromosomes in the African green monkey, suggesting that this monkey was ancestral to the other primates. Karyotype differences between these three primates and more primitive primates are predominantly translocations between chromosome types.

We can also compare chromosome patterns between species not as closely related. All mammals, for example, have identically banded X chromosomes. One section of human chromosome 1 that is alike in humans, apes, and monkeys is also remarkably similar to parts of chromosomes in cats and mice. A human even shares several chromosomal segments with a horse (**figure 16.13**), but our karyotype is much less like that of the aardvark, the most primitive placental mammal.

Chromosome band pattern similarities, obtained with stains, are not ideal measures of species relatedness because a band can contain many genes that differ from those within a band at a corresponding locus in another species's genome. DNA probes used as part of a FISH analysis are more precise because they mark particular genes. Direct correspondence of known gene order, or **synteny,** between species is better evidence

of close evolutionary relationships. For example, 11 genes are closely linked on the long arm of human chromosome 21, mouse chromosome 16, and on a chromosome called U10 in cows. However, several genes on human chromosome 3 are found near the human chromosome 21 counterpart in mice and cows. Perhaps a mammal ancestral to these three species had all of these genes together, and the genes dispersed to an additional chromosome in humans.

Comparing Proteins

Many different types of organisms use the same proteins, with only slight variations in amino acid sequence. The similarity of protein sequences is compelling evidence for descent from shared ancestors—that is, evolution. The similarities in amino acid sequences in human and chimpanzee proteins are astounding—many proteins are alike in 99 percent of their amino acids. Several proteins are virtually identical. When analyzing a gene's function, researchers routinely consult databases of known genes in many other organisms. Two of the most highly conserved proteins are cytochrome *c* and homeobox proteins.

Cytochrome *c*

One of the most ancient and well-studied proteins is cytochrome *c,* which helps to extract energy from nutrients in the mitochondria for use in the reactions of cellular

respiration. Twenty of 104 amino acids occupy identical positions in the cytochrome *c* of all eukaryotes. The more closely related two species are, the more alike their cytochrome *c* amino acid sequence is (**figure 16.14**). Human cytochrome *c,* for example, differs from horse cytochrome *c* by 12 amino acids, and from kangaroo cytochrome *c* by 8 amino acids. The human protein is identical to chimpanzee cytochrome *c.*

Homeobox Proteins

Another gene that has changed little across evolutionary time is a **homeobox** or HOX gene. It encodes a transcription factor that controls the order in which an embryo turns on genes. This cascade of gene action ultimately ensures that anatomical parts—whether a leg, petal, or segment of a larva—develop in the appropriate places. The highly conserved portion of a homeobox protein is a 60-amino-acid sequence called the homeodomain encoded by a 180-base DNA sequence called the homeobox. (Genes that include homeobox sequences are also termed *homeotic.*) In multicellular species, these genes organize body parts. Humans and most other vertebrates have 39 HOX genes in four clusters labeled A, B, C, and D. The genes have very few introns. Another intriguing aspect of HOX gene clusters is that the individual genes are expressed in a sequence, in developmental time or anatomical position, that mirrors their order on the chromosome.

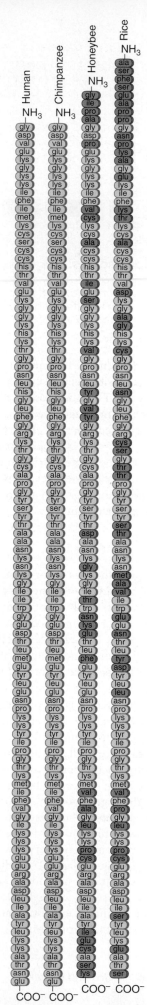

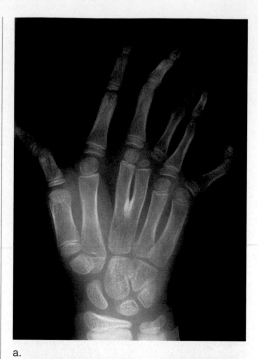

a.

Cytochrome *c* Evolution	
Organism	Number of amino acid differences from humans
Chimpanzee	0
Rhesus monkey	1
Rabbit	9
Cow	10
Pigeon	12
Bullfrog	20
Fruit fly	24
Wheat germ	37
Yeast	42

b.

Figure 16.14   Amino acid sequence similarities reflect evolutionary relatedness.   Similarities in amino acid sequence for the respiratory protein cytochrome *c* in humans and other species parallel the degree of species relatedness. **(a)** These chains show the differences in cytochrome *c* sequence among four species. Amino acids that differ from those in the human sequence are highlighted. **(b)** This chart compares the sequence differences in nine species for this highly conserved protein.

The terms *homeobox* and *homeodomain* derive from the homeotic mutants of the fruit fly *Drosophila melanogaster,* which have mixed-up body parts. *Antennapedia,* for example, has legs in place of its antennae; *proboscipedia* grows legs on its mouth. Geneticists have studied homeotic fruit flies for half a century. Researchers sequenced the fly homeobox gene in 1983 and then found it in frogs, mice, beetles, mosquitoes, slime molds, chickens, roundworms, corn, humans, petunias, and many other species. Because the homeobox protein is a transcription factor (see chapter 10), it controls the activities of other genes.

Mutations in homeobox genes cause human illnesses. In a form of leukemia, a homeobox mutation shifts certain white blood cell progenitors onto the wrong developmental pathway. The misguided cells retain the rapid cell division characteristic of progenitor cells, causing the cancer. DiGeorge syndrome (OMIM 188400) is also caused by a homeobox mutation. Although affected

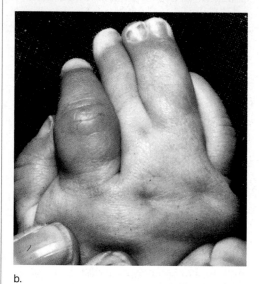

b.

Figure 16.15   A human *HOX* gene mutation causes synpolydactyly. Mutation in the *HOXD 13* gene disrupts development of fingers and toes, causing a very distinctive X-ray image **(a)** and phenotype **(b)**. The third and fourth fingers are partially fused with an extra digit within the webbed material.

individuals hardly sprout legs from their heads, as do *Antennapedia* flies, the missing thymus and parathyroid glands and abnormal ears, nose, mouth, and throat correspond to the sites of abnormalities in the flies. **Figure 16.15** shows another human disorder caused by a mutation in a HOX gene, synpolydactyly (OMIM 186000).

Experiments that transfer genes of one species into cells of another reveal how highly conserved the homeobox is, implying it is essential and ancient. If a mouse version of the *Antennapedia* gene is placed into the fertilized egg of a normal fly, the adult fly grows legs on its head, expressing the mouse gene as if it were the fly counterpart. The human version of the gene, placed into a mouse's fertilized egg, disrupts the adult mouse's head development. Homeotic genes and the proteins they encode, therefore, provide instructions for development of organisms whose bodies have many parts.

16.3 Molecular Clocks

A clock measures the passage of time by moving its hands through a certain degree of a circle in a specific and constant interval of time—a second, a minute, or an hour. Similarly, an informational molecule can be used as a "molecular clock" if its building blocks are replaced at a known and constant rate.

The similarity of nuclear DNA sequences in different species can help scientists estimate the time when the organisms diverged from a common ancestor, if they know the rate of base substitution mutation. For example, many nuclear genes studied in humans and chimpanzees differ in 5 percent of their bases, and substitutions occur at a rate of 1 percent per 1 million years. Therefore, 5 million years have presumably passed since the two species diverged. Mitochondrial DNA (mtDNA) sequences may also be tracked in molecular clock studies, as we will soon see.

Time scales based on fossil evidence and molecular clocks can be superimposed on evolutionary tree diagrams constructed from DNA or protein sequence data. However, evolutionary trees can become complex when a single set of data can be arranged into a large number of different tree configurations. A tree for 17 mammalian species, for example, can be constructed in 10,395 different ways! The sequence in which the data are entered into tree-building computer programs influences the tree's shape, which is vital to interpreting species relationships. With new sequence information, the tree possibilities change.

Parsimony analysis is a statistical method used to identify an evolutionary tree likely to represent what really happened. A computer connects all evolutionary tree sequence data using the fewest possible number of mutational events to account for observed DNA base sequence differences. For the 5-base sequence in **figure 16.16,** for example, the data can be arranged into two possible tree diagrams. Because mutations are rare events, the tree that requires the fewest mutations is more likely to reflect reality.

Neanderthals Revisited

Molecular clock data can provide clues to relationships among modern organisms and also fill the gaps in the fossil record. Consider our knowledge of Neanderthals.

Two molecular technologies—analyzing ancient DNA and using mtDNA clocks—indicate that Neanderthals were a side branch on our family tree and diverged from us more than half a million years ago. This is much farther back than the 300,000 years ago that the fossil evidence indicates.

In 1997, a graduate student, Matthias Krings, ground up a bit of arm bone from the original French Neanderthal skeleton. He then amplified several 100-base-pair-long pieces of mtDNA that do not encode protein and mutate very rapidly, perhaps because they do not affect the phenotype

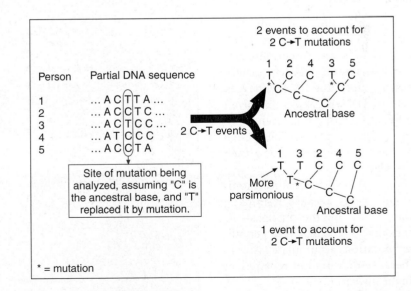

Figure 16.16 Parsimony analysis. Even a computer has trouble arranging DNA differences into an evolutionary tree showing species, population, or individual relationships. A parsimonious tree accounts for all data with the fewest number of mutations. Here, the two individuals who have a T in place of the ancestral C could have arisen in two mutational events or one, assuming that these individuals had a common ancestor. Since mutations are rare events, the more realistic scenario is one mutation.

and are therefore not under selective pressure. The DNA pieces were sequenced and compared to corresponding sequences from 986 modern *Homo sapiens*. The Neanderthal DNA differed at 26 positions. Not only is this three times the number of differences seen between pairs of the most unrelated modern humans, but the locations of the base differences were completely different from the places where modern genes vary (SNPs). Analysis of other Neanderthal bones supports the 1997 results. This genetic distinction suggests that it is highly unlikely that Neanderthals and modern humans ever interbred. In addition, sequences of mtDNA in the rib and leg bones of two Cro-Magnons from a cave in southern Italy, dating back 25,000 years, are unlike Neanderthal DNA, but very similar to DNA from modern humans. Extrapolations from mutation rates of genes in modern humans and chimps indicate the last shared ancestor between Neanderthals and humans lived from 690,000 to 550,000 years ago. Said paleontologist Tim White, "The Neanderthal vanished, leaving no trace of their genes in the modern human gene pool." This is what figure 16.8 indicates.

Tracking the Sexes: mtDNA and the Y Chromosome

To track the migrations of humanity, researchers consult two types of DNA: mitochondrial DNA and Y chromosome sequences. These sources of genetic material represent female and male lineages, respectively. MtDNA is passed exclusively from mothers to all offspring; Y chromosomes are unique to males. These DNA sequences have advantages beyond their gender specificity.

Mitochondrial DNA is ideal for monitoring recent events because it mutates faster than DNA in the nucleus—its sequences change by 2 to 3 percent per million years. Mutations accumulate faster in mtDNA because they are not repaired, as they are in the nucleus.

An advantage of using the Y chromosome is that much of it does not recombine. Crossing over, which it could only do with an X chromosome because there is no second Y, would break linkage from the past

generation and therefore make it impossible to trace relationships. As with comparisons of other DNA sequences, the logic behind mtDNA and Y chromosome studies is that the more alike the haplotypes (closely linked variants) between two individuals or groups, the more recently they shared an ancestor. Tree diagrams depict such data.

Information revealed in mtDNA and Y chromosome studies often mirrors historical records. For example, the Hazara people in Afghanistan have Y chromosome haplotypes that indicate that they are direct descendants of Genghis Khan. Their mtDNA haplotypes are also seen in people from eastern Eurasia. The interpretation, consistent with history, is that Genghis Khan's army picked up women from east Asia, whose DNA remained in the gene pool of today's Hazara. **Figure 16.17** is a map of parts of the world relevant to several sections of this chapter.

Genetic analysis of the Parsis, who live today in southeastern Pakistan, also supports history. They migrated originally from Iran in the 7th century A.D., stopped in

northwestern India, and two centuries later migrated to Pakistan. The Y chromosome haplotypes match those of groups in present-day Iran, rather than among neighbors in Pakistan, supporting the male origin in Iran. However, the mtDNA haplotypes match those in India. If the original Iranians included women, they didn't make it into the modern Parsis gene pool. More likely, men from Iran found female partners in India.

Out of Africa—The First Time

Theoretically, if a particular sequence of mtDNA could have mutated to yield the mtDNA sequences in modern humans, then that ancestral sequence may represent a very early human or humanlike female—a mitochondrial "Eve," or metaphorical first woman. **Figure 16.18** shows how one maternal line may have come to persist.

When might this theoretical first woman, the most recent female ancestor to us all, have lived? In the mid 1980s, graduate student Rebecca Cann at the University of Cali-

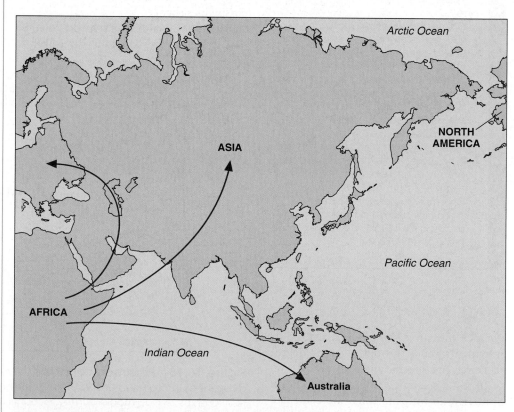

Figure 16.17   A map and DNA analysis can help trace human expansion patterns.

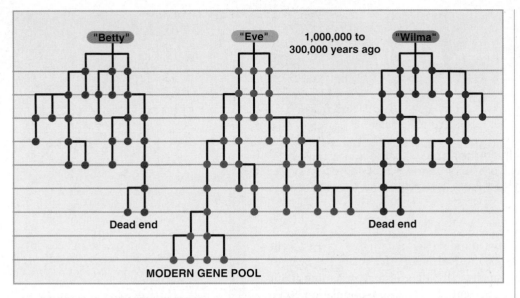

Figure 16.18 Mitochondrial Eve. According to the mitochondrial Eve hypothesis, modern mtDNA retains some sequences from a figurative first woman, "Eve," who lived in Africa 300,000 to 100,000 years ago. In this schematic illustration, the lines represent generations, and the circles, females. Lineages cease whenever a woman does not have a daughter to pass on the mtDNA.

fornia, Berkeley, with the late Allan Wilson and Mark Stoneking compared mtDNA sequences for protein encoding as well as noncoding DNA regions in a variety of people, including Africans, African Americans, Europeans, New Guineans, Australians, and others. They concluded from several methods that the hypothesized ancestral woman lived about 200,000 years ago, in Africa. In 2003, another research group analyzed mtDNA from 600 living East Africans, and arrived at 170,000 years ago for the beginning of the modern human line, which is remarkably close to the date of the *H. sapiens idaltu* fossils. The locations of fossil evidence, such as *H. sapiens idaltu* skulls, support an African origin, and Charles Darwin suggested it, too.

One way to reach this time estimate is by comparing how much the mtDNA sequence differs among modern humans to how much it differs between humans and chimpanzees. The differences in mtDNA sequences among contemporary humans amount to 1/25 the difference between humans and chimpanzees. The two species diverged about 5 million years ago, according to extrapolation from fossil and molecular evidence. Multiplying 1/25 by 5 million gives a value of 200,000 years ago, assuming that the mtDNA mutation rate is constant over time.

Where did Eve live? Mitochondrial DNA comparisons consistently find that African people have the most numerous and diverse mutations. The same is true for other regions of the genome. Therefore Africans have existed longer than other modern peoples, because it takes time for mutations to accumulate. In many evolutionary trees constructed by parsimony analysis, the individuals whose DNA sequences form the bases are from Africa. That is, gene variants in other modern human populations are subsets of an ancestral African genome.

The idea of mitochondrial Eve is part of the "out of Africa" view, or **replacement hypothesis** of human origins. It states that about 200,000 years ago, *H. sapiens* evolved from an *H. erectus* population in Africa. This may have occurred quickly, in small, isolated pockets, or gradually across a broader swath of the continent. However it happened, eventually descendants of these early *H. sapiens* expanded.

An alternate view, largely disproven, is the **multiregional hypothesis,** which maintains that human traits originated in several places, and *H. erectus* expanded well beyond Africa, mixing and sharing genes, gradually evolving into *H. sapiens* on a global scale, without isolated pockets of peoples. (Anthropologists prefer the word "expand" to "migrate" from Africa, because although people left, many also remained, and evolution continued in Africa.) The "out of Africa" and multiregionalism hypotheses were once so contentious that scientists fought over them at meetings and in print.

RELATING THE CONCEPTS

Does the existence of the Flores people argue for the replacement hypothesis or the multiregional hypothesis?

The African Slave Trade

Mitochondrial DNA has been used to track a much more recent out-of-Africa event: the slave trade. From the 15th through the 19th centuries, approximately 13 million Africans were captured and sent to the Americas, as slaves. Two million died en route, as did many others in the hard years to follow. Historical records show that about two-thirds of the slaves came from western Africa, about a third from west-central Africa, and most of the remainder from southeastern Africa.

European researchers compared haplotypes spanning a several hundred base sequence of mtDNA among 481 people of recent African ancestry in the Americas to DNA from 2,374 people from various parts of Africa. **Figure 16.19** clearly shows that the contributions from western and west-central Africa are echoed in today's African-Americans.

Although the mtDNA evidence is consistent with historical records, the researchers caution that it is probably not possible to trace any particular individual or family back to a specific place in Africa. Reasons range from the practical to the peculiarities of African populations:

- mtDNA samples from people in many parts of Africa today aren't available.

- Some of the haplotypes have been widespread about the African continent since prehistoric times, making them useless for tracing specific geographic areas.

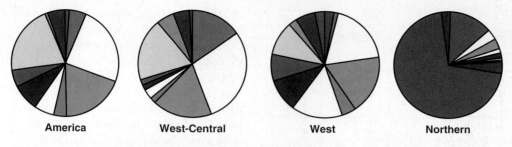

| America | West-Central | West | Northern |

Figure 16.19 The sources of slaves in the Americas. Comparison of mtDNA sequences in several modern populations supports historical records of slaves coming predominantly from west and west-central Africa. Each color represents a distinct set of mtDNA sequences (haplotypes).

- During the time of slave capture, there was a lot of migration in Africa, mixing up the haplotypes that might have defined particular groups.

Native American Origins

Identifying the origins of Native Americans is challenging because data from linguistics, archaeology, and genetics do not always agree. A study from 1987, for example, established three waves of migration across the Bering Strait land bridge that formed between Siberia and Alaska during low glacial periods. The basis was a comparison of many Native American languages, and subsequent grouping of the people into the "Eskimo-Aleut," who arrived 7,000 to 5,000 years ago, the "Na-Dene," who came 15,000 to 12,000 years ago, and the "Amerinds," who arrived about 33,000 years ago. But only a few sounds had been used to distinguish the languages, lumping several as Amerind that may actually have been distinct in their origins. In other words, the premise for grouping the people was faulty.

Since then, several genetic studies have countered the "three migration hypothesis," but even the genetic studies do not completely agree. One problem is that studying different parts of the human genome can yield different results, because DNA sequences change at different rates. Y chromosome analyses support one migration bringing in diverse genes—but from the Mongolian/Chinese border, not Siberia. The data suggest a continuous migration from 37,000 to 23,000 years ago. A study utilizing mtDNA also finds a single, long migration, from Mongolia, about 21,000 years ago. Yet another study of nine genes concurs with archaeological evidence that

the founding population that traveled from Asia to North America may have been as small as 80 individuals, representing about 1 percent of an ancestral population. But a trek from Mongolia was probably not the sole source of people in the Americas, according to mtDNA.

The mtDNA of modern Native Americans includes five major haplotypes, A through D and X. A through D account for 97 percent of Native Americans, who are of Asian origin. In contrast, haplotype X is not seen in Siberian people, but is found among certain European groups. It isn't clear whether the European contribution to the Native American gene pool arrived by crossing the Atlantic, or from migrating from the other direction across Asia, over the Bering strait.

Archaeological evidence suggests other scenarios for the peopling of the Americas. For example, a type of stone tool called a Clovis point is known in many American settlements, dating from 13,500 to 12,500 years ago. These tools are remarkably similar to tools from southwest France and northern Spain from 19,100 to 24,000 years ago, yet unlike stone tools in museums in Siberia, leading some anthropologists to envision expansion to the New World via the Atlantic ocean. This hypothesis might explain similar findings of European-style tools along the east coast of the United States, such as in Virginia and South Carolina. However, the similarity in tool shape could be a coincidence. Another hypothesis, based on archaeological evidence in a settlement from 14,500 years ago in Monte Verde, in southern Chile, is that people came along the west coast by boat, living off of the abundant marine mammals. They could have completed the long journey over several generations.

RELATING THE CONCEPTS

How might you use mtDNA or Y chromosome haplotypes on fossils of the Flores people to trace their origins?

Key Concepts

1. Molecular clocks apply mutation rates to time scales to estimate when two individuals or types of organisms most recently shared ancestors.
2. Different genes evolve at different rates. Parsimony analysis selects likely evolutionary trees from DNA data.
3. Mitochondrial DNA clocks trace maternal lineages, and Y chromosome sequences trace paternal lineages.
4. Molecular clocks have been used to examine the relationship of Neanderthals to modern humans and the origin and migrations of modern humans.

16.4 Eugenics

Fossil evidence, ancient DNA, and molecular clocks are useful in studying our past. We can control the future, to an extent, through reproductive choices that affect the gene pool. Some people try to control the genes in their offspring by seeking mates with high intelligence or certain physical characteristics. This idea was taken to an absurd extreme in a sperm bank in California where the donors are all Nobel Prize winners, and in a website advertising eggs donated by young women with high scores on college entrance exams. The ability to control reproductive choices raises many bioethical issues.

Eugenics is the control of individual human reproductive choices to achieve a societal goal. Sir Francis Galton coined the term, meaning "good in birth," in 1883 (**table 16.3**). He defined eugenics as "the science of improvement of the human race germplasm through better breeding." The 2,500-year-old caste system in India and the antimiscegenation laws in the United States that banned marriage between people of different races from 1930 to 1967 were clearly eugenic because they sought to control reproduction to change society.

Table 16.3

A Chronology of Eugenics

1883	Sir Francis Galton coins the term *eugenics.*
1889	Sir Francis Galton's writings are published in the book *Natural Inheritance.*
1896	Connecticut enacts law forbidding sex with a person who has epilepsy or is "feebleminded" or an "imbecile."
1904	Galton establishes the Eugenics Record Office at the University of London to keep family records.
1907	First eugenic law in the United States orders sterilization of institutionalized mentally retarded males and criminal males when experts recommend it.
1910	Eugenics Record Office founded in Cold Spring Harbor, New York, to collect family and institutional data.
1924	Immigration Act limits entry into the United States of "idiots, imbeciles, feebleminded, epileptics, insane persons," and restricts immigration to 7 percent of the U.S. population from a particular country according to the 1890 census—keeping out those from southern and eastern Europe.
1927	Supreme Court (*Buck vs. Bell*) upholds compulsory sterilization of the mentally retarded by a vote of 8 to 1, leading to many state laws.
1934	Eugenic sterilization law of Nazi Germany orders sterilization of individuals with conditions thought to be inherited, including epilepsy, schizophrenia, and blindness, depending upon rulings in Genetic Health Courts.
1939	Nazis begin killing 5,000 children with birth defects or mental retardation, then 70,000 "unfit" adults.
1956	U.S. state eugenic sterilization laws are repealed, but 58,000 people have already been sterilized.
1965	U.S. immigration laws reformed, lifting many restrictions.
1980s	California's Center for Germinal Choice is established, where Nobel Prize winners can deposit sperm to inseminate carefully chosen women.
1990s	Laws passed to prevent health insurance or employment discrimination based on genotype.
2000	Human genome sequenced.
2003	Many governments recommend certain genetic tests, and have legislation to prevent genetic discrimination. In the U.S., protective legislation is still in discussion.
2004	Genocide of black Africans occurs in Sudan.

Galton's ideas were popular for a time. Eugenics societies formed in several nations and attempted to practice his ideas in various ways. Creating incentives for reproduction among those considered superior constitutes positive eugenics. Interfering with reproduction among those judged inferior is negative eugenics.

One vocal supporter of the eugenics movement was Sir Ronald Aylmer Fisher. In 1930, he published a book, *The Genetical Theory of Natural Selection,* which connected the concepts of Charles Darwin and Gregor Mendel and listed the basic tenets of population genetics. Natural selection and Mendelian inheritance provided a framework for eugenics. The final five chapters of Fisher's otherwise highly regarded work tried to apply the principles of population genetics to human society. Fisher maintained that those at the top of a society tend to be "genetically infertile," producing fewer children than the less affluent classes. This, he claimed, was the reason why civilizations ultimately topple. He offered several practical suggestions to remedy this, including state monetary gifts to high-income families for each child born to them.

Early in the twentieth century, eugenics focused on maintaining purity. One prominent geneticist, Luther Burbank, realized the value of genetic diversity at the beginning of a eugenic effort. Known for selecting interesting plants and crossing them to breed plants with useful characteristics, such as less prickly cacti and a small-pitted plum, Burbank in 1906 applied his agricultural ideas to people. In a book called *The Training of the Human Plant,* he encouraged immigration to the United States so that advantageous combinations of traits would appear as the new Americans interbred. Burbank's plan ran into problems, however, at the selection stage, which allowed only those with "desirable" trait combinations to reproduce.

On the East Coast of the United States, Charles Davenport led the eugenics movement. In 1910, he established the Eugenics Record Office at Cold Spring Harbor, New York. There he headed a massive effort to compile data from institutions, prisons, circuses, and general society. In the rather simplistic view of genetics at the time, he attributed nearly every trait to a single gene. "Feeblemindedness," he thought, was inherited as an autosomal recessive trait. This was a catch-all phrase for a person with low intelligence (as measured on an IQ test) and such behavioral abnormalities as "criminality," "promiscuity," and "social dependency." In one famous case, a young woman named Carrie Buck was ordered to be sterilized when she, her mother, and her illegitimate infant daughter Vivian were declared feebleminded. Carrie had been raped by a relative of her foster parents, and was actually an average student. **Figure 16.20** shows the pedigree for Carrie Buck and her "inherited trait" of feeblemindedness.

Other nations practiced eugenics. From 1934 until 1976, the Swedish government forced certain individuals to be sterilized as part of a "scientific and modern way of changing society for the better," according

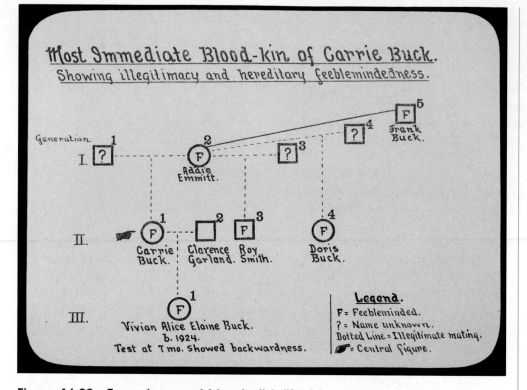

Figure 16.20 Eugenics sought to abolish "feeblemindedness." In 1927, 17-year-old Carrie Buck, of Charlottesville, stood trial for the crime of having a mother who lived in an asylum for the feebleminded, and having a daughter out-of-wedlock (following rape) also deemed feebleminded, as was Carrie herself, though she was a B student in school. Ruled Sir Oliver Wendell Holmes Jr., "three generations of imbeciles are enough." Carrie went down in history as the first person to be sterilized to prevent having another "socially inadequate offspring."

to one historian. At first, only mentally ill people were sterilized, but poor, single mothers were later included. Revelation of the Nazi atrocities did not halt eugenics in Sweden, but the women's movement in the 1970s pushed for an end to forced sterilizations.

In 1994, China passed the Maternal and Infant Health Care Law, which proposes "ensuring the quality of the newborn population" and forbids procreation between two people if physical exams show "genetic disease of a serious nature . . . that may totally or partially deprive the victim of the ability to live independently, that [is] highly possible to recur in generations to come, and that [is] medically considered inappropriate for reproduction." Such "genetic diseases" include mental retardation, mental illness, and seizures, conditions that are ill-defined in the law and are not necessarily inherited.

The Dor Yeshorim program in the orthodox Jewish community in New York City, described in chapter 15, may be considered eugenics. Founded in the early 1980s by a rabbi who had lost four children to Tay-Sachs disease, the program confidentially identifies carriers of genetic diseases in Ashkenazi populations. People use the information to prevent reproduction between certain individuals, but the goal is to prevent suffering, not to improve society.

Population genetic "biobanks," described in chapter 1, acquire genetic information on their citizens. These projects are not eugenic, however, because the information is used to improve health, not to make reproductive decisions. Eugenics, in contrast, uses such information to maximize the genetic contribution from those deemed desirable and minimize the contribution from those con-

sidered unacceptable. A major fallacy of eugenics is its subjectivity. Who decides which traits are desirable or superior?

Because genetic technologies may affect reproductive choices and can influence which alleles are passed to the next generation, modern genetics has sometimes been compared to eugenics. Medical genetics and eugenics differ in their overall goals. Eugenics aims to skew allele frequencies in future generations by allowing only people with certain "valuable" genotypes to reproduce, for the supposed benefit of the population as a whole. The goal of medical genetics, in contrast, is usually to skew allele frequencies in order to prevent suffering on a family level. Bioethics: Choices for the Future on page 325 presents one young man's view of prenatal testing.

One particularly frightening aspect of the eugenics movement early in the twentieth century was the vague nature of the traits considered hereditary and undesirable, such as "feeblemindedness." Now, as the human genome is analyzed, will eugenics resurge? Will we use new genetic information to choose the traits of the next generation? Many people fear that tests to identify carriers or disease susceptibility will be used eugenically, particularly in nations where the government does not underwrite health care. Laws are being implemented in many nations to prevent genetic discrimination. Let's hope that in addition to deciphering our genetic blueprints, we also learn how to apply that information wisely. Unlike other species, we have the ability to affect our own evolution.

Key Concepts

1. Eugenics is the control of individual human reproduction for societal goals, maximizing the genetic contribution of those deemed acceptable (positive eugenics) and minimizing the contribution from those considered unacceptable (negative eugenics).
2. Some people consider modern genetic screening practices eugenic, but genetic testing usually aims to prevent or alleviate human suffering.

Bioethics: Choices for the Future

Two Views of Neural Tube Defects

Genetic technologies permit people to make reproductive choices that can alter allele frequencies in populations. Identifying carriers of a recessive illness, who then may decide not to have children together, is one way to remove some disease-causing alleles from a population, by decreasing the number of homozygous recessive individuals. Screening pregnant women for fetal anomalies, then terminating affected pregnancies, also alters disease prevalence and, if the disorder has a genetic component, allele frequencies. This is the case for neural tube defects (NTDs), which are multifactorial.

An NTD forms at the end of the first month, when the embryo's neural tube does not completely close. If the opening is in the head, the condition is called anencephaly, and usually ends in miscarriage, stillbirth, or a newborn who dies within days. If the opening is in the spinal cord, the condition is called spina bifida. Usually the individual is paralyzed from the point of the lesion down, but can live into adulthood and have normal intelligence. Sometimes surgery can improve functioning in people with mild cases of spina bifida. People with spina bifida often also have hydrocephalus, or "water on the brain."

In 1992, the Centers for Disease Control and Prevention summarized studies indicating that taking the vitamin folic acid in pregnancy lowers the risk of NTD recurrence by 50 percent, from 3 to 4 percent to 1.5 to 2 percent. Women who had had an affected child began taking large doses of the vitamin in the months before conception. But when epidemiologists tried to monitor how well folic acid supplementation was working, they faced a problem—the prevalence values of NTDs were greatly underestimated. This happened because the statistics on NTD prevalence—vital to discovering whether folic acid was actually preventing the defect—included only newborns, stillborns, and older fetuses. Most reports did not account for pregnancies terminated following a prenatal diagnosis of an NTD. These pregnancies caused the underreporting of anencephaly by 60 to 70 percent, and of spina bifida, by 20 to 30 percent in some states. NTD screening and subsequent termination of affected pregnancies alters the allele frequencies by preventing causative genes from passing to new generations.

A Personal View

Blaine Deatherage-Newsom has a different view of population screening for neural tube defects because he has one (**figure 1**). Blaine was born in 1979 with spina bifida. Paralyzed from the armpits down, he has endured much physical pain, but he has also achieved a great deal. He put the question, "If we had the technology to eliminate disabilities from the population, would that be good public policy?" on the Internet—initiating a global discussion. His view on NTD screening is one we do not often hear:

> I was born with spina bifida and hydrocephalus. I hear that when parents have a test and find out that their unborn child has spina bifida, in more than 95 percent of the cases they choose to have an abortion. I also went to an exhibit at the Oregon Museum of Science and Industry several years ago where the exhibit described a child born with spina bifida and hydrocephalus, and . . . asked people to vote on whether the child should live or die. I voted that the child should live, but when I voted, the child was losing by quite a few votes.
>
> When these things happen, I get worried. I wonder if people are saying that they think the world would be a better place without me. I wonder if people just think the lives of people with disabilities are so full of misery and suffering that they think we would be better off dead. It's true that my life has suffering (especially when I'm having one of my 11 surgeries so far), but most of the time I am very happy and I like my life very much. My mom says she can't imagine the world without me, and she is convinced that everyone who has a chance to know me thinks that the world is a far better place because I'm in it.

Is eliminating disabilities good public policy? It depends on your point of view.

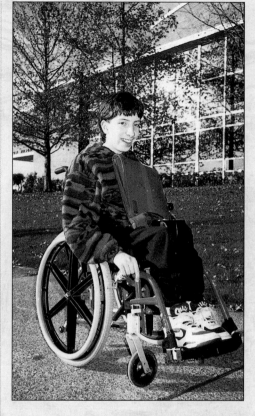

Figure 1 Blaine Deatherage-Newsom.

Excerpt by Blaine Deatherage-Newsom, "If we could eliminate disabilities from the population, should we? Results of a survey on the Internet." Reprinted by permission.

Summary

16.1 Human Origins

1. The first primates were rodentlike insectivores that lived about 60 million years ago. By 30 to 40 million years ago, monkeylike *Aegyptopithecus* lived. **Hominoids,** ancestral to apes and humans, lived 22 to 32 million years ago. They include *Dryopithecus* and other primates who began to walk upright.

2. **Hominids,** ancestral to humans only, appeared about 19 million years ago. These animals were more upright, dwelled on the plains, and had smaller brains than their forebears.

3. At least three types of hominids lived about 6 million years ago, shortly after the split from the chimp lineage.

4. The *Australopithecines* preceded and then coexisted with *Homo habilis,* who lived in caves, had strong family units, and used tools extensively. *Homo erectus* was a contemporary who outsurvived *H. habilis,* lived in societies, and used fire. *Homo sapiens idaltu* lived about 156,000 years ago, and looked like us.

5. Early *Homo sapiens* also included the Neanderthals and Cro-Magnons, although Neanderthals were a dead end. Modern humans appeared about 40,000 years ago, and culture was apparent by 14,000 years ago.

16.2 Molecular Evolution

6. Molecular evolution considers differences at the genome, chromosome, protein, or DNA sequence levels with mutation rates to estimate species relatedness. Animal models are possible because of similarities in DNA sequence among species.

7. Targeted comparative sequencing aligns corresponding DNA sequences in different species.

8. Humans and chimps share 98.7 percent of their protein-encoding gene sequences. Indels, introns, and repeats create genome differences between humans and chimps.

9. Single genes and differences in gene expression can account for great distinctions between chimps and humans.

10. The human genome shows many signs of past duplication.

11. Amplifying ancient DNA is difficult because contamination may occur.

12. Closely related species have similar chromosome banding patterns. Genes in the same order on chromosomes in different species are **syntenic.**

13. Cytochrome *c* and homeobox proteins are highly conserved.

16.3 Molecular Clocks

14. Gene sequence information from several species may be used to construct evolutionary tree diagrams, and a molecular clock based on the known mutation rate of the gene may then be applied. Different genes mutate at different rates. Molecular trees indicate when species diverged from shared ancestors.

15. Parsimony analysis selects the evolutionary trees requiring the fewest mutations, which are therefore the most likely.

16. Molecular clocks based on mtDNA date recent events through the maternal line because this DNA mutates faster than nuclear DNA. Y chromosome genes trace paternal lineage. Both types of evidence are used to study human origins and expansions.

16.4 Eugenics

17. **Eugenics** is the control of individual reproduction to serve a societal goal.

18. Positive eugenics encourages those deemed acceptable or superior to reproduce. Negative eugenics restricts reproduction of those considered inferior. Eugenics extends the concept of natural selection and Mendel's laws but does not translate well into practice.

19. Some aspects of genetic technology also affect reproductive choices and allele frequencies, but the goal is to alleviate or prevent suffering, not to change society.

Review Questions

1. What is the difference between a hominoid and a hominid?

2. Some anthropologists classify chimpanzees along with humans in genus *Homo.* How does this conflict with fossil evidence of the *Australopithecus* species?

3. Give an example of how a single gene difference can have a profound effect on the phenotypes of two species.

4. What is the evidence that *Australopithecus garhi* may have been a direct forebear of *Homo?*

5. Give an example of molecular evidence that is consistent with fossil or other evidence, and an example of molecular evidence that conflicts with other information.

6. How does the information provided by Y chromosome and mitochondrial DNA sequences differ from the information obtained from nuclear DNA sequences?

7. List three aspects of development, anatomy, or physiology that were important in human evolution.

8. Why are exons highly conserved, but introns are not?

9. Explain how indels could cause the divergence of our genome sequence from that of chimpanzees, yet not contribute to observable differences between the two species.

10. Protein-encoding genes have different mutation rates. How might this complicate the interpretation of targeted comparative sequencing experiments?

11. Cite two ways that humans and chimps can differ greatly at the genetic level, but still be very alike in terms of DNA sequence.

12. Why does comparing gene sequences offer more information for molecular evolution studies than comparing protein sequences?

13. Why can comparing the sequences of different genes or proteins lead to different conclusions about when two groups diverged from a common ancestor?

14. Why is comparing the DNA sequence of one gene a less accurate estimate of the evolutionary relationship between two

species than a DNA hybridization experiment that compares large portions of the two genomes?

15. Cite a limitation of comparing chromosome banding patterns to estimate species' relationships.

16. What types of information are needed to construct an evolutionary tree diagram? What assumptions are necessary? What are the limitations of these diagrams?

17. How can the human and chimp genomes be 99 percent alike in DNA sequence, yet still be different?

18. Give an example of how a single gene change can influence evolution.

19. Cite three examples of eugenic actions or policies.

20. Distinguish between positive and negative selection, and between positive and negative eugenics. How do selection and eugenics differ?

21. Science is based on hypotheses, which are explanations for observations that can be disproven as evidence accumulates. Choose an hypothesis mentioned in this chapter and discuss how evidence disproved it, sending scientists to revise their thinking.

22. How does the discovery of tools dating from 50,000 years ago in South Carolina refute or support other ideas about the peopling of the Americas?

Applied Questions

1. A geneticist aboard a federation starship is given the task of determining how closely related Humans, Klingons, Romulans, and Betazoids are. Each organism walks on two legs, lives in complex societies, uses tools and technologies, looks similar, and reproduces in the same manner. Each can interbreed with any of the others. The geneticist finds the following data:

   • Klingons and Romulans each have 44 chromosomes. Humans and Betazoids have 46 chromosomes. Human chromosomes 15 and 17 resemble part of the same large chromosome in Klingons and Romulans.

   • Humans and Klingons have 97 percent of their chromosome bands in common. Humans and Romulans have 98 percent of their chromosome bands in common, and Humans and Betazoids show 100 percent correspondence. Humans and Betazoids differ only by an extra segment on chromosome 11, which appears to be a duplication.

   • The cytochrome *c* amino acid sequence is identical in Humans and Betazoids, differs by one amino acid between Humans and Romulans, and differs by two amino acids between Humans and Klingons.

   • The gene for collagen contains 50 introns in Humans, 50 introns in Betazoids, 62 introns in Romulans, and 74 introns in Klingons.

   • Mitochondrial DNA analysis reveals many more individual differences between Klingons and Romulans than between Humans and Betazoids.

   a. Hypothesize the chromosomal aberrations that might explain the karyotypic differences among these four types of organisms.

   b. Which are our closest relatives among the Klingons, Romulans, and Betazoids? What is the evidence for this?

   c. Are Klingons, Romulans, Humans, and Betazoids distinct species? What information reveals this?

   d. Which of the evolutionary tree diagrams is consistent with the data?

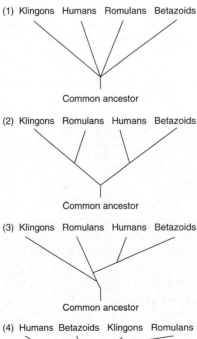

(1) Klingons Humans Romulans Betazoids

Common ancestor

(2) Klingons Romulans Humans Betazoids

Common ancestor

(3) Klingons Romulans Humans Betazoids

Common ancestor

(4) Humans Betazoids Klingons Romulans

Common ancestor

2. Give three examples of negative eugenic measures and three examples of positive eugenic measures.

3. A molecular anthropologist who is studying diabetes in native Americans feels that he can obtain information on why certain groups are prone to the disorder by analyzing genetic variants in small, isolated populations around the world. Do you think that the goal of understanding disease and alleviating suffering in one group of people justifies obtaining and studying the DNA of other people who have had little contact with cultures outside their own? Can you suggest a compromise intervention?

4. In 1997, law schools in two states reversed their affirmative action policies and began evaluating all applicants on an equal basis—that is, applying the same admittance requirements to all. In fall 1997, classrooms of new law students had few, if any, nonwhite faces. How was this action eugenic, and how was it not?

5. Several women have offered to be artificially inseminated with sperm from the Ice Man, the human who died 5,300 years ago and was found in the Alps. However, he had been castrated. If sperm could have been recovered, and a woman inseminated, what do you think the child would be like?

6. The Human Genome Diversity Project is sampling DNA from white blood cells collected from 1,000 people that represent 40 to 50 populations in order to learn about their differences. Suggest a plan for utilizing this information yet avoid eugenics.

7. Locate a disease-causing gene on a human chromosome, and consult figure 16.13 to

determine where it may have a counterpart in the horse genome.

8. Researchers select organisms to help them study versions of human diseases based on ease of cultivation. Mice, rats, fruit flies, and roundworms are examples of such "model" organisms. Do you think that when researchers choose model organisms, they should consider genome similarities? Cite a reason for your answer.

Web Activities

Visit the Online Learning Center (OLC) at www.mhhe.com/lewisgenetics7. Select **Student Edition, chapter 16,** and **Web Activities** to find the website links needed to complete the following activities.

9. Go to the website for the National Institutes of Health Intramural Sequencing Center. Under Scientific Projects, click Comparative Vertebrate Sequencing (first choice). Scroll down. Locate one where the red bars align for humans and chimps and/or another species. Read "gene name" at the bottom of the screen, and then look up this name on OMIM. Describe a human trait or condition that this gene confers.

10. Go to the Image Archive on the American Eugenics Movement website. Look at several images, and either find one that presents a genetic disorder and describe it, or find an image that presents biologically incorrect information, and explain the error.

11. Several websites promise to consult your Y chromosome or mitochondrial DNA and tell you where you came from. Some researchers have cautioned that these companies make false promises. Look at some of these websites, and identify limitations to the technologies used.

Case Studies and Research Results

12. In The Netherlands, the proposed Groningen Protocol presents guidelines for euthanasia of newborns who suffer from untreatable pain, or extreme deformities, or who require continual life support. The guidelines specifically mention disorders with genetic components, such as severe spina bifida and epidermolysis bullosa, which blisters the skin (see figure 2.12).

   Do you agree or disagree with the following justifications for newborn euthanasia? Cite reasons for your answers.

   a. Approving guidelines is a formality—hastening the death of a severely ill newborn is already practiced, and not only in The Netherlands.

   b. Having a government-sanctioned protocol will prevent abuse.

   c. The intent is to relieve suffering, and the procedure requires parental consent.

   d. Only 10 or fewer newborns will likely be euthanized in the country each year, out of a total population of 16 million.

   e. The euthanized infants would never have contributed genes to the next generation.

   f. The practice will be restricted to what the government considers hopeless cases.

13. Consider the following brain sizes:

Animal	Brain volume (cubic centimeters)
Homo sapiens sapiens (us)	1350
Homo sapiens idaltu	1450
Homo erectus	1000
Australopithecus	380
Chimpanzee	380
Homo floresiensis	380

   a. Explain how these data either support or refute the hypothesis that increasing brain size correlates to increased intelligence.

   b. Explain a limitation of the ways in which we learn about our ancestors or relatives that might account for confusing results when we compare traits.

14. For more than 20 million years, lice have lived on the skins of primates. Researchers compared a 1,525-base-pair sequence of mtDNA among modern varieties of lice, and, applying mutation rate, derived the following evolutionary tree. It depicts a split in the louse lineage, with one group of head and body lice living throughout the world, and another group of only head lice living in the Americas.

   a. What events in human evolution roughly correspond to the branch points in the louse evolutionary tree?

   b. What might be the significance of the similarity between the evolutionary trees for lice and humans?

   c. The researchers interpreted their findings to possibly indicate that lice were transferred from archaic humans to modern humans. What is the evidence for this hypothesis? What other types of evidence or background information might make it more convincing?

   d. What is a limitation of this research?

15. In the 1870s, a prison inspector and self-described sociologist named Richard Dugdale noticed that a disproportionate number of inmates at his facility in Ulster County, New York, were related. He began studying them, calling the family the "Jukes," although he kept records of their real names. Dugdale traced the family back seven generations to a son of Dutch settlers, a man named Max, a pioneer who lived off the land. Margaret, "the mother of criminals," as Dugdale would write in his 1877 book *The Jukes: A Study in Crime, Pauperism, Disease and Heredity,* married one of Max's sons, and the couple presumably gave rise to 540 of the 709 criminals on Dugdale's watch. Dugdale attributed the Jukes' less desirable

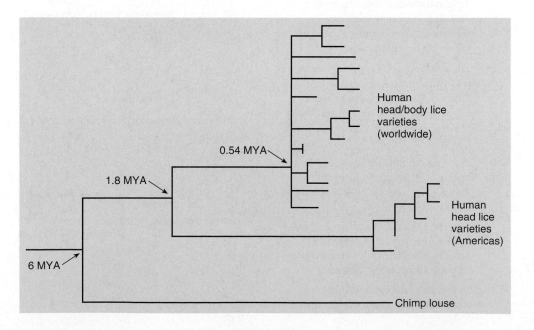

characteristics to heredity. He wrote, among many conclusions, that "harlotry may become a hereditary characteristic and be perpetuated without any specially favoring environment to call it into activity."

The Jukes study influenced social scientists to probe other families seemingly riddled with misfits—they were all caucasian, descended from colonial settlers, and poor. Poverty was not seen as an economic problem, but as a reflection of an inner, inborn, hard-to-define degeneracy, that if left unchecked would cost society dearly. Dugdale's book eventually became fodder for the fledgling eugenics movement. In 1911, researchers at the Eugenics Record Office in Cold Spring Harbor updated the Dugdale account, describing the Jukes' phenotype as "feeblemindedness, indolence, licentiousness, and dishonesty." Criminality, too, was considered an inherited trait. The evidence from the Jukes family and others was used to argue for compulsory sterilization of those deemed unfit. But the original research on the Jukes family was flawed, and its accuracy was never questioned. Less notorious Jukes family members served in respected professions, some even holding public office. The Jukes were vindicated in 2003, when archives at the State University of New York at Albany revealed the original names of the people in Dugdale's account; most were not even related. The Jukes family curse was more legend than fact.

a. What would have had to have happened to the original jailed Jukes family members or their descendants to be considered eugenic?

b. How could studies on one family harm others?

c. Cite an example of an idea based on eugenics today or in the recent past.

d. If you were a contemporary of Dugdale's, what type of evidence would you have sought to counter his ideas?

16. A Y chromosome haplotype consists of specific mutations for the SRY gene and genes called M96 and P29. Among modern Africans, there are three variants of this haplotype. Two of them are found only among Africans, but the third variant of the Y haplotype, called E3, is also seen in western Asia and in parts of Europe. Researchers examined specific subhaplotypes (variations of the variations) and found that one type, called E-M81, accounts for 80 percent of the Y chromosomes sampled in northwest Africa, diminishing sharply in incidence to the east, and not present in sub-Saharan Africa at all. That same haplotype is found in a small but significant percentage of the Y chromosomes in Spain and Portugal. Consult a map, and propose a scenario for this gene flow. What further information would be useful in reconstructing migration patterns?

Learn to apply the skills of a genetic counselor with this additional case found in the *Case Workbook in Human Genetics*:

Novelty seeking and ADHD

Genetics of Immunity

THE RETURN OF POLIO

For parents of young children in the early 1950s, late summer brought the terror of a fever that might explode into "infantile paralysis." Poliomyelitis is usually mild, but in about one percent of cases, the infecting virus invades spinal cord cells, causing paralysis. A huge effort to develop a vaccine succeeded in 1955 with an injected, inactivated form of the virus, and in 1962, with an oral vaccine. Schoolchildren lined up in school cafeterias to eat lumps of sugar stained pink with the vaccine.

A vaccine is an inactivated, or live but non-disease-causing, version of a pathogen, or part of a pathogen, that fools the human immune system into reacting as if infection has occurred. The immune system produces antibody proteins, and if the person encounters the actual pathogen, protection is swift. So successful was polio vaccination that epidemiologists predicted vanquishing the disease by 2000—but the date was pushed back to 2005 when outbreaks occurred in places where vaccination rates were not near 100 percent.

In 2003, disaster struck. In parts of Nigeria, religious leaders told people that polio vaccine could sterilize females and spread HIV. People stopped vaccinating their children—and polio came back. Then cases appeared in more than 12 nearby nations, the poliovirus genetically identical to the Nigerian strains.

The resurgence of polio illustrates the futility of considering a medical advance based on genetics in a vacuum. Even though the vaccine can, theoretically, forever banish this disease, fear and distrust prevented it from doing its job.

Polio has returned to several nations where parents stopped vaccinating their children in 2003, believing rumors that the vaccine causes infertility and spreads HIV infection. This child is receiving oral polio vaccine.

17.1 The Importance of Cell Surfaces

We share the planet with plants, microbes, fungi, and other animals. The human immune system has evolved in a way that keeps potentially harmful organisms out of our bodies. This system is a mobile army of about 2 trillion cells and the biochemicals they produce. Protection is based upon the ability of the immune system to recognize "foreign" or "nonself" surfaces, which include those of microorganisms such as bacteria and yeast; nonliving "infectious agents" such as viruses; and even tumor cells and transplanted cells. Then, the immune system carries out a highly coordinated, multipronged attack that includes both general responses and highly specific actions.

Pathogens

Bacteria are prokaryotic cells, which means that they lack membrane-bounded, complex organelles, but they are nonetheless cells (**figure 17.1**). Antibiotic drugs treat bacterial infections. Much of the action of the immune system is directed against viruses, which are much simpler than cells and straddle the boundary between the nonliving and the living. Few drugs can treat viral infections.

A **virus** is a single or double strand of RNA or DNA wrapped in a protein coat, and in some types, in an outer envelope, too. A virus can reproduce only if it enters and uses a host cell's energy resources, protein synthetic machinery, and secretion pathway. Still, it is a stunningly streamlined structure. A virus may have only a few protein-encoding genes, but many copies of the same protein can assemble to form an intricate covering, like the panes of glass in a greenhouse. Ebola virus, for example, has only seven types of proteins, but they assemble into a structure capable of reducing a human body to little more than a bag of blood and decomposed tissue (**figure 17.2a**). In contrast, the smallpox virus has more than 100 different types of proteins, and HIV is also complex (figure 17.2b).

Human chromosomes include viral DNA sequences that are vestiges of past infections, perhaps in distant ancestors. Many DNA viruses reproduce by inserting

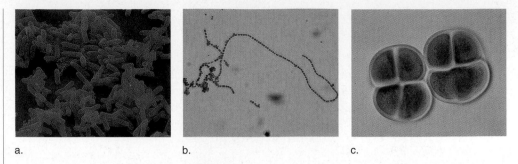
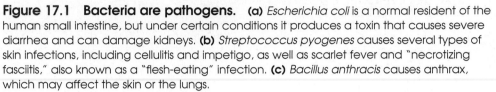

a.　　　　　b.　　　　　c.

Figure 17.1　Bacteria are pathogens.　(a) *Escherichia coli* is a normal resident of the human small intestine, but under certain conditions it produces a toxin that causes severe diarrhea and can damage kidneys. **(b)** *Streptococcus pyogenes* causes several types of skin infections, including cellulitis and impetigo, as well as scarlet fever and "necrotizing fasciitis," also known as a "flesh-eating" infection. **(c)** *Bacillus anthracis* causes anthrax, which may affect the skin or the lungs.

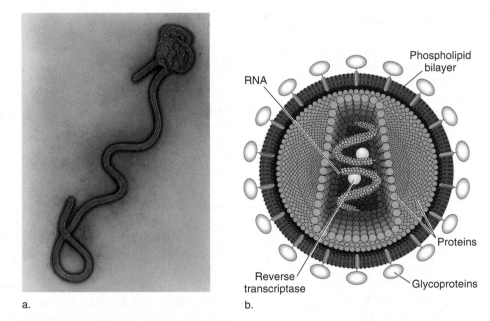

a.　　　　　b.

Figure 17.2　Virus structure.　Viruses are nucleic acids in protein coats. **(a)** Ebola virus is a single strand of RNA and just seven proteins. People become infected when they touch the body fluids of the infected. **(b)** HIV consists of RNA surrounded by several protein layers. Once inside a human cell, the virus uses a viral enzyme, reverse transcriptase, to make a DNA copy of its RNA. The virus then inserts this copy into the host cell's DNA. The infected cell not only dies, but produces and releases many viral particles.

DNA into the host cell's genetic material. In contrast, it takes several steps for an RNA virus to insert DNA into a human chromosome, because the RNA must first be copied into DNA by a viral enzyme called reverse transcriptase. The DNA that represents the RNA virus then inserts into the host cell's chromosome. (In some viruses, the invading DNA is replicated separately from a host chromosome.) Certain RNA viruses are called retroviruses because they transmit genetic information opposite the usual direction—from DNA to RNA to protein. HIV is a retrovirus.

Once viral DNA integrates into the host cell's DNA, it can either remain and replicate along with the host's DNA without causing harm, or it can take over and kill the cell. Viral genes direct the host cell to replicate viral DNA and then use it to manufacture viral proteins, at the expense of the cell's normal activities. The cell fills with viral DNA and protein, which assemble into new viruses. The cell bursts, releasing many new virus copies into the body.

Diverse viruses infect all types of organisms. They were discovered in tobacco plants, but also infect microorganisms, fungi, and, of course, animals. Their genetic material cannot repair itself, so the mutation rate may be high—which is one reason why we cannot develop an effective vaccine against HIV or the common cold, and why new influenza vaccines must be developed each year. Recall from chapter 10 that prions are infectious proteins. The immune system does not recognize infectious prions. This is because unlike viruses, prions are variants of proteins normally in the body that have the same amino acid sequence. They differ from harmless protein in their three-dimensional structure.

Genetic Control of Immunity

Genes that affect immunity may confer susceptibilities or resistances to certain infectious diseases, or raise the risk of developing an allergic or **autoimmune** condition, in which the immune system attacks an individual's own tissues. Most such effects are polygenic. For example, systemic lupus erythematosus is an autoimmune disorder caused by inheriting certain alleles of dozens of genes. OMIM entries list susceptibility genes, not causative ones. Interpreting such genotypes is more suggestive than diagnostic, and is more informative about a population than about a particular individual. For example, an allele of a gene called *PTPN22* is found in 1 in 6 Caucasians in the United States who do not have lupus, but in 1 in 4 who do have the disease. Lupus causes a distinctive facial rash in the shape of a butterfly, achy joints, extreme fatigue, fever, anemia, and hair loss.

A few types of single genes exert powerful effects on immunity. Certain classes of genes oversee immunity by encoding **antibodies** and **cytokines,** proteins that directly attack foreign antigens. An **antigen** is any molecule that elicits an immune response, and is usually a protein or carbohydrate. Genes specify the cell surface antigens that mark the body's cells as "self." They do this by encoding antigen proteins, or enzymes required to synthesize particular carbohydrates that are antigens.

Because genes control immunity, mutations can impair immune function, causing immune deficiencies, autoimmune disorders, allergies, and cancer. Understanding how genes control immunity makes it possible to enhance or redirect the system's ability to fight disease. We begin our look at normal immunity with some familiar examples of our personal cellular landscapes.

Blood Groups

Transplanting an organ as complex as a liver is risky. A far simpler type of transplant, although still very dependent on matching cell surfaces, is a blood transfusion. Using one person's blood to restore another's health was proposed centuries ago. To do so safely and successfully, however, it was necessary to understand the genetics of blood types.

ABO Blood Groups

The first transfusions, performed in the late 1600s, used lamb's blood. By the 1800s, physicians were trying to use human blood. Results were unpredictable—some recipients recovered, but others died. So poor was the success rate that, by the late 1800s, many nations banned transfusions.

Then Austrian physician Karl Landsteiner began investigating why transfusions sometimes worked and sometimes didn't. In 1901, he determined that human blood was of differing types, and only certain combinations were compatible. Specifically, he identified three types of blood that he called A, B, and O, and found that transfusing between types often led to disaster. In 1902, other researchers discovered the rare type AB. In 1910, identification of the ABO blood antigen locus (OMIM 110300) explained these four blood types and their incompatibilities (**figure 17.3**). Today we know of 26 blood group systems that include nearly 200 antigens.

Recall from chapter 5 that the *I* gene alleles encode enzymes that place antigens A, B, both A and B, or neither antigen on sugar chains on red blood cells (see table 5.1). Blood type incompatibility occurs when a person's immune system manufactures antibodies that attack the antigens his or her cells do not carry. A person with blood type A, for example, has antibodies against type B antigen. If he or she is transfused with type B blood, the anti-B antibodies clump the transfused red blood cells, blocking circulation and depriving tissues of oxygen. A person with type AB blood doesn't

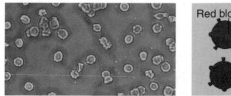
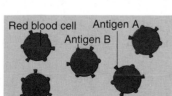

Compatible Blood Types (no clumping)

Donor	Recipient
O	O, A, B, AB
A	A, AB
B	B, AB
AB	AB

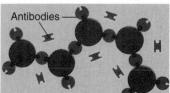

Incompatible Blood Types (clumping)

Donor	Recipient
A	B, O
B	A, O
AB	A, B, O

Figure 17.3 ABO blood types. Genetics explains blood incompatibilities.

manufacture antibodies against either antigen A or B, because if he or she did, the person's own blood would clump. Therefore, someone with type AB blood can receive any ABO blood type. Type O blood has neither A nor B antigens, so it cannot stimulate an immune response in a transfusion recipient; people with type O blood can therefore donate to anyone. However, the idea that a person with AB blood is a "universal recipient" and one with type O blood is a "universal donor" is more theoretical than practical, because antibodies to other donor blood antigens (for example, the Rh factor, discussed next) can cause slight incompatibilities. For this reason, blood is as closely matched as possible.

A person who receives mismatched blood quickly feels the effects—anxiety, difficulty breathing, facial flushing, headache, and severe pain in the neck, chest, and lower back. Red blood cells burst, releasing free hemoglobin that can damage the kidneys.

The Rh Factor

ABO blood type is often further differentiated by a $^+$ or $^-$, which refers to another blood group antigen called the Rh factor (OMIM 111700). Whether a person has the Rh factor (Rh$^+$) or not (Rh$^-$) is determined by a combination of alleles of three genes, designated *C, D,* and *E,* on chromosome 1. Each locus has two alleles: *C* or *c, D* or *d,* and *E* or *e.* A person whose red blood cells have at least one of these antigens is considered Rh$^+$. That is, the only Rh$^-$ genotype is *ccddee.* The Rh antigens were originally identified in rhesus monkeys, hence the name.

Rh type is important when an Rh$^+$ man and an Rh$^-$ woman conceive a child who is Rh$^+$. The pregnant woman's immune system reacts to the few fetal cells that enter her bloodstream by manufacturing antibodies against them (**figure 17.4**). Not enough antibodies form to harm the first fetus, but if she carries a second Rh$^+$ fetus, the woman's now plentiful antibodies attack the fetal blood supply. In the fetus, bilirubin, a breakdown product of red blood cells, accumulates, damaging the brain and turning the skin and whites of the eyes yellow. The fetal liver and spleen swell as they rapidly produce new red blood cells. If the fetus or newborn does not receive a transfusion of Rh$^-$ blood and have some of its Rh$^+$ blood removed, then the heart and blood vessels collapse and fatal respiratory distress sets in. Rh disease that progresses this far is called hydrops fetalis.

Fortunately, natural and medical protections make hydrops fetalis rare today, although exchange blood transfusions were once common. Determining parental ABO blood types indicates whether an immune reaction against the fetus of an Rh-incompatible couple will take place. If the woman has type O blood and the fetus is A or B, then her anti-A or anti-B antibodies attack the fetal blood cells in her circulation before her system has a chance to manufacture the anti-Rh antibodies. This blocks the anti-Rh reaction.

Obstetricians routinely determine a pregnant woman's blood type. If she and her partner are Rh incompatible, doctors inject a substance called RhoGAM during pregnancy and after the birth. RhoGAM is actually antibody against the Rh antigen. RhoGAM covers antigens on fetal blood cells in the woman's circulation so that she does not manufacture anti-Rh antibodies. However, events other than pregnancy and childbirth can expose an Rh$^-$ woman's system to Rh$^+$ cells, placing even her first child at risk. These include amniocentesis, a blood transfusion, an ectopic (tubal) pregnancy, a miscarriage, or an abortion.

Other Blood Groups

The second blood group discovered, in 1927, was the MN system; a third allele, *S,* was identified 20 years later. The three alleles are codominant, combining to form six different genotypes and phenotypes. The MNS antigens bind to two specific glycoproteins on red blood cell surfaces.

Another blood-type determining gene is called Lewis (OMIM 111100). It encodes an enzyme, fucosyltransferase (FUT3), that adds an antigen to the sugar fucose, which the product of the *H* gene then places on red blood cells. (Recall from section 5.1 that the *H* gene is necessary for ABO expression.) Individuals with genotype *LeLe* or *Lele* have the Lewis antigen on red blood cell plasma membranes and in saliva,

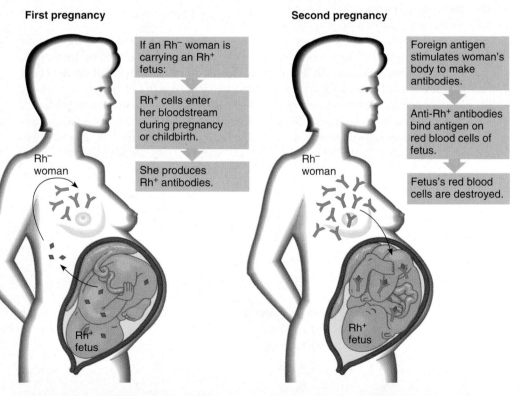

First pregnancy

If an Rh$^-$ woman is carrying an Rh$^+$ fetus:

Rh$^+$ cells enter her bloodstream during pregnancy or childbirth.

She produces Rh$^+$ antibodies.

Rh$^-$ woman

Rh$^+$ fetus

Second pregnancy

Foreign antigen stimulates woman's body to make antibodies.

Anti-Rh$^+$ antibodies bind antigen on red blood cells of fetus.

Fetus's red blood cells are destroyed.

Rh$^-$ woman

Rh$^+$ fetus

Figure 17.4   Rh incompatibility. Fetal cells entering the pregnant woman's bloodstream can stimulate her immune system to make anti-Rh antibodies, if the fetus is Rh$^+$ and she is Rh$^-$. A drug called RhoGAM prevents attacks on subsequent fetuses.

whereas *lele* people do not produce the antigen. Another interesting gene that affects the blood is the secretor gene (*FUT2*, OMIM 182100). People who have the dominant allele *Se* secrete the A, B, and H antigens in body fluids, including semen, saliva, tears, and mucus.

Cell surfaces are dotted with many molecules other than those that confer blood type. Many of these protein surface features are encoded by genes that are part of a 6-million-base-long cluster on the short arm of chromosome 6 called the **major histocompatibility complex** (MHC). The MHC includes about 70 genes.

The Human Leukocyte Antigens

The genes of the MHC are classified into three functional groups. Class III MHC genes encode proteins that are in blood plasma (the liquid portion of blood) and that carry out some of the nonspecific immune functions. Class I and II genes of the MHC encode the **human leukocyte antigens** (HLA), first studied in leukocytes (white blood cells). The HLA proteins link to sugars, forming branchlike glycoproteins that emanate from cell surfaces. Some HLA glycoproteins latch onto bacterial and viral proteins, displaying them like badges to alert other immune system cells. This action, called antigen processing, is often the first step in an immune response. The cell that displays the foreign antigen is called an **antigen-presenting cell. Figure 17.5** shows how a large cell called a macrophage displays bacterial antigens. Certain white blood cells called T cells (or T lymphocytes) are also antigen-presenting cells. Class I and II HLA proteins differ in the types of immune system cells they alert.

All cells with nuclei (that is, all cells except red blood cells) have some HLA antigens that identify them as "self," or belonging to the same individual. In addition to these common HLA markers are more specific markers that distinguish particular tissue types. Class I includes three genes, *A*, *B*, and *C*, that vary greatly and are found on all cell types, and three other genes, *E*, *F*, and *G*, that are more restricted in their distribution. Class II includes three major genes whose encoded proteins are found mostly on antigen-presenting cells.

Because the HLA classes consist of several genes that have many alleles, individuals have an overall HLA "type." Only 2 in every 20,000 unrelated people match for the six major HLA genes by chance. When transplant physicians attempt to match donor tissue to a potential recipient, they determine how alike the two individuals are at these six loci. Usually at least four of the genes must match for a transplant to have a reasonable chance of success. Before DNA profiling, HLA typing was the predominant type of blood test used in forensic and paternity cases to rule out involvement of certain individuals. However, HLA genotyping has become very complex because hundreds of alleles are now known, and HLA genotype and disease associations differ in different populations.

About 50 percent of the genetic influence on immunity stems from HLA genes. However, a few disorders are very strongly associated with inheriting particular HLA types. This is the case for ankylosing spondylitis, which inflames and deforms vertebrae. A person with either of two particular subtypes of an HLA antigen called B27 is 100 times as likely to develop the condition as someone who lacks either form of the antigen. HLA-associated risks are not absolute. More than 90 percent of people who suffer from ankylosing spondylitis have the B27 antigen, which occurs in only 5 percent of the general population. However, 10 percent of people

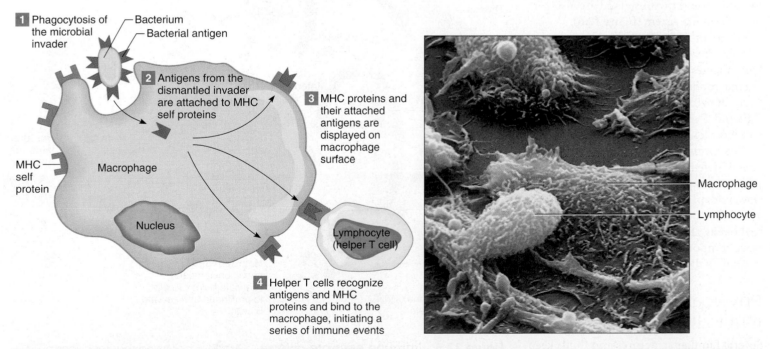

Figure 17.5 Macrophages are antigen-presenting cells. A macrophage engulfs a bacterium, then displays foreign antigens on its surface, held in place by major histocompatibility complex (MHC) self proteins. This event sets into motion many immune reactions.

who have ankylosing spondylitis do *not* have the B27 antigen, and some people who have the antigen never develop the disease. **Table 171.** lists some HLA type-disease associations.

Key Concepts

1. The immune system consists of cells and biochemicals that distinguish self from nonself antigens.
2. Pathogens include microbes and infectious agents, such as viruses, which take over a host cell's protein synthesis machinery to reproduce.
3. Blood types result from self antigen patterns on red blood cells. HLA cell surface proteins establish self and display foreign antigens.

17.2 The Human Immune System

The immune system has macroscopic (large-scale) and microscopic (small-scale) components.

On a macroscopic level, the immune system includes a network of vessels called lymphatics. These transport lymph, a watery fluid, to bean-shaped structures throughout the body called lymph nodes. The spleen and thymus gland are also part of the immune system (**figure 17.6**).

On a microscopic level, the immune system consists of white blood cells called lymphocytes and the wandering, scavenging macrophages that capture and degrade bacteria, viruses, and cellular debris. **B cells** and **T cells** are the two major types of lymphocytes.

The immune response consists of two lines of defense—an immediate generalized **innate immunity,** and a more specific, slower **adaptive immunity.** These defenses act after various physical barriers block pathogens. **Figure 17.7** summarizes the basic components of immunity, discussed in detail in the following sections.

Physical Barriers and the Innate Immune Response

Several familiar structures and fluids keep pathogens from entering the body in the innate immune response. Unbroken skin

Table 17.1

HLA-Disease Associations

Condition	Description	Relative Risk for Certain HLA Type*
Narcolepsy	Suddenly falling asleep	264
Ankylosing spondylitis	Inflamed and deformed vertebrae	77
Reiter's disease	Inflamed joints, eyes, and urinary tract	37
Dermatitis herpetiformia	Burning, itchy skin lesions	16
Psoriasis	Scaly skin lesions	8
Autoimmune hepatitis	Inflamed liver	7.4
Type 1 diabetes	Inability to produce insulin	5–7
Graves disease	Malfunction of thyroid gland	5.5
Celiac disease	Severe diarrhea	5.1
Myasthenia gravis	Fluctuating weakness of voluntary muscles	4.2
Rheumatoid arthritis	Severely inflamed joints	3–6
Multiple sclerosis	Degeneration of brain and spinal cord, producing weakness and poor coordination	2.7
Systemic lupus erythematosus	Facial rash, high persistent fever, destruction of heart, brain, kidneys	2

*Recall from chapter 1 that a relative risk of greater than 1 indicates a greater risk than that of the general population.

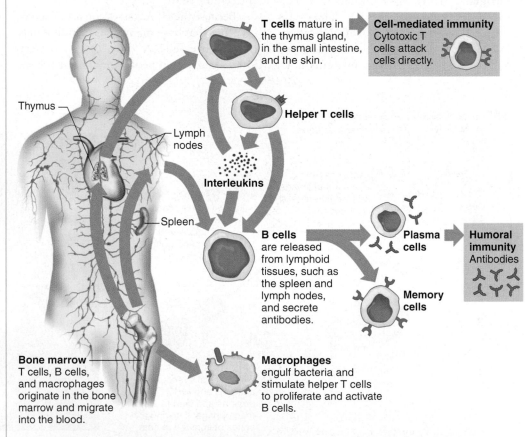

T cells mature in the thymus gland, in the small intestine, and the skin.

Cell-mediated immunity Cytotoxic T cells attack cells directly.

Helper T cells

Interleukins

B cells are released from lymphoid tissues, such as the spleen and lymph nodes, and secrete antibodies.

Plasma cells

Humoral immunity Antibodies

Memory cells

Bone marrow T cells, B cells, and macrophages originate in the bone marrow and migrate into the blood.

Macrophages engulf bacteria and stimulate helper T cells to proliferate and activate B cells.

Thymus

Lymph nodes

Spleen

Figure 17.6   Immune cells are diverse. T cells, B cells, and macrophages build an overall immune response. All three types of cells originate in the bone marrow and circulate in the blood.

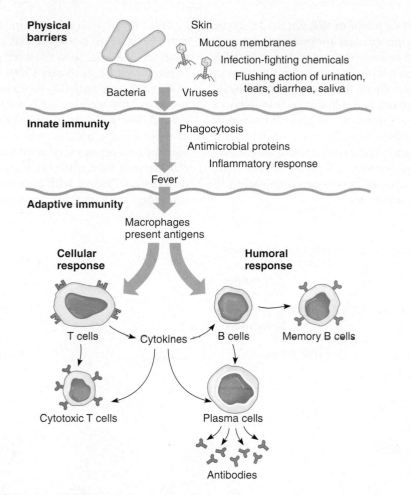

Physical barriers

Skin
Mucous membranes
Infection-fighting chemicals
Flushing action of urination, tears, diarrhea, saliva

Bacteria   Viruses

Innate immunity

Phagocytosis
Antimicrobial proteins
Inflammatory response

Fever

Adaptive immunity

Macrophages present antigens

Cellular response   **Humoral response**

T cells → Cytokines → B cells → Memory B cells

Cytotoxic T cells   Plasma cells

Antibodies

Figure 17.7   Levels of immune protection.   Disease-causing organisms and viruses (pathogens) first must breach physical barriers, then nonspecific cells and molecules attack in the innate immune response. If this is ineffective, the adaptive immune response begins: Antigen-presenting cells stimulate T cells to produce cytokines, which activate B cells to differentiate into plasma cells, which secrete antibodies. Once activated, these specific cells "remember" the pathogen, allowing faster responses to subsequent attacks.

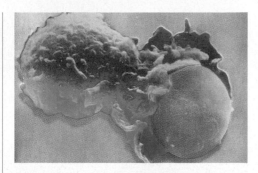

Figure 17.8   Nature's garbage collectors.   A human phagocyte engulfs a yeast cell.

and mucous membranes such as the lining inside the mouth are part of this first line of defense, as are earwax and the waving cilia that push debris and pathogens up and out of the respiratory tract. Most microbes that make it to the stomach perish in a vat of churning acid—though a notable exception is the bacterium that causes peptic ulcers. Other microbes are flushed out in diarrhea. These barriers are nonspecific—they keep out anything foreign, not just particular pathogens.

If a pathogen breaches these physical barriers, innate immunity provides a rapid, broad defense. The term *innate* refers to the fact that these general defenses are in the body, ready to function should infection begin. A process called **inflammation** is a central part of the innate immune response.

Inflammation creates a hostile environment for certain types of pathogens at an injury site, sending in cells that engulf and destroy them. Such cells are called phagocytes, and their engulfing action is phagocytosis (**figure 17.8**). Some blood cell types, such as neutrophils, are phagocytes, as are the large, wandering macrophages. At the same time during inflammation, plasma accumulates, which dilutes toxins and brings in antimicrobial chemicals. Increased blood flow with inflammation warms the area, turning it swollen and red. The person may not be very comfortable, but often, the pathogen does not survive. Inflammation at the site of an injury can prevent infection.

Three major classes of proteins participate in the innate immune response—the complement system, collectins, and cyto-

kines. Mutations in the genes that encode these proteins can produce disorders that increase susceptibility to infection. Yet, some mutations have no effect; perhaps other proteins provide the function.

The complement system consists of plasma proteins that assist, or complement, several of the body's other defense mechanisms. Some complement proteins puncture bacterial plasma membranes, bursting the cells. Others dismantle viruses. Yet other complement proteins assist inflammation by triggering the release of histamine from mast cells, another type of immune system cell that is involved in allergies. Histamine dilates blood vessels, enabling fluid to rush to the infected or injured area. Still other complement proteins attract phagocytes to an injury site.

Collectins are proteins that broadly protect against bacteria, yeasts, and some viruses by detecting slight differences from human cells in their surfaces. Groups of human collectins correspond to the surfaces of different types of pathogens, such as the distinctive sugars on infecting yeast, the linked sugars and lipids of certain bacteria, and the surface features of some RNA viruses.

Cytokines play many roles in immunity. As part of the innate immune response, cytokines called interferons alert other components of the immune system to the presence of cells infected with viruses. These cells are then destroyed, which limits the spread of infection. Interleukins are cytokines that cause fever, temporarily triggering a higher body temperature that directly kills some infecting bacteria and viruses. Fever also counters microbial growth indirectly, because higher body temperature reduces the iron level in the blood. Bacteria and fungi require more iron as the body temperature rises; therefore, a fever-ridden body stops their growth. Phagocytes also attack more vigorously when the temperature rises. Tumor necrosis factor is another type of cytokine that activates other protective biochemicals, destroys certain bacterial toxins, and also attacks cancer cells. Many of the aches and pains we experience from an infection are actually due to the immune response, not directly to the actions of the pathogens.

The Adaptive Immune Response

Adaptive immunity must be stimulated into action. It may take days to respond, compared to minutes for innate immunity. Adaptive immunity is highly specific and directed.

B cells and T cells carry out adaptive immunity. B cells produce antibodies in response to activation by T cells in the **humoral immune response.** ("Humor" means fluid; antibodies are carried in fluids.) T cells produce cytokines and activate other cells in the **cellular immune response.** B and T cells differentiate in the bone marrow and migrate to the lymph nodes, spleen, and thymus gland, as well as circulate in the blood and tissue fluid.

The adaptive arm of the immune system has three basic characteristics. It is *diverse*, vanquishing many types of pathogens. It is *specific*, distinguishing the cells and molecules that cause disease from those that are harmless. The immune system also *remembers*, responding faster to a subsequent encounter with a foreign antigen than it did the first time. The first assault initiates a primary immune response. The second assault, based on the system's "memory," is a secondary immune response. This is why we get some infections, such as chickenpox, only once. However, upper respiratory infections and influenza recur because the causative viruses mutate, presenting a different face to our immune systems each season.

The Humoral Immune Response— B Cells and Antibodies

An antibody response begins when an antigen-presenting macrophage activates a T cell. This cell in turn contacts a B cell that

has surface receptors that can bind the type of foreign antigen the macrophage presents. The immune system has so many B cells, each with different combinations of surface antigens, that there is almost always one or more available that corresponds to a particular foreign antigen. Turnover of these cells is high. Each day, millions of B cells perish in the lymph nodes and spleen, while millions more form in the bone marrow, each with a unique combination of surface molecules.

Once the activated T cell finds a B cell match, it releases cytokines that stimulate the B cell to divide. Soon the B cell gives rise to two types of cells (**figure 17.9**). The first, plasma cells, are antibody factories, secreting up to 2,000 identical antibodies each per second into the bloodstream at the height of their few-day lifespan. These cells provide the primary immune response. Plasma cells derived from different B cells secrete different antibodies, with each type corresponding to a specific portion of the

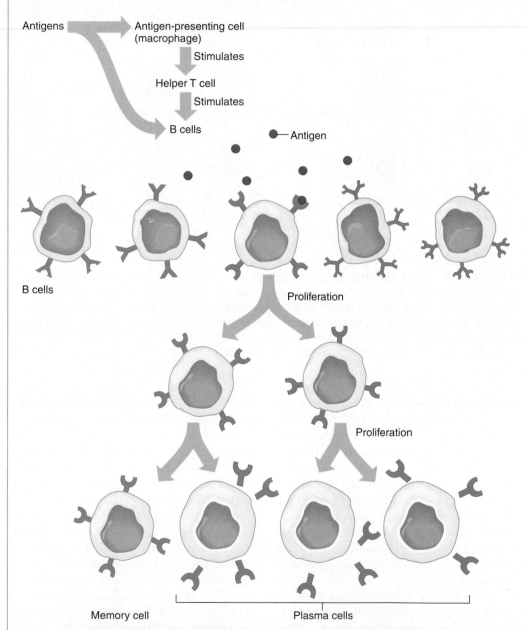

Figure 17.9 Production of antibodies. In the humoral immune response, B cells proliferate and mature into antibody-secreting plasma cells. Note that only the B cell that binds the antigen proliferates; its descendants may develop into memory cells or plasma cells. Plasma cells outnumber memory cells.

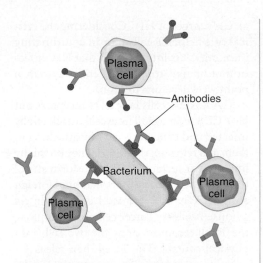

Figure 17.10 An immune response recognizes many targets. A humoral immune response is polyclonal, which means that different plasma cells produce antibody proteins that recognize and bind to different features of a foreign cell's surface.

pathogen in what is called a polyclonal antibody response (**figure 17.10**). This response is like hitting a person in different parts of the body. The second type of B cell descendant, memory cells, are far fewer and usually dormant. They respond to the foreign antigen faster and with more force should it appear again. This is a secondary immune response.

Antibodies are constructed of several polypeptides and are therefore encoded by several genes. The simplest antibody molecule is four polypeptide chains connected by disulfide (sulfur-sulfur) bonds, forming a shape like the letter Y (**figure 17.11**). A large antibody molecule might consist of three, four, or five such Ys joined.

In a Y-shaped antibody subunit, the two longer polypeptides are called heavy chains, and the other two light chains. The lower portion of each chain is an amino acid sequence that is very similar in all antibody molecules, even in different species. These areas are called constant regions. The amino acid sequence of the upper portions of each polypeptide chain, the variable regions, can differ greatly among antibodies.

Antibodies can bind certain antigens because of the three-dimensional shapes of the tips of the variable regions. These specialized ends are called antigen binding sites, and the parts that actually contact the antigen are called idiotypes. The parts of the antigens that idiotypes bind are epitopes. An antibody contorts to form a pocket around the antigen.

Antibodies have several functions. Antibody-antigen binding may inactivate a pathogen or neutralize the toxin it produces. Antibodies can also clump pathogens, making them more visible to macrophages, which then destroy them. Antibodies also activate complement, extending the innate immune response.

Antibodies are of five major types, distinguished by location and function (**table 17.2**). (Antibodies are also called immunoglobulins, abbreviated *Ig.*) Different antibody types predominate in different stages of an infection.

The human body can manufacture seemingly limitless varieties of antibodies, though the genome has a limited number of antibody genes. This great diversity is possible because parts of different antibody genes combine. During the early development of B cells, sections of their antibody genes move to other chromosomal locations, creating new genetic instructions for antibodies. In this way, 200 genes generate about 100 trillion different antibody types. Because of this tremendous diversity, a human body can respond to nearly any infection. A single stimulated B cell is said to give rise to a clone of plasma and memory cells because all of its descendants express the same antibody gene combinations.

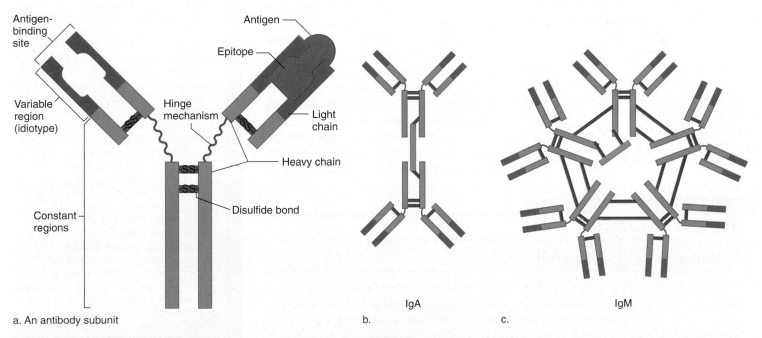

a. An antibody subunit
b.
IgA
c.
IgM

Figure 17.11 Antibody structure. The simplest antibody molecule **(a)** consists of four polypeptide chains, two heavy and two light, joined by two sulfur atoms that form a disulfide bond. Part of each polypeptide chain has a constant sequence of amino acids, and the remainder varies. The tops of the Y-shaped molecules form antigen binding sites. **(b)** IgA consists of two Y-shaped subunits, and IgM **(c)** consists of five subunits.

Table 17.2

Types of Antibodies

Type*	Location	Functions
IgA	Milk, saliva, urine, and tears; respiratory and digestive secretions	Protects against pathogens at points of entry into body
IgD	On B cells in blood	Stimulates B cells to make other types of antibodies, particularly in infants
IgE	In secretions with IgA and in mast cells in tissues	Acts as receptor for antigens that cause mast cells to secrete allergy mediators
IgG	Blood plasma and tissue fluid; passes to fetus	Protects against bacteria, viruses, and toxins, especially in secondary immune response
IgM	Blood plasma	Fights bacteria in primary immune response; includes anti-A and anti-B antibodies of ABO blood groups

*The letters *A, D, E, G,* and *M* refer to the specific conformation of heavy chains characteristic of each class of antibody.

The Cellular Immune Response— T cells and Cytokines

T cells provide the cellular immune response, so-called because the cells themselves travel to where they act, unlike B cells, which secrete antibodies into the bloodstream. T cells descend from stem cells in the bone marrow, then travel to the thymus gland ("T" refers to thymus). As the immature T cells, called thymocytes, migrate toward the interior of the thymus, they display diverse cell surface receptors. Then selection happens. As the wandering thymocytes touch lining cells in the gland that are studded with "self" antigens, thymocytes that do not attack the lining cells begin maturing into T cells, whereas those that harm the lining cells die by apoptosis—in great numbers. Gradually, T-cells-to-be that recognize self are selected and persist; those that can harm the body are destroyed.

Several types of T cells are distinguished by the types and patterns of receptors on their surfaces, and by their functions. Helper T cells recognize foreign antigens on macrophages, stimulate B cells to produce antibodies, secrete cytokines, and activate another type of T cell called a cytotoxic T cell, also called a killer T cell. Certain T cells may help to suppress an immune response when it is no longer required. The cytokines that helper T cells secrete include interleukins, interferons, tumor necrosis factor, and colony stimulating factors, which stimulate white blood cells in bone marrow to mature (**table 17.3**). Cytokines interact with and signal each other, sometimes in complex cascades.

Distinctive surfaces distinguish subsets of helper T cells. Certain antigens called cluster-of-differentiation antigens, or CD antigens, enable T cells to recognize foreign antigens displayed on macrophages. One such cell type, called a CD4 helper T cell, is an early target of HIV. Considering the critical role helper T cells play in coordinating immunity, it is little wonder that HIV infection ultimately topples the entire system, a point we will return to soon.

Cytotoxic T cells lack CD4 receptors but have CD8 receptors. These cells attack virally infected and cancerous cells by attaching to them and releasing chemicals. They do this by linking two surface peptides to form structures called T cell receptors that bind foreign antigens. When a cytotoxic T cell encounters a nonself cell—a cancer cell, for example— the T cell receptors draw the two cells into physical contact. The T cell then releases a protein called perforin, which pierces the cancer cell's plasma membrane, killing it (**figure 17.12**). Cytotoxic T cell receptors also attract body cells that are covered with certain viruses, destroying the cells before the viruses on them can enter, replicate, and spread the infection. Cytotoxic T cells continually monitor body cells, recognizing and eliminating virally infected and tumor cells.

Table 17.4 summarizes types of immune system cells.

Table 17.3

Types of Cytokines

Cytokine	Function
Colony stimulating factors	Stimulate bone marrow to produce lymphocytes
Interferons	Block viral replication, stimulate macrophages to engulf viruses, stimulate B cells to produce antibodies, attack cancer cells
Interleukins	Control lymphocyte differentiation and growth, cause fever that accompanies bacterial infection
Tumor necrosis factor	Stops tumor growth, releases growth factors, stimulates lymphocyte differentiation, dismantles bacterial toxins

Key Concepts

1. The immune system consists of physical barriers; an innate immune response of inflammation, phagocytosis, complement, collectins, and cytokines; and a more directed adaptive immune response that is diverse, specific, and remembers. It has two components.
2. In the humoral immune response, stimulated B cells divide and differentiate into plasma cells and memory cells. A plasma cell secretes abundant antibodies of a single type. Antibodies are made of Y-shaped polypeptides, each consisting of two light and two heavy chains. Each chain consists of a constant and a variable region, and the tips of the Y form an antigen binding site with a specific idiotype. Antibodies make foreign antigens more visible to macrophages and stimulate complement. Shuffling gene pieces generates great antibody diversity.
3. In the cellular immune response, helper T cells stimulate B cells to manufacture antibodies and cytotoxic T cells to secrete cytokines. Using T cell receptors, cytotoxic T cells bind to nonself cells and virus-covered cells and burst them.

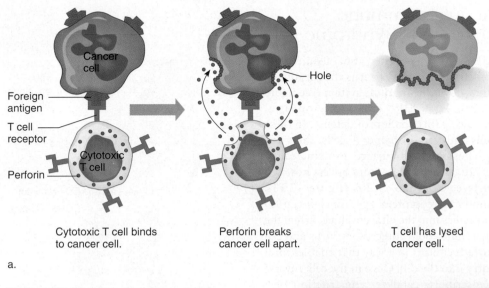

Cytotoxic T cell binds to cancer cell.

Perforin breaks cancer cell apart.

T cell has lysed cancer cell.

a.

Figure 17.12 Death of a cancer cell.
(a) A cytotoxic T cell binds to a cancer cell and injects perforin, a protein that pierces the cancer cell's plasma membrane. The cancer cell dies, leaving debris that macrophages clear away. **(b)** The smaller cell is a cytotoxic T cell, which homes in on the surface of the large cancer cell above it. The cytotoxic T cell will shatter the cancer cell, leaving scattered fibers.

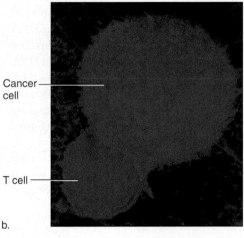

Cancer cell

T cell

b.

Table 17.4

Types of Immune System Cells

Cell Type	Function
Macrophage	Presents antigens
	Performs phagocytosis
Mast cell	Releases histamine in inflammation
	Releases allergy mediators
B cell	Matures into antibody-producing plasma cell or into memory cell
T cells	
Helper	Recognizes nonself antigens presented on macrophages
	Stimulates B cells to produce antibodies
	Secretes cytokines
	Activates cytotoxic T cells
Cytotoxic	Attacks cancer cells and cells infected with viruses upon recognizing antigens
Natural killer	Attacks cancer cells and cells infected with viruses without recognizing antigens; activates other white blood cells
Suppressor	Inhibits antibody production

17.3 Abnormal Immunity

The immune system continually adapts to environmental change. Because the immune response is so diverse, its breakdown affects health in many ways. Immune system malfunction may be inherited or acquired, and immunity may be too weak, too strong, or misdirected.

Inherited Immune Deficiencies

There are more than 20 types of inherited immune deficiencies, affecting both innate and adaptive immunity. In chronic granulomatous disease, neutrophils can engulf bacteria, but, due to deficiency of an enzyme called oxidase, they cannot produce the activated oxygen compounds that kill bacteria. Because this enzyme is made of four polypeptide chains, four genes encode it, and there are four ways to inherit the disease, all X-linked. A very rare autosomal recessive form is caused by a defect in the vacuole that encloses bacteria. Antibiotics and gamma interferon are used to prevent bacterial infections in these patients, and the disease can be cured with a bone marrow or an umbilical cord stem cell transplant.

Mutations in genes that encode cytokines or T cell receptors impair cellular immunity, which primarily targets viruses and cancer cells. But because T cells activate the B cells that manufacture antibodies, abnormal cellular immunity (T cell function) disrupts humoral immunity (B cell function). Mutations in the genes that encode antibody segments, that control how the segments join, or that direct maturation of B cells mostly impair immunity against bacterial infection. Inherited immune deficiency can also result from defective B cells. In one condition, B cells lack B cell linker protein, which normally signals the cells to mature into plasma cells. A person with this type of immune deficiency is highly vulnerable to certain bacterial infections, particularly of the ears and sinuses.

Severe combined immune deficiencies (SCID) affect both humoral and cellular immunity. About half of SCID cases are X-linked. In a less severe form, the individual lacks B cells but has T cells. Before antibiotic drugs became available, individuals with this form of SCID died before age 10 of overwhelming bacterial infection. In a more severe form of X-linked SCID, lack of B and T cells causes death by 18 months of age, usually of severe and diverse infections. Gene therapy for X-linked SCID is effective, but can cause leukemia. Ironically, an autosomal recessive form of SCID, called adenosine deaminase deficiency, became the first illness successfully treated with gene therapy. Both are discussed further in chapter 20.

A young man named David Vetter taught the world about the difficulty of life without immunity years before AIDS made it commonplace. David had an autosomal recessive form of SCID that caused him to be born without a thymus gland. His T cells could not mature and activate B cells, leaving him defenseless in a germ-filled world. Born in Texas in 1971, David spent his short life in a vinyl bubble, awaiting a treatment that never came. As he reached adolescence, David wanted to leave his bubble. A bone marrow transplant was unsuccessful—soon afterward, David began vomiting and developed diarrhea, both signs of infection. David left the bubble, but died within days of a massive infection.

Table 17.5 lists some inherited immune deficiencies.

Acquired Immune Deficiency Syndrome

AIDS is not inherited, but acquired by infection with HIV, a virus that gradually shuts down the immune system (see figure 17.2*b*). First, HIV enters macrophages, impairing this first line of defense. In these cells and later in helper T cells, the virus adheres with its surface protein, called gp120, to two coreceptors on the host cell surface, CD4 and CCR5 (**figure 17.13**). Another glycoprotein, gp41, anchors gp120 molecules into the viral envelope. When the virus binds both coreceptors, virus and cell surface contort in a way that enables viral entry into the cell. Once in the cell, reverse transcriptase catalyzes construction of a DNA strand complementary to the viral RNA, which replicates to form a DNA double helix. This enters the nucleus and inserts into a chromosome. The viral DNA sequences are transcribed and translated, and the cell fills with viral pieces, which are assembled into complete new viral particles that eventually bud from the cell (**figure 17.14**).

Once helper T cells start to die at a high rate, bacterial infections begin, because B cells aren't activated to produce antibodies.

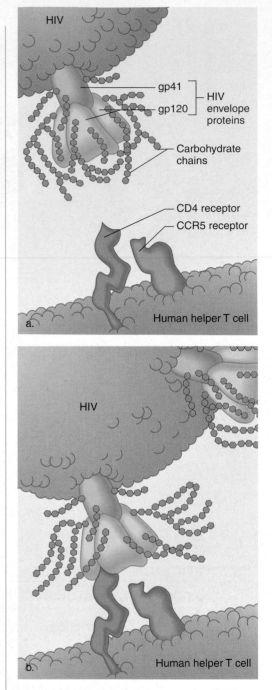

Figure 17.13   HIV binds to a helper T cell. **(a)** The part of HIV that binds to helper T cells is called gp120 (gp stands for glycoprotein). **(b)** When the carbohydrate chains that shield the protein portion of gp120 move aside as they approach the cell surface, the viral molecule can bind to a CD4 receptor. Binding to a receptor called CCR5 is also necessary for HIV to dock at a helper T cell. Once bound to the cell surface, the viral envelope fuses with the plasma membrane, enabling the virus to enter. A few lucky individuals lack CCR5 and thus cannot be infected by HIV. New types of drugs to fight HIV infection block the steps of viral entry. (The size of HIV here is greatly exaggerated.)

Table 17.5

Inherited Immune Deficiencies

Disease	OMIM	Inheritance*	Defect
Chronic granulomatous disease	306400	ar, AD, xlr	Abnormal phagocytes can't kill engulfed bacteria
Immune defect due absence of thymus	242700	ar	No thymus, no T cells
Neutrophil immunodeficiency syndrome	608203	ar	Deficiencies of T cells, B cells, and neutrophils
SCID			
Adenosine deaminase deficiency	102700	ar	No T or B cells
Adenosine deaminase deficiency with sensitivity to ionizing radiation	602450	ar	No T, B, or natural killer cells
IL-2 receptor mutation	300400	xlr	No T, B, or natural killer cells

*ar = autosomal recessive
AD = autosomal dominant
xlr = X-linked recessive

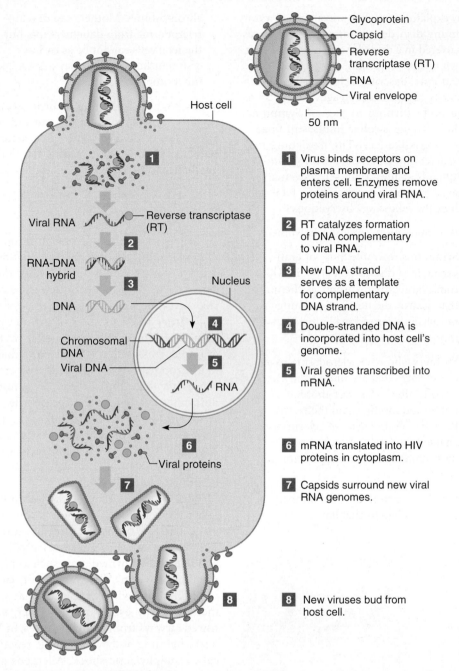

able nature has important clinical implications. Combining drugs with different actions is the most effective way to slow the disease process, so that AIDS becomes a chronic, lifelong, but treatable illness, instead of a killer.

Three types of drugs have cut the death rate from AIDS dramatically. Reverse transcriptase inhibitors block the copying of viral RNA into DNA. Protease inhibitors block the trimming of certain viral proteins, which is required for new viral particles to assemble. Entry inhibitors block the ability of the virus to bind to a cell, fuse with the plasma membrane, and enter.

Clues to developing new drugs to treat AIDS come from people at high risk who resist infection. For example, researchers have identified four receptors or the molecules that bind to them that mutation alters in ways that keep HIV out of cells. To find these receptors, researchers scrutinized the DNA of people who had unprotected sex with many partners, and people with hemophilia who had received HIV-tainted blood in the 1980s—but who were not infected. Some of them were homozygous recessive for a 32-base deletion in the CCR5 gene. Their CCR5 coreceptors were too stunted, thanks to a premature "stop" codon, to reach the cell's surface (see figure 17.13). Like a ferry arriving at shore to find no dock, HIV has nowhere to bind on the cells of these fortunate individuals. Heterozygotes, with one copy of the deletion, can become infected with HIV, but they remain healthy for several years longer than people who do not have the deletion. Curiously, the same *CCR5* mutation may have enabled people to survive plague in Europe during the middle ages. Apparently both the virus that causes AIDS and the bacterium that causes plague use the same portal into a human cell.

The diagram labels and numbered steps:

Glycoprotein
Capsid
Reverse transcriptase (RT)
RNA
Viral envelope

50 nm

Host cell

Viral RNA — Reverse transcriptase (RT)

RNA-DNA hybrid

DNA

Nucleus

Chromosomal DNA

Viral DNA

RNA

Viral proteins

1 Virus binds receptors on plasma membrane and enters cell. Enzymes remove proteins around viral RNA.

2 RT catalyzes formation of DNA complementary to viral RNA.

3 New DNA strand serves as a template for complementary DNA strand.

4 Double-stranded DNA is incorporated into host cell's genome.

5 Viral genes transcribed into mRNA.

6 mRNA translated into HIV proteins in cytoplasm.

7 Capsids surround new viral RNA genomes.

8 New viruses bud from host cell.

Figure 17.14 How HIV infects. HIV integrates into the host chromosome, then commandeers transcription and translation, ultimately producing more virus particles.

Much later in infection, HIV variants arise that can bind to a receptor called CXCR4 on cytotoxic T cells, killing them. Loss of these cells renders the body very vulnerable to viral infections and cancer.

HIV has an advantage over the human immune system because it replicates quickly, changes quickly, and can hide. The virus is very prone to mutation, both because it cannot repair replication errors, and because those errors happen frequently—1 per every 5,000 or so bases—because of the "sloppiness" of reverse transcriptase. The immune system simply cannot keep up; antibodies against one viral variant are useless against the next. For several years, the bone marrow produces 2 billion new T and B cells a day to counter the million to billion new HIV particles that bud daily from cells.

So genetically diverse is the population of HIV in a human host that, within days of the initial infection, variants arise that resist the drugs used to treat AIDS. HIV's change-

Autoimmunity

Autoimmunity is a reaction that occurs when the immune system produces antibodies, called **autoantibodies,** that attack the body's own healthy tissues. The signs and symptoms of autoimmune disorders reflect the cell types under attack. For example, autoimmune ulcerative colitis affects colon cells, causing severe abdominal pain.

A mutation in a single gene can cause varied symptoms of an autoimmune disorder. For example, a mutation in a gene on chromosome 21q causes autoimmune polyendocrinopathy syndrome type I (OMIM 240300). In this condition, malfunction of various endocrine glands occurs in a sequence. Children under 5 develop candidiasis, a fungal infection (not an autoimmune disorder). By age 10, the individual's parathyroid glands begin to fail, affecting calcium metabolism. By age 15, most affected individuals also develop Addison disease, reflecting a deficiency in adrenal gland hormones. Other associated conditions include thyroid deficiency, diabetes mellitus, vitiligo (skin whitening), and alopecia (hair loss).

The symptoms of many autoimmune conditions can arise by other mechanisms, making diagnosis difficult. Hemolytic anemia, for example, may be autoimmune, inherited, or a reaction to toxin exposure.

Autoimmunity may arise in several ways:

- A virus replicating within a cell incorporates proteins from the cell's surface onto its own. When the immune system "learns" the surface of the virus to destroy it, it also learns to attack human cells that normally bear the protein.

- Some thymocytes that should have died in the thymus somehow escape the massive die-off, persisting to attack "self" tissue later on.

- A nonself antigen coincidentally resembles a self antigen, and the immune system attacks both. In rheumatic fever, for example, antigens on heart valve cells resemble those on *Streptococcus* bacteria; antibodies produced to fight a strep throat also attack the heart valve cells.

Some disorders thought to be autoimmune may in fact have a more bizarre cause—fetal cells persisting in a woman's circulation, even decades after her offspring has grown up! In response to an as yet unknown trigger, the fetal cells, perhaps "hiding" in a tissue such as skin, emerge, stimulating antibody production and symptoms in the mother. This mechanism, called microchimerism ("small mosaic"),

may explain the higher prevalence of autoimmune disorders among women. It was discovered in a disorder called scleroderma, which means "hard skin" (**figure 17.15**).

Patients describe scleroderma, which typically begins between ages 45 and 55, as "the body turning to stone." Symptoms include fatigue, swollen joints, stiff fingers, and a masklike face. The hardening may also affect blood vessels, the lungs, and the esophagus. Clues that scleroderma is a delayed response to persisting fetal cells include the following observations:

- It is much more common in women.

- Symptoms resemble those of graft-versus-host disease (GVHD), in which transplanted tissue produces chemicals that destroy the host's body. Antigens on cells in scleroderma lesions match those that cause GVHD.

- Mothers who have scleroderma have cell surfaces that are more similar to those of their sons than those of unaffected mothers and their sons. Perhaps the similarity of cell surfaces enabled the fetal cells to escape destruction by the woman's immune system.

- Skin lesions from affected mothers of sons include cells that have Y

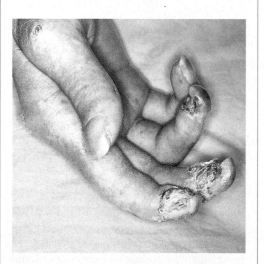

Figure 17.15 An autoimmune disorder—maybe. Scleroderma hardens the skin. Some cases appear to be caused by a long-delayed immune response to cells retained from a fetus decades earlier!

chromosomes. Mothers can develop scleroderma from daughters too, but the fetal cells cannot be as easily distinguished because they are XX, like the mothers' cells.

It's possible that other disorders traditionally considered autoimmune and that are more prevalent in women may actually reflect an immune system response to lingering fetal cells.

Allergies

An allergy is an immune system response to a substance, called an allergen, that does not actually present a threat. Many allergens are particles small enough to be carried in the air and enter a person's respiratory tract. The size of the allergen may determine the type of allergy. For example, grass pollen is large and remains in the upper respiratory tract, where it causes hay fever. But allergens from house dust mites, cat dander, and cockroaches are small enough to infiltrate the lungs, triggering asthma (**figure 17.16**). Asthma is a chronic disease in which contractions of the airways, inflammation, and accumulation of mucus block air flow.

Both humoral and cellular immunity take part in an allergic response. Antibodies of class IgE bind to mast cells, sending signals that cause the mast cells to open and release allergy mediators such as histamine and heparin. Allergy mediators cause inflammation, with symptoms that may include runny eyes from hay fever, narrowed airways from asthma, rashes, or the overwhelming bodywide allergic reaction called anaphylactic shock. Allergens also activate a class of helper T cells that produce a particular mix of cytokines whose genes are clustered on chromosome 5q. Regions of chromosomes 12q and 17q have genes that control IgE production.

The fact that allergies have become very common only during the past century suggests a much stronger environmental than genetic component. Still, people inherit susceptibilities to allergy. Twin studies of various allergies reveal about a 75 percent concordance, and isolated populations with a great deal of inbreeding tend to have a high prevalence of certain allergies.

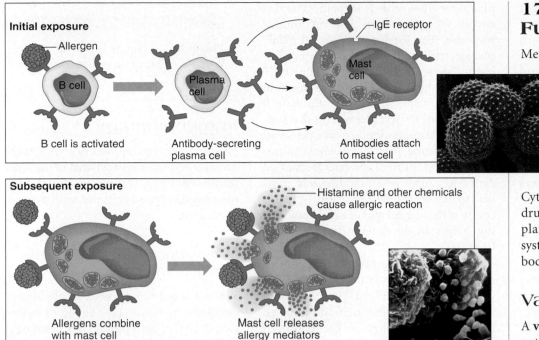

Initial exposure

Allergen

B cell

B cell is activated

Plasma cell

Antibody-secreting plasma cell

IgE receptor

Mast cell

Antibodies attach to mast cell

Subsequent exposure

Histamine and other chemicals cause allergic reaction

Allergens combine with mast cell

Mast cell releases allergy mediators

Figure 17.16 Allergy. In an allergic reaction, an allergen such as pollen (upper inset) activates B cells, which divide and give rise to antibody-secreting plasma cells. The antibodies attach to mast cells. When the person encounters allergens again, the allergens combine with the antibodies on the mast cells, which then burst (lower inset), releasing the chemicals that cause itchy eyes and a runny nose.

The allergies that people suffer today may be a holdover of an immune function that was important in the past. Evidence for this hypothesis is that people with allergies have higher levels of white blood cells called eosinophils than do others, and these cells fight parasitic infections that are no longer common. In a more general sense, because allergies are more common in developed nations and have become more prevalent since the introduction of antibiotic drugs, some researchers hypothesize that allergies may result from a childhood relatively free of infection, compared to times past—almost as if the immune system is reacting to being underutilized. This idea that allergies stem from an environment too clean to have stimulated the immune system very much is called the hygiene hypothesis.

Key Concepts

1. Inherited immune deficiencies affect innate and adaptive immunity.
2. HIV replicates very rapidly, and T cell production matches it until the immune response is overwhelmed. HIV is a retrovirus that injects its RNA into host cells by binding coreceptors. Reverse transcriptase then copies viral RNA into DNA. HIV uses the cell's protein synthesis machinery to mass produce itself; then the cell releases virus. HIV continually mutates, becoming resistant to drugs.
3. In autoimmune disorders, autoantibodies attack healthy tissue. These conditions may be caused by a virus that borrows a self antigen, T cells that never recognize self, or healthy cells bearing antigens that resemble nonself antigens. Some conditions considered autoimmune may reflect response to retained fetal cells.
4. An overly sensitive immune system causes allergies.
5. Allergens bind to IgE antibodies on mast cells, which release allergy mediators. A subset of helper T cells secretes cytokines that contribute to allergy symptoms.

17.4 Altering Immune Function

Medical technology can alter or augment immune system functions in various ways. Vaccines trick the immune system into acting early. Antibiotic drugs, which are substances derived from organisms such as fungi and soil bacteria, have been used for decades to assist an immune response. Cytokines and altered antibodies are used as drugs to treat a variety of conditions. Transplants require suppression of the immune system so that the body will accept a nonself body part.

Vaccines

A **vaccine** is an inactive or partial form of a pathogen that stimulates the immune system to alert B cells to produce antibodies. When the person then encounters the pathogen, a secondary immune response ensues, even before symptoms arise. Vaccines consisting of entire viruses or bacteria can, rarely, cause illness if they mutate to a pathogenic form. This is a risk of the original smallpox vaccine. A safer vaccine uses only the part of the pathogen's surface that elicits an immune response. Vaccines against different illnesses can be combined into one injection, or the genes encoding antigens from several pathogens can be inserted into a harmless virus and delivered as a "super vaccine."

Vaccine technology dates back to the eleventh century in China. Because people observed that those who recovered from smallpox never got it again, they crushed scabs into a powder that they inhaled or rubbed into pricked skin. In 1796, the wife of a British ambassador to Turkey witnessed the Chinese method of vaccination and mentioned it to English country physician Edward Jenner. Intrigued, Jenner was vaccinated the Chinese way, and then thought of a different approach.

It was widely known that people who milked cows contracted a mild illness called cowpox, but did not get smallpox. The cows became ill from infected horses. Since the virus seemed to jump from one species to another, Jenner wondered whether exposing a healthy person to cowpox lesions might protect against smallpox. A slightly different virus causes cowpox, but Jenner's approach worked, leading to development of the first vaccine (the word comes from the Latin *vacca,* for "cow"). Jenner tried his first vaccine on a volunteer, 8-year-old James Phipps. Jenner dipped a needle in pus oozing from a small cowpox sore on a milkmaid named Sarah Nelmes, then scratched the boy's arm with it. He then exposed the boy to people with smallpox. Young James never became ill. Eventually, improved versions of Jenner's smallpox vaccine eradicated a disease that once killed millions. (**figure 17.17**). By the 1970s, vaccination became unnecessary. However, several nations have resumed smallpox vaccination, as section 17.5 discusses.

People still receive most vaccines as injections, but new delivery methods include nasal sprays and genetically modified fruits and vegetables. A banana as a vaccine makes sense in theory, but in practice it is proving difficult to obtain a uniform product. Edible plants are given genes from pathogens that encode the antigens that evoke an immune response. The foreign antigens stimulate phagocytes beneath the small intestinal lining to "present" the antigens to nearby T cells. From here, the antigens are passed to the bloodstream, where they stimulate B cells to divide to yield plasma cells that produce IgA. These antibodies coat the small intestinal lining, protecting against pathogens in food. Current research focuses on converting plant-based vaccines into powders so that doses can be regulated—but this counters the original goal of easily immunizing babies in developing countries with bananas.

Whatever the form of vaccine, it is important that a substantial proportion of a population be vaccinated. This establishes "herd immunity"—that is, if unvaccinated people are rare, then if the pathogen appears, it does not spread, because so many people are protected. If the population includes pockets of unvaccinated individuals, the disease can spread.

Immunotherapy

Immunotherapy amplifies or redirects the immune response. It originated in the nineteenth century to treat disease. Today, a few immunotherapies are in use, with more in clinical trials.

Monoclonal Antibodies Boost Humoral Immunity

When a B cell recognizes a single foreign antigen, it manufactures a single, or monoclonal, type of antibody. A large amount of a single antibody type could target a particular pathogen or cancer cell because of the antibody's great specificity.

In 1975, British researchers Cesar Milstein and George Köhler devised **monoclonal antibody (MAb) technology,** which mass-produces a single B cell, preserving its specificity and amplifying its antibody type. First, they injected a mouse with a sheep's red blood cells (**figure 17.18**). They then isolated a single B cell from the mouse's spleen and fused it with a cancerous white blood cell from a mouse. The fused cell, called a hybridoma, had a valuable pair of talents. Like the B cell, it produced large amounts of a single antibody type. Like the cancer cell, it divided continuously.

Today MAbs are made to more closely resemble natural human antibodies because the original mouse preparations caused allergic reactions. MAbs are used in basic research, veterinary and human health care, agriculture, forestry, and forensics. MAbs are used to diagnose everything from strep throat to turf grass disease. In a home pregnancy test, a woman places drops of her urine onto a paper strip containing a MAb that binds hCG, the "pregnancy" hormone. The color changes if the MAb binds its target. In cancer diagnosis, if a MAb attached to a fluorescent dye and injected into a patient or applied to a sample of tissue or body fluid binds its target—an antigen found mostly or only on cancer cells—a scanning technology or fluorescence micro-

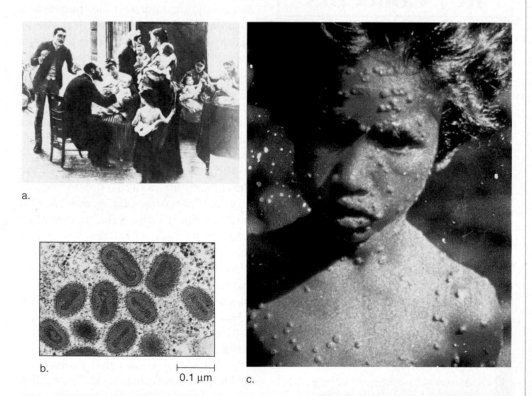

Figure 17.17 Smallpox: Gone? (a) Edward Jenner invented the modern version of a smallpox **(b)** vaccine in 1798. This boy **(c)** is one of the last victims of smallpox, which has not naturally infected a human since 1977. Because many doctors are unfamiliar with smallpox, and people are no longer vaccinated, an outbreak would be a major health disaster.

a.

b.

0.1 μm

c.

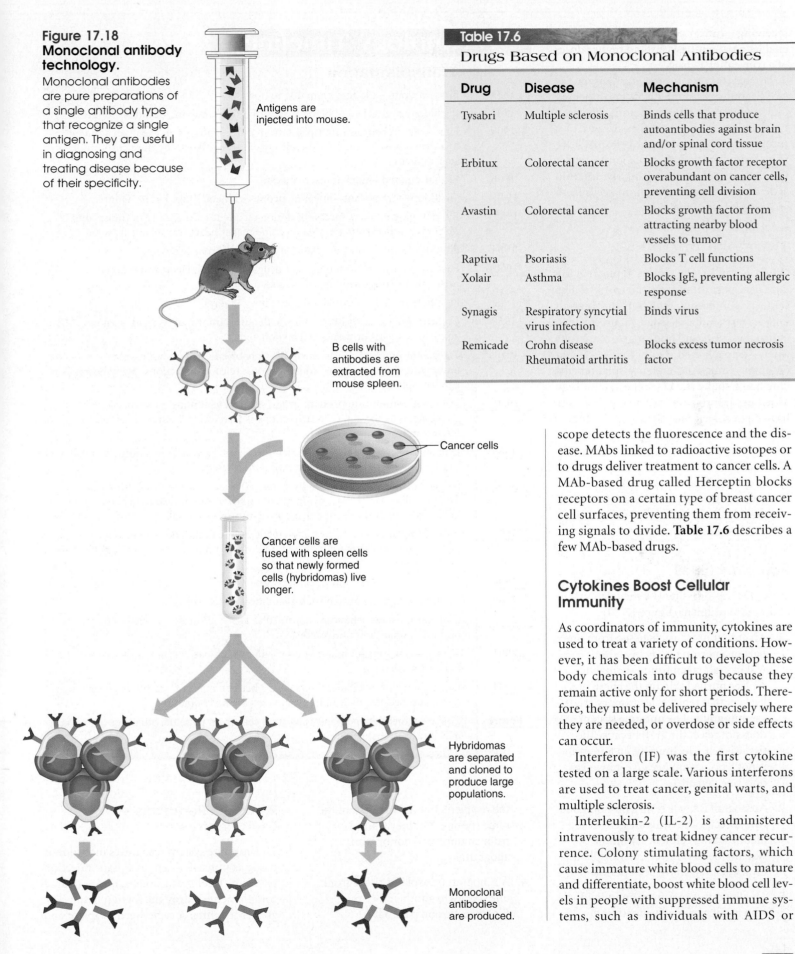

Figure 17.18
Monoclonal antibody technology.
Monoclonal antibodies are pure preparations of a single antibody type that recognize a single antigen. They are useful in diagnosing and treating disease because of their specificity.

Antigens are injected into mouse.

B cells with antibodies are extracted from mouse spleen.

Cancer cells

Cancer cells are fused with spleen cells so that newly formed cells (hybridomas) live longer.

Hybridomas are separated and cloned to produce large populations.

Monoclonal antibodies are produced.

Table 17.6
Drugs Based on Monoclonal Antibodies

Drug	Disease	Mechanism
Tysabri	Multiple sclerosis	Binds cells that produce autoantibodies against brain and/or spinal cord tissue
Erbitux	Colorectal cancer	Blocks growth factor receptor overabundant on cancer cells, preventing cell division
Avastin	Colorectal cancer	Blocks growth factor from attracting nearby blood vessels to tumor
Raptiva	Psoriasis	Blocks T cell functions
Xolair	Asthma	Blocks IgE, preventing allergic response
Synagis	Respiratory syncytial virus infection	Binds virus
Remicade	Crohn disease Rheumatoid arthritis	Blocks excess tumor necrosis factor

scope detects the fluorescence and the disease. MAbs linked to radioactive isotopes or to drugs deliver treatment to cancer cells. A MAb-based drug called Herceptin blocks receptors on a certain type of breast cancer cell surfaces, preventing them from receiving signals to divide. **Table 17.6** describes a few MAb-based drugs.

Cytokines Boost Cellular Immunity

As coordinators of immunity, cytokines are used to treat a variety of conditions. However, it has been difficult to develop these body chemicals into drugs because they remain active only for short periods. Therefore, they must be delivered precisely where they are needed, or overdose or side effects can occur.

Interferon (IF) was the first cytokine tested on a large scale. Various interferons are used to treat cancer, genital warts, and multiple sclerosis.

Interleukin-2 (IL-2) is administered intravenously to treat kidney cancer recurrence. Colony stimulating factors, which cause immature white blood cells to mature and differentiate, boost white blood cell levels in people with suppressed immune systems, such as individuals with AIDS or receiving cancer chemotherapy. This enables a patient to withstand higher doses of a conventional drug.

Because excess tumor necrosis factor (TNF) underlies some disorders, blocking its activity treats some conditions. The drug Enbrel, for example, consists of part of a receptor for TNF. Taking it prevents TNF from binding to cells that line joints, relieving arthritis. Excess TNF in rheumatoid arthritis prevents the joint lining cells from secreting lubricating fluid.

Transplants

When a car breaks down, replacing the damaged part often fixes the trouble. The same is sometimes true for the human body. Hearts, kidneys, livers, lungs, corneas, pancreases, skin, and bone marrow are routinely transplanted, sometimes several organs at a time. Although transplant medicine had a shaky start (see the Technology Timeline: Transplantation), many problems have been solved (see figure 17.1). Today, thousands of transplants are performed annually and recipients gain years of life. The challenge to successful transplantation lies in genetics because individual inherited differences in cell surfaces determine whether the body will accept tissue from a particular donor.

Transplant Types

Transplants are classified by the relationship of donor to recipient (**figure 17.19**):

1. An autograft transfers tissue from one part of a person's body to another. A skin graft taken from the thigh to replace burned skin on the chest, or a leg vein that replaces a coronary artery, are autografts. The immune system does not reject the graft because the tissue is self. (Technically, an autograft is not a transplant because it involves only one person.)

2. An isograft is tissue from a monozygotic twin. Because the twins are genetically identical, the recipient's immune system does not reject the transplant. Ovary isografts have been performed.

3. An allograft comes from an individual who is not genetically identical to the recipient, but is a member of the same species. A kidney transplant from an unrelated donor is an allograft.

4. A xenograft transplants tissue from one species to another. (See the Bioethics Box on page 349.)

Technology Timeline

Transplantation

1899	First allograft—a kidney from dog to dog.
1902	Pig kidney is attached to blood vessels of a woman dying of kidney failure.
1905	First successful corneal transplant, from a boy who lost an eye in an accident to a man whose cornea was chemically damaged. Works because cornea cells lack antigens.
1906	First attempted kidney transplant fails.
1940s	First kidney transplants on young people with end-stage kidney failure.
1950s	Blood typing predicts success of donor-recipient pairs for organ transplants. Invention of heart-lung bypass machine makes heart transplants feasible.
1954	First successful organ transplant, of a kidney from a MZ twin.
1960s	First effective immunosuppressant drugs developed. Heart transplants performed in dogs with mixed success.
1967	First human heart transplant. Patient lives 18 days.
1968	Uniform Anatomical Gift Act passes. Requires informed consent from next of kin before organs or tissues can be donated.
1970s	Transplant problems: they extend life only briefly and do not correct underlying disease; surgical complications; rejection reactions. Many hospitals ban transplants.
1980s	Improved immunosuppressant drugs, surgical techniques, and tissue matching, plus the ability to strip antigens from donor tissue, reawaken interest in transplants.
1984	Doctors transplant a baboon's heart into "Baby Fae," who was born with half a heart. She lives 20 days before rejecting the xenograft.
1992	Surgeons transplant a baboon liver into a 35-year-old man with hepatitis. The man lives for 71 days, dying of an unrelated cause. Researchers realize lower doses of anti-rejection drugs may improve survival.
1997	Pig cell implants used to treat pancreatic failure and Parkinson disease. Pig liver used to maintain liver function for six hours as young man awaited a human liver.
1998	Transplants of hands and forearms begin.
2000	Cloning of pigs brings xenotransplantation closer to reality.
2003	Cloned mini-pigs genetically modified to lack cell surface molecules that provoke human immune response.
2003	DNA gene expression microarrays predict which patients are likely to reject kidney transplants.
2004	Researchers discover that infants have a better chance of accepting a heart transplant because their immune systems are not as mature.
Future	Tissue engineering, perhaps using stem cells from patients, others, or embryos, may replace transplantation.

Rejection Reactions— Or Acceptance

The immune system recognizes most donor tissue as nonself. Then, in a tissue rejection reaction, T cells, antibodies, and activated complement destroy the foreign tissue. The greater the difference between recipient and

Pig Parts

In 1902, a German medical journal reported an astonishing experiment. A physician, Emmerich Ullman, had attached the blood vessels of a patient dying of kidney failure to a pig's kidney set up by her bedside. The experiment failed when the patient's immune system rejected the attachment almost immediately.

Nearly a century later, in 1997, an eerily similar experiment took place. Robert Pennington, a 19-year-old suffering from acute liver failure and desperately needing a transplant, survived for six and a half hours with his blood circulating outside of his body through a living liver removed from a 15-week-old, 118-pound pig named Sweetie Pie. The pig liver served as a bridge until a human liver became available. But Sweetie Pie was no ordinary pig. She had been genetically modified and bred so that her cells displayed a human protein that controlled rejection of tissue transplanted from an animal of another species. Because of this slight but key bit of added humanity, plus immunosuppressant drugs, Pennington's body tolerated the pig liver's help for the few crucial hours. Baboons have also been used as sources of transplant organs (**figure 1**).

Successful xenotransplants would help alleviate the organ shortage. However, some people object to raising animals to use their organs as transplants because it requires killing the donors. One researcher counters such protests by comparing the use of animal organs to eating them.

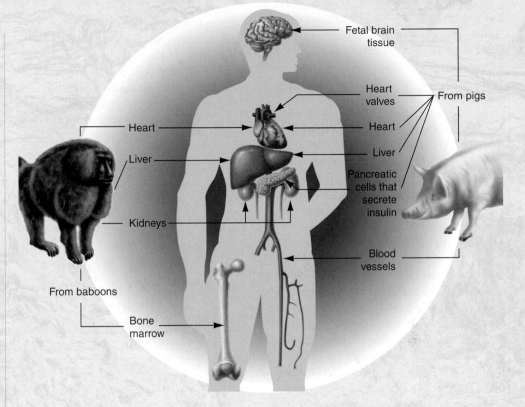

Figure 1   Baboons and pigs can provide tissues and organs for transplant.

A possible danger of xenotransplants is that people may acquire viruses from the donor organs. Viruses can "jump" species, and the outcome in the new host is unpredictable. For example, a virus called PERV—for "porcine endogenous retrovirus"—can infect human cells in culture. However, several dozen patients who received implants of pig tissue did not show evidence of PERV years later. That study, though, looked only at blood. We still do not know what effect pig viruses can have on a human body. Because many viral infections take years to cause symptoms, introducing a new infectious disease in the future could be the trade-off for using xenotransplants to solve the current organ shortage.

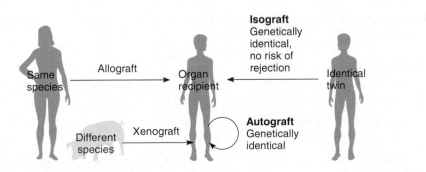

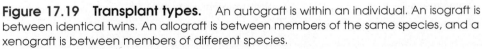

Figure 17.19   Transplant types.   An autograft is within an individual. An isograft is between identical twins. An allograft is between members of the same species, and a xenograft is between members of different species.

donor cell surfaces, the more rapid and severe the rejection reaction. An extreme example is the hyperacute rejection reaction against tissue transplanted from another species—the donor tissue is usually destroyed in minutes as blood vessels blacken and cut off the blood supply.

Physicians use several approaches to dampen rejection so that a transplant recipient can survive. These include closely matching the HLA types of donor and recipient and stripping donor tissue of antigens. Newer immunosuppressive drugs inhibit production of the antibodies and T cells that specifically attack transplanted tissue, while sparing other components of the immune system. Experiments on transplanted tissues using gene expression microarrays reveal at least three types of rejection not otherwise obvious. Such profiling will likely be used to avoid rejection reactions.

Rejection is not the only problem that can arise from an organ transplant. Graft-versus-host disease develops sometimes when bone marrow transplants are used to correct certain blood deficiencies and cancers. The transplanted bone marrow, which is actually part of the donor's immune system, attacks the recipient—its new body—as foreign. Symptoms include rash, abdominal pain, nausea, vomiting, hair loss, and jaundice.

Sometimes a problem arises if a bone marrow transplant to treat cancer is too closely matched to the recipient. This may at first seem illogical, but what happens is that if the cancer returns with the same cell surfaces as it had earlier, the patient's new bone marrow is so similar to the old marrow that it is equally unable to fight the cancer. The best tissue for transplant may be highly similar, but not identical, to the recipient's tissues. That is, the donor bone marrow should be different enough to control the cancer, but not so different that rejection occurs.

17.5 A Genomic View of Immunity—The Pathogen's Perspective

Immunity against infectious disease involves interactions of two genomes—ours and the pathogen's. At the same time that human genome information is revealing how the immune system halts infectious disease, information coming from the sequencing of the genomes of pathogens is also useful. **Table 17.7** lists some of the pathogens whose genomes have been sequenced.

Knowing the DNA sequence of a pathogen's genome, or the sequences of key genes, can reveal exactly how that organism causes illness in humans, which can suggest new treatment strategies. The sequence for *Streptococcus pneumoniae*, for example, revealed instructions for a huge protein that enables the bacterium to adhere to human cells. Pharmaceutical researchers can now synthesize potential drugs that dismantle this previously unknown adhesion protein.

Pathogen genome information is also used to protect against infection, in an approach called reverse vaccinology. Instead of culturing hard-to-grow pathogens in the laboratory, researchers scan genome sequences and identify antigens that will evoke a protective response from the human immune system. This strategy enabled researchers to rapidly develop experimental vaccines against severe acquired respiratory syndrome (SARS).

Crowd Diseases

History provides clues to the complex and ever-changing relationships between humans and our pathogens. Because adap-

Key Concepts

1. Vaccines are disabled pathogens or their parts that elicit an immune response, protecting against infection by the active pathogen.
2. Immunotherapy uses immune system components to fight disease. B cells fused with cancer cells produce MAbs that target specific antigens. Cytokines boost immune function and destroy cancer cells.
3. Autografts transfer tissue from one part of a person's body to another; isografts are between identical twins; allografts are between members of the same species; and a xenograft is a cross-species transplant.
4. Allografts can cause tissue rejection reactions, and xenografts can set off hyperacute rejection. In graft-versus-host disease, transplanted bone marrow rejects the recipient's tissues.

Table 17.7

Pathogens with Sequenced Genomes

Pathogen	Human Disease
Bacterial	
Borrelia burgdorferi	Lyme disease
Brucella suis	Fever (infertility in other animals)
Campylobacter jejuni	Food poisoning
Clostridium perfringens	Food poisoning
Enterococcus faecalis	Urinary tract, wound, intestinal, and heart infections
Listeria monocytogenes	Lethal infection in newborns
Mycobacterium tuberculosis	Tuberculosis
Neisseria meningitides	Meningitis and septicemia (brain membrane inflammation and blood poisoning)
Streptococcus pyogenes	Puerperal fever, scarlet fever, pharyngitis, impetigo, cellulitis, "flesh-eating bacteria"
Treponema pallidum	Syphilis
Vibrio cholerae	Cholera
Yersinia pestis	Plague
Nonbacterial	
Brugia malayi (a worm)	Elephantiasis (grossly enlarged lymph nodes)
Entamoeba histolytica	Intestinal infection
Plasmodium falciparum	Malaria
Schistosoma mansoni	Schistosomiasis
Toxoplasma gondii	Birth defects, opportunistic infection in AIDS
Trypanosoma brucei	African sleeping sickness
Trypanosoma cruzi	Chagas disease

tive immunity responds to an environmental stimulus, epidemics often followed the introduction of a pathogen into a population that had not encountered it before.

When Europeans first explored the New World, they inadvertently brought bacteria and viruses to which their immune systems had adapted. The immune systems of Native Americans, however, had never encountered these pathogens. Many people died. Smallpox decimated the Aztec population in Mexico from 20 million in 1519, when conquistador Hernán Cortés arrived from Spain, to 10 million by 1521, when Cortés returned. By 1618, the Aztec nation had fallen to 1.6 million. The Incas in Peru and northern populations were also dying of smallpox. When explorers visited what is now the southeast United States, they found abandoned towns where natives had died from smallpox, measles, pertussis, typhus, and influenza.

The diseases that so easily killed Native Americans are known as "crowd" diseases, because they arose with the spread of agriculture and urbanization and affect many people. Crowd diseases swept Europe and Asia as expanding trade routes spread bacteria and viruses along with silk and spices. More recently, air travel has spread crowd diseases, such as SARS.

Crowd diseases tend to pass from conquerors who live in large, intercommunicating societies to smaller, more isolated and more susceptible populations, and not vice versa. When Columbus arrived in the New World, the large populations of Europe and Asia had existed far longer than American settlements. In Europe and Asia, infectious diseases had time to become established and for human populations to adapt to them. In contrast, an unfamiliar infectious disease can quickly wipe out an isolated tribe, leaving no one behind to give the illness to new invaders. We may never even know of them.

Most crowd diseases vanish quickly, for several reasons: vaccines or treatments may stop transmission; people may alter their behaviors to avoid contracting the infection; or the disease may kill before individuals can pass it on. Sometimes, we don't know why a disease vanishes or becomes milder.

We may be able to treat and control newly evolving infectious diseases one at a time, with new drugs and vaccines. But the mutation process that continually spawns new genetic variants in microbe populations—resulting in evolution—means that new infectious diseases will continue to arise, and old ones to return or ravage new populations.

Bioweapons

It may seem incomprehensible that anyone would ever use pathogens to intentionally harm people, but it is a sad fact of history—and the present—that such bioweapons exist. Biological weapons have been around since medieval warriors catapulted plague-ridden corpses over city walls to kill the inhabitants. During the French and Indian War, the British gave Native Americans blankets intentionally contaminated with secretions from smallpox victims. Although international law banned "germ warfare" in 1925, from 1932 until 1942 Japan field-tested bacterial bioweapons in rural China, killing thousands.

In 1973, the Soviet Union established an organization called Biopreparat. Thousands of workers in 50 facilities prepared anthrax bombs and other bioweapons under the guise of manufacturing legitimate drugs, vaccines, and veterinary products. Soviet bioweapons were even more lethal than their natural counterparts. Plague bacteria, for example, were genetically modified to resist sixteen antibiotics and to manufacture a protein that strips nerve cells of their fatty coats, adding paralysis to the list of natural symptoms.

Anthrax is another bacterial infection that has left its mark on the history of bioweaponry. In 1979, an accident occurred in a Soviet city then called Sverdlovsk. At Military Compound Number 19, a miscommunication among shift workers in charge of changing safety air filters caused the release of a cloud of dried anthrax spores over the city. Within weeks, more than 100 people died of anthrax. They were mostly young, healthy men who were outside on that Friday night and breathed in enough anthrax spores to cause the inhaled form of the illness. The government officially announced that the deaths were due to eating infected meat. Then workers sprayed water everywhere, reaerosolizing the spores and causing more infections. The deadliness of inhalation anthrax is why health officials were so concerned when spores were mailed to a senator, media representatives, and others in 2001 in the U.S. Several people died.

The symptoms of inhalation anthrax result from a toxin that consists of three proteins. One protein forms a barrel-like structure that binds to macrophages and admits the other two components of the toxin. One of these components overloads signal transduction and impairs the cell's ability to function as a phagocyte. The other toxin component breaks open macrophages, which release tumor necrosis factor and interleukins. Early symptoms of inhalation anthrax resemble influenza, but the victim rapidly suffers respiratory collapse.

In September 1992, then-president Boris Yeltsin officially halted bioweapon research in the former Soviet Union. Twenty years earlier, political leaders in London, Moscow, and Washington had signed the Biological Weapons Convention, an effort to prevent bioterrorism. Its protocols are being strengthened today.

In the United States, a small-scale bioweapons effort began in 1942. A facility at Fort Detrick in Frederick, Maryland, stored 5,000 bombs loaded with anthrax spores; a production facility for the bombs was located in Terre Haute, Indiana; and Mississippi and Utah had test sites. President Richard Nixon halted the program in 1969; he thought that conventional and nuclear weapons were a sufficient deterrent and defense.

Today bioweapons are again a threat. Such weapons have come far since the Japanese dropped porcelain containers of plague-ridden fleas over China. Today's bioterrorists not only know how to grow and dry pathogens, but how to control particle size to ease infection. In addition, genetic modification can alter the characteristics of a virus or bacterium intended for use as a weapon, making it even deadlier, or targeting specific types of victims.

Key Concepts

1. Knowing the genome sequence of a pathogen can reveal how it evades the human immune system.
2. Crowd diseases happen when infectious agents are introduced into a population that hasn't encountered them before.
3. Bioterrorism is the use of pathogens— either in their natural state or genetically manipulated—to kill people.

Summary

17.1 The Importance of Cell Surfaces

1. The cells and biochemicals of the immune system distinguish self from nonself, protecting the body against infections and cancer.

2. Most genetic effects on immunity are polygenic, but a few single genes have significant effects.

3. Patterns of cell surface proteins and glycoproteins determine blood types. An **antigen** is a molecule that elicits an immune response. A blood incompatibility occurs if a blood recipient manufactures **antibodies** against antigens in donor blood. Blood type systems include ABO and Rh.

4. **HLA** genes are closely linked on chromosome 6 and encode cell surface antigens that present foreign antigens to the immune system.

17.2 The Human Immune System

5. If a pathogen breaches physical barriers, the **innate immune response** produces the redness and swelling of inflammation, plus complement, collectins, and cytokines. The response is broad and general.

6. The **adaptive immune response** is slower, specific, and has memory. This response is both humoral and cellular.

7. The **humoral immune response** begins when macrophages display foreign antigens near HLA antigens. This activates **T cells,** which activate **B cells.** The B cells, in turn, give rise to plasma cells and secrete specific antibodies. Some B cells give rise to memory cells.

8. An antibody is Y-shaped and made up of four polypeptide chains, two heavy and two light. Each antibody molecule has regions of constant amino acid sequence and regions of variable sequence.

9. The tips of the Y of each subunit form antigen binding sites, which include the more specific idiotypes that bind foreign antigens at their epitopes.

10. Antibodies bind antigens to form immune complexes large enough for other immune system components to detect and destroy. Antibody genes are rearranged during early B cell development, providing instructions to produce a great variety of antibodies.

11. T cells carry out the **cellular immune response.** Their precursors, called thymocytes, are selected in the thymus to recognize self. Helper T cells secrete cytokines that activate other T cells and B cells. A helper T cell's CD4 antigen binds macrophages that present foreign antigens. Cytotoxic T cells release biochemicals that bore into and kill bacteria and also destroy cells covered with viruses.

17.3 Abnormal Immunity

12. Mutations in antibody or cytokine genes, or in genes encoding T cell receptors, cause inherited immune deficiencies. Severe combined immune deficiencies affect both branches of the immune system.

13. HIV binds to the coreceptors CD4 and CCR5 on macrophages and helper T cells, and, later in infection, triggers apoptosis of cytotoxic T cells. As HIV replicates, it mutates, evading immune attack. Falling CD4 helper T cell numbers allow opportunistic infections and cancers to flourish. People who cannot produce a complete CCR5 protein resist HIV infection.

14. In an **autoimmune disease,** the body manufactures **autoantibodies** against its own cells. Autoimmunity may result from a virus that incorporates and displays a self antigen, from bacteria or cancer cells that have antigens that resemble self antigens, from unselected T cells, or from lingering fetal cells.

15. In susceptible individuals, allergens stimulate IgE antibodies to bind to mast cells, which causes the cells to release allergy mediators. Certain helper T cells release selected cytokines. Allergies may be a holdover of past immune function.

17.4 Altering Immune Function

16. A **vaccine** presents a disabled pathogen, or part of one, to elicit a primary immune response.

17. Immunotherapy enhances or redirects immune function. **Monoclonal antibodies** are useful in diagnosing and treating some diseases because of their abundance and specificity. To create MAbs, individual activated B cells are fused with cancer cells to form hybridomas. Cytokines are used to treat various conditions.

18. Transplant types include autografts (within oneself), isografts (between identical twins), allografts (within a species), and xenografts (between species). A tissue rejection reaction occurs if donor tissue is too unlike recipient tissue.

17.5 A Genomic View of Immunity—The Pathogen's Perspective

19. Learning the genome sequences of pathogens can reveal how they infect, which provides clues to developing new treatments.

20. Crowd diseases spread rapidly through a population that has had no prior exposure, passed from members of a population that have had time to adapt to the pathogen.

21. Throughout history, people have used bacteria and viruses as weapons.

Review Questions

1. Match the cell type to the type of biochemical it produces.

   1. mast cell
   2. T cell
   3. B cell
   4. macrophage
   5. all cells with nuclei
   6. antigen-presenting cell

   a. antibodies
   b. HLA class II genes
   c. interleukin
   d. histamine
   e. interferon
   f. heparin
   g. tumor necrosis factor
   h. HLA class I *A, B,* and *C* genes

2. Distinguish between
   a. a T cell and a B cell.
   b. innate and adaptive immunity.
   c. a primary and secondary immune response.
   d. a cellular and humoral immune response.
   e. an autoimmune condition and an allergy.
   f. an inherited and acquired immune deficiency.

3. What is the physical basis of a blood type? of blood incompatibility?

4. What would be the consequences of lacking
   a. helper T cells?
   b. cytotoxic T cells?
   c. B cells?
   d. macrophages?

5. State the function of each of the following immune system biochemicals:
   a. complement proteins
   b. collectins
   c. antibodies
   d. cytokines

6. Which components of the human immune response explain why we experience the same symptoms of an upper respiratory infection (a "cold") when many different types of viruses can cause these conditions?

7. What does HIV do to the human immune system?

8. What are the dangers of a bone marrow transplant being too different from the recipient's tissues? too similar?

9. Cite three reasons why developing a vaccine against HIV infection has been challenging.

10. It was once said that thymocytes are "educated" in the thymus, meaning that immature T cells are somehow "taught" to recognize self cell surfaces and refrain from attack. This is not exactly what happens. Why?

11. What part do antibodies play in allergic reactions and in autoimmune disorders?

12. How do each of the following illnesses disturb immunity?
    a. graft-versus-host disease
    b. SCID
    c. scleroderma
    d. AIDS
    e. hayfever

13. Why is a deficiency of T cells more dangerous than a deficiency of B cells?

14. What do a plasma cell and a memory cell descended from the same B cell have in common? How do they differ?

15. Why is a polyclonal antibody response valuable in the body, but a monoclonal antibody valuable as a diagnostic tool?

Applied Questions

1. A man is flown to an emergency room of a major medical center, near death after massive blood loss in a car accident. There isn't time to match blood types, so the physician orders type O negative blood. Why did she order this type of blood?

2. Rasmussen's encephalitis causes children to have 100 or more seizures a day. Affected children have antibodies that attack brain cell receptors that normally receive nervous system biochemicals. Is this condition most likely an inherited immune deficiency, an adaptive immune deficiency, an autoimmune disorder, or an allergy? State a reason for your answer.

3. Allergy to a protein in peanuts can cause anaphylactic shock. An experimental vaccine consists of the gene encoding this protein, wrapped in an edible carbohydrate, so it can be eaten and stimulate production of protective antibodies in the small intestine. When this vaccine was fed to rats who have a peanut allergy, their blood showed lowered levels of IgE, but increased levels of IgG. Also, their bowel movements contained higher than usual levels of IgA. Is the vaccine working? How can you tell?

4. In people with a certain HLA genotype, a protein in their joints resembles an antigen on the bacterium that causes Lyme disease, an infection transmitted in a tick bite that causes flulike symptoms followed by arthritis (joint inflammation). When these individuals become infected, their immune systems attack not only the bacteria, but also their joints. Explain why antibiotic therapy helps treat the early phase of the disease, but not the arthritis.

5. Even in overwhelmingly deadly infectious diseases, such as bubonic plague and Ebola hemorrhagic fever, a small percentage of the human population survives. Suggest two mechanisms based on immune system functioning that can account for their survival.

6. A person exposed for the first time to Coxsackie virus develops a painful sore throat. How is the immune system alerted to the exposure to the virus? When the person encounters the virus again, why doesn't she develop symptoms?

7. A young woman who has aplastic anemia will soon die as her lymphocyte levels drop sharply. What type of cytokine might help her?

8. In Robin Cook's novel *Chromosome Six,* a geneticist places a portion of human chromosome 6 into fertilized ova from bonobos (pygmy chimps). The bonobos that result are used to provide organs for transplant into specific individuals. Explain how this technique would work.

9. Suggest ways that local, state, and federal governments can prepare to handle a bioterrorism attack.

10. Is the heritability of SCID likely to be higher or lower than that for an allergy? Why?

Web Activities

Visit the Online Learning Center (OLC) at www.mhhe.com/lewisgenetics7. Select **Student Edition, chapter 17,** and **Web Activities** to find the website links needed to complete the following activities.

11. Many websites describe products (food supplements) that supposedly "boost" immune system function. Very often, the descriptions are vague, use meaningless jargon or buzzwords, or contain misinformation. Locate such a website and identify claims that are unclear, deceptive, or incorrect. Alternatively, identify a claim that *is* consistent with the description of immune system function in this chapter.

12. Go to the Blood Book website. This site explains human blood groups; use it to try to answer the following questions.

    a. Which blood group is capable of being the most diverse in a population?

    b. Why has classification of human blood types been so confusing?

    c. In predicting blood types of future offspring, why would it be important to know whether the genes encoding different blood group antigens, or the enzymes that make their synthesis possible, are linked or unlinked?

Case Studies and Research Results

13. The following couples each have one child and are expecting their second. Consult their Rh genotypes, and determine whether each second pregnancy will definitely have an Rh incompatability, will definitely *not* have an incompatibility, or might have one. (*Hint:* Determine whether the woman is Rh⁻, and the likelihood that the fetus is Rh⁺.)

    a. Rachel is genotype *CcDdee* and Ross is *ccddee.*

    b. Monica is *ccddee* and Chandler is *CCDDEE.*

    c. Elaine and Jerry are both *ccddee.*

    d. Gwen is *ccddee* and Gavin is *ccDdEe.*

14. African Americans are at higher risk of developing conditions associated with a too vigorous inflammatory response (heart attack, diabetes, stroke, and kidney disease) than people in other populations. To see whether this tendency is inherited, researchers examined variants of several genes whose protein products regulate cytokines (interleukins and tumor necrosis factor). The investigators compiled the genetic information for 179 African American women and 396 white women who delivered healthy babies at a Boston hospital between 1997 and 2001. The African Americans were much more likely to have genetic profiles indicating increased inflammation than the white women.

    a. From this information, what would you conclude about the causes of the increased risk for inflammatory conditions among African Americans?

    b. What other factors besides an association with skin color might contribute to the increased risk?

    c. Why might certain alleles be more common in one population compared to another?

    d. Design an experiment to test whether some other factor, such as economic status or whether one lives in a city or rural area, contributes to elevated risk of developing inflammatory disorders.

15. State whether each of the following situations involves an autograft, an isograft, an allograft, or a xenograft.

    a. A man donates part of his liver to help his daughter, who has a damaged liver due to cystic fibrosis.

    b. A woman with infertility receives an ovary transplant from her identical twin sister.

    c. A man receives a heart valve from a pig.

    d. A woman who has had a breast removed has a new breast built using fatty tissue from her thigh.

16. Mark and Louise are planning to have their first child, but they are concerned because they think that they have an Rh incompatibility. He is Rh⁻ and she is Rh⁺. Will there be a problem? Why or why not?

Learn to apply the skills of genetic counselor with this additional case found in the *Case Workbook in Human Genetics:*

Stiff person syndrome

A vanishing twin helps an athlete

CHAPTER

18

The Genetics of Cancer

CHAPTER CONTENTS

MICROARRAYS ILLUMINATE THYROID CANCER

I never thought I would care very much about the cells composing my thyroid gland. That changed on August 4, 1993, when my physician, looking at me from across a room, said, "What's that lump in your neck?"

Soon after, a specialist stuck six thin needles into the lump to sample thyroid cells for testing, telling me that 99 percent of thyroid lumps are not cancerous. However, when he approached with a seventh needle for a sample for a study on something called *p53*, I began to worry, because *p53* is a gene associated with cancer. When the specialist called early on a Monday morning, I knew I was among the unlucky 1 percent, because doctors never deliver bad news at the start of a weekend. I had papillary thyroid cancer, which accounts for 80 percent of cases and is easily treated with surgery and radioactive iodine. But when I was on the operating table, the surgeon did not think the lesion looked like a papillary tumor. Off it went to the pathology lab, while I waited on the table, adding to the danger of my surgery. The results: I had two tumors, one papillary, one follicular. I was successfully treated.

Had I developed thyroid cancer today, I might not have had to wait on an operating table while a pathologist examined my cells for the telltale distinctions between tumor types. DNA microarrays can now highlight five key genes that are expressed differently in papillary and follicular thyroid cancer. My physicians would have known, before surgery, the genetic nature of my tumors. This approach is very valuable for cancer in which treatment differs depending upon the genetic profiles of the cells, such as breast or prostate cancers.

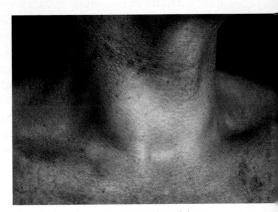

I had no symptoms of my thyroid cancer, and did not even notice the subtle, yet egg-sized swelling in my neck.

Cancer has been part of human existence for eons. Egyptian mummies from 3000 B.C. show evidence of cancerous tumors, and by 1600 B.C., the Egyptians were clearly attempting to treat cancer. Papyruses illustrate them cutting or burning off growths, and using more inventive treatments for less obvious tumors. A remedy for uterine cancer, for example, introduced fresh ground dates mixed with pig's brain into the vagina!

By 300 B.C., Hippocrates had described several types of tumors, and coined the term "cancer" to describe the crablike shape of a tumor invading normal tissue. He attributed cancer to a buildup of black bile; others blamed it on fermenting lymph, injury, irritation, or simply "melancholia." Today we know that the collection of diseases called cancer reflects a profound derangement of the cell cycle. Sequences of mutations in somatic cells underlie the progression of cancer as it invades and spreads.

Cancer has or will affect one in three of us. Diagnosis and treatment are becoming increasingly individualized, thanks largely to a genomic approach to describing cancer cells.

18.1 Cancer Is Genetic, But Usually Not Inherited

Cancer is a complication of being a many-celled organism. Our specialized cells must follow a schedule of mitosis—the cell cycle—so that organs and other body parts either grow appropriately during childhood, stay a particular size and shape in an adult, or repair damage by replacing tissue. If a cell in solid tissue escapes normal controls on its division rate, it forms a growth called a tumor (**figure 18.1**). In the blood, such a cell divides to take over the population of blood cells. A tumor is benign if it grows in place but does not spread into, or "invade," surrounding tissue; a tumor is cancerous, or malignant, if it infiltrates nearby tissue. A malignant tumor also sends parts of itself into the bloodstream or lymphatic vessels, either of which transports it to other areas, where the cancer cells "seed" the formation of new tumors. The process of spreading is termed **metastasis,** which means "not standing still."

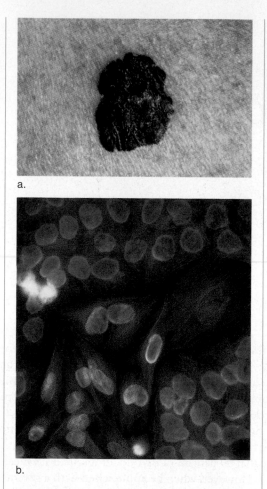

a.

b.

Figure 18.1 Cancer cells stand out. **(a)** A melanoma is a cancer of the pigment-producing cells (melanocytes) in the skin. It may have any or all of four characteristics, abbreviated ABCD: it is **a**symmetric, has **b**orders that are irregular, **c**olor variations, and a **d**iameter of more than 5 millimeters. **(b)** Stains and dyes can reveal cancer at the cellular level. These melanoma cells stain orange. The different staining characteristics of cancer cells reflect differences in gene expression patterns between the normal and cancerous states.

Cancer is a group of disorders that arise from alterations in genes. Only about 10 percent of cases are inherited as single-gene disorders, in which the faulty instructions are in every cell. More often, mutations in cancer-causing genes occur in a few somatic cells over a lifetime. Cancer is thus a genetic disease at the cellular level, but not at the whole-body level. That is, usually cancer is not inherited from a parent, but results from a somatic mutation in the affected person's cell.

Probably combinations of particular gene variants sum to increase the risk of cancer, which may explain how cancer can "run in families" yet not follow a Mendelian pattern of inheritance. Cancer often takes years to develop, as a sequence of genes mutate in the affected tissue. Then, the cells whose mutations enable them to divide more often than others gradually take over the tissue. At the same time, changes at the gene expression level fuel the disease process. Even though a cancer may not spread for years, mutations or changes in gene expression that indicate that it will do so can occur early in the course of illness.

Cancer wasn't always considered a genetic phenomenon. When President Richard Nixon declared a "war on cancer" in 1971, the targets were radiation, viruses, and chemicals. These agents cause cancer, but they do so by interfering with the precise genetic controls of cell division.

Researchers first discovered genes that could cause cancer in 1976, but there had been earlier hints in that most substances known to be carcinogens (causing cancer) are also mutagens (damaging DNA). Did the genetic change cause the cancer? A second line of evidence came from families in which colon or breast cancer was so prevalent that it fit the inheritance pattern of a Mendelian trait.

RELATING THE CONCEPTS

Why was the author not concerned about passing on her thyroid cancer to her children?

From Single Mutations to Sweeping Changes in Gene Expression

In the 1980s and 1990s, searches for cancer-causing genes began with rare families that had many young members who had the same type of cancer. Researchers then identified parts of the genome that only the affected individuals shared, such as a chromosomal aberration or a unique DNA sequence. Then the search focused on specific genes in the identified region whose protein products could affect cell cycle control. This approach led to the discovery of

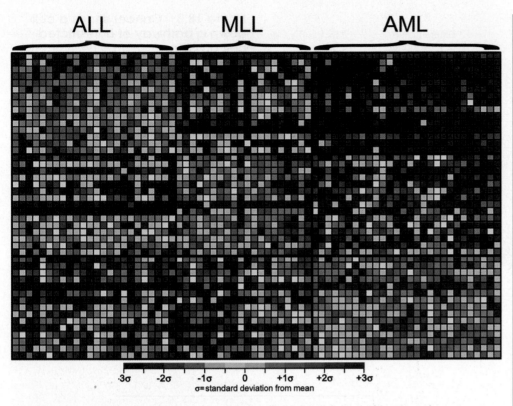

ALL   MLL   AML

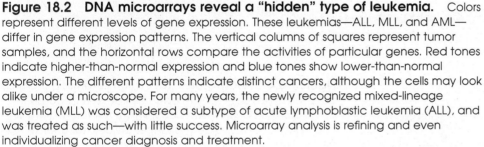

-3σ   -2σ   -1σ   0   +1σ   +2σ   +3σ
σ = standard deviation from mean

Figure 18.2 DNA microarrays reveal a "hidden" type of leukemia. Colors represent different levels of gene expression. These leukemias—ALL, MLL, and AML—differ in gene expression patterns. The vertical columns of squares represent tumor samples, and the horizontal rows compare the activities of particular genes. Red tones indicate higher-than-normal expression and blue tones show lower-than-normal expression. The different patterns indicate distinct cancers, although the cells may look alike under a microscope. For many years, the newly recognized mixed-lineage leukemia (MLL) was considered a subtype of acute lymphoblastic leukemia (ALL), and was treated as such—with little success. Microarray analysis is refining and even individualizing cancer diagnosis and treatment.

more than 100 **oncogenes,** which cause cancer when they are inappropriately activated, and of more than 30 **tumor suppressor genes,** whose deletion or inactivation causes cancer.

A genomics-based approach to understanding cancer uses DNA microarrays that highlight the expression differences among thousands of genes. At first this approach was used to supplement cancer cell distinctions that could be seen under a microscope, such as nucleus shape. Microarray analysis has become so refined that it can reveal previously unrecognized cancer subtypes. **Figure 18.2** highlights how microarrays revealed a new type of leukemia, discussed in the opener to chapter 11. In the future, DNA microarray gene expression panels will provide "molecular portraits" of many cancers.

Loss of Cell Cycle Control

Cancer is a consequence of cell cycle disruption. It begins when a cell divides more frequently, or more times, than the normal cell type it descended from (**figure 18.3**).

The timing, rate, and number of mitoses a cell undergoes depend on protein growth factors and signaling molecules from outside the cell, and on transcription factors from within. Because these biochemicals are under genetic control, so is the cell cycle. Cancer cells probably arise often, because mitoses are so frequent that an occasional cell escapes control. However, the immune system destroys most cancer cells after recognizing tumor-specific antigens on their surfaces.

The discovery of the checkpoints that control the cell cycle revealed how cancer can begin (see figure 2.18). A mutation in a gene that normally halts or slows the cell cycle can lift the constraint, leading to inappropriate mitosis. Failure to pause long enough to repair DNA can allow a mutation in an oncogene or tumor suppressor gene to persist.

Loss of control over telomere length may also contribute to cancer by affecting the cell cycle. Recall from figure 2.19 that telomeres, or chromosome tips, protect chromosomes from breaking. Human telomeres consist of the DNA sequence TTAGGG repeated thousands of times. The repeats are normally lost as a cell matures, at the rate of about 15 to 40 nucleotides per cell division. The more specialized a cell, the shorter its telomeres. The chromosomes in skin, nerve, and muscle cells, for example, have short telomeres. Chromosomes in a sperm cell or oocyte, however, have long telomeres. This makes sense—as the precursors of a new organism, gametes must retain the capacity to divide many times.

Gametes keep their telomeres long, thanks to an enzyme, telomerase, that is a complex of RNA and protein. Part of the RNA—the sequence AAUCCC—serves as a template for the 6-DNA-base repeat that builds telomeres (**figure 18.4**). Telomerase moves down a chromosome tip like a zipper, adding six "teeth" at a time.

In normal, specialized cells, telomerase is turned off, and telomeres shrink, signaling a halt to cell division when they reach a certain size. In cancer cells, telomerase is turned back on. Telomeres extend, and this releases the normal brake on rapid cell division. As daughter cells of the original abnormal cell continue to divide uncontrollably, a tumor forms, grows, and may spread. Usually, the longer the telomeres in cancer cells, the more advanced the disease. However, turning on telomerase production in a cell is not sufficient in itself to cause cancer. Many things must go wrong for cancer to begin.

Inherited Versus Sporadic Cancer

Most cancers are isolated, or sporadic, which means that the causative mutation occurs only in cells of the affected tissue. This is a **somatic mutation,** because it

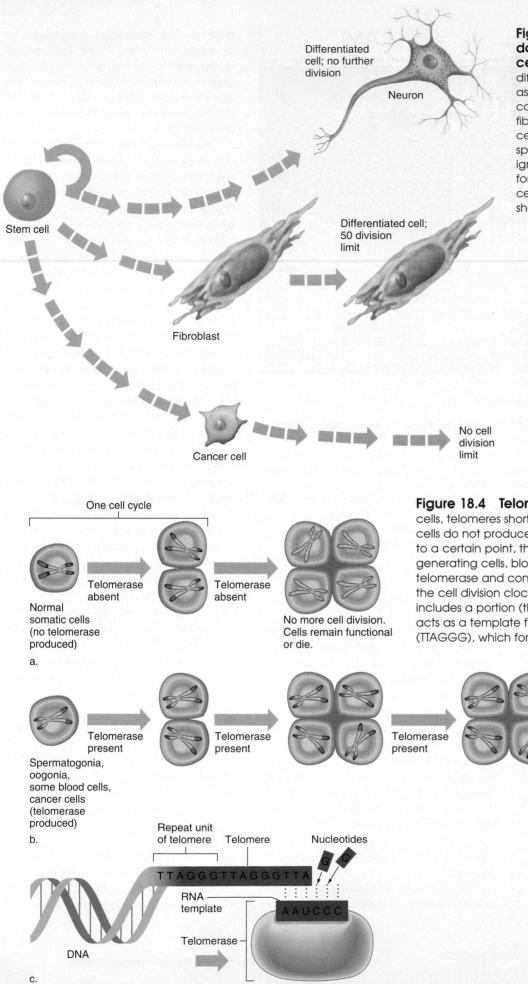

Figure 18.3 Cancer sends a cell down a pathway of unrestricted cell division. Cells may be terminally differentiated and no longer divide, such as a neuron, or differentiated yet still capable of limited cell division, such as a fibroblast (connective tissue cell). Cancer cells either lose specializations or never specialize and divide unceasingly, ignoring the 50-or-so division "Hayflick limit for cultured cells." (Arrows represent some cell divisions; not all daughter cells are shown.)

Differentiated cell; no further division

Neuron

Stem cell

Differentiated cell; 50 division limit

Fibroblast

Cancer cell

No cell division limit

Figure 18.4 Telomeres. **(a)** In normal somatic (nonsex) cells, telomeres shorten with each cell division because the cells do not produce telomerase. When the telomeres shrink to a certain point, the cell no longer divides. **(b)** Sperm-generating cells, blood cells, and cancer cells produce telomerase and continually extend their telomeres, resetting the cell division clock. **(c)** Telomerase contains RNA, which includes a portion (the nucleotide sequence AAUCCC) that acts as a template for the repeated DNA sequence (TTAGGG), which forms the telomere.

One cell cycle

Normal somatic cells (no telomerase produced)

Telomerase absent

Telomerase absent

No more cell division. Cells remain functional or die.

a.

Spermatogonia, oogonia, some blood cells, cancer cells (telomerase produced)

Telomerase present

Telomerase present

Telomerase present

Cells continue dividing under influence of telomerase.

b.

Repeat unit of telomere

Telomere

Nucleotides

G G

TTAGGGTTAGGGTTA

RNA template

AAUCCC

Telomerase

DNA

c.

occurs in somatic (nonsex) cells. A sporadic cancer may result from a single dominant mutation, or from two recessive mutations in the same gene. The cell harboring the mutation loses control of its cell cycle, divides continuously, and forms a tumor. Susceptibility to developing a sporadic cancer is *not* directly passed on to future generations because the gametes do not carry the mutant allele or alleles. In contrast, a **germline,** or inherited, cancer susceptibility *is* directly passed to future generations because it is in every cell, including gametes. Cancer develops when a second mutation occurs in a somatic cell in the affected body part (**figure 18.5**).

Germline mutations may explain why some heavy smokers develop lung cancer, but many do not; the unlucky ones may have inherited a susceptibility allele in every cell. Years of exposing lung tissue to the carcinogens in smoke eventually cause a mutation in a tumor suppressor gene or oncogene of a lung cell, giving it a proliferative advantage. Without the susceptibility gene, two such somatic mutations are nec-

essary to trigger the cancer. This, too, can be the result of an environmental insult, but it takes longer for two events to occur than one. Germline cancers are rare, but have high penetrance and tend to strike earlier in life than sporadic cancers.

Key Concepts

1. Cancer is genetic, but not necessarily inherited.
2. Single genes (oncogenes and tumor suppressors), when mutant, can cause cancer. New diagnostic approaches monitor expression of many genes.
3. Cancer is caused by a loss of cell division control. Implicated genes encode growth factors, transcription factors, or telomerase.
4. Most mutations leading to cancer occur in somatic cells.
5. Cancer may develop when an environmental trigger causes mutations in a somatic cell (sporadic) or when a somatic mutation compounds an inherited susceptibility (germline).

18.2 Characteristics of Cancer Cells

Cell division is rigorously controlled. Whether a cell divides or stops dividing and whether it differentiates depend upon signals from surrounding cells. A cancer cell simply stops "listening" to those signals.

Cancer cells can divide continuously if given sufficient nutrients and space. Cervical cancer cells of a woman named Henrietta Lacks, who died in 1951, vividly illustrate the hardiness of these cells. Her cells persist today as standard cultures in many research laboratories. These "HeLa" cells divide so vigorously that when they contaminate cultures of other cells they soon take over.

Cancer cells divide more frequently or more times than the cells from which they arose. Some cells normally divide frequently, and others rarely (**table 18.1**). Even the fastest-dividing cancer cells, which complete mitosis every 18 to 24 hours, do not divide as often as some cells in a normal human embryo do. A cancerous tumor eventually grows faster than surrounding tissue because a greater proportion of its cells is dividing.

Some cancers grow alarmingly fast. The smallest detectable fast-growing tumor is half a centimeter in diameter and can contain a billion cells. These cells divide to produce

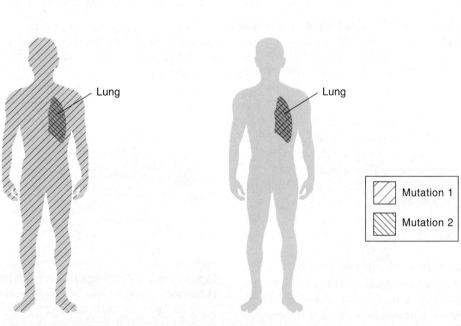

a. Germline (inherited) cancer   b. Sporadic cancer

Mutation 1
Mutation 2

Figure 18.5   Germline versus sporadic cancer.   (a) In germline cancer, every cell has one gene variant that increases cancer susceptibility, and a second mutation occurs in a cell of the affected tissue. This type of predisposition to cancer is a Mendelian trait. **(b)** A sporadic cancer forms when a dominant mutation occurs in a somatic cell or two recessive mutations occur in the same gene in the same somatic cell. An environmental factor, such as exposure to radiation or a chemical, can cause the somatic mutations that cause cancer.

Table 18.1

Cell Division Rates for Normal and Cancer Cells

Cell Type	Hours Between Divisions
Normal Cells	
Bone marrow precursor cells	18
Lining cells of large intestine	39
Lining cells of rectum	48
Fertilized ovum	36–60
Cancer Cells	
Stomach	72
Acute myeloblastic leukemia	80–84
Chronic myeloid leukemia	120
Lung	196–260

a million or so new cells in an hour. If 99 percent of the tumor's cells are destroyed, 10 million are left to proliferate. Other cancers develop over years. A tumor grows more slowly at first, because fewer cells divide. By the time the tumor is the size of a pea—when it is usually detectable—billions of cells are actively dividing.

A cancer cell looks different from a normal cell. It is rounder because it does not adhere to surrounding normal cells as strongly as other cells do. Also, because the plasma membrane is more fluid, different substances cross it. A cancer cell's surface may sport different antigens than are on other cells or different numbers of antigens that are also on normal cells. The "prostate specific antigen" (PSA) blood test that indicates increased risk of prostate cancer, for example, detects elevated levels of this protein that may come from cancer cells.

When a cancer cell divides, both daughter cells are cancerous, since they inherit the altered cell cycle control. Therefore, cancer is said to be heritable because it is passed from parent cell to daughter cell. A cancer is also transplantable. If a cancer cell is injected into a healthy animal of the same species, it will proliferate there.

A cancer cell is **dedifferentiated,** which means that it is less specialized than the normal cell types near it that it might have descended from. A skin cancer cell, for example, is rounder and softer than the flattened, scaly, healthy skin cells above it in the epidermis, more like a stem cell in both appearance and proliferative capacity. Cancer cell growth is also unusual. Normal cells placed in a container divide to form a single layer; cancer cells pile up on one another. In an organism, this pileup would produce a tumor. Cancer cells that grow all over one another lack contact inhibition—they do not stop dividing when they crowd other cells.

Cancer cells have surface structures that enable them to squeeze into any space, a property called invasiveness (**figure 18.6**).

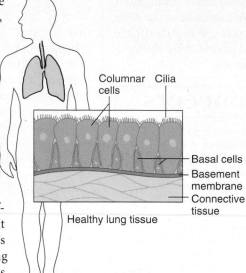

Healthy lung tissue

Columnar cells — Cilia

Basal cells
Basement membrane
Connective tissue

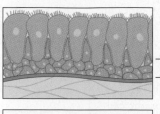
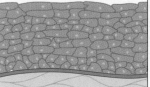

Basal cells proliferate (1 year after smoking starts)

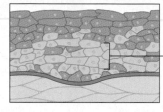

Cilia and columnar cells destroyed. Squamous or "flattened" cells (5 years after smoking starts)

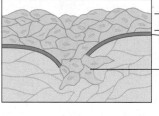

Cancer cells with atypical nuclei (8 years after smoking starts)

Cancer cells with atypical nuclei
Basement membrane

Early cancerous invasion (20–22 years after smoking starts; first symptoms)

Figure 18.6 Cancers take many years to spread. Lung cancer due to smoking begins with irritation of the lining tissue in respiratory tubes (bronchial epithelium). Ciliated cells die (but can be restored if smoking ceases), basal cells divide, and then, if the irritation continues, cancerous changes may appear.

They anchor themselves to tissue boundaries, called basement membranes, where they secrete enzymes that cut paths through healthy tissue. Unlike a benign tumor, an invasive malignant tumor grows irregularly, sending tentacles in all directions.

Cancer cells eventually reach the bloodstream or lymphatic vessels, which take them to other parts of the body—unless treatment stops the disease process. The traveling cancer cells settle into new sites. Once they've grown to the size of a pinhead, interior cancer cells respond to the oxygen-poor environment by secreting a protein, called vascular endothelial growth factor, that stimulates nearby capillaries (the tiniest blood vessels) to sprout new branches that extend toward the tumor, bringing in oxygen and nutrients and removing wastes. This growth of new capillaries is called angiogenesis. Capillaries may snake into and out of the tumor (**figure 18.7**). Cancer

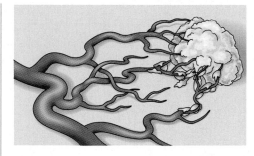

Figure 18.7 Angiogenesis nurtures a tumor. Cells starved for oxygen deep within a tumor secrete vascular endothelial growth factor, which stimulates nearby capillaries to extend branches toward a tumor.

cells wrap around the blood vessels and creep out upon this scaffolding, invading nearby tissue. Cancer cells may also secrete hormones that encourage their own growth. This is a new ability because the

Table 18.2

Characteristics of Cancer Cells

Oilier, less adherent

Loss of cell cycle control

Heritable

Transplantable

Dedifferentiated

Lack contact inhibition

Induce local blood vessel formation (angiogenesis)

Invasive

Increased mutation rate

Can spread (metastasize)

cells they descend from do not produce these hormones.

Once cancer cells move to a new body part, the disease has metastasized. The DNA of secondary tumor cells often mutates, and chromosomes may break or rearrange. Many cancer cells are aneuploid. The metastasized cancer thus becomes a new genetic entity that may resist treatments that were effective against most cells of the original tumor. Because gene expression patterns associated with metastasis are detectable early, new cancer treatments may actually prevent metastasis.

Table 18.2 summarizes the characteristics of cancer cells.

Key Concepts

1. Cancer occurs when cells divide faster or more times than normal.
2. Cancer cells are heritable, transplantable, and dedifferentiated. They lack contact inhibition, cutting through basement membranes.
3. A cancerous growth is invasive and can metastasize and stimulate angiogenesis, spreading further.

18.3 Origins of Cancer Cells

Mutations cause the changes that turn a cell cancerous, but that is only a first step. The degree to which the cell in which cancer begins is specialized, and the location of that cell in a tissue, influence whether or not disease develops.

Most cancer cells are more specialized than stem cells but considerably less specialized than the differentiated cells near them in a tissue. From which does the cancer cell arise? Does the cancer cell descend from a stem cell that yields slightly differentiated daughter cells that retain the capacity to self-renew, or does a cancer cell arise from a specialized cell that loses some of its features and can divide? The situation can be compared to a person in his underwear. Is he dressing (a stem cell's daughter becoming partially specialized?) or undressing (a differentiated cell's daughter shedding specializations?)?

The end results of "dressing" a stem cell or "undressing" a differentiated cell are the same—cancer—but the pathways to get there are opposite. Designing experiments to demonstrate this distinction has been very challenging. Researchers have identified signaling molecules that enable a cell to dedifferentiate, but haven't yet implicated them in causing cancer. On the other hand, experiments can trace certain stem cells, called **cancer stem cells,** that veer from normal development to produce both cancer cells and abnormal specialized cells. Cancer stem cells are found in cancers of the brain, blood, and epithelium (lining tissues such as those in breast, colon, and prostate cancers).

Recall from chapter 2 that stem cells self-renew as well as produce progenitor cells that in turn give rise to daughter cells that differentiate (see figures 2.23 and 2.24). **Figure 18.8** illustrates how cancer stem cells may cause brain tumors. As cancer stem cells give rise to progenitors and then differentiated cells (neurons, astrocytes, and oligodendrocytes), a cell surface molecule called CD133 is lost (designated CD133$^-$) at the late progenitor stage. In contrast, cancer cells retain the molecule (designated CD133$^+$). Some progenitor cells that descend from a cancer stem cell can relentlessly divide, and they ultimately accumulate, forming a brain tumor.

If the cancer stem cell theory of cancer origin is the "putting underwear on" model, then dedifferentiation is the "taking outerwear off" counterpart. Specialized cells may lose some of their distinguishing characteristics as mutations occur when they divide, or they may begin to express "stemness" genes that override signals to differentiate **(figure 18.9).** Whatever the mechanism, the result is dedifferentiation. So far experiments have not captured the exact moment when a cell both loses specializations and becomes able to continually divide. However, researchers have identified a biochemical, named "reversine," that can stimulate differentiated cells to divide and give rise to progenitor cells in mice. Therefore, reversine may play a role in the dedifferentiation of cancer cells.

Another possible origin of cancer may be a loss of balance at the tissue level in favor of cells that can divide continually or frequently—like a population growing faster if more of its members are of reproductive age. Consider a tissue that is 5 percent stem cells, 10 percent progenitors, and 85 percent differentiated cells. If a mutation, over time, shifts the balance in a way that creates more stem and progenitor cells, the extra cells pile up, and a tumor forms **(figure 18.10).**

Uncontrolled tissue repair may cause cancer **(figure 18.11).** If too many cells divide to fill in the space left by injured tissue, and those cells keep dividing, an abnormal growth may result. It is a little like ordering too many bricks to build a wall and using the extra to form a lump extending outward.

With so many millions of cells undergoing so many error-prone DNA replications, and so many ways that cancer can arise, it perhaps isn't surprising that cancer is so common. Yet most of the time, the immune system vanquishes a cancer before it progresses very far. **Table 18.3** reviews possible origins of cancer at the cellular level.

Key Concepts

1. Cancer stem cells are cells that veer from normal development to produce both cancer cells and abnormal specialized cells.
2. Dedifferentiation might occur through mutation or overexpression of "stemness" genes.
3. Upsetting the balance of stem and progenitor to differentiated cells can cause cancer as excess, fast-dividing cells accumulate.

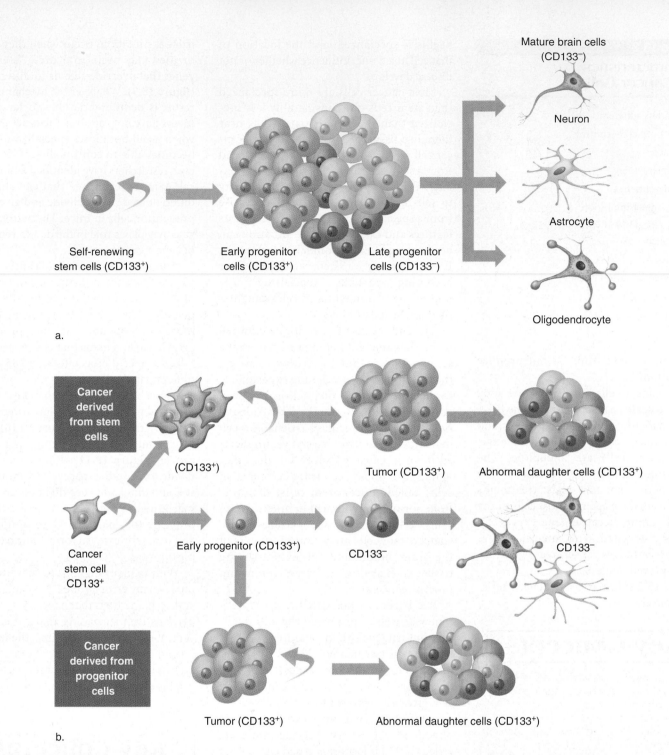

a.

b.

Figure 18.8 Cancer stem cells. (a) In the developing brain, stem cells divide to self-renew and give rise to early progenitor cells, which in turn divide to yield late progenitor cells. These late progenitor cells lose the CD133 cell surface marker, and divide to give rise to daughter cells that specialize as neurons or two types of supportive cells, astrocytes or oligodendrocytes. **(b)** A cancer stem cell can divide to self-renew and give rise to a cancer cell, which in turn can also spawn abnormal daughter cells. Some early progenitors give rise to normal differentiated cells. Sometimes the cancer-causing mutations occur in the cancer stem cell-derived early progenitor cell. In this case, the early progenitors form the tumor, which may spawn some abnormal daughter cells. Note that stem cells, cancer stem cells, progenitor cells, and abnormal daughter cells all have the CD133$^+$ marker, but the differentiated cells do not.

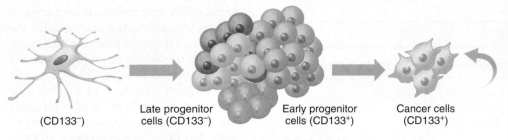

(CD133⁻) Late progenitor cells (CD133⁻)   Early progenitor cells (CD133⁺)   Cancer cells (CD133⁺)

Figure 18.9 Dedifferentiation reverses specialization. Mutations in a differentiated cell could reactivate latent "stemness" genes, giving the cell greater capacity to divide while causing it to lose some of its specializations. These are two of the defining characteristics of cancer.

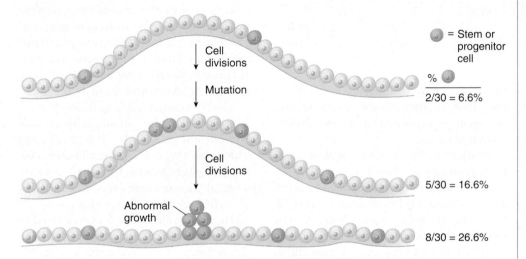

Figure 18.10 Shifting the balance in a tissue towards cells that divide. If a mutation renders a differentiated cell able to divide to yield other cells that frequently divide, then over time these cells may take over, forming an abnormal growth.

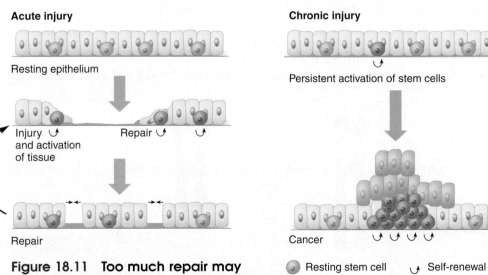

Figure 18.11 Too much repair may trigger tumor formation. If epithelium is occasionally damaged, resting stem cells can become activated and divide to fill in the tissue. If injury is chronic, the persistent activation of stem cells to renew the tissue can veer out of control, fueling an abnormal growth.

Table 18.3
Origins of Cancer Cells

1. Cancer stem cells
2. Dedifferentiation and reacquisition of self-renewal
3. Increase in proportion of stem and/or progenitor cells in tissue
4. Out-of-control tissue repair

18.4 Genes Associated with Cancer

Most mutations that cause cancer are in oncogenes or tumor suppressor genes (**table 18.4**). A third category includes mismatch mutations in DNA repair genes (see section 12.6) that allow other mutations to persist unfixed and to accumulate. When such mutations activate oncogenes or inactivate tumor suppressor genes, cancer results. DNA repair disorders are often inherited in a Mendelian fashion, and are quite rare. They tend to cause diverse and widespread tumors.

Oncogenes

Genes that normally trigger cell division when it is appropriate are called **proto-oncogenes.** They are active where and when high rates of cell division are necessary, such as in a wound or in an embryo. When proto-oncogenes are turned on at the wrong time or place, they function as oncogenes ("onco" means cancer). Abnormal activation of a proto-oncogene into an oncogene may be the result of a mutation or a change in expression of the wild type gene. A single base change in a proto-oncogene causes bladder cancer, for example. Alternatively, a proto-oncogene may be moved near a gene that is highly expressed; then it, too, is rapidly or frequently transcribed. For example, a human proto-oncogene is normally activated in cells at the site of a wound, where it stimulates production of growth factors that cause mitosis to fill the damaged area in with new cells. When that proto-oncogene is activated at a site other than a wound—as an oncogene—it still hikes growth factor production and stimulates mitosis. However, because the

Table 18.4

Types of Cancer-Causing Genes

Type of Gene	Mechanism of Carcinogenesis	Mode of Inheritance
Oncogene	Actively promotes cancer. Normal version, a proto-oncogene, controls cell cycle. Oncogene activates cell division at inappropriate time or place.	Recessive
Tumor suppressor gene mutation	Mutation removes normal suppression of cell division.	Dominant
DNA repair gene mutation	Effect indirect. Faulty DNA repair gene allows many mutations to accumulate, some in proto-oncogenes and tumor suppressor genes.	

site of the action is not damaged tissue, the new cells form a tumor.

Some proto-oncogenes encode transcription factors that, as oncogenes, are too highly expressed. (Recall from chapter 10 that transcription factors bind to specific genes and activate transcription.) The products of these activated genes then contribute to the cancer cell's characteristics. Oncogenes may also block apoptosis. As a result, damaged cells do not die, but divide.

Increased Expression in a New Location

A proto-oncogene can be transformed into an oncogene when it is placed next to a gene that boosts its expression. A virus infecting a cell, for example, may insert DNA next to a proto-oncogene. When the viral DNA is rapidly transcribed, the adjacent proto-oncogene (now an oncogene) is also rapidly transcribed. Increased production of the oncogene's protein product then switches on genes that promote mitosis, triggering the cascade of changes that leads to cancer. Kaposi sarcoma and acute T cell leukemia are human cancers caused by viruses.

A proto-oncogene can also be activated when it is moved next to a very active gene that is a normal part of the genome. This can happen when a chromosome is inverted or translocated. For example, a cancer of the parathyroid glands in the neck is associated with an inversion on chromosome 11, which places a proto-oncogene next to a DNA sequence that controls transcription

of the parathyroid hormone gene. When the gland synthesizes the hormone, the oncogene is expressed. The cells divide, forming a tumor.

Ironically, the immune system contributes to cancer when a translocation or inversion places a proto-oncogene next to an antibody gene. Recall from chapter 17 that antibody genes normally move into novel combinations when a B cell is stimulated and they are very actively transcribed. Cancers associated with viral infections, such as liver cancer following hepatitis, may be caused when proto-oncogenes are mistakenly activated along with antibody genes. Similarly, in Burkitt lymphoma, a cancer common in Africa, a large tumor develops from lymph glands near the jaw. People with Burkitt lymphoma are infected with the Epstein-Barr virus, which stimulates specific

chromosome movements in maturing B cells to assemble antibodies against the virus. A translocation places a proto-oncogene on chromosome 8 next to an antibody gene on chromosome 14. The oncogene is overexpressed, and the cell division rate increases. Tumor cells of Burkitt lymphoma patients reveal the characteristic chromosome 8 to 14 translocation (**figure 18.12**).

Fusion Proteins with New Functions

Oncogenes are also activated when a proto-oncogene moves next to another gene, and the gene pair is transcribed and translated together, as if they were one gene. The double gene product, called a **fusion protein,** activates or lifts control of cell division.

The first cancer-causing fusion protein was discovered in patients with chronic myeloid leukemia (CML). It is an old story with an exciting new chapter added in 2002.

Most patients with CML have a small, unusual chromosome called the Philadelphia chromosome (Ph[1]), which consists of the tip of chromosome 9 translocated to chromosome 22. Discovering this cancer-chromosome link was quite a feat. On August 13, 1958, two men entered hospitals in Philadelphia and reported weeks of fatigue. Each had very high white blood cell counts and were diagnosed with CML. Too many immature white blood cells were crowding out the healthy cells. The men's blood samples eventually fell into the hands of University of Pennsylvania assistant professor Peter Nowell and graduate student

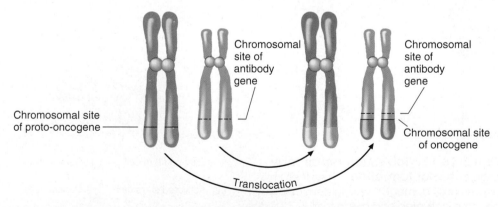

Figure 18.12 A translocation that causes cancer. The cause of Burkitt lymphoma is translocation of a proto-oncogene on chromosome 8 to chromosome 14, next to a highly expressed antibody gene. Overexpression of the translocated proto-oncogene, now an oncogene, triggers the molecular and cellular changes of cancer.

David Hungerford, who found the telltale Philadelphia chromosome.

With refinements in chromosome banding, important details emerged as to how the unusual chromosome causes cancer. In 1972, Janet Rowley at the University of Chicago used new stains that distinguished AT-rich from GC-rich chromosome regions to tell that Ph[1] is a translocated chromosome. By 1984, researchers had homed in on the two genes juxtaposed in the 9 to 22 translocation.

One gene from chromosome 9 is called the Abelson oncogene (*abl*), and the other gene, from chromosome 22, is called the breakpoint cluster region (*bcr*). Because the translocation is reciprocal, swapping parts of two chromosomes, two different fusion genes form. The *bcr-abl* fusion gene is part of the Philadelphia chromosome, and it causes CML. The encoded fusion protein, called the BCR-ABL oncoprotein, is a form of the enzyme tyrosine kinase, which is the normal product of the *abl* gene. The cancer-causing form of tyrosine kinase is active for too long, which sends signals into the cell, stimulating it to divide for too long. (The other fusion gene does not affect health.)

The discovery that a fusion oncoprotein sets into motion the cellular changes that cause CML led directly to development of a very effective drug, Gleevec. This drug is a small molecule that nestles into the pocket on the tyrosine kinase that must bind ATP to stimulate cell division (**figure 18.13**). With ATP binding blocked, cancer cells cease dividing.

A different fusion oncoprotein causes acute promyelocytic leukemia. (Leukemias differ by the type of white blood cell affected.) A translocation between chromosomes 15 and 17 brings together a gene coding for the retinoic acid cell surface receptor and an oncogene called *myl*. The fusion protein functions as a transcription factor, which when overexpressed causes cancer. The nature of this fusion protein explained an interesting clinical observation—some patients who receive retinoid (vitamin A-based) drugs recover. Their immature, dedifferentiated cancer cells, apparently stuck in an early stage of development where they divide frequently, suddenly differentiate, mature, and then die. Perhaps the cancer-causing fusion protein prevents affected white blood cells from getting

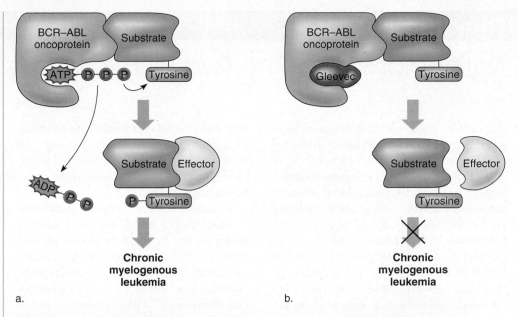

Figure 18.13 How Gleevec treats chronic myelogenous leukemia. In CML, a translocation forms the fusion oncoprotein BCR-ABL, which functions as a tyrosine kinase. A tyrosine on a substrate molecule picks up a phosphate from the ATP nestled in the oncoprotein, making the substrate able to bind to another protein, called an effector, that triggers runaway cell division **(a)**. Gleevec replaces the ATP **(b)**. Without phosphorylation of the tyrosine on the substrate, division of the abnormal cells stops. As cancer progresses, some cells undergo mutations that make the shape of their pockets unable to bind the drug. Newer drugs can replace Gleevec once the cancer becomes resistant.

enough retinoids to specialize, locking them in an embryoniclike, rapidly dividing state. Supplying extra retinoids allows the cells to continue along their normal developmental pathway.

Receiving a Too-Strong Division Signal

In about 25 percent of women with breast cancer, the affected cells have 1 to 2 million copies of a cell surface protein called Her-2/neu that is the product of an oncogene. In breast cells of other women, the number of these proteins is only 20,000 to 100,000.

The Her-2/neu proteins are receptors for epidermal growth factor. The receptors traverse the plasma membrane, extending outside the cell into the extracellular matrix and also dipping into the cytoplasm. They function as a tyrosine kinase, as is the case for CML. When the growth factor binds to the tyrosine of the receptor, it picks up a phosphate group, and this sends a signal into the cell that activates transcription of genes that stimulate cell division. The cause of Her-2/neu breast cancer is clear: Too

many tyrosine kinase receptors send too many signals to divide.

Her-2/neu breast cancer usually strikes early in adulthood and spreads quickly. However, a monoclonal antibody-based drug called Herceptin binds to the receptors, blocking the signal to divide. Interestingly, Herceptin works when the extra receptors arise from multiple copies of the gene, rather than from extra transcription of a single *Her-2/neu* gene.

Tumor Suppressors

Some cancers result from loss of a gene that normally suppresses tumor formation in response to growth-inhibiting signals. Whereas oncogene activation is usually associated with a point mutation, chromosomal translocation or inversion, and a gain of function, a tumor suppressor gene mutation that causes cancer is usually a deletion that removes a function. DNA viruses that are associated with a high risk of developing particular cancers interact with the normal products of tumor suppressor genes.

Retinoblastoma—The Two-Hit Hypothesis

Our current understanding of tumor-suppressing genes began with observations by Alfred Knudson. He studied retinoblastoma (RB), a rare childhood eye cancer. Distinct tumors, representing individual original cancerous cells, develop in the eye (**figure 1**). Sometimes RB affects one eye, and sometimes both. Knudson examined the medical records of 48 children with RB admitted to M.D. Anderson Hospital in Houston between 1944 and 1969. He recorded the following information for each child:

1. Whether one eye or two were affected

2. How old the child was at the time of diagnosis

3. Whether any other relatives had RB

4. Sex

5. Number of tumors per eye

The fact that RB occurred in boys and girls told Knudson that any genetic control was autosomal. Pooling data from families with more than one case of RB revealed that approximately 50 percent of the children of an affected parent were also affected, suggesting dominant inheritance. Knudson also noted that in some families, a child

with two affected eyes would have an affected grandparent, but both parents had healthy eyes. A picture of autosomal dominant inheritance with incomplete penetrance began to emerge.

Knudson, however, proposed a different explanation: An initial, inherited recessive mutation had to be followed by a second, somatic mutation in the eye to trigger tumor formation. This idea became known as the "two-hit hypothesis" of cancer causation. Occasionally the second mutation would not occur, and this would result in an unaffected generation between affected ones.

Two mutations explained another observation Knudson gleaned from medical records. Children with tumors in both eyes become affected much earlier than children with tumors in only one eye—generally before the age of five. This would make sense if a hereditary, bilateral (two-eye) form of RB requires a germline mutation followed by a somatic mutation, but a nonhereditary, unilateral (one-eye) form results from two somatic mutations in the same gene in the same cell. That is, in inherited RB, a newborn is halfway on the road to tumor development—just one somatic mutation in the eye is needed. The unilateral, noninherited form appears later in childhood because it

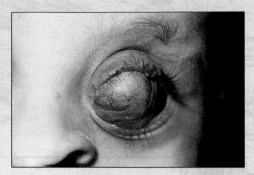

Figure 1 Retinoblastoma. In inherited retinoblastoma, all of the person's cells are heterozygous for a mutation in the *RB* gene. A second mutation, occurring in the original unmutated allele in cone cells in the retina, releases controls on mitosis, and a tumor develops.

takes longer for two somatic mutations in the same gene to occur in the same cell. Knudson used mathematics to show that the average number of tumors per eye—three—was consistent with two "hits."

Although it would be another 15 years before researchers identified the *RB* gene on chromosome 13, and longer still before its role in controlling the cell cycle was identified, Knudson's insights paved the way for those discoveries and for recognizing the widespread action of tumor suppressors in general.

Wilms' tumor is a cancer that develops from loss of tumor suppression. When a gene that normally halts mitosis in the rapidly developing kidney tubules in the fetus is absent, a child's kidney retains pockets of cells dividing as frequently as if they were still in the fetus, forming a tumor. Among the best-studied tumor suppressor genes are the retinoblastoma gene, the *p53* gene, and *BRCA1*.

Retinoblastoma (RB) and the Two-Hit Hypothesis

Alfred Knudson proposed the role of tumor-suppressing genes in causing cancer in 1971, as the intriguing story of a rare

childhood eye tumor called retinoblastoma (OMIM 180200) unfolded (**Reading 18.1**). Retinoblastoma has a long history. In 1597, a Dutch anatomist clinically described the eye cancer as a growth "the size of two fists." In 1886, researchers identified inherited cases. At that time, a child would survive the cancer only if the affected eye was removed. Today, children with an affected parent or sibling can be monitored from birth, so that noninvasive treatment can begin early. Full recovery is common. Often the first abnormal sign is an unusual gray area that appears in an eye in a photograph—the tumor reflects light differently than unaffected parts of the eye.

About half of the 1 in 20,000 infants who develop RB inherit susceptibility to the disorder: They harbor one germline mutant allele for the *RB* gene in each of their cells. Cancer develops in any somatic cell where the second copy of the *RB* gene mutates. Therefore, inherited retinoblastoma requires two point mutations or deletions, one germline and one somatic. In sporadic (noninherited) cases, two somatic mutations occur in the *RB* gene. Either way, RB usually starts in a cone cell of the retina, which provides color vision.

Many children with RB have deletions in the same region of the long arm of chromosome 13, which led researchers to the cancer-causing gene. In 1987, they found the gene altered in RB and identified its protein product, linking the cancer to control of the cell cycle. The 928-amino-acid-long protein normally binds transcription factors so that they cannot activate genes that carry out mitosis. It normally halts the cell cycle at G_1. When the *RB* gene is mutant or missing, the hold on the transcription factor is released, and cell division ensues.

Mutations in the *RB* gene cause other cancers. Children successfully treated for retinoblastoma often develop bone cancer as teens or bladder cancer as adults. Mutant *RB* genes have been found in the cells of patients with breast, lung, or prostate cancers, or acute myeloid leukemia, who never had the eye tumors. These other cancers may be caused by expression of the same genetic defect in different tissues.

p53 Normally Prevents Many Cancers

Another single gene that causes a variety of cancers when mutant is *p53*. Recall from chapter 12 that the p53 protein transcription factor "decides" whether a cell repairs DNA replication errors or dies by apoptosis. If a cell loses a *p53* gene, or if the gene mutates and malfunctions, a cell with damaged DNA is permitted to divide, and cancer may be the result.

More than half of human cancers involve a point mutation or deletion in the *p53* gene. The precise locations and types of mutations—a transition (purine to purine or pyrimidine to pyrimidine) or transversion (purine to pyrimidine or vice versa)—are different in different cancers.

Mutational analysis and epidemiological observations reveal that p53 protein may be a genetic mediator between environmental insults and development of cancer (**figure 18.14**). Consider a type of liver cancer prevalent in populations in southern Africa and Qidong, China. These two groups have in common exposure to the hepatitis B virus and to a food contaminant called aflatoxin B1. Most of the people with the liver cancer have the same point mutation in the *p53* gene that substitutes a T for a G. Could the food toxin, hepatitis virus, or both cause the mutation? Alternatively, the virus or toxin might trigger expression of an existing *p53* gene variant.

In most *p53*-related cancers, mutations occur only in somatic cells. However, in the germline condition Li-Fraumeni syndrome (OMIM 151623), family members who inherit a mutation in the *p53* gene have a very high risk of developing cancer—50 percent by age 30, and 90 percent by age 70. The risk of breast cancer is near 100 percent (see In Their Own Words on page 368). A somatic mutation in the affected tissue is necessary for cancer to develop.

BRCA1—A Genetic Counseling Challenge

Breast cancer that recurs in families can reflect the inheritance of a germline mutation (familial), or be caused by two-hit somatic mutations that happen more than once in a family due to chance (sporadic). Familial breast cancer touches on many of the complications of Mendel's laws discussed in chapter 5—multiple alleles, incomplete penetrance, variable expressivity, environmental influences, genetic heterogeneity, and polygenic inheritance and epistasis.

Only 5 percent of breast cancers are familial, and of these, 15 to 20 percent are caused by mutations in the genes *BRCA1* or *BRCA2*. Because breast cancer is so common, some women seeking genetic counseling may have sporadic cases that appear to be inherited—or the reverse. **Table 18.5** highlights some of the challenges encountered in genetic counseling for breast cancer.

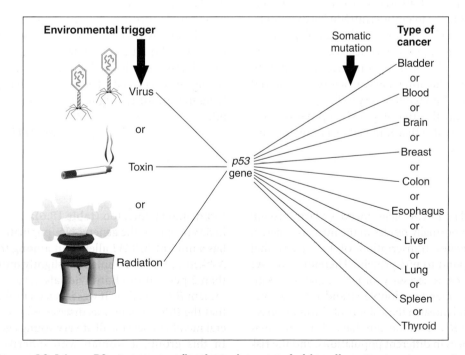

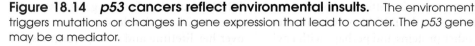

Figure 18.14 *p53* cancers reflect environmental insults. The environment triggers mutations or changes in gene expression that lead to cancer. The *p53* gene may be a mediator.

Table 18.5

The Complexities of Providing Genetic Counseling for Familial Breast Cancer

1. Many mutations and polymorphisms are known in breast cancer genes.

2. When more than one case occurs in a family, it can be either familial or sporadic. A woman who does not have a *BRCA1* or *BRCA2* mutation can still develop breast cancer. A woman with no affected relatives can have a *BRCA1* or *BRCA2* mutation.

3. *BRCA1* and *BRCA2* are incompletely penetrant—that is, inheriting a disease-causing allele does not always mean developing cancer.

4. The risk associated with *BRCA1* or *BRCA2* mutations varies depending upon interactions with other genes as well as environmental exposures.

Key Concepts

1. Proto-oncogenes normally control the cell cycle. They can become oncogenes when they mutate, when they move next to a gene that is highly expressed, or when they are transcribed and translated along with another gene, resulting in a fusion protein that triggers cancer.
2. Mutations in tumor suppressor genes usually are deletions that cause a cell to ignore extracellular constraints on cell division.

18.5 A Series of Genetic Changes Causes Some Cancers

Most cancers reflect the interplay of several genes. Certain single genes, when mutant, can have a generalized effect, causing a chromosomal instability that sets the stage for cancer. This is the case for a gene called *BUB1B*. Mutations in this gene cause mosaic variegated aneuploidy (OMIM 257300), in which about a quarter of a person's cells have missing or extra chromosomes. The autosomal recessive condition causes various types of childhood cancer. The functional gene apparently ensures that the correct number of chromosomes is distributed to daughter cells as mitosis completes.

Genes that guide a cell towards the cancerous state when mutant are sometimes considered in two broad categories, based on their effects. "Gatekeeper" genes control mitosis and apoptosis, which must be in balance to maintain the number of cells forming the affected tissue. Their effect is direct. "Caretaker" genes, in contrast, control the mutation rates of gatekeepers, and may have the overall effect, when mutant, of destabilizing the genome.

Some cancers are the culmination of a series of changes in several specific genes, involving gatekeepers and caretakers. To identify the steps, researchers examine tumor cell DNA from people in various stages of the same type of cancer. The older the tumor, the more genetic changes accumulate. Therefore, a mutation present in all stages acts early in the disease process, whereas a mutation seen only in the tumor cells of sicker people functions late in the process. Each step provides a potential point of treatment. Following is a closer look at two types of cancer that reflect a series of genetic changes.

A Rapidly Growing Brain Tumor

Astrocytomas, the most common types of brain tumors, affect cells called astrocytes. These tumors grow quickly. The man whose brain is shown in **figure 18.15** died just three months after noticing twitching in an eye. During that time, a series of single-gene and chromosomal changes occurred. Loss of both *p53* alleles came early because this change appears in many early-stage tumor cells, as well as in later ones.

By the time an astrocytoma has grown into a small tumor, another genetic change is apparent—loss of both alleles of several genes on chromosome 9. Some of the missing genes encode interferons, so the loss probably disrupts immune protection against the developing cancer. Two other deleted genes encode tumor suppressors.

At least two additional mutations speed the tumor's growth. First, an oncogene on chromosome 7 is activated, overexpressing a gene that encodes a cell surface receptor for a growth factor. The cancer cells bear too many growth factor receptors and receive too many messages to divide. Finally, the cancer cells lose one or even both copies of chromosome 10. This is a final change, because it is seen in all end-stage tumors, but not in early ones.

Colon Cancer

Colon (large intestine) cancer does not usually occur in families with the frequency or pattern expected of a single-gene disorder. However, when family members with noncancerous growths (polyps) in the colon are considered with those who have colon cancer, a Mendelian pattern emerges. Five percent of colon cancer cases are inherited. One in 5,000 people in the United States has precancerous colon polyps, a condition called familial adenomatous polyposis (FAP; OMIM 175100).

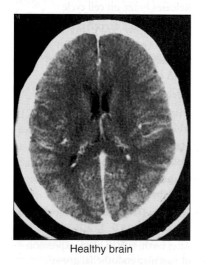

Healthy brain

Loss or mutation of both *p53* tumor suppressor alleles → Loss of chromosome 9 genes encoding interferons and tumor suppressors → Oncogene activation increases growth signals → Loss of chromosome 10, role unknown

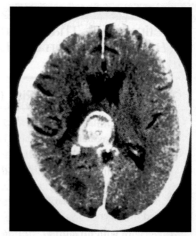

Astrocytoma

Figure 18.15 **Several genes can contribute to a cancer.** A series of genetic changes transforms normal astrocytes, which support nerve cells in the brain, into a rapidly growing cancer.

FAP begins in early childhood with tiny colon polyps, often hundreds, that progress over many years to colon cancer. Colon lining cells typically live three days. In FAP, they fail to die on schedule and instead build up, forming polyps. Connecting FAP to the development of colon cancer enabled researchers to view the stepwise progression of a cancer (**figure 18.16**). Both oncogenes and tumor suppressors take part.

The study of the hereditary nature of some colon cancers began at the University of Utah in Salt Lake City in the fall of 1947, when young professor Eldon Gardner stated that he thought cancer might be inherited. A student, Eugene Robertson, excitedly told the class that he knew of a family in which a grandmother, her three children, and three grandchildren had colon cancer.

Intrigued, Gardner delved into the family's records and began interviewing relatives. He eventually found 51 family members and arranged for each to be examined with a colonoscope, a lit instrument passed into the rectum to view the wall of the colon. The colons of 6 of the 51 people were riddled with the gobletlike precancerous polyps, although none of the 6 had symptoms. Removal of the affected tissue probably saved their lives.

In the years that followed, researchers identified other families with more than one case of colon polyps. Individuals with only polyps were diagnosed with FAP. If a person with colon polyps had cancer elsewhere, extra teeth, and pigment patches in the eye, the condition was called Gardner syndrome, named for the professor.

Researchers identified the chromosomal defect that causes Gardner syndrome in 1985 with the help of a 42-year-old man at the Roswell Park Cancer Institute in Buffalo, New York. He had several problems—no gallbladder, an incomplete liver, an abnormal kidney, mental retardation, and Gardner syndrome. To a geneticist, a seemingly unrelated combination of symptoms suggests a chromosomal abnormality affecting several genes. Sure enough, the man's karyotype revealed a small deletion in the long arm of chromosome 5. This was the first piece to the puzzle of colon cancer. Today, intestinal cells in stool samples can be tested for absence of a protein called APC, indicating the first step in this particular route to colon cancer.

The deletion in chromosome 5 detected in the man from Buffalo in 1985 removed the *APC* gene. This is the main "gatekeeper" for this type of colon cancer, and is the first step depicted in figure 18.16. Normally APC protein binds to another protein, b-catenin, causing a phosphate to be added to it. The phosphorylation prevents b-catenin from acting. But when the *APC* gene is deleted, b-catenin isn't silenced, and instead it enters the nucleus and activates genes that promote mitosis. The cell becomes unable to stop dividing. A tumor forms, but it is not yet malignant. Other pathways, such as those controlled by the genes *TGF* and *p53*, push the abnormal cells to become cancerous. *TGF* normally inhibits mitosis, and *p53* normally sends cells to a fate of apoptosis. *PRL-3* is a gene that acts late in the process, enabling the

cancer to spread. Several caretaker genes affect the expression of the gatekeepers, so the overall picture is quite complex.

18.6 Environmental Causes of Cancer

Environmental factors can contribute to cancer by mutating genes that control the processes that go awry in cancer, such as the cell cycle, apoptosis, or DNA repair. It would take fewer such insults to result in cancer in an individual who has inherited a susceptibility gene, but it can also happen in somatic tissue in anyone.

Looking at cancer at a population level reveals the dual roles of genetics and the environment. Researchers in Iceland consulted their national cancer registry of 32,000 patients, which tracks the health of relatives over the past 50 years. Six cancers were about twice as likely to occur in blood relatives of affected individuals as in people without affected relatives—prostate, stomach, lung, colon, kidney, and bladder. These may reflect inheritance of specific susceptibility genes. On the other hand, cancers of the stomach, colon, and lung were more common in the spouses of affected individuals—and these can be explained by sharing an environment (diet for the stomach and colon cancers, air pollution and smoke for the lung cancer).

On an individual level, one way to lower the chance of developing cancer is to avoid certain high-risk environmental factors, such as cigarette smoking and excess sun exposure. A more active approach to minimize environmental influences on cancer risk is chemoprevention, which is taking certain nutrients, plant extracts, or drugs. Promising "chemopreventatives" include folic acid, vitamins D and E, selenium, compounds

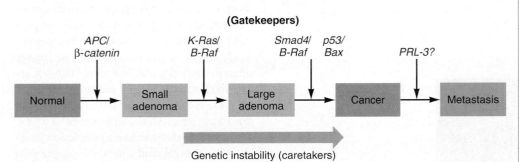

Figure 18.16 Several mutations contribute to FAP colon cancer. Cells lining the colon divide more frequently when the *APC* gene on chromosome 5q undergoes a point mutation, causing small benign tumors (adenomas) to form. Activation of certain oncogenes, such as *K-Ras* and *B-Raf*, fuel growth of the adenomas. Mutations in *p53* and other genes push the adenoma cells to become cancerous. Finally, mutations in a gene called *PRL-3* trigger metastasis. Caretaker genes cause genetic instability that contributes to the disease process.

Considering Carcinogens

Determining precisely how an environmental factor such as diet affects cancer risk can be complicated. Consider the cruciferous vegetables, such as broccoli and Brussels sprouts, which are associated with decreased risk of developing colon cancer. Experiments show that these vegetables release compounds called glucosinolates, which in turn activate "xenobiotic metabolizing enzymes" that detoxify carcinogenic products of cooked meat called heterocyclic aromatic amines. With a vegetable-poor, meaty diet,

heterocyclic amines accumulate. They cross the lining of the digestive tract and circulate to the liver, where enzymes metabolize them into compounds that cause the mutations associated with colon cancer (**figure 18.17**).

Environmental exposures to carcinogens—in the workplace, home, or outdoors—can also raise cancer risk. Chemical carcinogens were recognized as long ago as 1775, when British physician Sir Percival Potts suggested that the high rate of skin cancer in the scrotums of chimney sweeps in London was due to their exposure to a chemical in soot. Since then, epidemiological studies have identified many chemicals as possibly causing cancer in certain populations (**table 18.8**). However, most studies

reveal correlations rather than cause-and-effect relationships. In the strongest cases, genetic or biochemical evidence explains the observed environmental connection.

Methods to Study Cancer-Environment Links

Epidemiologists use different statistical tools to establish links between environmental exposures and cancer. Connections are strengthened when different types of investigations yield consistent results. This is true, for example, of the association between eating whole grain cereals and reduced incidence of colorectal cancer.

Epidemiological studies of cancer causation compare people in different ways. A **population study** compares the incidence of a type of cancer among very different groups of people. If the incidence differs, then some distinction among the populations may be responsible. For example, an oft-mentioned study from 1922 found that primitive societies have much lower rates of many cancers than more developed societies. The study attributed the lack of cancer to the high level of physical activity among the primitive peoples—but diet might also have explained the difference.

Population studies often have too many variables to clearly establish cause and effect. Consider the very high incidence of breast cancer on Long Island, New York. One hypothesis attributes the mini-epidemic to pesticide exposure, but this population also has a high frequency of *BRCA1* mutations among its Ashkenazi citizens. Sociological factors come into play, too. In this population, women have frequent mammograms starting at a young age. As a result, the percentage of the population with recognized early stages of the disease may be higher than in other populations where women are less likely to have regular mammograms. All of these factors may contribute to the high breast cancer incidence in this area.

More informative than a population study is a **case-control study**, in which people with a type of cancer are matched with healthy individuals for age, sex, and other characteristics. Then researchers look for differences between the pairs. If, for example, the cancer patients had extensive dental X rays at a young age but the control group didn't, X-ray exposure may be a causal factor. The problem

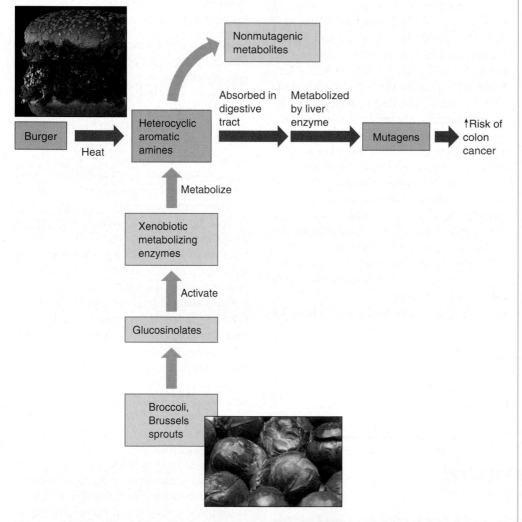

Figure 18.17 One way that cruciferous vegetables lower cancer risk.
Compounds called heterocyclic aromatic amines form in cooking meat, are absorbed into the digestive tract, and are metabolized by a liver enzyme into mutagens, which may cause colon cancer. Broccoli and Brussels sprouts produce glucosinolates, which activate xenobiotic metabolizing enzymes that block part of the pathway that leads to production of the mutagens.

Table 18.8

Increase in Death Rates for Certain Cancers in Particular Geographical Areas in the United States

Cancer Type	Region	Possible Explanation
Breast	Northeast	BRCA1 mutations, greater lifetime exposure to estrogens (early menstruation, late menopause, older age of first birth, exposure to pesticides)
Colon	Northeast	Dietary factors, diagnostic radiation exposure
Lung	White men in south, white women in west, blacks in northern cities	Changes in regional trends in cigarette smoking
Lung	Men in southern coastal areas	Asbestos exposure while working in shipyards during World War II
Mouth, throat	Women in rural south	Smokeless tobacco
Esophagus	Washington, D.C., coastal South Carolina	Alcohol and tobacco, dietary deficiencies of fruits and vegetables

with this type of study is that much of the information is often based on recall, people make mistakes, and not all relevant factors are identified and taken into account.

The most informative type of epidemiological investigation is a **prospective study,** in which two or more groups of people follow a specified activity plan, such as a dietary regimen, and are checked periodically for cancer. By looking ahead, the investigator has more control over the activities and can verify information. However, a limitation of this type of cancer study is that cancer usually takes many years to appear and progress.

Once epidemiological studies indicate a correlation, a biological explanation is necessary to draw conclusions or even suggest further studies. For example, finding lower cancer rates among vegetarians might be explained by the observation that certain vegetables contain antioxidant compounds, which deactivate the free radicals that can damage DNA, thereby preventing mutations.

18.7 Human Genome Data Tailor Diagnosis and Treatment

Estimating the risk that a certain type of cancer will occur in a particular individual is only possible for a few disorders that are inherited in a Mendelian fashion through known genes. More often, discovery of cancer follows a screening test such as mammography or high levels of prostate specific antigen in the bloodstream, or after symptoms occur or a person feels a lump. Then treatments begin—and there are usually many options.

The oldest cancer treatment is surgery—it prevents invasiveness by removing the tumor. Radiation and chemotherapy kill all cells that divide rapidly. Unfortunately, this also affects healthy cells in the digestive tract, hair follicles, and bone marrow, causing nausea, hair loss, great fatigue, and susceptibility to infection. Patients receive several other drugs to help them tolerate the side effects, including colony stimulating factors to replenish bone marrow. These other drugs enable patients to withstand higher and more effective doses of chemotherapy.

Several newer types of cancer drugs affect cancer cell characteristics or activities other than hiked division rate. Some treatments stimulate cells to regain specialized characteristics, such as drugs based on retinoic acid. Another approach is to inhibit telomerase, which prevents cancer cells from elongating their telomeres and continually dividing. In yet another approach, angiogenesis inhibitors rob a cancer of its blood supply. Yet other treatment approaches induce apoptosis, which counters the runaway cell division of cancer.

The evolution of diagnosis and treatments for breast cancer illustrates how genetic and genomic information are refining how physicians manage these diseases. The earliest treatments simply removed or destroyed the affected tissue (**table 18.9**). Then physicians began to determine whether tumor cells have receptors for estrogen or progesterone, two hormones. Women with estrogen receptor-positive tumors typically begin a several-year course either of a drug that blocks these receptors from receiving signals to divide, or a drug that inhibits an enzyme called aromatase, required to produce estrogen.

Determining estrogen receptor status is subtyping by phenotype. With the discovery of single genes that cause cancer, diagnosis began to include genotyping, too; a woman might have BRCA1 or Her-2/neu breast cancer. Increasingly, cancer diagnosis will be based on DNA microarrays that scan both genotype and gene expression patterns,

Table 18.9

Evolution of Treatments for Breast Cancer

Strategy	Examples
Remove or destroy cancerous tissue	Surgery, radiation, chemotherapy
Use phenotype to select drug	Estrogen receptor-positive women take a selective estrogen receptor modulator or an aromatase inhibitor or both
Use genotype to select drug	Women with Her-2/neu-positive cancers take Herceptin (monoclonal antibody)
Genomic level	Gene expression profile on DNA microarray used to guide drug choice; 70-gene signature predicts metastasis

enabling physicians to match a particular patient to the treatments most likely to work right from the start, or predict metastasis. Genomic analyses will also identify patients likely to suffer side effects from particular drugs.

The limitation of any cancer treatment, old or new, is defined by the strength of the enemy. Cancer cells are incredibly abundant and ever-changing. Surgery followed by a barrage of drugs and radiation can slow the course of the disease, but all it takes is a few escaped cancer cells—called micrometastases—to sow the seeds of a future tumor. The DNA of cancer cells mutates in ways that enable the cells to pump out any drug sent into them. In addition, cancer cells have redundancies, so that if a drug shuts down angiogenesis or invasiveness, the cell completes the task another way. Although cancer treatments can completely cure the illness, it is more likely that

they kill enough cancer cells, and sufficiently slow the spread, that it takes the remainder of a lifetime for the tumors to grow back. In this way, cancer becomes a chronic, manageable condition.

Cancer may even be, to some extent, preventable. Kari Stefansson, head of the Icelandic biobank effort, says, "Genetic factors contribute to the risk of specific cancers, but also, certain types of cancer can be looked upon collectively as broad, complex phenotypes. Lifestyle and environmental factors play a very significant role in the development of cancer, and are things we may all be able to do something about today."

RELATING THE CONCEPTS

How is the application of gene expression profiling in diagnosing thyroid cancer similar to its use in breast cancer?

Key Concepts

1. Lower cancer risk is associated with eating more fruits, vegetables, and whole grain cereals.
2. Treatments for cancer target the characteristics of cancer cells. Surgery removes tumors. Chemotherapy and radiation nonselectively destroy rapidly dividing cells.
3. Newer treatments target receptors on cancer cells, block telomerase, stimulate differentiation, or attack a tumor's blood supply.
4. Diagnosis and treatment of cancer will increasingly consider genomic information.

Summary

18.1 Cancer Is Genetic, But Usually Not Inherited

1. Cancer is a genetically dictated loss of cell cycle control, creating a population of highly proliferative cells that outgrows and overwhelms surrounding tissue.

2. Sporadic cancers result from **somatic mutations.** They are more common than germline cancers, which are caused by a **germline** mutation plus a somatic mutation in affected tissue. Cancer may also be polygenic. Changing gene expression patterns also contribute to cancer, and can be used to distinguish types.

3. Mutations in genes that encode or control transcription factors, cell cycle checkpoint proteins, growth factors, repair proteins, or telomerase may disrupt the cell cycle sufficiently to cause cancer.

18.2 Characteristics of Cancer Cells

4. A tumor cell divides more frequently or more times than cells surrounding it, has altered surface properties, loses the specializations of the cell type it arose from, and produces daughter cells like itself.

5. A malignant tumor infiltrates nearby tissues and can **metastasize** by attaching to basement membranes and secreting enzymes that penetrate tissues and open a route to the bloodstream. From there, a cancer cell can travel, establishing secondary tumors.

18.3 Origins of Cancer Cells

6. Cell specialization and position within a tissue are important determinants of whether cancer begins.

7. **Cancer stem cells** can divide to yield cancer cells as well as abnormally differentiated cells.

8. A cell that dedifferentiates and/or turns on expression of "stemness" genes can begin a cancer.

9. A mutation that endows a cell with the ability to divide continually can alter the percentages of cells in a tissue that can divide, resulting in an abnormal growth.

10. Chronic repair of tissue damage can provoke stem cells into producing an abnormal growth.

18.4 Genes Associated with Cancer

11. Cancer is often the result of a series of genetic changes involving the activation of **proto-oncogenes** to **oncogenes,** and the inactivation of **tumor suppressor** genes. Mutations in DNA repair genes can cause cancer by increasing the mutation rate.

12. Proto-oncogenes normally promote controlled cell growth, but are overexpressed because of a point mutation, placement next to a highly expressed gene, or transcription and translation with another gene, producing a **fusion protein.** Oncogenes may also be overexpressed growth factor receptors.

13. A tumor suppressor is a gene that normally enables a cell to respond to factors that limit its division. Tumor suppressor genes include *RB, p53,* and *BRCA1.*

18.5 A Series of Genetic Changes Causes Some Cancers

14. Many cancers result from "two hits" or mutations, but some entail a longer series of genetic changes.

15. To decipher the gene action sequences that result in cancer, researchers examine the mutations in cells from patients at various stages of the same type of cancer. Those mutations present at all stages of the cancer are the first to occur.

16. Astrocytoma and FAP are two cancers that require several mutations to develop.

18.6 Environmental Causes of Cancer

17. **Population, case-control,** and **prospective studies** can reveal correlations between environmental exposures and the development of certain cancers, but usually cannot establish cause and effect. Biochemical and/or genetic evidence is important to explain epidemiological observations.

18.7 Human Genome Data Tailor Diagnosis and Treatment

18. Traditional cancer treatments are surgery, radiation, and chemotherapy. Newer approaches based on molecular biology include blocking hormone receptors, stimulating cell specialization, blocking telomerase, and inhibiting angiogenesis. A genomic approach identifies differences in gene expression that define cancer subtypes.

Review Questions

1. Cite three reasons why cancer may not follow a Mendelian pattern, but nevertheless involves abnormal gene function.

2. How can detecting single alleles or gene expression patterns improve cancer diagnosis and/or treatment?

3. Why don't all cancers of the same cell type respond to the same drug?

4. How is the cell cycle controlled from outside and inside the cell?

5. What is inaccurate about the statement that "cancer cells are the fastest dividing cells in the body?"

6. Is any cell with long telomeres a cancer cell?

7. What would be the value of knowing whether a person's cancer is sporadic or inherited?

8. List four characteristics of cancer cells.

9. Describe four ways that cancer might originate at the cellular level.

10. How can the same cancer be associated with deletions as well as translocations of genetic material?

11. How would mutations in genes that encode the following proteins lead to cancer?

    a. a transcription factor

    b. the p53 protein

    c. the retinoblastoma protein

    d. the *myl* oncogene's protein

    e. a repair enzyme

    f. the APC protein

    g. CHEK2 protein

12. Three percent of all cancer cells have chromosome rearrangements. What other type of genetic change might be present in a cancer cell?

13. Distinguish among the following types of studies:

    a. population

    b. case-control

    c. prospective

14. List four new strategies for treating cancer, and explain how they work.

Applied Questions

1. An individual can develop breast cancer by inheriting a germline mutation, then undergoing a second mutation in a breast cell; or by undergoing two mutations in a breast cell, one in each copy of a tumor suppressor gene. Cite another type of cancer, discussed in the chapter, that can arise in these two ways.

2. For women under 55 with moderate-sized breast tumors that have not spread to the lymph nodes, treatment is surgery, then radiation and chemotherapy—but 60 percent of these women are cured with surgery alone, and would not benefit further from more treatment. Suggest how this treatment might be better targeted to selected women.

3. Humans missing both *p53* alleles in all cells are unknown. People with p53-related cancers either have a germline mutation and a somatic mutation in affected tissue, or two somatic mutations in the tissue. Experiments show that mice missing both copies of their *p53* genes die as embryos, with massive brain abnormalities.

    a. Why don't we see people with two missing or mutant *p53* alleles in all cells?

    b. Under what circumstances might a human with two mutant *p53* alleles be conceived?

4. Von Hippel-Lindau disease is an inherited cancer syndrome. The responsible mutation lifts control over the transcription of certain genes, which, when overexpressed, cause tumors to form in the kidneys, adrenal glands, and blood vessels. Is the von Hippel-Lindau gene an oncogene or a tumor suppressor? Cite a reason for your answer.

5. Is the gene that causes mosaic variegated aneuploidy an oncogene or a tumor suppressor? Is it a gatekeeper or a caretaker gene? Cite reasons for your answers.

6. A tumor is removed from a mouse and broken up into cells. Each cell is injected into a different mouse. Although all the mice used in the experiment are genetically identical and raised in the same environment, the animals develop cancers with different rates of metastasis. Some mice die quickly, some linger, and others recover. What do these results indicate about the characteristics of the original tumor cells?

7. Colon, breast, and stomach cancers can be prevented by removing the affected organ. Why is this approach not possible for chronic myeloid leukemia?

8. A vegetarian develops pancreatic cancer and wants to sue the nutritionist who suggested she follow a vegetarian diet. Is her complaint justified? Why or why not?

9. Iron foundry workers in Finland and coke oven workers in Poland have high exposures to polycyclic aromatic hydrocarbons, and they tend to develop cancers caused by mutations in the *p53* gene. What information would help determine whether the chemical exposure causes the mutation and whether the mutation causes the cancers?

10. MammaPrint is a DNA microarray-based test of the expression of 70 genes implicated in breast cancer. Certain patterns are significantly more common in cancers that spread, creating a "signature" that doctors can use to guide treatment decisions. Cite an advantage and a shortcoming of this test.

11. The discovery of cancer stem cells suggests a new type of treatment—develop a drug that stops self-renewal. Explain how such a drug might work, and what an adverse effect might be.

12. A project at the Sanger Centre in the United Kingdom is screening 1,500 types of cancer cells growing in culture to detect homozygous deletions. Will this approach identify oncogenes or tumor suppressor genes? Cite a reason for your answer.

13. A mutation in a gene called *FLT3*, which encodes a tyrosine kinase receptor, causes acute myelogenous leukemia, which has a five-year survival rate of 20 percent. A new drug blocks the receptor on white blood cells. Explain how it works.

Web Activity

Visit the Online Learning Center (OLC) at www.mhhe.com/lewisgenetics7. Select **Student Edition, chapter 18,** and **Web Activities** to find the website link needed to complete the following activity.

14. Go to the Cancer Quest website. Click on one oncogene and one tumor suppressor, and describe how, when mutant, they cause cancer.

Case Studies and Research Results

15. Elsie finds a small lump in her breast and goes to her physician, who takes a medical and family history. She mentions that her father died of brain cancer, a cousin had leukemia, and her older sister was just diagnosed with a tumor of connective tissue. The doctor assures her that the family cancer history doesn't raise the risk that her breast lump is cancerous, because the other cancers were not in the breast. Is the doctor correct?

16. Marcy goes to see a genetic counselor because her sister developed breast cancer at age 52, and her husband's sister at age 34. Marcy's mother was diagnosed with breast cancer at age 68. With these three cases in her family, she is convinced she has a mutant *BRCA1* gene and will suffer the same fate.

    a. What questions should the genetic counselor ask Marcy?

    b. Do you think she would benefit from a *BRCA1* gene test?

    c. What complications might arise from such testing?

17. Researchers compared the expression of a gene that encodes a protein called Pirh2 in lung tumor cells and noncancerous cells from humans and mice. Expression was significantly higher in the cancer cells than in the noncancerous cells of both humans and mice. Additional laboratory experiments showed that Pirh2 protein attaches ubiquitin (see figure 10.21) to p53 protein.

    a. Is *Pirh2* an oncogene or a tumor suppressor? How can you tell the difference?

    b. Explain how overexpression of the *Pirh2* gene causes lung cancer.

    c. Why needn't the *p53* gene be mutant for this cancer to develop?

    d. Drugs already exist that target proteasomes, the cellular garbage dumps for ubiquitin-tagged proteins. What is a risk of using such a drug to treat cancer?

18. The Aral Sea in central Asia, about equidistant from Kazakhstan, Turkmenistan, and Uzbekistan, is one of the most polluted areas on the planet. Much of the sea has dried up, leaving spotty pools among salty sediments laced with toxic by-products of pesticides. Cancer rates are 3 to 5 times greater than elsewhere, and more than 80 percent of the cases are cancer of the esophagus. What information would indicate whether or not these cancers are likely to be passed to future generations genetically?

Learn to apply the skills of a genetic counselor with these additional cases found in the *Case Workbook in Human Genetics:*

Li-Fraumeni syndrome

Multiple endocrine neoplasia

Cancer stem cells

CHAPTER

19

Genetic Technologies: Amplifying, Modifying, and Monitoring DNA

CHAPTER CONTENTS

IMPROVING PIG MANURE

Pig manure presents a serious environmental problem. The animals do not have an enzyme that would enable them to extract the mineral nutrient phosphorus from a compound called phytate in grain, so they are given dietary phosphorus supplements. As a result, their manure is full of phosphorus. The element washes into natural waters, contributing to fish kills, oxygen depletion in aquatic ecosystems, algal blooms, and even the greenhouse effect. But biotechnology may have solved the "pig poop" problem.

In the past, pig raisers have tried various approaches to keep their animals healthy and the environment clean. Efforts included feeding animal by-products from which the pigs can extract more phosphorus, and giving supplements of the enzyme phytase, which liberates phosphorus from phytate. But consuming animal by-products can introduce prion diseases, and giving phytase before each meal is costly. A "phytase transgenic pig," however, is genetically modified to secrete bacterial phytase in its saliva, which enables it to excrete low-phosphorus manure.

A transgenic organism has a genetic change in each of its cells, often the addition of a gene from a different species. The transgenic pig has a phytase gene from the bacterium *E. coli*. Its manure has 75 percent less phosphorus than normal pig excrement. Says one researcher, "These pigs offer a unique biological approach to the management of phosphorus nutrition and environmental pollution in the pork industry."

Transgenic pigs given a bacterial digestive enzyme excrete genetically modified, less-polluting manure. The author poses amidst a pile of the nonmodified material at the University of Georgia.

19.1 Patenting DNA

Biotechnology is the use or alteration of cells or biological molecules for specific applications. It is an ancient art as well as a modern science. Using yeast to ferment fruit or produce wine are biotechnologies, as is extracting biochemicals from organisms.

The popular terms "genetic engineering" and "genetic modification" refer broadly to any biotechnology that manipulates genetic material. This includes altering the DNA of an organism to suppress or enhance the activities of its own genes, as well as combining the genetic material of different species. Organisms that harbor DNA from other species are termed **transgenic** and their DNA is called **recombinant DNA.** Creating them is possible because all life uses the same genetic code (**figure 19.1**). It is this mixing of DNA from different species that some people object to as being unnatural. In fact, DNA moves and mixes between species in nature—bacteria do it, and it is why we have viral DNA sequences in our chromosomes. But human-instituted genetic modification usually endows organisms with traits they would probably not acquire naturally, such as pigs with low-phosphorus manure, tomatoes that grow in salt water, and bacteria that synthesize human insulin.

Transgenic organisms raise legal questions. Is a corn plant that manufactures a protein naturally found only in green beans an invention deserving of patent protection? To qualify for such protection, a transgenic organism must be new, useful, and not obvious (see Technology Timeline). Patent law has evolved to keep up with modern biotechnology. In the 1980s, when sequencing a gene was painstakingly slow, only a few genes were patented. Then, in the mid-1990s, with faster sequencing technology and shortcuts to finding the protein-encoding portions of the genome, the U.S. National Institutes of Health and biotech companies began seeking patent protection for thousands of short DNA sequences, even if their functions weren't known. Someday, they would be, and the patent holders then would stand to make money. Because of the flood of applications, the U.S. Patent and Trademark Office began to tighten requirements for utility. Today, although entire genomes are sequenced in the time it once took to decipher a single gene, a DNA sequence alone does not warrant patent protection. The described molecule must be useful as a tool for research or as a novel or improved product.

Despite the increasing stringency of patent requirements, problems still arise concerning the status of DNA sequences. A biotechnology company in the United States, for example, holds a patent on the *BRCA1* breast cancer gene that includes any diagnostic tests based on the DNA sequence. That company's tests, however, do not cover all mutations in the gene. A French physician working with a family that has a unique large deletion is challenging the patent, because to be tested, her patients must pay a high licensing fee to the U.S. company that "owns" the gene sequence.

The influx of human genome information continues to complicate patenting. One problem is redundancy. For the same gene, it is possible to patent the entire sequence (termed genomic DNA), or just the protein-

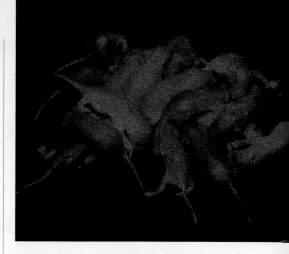

Figure 19.1 The universality of the genetic code makes biotechnology possible. These transgenic mice contain the gene encoding a jellyfish's green fluorescent protein (GFP). Researchers use GFP to mark genes of interest to follow their expression. The GFP mice glow less greenly as they mature and more hair covers the skin. The non-green mice are not genetically modified.

Technology Timeline

Patenting Life and Genes

1790	U.S. patent act is enacted. An invention must be new, useful, and not obvious to earn a patent.
1873	Louis Pasteur is awarded first patent on a life form, for yeast used in industrial processes.
1930	New plant variants can be patented.
1980	First patent is awarded on a genetically modified organism, a bacterium given four plasmids (DNA rings) that enable it to metabolize components of crude oil. The plasmids are naturally occurring, but do not all occur naturally in a single type of bacterium.
1988	First patent is awarded for a transgenic organism, a mouse that manufactures human protein in its milk. Harvard University granted patent for "OncoMouse" transgenic for human cancer.
1992	Biotechnology company is awarded a broad patent covering all forms of transgenic cotton. Groups concerned that this will limit the rights of subsistence farmers contest the patent several times.
1996–1999	Companies patent partial gene sequences and certain disease-causing genes as the basis for developing specific medical tests.
2000	With gene and genome discoveries pouring into the Patent and Trademark Office, requirements tightened for showing utility of a DNA sequence.
2003	Attempts to enforce patents on nonprotein-encoding parts of the human genome anger researchers who support open access to the information.
2005	Read about patents relevant to genetics, genomics, and biotechnology at http://dnapatents.georgetown.edu/

encoding exons. A researcher can also patent a gene variant, such as a sequence containing a SNP or mutation. A company or researcher developing a tool or test based on a particular gene or its encoded protein might infringe upon several patents that are based on essentially the same information. For example, it is unclear how patents would cover exons common to different genes. Now, as genetics shifts from a gene-by-gene focus to analyzing expression patterns of suites of interacting genes, patent law will have to once again adjust to keep up with scientific developments.

RELATING THE CONCEPTS

A company has developed a biotechnology that kills bacteria responsible for an intestinal illness in people who drink milk from cows that harbor the bacteria. Do you think that this invention infringes upon the patent for treating pig excrement described in the chapter opener? Cite a reason for your answer.

Key Concepts

1. Biotechnology is the use or modification of cells or biological molecules for a specific application.
2. Patent law regarding DNA has evolved with the technology since the 1970s, and is still changing.

19.2 Amplifying DNA

Some forensic and medical tests require many copies of a specific DNA sequence from a small initial sample. In one case, investigators collected a minute smear of brain tissue on a car fender and had its DNA extracted, sequenced, and compared to DNA from three unidentified headless corpses in a city morgue. In another example, at pharmaceutical companies, bacteria harboring human genes that they use to produce protein-based drugs occupy huge containers called bioreactors.

Mass-producing a DNA sequence is called nucleic acid amplification. Technologies that amplify DNA were invented in the 1970s and 1980s. The best known is the **polymerase chain reaction** (PCR), which amplifies but does not change the initial DNA sequence. PCR was used to identify the brain material on the car fender. Producing drugs in bacteria uses another approach, recombinant DNA technology, that amplifies DNA that includes sequences from other types of organisms. PCR is done on molecules; recombinant DNA technology works in cells. Recombinant DNA technology is addressed in the next section.

PCR is based on the natural process of DNA replication. Recall from chapter 9 that every time a cell divides, it replicates all of its DNA. PCR uses DNA polymerase to rapidly replicate a specific DNA sequence millions of times.

Applications of PCR are eclectic (**Table 19.1**). In forensics, it is used routinely to amplify DNA sequences that are profiled to establish blood relationships, to identify remains, and to help convict criminals or exonerate the falsely accused. When used to amplify the nucleic acids of pathogens, PCR is useful in agriculture, veterinary medicine, environmental science, and human health care. In genetics, PCR is both a crucial laboratory tool to identify genes and it is a component of many diagnostic tests.

PCR was born in the mind of Kary Mullis on a moonlit night in northern California in 1983. As he drove the hills, Mullis was thinking about the precision and power of DNA replication, and a way to tap into that power popped into his mind. He excitedly explained his idea to his girlfriend and then went home to think it through. "It was difficult for me to sleep with deoxyribonuclear bombs exploding in my brain," he wrote much later.

The idea behind PCR was so simple that Mullis had trouble convincing his superiors at Cetus Corporation that he was really onto something. He spent the next year using the technique to amplify a well-studied gene. One by one, other researchers glimpsed Mullis's vision of that starry night. After winning over his colleagues at Cetus, Mullis published a landmark 1985 paper and filed patent applications, launching the field of nucleic acid amplification. Mullis received only a $10,000 bonus from Cetus for his invention, which the company sold to another for $300 million. Mullis did, however, win a Nobel prize.

Table 19.1

Uses of PCR

PCR has been used to amplify DNA from:

- a cremated man, from skin cells left in his electric shaver, to diagnose an inherited disease in his children.
- human tissue from the site of the World Trade Center in the days following September 11, 2001, to identify victims.
- a preserved quagga (a relative of the zebra) and a marsupial wolf, both extinct.
- microorganisms that cannot be cultured for study.
- the brain of a 7,000-year-old human mummy.
- the digestive tracts of carnivores, to reveal food web interactions.
- roadkills and carcasses washed ashore, to identify locally threatened species.
- products illegally made from endangered species, such as powdered rhinoceros horn, sold as an aphrodisiac.
- genetically altered bacteria that are released in field tests, to follow their dispersion.
- one cell of an 8-celled human embryo to detect a disease-related genotype.
- poached moose meat in hamburger.
- remains in Jesse James's grave, to make a positive identification.
- the guts of genital crab lice on a rape victim, which matched the DNA of the suspect.
- dried semen on a blue dress belonging to a White House intern.
- fur from Snowball, a cat that linked a murder suspect to a crime.

PCR rapidly replicates a selected sequence of DNA in a test tube (**figure 19.2**). The requirements include:

1. Knowing parts of a target DNA sequence to be amplified.

2. Two types of lab-made, single-stranded, short pieces of DNA called primers. These are complementary in sequence to opposite ends of the target sequence.

3. A large supply of the four types of DNA nucleotide building blocks.

4. Taq1, a DNA polymerase produced by *Thermus aquaticus*, a microbe that inhabits hot springs. This enzyme is adapted to its host's hot surroundings and makes PCR easy because it does not fall apart when DNA is heated, as most proteins do.

In the first step of PCR, heat is used to separate the two strands of the target DNA. Next, the two short DNA primers and Taq1 DNA polymerase are added. The temperature is lowered. Primers bind by complementary base pairing to the separated target strands. In the third step, the Taq1 DNA polymerase adds bases to the primers and builds a sequence complementary to the target sequence. The newly synthesized strands then act as templates in the next round of replication, which is initiated immediately by raising the temperature. All of this is done in an automated device called a thermal cycler that controls the key temperature changes. The heat-resistant DNA polymerase is crucial to the process.

The pieces of identical DNA accumulate exponentially. The number of amplified pieces of DNA equals 2^n, where n equals the number of temperature cycles. After just 20 cycles, 1 million copies of the original sequence have accumulated in the test tube.

PCR's greatest strength is that it works on crude samples of rare, old, and minute sequences, as Reading 9.2 describes. PCR's greatest weakness, ironically, is its exquisite sensitivity. A blood sample submitted for diagnosis of an infection, if contaminated by leftover DNA from a previous test, or a stray eyelash from the person running the reaction, can yield a false result.

The invention of PCR inspired other nucleic acid amplification techniques. One is transcription-mediated amplification, which copies target DNA into RNA and then uses RNA polymerase to amplify the RNA. This procedure doesn't require temperature shifts, and it generates 100 to 1,000 copies per cycle, compared to PCR's doubling, and can yield 10 billion copies of a selected sequence in a half hour!

Nucleic acid amplification has improved diagnostic testing. For example, U.S. laboratories began using it in 1999 to test donated blood for RNA from HIV. Past techniques detected either a person's antibody response to infection, or antigens unique to HIV's surface. These approaches are not sensitive enough to detect infection until it's already been present for weeks. Of more than 37 million blood samples tested for HIV RNA, nucleic acid amplification found 12; antigen testing found only 2.

Key Concepts

1. PCR rapidly replicates a small part of an organism's genome.
2. PCR has many uses.
3. Other nucleic acid amplification technologies followed PCR.

19.3 Modifying DNA

Recombinant DNA technology adds genes from one type of organism to the genome of another. It was the first gene modification biotechnology, and was initially done in bacteria. When bacteria bearing recombinant DNA divide, they yield many copies of the "foreign" DNA, and under proper conditions they produce many copies of the protein that the foreign DNA specifies. Recombinant DNA technology is also known as gene cloning. "Cloning" in this context refers to making many copies of a specific DNA sequence.

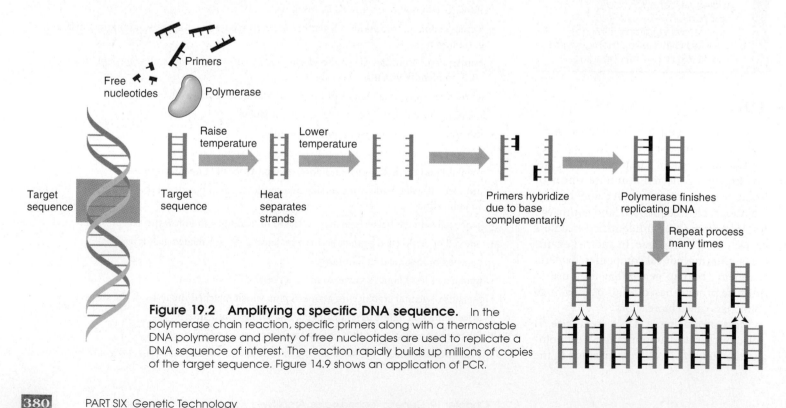

Figure 19.2 Amplifying a specific DNA sequence. In the polymerase chain reaction, specific primers along with a thermostable DNA polymerase and plenty of free nucleotides are used to replicate a DNA sequence of interest. The reaction rapidly builds up millions of copies of the target sequence. Figure 14.9 shows an application of PCR.

Free nucleotides   Primers   Polymerase

Target sequence   Target sequence   Raise temperature   Heat separates strands   Lower temperature   Primers hybridize due to base complementarity   Polymerase finishes replicating DNA   Repeat process many times

Recombinant DNA Technology and Transgenic Organisms

Researchers first began pondering the potential uses and risks of mixing DNA from different species in the 1970s.

In February 1975, 140 molecular biologists convened at Asilomar, a conference center on California's Monterey Peninsula, to discuss the safety and implications of a new type of experiment. Investigators had found a simple way to combine the genes of two species, and they were concerned about the safety of experiments requiring the use of a cancer-causing virus, and about where the field was headed. They discussed restricting the types of organisms used in recombinant DNA research and explored ways to prevent escape of a resulting organism from the laboratory. The guidelines drawn up at Asilomar outlined measures of "physical containment," such as using specialized hoods and airflow systems that would keep the organisms inside the laboratory, and "biological containment," such as weakening organisms so that they could not survive outside the laboratory.

A decade after the Asilomar meeting, many members of the original group reconvened at the meeting site to assess progress in the field. Nearly all agreed on two points: Recombinant DNA technology was safer than expected, and the technology had spread to industry more swiftly and in more diverse ways than anyone had imagined. At a meeting twenty-five years after the original event, attendees concluded that biotechnology had become so commercialized that the open atmosphere of the original gathering was no longer possible.

Recombinant DNA-based products have been slow to reach the marketplace because of the high cost of the research and the long time it takes to develop any new drug. Today, several dozen such drugs are available, and more are in the pipeline. Recombinant DNA research initially focused on direct gene products such as peptides and proteins with therapeutic actions, such as insulin, growth hormone, and clotting factors. However, the technology can target other biochemicals, such as carbohydrates and lipids, by affecting the genes that encode enzymes required to synthesize them.

RELATING THE CONCEPTS

How does the genetic manipulation of pigs to alter their excrement differ from the earliest use of recombinant DNA technology, which enabled bacteria to produce human proteins?

Constructing Recombinant DNA Molecules—An Overview

Manufacturing recombinant DNA molecules requires restriction enzymes that cut donor and recipient DNA at the same sequence; DNA to carry the donor DNA (called cloning vectors); and recipient cells (bacteria or other cultured single cells).

After inserting donor DNA into vectors, the procedure requires several steps:

- Selecting cells where the genetic material includes foreign DNA

- Selecting cells that received the gene of interest

- Stimulating transcription of the foreign gene and translation of its protein product

- Collecting and purifying the desired protein

The natural function of restriction enzymes is to protect bacteria by cutting and thereby inactivating the DNA of infecting viruses. Protective methyl (CH_3) groups shield the bacterium's own DNA from its restriction enzymes. Bacteria have hundreds of types of restriction enzymes. Some cut DNA at particular sequences of four, five, or six bases. These targets are symmetrical—the recognized sequence reads the same, from the 5′ to 3′ direction, on both strands of the DNA. For example, the restriction enzyme EcoR1, shown in **figure 19.3,** cuts at the sequence GAATTC. The

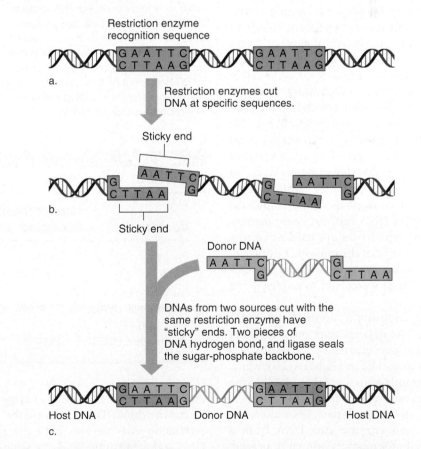

Figure 19.3 Recombining DNA. A restriction enzyme makes "sticky ends" in DNA by cutting it at specific sequences. **(a)** The enzyme EcoR1 cuts the sequence GAATTC between G and A. **(b)** This staggered cutting pattern produces "sticky ends" of sequence AATT. The ends attract through complementary base pairing. **(c)** DNA from two sources is cut with the same restriction enzyme. Pieces join, forming recombinant DNA molecules.

complementary sequence on the other strand is CTTAAG, which, read backwards, is GAATTC. (You can try this with other sequences to see that it rarely works this way!) In the English language, this type of symmetry is called a palindrome, referring to a sequence of letters that reads the same in both directions, such as "Madam, I'm Adam." Palindromic sequences in DNA reflect the sequences on two strands.

The cutting action of some restriction enzymes on double-stranded DNA creates single-stranded extensions. They are called "sticky ends" because they are complementary to each other and as a result form hydrogen bonds as their bases pair. Restriction enzymes work as molecular scissors in creating recombinant DNA molecules because they cut at the same sequence in any DNA source. That is, the same sticky ends result from the same restriction enzyme, whether the DNA is from a mockingbird or a maple.

Another natural "tool" used in recombinant DNA technology is a cloning vector. This structure carries DNA from the cells of one species into the cells of another. A vector can be any piece of DNA in which other DNA can insert for transfer into the cell of another organism. A commonly used type of vector is a **plasmid,** which is a small circle of double-stranded DNA that occurs naturally in some bacteria, yeasts, plant cells, and other types of organisms (**figure 19.4**). Viruses that infect bacteria, called bacteriophages, provide another type of vector. Bacteriophages are manipulated to transport DNA but not cause disease. Disabled retroviruses are used as vectors too, as are artificial chromosomes from bacteria and yeast. Bacterial artificial chromosomes (BACs) were used to sequence the human genome.

When choosing a cloning vector, size matters. The desired gene must be short enough to insert into the vector. Gene size is typically measured in kilobases (kb), which are thousands of bases. **Table 19.2** lists the capacities of several cloning vectors.

To create a recombinant DNA molecule, a restriction enzyme cuts DNA from a donor cell at sequences known to bracket the gene of interest (**figure 19.5**). The enzyme leaves single-stranded ends on the cut DNA, each bearing a characteristic base sequence. Next, a plasmid is isolated and

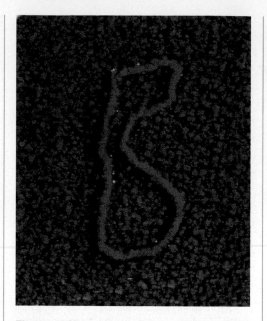

Figure 19.4 Plasmids. Plasmids are small circles of DNA found naturally in the cells of some organisms. A plasmid can replicate independent of the host genetic material, including any other DNA inserted into it. For this reason, plasmids make excellent cloning vectors—structures that carry DNA from the cells of one species into the cells of another. They naturally enter about 1 in 1,000 bacterial cells and include several restriction enzyme cutting sites and some antibiotic resistance genes.

Table 19.2

Cloning Vectors

Vector	Size of Insert Accepted (kb)
Plasmid	up to 15
Bacteriophage	up to 90
Bacterial artificial chromosome (BAC)	100–500
Yeast artificial chromosome (YAC)	250–2,000

cut with the same restriction enzyme used to cut the donor DNA. Because the same restriction enzyme cuts both the donor DNA and the plasmid DNA, the same complementary single-stranded base sequences extend from the cut ends of each. When the cut plasmid and the donor DNA are mixed, the single-stranded sticky ends of some

plasmids base pair with the sticky ends of the donor DNA. The result is a recombinant DNA molecule, such as a plasmid carrying the human insulin gene. The plasmid and its human gene can now be transferred into a cell, typically bacterial.

The following parts of this section discuss the steps of creating recombinant DNA molecules.

Isolating the Gene of Interest

Constructing recombinant DNA molecules usually begins by cutting all of the DNA of the donor cell. This DNA, which includes non-protein-encoding sequences, is termed genomic DNA. Researchers assemble collections of recombinant bacteria (or other single cells) that harbor pieces of a genome. By using several copies of a genome, the pieces overlap where sequences align. Such a collection is called a **genomic library.** For each application, such as using a human protein as a drug, a particular piece of DNA must be identified and isolated from a genomic library. There are several ways to do this "needle in a haystack" type of search.

A synthetic piece of DNA that is complementary to part of the template strand of the gene in question can be linked to a label, such as a radioactive or fluorescent molecule. This labeled gene fragment is called a **DNA probe.** It emits a signal when it binds to its complement in a cell that contains a recombinant plasmid. DNA probes can also be made using genes of similar sequence from other species—they will bind the human version of the gene. Using such a probe is a little like mistakenly using hipropotamus to search for hippopotamus on the Internet. You'd probably still come up with a hippo.

A genomic library contains too much information for a researcher seeking a particular protein-encoding gene—it may also contain introns, the genes that encode rRNAs and tRNAs, and many repeated sequences. A shortcut is to use another type of library, called a complementary DNA, or **cDNA library,** that represents only protein-encoding genes. A cDNA library is made from the mRNAs in a differentiated cell, which represent the proteins manufactured there. For example, a muscle cell has abundant mRNA that encodes contractile proteins, whereas a fibroblast has many mRNAs that represent connective tissue proteins.

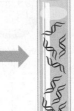

DNA isolated from donor cell (animal or plant)

A specific restriction enzyme fragments donor DNA

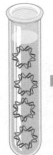

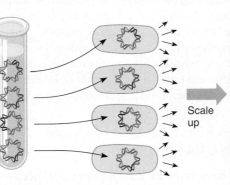

Plasmid isolated from bacterium

The same restriction enzyme that fragmented donor DNA is also used to open plasmid DNA

Donor and plasmid DNA are mixed; "sticky ends" of donor DNA hydrogen bond with sticky ends of plasmid DNA fragment; recombinant molecule is sealed with ligase

Modified plasmids (recombinant DNA) are introduced into bacteria

Bioreactor

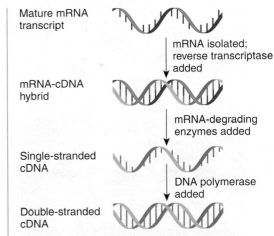

Bacteria divide and clone the gene from donor cells that were spliced into the plasmids

Scale up

Drug is produced

Figure 19.5 Recombinant DNA. To construct a recombinant DNA molecule, DNA isolated from a donor cell and a plasmid are cut with the same restriction enzyme and mixed. Some of the sticky ends from the donor DNA hydrogen bond with the sticky ends of the plasmid DNA, forming recombinant DNA molecules. When such a modified plasmid is introduced into a bacterium, it is mass produced as the bacterium divides.

To make a cDNA library, researchers first extract the mRNAs from cells. Then, these RNAs are used to construct complementary or "c" DNA strands using reverse transcriptase, DNA nucleotide triphosphates, and DNA polymerase (**figure 19.6**). Reverse transcriptase synthesizes DNA complementary to RNA. DNA polymerase and the nucleotides then can synthesize the complementary strand to the single-stranded cDNA to form a double-stranded DNA. Different cell types yield different cDNA collections, or libraries, that reflect which genes are expressed. They do not, however, reveal protein abundance, because in a cell mRNA molecules are transcribed and degraded at different rates.

A specific cDNA can be taken from a cDNA library and used as a probe to isolate the original gene of interest from the genomic library. If the goal is to harness the gene and eventually collect its protein product, then the genomic version is useful, because it includes control regions such as promoters. Once a gene of interest is transferred to a cell where it can be transcribed into mRNA and that RNA can be translated,

the protein is collected. Such cells are typically grown in devices called bioreactors, with nutrients sent in and wastes removed. A researcher collects the desired product from the medium the cells are growing in.

Selecting Recombinant DNA Molecules

Much of the effort in recombinant DNA technology entails identifying and separating cells that contain the gene of interest, once the foreign DNA is inserted into the vector. Three types of recipient cells can result:

1. Cells that lack plasmids

2. Cells that contain plasmids that do not contain a foreign gene

3. Cells that contain plasmids that have picked up a foreign gene (the goal)

The procedure is set up to distinguish bacteria that have taken up recombinant plasmids from those that have not taken up plasmids or that have admitted plasmids that do not carry foreign DNA. The two-step strategy uses a plasmid that includes a gene

Mature mRNA transcript

mRNA isolated; reverse transcriptase added

mRNA-cDNA hybrid

mRNA-degrading enzymes added

Single-stranded cDNA

DNA polymerase added

Double-stranded cDNA

Figure 19.6 Copying DNA from RNA. Researchers make cDNA from mRNA using reverse transcriptase, an enzyme from a retrovirus. A cDNA version of a gene includes the codons for a mature mRNA, but not sequences corresponding to promoters and introns. Labeled cDNAs are used as probes to locate genes in genomic libraries.

that imparts resistance to an antibiotic drug, such as ampicillin in **figure 19.7**. First, human and plasmid DNA are cut with the same restriction enzymes and mixed. Hopefully, some human DNA fragments insert into the plasmids. The plasmids are closed up with ligase (the enzyme that glues the sugar-phosphate backbone when DNA replicates), and transferred to bacterial cells. Ampicillin is applied, and only cells harboring plasmids survive. Cells divide, forming visible colonies on a plate. The second manipulation is to include in the plasmid a gene, called *lacZ*, that encodes the enzyme β galactosidase. When the substrate of the reaction that β galactosidase catalyzes, called X-Gal, is applied to cultured cells, a reaction ensues and a blue color

appears. If inserting human DNA interrupts the *lacZ* gene, then the reaction cannot proceed and the colony doesn't appear blue and is easily distinguished (**figure 19.8**).

When cells containing the recombinant plasmid divide, so does the plasmid. Within hours, the original cell gives rise to many harboring the recombinant plasmid. The enzymes, ribosomes, energy molecules, and factors necessary for protein synthesis transcribe and translate the plasmid DNA and its foreign gene, producing the desired protein.

Products from Recombinant DNA Technology

In basic research, recombinant DNA technology provides a way to isolate individual genes from complex organisms and observe their functions on the molecular level. Recombinant DNA has many practical uses, too. The first was to mass-produce protein-based drugs, and the first such drug was human insulin.

Before 1982, people with type 1 diabetes mellitus obtained the insulin that they

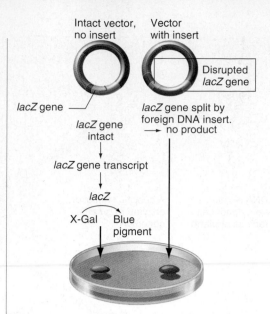

Figure 19.8 Selecting recombinant DNA molecules step 2: Identifying DNA insertion. Exposure to X-Gal distinguishes bacteria that have taken up plasmids in which inserted foreign (human) DNA splits the *lacZ* gene. The desired recombinant DNA-bearing bacterial colonies do not turn blue in the presence of X-Gal.

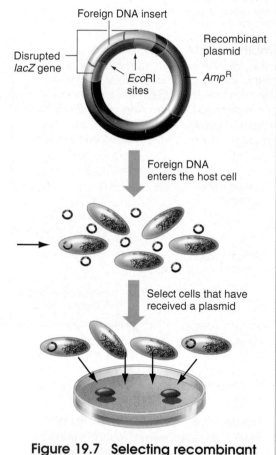

Figure 19.7 Selecting recombinant DNA molecules step 1: Taking up a plasmid. Recombinant plasmids include restriction enzyme cutting sites (EcoR1 is a restriction enzyme), an antibiotic resistance gene (to ampicillin here), foreign DNA, and a gene called *lacZ* that imparts a blue color to the growth medium when its encoded protein is produced in the presence of substrate. Bacterial cells that harbor plasmids can grow in the presence of the antibiotic.

Table 19.3

Drugs Produced Using Recombinant DNA Technology

Drug	Use
Atrial natriuretic peptide	Dilates blood vessels, promotes urination
Colony stimulating factors	Help restore bone marrow after marrow transplant; restore blood cells following cancer chemotherapy
Deoxyribonuclease (DNase)	Thins secretions in lungs of people with cystic fibrosis
Epidermal growth factor	Accelerates healing of wounds and burns; treats gastric ulcers
Erythropoietin (EPO)	Stimulates production of red blood cells in cancer patients
Factor VIII	Promotes blood clotting in treatment of hemophilia
Glucocerebrosidase	Gaucher disease
Human growth hormone	Promotes growth of muscle and bone in people with very short stature due to hormone deficiency
Insulin	Allows cells to take up glucose in treatment of type 1 diabetes
Interferons	Genital warts, hairy cell leukemia, hepatitis C and B, Kaposi sarcoma, multiple sclerosis
Interleukin-2	Kidney cancer
Lung surfactant protein	Helps lung alveoli to inflate in infants with respiratory distress syndrome
Renin inhibitor	Lowers blood pressure
Somatostatin	Decreases growth in muscle and bone in pituitary giants
Superoxide dismutase	Prevents further damage to heart muscle after heart attack
Tissue plasminogen activator	Dissolves blood clots in treatment of heart attacks, stroke, and pulmonary embolism

The Ethics of Using a Recombinant Drug: EPO

EPO is a hormone produced in the kidneys that is a 165-amino-acid protein plus four carbohydrate chains. When the oxygen level in the blood is too low, cells in the kidneys produce EPO, which travels to the bone marrow and binds to receptors on cells that give rise to red blood cell progenitors. Soon, more red blood cells enter the circulation, carrying more oxygen to the tissues (**figure 1**).

The value of EPO as a drug became evident after the invention of hemodialysis to treat kidney failure in 1961. This otherwise highly successful treatment also causes severe anemia, because dialysis removes EPO from the blood. Counteracting the anemia required boosting EPO levels. In 1970, the U.S. government sought ways to mass-produce EPO. But levels of EPO in human plasma are too low to pool from donors. A more likely potential source was people suffering from disorders, such as aplastic anemia and hookworm infection, that caused them to secrete large amounts of EPO in their urine.

In the 1970s, the U.S. government obtained EPO from South American farmers with hookworm infections and Japanese aplastic anemia patients. But was it ethical to obtain a scarce substance from the poor to treat the comparatively wealthy? Then AIDS came. Biochemicals from human body fluids were no longer safe.

Recombinant DNA technology solved the EPO problem. The hormone is produced in hamster kidney cells, which can attach EPO's four carbohydrate groups. Today, EPO is sold under various names to treat anemia in dialysis and AIDS patients and is also given with cancer chemotherapy to avoid the need for transfusions.

Recombinant EPO found an unexpected market among Jehovah's Witnesses, whose religion forbids transfused blood, which they believe destroys the soul. They can, however, use EPO, because it has been declared a product of recent technology, rather than blood. Some Jehovah's Witnesses have used it before surgery to boost red blood cell supplies, or afterward to compensate for blood loss.

EPO's ability to increase the oxygen-carrying capacity of blood, and thereby to increase physical endurance, has attracted the attention of competitive athletes. Training at high altitudes increases endurance, and the reason is EPO. The scarcer oxygen in the air at high altitudes stimulates the kidneys to produce EPO, which stimulates production of more red blood cells. Athletes have attempted to reproduce this effect by abusing EPO. A few Dutch bicyclists developed dangerous blood clots in their legs from taking overly high, and medically unnecessary, doses. It is now routine to screen Olympic athletes for EPO abuse.

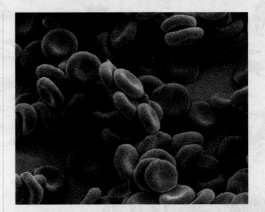

Figure 1 These red blood cells are mass-produced in a patient treated with erythropoietin (EPO).

One large Scandinavian family, however, gets its extra EPO naturally. They have an inherited condition, benign erythrocytosis (OMIM 263400), in which they overproduce the hormone. One of them won a gold medal for skiing in the Olympics. However, athletes who abuse EPO raise the issue of how to control the use of an otherwise valuable drug derived from biotechnology.

EPO is not the only abused biotech-derived drug. Body builders abuse a recombinant version of the hormone somatotropin, used to treat AIDS wasting syndrome, and the Internet is full of ads for recombinant human growth hormone.

---

injected daily from pancreases removed from cattle in slaughterhouses. Cattle insulin is so similar to the human peptide, different in only 2 of its 51 amino acids, that most people with diabetes could use it. However, about 1 in 20 patients is allergic to cow insulin because of the slight chemical difference. Until recombinant DNA technology was developed, the allergic patients had to use expensive combinations of insulin from other animals or human cadavers. A person with diabetes can now purchase "Humulin," the human protein made in *E. coli*, at a local drugstore. **Table 19.3** lists some drugs produced using recombinant DNA technology. Each year a few new drugs join the list.

Drugs developed using recombinant DNA technology must compete in the marketplace with conventional products. Deciding whether a recombinant drug is preferable to an existing similar drug is often a matter of economics, marketing, and common sense. For example, interferon β-1b helps some people with a certain type of multiple sclerosis, but it costs more than $20,000 per year per patient. British researchers calculated what it would cost to treat the nation's patients who would benefit from the drug, and then determined how else the funds could be spent. They concluded that more people would be served if the money were spent on improved supportive care for many rather than on a costly new treatment for a few.

Tissue plasminogen activator (tPA), a recombinant clot-busting drug developed in the mid-1980s, also has cheaper alternatives. If injected within four hours of a heart attack, tPA dramatically limits damage to the heart muscle by restoring blood flow. It costs $2,200 a shot. An older drug, streptokinase, is extracted from unaltered bacteria and is nearly as effective, at $300 per injection. tPa is very valuable for patients who have already had streptokinase and could have an allergic reaction if they were to use it again. Bioethics: Choices for the Future considers another drug derived from recombinant DNA technology, erythropoietin (EPO).

An application of recombinant DNA technology in the textile industry is a new source of indigo, the dye used to make blue jeans blue. The dye originally came from mollusks and fermented leaves of the European woad plant or Asian indigo plant. The 1883 discovery of indigo's chemical structure led to the invention of a synthetic process to produce the dye using coal-tar. That method has dominated the industry, but it releases toxic by-products.

In 1983, microbiologists discovered that *E. coli,* with a little help, can produce indigo. The bacterium converts glucose to the amino acid tryptophan, which then forms indole, a precursor to indigo. By learning the steps and offshoots of this biochemical pathway, researchers altered bacteria to suppress the alternative pathways for metabolizing glucose, allowing the cells to synthesize excess tryptophan. When the *E. coli* cells incorporate genes from another bacterial species, they extend the biochemical pathway all the way to produce indigo. The result: common bacteria that manufacture the blue dye of denim jeans from glucose, a simple sugar.

Delivering DNA in Plants and Animals

When recombinant DNA technology is applied to cells of multicellular organisms, individuals then must develop and be bred to yield homozygotes for desired recessive traits. Dominant traits would be present in the initial generation. The added or altered characteristic in the multicellular transgenic organism is a visible trait or an unusual secretion.

A transgenic plant is easier to create than a transgenic animal because it can be derived from somatic cells. (This makes sense. You can grow a new plant from a cutting; animal development does not work this way.) Different vectors and gene transfer techniques are sometimes used in plants because their cell walls, not present in animal cells, are difficult to penetrate. Some manipulations are done on plant cells that have had their cell walls removed. These denuded plant cells are called protoplasts.

A commonly used plant vector is a **Ti plasmid** (for "tumor-inducing"), which occurs naturally in the bacterium *Agrobacterium tumefaciens* (**figure 19.9**). A *Ti* plasmid normally causes a tumorlike growth;

researchers remove the genes controlling this process. For example, a gene from the bacterium *Bacillus thuringiensis (bt)* specifies a protein that destroys the stomach linings of certain insect larvae. When the *bt* gene is introduced into corn cells via a *Ti* plasmid, the cells regenerate corn plants that produce their own insecticide. More than two-thirds of the corn plants grown in the United States are transgenic for the *bt* insecticide gene.

The same gene is used to protect trees such as larch, white spruce, and pioneer elm from gypsy moth and forest tent caterpillars. Use of *bt* protein as a natural insecticide isn't new—organic farmers have been

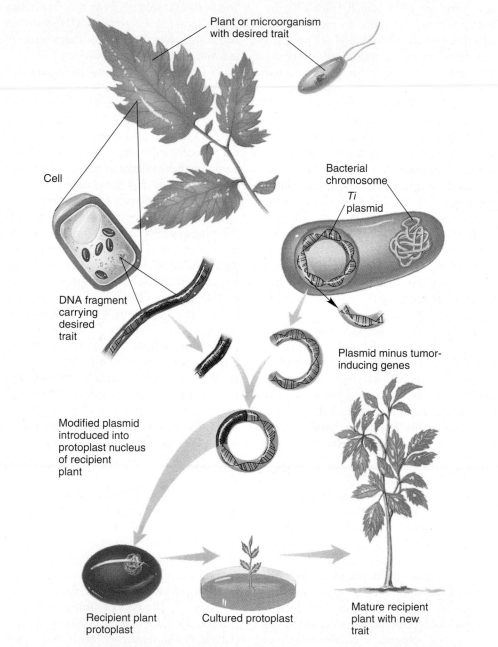

Figure 19.9 Producing a transgenic plant. A DNA fragment carrying the desired gene—conferring resistance to an herbicide, for example—is isolated from its natural source and spliced into a *Ti* plasmid with the tumor-inducing genes removed. The plasmid incorporating the foreign DNA then invades a cell of the recipient plant, entering the nucleus and integrating into the plant's DNA. Finally, in culture, the cell is regenerated into a mature transgenic plant that expresses the desired trait and passes it on to its progeny. A breeding step may be necessary to obtain plants homozygous for a recessive trait.

using the protein for years, applying unmodified bacteria directly to plants. A transgenic plant's leaves, which are the parts that many insects eat, contain the *bt* insecticide. Transgenic plants are also created whose seeds store proteins useful as drugs, or whose roots secrete useful substances.

Table 19.4 lists some useful transgenic plants, and **table 19.5** describes some agricultural challenges that transgenic crops can address.

Transgenic plants acquire their new capabilities quickly, in one generation. In contrast, obtaining crops with valuable characteristics using traditional agriculture may require breeding plants through several generations of specific genetic crosses. The recombinant DNA route is also more precise than traditional breeding because it deals with one characteristic at a time, rather than having to follow traits that reassort during meiosis in traditional agriculture.

The creation of a transgenic plant follows these general steps:

1. A gene that confers an agriculturally useful characteristic is isolated and inserted into a cloning vector.

2. The recombinant vector is selected and delivered into protoplasts.

3. Whole plants are regenerated from genetically altered cells, perpetuating the change in all cells.

4. If the trait is recessive, crosses are set up to generate homozygous recessive plants.

5. The transgenic plants are tested in the laboratory to see if the desired protein is manufactured, and if the introduced gene (a transgene) is passed on to the next generation.

6. The plants are field-tested to see if the desired trait persists and if the crops affect the ecosystem.

Table 19.4
Genetically Modified Crops

Altered Plant	Effect
Rice with beta carotene and extra iron	Added nutritional value
Canola with high-laurate oil	Can be grown domestically; less costly than importing palm and coconut oils
Delayed ripening tomato	Extended shelf life
Herbicide-resistant cotton	Herbicide kills weeds without harming crop
Minipeppers	Improved flavor, fewer seeds
Bananas resistant to fungal infection	Extended shelf life
Delayed-ripening bananas and pineapples	Extended shelf life
Elongated sweet pepper	Improved flavor, easier to slice
Altered cotton fiber	"Plasticized" fabric
Altered paper pulp trees	Paper component (lignin) easier to process
High-starch potatoes	Absorb less oil when fried
Pest-resistant corn	Can resist European corn borer
Seedless minimelons	Single serving size
Sweet peas and peppers	Retain sweetness longer
Sugarcane with corn gene	Resists bacterial and fungal toxins

Table 19.5
Transgenic Approaches to Agricultural Challenges

Challenge	Possible Solution
Frost damages crops	Spray crops with bacteria genetically altered to lack surface proteins that promote ice crystallization. Bacteria can also be manipulated to stimulate ice crystallization, then used to increase snow buildup at winter sports facilities.
Herbicides and pesticides damage crops	Isolate genes from an organism not affected by the chemical and insert it into the genome of a crop plant.
Crops need costly nitrogen fertilizer because atmospheric nitrogen is not biologically usable	Short-term: Genetically manipulate nitrogen-fixing *Rhizobium* bacteria to overproduce enzymes that convert atmospheric nitrogen to a biologically usable form in root nodules of legumes. Alter *Rhizobium* to colonize a wider variety of plants. Long-term: Transfer *Rhizobium* nitrogen-fixation genes into plant cells and regenerate transgenic plants.
A plant food is low in a particular amino acid	Transfer gene from another species that controls production of a protein rich in the amino acid the crop plant lacks.
A virus destroys a crop	Genetically alter crop plant to manufacture a protein on its cell surface normally found on the virus's surface. Plant becomes immune to virus.
Public concern about the safety of synthetic pesticides	Stimulate *Bacillus thuringiensis* to overproduce its natural pesticide, which destroys insects' stomach linings. Transfer *B. thuringiensis* bioinsecticide gene to crop plant.

Creating transgenic animals is more challenging than working with plants because of fundamental differences in early development. Researchers use several techniques to insert DNA into animal cells. Chemicals are used to open transient holes in plasma membranes, admitting DNA. Liposomes are fatty bubbles that can carry DNA into cells as plasma membranes envelop them. In electroporation, a brief jolt of electricity opens transient holes in plasma membranes that may permit foreign DNA to enter. DNA is also injected into cells using microscopic needles. This is called microinjection.

Another way to introduce DNA into cells is particle bombardment. A gunlike device shoots tiny metal particles, usually gold or tungsten, coated with foreign DNA. When aimed at target cells, some of the projectiles enter. This technique is used to deliver DNA to plant cells, too.

As in plant cells, once foreign DNA is introduced into an animal cell, it must enter the nucleus, replicate along with the cell's own DNA, and be transmitted when the cell divides. Finally, an organism must be regenerated from the altered cell. If the trait is dominant, the transgenic organism must express it in the appropriate tissues at the right time in development. If the trait is recessive, crosses between heterozygotes may be necessary to yield homozygotes that express the trait. Then the organisms must pass the characteristic on to the next generation.

Figure 19.10 illustrates transgenic silkworms that have been turned into a factory for a human protein, collagen. The silkworm *Bombyx mori* has been cultivated and its silk coveted for nearly five thousand years. About half of a larva's body is devoted to silk production. Two silk glands secrete threads consisting of linked proteins called fibroins. The insect pulls the strands out of its hindquarters in an elaborate figure-eight pattern, and once these strands hit the air, they harden into silk. Because of unusual mitotic cell division in which DNA replicates but cells do not divide, each cell in a silk gland has 400,000 copies of the genome, enabling the silkworm to build its cocoon rapidly.

To take advantage of the silkworm's mass-production capabilities, researchers placed recombinant DNA into a BAC vector, which they introduced in a transposon

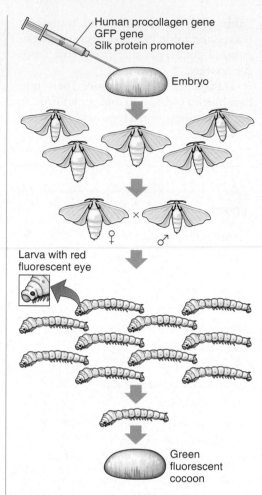

Human procollagen gene
GFP gene
Silk protein promoter

Embryo

♀ × ♂

Larva with red fluorescent eye

Green fluorescent cocoon

Figure 19.10 Silk with human collagen, courtesy of transgenic silkworms. Early embryos receive injections of a human procollagen gene linked to a silk protein promoter, as well as to the green fluorescent protein (GFP) gene. The transgenic silkworms also have a red fluorescent protein gene linked to a promoter that causes it to be expressed in larval eyes. The affected larvae with red eyes indicate individuals that are homozygous for the transgenes, enabling their separation before they actually make their cocoons.

(a "jumping gene") naturally found in many insect cells. Then they injected the loaded vector into early embryo cells. The recombinant DNA included a promoter from a fibroin gene, so that when a larva got the hormonal urge to make silk, it would make silk and also human collagen. To mark the proteins, the investigators used GFP—from the same jellyfish gene that lights up the mice in figure 19.1. As an added guide, the recombinant DNA also included a gene for red fluorescent protein, but with a promoter that causes it to be expressed in the eye. The researchers simply selected red-eyed larvae for collagen-silk harvesting, then separated the collagen from the silk of the cocoons. Collagen is used in cosmetics and in tissue engineering.

Bioremediation

Recombinant DNA technology and transgenic organisms provide processes as well as products. In bioremediation, bacteria or plants with the ability to detoxify certain pollutants are released or grown in a particular area to cleanse the environment. Natural selection has sculpted such organisms, perhaps as adaptations that render them unpalatable to predators. Bioremediation uses genes that enable an organism to metabolize a substance that, to another species, is a toxin. The technology uses unaltered organisms, and also transfers "detox" genes to other species so that the protein products can more easily penetrate a polluted area.

Nature offers many organisms with interesting tastes. One type of tree that grows in a tropical rainforest on an island near Australia, for example, accumulates so much nickel from soil that slashing its bark releases a bright green latex ooze. This tree can be used to clean up nickel-contaminated soil.

Bioremediation can tap the metabolisms of transgenic microorganisms, sending them into plants whose roots then distribute the detox proteins in the soil. For example, transgenic yellow poplar trees can thrive in mercury-tainted soil if they have a bacterial gene that encodes an enzyme, mercuric reductase, that converts a highly toxic form of mercury in soil to a less toxic gas. The tree's leaves then release the gas.

Bioremediation cleans up munitions dumps from wars. One application uses bacteria that normally break down dinitrotoluene—better known as TNT, the major ingredient in dynamite and land mines. The enzyme that provides this capability is linked to the GFP gene (see figure 19.1). When the bacteria are spread in a contaminated area, they glow near land mines, revealing the locations much more specifically than a metal detector could. Once the land mines are removed, the bacteria die as their food vanishes.

Gene Targeting

Transgenic technology is imprecise because it does not introduce DNA to a particular chromosomal locus. Instead, the transgene can disrupt another gene, or come under another gene's control sequence. Even if a transgene inserts into a chromosome and is expressed, the host's version of the same gene may overshadow the transgene's effect.

A more precise genetic modification, gene targeting, uses a natural process called **homologous recombination,** in which a DNA sequence replaces a similar or identical sequence in a host chromosome. The technique was developed in the late 1980s by introducing an inactivated gene into a mouse cell's nucleus, thereby "knocking out" function of the gene it replaced. By observing what happens (or doesn't happen) when a gene doesn't function, researchers could deduce the gene's normal function as well as learn more about associated inherited disease. A variation on gene targeting exchanges genes that have an altered function, producing a "knockin."

Gene targeting in mammals entails genetic modification plus complex developmental manipulations because the technique does not work on fertilized ova; instead, it is done in embryonic stem (ES) cells (see figure 2.25). Most gene targeting research uses mice because their embryo cells are easiest to manipulate. The first attempts followed transmission of coat color genes, so that results would be easy to see. In one scheme, the knockout mice were white (**figure 19.11**).

To begin, researchers used electroporation or microinjection to deliver an inactivated pigment gene into a pigmented mouse ES cell. Next, altered ES cells were injected into early embryos from colored mice. The embryos were implanted into surrogate mothers, where they developed into individuals that had some cells bearing the targeted gene. The newborn mice were mosaics, with patches of tissue whose cells contained the inactivated pigment gene. When the mosaic mice were mated to each other, some of the pups developed from a sperm and egg that each had the knocked-out gene. These rare homozygotes were easily distinguished because they were white, while their siblings were pigmented.

Gene-Targeted Mice as Models

Gene targeting is very useful in developing animal models of human genetic diseases. First, researchers identify the model's version of a human disease-causing allele. Then they transfer a corresponding human mutant allele to mouse ES cells and follow the steps previously outlined to breed a homozygous animal. The mouse is a knockout if gene function is gone, and a knockin if it makes the human protein.

Knockout or knockin mice are valuable models of human genetic disease because they provide a controllable test population. Consider severe combined immune deficiency due to adenosine deaminase deficiency

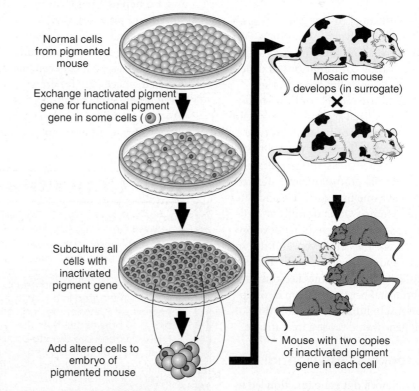

Figure 19.11 Gene targeting. Inactivated pigment-encoding genes are inserted into mouse embryonic stem (ES) cells, where they "knock out" functional pigment-encoding alleles. The modified ES cells are cultured and injected into early mouse embryos. Mosaic mice develop, with some cells heterozygous for the inactivated allele. These mice are bred to each other, and, if all goes well, yield some offspring homozygous for the knocked-out allele. It's easy to tell which mouse this is in the pigment gene example, but gene targeting is particularly valuable for revealing unknown gene functions by inactivating targeted alleles.

(OMIM 102700). The phenotype varies depending on the environment. A child raised in isolation might be relatively healthy; a child out among others would likely have frequent infections. This form of immune deficiency is so rare that it is difficult to find enough individuals to conduct a study. Instead, genetically identical mice with knocked-out ADA genes can be exposed to particular infectious agents or undergo treatments such as gene therapy.

Knockout mice are also helpful as models of sickle cell disease. Recall from chapter 11 that the human globin genes occur in two clusters, on different chromosomes. One research group knocked out the mouse's beta globin genes, and another knocked out the alpha genes—then each research team heard about the other. They joined forces, knocked out all the mouse globin genes, and knocked in their human counterparts. The mice make human sickle cell hemoglobin and have the same symptoms as humans with sickle cell disease. They are used to test new drugs and gene therapies.

Animals with knocked-out genes are also useful in studying polygenic disorders. For example, researchers are studying atherosclerosis by inactivating combinations of genes whose products oversee lipid metabolism.

Knockout mouse models of human disease also reveal prenatal aspects of the condition. People with neurofibromatosis type 1 have benign tumors beneath the skin and distinctive pigmented areas of skin. The condition is autosomal dominant, and affected individuals have one mutant allele and one normal one. Homozygous dominant individuals are so severely affected that development ceases in the embryo, ending in spontaneous abortion. Mouse embryos with both neurofibromin genes knocked out have severely abnormal hearts and stop developing as embryos. In humans, a double dose of the mutant gene would cause a miscarriage.

When Knockouts Are Normal

The ability to knock out gene function led to surprises when animals missing supposedly vital genes were much healthier than expected. This was the case for mice lacking a gene that encodes a type of collagen. Scientists thought that type X collagen promoted the normal growth and development of long bones in both mice and humans. Mutations

in collagen genes cause a variety of syndromes in humans (see figure 12.3). Yet mice with knocked-out type X collagen genes had normal skeletons! How could this be?

Gene targeting experiments suggest that the importance of a gene's product must be considered in the context of the entire genome and organism. A broader view of interacting genes presents several possible explanations for healthy knockout mice:

1. Other genes that encode the same or similar proteins as the knocked-out gene may replace its function, so that disabling one gene does not affect the phenotype. The genome is redundant for important functions.

2. An absent protein in a knockout may not alter the phenotype, though an abnormal protein might.

3. The knocked-out gene may not do what we thought, and it may even have no function at all.

4. The knocked-out gene may function under different circumstances than the experiment provides. Various environmental challenges may be required to reveal the gene's function. For example, type X collagen, rather than being necessary for growth and development, may repair fractures. The protein would therefore usually be unnecessary in mouse embryos and newborns.

Key Concepts

1. In gene targeting in mice, a gene inserted into an ES cell recombines at the chromosomal site where it normally resides. The ES cell is then incorporated into a developing embryo from another individual, which is implanted into a surrogate. Animals with phenotypes indicating that they harbor cells with the targeted gene are bred to each other to yield homozygotes.
2. Swapping an inactivated allele for a gene of interest produces a knockout mouse, and replacing a gene with another that has an altered function creates a knockin mouse.
3. These animals model human disease. They can reveal that a gene does not function as we thought.

19.4 Monitoring Gene Function

A genome is like an orchestra. Just as not all of the instruments play with the same intensity at every moment, not all genes are expressed continually at the same levels. Before the field of genomics began in the 1990s, the study of genetics proceeded one gene at a time, like hearing the separate contributions of a violin, a viola, and a flute. Many genetic investigations today, in contrast, track the crescendos of gene activity that parallel events in an organism's life. This new view has introduced the element of time to genetic analysis. Unlike the gene maps of old, which ordered genes on chromosomes, new types of maps reveal the orders of events in unfolding programs of development and response to the environment.

DNA microarrays—or "gene chips"—are devices that reveal which genes are active and which are inactive in a tissue sample. The creativity of the technique lies in choosing the samples and deciding which genes to consider. We've already encountered applications in cancer diagnosis. The opening essay for chapter 11 and figure 18.2 describe how DNA microarrays led researchers to discover a type of leukemia that hadn't been distinguishable based on symptoms or the characteristics of the affected cells. Chapter 1 introduced the two major uses of DNA microarrays: **gene expression profiling** that indicates which genes are transcribed, and **DNA variation screening,** which detects mutations or SNPs (single nucleotide polymorphisms).

Gene Expression Microarrays Track the Aftermath of Spinal Cord Injury

Consider a specific example to illustrate the basic steps in creating a DNA microarray to assess gene expression—spinal cord injury. Researchers knew that in the hours after such a devastating injury, immune system cells and inflammatory biochemicals flood the area, but it took gene expression profiling to reveal just how fast healing begins.

A microarray is a piece of glass or plastic that is about 1.5 centimeters square—smaller than a postage stamp. Many small pieces of DNA (oligonucleotides) of known sequence are bound to one surface, in a grid pattern. The researcher records the position of each DNA piece within the grid. Typically a sample from an abnormal situation (such as disease, injury, or environmental exposure) is compared to a normal control. **Figure 19.12** compares cerebrospinal fluid (CSF; the liquid that bathes the spinal cord) from an injured person (sample A) to fluid from a healthy person (sample B). Messenger RNAs are extracted from the samples and cDNAs made (see figure 19.6). The cDNAs from the abnormal sample are labeled with a red fluorescent dye, and the cDNAs from the control sample are labeled with a green fluorescent dye. These labeled DNAs are then used as identification probes by incubating them with the microarray, which displays thousands of genes that the researchers expect to be involved in a spinal cord injury—or perhaps the entire human genome, to avoid the problem of not knowing what to look for.

Probes that find and hybridize with (bind) their complements on the grid fluoresce in place. A laser scanner then detects and converts the results to a colored image. Each position on the microarray can bind probes from both samples, either, or neither. The scanner also detects fluorescence intensities, which provides information on how strongly the gene is expressed. Then a computer interprets the pattern of gene expression, which may or may not make visual sense to the researcher—microarray experiments often yield surprise results. For the spinal cord example, the visual data mean the following:

- Red indicates a gene expressed in CSF only when the spinal cord is intact.

- Green indicates a gene expressed in CSF only when the spinal cord is injured (and presumably leaking inflammatory molecules).

- Yellow indicates positions where both red- and green-bound dyes fluoresce. These represent genes that are expressed whether or not the spinal cord has been injured.

- **Black,** or a lack of fluorescence, corresponds to DNA sequences that are not expressed in CSF.

A computer analyzes the color and intensity pattern, which provides a snapshot of gene expression that follows spinal cord injury. The technique is even more powerful when repeated at different times after injury. When researchers at the University of Florida did exactly that on injured rats, they discovered

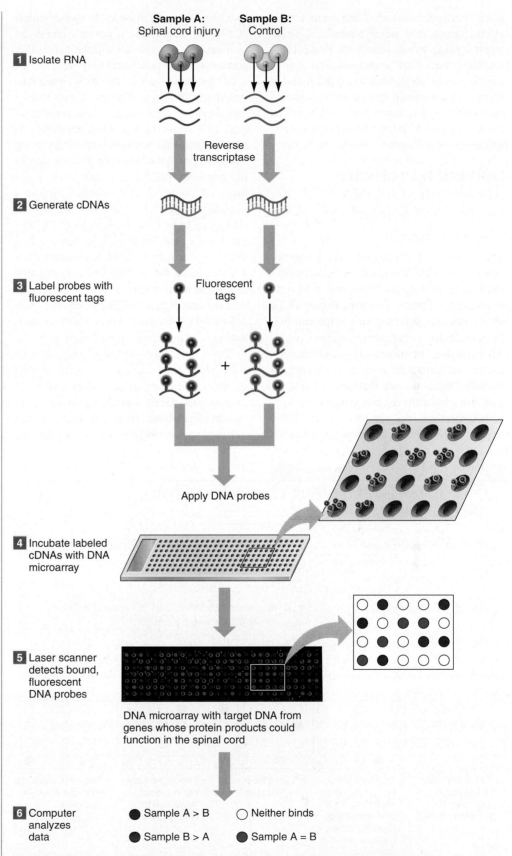

Figure 19.12 A DNA microarray experiment reveals gene expression in response to spinal cord injury.

genes expressed just after the event whose participation they never suspected. Their microarrays, summarized in **table 19.6**, revealed waves of expression of genes involved in healing. Analysis on the first day indicated activation of the same suite of genes whose protein products heal injury to the deep layer of skin—a total surprise that suggests new points for drugs to intervene.

Solving a Problem: Interpreting a DNA Sequence Variation Microarray

The second major type of DNA microarray experiment, a DNA sequence variation analysis, screens mutations, SNPs, and wild type sequence for a particular gene. **Figure 19.13** shows possible patterns for comparing two individuals for a single-gene recessive disorder. Each person's microarray would have two fluorescing spots if he or she is a heterozygote (because there are two different alleles), or just one if he or she is a homozygote.

Interpreting the results of a DNA sequence variation analysis depends upon the specific nature of a disorder. In cystic fibrosis (CF), for example, a person could be homozygous recessive for a mutant allele that confers symptoms so mild they were attributed to recurrent respiratory infection. Another scenario is different alleles in two heterozygotes combining to cause severe illness in a child. This is what happened to Monica and Bill. Routine screening during Monica's pregnancy revealed that she carries the most common CF allele in many populations, *ΔF508*, which is associated with severe disease (**figure 19.14a**). Bill was then tested and found to be a carrier, too (figure 19.14b), but for the rarer allele *G542X*. Monica had amniocentesis, and a DNA microarray constructed using the sampled fetal cells revealed the pattern in figure 19.14c—with both mutant alleles present. The computer then consulted a database of known allele combinations and predicted a poor prognosis.

DNA sequence variation analysis also uses "SNP chips" that cover selected regions of more than one gene and can identify disease-associated variations over wide swaths of a genome. Such tests might predict whether variants in genes other than the one that causes CF could affect the phenotype of Monica and Bill's offspring, such as which respiratory infections the child is most likely to contract. SNP patterns can also be correlated to drug responses, guiding treatment.

Yet another application of DNA microarray technology is **microarray comparative genomic hybridization,** which tracks deletions and amplifications of specific DNA sequences between samples. One specific use is in monitoring cancer progression in an individual—such large-scale changes tend to accumulate with time. In evolutionary studies, this technique is used to flag DNA sequences that are present in more copies in the human genome than in the great apes, suggesting regions of the genome that may have contributed to our differences from our closest relatives (see Reading 16.1). Microarrays are also used to search among proteins and small molecules to reveal the molecular choreographies that underlie life.

Table 19.7 reviews the major uses of DNA microarray technology to monitor phenotypes and genotypes. It is rapidly becoming as important in the twenty-first century as gene maps were in the twentieth.

Table 19.6	
Gene Expression Profiling Chronicles Repair After Spinal Cord Injury	
Time After Injury (rats)	**Type of Increased Gene Expression**
Day 1	Protective genes to preserve remaining tissue
Day 3	Growth, repair, cell division
Day 10	Repair of ground substance (part of connective tissues) Angiogenesis (blood vessel extension)
Days 30–90	Blood vessels mature New type of ground substance associated with healing

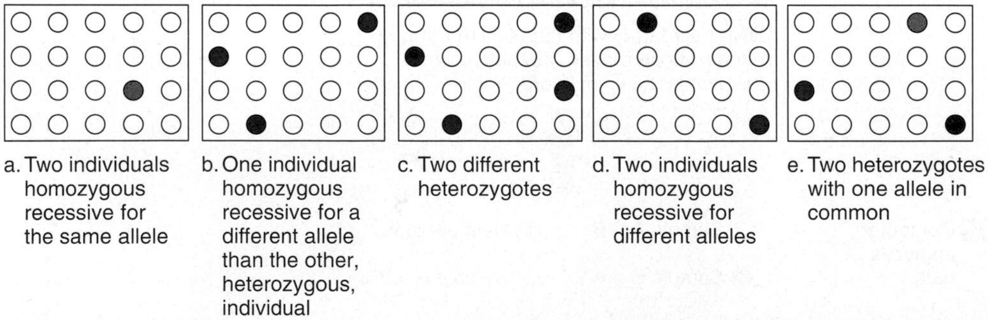

a. Two individuals homozygous recessive for the same allele

b. One individual homozygous recessive for a different allele than the other, heterozygous, individual

c. Two different heterozygotes

d. Two individuals homozygous recessive for different alleles

e. Two heterozygotes with one allele in common

Figure 19.13 DNA sequence variation analysis reveals alleles. Heterozygotes have two fluorescent spots, and homozygotes have one. Target DNAs that two differently labeled DNAs share fluoresce yellow.

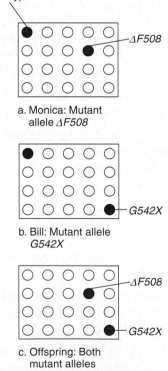

a. Monica: Mutant allele *ΔF508*

b. Bill: Mutant allele *G542X*

c. Offspring: Both mutant alleles

Figure 19.14 Microarrays show inheritance of cystic fibrosis. Bill and Monica's offspring was unfortunate enough to receive rare CF alleles from both heterozygous parents.

Table 19.7

Types of DNA Microarray Studies

DNA Microarray Application	Type of Information Revealed
Gene expression profiling	Which genes are transcribed and which are not under defined conditions
DNA variation screening	Which mutations or SNPs are present in an individual
Microarray comparative genomic hybridization	Deletions and amplifications of DNA sequences between samples (cells or species)

Summary

19.1 Patenting DNA

1. **Biotechnology** is the alteration of cells or biochemicals to provide a useful product. It includes extracting natural products, altering an organism's genetic material, and combining DNA from different species.

2. A **transgenic** organism has DNA from a different species in its genome. **Recombinant DNA** comes from more than one type of organism. Both are possible because of the universality of the genetic code.

3. Patented DNA must be useful.

19.2 Amplifying DNA

4. Nucleic acid amplification, such as **PCR**, uses the power and specificity of DNA replication enzymes to selectively mass-produce DNA sequences.

5. In PCR, primers corresponding to a DNA sequence of interest direct polymerization of supplied nucleotides to make many copies.

19.3 Modifying DNA

6. **Recombinant DNA technology** is used to mass-produce proteins in bacteria or other single cells. Begun hesitantly in 1975, the

technology has matured into a valuable method to produce proteins.

7. Constructing a recombinant DNA molecule begins when **restriction enzymes** cut both the gene of interest and a **cloning vector** at a short palindromic sequence, creating complementary "sticky ends." The cut foreign DNA and vector DNA are mixed, and vectors that pick up foreign DNA are selected.

8. **Genomic libraries** consist of recombinant cells containing fragments of a foreign genome. **DNA probes** are used to select genes of interest from genomic libraries. DNA probes may be synthetic, taken from another species, or a **cDNA**, which is reverse transcribed from mRNA.

9. Genes conferring antibiotic resistance and color changes in growth media are used to select cells harboring recombinant DNA. Useful proteins are isolated and purified.

10. To produce a multicellular transgenic organism, a gamete in an animal or plant, somatic plant cell, or early embryo cell receives a foreign gene via a *Ti* plasmid (some plants) or other technique. The organism develops, including the change in each cell and passing it to the next generation. Heterozygotes for the

transgene are then bred to yield homozygotes.

11. Recombinant DNA technology provides novel drugs, textiles, foods, and bioremediation.

12. **Gene targeting** uses attraction of a DNA sequence for its complement (**homologous recombination**) to precisely exchange genes. It is done on an ES cell, which is inserted into an embryo and transferred to a surrogate mother. Heterozygotes are bred to yield homozygotes.

13. Knockouts inactivate a gene. Knockins replace a gene. Knockout mice model human disease or may reveal that a gene product is not vital.

19.4 Monitoring Gene Function

14. **DNA microarrays** are devices that hold DNA pieces to which fluorescently labeled DNA probes from samples are applied.

15. There are three major uses of DNA microarrays. cDNAs are applied in **gene expression profiling**. **DNA variation screening** detects mutations and SNPs. **Microarray comparative genomic hybridization** tracks deletions and amplifications.

Review Questions

1. Cite three examples of a DNA sequence that meets requirements for patentability.

2. How are PCR and recombinant DNA technology similar, and how do they differ?

3. Describe the roles of each of the following tools in a biotechnology:

   a. restriction enzymes

   b. embryonic stem cells

   c. cloning vectors

4. How do researchers use antibiotics to select cells containing recombinant DNA?

5. List the components of an experiment to produce recombinant human insulin in *E. coli* cells.

6. Why would recombinant DNA technology be restricted if the genetic code were not universal?

7. What is an advantage of a drug produced using recombinant DNA technology compared to one extracted from natural sources?

8. Describe three ways to insert foreign DNA into cells.

9. Why isn't transgenic technology as precise as gene targeting?

10. What do gene targeting and RNA interference (see chapter 11) have in common?

11. What is the difference between the types of information obtained in a gene expression profile and a DNA sequence variation analysis?

12. How does the information from a DNA microarray differ from the information in the haplotype of figure 5.17?

Applied Questions

1. To diagnose a rare form of encephalitis (brain inflammation), a researcher needs a million copies of a viral gene. She decides to use PCR on a sample of the patient's cerebrospinal fluid. If one cycle takes two minutes, how long will it take to obtain a millionfold amplification?

2. HIV infection was formerly diagnosed by detecting antibodies in a person's blood or documenting a decline in the number of the type of white blood cells that HIV initially infects. Why is PCR detection of HIV more sensitive?

3. Genetic modification endows organisms with novel abilities. From the following three lists (choose one item from each list), devise an experiment to produce a particular protein, and suggest its use.

Organism	Biological Fluid	Protein Product
pig	milk	human beta globin chains
cow	semen	human collagen
goat	silk	human EPO
chicken	egg white	human tPA
aspen tree	sap	human interferon
silkworm	blood plasma	jellyfish GFP
rabbit	honey	human clotting factor
mouse	saliva	alpha-1-antitrypsin

4. Collagen is a connective tissue protein that is used in skincare products, shampoo, desserts, and in artificial skin. For many years it was obtained from the hooves and hides of cows collected from slaughterhouses. Human collagen can be manufactured in transgenic mice. Describe the advantages of the mouse system for obtaining collagen.

5. A woman was outraged to read a magazine article about certain genetic modification experiments. She wrote to this author, "Scientists have produced mice with human brains, mice with human blood flowing through their veins, and pigs with human cell surfaces. I am a big Mickey Mouse fan, but this is too much!" She is referring to:

   - Mice with spinal cord injuries that receive implants of human neural stem cells. They regain movement.
   - Mice with human beta globin genes that are models of sickle cell disease.
   - Pigs that have some human proteins on their cell surfaces in attempts to make their organs more compatible for transplantation to humans.

   a. Many people are alarmed at genetic modification experiments. What is your opinion about the utility and bioethics of each of these examples of genetic modification? (The first example is actually a cell implant, not genetic modification. But the principle of mixing material from two species is similar.)

   b. Whose responsibility do you think it is to present accurate information on genetic manipulations: journalists, scientists, or the marketers of affected products?

   c. There was no public outcry over the development of human insulin produced in bacterial cells and used to treat diabetes. Yet some people object to mixing DNA from different species in agricultural biotechnology. Why do you think that the same general technique is perceived as beneficial in one situation, yet a threat in another?

6. A human oncogene called *ras* is inserted into mice, creating transgenic animals that develop a variety of tumors. Why are mouse cells able to transcribe and translate human genes?

7. A healthy knockout mouse cannot manufacture what was thought to be a vital enzyme. Suggest three possible explanations for this finding.

8. In a mouse model of a human condition called "urge syndrome," in which the feeling of impending urination occurs frequently, researchers inactivate a gene encoding nitric oxide synthase, which produces nitric oxide (NO). NO is the neurotransmitter that controls muscle contraction in the bladder. Name a biotechnology that could accomplish this.

9. Mouse models for cystic fibrosis have been developed by inserting a human transgene and by gene targeting to inactivate the mouse counterpart of the alleles that cause the disorder. How do these methods differ? Which method do you think produces a more accurate model?

10. In a DNA microarray experiment, the researcher selects the DNA pieces that are attached to the grid. For example, to study an injury, he or she might choose genes known to be involved in the inflammatory response. How might this approach be limited?

11. Devise an experiment using DNA microarrays to determine whether men and women have different hormonal responses to watching an emotional film. (A hormone is a type of messenger molecule that is carried in the blood).

Web Activities

Visit the Online Learning Center (OLC) at www.mhhe.com/lewisgenetics7. Select **Student Edition, chapter 19,** and **Web Activities** to find the website links needed to complete the following activities.

12. Use the Web to identify three drugs made using recombinant DNA technology, and list the conditions they are used to treat.

13. About how many DNA patents are currently being evaluated at the U.S. Patent and Trademark office, and what percentage of them are for DNA from the human genome?

Case Studies

14. In the 1990s, Australian researcher Malcolm Simons filed many patent applications on the use of non-protein-encoding parts of the human genome to predict the risk of developing certain diseases. At the time, research interest focused on the protein-encoding parts of the genome—many researchers called the rest "junk." But towards the turn of the century, research correlating SNP patterns in the "junk" to disease risk became a top priority at many biotech companies, and researchers there encountered Simons's patents. He is now asking researchers to pay to use his idea. Many geneticists denounced Simons's actions as counter to the spirit of open access to human genome information. Do you support or object to Simons's restricting access to the DNA sequences he predicted would have clinical utility?

15. Nancy is a transgenic sheep who produces human alpha-1-antitrypsin (AAT) in her milk. This protein, normally found in blood serum, enables the microscopic air sacs in the lungs to inflate. Without it, inherited emphysema results—a severe illness that usually kills humans by early adulthood. Donated blood cannot yield enough AAT to help the thousands who need it. Describe the steps taken to enable Nancy to secrete human AAT in her milk.

16. To investigate causes of acne, researchers used DNA microarrays that are studded with DNA sequences that cover the entire human genome. Samples came from facial skin of people with flawless complexions and from people with severe acne. In the following simplified portion of a DNA microarray, one sample is labeled green and comes from healthy skin; a second sample is labeled red and represents skin with acne. Sites on the microarray where both probes bind fluoresce yellow. The genes are indicated by letter and number.

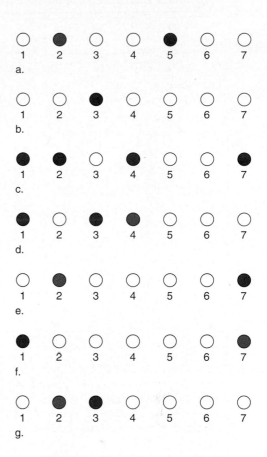

a. Which genes are expressed in skin whether or not a person has acne?

b. Which genes are expressed only when acne develops?

c. List three DNA pieces that correspond to genes that are not expressed in skin.

d. How would you use microarrays to trace changes in gene expression as acne begins and worsens?

e. Design a microarray experiment to explore gene expression in response to sunburn.

Learn to apply the skills of a genetic counselor with additional cases found in the *Case Workbook of Human Genetics.*

Hemophilia A and B
Infertility drugs
Transgenic tobacco

CHAPTER

20

Genetic Testing, Genetic Counseling, and Gene Therapy

GENE THERAPY FOR CANAVAN DISEASE

Ilyce and Mike Randell, of Buffalo Grove, Illinois, were worried. Their 5-month-old son, Max, born in October, 1997, didn't seem able to do what the baby books said he should—he could not hold up his head, roll over, reach for objects, sleep more than an hour, and was unresponsive to those around him. When doctors diagnosed Canavan disease (OMIM 271900), which robs brain neurons of their fatty sheaths so that the brain slowly degenerates, they suggested that the best place for the golden-haired boy would be a nursing home, to live out his expected two years. Instead, Max went on to become the youngest person to receive gene therapy for a degenerative brain disease. Today he plays with brother Alex, attends school, gets around in a wheelchair and communicates by using a computer.

Max received his first gene therapy, to test safety, in 1998, and then a more intensive version in 2001. Ninety billion viral particles that each carried a corrected copy of the gene that is mutant in Canavan disease were delivered through six holes drilled into his skull. Although his progress is slight—his head and neck are stronger, he sees better, and has more control over his limbs—he may be paving the way for delivery of healing genes for much more common brain disorders, such as Parkinson disease and Alzheimer disease. Only about 4,000 children in the world have Canavan disease, and they do not live more than 10 years.

The news most often mentions gene therapy failures. But Ilyce Randell tells Max's tale whenever a reporter asks. "I cannot even begin to describe the joy we feel just seeing him regain even the slightest bit of functional mobility; he just beams with pride when his body does what he wants it to."

Max Randell's experience has been a gene therapy success story.

20.1 Genetic Testing and Counseling

Living with a genetic disease or treating its symptoms takes two basic steps: identifying the condition and, with an understanding of what goes awry, taking action to counter or mask expression of the mutant genes. Genetic counselors are health care professionals who help families along this journey.

Genetic Testing

If Max Randell had been born today, newborn screening tests would have likely detected the mutant genes that cause Canavan disease. Such newborn screening is one example of the growing incorporation of genetic testing into medical practice. Some of the tests and technologies aren't actually new, but what is new is the scope, and the economic and ethical issues that arise with genetic testing at the population level.

Technology to detect and treat inborn errors of metabolism started with phenylketonuria (PKU). Beginning in 1961, the Guthrie test sampled blood from a newborn's heel and tested for the amino acid buildup of PKU. In 1963, a specialized diet (legally termed a "medical food" so that insurance will cover the high cost) became available, with dramatic positive results. Today, a technique called tandem mass spectrometry is used to identify the chemical imbalances of many inborn errors. Then the polymerase chain reaction is used on blood samples to amplify specific mutations so that they can be identified.

Increasing use of tandem mass spectrometry to simultaneously detect many inborn errors of metabolism is widening the availability of tests for certain rare conditions. For many years, in the United States, the March of Dimes recommended such newborn screening for the eight conditions listed in **table 20.1,** plus congenital hearing loss. States could provide whatever number of tests they liked, which meant that a child with a certain disorder would be diagnosed and possibly treated in one state, but not another. In 2005, following recommendations of the American College of Medical Genetics, the U.S. government's Maternal and Child Health Bureau mandated testing for 29 conditions that are treatable and recommended testing for an additional 25 conditions that are currently not treatable. Testing for these 25 disorders can help a physician narrow down a diagnosis of another condition by ruling out possibilities.

Selecting which disorders to test for illustrates the ethical and economic issues of newborn testing. The advantage of early detection is clear for a treatable condition, such as PKU or maple syrup urine disease, described in Reading 2.1—preventing symptoms and enabling normal development. But what might be the consequences of revealing a serious disorder without the ability to prevent symptoms or halt the disease's progression? Could testing cause more harm than good?

Even though inborn errors of metabolism are very rare, testing for them is not costly, because a battery of tests requires only one blood sample, which is routinely taken anyway. Screening for 50 conditions using tandem mass spectrometry today costs less than the price of a pair of running shoes. Yet the economics can be viewed another way, by considering population statistics. Each year in the United States, 4 million babies are born, and 0.1 percent of them have one of the 54 disorders that are most commonly detected with newborn screening. Health officials in some states argue that the cost of testing so many to detect so few is not practical. Supporters of

Table 20.1

Newborn Screening Tests

Disease	Incidence	Symptoms	Treatment
Biotinidase deficiency	1/70,000 Rare in blacks or Asians	Convulsions; hair, hearing, and vision loss; developmental abnormalities, coma, sometimes death	Most physical symptoms reversed by oral biotin
Maple syrup urine disease	1/250,000–300,000 More common in blacks and Asians	Lethargy, mental retardation, sweet-smelling urine, irritability, vomiting, coma, death by one month	Diet very low in overproduced amino acids
Congenital adrenal hyperplasia	1/12,000 whites 1/15,000 Jews 1/680 Yupik eskimos	Masculinized female genitalia, dehydration, precocious puberty in males, accelerated growth, short stature, ambiguous sex characteristics	Hormone replacement, surgery
Congenital hypothyroidism	1/3,600–5,000 whites Rare in blacks, more common in Hispanics	Mental retardation, growth failure, hearing loss, underactive thyroid, neurological impairment	Hormone replacement
Galactosemia	1/60,000–80,000	Muscle weakness, cerebral palsy, seizures, mental retardation, cataracts, liver disease	Galactose-free diet
Homocystinuria	1/50,000–150,000	Blood clots, thin bones, mental retardation, seizures, muscle weakness, mental disturbances	Low-methionine, high-cysteine diet, drugs
Phenylketonuria (PKU)	1/10,000–25,000	Mental retardation	Low-phenylalanine diet
Sickle cell and other hemoglobinopathies	1/400 U.S. blacks	Joint pain, severe infection, leg ulcers, developmental delay	Prophylactic antibiotics

testing counter that the costs saved by being able to treat identified children justifies the expense of populationwide testing.

Newborns are not the only individuals offered genetic testing, and not all genetic tests are for disorders as rare as those detected in newborns. Section 13.2 describes several tests and techniques that provide clues to the genetic health of a fetus. Throughout life, genetic tests are often part of confirming a diagnosis, or excluding one otherwise based on observing symptoms. Chapter 18 discusses how gene expression profiling is impacting the diagnosis and treatment of cancer. Many genetic tests are also being used to predict elevated risks and inherited susceptibilities to a variety of illnesses. In the future, such tests will increasingly provide information on environmental factors that pose a health threat to individuals with particular genetic backgrounds. **Table 20.2** summarizes some uses of genetic tests.

Genetic Counseling

Until recently, most medical professionals received very limited training in genetics. Since the 1970s, genetic counselors have led the way in explaining single-gene disease inheritance patterns and recurrence risks to patients. Today, their role is expanding to embrace multifactorial disorders, such as cancer and cardiovascular disease.

The knowledge that genetic counselors share with their patients reflects what you have read so far in this book, presented in a personalized manner. A genetic counselor might explain Mendel's laws, but substitute the particular family's situation for the pea plant experiments.

A genetic counseling session begins with a discussion of the family's health history. Using a computer program or pencil and paper, the counselor derives a pedigree, then explains the risks of recurrence for particular family members for their particular inherited illness (**figure 20.1**). The counselor provides detailed information on the condition and refers the family to support groups. If a couple wants to have a biological child who does not have the illness, a discussion of assisted reproductive technologies (see chapter 21) might be in order.

A large part of the genetic counselor's job is to determine when specific biochemical, gene, or chromosome tests are appropriate, and to arrange for people to take the tests. The counselor interprets test results and helps the patient or family choose among medical options. Genetic counselors are also often asked to provide information on drugs that can cause birth defects, although this really is related to development, not genetics.

People seek genetic counseling for two major reasons: prenatal diagnosis, and a disease in the family. Prenatal genetic counseling typically presents population (empiric) and family-based risks, explains tests, and discusses whether the benefits of testing outweigh the risks. The couple or woman decide whether amniocentesis, chorionic villus sampling, maternal serum screening, or no testing is appropriate for them. A genetic counselor explains that tests do not guarantee a healthy baby. For example, many people assume that amniocentesis checks every gene, when it actually detects only large-scale chromosome aberrations. Single-gene tests must be requested separately. If a test reveals that the fetus has a serious medical condition, the counselor discusses possible outcomes, treatment plans, and the options of ending the pregnancy or treating a newborn.

Counseling when there is an inherited disease in a family is another matter. For recessive disorders, the affected individual is usually a child, whose condition often came as quite a surprise. Often the hardest information to communicate is the risk to subsequent children. Many people think that if one child is affected with an autosomal recessive condition, then the next three will be healthy, rather than realizing that each offspring has a 1 in 4 chance of inheriting

---

Table 20.2

Types of Genetic Tests

Type of Test	Information Provided	Example
Carrier screen	Identifies heterozygotes—people with one copy of a mutant gene	The healthy sibling of a child with CF is tested—chance of being a carrier is 2/3.
Prenatal test	Detects mutant gene in a fetus for a condition present in a family	A couple who know they are carriers of Tay-Sachs disease have a fetus tested.
Prenatal screen	Tests embryos or fetuses from a population for increased risk of a condition, not based on family history	A pregnant woman's blood is tested for elevated level of a protein indicating increased risk for a neural tube defect.
Newborn screen	Populationwide testing for several treatable inborn errors of metabolism	A child with identified sickle cell disease genes at birth can ease or delay symptoms with antibiotics.
Diagnostic test	Confirms diagnosis based on symptoms	A child with "failure to thrive" and frequent lung infections is tested for mutant alleles for CF.
Predisposition test	Detects allele(s) associated with an illness, but not absolutely diagnostic of it	A young Jewish woman with a strong family history of breast cancer has a mutant *BRCA1* allele, giving her an 85 percent lifetime risk of developing the condition.
Predictive test	Detects highly penetrant mutation with adult onset in an individual at high risk based on family history	A healthy person is tested for the Huntington disease mutation because a parent has the condition.

Reasons to seek genetic counseling:

"Advanced maternal age"

Elevated risk of single gene disorder

Family history of multifactorial disorder

Family history of cancer

Genetic counseling sessions:

Family history

Pedigree construction

Information provided on specific disorders, modes of inheritance, tests to identify at-risk family members

Testing arranged, discussion of results

Links to support groups, appropriate services

Follow-up contact

Figure 20.1   The genetic counseling process.

the illness. Counseling for subsequent pregnancies requires great sensitivity and sometimes a little bit of mind reading. Many people will not terminate a pregnancy when the fetus has a condition that already affects their living child, yet some will see that as the kindest option. Genetic counselors must respect these feelings, and tailor the options discussed accordingly. Counseling for adult-onset disorders may include helping a patient decide whether to take a predictive test, such as that for Huntington disease. Predictive tests are introducing a new type of patient, the "genetically unwell"—those with mutant genes but no symptoms.

RELATING THE CONCEPTS

Max Randell has Canavan disease and his brother Alex is a carrier. If the parents have another child, what is the chance that he or she will be free of the disease-causing allele? Canavan disease is autosomal recessive.

Until recently, genetic counseling was "nondirective," meaning that the practitioner did not offer an opinion or suggest a course of action, but only presented options. That approach is changing as the field moves from analyzing single-gene disorders to considering inherited susceptibilities to more common illnesses that are more likely to have treatments. A recent definition of the role of the genetic counselor is "shared deliberation and decision making between the counselor and the client."

In some cases it can be challenging to provide nondirective counseling. One such situation is when a couple has the same autosomal recessive form of blindness or deafness, but want to have children, even though they know that the children will be blind or deaf, because they, too, will be homozygous recessive. In one case, when a genetic counselor suggested adoption or intrauterine insemination by donor, the couple considered the advice a value judgment on their choice to have a child.

The United States has about 2,200 genetic counselors with masters degrees in the field. Genetic counselors work in medical centers, clinics, hospitals, biotechnology companies, pharmaceutical companies, research and diagnostic testing laboratories, and medical practices.

Because there are so few genetic counselors, and most of them practice in urban areas, access to their services is limited. Finding a genetic counselor with a specific expertise is difficult. For example, only 400 genetic counselors in the United States are specially trained in cancer genetic counseling. Due to the shortage of counselors and demand for their services, sometimes other types of health care professionals, such as physicians, nurses, social workers, and PhD geneticists, provide counseling. One survey found that dietitians, physical therapists, psychologists, and speech-language pathologists regularly discuss genetics with their patients, who regularly ask questions about heredity.

Other stand-ins for trained genetic counselors can be less effective. For example, some companies that provide genetic tests for cancer risk train licensed practical nurses to counsel patients, or use "virtual" genetic counseling, in which an interactive computer program calculates risks and provides basic information. In one study, women seeking genetic counseling for possibly carrying either the *BRCA1* or *BRCA2* genes that cause breast cancer had virtual counseling. Although the experience increased knowledge, especially for low-risk women who had overestimated their danger, it did not lower anxiety nearly as well as a real genetic counselor with expertise in cancer genetics.

Other medical professionals sometimes lack the unique combination of skills that a genetic counselor offers. For example, a survey of mothers of children with trisomy 21 Down syndrome found many instances of physicians without appropriate training in psychology presenting only negative facts to new parents, causing much distress. Said one mother, "The doctor flat out told my husband that this could have been prevented or discontinued at an earlier stage of the pregnancy." Another woman overheard her doctor call her son an "FLK," which is medical slang for "funny-looking kid."

Genetic counselors are also concerned that direct-to-consumer marketing of genetic tests without accompanying advice and interpretation can lead to misinformed health care consumers. For example, one website offers a test for hereditary hemochromatosis (OMIM 235200) that detects the mutation that causes most cases. This disorder, which causes "iron overload," is discussed in section 20.2. A client mails in a cheek brush sample, and results are sent back by mail. Although the website cautions that the results can only confirm other evidence of iron overload, and offers genetic counseling, a person could receive news of having one or two mutations without realizing that the gene's penetrance is low. That is, many people with two mutations do not ever develop symptoms!

As genetic testing becomes more common as we learn more about the human genome, the need will increase for genetic counselors, or other genetics-savvy professionals, to help families navigate through the new information.

Genetic Privacy

Genetic testing provides information that can have effects beyond the individual. People are very concerned about who should have access to such information. Physicians must weigh the risks and benefits of keeping medical information confidential when to do so could harm others. Consider the following true cases:

- A couple suffered several miscarriages, then had a child with multiple problems. A chromosome check found that the father carried a translocation. The man's siblings may also have carried the translocated chromosome, but the couple did not want to tell anyone the test result.

- Mr. and Mrs. Gold knew that their mildly retarded son had a chromosomal deletion. A questionnaire from the special education department in their school district asked if he'd ever had a chromosome test. If the parents answered yes, their child may have been stigmatized. If they answered no, he may not have gotten appropriate support. They did not know what to do.

- A subway driver had familial hypercholesterolemia. Although he had chest pains and high blood pressure and serum cholesterol, he had never had a heart attack. He knew that he could suddenly die, but he wouldn't tell the transportation department because he was retiring in a few months. The genetic counselor knew the diagnosis.

These cases are from the 1980s and 1990s, before many genetic tests were available. Today, the question of genetic privacy is arising more and more with cancer susceptibility, because knowledge can warn a person to have specific medical tests or change lifestyle habits. For example, one physician did not know what to do when testing revealed that her patient had inherited a *BRCA1* mutation, but the patient refused to share the information with her sister, who has a 50 percent chance of having inherited it, too. To keep quiet denies the sister the chance of early detection; to tell her breaks the doctor's confidentiality with her patient. Did the doctor have a "duty to warn" the patient's sister?

Medical decision making on the "duty to warn" often refers to legal precedents, specifically a 1976 case, *Tarasoff versus the Regents of the University of California*. A psychiatrist did not warn a young woman that her parent, who was the patient, had threatened to kill her. The woman was indeed murdered, and the California Supreme Court decided that the doctor should have warned the victim. Since then, several cases of "duty to warn" about genetic disease risks have gone through the courts (**table 20.3**).

During the Clinton Administration, a presidential commission drew up guidelines for the "duty to warn" dilemma. According to this report, a physician should disclose information if

1. harm from keeping confidentiality outweighs the harm from breaching it.

2. relative(s) at risk can be identified.

3. failure to warn places the person at great risk of harm.

Medical organizations have had their say on the "duty to warn" situation too. The American Medical Association advises physicians to discuss with patients, before or at the time of genetic testing, the situations in which the doctor feels he or she should notify relatives. This discussion should be part of the informed consent process. Regulations are stricter in Canada, where physicians have a "duty to rescue" when breaking a confidence can help someone. In the United States, however, the idea of mandating "duty to warn" has met with much opposition. The National Alliance for Breast Cancer Organizations and other consumer groups have expressed concern that such a practice could discourage people from taking genetic tests, as well as place too great a demand on physicians to counsel relatives of their patients. But from a legal standpoint, breaching confidentiality in the

Table 20.3	
Legal Cases on "Duty to Warn" of Cancer Risk	
Case	**Details**
Pate v. Threlkel (1995)	Daughter sued doctor for not informing her that her mother had autosomal dominant medullary thyroid cancer, the most serious type. Heidi Pate was diagnosed three years after her mother; her case was advanced. Early detection has a higher chance of cure. The Supreme Court of Florida ruled in her favor.
Safer v. Estate of Pack (1996)	Daughter sued doctor for not informing family members when her father was diagnosed with colon cancer 30 years earlier that it is hereditary and occurs by age 40. By the time of the daughter's diagnosis, the disease was advanced. New Jersey Supreme Court ruled in doctor's favor because plaintiff had indeed been tested as a child, but her mother had not told her of the family history.
Molloy v. Meier (2004)	Mother sued physician for not informing her that fragile X syndrome, diagnosed 10 years earlier in her daughter, could affect half-siblings. Mother would not have had other children if she'd known. Decision pending.

United States goes against the HIPAA (Health Insurance Portability and Accountability Act of 1996) regulation that maintains privacy of medical records.

Perhaps the solution to the duty-to-warn dilemma is to tackle this issue on a case-by-case basis, rather than using legal precedents that did not apply to genetic disease, or legislating new guidelines. The biological and psychological aspects of inherited disease or disease susceptibilities must be taken into account. These include the penetrance of a disease-causing mutation, the degree of susceptibility of a particular individual, whether treatment is available, and feelings of guilt about passing a health-related gene to children, not inheriting a condition when a sibling has, or surviving a condition when relatives do not.

Key Concepts

1. Newborn screening is using tandem mass spectrometry to expand coverage to dozens of inborn errors of metabolism.
2. Genetic tests are becoming an increasing part of diagnostic medicine, for rare and common disorders.
3. A genetic counselor provides information to individuals, couples expecting children, and families on modes of inheritance, recurrence risks, genetic tests, and treatments. The counselor helps the clients make decisions while being sensitive to individual choices.
4. The shortage of genetic counselors has led other health care professionals to provide the service, and has resulted in the use of computer programs to dispense information and advice, and direct-to-consumer marketing of genetic tests.
5. Medical organizations consider legal precedents as well as the specific circumstances of genetic testing to decide when to break a confidence concerning a test result.

20.2 Treating Genetic Disease

In parallel to the implementation of genetic testing has been development of new types of treatments for genetic disease. However, while genetic testing is so far along that in some settings it is required by law or available over the Internet, genetic disease treatments are mostly still experimental.

Treatment for genetic diseases has evolved in three phases: (1) replacing missing proteins with material from donors, (2) obtaining pure proteins using recombinant DNA technology, and (3) delivering replacement genes to correct the problem at its source, called **gene therapy.** In Their Own Words on page 403 tells the story of an early recipient of gene therapy, and section 20.3 describes other cases.

More than a thousand clinical trials of gene therapies have been conducted since 1990. As the new millennium dawned, researchers had expected that the sequencing of the human genome would accelerate the pace of gene therapy development. Instead, new information about the complexity of how genes interact, and a few cases where the experimental treatment harmed the patient, have led to a reevaluation of the idea that we can augment or replace a gene with predictable effects.

Alleviating or even preventing symptoms of some genetic disorders is possible at the phenotypic and/or genotypic levels.

Treating the Phenotype

Preventing a disease-associated phenotype can be as straightforward as supplying a missing protein **(table 20.4).** A child with cystic fibrosis sprinkles powdered digestive enzymes from cows onto applesauce, which she eats before each meal to replace the enzymes her clogged pancreas cannot secrete. A boy with hemophilia receives a clotting factor. Even wearing eyeglasses is a way of altering the expression of one's inheritance.

An inherited illness with an unusual phenotypic treatment is hereditary hemochromatosis. Because HH results in "iron overload," the treatment is to periodically remove blood, removing iron, which is part of the hemoglobin molecule (see figure 11.3).

In the United States, 1.5 million people have this autosomal recessive condition, and 32 million people—1 in 8—carry a

Table 20.4

Enzyme Replacement Therapy

Disease	OMIM	Enzyme	Mode*	Symptoms
Fabry disease	301500	alpha galactosidase	xlr	Skin lesions, abdominal pain, heart problems, kidney failure, symptoms begin any time
Gaucher disease	230800	glucocerebrosidase	ar	Enlarged spleen, bone lesions, skin pigmentation, symptoms begin any time
Hunter disease (mucopolysaccharidosis I H/S)	309900	iduronate sulfatase	xlr	Dwarfism, abnormal facial features, enlarged liver and spleen, heart problems, deafness, urination of unusual metabolites, mild and severe forms
Hurler-Scheie disease (mucopolysaccharidosis II)	607015	alpha-L-iduronidase	ar	Short stature, corneal clouding, stiff joints, umbilical hernia, enlarged liver and spleen, mental retardation, onset 3 to 8 years, survival to adulthood
Pompe disease	232300	acid alpha-1,4-glucosidase	ar	Heart problems, flaccid muscles; infantile, juvenile, and adult forms

*xlr = X-linked recessive
ar = autosomal recessive

Living with Hemophilia

Don Miller was born in 1949 and is semi-retired from running the math library at the University of Pittsburgh. Today he has a sheep farm. On June 1, 1999, he was the first hemophilia patient to receive a disabled virus that delivered a functional gene for clotting factor VIII to his bloodstream. He described his life with hemophilia and his treatment. It worked—today his disease is under control.

The hemophilia was discovered when I was circumcised, and I almost bled to death, but the doctors weren't really sure until I was about 18 months old. No one where I was born was familiar with it.

When I was three, I fell out of my crib and I was black and blue from my waist to the top of my head. The only treatment then was whole blood replacement. So I learned not to play sports. A minor sprain would take a week or two to heal. One time I fell at my grandmother's house and had a 1-inch-long cut on the back of my leg. It took five weeks to stop bleeding, just leaking real slowly. I didn't need whole blood replacement, but if I moved a little the wrong way, it would open and bleed again.

I had transfusions as seldom as I could. The doctors always tried not to infuse me until it was necessary. Of course there was no AIDS then, but there were problems with transmitting hepatitis through blood transfusions, and other blood-borne diseases. All that whole blood can kill you from kidney failure. When I was nine or ten I went to the hospital for intestinal polyps. I was operated on and they told me I'd have a 10 percent chance of pulling through. I met other kids there with hemophilia who died from kidney failure due to the amount of fluid from all the transfusions. Once a year I went to the hospital for blood tests. Some years I went more often than that. Most of the time I would just lay there and bleed. My joints don't work from all the bleeding.

By the time I got married at age 20, treatment had progressed to gamma globulin from plasma. By then I was receiving gamma globulin from donated plasma and small volumes of cryoprecipitate, which is the factor VIII clotting protein that my body cannot produce pooled from many donors. We decided not to have children because that would end the hemophilia in the family.

I'm one of the oldest patients at the Pittsburgh Hemophilia Center. I was HIV negative, and over age 25, which is what they want. By that age a lot of people with hemophilia are HIV positive, because they lived through the time period when we had no choice but to use pooled cryoprecipitate.

I took so little cryoprecipitate that I wasn't exposed to very much. And, I had the time. The gene therapy protocol involves showing up three times a week.

The treatment is three infusions, one a day for three days, on an outpatient basis. So far there have been no side effects. Once the gene therapy is perfected, it will be a three-day treatment. A dosage study will follow this one, which is just for safety. Animal studies showed it's best given over three days. I go in once a week to be sure there is no adverse reaction. They hope it will be a one-time treatment. The virus will lodge in the liver and keep replicating.

In the eight weeks before the infusion, I used eight doses of factor. In the fourteen weeks since then, I've used three. Incidents that used to require treatment no longer do. As long as I don't let myself feel stressed, I don't have spontaneous bleeding. I've had two nosebleeds that stopped within minutes without treatment, with only a trace of blood on the handkerchief, as opposed to hours of dripping.

I'm somewhat more active, but fifty years of wear and tear won't be healed by this gene therapy. Two of the treatments I required started from overdoing activity, so now I'm trying to find the middle ground.

Don Miller

---

mutant allele for the HH gene. It is most common among those of Irish, Scottish, or British descent. In HH, cells in the small intestine absorb too much iron from food. Over many years, the excess iron is deposited throughout the body, causing various symptoms and secondary conditions. The liver develops cirrhosis (scarring) and sometimes cancer; the heart may fail or beat irregularly; an iron-loaded pancreas may cause diabetes; joints become arthritic; and the skin darkens. Early signs and symptoms include chronic fatigue, infection, hair loss, infertility, muscle pain, and feeling cold.

Diagnosis requires a blood test to detect the telltale blood-level increase in ferritin, a protein that carries iron, and a liver biopsy. Determining the genotype alone is not sufficient for diagnosis because the penetrance is very low. That is, although most people with iron overload have mutations in the HH gene, only a small percentage of people with a homozygous recessive genotype actually have symptoms. More men than women develop HH symptoms, because a woman loses some blood each month when she menstruates. Once women pass the age of menopause, the sex ratio equalizes. Having blood removed every few months, however, is a simple way to keep the body's iron levels down. (The blood is discarded.)

Gene Therapy

Altering genes to treat an inherited disorder theoretically can provide a longer-lasting effect than treating symptoms, but this is much easier said than done. The first gene therapy efforts focused on inherited disorders that researchers knew the most about, even though the conditions are very rare, so that treatment strategies could be developed with considerable knowledge of the disease mechanism. With the increasing understanding of human genes and their functions made possible by knowing the human genome sequence, gene therapy efforts are targeting more common illnesses, such as heart disease and cancers. **Tables 20.5** and **20.6** list some general requirements and concerns related to gene therapy.

Gene therapy approaches vary in the way that healing genes are delivered and to which cells they are sent. Gene therapy approaches also vary in invasiveness (**figure 20.2**). The first gene therapy on Ashanthi DeSilva (Section 20.3) altered cells outside her body and then infused the corrected cells. This approach is called *ex vivo* **gene therapy.** In *in situ* **gene therapy,** the healthy gene plus the DNA that delivers it (the vector) is injected into a very localized and accessible body part, such as a single melanoma skin cancer. In the most invasive approach, *in vivo* ("in the living body") **gene therapy,** the vector is introduced directly into the body.

Germline Versus Somatic Gene Therapy

Researchers distinguish two general types of gene therapy, depending upon whether it affects gametes or fertilized ova, or somatic tissue.

Germline gene therapy (also known as heritable gene therapy) alters the DNA of a gamete or fertilized ovum. As a result, all cells of the individual have the change. Germline gene therapy is heritable—it passes to offspring. It is not being done in humans, although it is done to create the transgenic organisms discussed in chapter 19.

Somatic gene therapy corrects only the cells that an illness affects. It is nonheritable: A recipient does not pass the genetic correction to offspring. An example of somatic gene therapy is clearing lungs congested from cystic fibrosis with a nasal spray containing functional CFTR genes. The treatment doesn't alter the gametes (sperm or oocytes), so a treated person could not pass a normal CFTR allele to offspring.

Gene Delivery

Researchers obtain therapeutic genes using the recombinant DNA and polymerase chain reaction technologies described in

Table 20.5

Gene Therapy Concerns

Scientific	Bioethical
1. Which cells should be treated, and how?	1. Does the participant in a gene therapy trial truly understand the risks?
2. What proportion of the targeted cell population must be corrected to alleviate or halt progression of symptoms?	2. If a gene therapy is effective, how will recipients be selected, assuming it is expensive at first?
3. Is overexpression of the therapeutic gene dangerous?	3. Should rare or more common disorders be the focus of gene therapy research and clinical trials?
4. Is it dangerous if the altered gene enters cells other than the intended ones?	4. What effect should deaths among volunteers have on research efforts?
5. How long will the affected cells function?	5. Should clinical trials be halted if the delivered gene enters the germline?
6. Will the immune system attack the introduced cells?	
7. Does the targeted DNA sequence occur in more than one gene?	

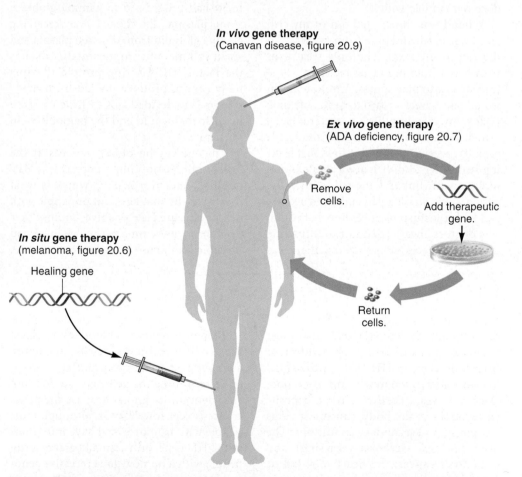

Figure 20.2 **Gene therapy invasiveness.** Therapeutic genes are delivered to cells removed from the body that are then returned (*ex vivo* gene therapy); delivered directly to an accessible body part such as skin (*in situ* gene therapy); or delivered directly to an interior body part, such as through the skull for Canavan disease or to an artery leading to the liver (*in vivo* gene therapy).

chapter 19. In the future, researchers and, someday, clinicians may deliver "artificial" genes synthesized on microchips using digital technology. This new approach, called picoarray gene synthesis, may reduce the cost of gene therapy by a hundredfold or more.

Researchers use physical, chemical, and biological methods to send DNA into cells. Physical methods include electroporation, microinjection, and particle bombardment. Chemical methods include liposomes that enclose the gene cargo and other types of lipid that carry DNA across the plasma membrane. The lipid carrier can penetrate the plasma membrane that DNA alone cannot cross. However, lipid-based methods often fail to deliver a sufficient payload, and gene expression is transient.

One way to improve lipid-mediated gene delivery is to link the lipid to a peptide that carries the gene of interest and also binds it to a specific integrin on the target cell. (Recall from chapter 2 that an integrin is a cellular adhesion molecule.) Another strategy is to alter liposome surfaces so that they resemble viruses that can enter particular cell types.

Biological approaches to gene transfer utilize a vector, such as a viral genome (**table 20.7**). Researchers remove the viral genes that cause symptoms or alert the immune system and add the corrective gene. Different viral vectors are useful for different types of experiments. A certain virus may transfer its cargo with great efficiency but carry only a short DNA sequence. Another virus might carry a large piece of DNA but send it to many cell types, causing side effects. Still another virus may not infect enough cells to alleviate symptoms. Some retroviruses have limited use because they infect only dividing cells.

Researchers target some viral vectors to cells that are normally infected. For example, adenoviruses that transport CFTR genes to the airway passages of people with cystic fibrosis normally infect lung tissue. By adding portions of other viruses, researchers can redirect a virus to infect a certain cell type. Adeno-associated virus (AAV), for example, infects many cell types, but adding a promoter from a parvovirus gene called *B19* restricts infectivity to red blood cell progenitors in bone marrow. Add a human gene that encodes a protein normally found in red blood cells, and the entire vector can treat an inherited disorder of blood, such as sickle cell disease (**figure 20.3**).

Sites of Gene Therapy

Several somatic gene therapies are in clinical trials, targeting several different tissues (**figure 20.4**). Gene delivery may be directly to the affected tissue, or into cells that can produce the needed protein and divide. Researchers are increasingly turning to stem and progenitor cells, because these cells can divide as well as travel and even change the fates of their daughter cells. Therefore, gene therapies can target stem or progenitor cells that are part of an organ, or such cells in the bone marrow. The altered cells can then migrate to other body parts and either specialize into the needed cell types, or fuse with cells there, providing them with a functional copy of the gene. Duchenne muscular dystrophy may be treated in this way, taking advantage of the ability of certain stem cells in the blood, under certain conditions, to travel to muscle tissue and give rise to immature muscle cells.

Table 20.7

Viral Vectors Used in Gene Therapy

Vector	Characteristics	Applications
Adeno-associated virus (AAV)	Integrates into specific chromosomal site Long-term expression Nontoxic Infects dividing and nondividing cells Carries small genes	Cystic fibrosis Sickle cell disease Thalassemias Canavan disease
Adenovirus (AV)	Large virus, carries large genes Transient expression Evokes immune response Infects dividing and nondividing cells	Cystic fibrosis Hereditary emphysema
Herpes	Long-term expression Infects neuroglia	Brain tumors
Retrovirus	Stable but imprecise integration Long-term expression Most types infect only dividing cells Nontoxic	Gaucher disease HIV infection Cancers SCID

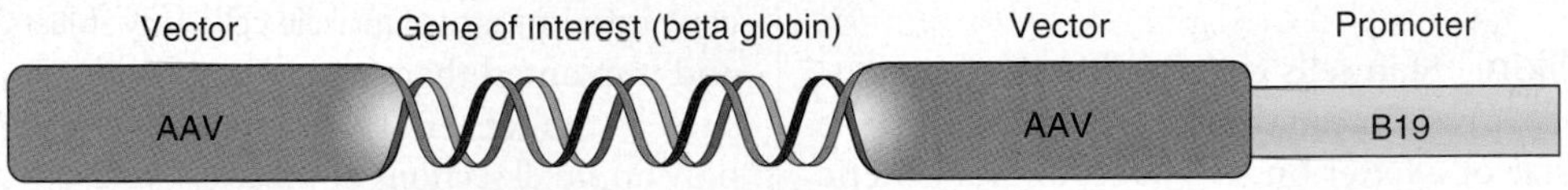

Vector — Gene of interest (beta globin) — Vector — Promoter

AAV — AAV — B19

Figure 20.3 Correcting defects in red blood cells. A promoter from a parvovirus B19 gene directs the adeno-associated virus (AAV) genome harboring a human beta globin gene to erythroid progenitor cells, which give rise to cells that mature into red blood cells filled with hemoglobin.

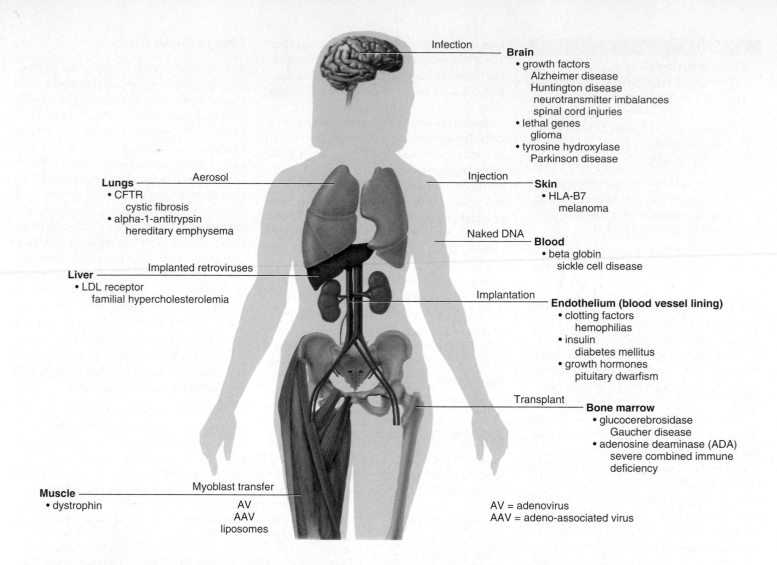

Figure 20.4   Gene therapy sites.   Beneath the label for each site are listed the targeted protein (•) and then the disease.

Following are descriptions of some gene therapy targets under investigation.

Endothelium   Endothelium forms capillaries. Genetically altered endothelium can secrete a needed protein directly into the bloodstream, such as insulin for a person with diabetes or a clotting factor for someone with hemophilia. Endothelium is implanted with collagen to provide support and angiogenesis factors to stimulate capillary growth.

Skin   Skin cells grow well in the laboratory. A person can donate a patch of skin the size of a letter on this page; after a genetic manipulation, the sample can grow to the size of a bathmat within just three weeks, and the skin can then be grafted back onto the person. Skin grafts can be genetically modified to secrete therapeutic proteins.

Muscle   Muscle tissue is a good target for gene therapy because it comprises about half of the body's mass, is easily accessible, and is near a blood supply. However, it is a challenge to correct enough muscle cells to alleviate symptoms. Consider gene therapy for Duchenne muscular dystrophy (DMD) (see figure 2.1). An early milestone was to cut the dystrophin gene—about 3 million bases—down to a size small enough to deliver to cells. Researchers then sent the shortened gene into immature muscle cells (myoblasts) and implanted them at easily testable sites, such as the toes. But the approach worked only on small sections of muscle.

Liver   This largest organ in the body is an important candidate for gene therapy because it has many functions and can regenerate. An implant of corrected cells can take over liver function. For the inborn error of metabolism tyrosinemia (OMIM 276700), for example, only 5 percent of the liver's 10 trillion cells would need to be corrected.

Gene therapy to deliver the gene that encodes the LDL receptor can treat familial hypercholesterolemia, (see figure 5.2). Recall that when liver cells lack LDL receptors, cholesterol accumulates on artery interiors. Heterozygotes have half the normal number of LDL receptors and suffer heart attacks in early or mid-adulthood. Homozygotes die in childhood. Genetically altering liver cells to produce more LDL receptors can reverse the effects of FH. One young woman heterozygous for FH had 15 percent of her liver removed. The cells were given functional LDL receptor genes, and then redelivered through a major liver vein.

Eighteen months later, the grafted liver cells bore more LDL receptors, and the woman's serum cholesterol levels had improved.

Lungs The respiratory tract is easily accessed with an aerosol spray, eliminating the need to remove, treat, and reimplant cells. Several aerosols to treat cystic fibrosis attempt to replace the defective gene, but so far the correction is short-lived and localized. Another lung disorder that may be treatable with gene therapy is alpha-1-antitrypsin (AAT) deficiency (OMIM 107400), which causes hereditary emphysema. Absence of the enzyme AAT enables levels of another enzyme, elastase, to accumulate and destroy lung tissue. White blood cells in the lungs normally produce elastase to destroy infecting bacteria, but in AAT deficiency, elastase levels rise too high. Delivering the AAT gene may normalize levels of both enzymes.

Nervous Tissue Gene therapy on nervous tissue isn't restricted to correcting inherited diseases. Many common, non-inherited illnesses and injuries affect the nervous system too, including seizures, strokes, and spinal cord injuries. If a protein could be used to correct an abnormal situation, then cell implants or gene delivery can possibly heal. However, neurons are difficult targets for gene therapy because they do not divide. Gene therapy efforts can alter other cell types, such as fibroblasts (connective tissue cells), to secrete nerve growth factors or manufacture the enzymes necessary to produce certain neurotransmitters. Then the altered cells can be implanted.

Cancer About half of current gene therapy trials target cancer. In general, these approaches enable cancer cells, or their neighbors, to produce proteins that dampen oncogene expression, bolster tumor suppression, strengthen or redirect the immune response, or induce apoptosis. Two strategies are suicide gene therapy and manipulation of the immune response to create cancer vaccines.

Viruses are used to treat a type of brain tumor, called a glioma, that affects a type of neuroglial cell that supports and interacts with neurons. Unlike neurons, neuroglia can divide. Cancerous neuroglia divide very fast, usually causing death within a year. Researchers reasoned that an agent directed against only the dividing cells might halt the cancer. One candidate was a "suicide" gene from the herpes simplex virus. In the presence of a certain drug, activation of the gene kills the cell that contains it. Would cancer cells infected with a virus carrying this gene self-destruct?

Figure 20.5 illustrates the herpes suicide gene therapy system. First, mouse fibroblasts are infected with a retrovirus or adenovirus vector that contains a herpes gene that encodes an enzyme called thymidine kinase. Any cell that produces thymidine kinase is susceptible to the anti-herpes drug ganciclovir. Because a retrovirus can only infect dividing cells, it does not harm nondividing, healthy brain neurons, but does enter cancerous neuroglia. The modified mouse fibroblasts are injected into a person's brain tumor through a hole drilled in the skull. The implanted cells release viruses, which infect neighboring tumor cells, which then produce thymidine kinase. When the patient takes ganciclovir, the drug is changed into a toxin that kills the cancer cells as well as nearby cells in what is called a "bystander effect." A few people have improved with this treatment, but immune response to the vectors may limit use.

Cancer vaccines enable tumor cells to produce immune system biochemicals or mark tumor cells so that the immune system recognizes them more easily.

Melanoma, a skin cancer, is a candidate for cancer vaccine treatment because it is easily accessible. In one group of experiments, melanoma cells were removed, given genes encoding interleukins, and reimplanted. The genes were expressed, evoking an immune response, and the tumors shrank. In another strategy, researchers inject liposomes bearing genes encoding HLA proteins directly into tumors. **Figure 20.6** describes this approach for an HLA protein called HLA-B7. This protein, when displayed on tumor cell surfaces, stimulates the immune system to respond to the tumor as if it were foreign tissue.

RELATING THE CONCEPTS

1. Does Max Randell's gene therapy alter his phenotype, genotype, or both?
2. Is his gene therapy germline or somatic?

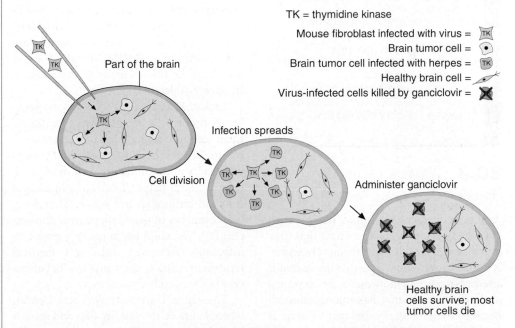

TK = thymidine kinase
Mouse fibroblast infected with virus =
Brain tumor cell =
Brain tumor cell infected with herpes =
Healthy brain cell =
Virus-infected cells killed by ganciclovir =

Part of the brain

Infection spreads

Cell division

Administer ganciclovir

Healthy brain cells survive; most tumor cells die

Figure 20.5   A herpes virus attacks cancer. Mouse fibroblasts harboring a thymidine kinase gene from the herpes simplex virus, in a retrovirus vector, are implanted near the site of a brain tumor. The altered viruses infect the rapidly dividing tumor cells. When the patient takes the antiviral drug ganciclovir, the cells producing thymidine kinase are selectively killed, providing a gene-based cancer treatment from within.

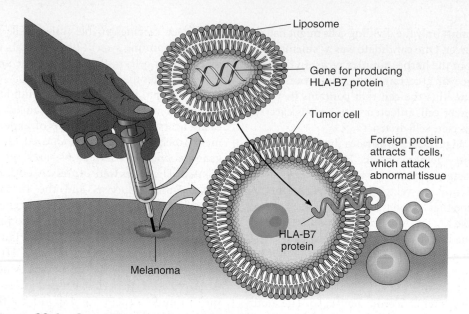

Figure 20.6   Gene therapy for melanoma.   A gene that encodes a cell surface protein that attracts the immune system's tumor-killing T cells is injected directly into a melanoma tumor. (The size of the liposome relative to the tumor cell is greatly exaggerated.)

Labels in figure:
Liposome
Gene for producing HLA-B7 protein
Tumor cell
Foreign protein attracts T cells, which attack abnormal tissue
HLA-B7 protein
Melanoma

Key Concepts

1. Protein-based therapies replace gene products and treat the phenotype.
2. Gene therapies replace malfunctioning or absent genes.
3. Gene therapy may be *in situ, ex vivo,* or *in vivo.*
4. Germline gene therapy targets gametes or fertilized ova and is heritable.
5. Vectors for delivering genes include liposomes and viruses.
6. Somatic gene therapy targets various types of somatic tissue as well as cancer cells and is not heritable.

20.3 Three Gene Therapies

Any new medical treatment begins with courageous volunteers who know that they may risk their health. Gene therapy, however, is unlike conventional drug therapy in that it attempts to alter an individual's genotype in a part of the body that has malfunctioned. Because the potentially therapeutic gene is usually delivered with other DNA, and it may be taken up by cell types other than those affected in the disease, reactions are unpredictable. Following is a look at some of the volunteers who pioneered gene therapy.

An Early Success

For the first few years of her life, Laura Cay Boren didn't know what it was like to feel well (**figure 20.7a**). From her birth in July 1982, she fought infection after infection. Colds rapidly became pneumonia, and routine vaccines caused severe abscesses. In February 1983, doctors identified Laura's problem—severe combined immune deficiency (SCID) due to adenosine deaminase (ADA) deficiency (OMIM 608958). She had inherited the inborn error of metabolism from carrier parents.

Lack of ADA blocks a biochemical pathway that normally breaks down a metabolic toxin into uric acid, which is then excreted (figure 20.7b). The substance that ADA normally acts upon builds up and destroys T cells. Without helper T cells to stimulate B cells, no antibodies are made. Therefore, both branches of the adaptive immune system fail. The child becomes very prone to infections and cancer, and despite medical treatment, usually does not live beyond a year in the outside environment.

The Duke University Medical Center, where Laura celebrated her first and second birthdays, became her second home. In 1983 and 1984, she received bone marrow transplants from her father, which temporarily bolstered her immunity. Red blood cell transfusions also helped, but Laura was still spending more time in the hospital than out. By the end of 1985, she was gravely ill. She had to be fed through a tube, and repeated infection had severely damaged her lungs. Then Laura was chosen to participate in a trial for a new treatment. She had been second in line to a boy who died just before beginning treatment. In the spring of 1986, Laura received her first injection of PEG-ADA. This is the missing enzyme, ADA, taken from a cow and stabilized with polyethylene glycol (PEG) chains (PEG is the major ingredient in antifreeze).

Previous enzyme replacement without PEG didn't work, because what remained of the immune system destroyed the injected, unaltered enzyme. Patients needed frequent doses, which provoked the immune system further, causing severe allergic reactions. Laura's physicians hoped that adding PEG would keep ADA in her blood long enough to work.

Laura began responding to PEG-ADA almost immediately, and within hours, her enzyme level increased twentyfold. After three months, toxins no longer showed up in her blood, but her immunity was still suppressed. After six months, though, Laura's immune function neared normal for the first time ever—and stayed that way, with weekly doses of PEG-ADA. Her life changed dramatically as she ventured beyond the hospital's germ-free rooms. By summer 1988, she could finally play with other children without fear of infection. She began first grade in fall 1989, but had to repeat the year—she had spent her time socializing!

PEG-ADA revolutionized treatment of this form of SCID, but it replaced the protein, not the gene. The gene therapy approach began on September 14, 1990, at 12:52 P.M. Four-year-old Ashanthi DeSilva sat up in bed at the National Institute of Health in Bethesda, Maryland, and began receiving her own white blood cells intravenously. Earlier, doctors had removed the cells and inserted functioning ADA genes. The gene delivery worked, but did not alter enough cells to restore immunity, so it had to be repeated, or PEG-ADA given at intervals. However, Ashanthi is now healthy, and she tells her story at scientific meetings (see In Their Own Words on page 409).

A longer-lasting treatment could result from altering progenitor cells (see figure 2.24). The progenitor of T cells accounts for only one in several billion bone marrow cells.

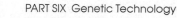

The First Gene Therapy Patient

In the late 1980s, the DeSilva's did not think their little girl, Ashanthi ("Ashi"), would survive. She suffered near-continual coughs and colds, and was so fatigued that she could walk only a few steps before becoming winded, her father Raj recalls. "We took her to so many doctors that I stopped counting. One doctor after another would say it was asthma, an allergy, or bronchitis."

Raj's brother, an immunologist, suggested the blood tests that would eventually reveal Ashi's underlying problem—severe combined immune deficiency due to adenosine deaminase (ADA) deficiency. Although unlucky in inheriting a disease, Ashi was lucky in that it was a condition so well understood that it was first in line for gene therapy. Through a series of physician contacts, Ashi became the first recipient.

The medical team at the National Institute of Health—W. French Anderson, Kenneth Culver, and Michael Blaese—had spent years planning the gene therapy, and were fairly certain that it would work. Within weeks following the therapy, Ashi began to make her own, functional T cells. Although she required further treatments, today she is well and excited about her future, anticipating a career in the music industry after college.

Over the years, she has championed gene therapy at biomedical conferences. The photo shows her at a meeting when she was 17, where she introduced Dr. Blaese: "Our duty on Earth is to help others. I thank you from the bottom of my heart for all you have enabled me to do."

Gene therapy has hit snags in recent years, but overall has had an excellent track record. Says Dr. Blaese, "You have to consider the context. In the years since the first patient, there has been one death and two malignancies. Compare that to the first 100 heart transplants, where only one person lived more than a year. Gene therapy has had a remarkable safety record, yet there are still problems."

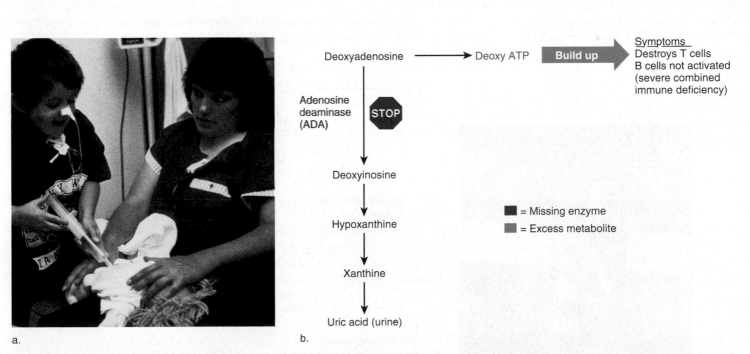

Figure 20.7 Correcting ADA deficiency. (a) Laura Cay Boren spent much of her life in hospitals until she received the enzyme that her body lacks, adenosine deaminase (ADA). Here, she pretends to inject her doll as her mother looks on. Today, gene therapy is possible using cord blood stem cells. (b) ADA deficiency causes deoxy ATP to build up, destroying T cells, which therefore cannot stimulate B cells to secrete antibodies. The result is severe combined immune deficiency (SCID).

Umbilical cord blood was a more plentiful source. If fetuses who had inherited ADA deficiency could be identified, then stem cells could be extracted from their cord blood at birth, given ADA genes, and reinfused.

Crystal and Leonard Gobea had already lost a five-month-old baby to ADA deficiency when amniocentesis revealed that their second fetus was affected. They and two other couples were asked to participate in an experiment. Andrew Gobea and the other two babies received their own bolstered cord blood cells on the fourth day after birth, along with PEG-ADA to prevent symptoms in case the gene therapy did not work right away. T cells carrying normal ADA genes gradually appeared in their blood. By the summer of 1995, the three toddlers each had about 3 in 100 T cells carrying the ADA gene, and they continued to improve.

A few years after the three children with ADA deficiency were treated, another gene therapy trial for a severe combined immune deficiency (SCID) began in France. Nine male infants with a type of X-linked SCID had T cell progenitors removed and given the gene they were missing, which encodes part of a cytokine receptor. The researchers hoped that restoring cells' ability to recognize cytokines would cure the immune deficiency. It did, but since 2003, three of the boys have developed a rare leukemia. The retrovirus that delivered the therapeutic gene had inserted into a proto-oncogene. The boys were successfully treated for the leukemia, but this very unexpected side effect initially stalled many gene therapy trials. It was not the first time such an experiment had a tragic outcome. First came Jesse Gelsinger.

A Major Setback

Eighteen-year-old Jesse Gelsinger died in September 1999, days after receiving gene therapy, due to an overwhelming immune system reaction against the DNA used to introduce the therapeutic gene.

He had ornithine transcarbamylase deficiency (OTC) (OMIM 311250). In this X-linked recessive disorder, one of five enzymes required to break down amino acids liberated from dietary proteins is absent **(figure 20.8)**. The nitrogen released from the amino acids combines with hydrogens to form ammonia (NH_3), which rapidly accumulates in the bloodstream and travels to the brain, with devastating effects. The condition usually causes irreversible coma within 72 hours of birth. Half of affected babies die within a month, and another quarter by age five. The survivors can control their symptoms by following a special low-protein diet and taking drugs that bind ammonia.

Jesse wasn't diagnosed until he was two, because he was a mosaic—some of his cells could produce the enzyme, so his symptoms were milder. When he went into a coma in December 1998 after missing a few days of his medications, he and his father considered whether he should volunteer for a gene therapy trial they had read about. The researchers would not accept Jesse until he turned 18, so the next summer, four days after his birthday, Jesse underwent testing at the University of Pennsylvania, and was admitted to the trial. He was jubilant. He knew he might not directly benefit, but he had wanted to try to help babies who die of the condition. A bioethics committee had advised that the experimental treatment could not be tried on newborns because the parents would be too distraught to give informed consent to an untried medical procedure. Instead, volunteers were affected males and carrier females. Said Jesse at the time, "What's the worst that can happen to me? I die, and it's for the babies."

The gene therapy was an adenovirus with a functional human OTC gene inserted. This virus had already been used safely in about a quarter of the gene therapy experiments done since 1990. It is a disabled virus, with the genes that enable it to replicate and cause respiratory symptoms removed. Three groups of six patients each were to receive three different doses, to identify the lowest dose that would fight the OTC deficiency without side effects.

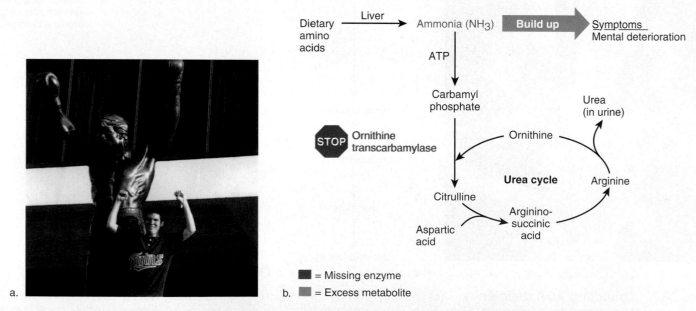

Figure 20.8 A brave example. **(a)** Jesse Gelsinger received gene therapy for an inborn error of metabolism in September 1999. He died four days later from an overwhelming immune response. **(b)** Lack of ornithine transcarbamylase causes ammonia to accumulate, which is toxic to the brain.

Jesse entered the hospital on Monday, September 13, after the 17 others in the trial had been treated and suffered nothing worse than fever and minor aches. Several billion altered viruses were placed in an artery leading into his liver. That night, Jesse developed a high fever—still not unusual. But by morning, the whites of his eyes were yellow, indicating that his liver was struggling to dismantle the hemoglobin released from burst red blood cells. A flood of hemoglobin meant a flood of protein, so the ammonia level in his liver soon skyrocketed, reaching 10 times normal levels by mid-afternoon. Jesse became disoriented, then comatose. By Wednesday, doctors had controlled the ammonia buildup, but his lungs began to fail, and Jesse was placed on a ventilator. Thursday, vital organs began to fail, and by Friday, he was brain dead. His dedicated and devastated medical team stood by as his father turned off life support, and Jesse died.

At a public hearing in December, doctors explained that the autopsy showed that Jesse had had a parvovirus infection, which may have led his immune system to attack the adenovirus. In the liver, the adenovirus had targeted not the hepatocytes as expected, but macrophages that function as sentries for the immune system. In response, interleukins flooded his body, and inflammation raged. Although parents of children with OTC who spoke at the hearing implored government officials to continue the research, the death of Jesse Gelsinger led to suspension of several gene therapy trials. The death drew particular attention to safety because, unlike most other volunteers, Jesse had not been very ill.

In the months following his death, government health officials identified several procedural factors that may have contributed to the tragedy: failure to report all side effects in a timely fashion, failure to fill out the proper eligibility forms, and inadequately documented informed consent. Also, more careful attention should have been paid to screening potential participants for underlying medical conditions. Extreme underreporting of adverse side effects was found in other gene therapy trials, but this might have reflected the fact that researchers attributed many adverse effects to the underlying disease, and not to the experimental treatment.

A Success in the Making

Efforts begun in 1995 to treat Canavan disease continued, despite Jesse Gelsinger's fate. Canavan disease is an ideal candidate for gene therapy for several reasons:

1. The gene and protein are well known.

2. There is a window of time when affected children are healthy enough to be treated.

3. Only the brain is affected.

4. Brain scans can monitor response to treatment.

5. No treatment exists.

Canavan disease disrupts the interaction between neurons and neighboring oligodendrocytes, which produce the fatty myelin that coats neurons, enabling them to transmit impulses fast enough for the brain to function (**figure 20.9**). Specifically, neurons normally release N-acetylaspartate (NAA),

Symptoms: Brain atrophy, kills oligodendrocytes

Build up

■ = Missing enzyme
■ = Excess metabolite

N-acetylaspartate (NAA)

Aspartoacylase

STOP

Myelin

Neuron

Aspartoacylase gene

Oligodendrocyte

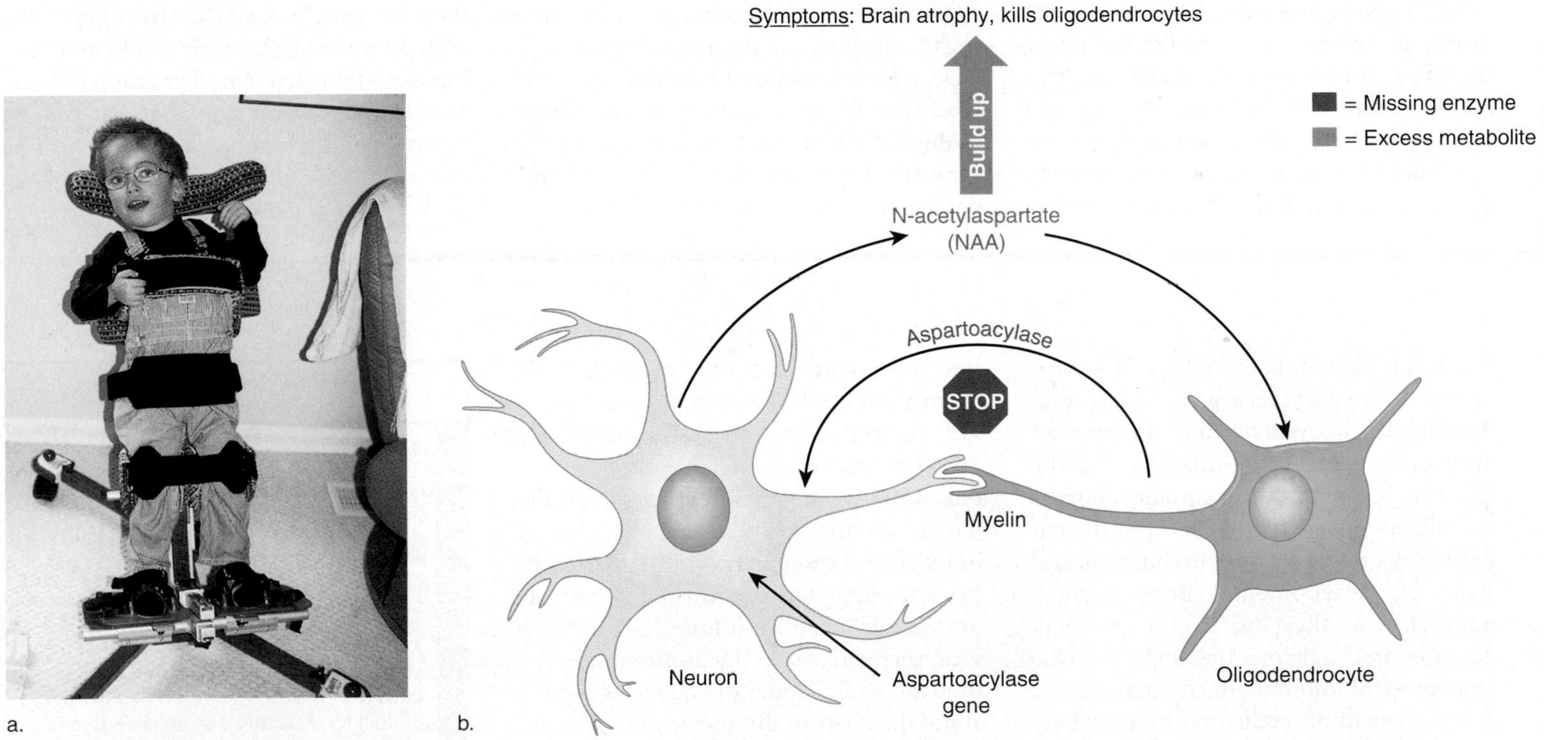

a.

b.

Figure 20.9 Canavan disease. (a) Max Randell, shown here at two-and-a-half years old, is battling Canavan disease. He underwent gene therapy that delivered viruses carrying a functional gene directly into his brain at six sites. Within ten days, he began to improve. **(b)** In Canavan disease, stripping of the lipid layer on brain neurons occurs because oligodendrocytes lack an enzyme that enables them to break down N-acetylaspartate, which neurons produce. Buildup of N-acetylaspartate eventually destroys the oligodendrocytes, so that the neurons lack myelin. Gene therapy enables the neurons to secrete the enzyme (rather than the oligodendrocytes), ultimately restoring the fatty covering that makes nerve transmission possible in the brain.

Canavan Disease: Patients Versus Patents

When Debbie Greenberg gave birth to Jonathan in 1981, she and her husband Dan had no idea that they would one day be leading the first effort to challenge how a researcher and a hospital obtained a patent on a gene. Like Max Randell's parents, Debbie and Dan were each carriers of Canavan disease, and Jonathan was affected. Although Jonathan lived 11 years, his brain never developed past infancy. The couple had an affected daughter, Amy, a few years after Jonathan was born, and three healthy children.

Shortly after Jonathan's diagnosis, the Greenbergs started the Canavan Foundation, which established a tissue bank that stored blood, urine, and autopsy tissue from affected children. In 1987, the Greenbergs met Dr. Reuben Matalon at a Tay-Sachs disease screening event in Chicago, and convinced him to begin a search for the Canavan gene. The Greenbergs helped to collect tissue from families from all over the world, which was critical to Dr. Matalon's success in identifying the gene and the causative mutation in 1993, when he was working at Miami Children's Hospital.

Finding the gene made it possible to detect the mutation, which could be used to confirm diagnoses and detect carriers, and in prenatal diagnosis. By 1996, the Canavan Foundation was offering free testing. But unknown to the members of the organization who had donated their childrens' tissues for the gene search, Dr. Matalon and Miami Children's Hospital had filed for a patent on their discovery. The U.S. Patent and Trademark office granted invention number 5,679,635—the Canavan gene—in 1997. A year later, the American College of Obstetricians and Gynecologists advised their physician members to offer carrier testing for Canavan disease to Ashkenazi Jewish patients, because 1 in 40 such women is a carrier. Identifying couples in which both people are carriers would give them the option of avoiding giving birth to affected children, a strategy that has reduced the number of children born with the similar Tay-Sachs disease to nearly zero. That same year, Miami Children's Hospital began to exercise its patent rights by requiring that doctors and diagnostic laboratories charge for a Canavan test. Suddenly, families whose donations—both monetary and biological—had made the discovery of the gene possible had to pay for carrier and prenatal tests. They were outraged.

On November 30, 2000, a group of parents and three nonprofit organizations filed suit in Chicago against Dr. Reuben Matalon and Miami Children's Hospital. The suit does not challenge the patent, but does challenge the way in which it was obtained—in secret, they claim. They wish to recover earnings from the gene test to be turned over to the families who had to pay to offset licensing fees. In 2003, U.S. district judge Federico Morena upheld the rights of the parents to sue, calling the case "a tale of successful research collaboration gone sour."

The Greenbergs' fight against a disease gene patent not only is a legal precedent, but also sounded a warning bell to other patient groups. As a direct result of the Canavan case, Sharon and Patrick Terry, of Sharon, Massachusetts, started a support group and tissue bank for the disorder their child has—pseudoxanthoma elasticum (PXE) (OMIM264800), which causes connective tissue to calcify. Like the Greenbergs, the PXE parents supplied their childrens' tissue, but they stipulated that their group be listed as a coinventor on any gene patents. By doing this, the families hope they will be able to retain control over the fate of the gene that they helped discover.

---

which is broken down into harmless compounds by an enzyme, aspartoacylase, that the oligodendrocytes produce. In Canavan disease, the enzyme is missing, and the resulting NAA buildup eventually destroys the oligodendrocytes. Without sufficient myelin, the neurons cease to function, and signs of developmental delay become noticeable, as they did in Max Randell, described in the chapter opening essay. Due to a powerful founder effect, Canavan disease is seen almost exclusively in the Ashkenazi Jewish population. Bioethics: Choices for the Future relates the bitter battle over access to genetic tests for Canavan disease.

The first attempts at gene therapy for Canavan disease introduced the needed gene in a liposome, through holes bored into the skull. The first recipient, 18-month-old Lindsay Karlin, regained some skills for awhile. Previously, she could barely open her eyes and did not interact with anyone. But three months after the therapy, she looked around, moved, and vocalized. A brain scan showed neuron myelination in regions where it had vanished. Lindsay was not treated again until June 2001, when a viral vector replaced the liposomes. In the interim, while regulatory agencies argued about the safety of this therapy for a disease that had no other treatment, Lindsay lost some of the gains from the first round, such as being able to hold her head up. But the gene therapy does appear to be working, in Lindsay, Max, and other children participating in current clinical trials.

Key Concepts

1. ADA deficiency was the first disorder treated with gene therapy. It began by replacing the missing enzyme, progressed to genetically altering mature white blood cells in ill children, and finally moved to infusing genetically modified umbilical cord stem cells into newborns.
2. Gene therapy for OTC deficiency led to a death due to a severe immune reaction to the virus used to deliver the genes.
3. For Canavan disease, gene therapy may provide the only treatment.

20.4 A Closer Look: Treating Sickle Cell Disease

Sometimes effective treatments are suggested by the peculiarities of a specific disorder. This is the case for sickle cell disease, perhaps the best-studied inherited illness. We know the precise mutation that causes the disease, how the phenotype arises in response to locally oxygen-poor conditions in the bloodstream, and how the globin genes are regulated in the embryo, fetus, and newborn (see figures 11.2 and 12.1). Knowledge of this developmental regulation has led to a unique way to prevent sickling, using a drug called hydroxyurea.

The story began with the observation that people who have "hereditary persistence of fetal hemoglobin" (OMIM 142470, HPFH) are healthy or have very mild anemia, indicating that fetal hemoglobin doesn't harm an adult. Could "turning on" fetal hemoglobin cure sickle cell disease by replacing the mutant beta chains with normal gamma (fetal) globin chains?

In adults, the gamma genes are normally silenced with methyl groups. A drug that removes the methyl groups might expose the gamma genes, enabling their expression. In 1982, researchers found that a drug called 5-azacytidine indeed removed the methyl groups and raised the proportion of fetal hemoglobin in the blood. But this drug causes cancer and is also toxic to ("stresses") red blood cells by producing reactive oxygen molecules. Which effect stimulates production of fetal hemoglobin, removing methyls or adding stress? Researchers found that an existing, safer drug, hydroxyurea, stresses red blood cell progenitors. Could it also turn on fetal globin genes? From 1984 until 1992, experiments in monkeys, healthy people, and sickle cell disease patients showed that hydroxyurea raises fetal hemoglobin levels, halving the number of crises and greatly cutting the need for transfusions in patients.

The way hydroxyurea works involves biology, chemistry, and physics. First, the drug increases the proportion of gamma globin molecules in the blood. Some of the gamma globins bind to mutant beta globins, preventing them from forming the circulation-strangling sheets. This prolongs the time it takes for the mutant beta globin chains to join, simply because gamma chains are now

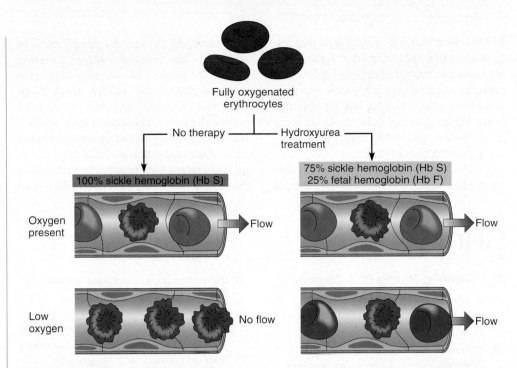

Figure 20.10 Reactivating globin genes. Hydroxyurea stimulates production of fetal hemoglobin (Hb F), which dilutes sickled hemoglobin (Hb S). With fewer polymerized hemoglobin molecules, the red blood cells do not bend out of shape as much and can reach the lungs. Oxygen restores the cells' shapes, averting a sickling crisis.

in the mix (**figure 20.10**). The delay is long enough for a red blood cell to return to the lungs. Once the cell picks up oxygen, sickling cannot occur. By slowing the sickling process, hydroxyurea corrects a devastating phenotype.

Sickle cell disease can be cured with a bone marrow transplant, but the procedure has a 10 percent mortality rate. Experimental gene therapy uses AAV as a vector.

Key Concepts

1. The more we know about a genetic disease, the more treatment options we can develop.
2. A unique approach that reactivates fetal hemoglobin genes treats sickle cell disease.

20.5 Perspective: A Slow Start, But Great Promise

Medical genetics is no longer just the realm of a few physician specialists, genetic counselors, and the families that have conditions so rare that they are called "orphan" diseases. As we begin to understand precisely how genes control the functioning of the human body, genetic testing, counseling, and therapy will likely become applied to common as well as rare disorders. But it has been a long time in coming.

When the age of gene therapy dawned in the 1990s, expectations were high—and for good reason. Work in the 1980s had clearly shown abundant, pure, human biochemicals, useful as drugs, could come from transgenic organisms. It was a matter of time, researchers speculated, before genetic altering of our own somatic tissue would treat a variety of ills.

In reality, gene therapy progress has been painstakingly slow. Boys with Duchenne muscular dystrophy who receive myoblasts with healthy dystrophin genes do not walk again. Instead, they might be able to wiggle a toe for a time. People with cystic fibrosis who inhale viruses bearing the CFTR gene do not permanently breathe easier, but might feel relief for a few weeks. And we now know that delivered genes do not always go where intended.

The sequencing of the human genome has not provided a list of new gene defects to correct—many of the disease-causing genes were already known—but instead has revealed a complexity to genome structure and function that will impact gene therapy. Consider the fact that the same exon sequence can be part of different genes. Targeting an exon because it is part of one gene may affect others, healing one set of symptoms while causing others. As genome researchers continue to identify gene functions, this risk should lessen.

Discovery of RNA interference, discussed in chapter 11, may also complicate gene therapy. Correcting a genetic mistake in the nucleus may not counter a disease phenotype because of what may happen in the cytoplasm—the mRNA transcribed from the delivered gene may be silenced before the needed protein is even synthesized. Yet another area of uncertainty is the issue of somatic versus germline gene therapy. A corrected gene targeted to a particular tissue may find its way, in the circulation, to the reproductive tract, enter a gamete, and thereby affect the next generation.

Despite these drawbacks—real and theoretical—at a molecular and cellular level, gene therapy *is* working. The patients with muscular dystrophy, cystic fibrosis, and SCID have cells that have accepted and expressed therapeutic genes. The challenge now is to find just the right vector to deliver a sustained, targeted, and safe genetic correction.

Summary

20.1 Genetic Testing and Counseling

1. Newborns are routinely screened for several inborn errors of metabolism, some of which are treatable.

2. Other genetic tests include prenatal diagnosis, cancer susceptibility tests, and predictive testing for genetic disease.

3. **Genetic counselors** provide information on inheritance patterns, disease risks and symptoms, and available tests and treatments.

4. Prenatal counseling and counseling a family coping with a particular disease pose different challenges.

5. Because of the limited availability of genetic counselors, sometimes other professionals or interactive computer programs provide information on genetic testing.

6. Genetic testing raises privacy issues when physicians must decide when it is appropriate to breach confidentiality about a test result.

20.2 Treating Genetic Disease

7. Protein supplementation, from donors and then from recombinant DNA technology, preceded **gene therapy,** which replaces malfunctioning genes.

8. Hereditary hemochromatosis is treated by having blood removed.

9. **Ex vivo** gene therapy is applied to cells outside the body that are then reimplanted or reinfused into the patient. **In situ gene therapy** occurs directly on accessible body parts. **In vivo gene therapy** is applied in the body.

10. **Germline gene therapy** affects gametes or fertilized ova, affects all cells of an individual, and is transmitted to future generations. It is not performed in humans.

11. **Somatic gene therapy** affects somatic tissue and is not passed to offspring.

12. Gene therapy delivers new genes and encourages production of a needed substance at appropriate times and in therapeutic (not toxic) amounts.

13. Several types of vectors are used to deliver therapeutic genes, including liposomes and viral genomes.

14. Some gene therapies target stem or progenitor cells, because they can divide and move.

15. There are many sites of somatic gene therapy, including cancers.

20.3 Three Gene Therapies

16. Gene therapy for ADA deficiency began with enzyme replacement, then gene therapy in white blood cells, then gene therapy in progenitor cells that could better replace affected cells.

17. Children developed leukemia from gene therapy to treat SCID. A death in a gene therapy trial for OTC deficiency was due to an immune system response to the viral genome used to introduce the gene.

18. Gene therapy for Canavan disease enables brain neurons to produce a missing enzyme.

20.4 A Closer Look: Treating Sickle Cell Disease

19. Drugs to reactivate fetal globin genes can treat sickle cell disease.

20.5 Perspective: A Slow Start, But Great Promise

20. Development of gene therapy has been slower than anticipated because of the unexpected complexities of gene interactions and the challenge of adequately targeting and sustaining therapeutic effects.

Review Questions

1. Why is newborn screening economically feasible?

2. Using information from this or other chapters, or the Internet, cite genetic tests given to a newborn, young adult, and middle-aged person (three different tests).

3. Describe what a genetic counselor does.

4. What are the advantages and disadvantages of "virtual" genetic counseling (using an interactive computer program rather than a human)?

5. What factors do you think are important in deciding whether or not to provide populationwide newborn screening for a particular inherited disorder?

6. What are the three stages of the evolution of treatments for single-gene disorders?

7. Why is the removal of blood in people with hereditary hemochromatosis not gene therapy?

8. Explain the differences among *ex vivo, in situ,* and *in vivo* gene therapies. Give an example of each.

9. Would somatic gene therapy or germline gene therapy have the potential to affect evolution? Explain your answer.

10. What factors would a researcher consider in selecting a viral vector for gene therapy?

11. Gene therapies for Duchenne muscular dystrophy have used AV, AAV, and liposomes. Explain how each approach works.

12. Why is it easier to "fix" a liver with gene therapy than to treat a muscle disease?

13. Why is a bone marrow transplant from a healthy child to a sibling with sickle cell disease technically not gene therapy, while removing an affected child's bone marrow and replacing the mutant gene with a normal one in particular cells is gene therapy?

14. Explain how gene therapy can target cancer.

15. Select a gene therapy discussed in the chapter and explain how it works.

16. Explain how hydroxyurea can be used to prevent symptoms of sickle cell disease.

Applied Questions

1. Why would the American College of Medical Genetics ask that the government mandate testing for inborn errors of metabolism that do not have treatments?

2. What are the issues that a physician faces in deciding whether to violate confidentiality with a patient's genetic test results?

3. A lentivirus is a rare type of retrovirus that can infect nondividing cells, therefore widening its applicability as a gene therapy vector. HIV is a lentivirus that is being evaluated in a disabled form as a vector for gene therapy. What would have to be done to it to make this feasible?

4. Researchers have discovered that red blood cells from people with sickle cell disease hold oxygen longer when exposed to nitric oxide in a test tube. How can this observation be used to develop a new treatment for sickle cell disease?

5. Parkinson disease is a movement disorder in which neurons in a part of the brain called the substantia nigra can no longer produce the neurotransmitter dopamine. This neurotransmitter is not a protein. What are two difficulties in developing gene therapy for Parkinson disease?

6. Create a gene therapy by combining items from the three lists below. Describe the condition to be treated, and how a gene therapy might correct the symptoms.

Cell Type	Vector	Disease Target
fibroblast	AV	Duchenne muscular dystrophy
skin cell	AAV	Alzheimer disease
neuroglial cell	retrovirus	sickle cell disease
red blood cell progenitor	liposome	cystic fibrosis
myoblast	herpes virus naked DNA	glioma melanoma

7. Genes can be transferred into the cells that form hair follicles. Would gene therapy to treat baldness most likely be *ex vivo, in situ,* or *in vivo?* Cite a reason for your answer.

8. Why might a gene therapy for Canavan disease be more likely to pass requirements of a bioethics review board than the trial that Jesse Gelsinger took part in?

Web Activities

Visit the Online Learning Center (OLC) at www.mhhe.com/lewisgenetics7. Select **Student Edition, chapter 20,** and **Web Activities** to find the website links needed to complete the following activities.

9. Use OMIM to identify a disease that might be treated with gene therapy, and describe how the therapy would work.

10. Consult a website dealing with newborn screening, and list three inherited conditions that can be detected as well as treated.

Case Studies and Research Results

11. How would you, as a genetic counselor, handle the following situations (all real)? What would you tell the patients, and what tests would you suggest? (You might have to consult past chapters for specific information.)

   a. A couple in their early forties is expecting their first child. Amniocentesis indicates that the fetus is XXX, which might never have been noticed without the test. When they learn of the abnormality, the couple asks to terminate the pregnancy.

   b. A 25-year-old woman gives birth to a baby with trisomy 21 Down syndrome. She and her husband are shocked—they thought that this could only happen to a woman over the age of 35.

   c. Two people of normal height have a child with achondroplastic dwarfism, an autosomal dominant trait. They are concerned that subsequent children will also have the condition.

   d. A newborn has a medical condition not associated with any known gene mutation or chromosomal aberration. The parents want to sue the genetics department of the medical center because the amniocentesis did not indicate a problem.

12. Two women are in the hospital suffering from emphysema. Linda, 20 years old, does not smoke. She has battled the condition all her life—she has an inherited form of emphysema called alpha-1-antitrypsin deficiency. She knows that she is unlikely to see her thirtieth birthday. Linda's roommate in the hospital, Bernice, is also struggling to breathe with emphysema. She is 58 years old and developed the condition from smoking since age 16. A pair of lungs becomes available for transplant, and they match the tissue types of both Linda and Bernice. If Linda has the transplant, the new lungs will eventually become diseased, because her underlying enzyme deficiency is still present. Still, she could gain a decade of life. If the lungs go to Bernice, they would likely stay healthy, as long as she does not smoke —which she is not certain she can do. What criteria should the transplant and bioethics teams consider to decide who receives the lungs?

13. Three-year-old Tawny Fitzgerald has been to the emergency department repeatedly for broken bones. At the last visit, a nurse questioned Tawny's parents, Donald and Rebecca, about possible child abuse. No charges were filed—the child just appeared to be clumsy. Then Tawny's brother Winston was born. When he was six months old, Donald found him screaming

in pain one morning. A trip to the hospital revealed a broken arm. This time, a social worker was sent to the Fitzgerald home. Donald and Rebecca were interviewed in great depth and advised to find a lawyer. A relative in medical school suggested that they have the children examined for osteogenesis imperfecta, also known as "brittle bone disease."

Consult OMIM and list the facts about a form of this condition that could affect both sexes, with carrier parents. If you were the genetic counselor hired to help this couple, what would you ask them, and tell them, to help them deal with the legal and social services authorities who might need a biology lesson?

14. Jill and Scott S. had thought six-month-old Hannah was developing just fine until Scott's sister, a pediatrician, noticed that the baby's abdomen was swollen and hard. Knowing that the underlying enlarged liver and spleen could indicate an inborn error of metabolism, Scott's sister suggested the child undergo several tests. She had inherited sphingomyelin lipidosis, also known as Niemann-Pick disease type A (OMIM 257200). Both parents were carriers, but Jill had tested negative when she took a Jewish genetic disease panel during her pregnancy because her particular mutation was very rare and not included in the test panel. Happily, Hannah went on to become the first child with Niemann-Pick disease to survive treatment with a transplant of umbilical cord blood cells from a donor. As of her first birthday, she was catching up developmentally and was more alert than she would have been without the treatment. Monocytes, a type of white blood cell, from the cord blood traveled to her brain and manufactured the deficient enzyme. Dietary therapy does not work for this condition because the enzyme cannot cross from the blood to the brain. Monocytes, however, can enter the brain.

a. Did Hannah's treatment alter her phenotype, genotype, or both?

b. Why did the transplant have to come from donated cord blood, and not from Hannah's own, which had been stored?

c. If you were the genetic counselor, what advice would you give this couple if they conceive again?

15. A survey conducted at Harvard Medical School of 1,250 mothers of children with Down syndrome found several instances when physicians delivered the diagnosis in extremely negative and upsetting language. The researcher offers suggestions on how doctors can be more sensitive in this situation, but concludes that only a physician should deliver such news. Suggest an alternative.

Learn to apply the skills of a genetic counselor with these additional cases found in the *Case Workbook in Human Genetics:*

Gene doping

Hemophilia B

MCAD deficiency

Rheumatoid arthritis gene therapy

CHAPTER

21

Reproductive Technologies

POSTMORTEM SPERM RETRIEVAL

Bruce and Gaby Vernoff, in their early thirties, had delayed becoming parents, confident that their good health would make pregnancy possible later. But then Bruce suddenly died of an allergic reaction to a medication. Because she knew how much he had wanted to be a father, Gaby requested that physicians take some of Bruce's sperm after his death. Thirty hours after Bruce died, the medical examiner collected a sperm sample and sent it to the California Cryobank (a sperm bank), where it lay deeply frozen for more than a year. In the summer of 1978, Cappy Rothman, medical director of the bank, used the defrosted sperm to fertilize one of Gaby's oocytes. On March 17, their daughter was born. It was the first case of postmortem sperm retrieval in which the father did not actively participate in the decision. In other cases, the man was dying from cancer and had time to state his wishes to be a father posthumously. More recently, servicemen in the Gulf War in 1990–91 and in Operation Iraqi Freedom in 2003–5, fearing infertility from exposure to chemical or biological weapons, took advantage when sperm banks offered discounted sperm preservation to the military.

Postmortem sperm retrieval raises legal and ethical issues. In another case, in 1995, a woman conceived twins with her husband's consent sixteen months after her husband had died of leukemia at age 30. But the Social Security Administration refused to provide survivor benefits to their daughters, claiming that the father was not a father, but a sperm donor. The Massachusetts Superior Court reversed this decision. Because postmortem sperm retrieval, like other assisted reproductive technologies, is not regulated at the federal level in the United States, bioethicists have identified situations to avoid:

- Someone other than a spouse wishing to use the sperm
- A too-hasty decision based on grief
- Use of the sperm for monetary gain

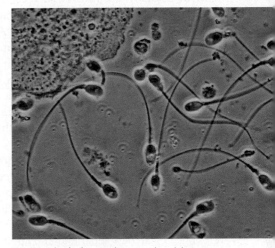

Under what circumstances should sperm from a deceased soldier be used to conceive a child?

A couple in search of an oocyte donor advertises in a college newspaper seeking an attractive young woman from an athletic family. A cancer patient stores her oocytes before undergoing treatment. Two years later, she has several of them fertilized in a laboratory dish with her partner's sperm, and has a cleavage embryo implanted in her uterus. She survives the cancer and becomes a mother. A man paralyzed from the waist down has sperm removed and injected into his partner's oocyte. He, too, becomes a parent when he thought he never would.

Lisa and Jack Nash sought to have a child for a different reason. Their daughter Molly, born on July 4, 1994, had Fanconi anemia (OMIM 227650), an autosomal recessive condition that would destroy her bone marrow, severely impairing her immunity. An umbilical cord stem cell transplant from a sibling could likely cure her, but Molly had no siblings. Nor did her parents wish to have another child with a 1 in 4 chance of also inheriting the disorder, as Mendel's first law dictates. Technology offered another solution.

In late 1999, researchers at the Reproductive Genetics Institute at Illinois Medical Center mixed Jack's sperm with Lisa's oocytes in a laboratory dish. After allowing 15 of the fertilized ova to develop to the 8-cell stage, researchers separated and applied DNA probes to one cell from each embryo. A cell that had wild type Fanconi anemia alleles and that matched Molly's human

Figure 21.1   Special siblings.
Adam Nash was conceived and selected to save his sister Molly's life. But he is also a much-loved sibling and son. Several other families have followed the Nashes' example in conceiving one child to help another. Criticism has faded.

leukocyte antigen (HLA) type was identified and its 7-celled remainder implanted into Lisa's uterus. Adam was born in late summer, and a month later, physicians infused his umbilical cord stem cells into his sister, saving her life (**figure 21.1**).

Increased knowledge of how the genomes of two individuals come together and interact has spawned several novel ways to have children. **Assisted reproductive technologies** replace the source of a male or female gamete, aid fertilization, or provide a uterus. These procedures, developed to treat infertility, are becoming a part of genetic screening, a role that will likely grow with the influx of human genome information.

21.1 Infertility and Subfertility

Infertility is the inability to conceive a child after a year of frequent intercourse without the use of contraceptives. Some specialists use the term *subfertility* to distinguish those individuals and couples who can conceive unaided, but for whom this may take longer than usual. On a more personal level, infertility is a seemingly endless monthly cycle of raised hopes and crushing despair. In addition, as a woman ages, the incidence of pregnancy-related problems rises, including chromosomal anomalies, fetal deaths, premature births, and low-birthweight babies. For most conditions, the man's age does not raise the risk of pregnancy complications.

Physicians who specialize in infertility treatment can identify a physical cause in 90 percent of cases. Of these, 30 percent of the time the problem is primarily in the male, and 60 percent of the time it is primarily in the female. However, for cases in which a physical problem is not obvious, the cause is usually a mutation or chromosomal aberration that impairs fertility in the male. The statistics are somewhat unclear, because in 20 percent of the 90 percent, both partners have a medical condition that could contribute to infertility or subfertility. A common combination, for example, is a woman with an irregular menstrual cycle and a man with a low sperm count.

One in six couples has difficulty in conceiving or giving birth to children. **Table 21.1** summarizes causes of subfertility and infertility.

Male Infertility

Infertility in the male is easier to detect but sometimes harder to treat than female infertility. One in 25 men is infertile. Some men have difficulty fathering a child because they produce fewer than the average 120 million sperm cells per milliliter of ejaculate, a condition called oligospermia that has several causes. If a low sperm count is due to a hormonal imbalance, administering the appropriate hormones may boost sperm output. Sometimes a man's immune system produces IgA antibodies that cover the sperm and prevent them from binding to oocytes. Male infertility can also be due to a varicose vein in the scrotum. This enlarged vein produces too much heat near developing sperm, and they cannot mature. Surgery can remove a scrotal varicose vein.

Most cases of male infertility are genetic. About a third of infertile men have small deletions of the Y chromosome that remove the only copies of key genes whose products control spermatogenesis. Other genetic causes of male infertility include mutations in genes that encode androgen receptors or protein fertility hormones, or that regulate sperm development or motility.

For many men with low sperm counts, fatherhood is just a matter of time: They are subfertile, not infertile. If an ejaculate contains at least 60 million sperm cells, fertilization is likely eventually. To speed conception, a man with a low sperm count can donate several semen samples over a period of weeks at a fertility clinic. The samples are kept in cold storage, then pooled. Some of the seminal fluid is withdrawn to leave a sperm cell concentrate, which is then placed in the woman's reproductive tract. It isn't very romantic, but it is highly effective at achieving pregnancy.

Sperm quality is more important than quantity. Sperm cells that are unable to move—a common problem—or are shaped abnormally, cannot reach an oocyte. However, the genetic package of an immobile or abnormally shaped sperm cell can be injected into an oocyte (see **figures 21.2** and 21.4). If the cause of male infertility is hormonal, however, replacing the absent hormones can sometimes make sperm move.

Hampered sperm motility is also associated with the presence of white blood cells

Table 21.1

Causes of Subfertility and Infertility

Men

Problem	Possible Causes	Treatments
Low sperm count	Hormone imbalance, varicose vein in scrotum, possibly environmental pollutants	Hormone therapy, surgery, avoiding excessive heat
	Drugs (cocaine, marijuana, lead, arsenic, some steroids and antibiotics, chemotherapy)	
	Oxidative damage	
	Y chromosome gene deletions	
Immobile sperm	Abnormal sperm shape	Intracytoplasmic sperm injection
	Infection	Antibiotics
	Malfunctioning prostate	Hormones
	Deficient apoptosis	
Antibodies against sperm	Problem in immune system	

Women

Problem	Possible Causes	Treatment
Erratic ovulation	Pituitary or ovarian tumor	Surgery
	Underactive thyroid	Hormone therapy
	Polycystic ovary syndrome	Oral contraceptives
Antisperm secretions	Unknown	Acid or alkaline douche, estrogen therapy
Blocked uterine tubes	Infection caused by IUD, abortion, or by sexually transmitted disease	Laparotomy, oocyte removed from ovary and placed in uterus
Endometriosis	Unknown	Hormones, laparotomy, drugs

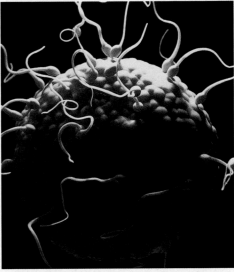

a.

b.

Figure 21.2 Sperm shape and motility are important. **(a)** Healthy sperm in action. **(b)** A misshapen sperm cannot fertilize an oocyte.

in semen. The blood cells produce reactive oxygen species, which are toxic molecules that bind sperm cell plasma membranes and destroy enzymes essential for the reactions within the sperm cell that generate ATP. With too little ATP to supply energy, sperm cannot move effectively. Fertility declines. Clinical trials are underway to test antioxidants to treat this form of male infertility.

Faulty apoptosis (programmed cell death) can also cause male infertility. Apoptosis selectively kills abnormally shaped sperm. Studies show that men with high percentages of abnormally shaped sperm often have cell surface molecules that indicate impaired apoptosis.

Female Infertility

Female infertility can be caused by abnormalities in any part of the reproductive system (**figure 21.3**). Many women with subfertility or infertility have irregular menstrual cycles, making it difficult to pinpoint when conception is most likely. In an average menstrual cycle of 28 days, ovulation usually occurs around the 14th day after menstruation begins, and this is when a woman is most likely to conceive.

For a woman with regular menstrual cycles who is under 30 years old and not using birth control, pregnancy typically happens within three or four months. A woman with irregular menstrual periods can use an ovulation predictor test, which

detects a peak in the level of leutinizing hormone that precedes ovulation by a few hours. Another way to detect the onset of ovulation is to record body temperature each morning using a digital thermometer with subdivisions of hundredths of a degree Fahrenheit, which can indicate the 0.4 to 0.6 rise in temperature that occurs when ovulation starts. This is when a woman is at her most fertile. She can then time intercourse for when she is most likely to conceive. Sperm can fertilize oocytes if they have been in the woman's body for up to five days before ovulation, but can fertilize for only a short time after ovulation.

The hormonal imbalance that usually underlies irregular ovulation has various

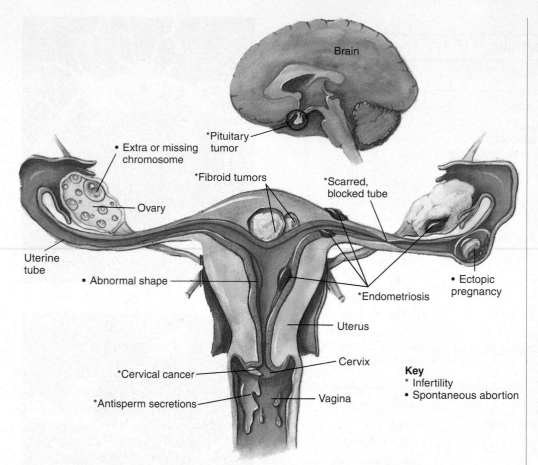

Figure 21.3   Sites of reproductive problems in the human female.

Labels on figure: Brain; *Pituitary tumor; • Extra or missing chromosome; *Fibroid tumors; *Scarred, blocked tube; Ovary; Uterine tube; • Abnormal shape; *Endometriosis; • Ectopic pregnancy; Uterus; *Cervical cancer; Cervix; *Antisperm secretions; Vagina; **Key** * Infertility • Spontaneous abortion

causes—a tumor in the ovary or in the pituitary gland in the brain that hormonally controls the reproductive system, an underactive thyroid gland, or use of steroid-based drugs such as cortisone. Sometimes a woman produces too much prolactin, the hormone that normally promotes milk production and suppresses ovulation in new mothers. If prolactin is abundant in a non-pregnant woman, she will not ovulate.

Fertility drugs can stimulate ovulation, but they can also cause women to "super-ovulate," producing more than one oocyte each month. A commonly used drug, clomiphene, raises the chance of having twins from 1 to 2 percent to 4 to 6 percent. If a woman's ovaries are completely inactive or absent (due to a birth defect or surgery), she can become pregnant only if she uses a donor oocyte.

A common cause of female infertility is blocked uterine tubes. Fertilization usually occurs in open tubes. Blockage can prevent sperm from reaching the oocyte, or entrap a fertilized ovum, keeping it from descending into the uterus. The embryo begins developing in the tube and if it is not removed and continues to enlarge, the tube can burst and the woman can die. This "tubal pregnancy" is called an ectopic pregnancy.

Uterine tubes can also be blocked due to a birth defect or, more likely, from an infection such as pelvic inflammatory disease. A woman may not know she has blocked uterine tubes until she has difficulty conceiving and medical tests uncover the problem. Surgery can sometimes open blocked uterine tubes.

Excess tissue growing in the uterine lining may make it inhospitable to an embryo. This tissue can include benign tumors, called fibroids, or areas of thickened lining that reflect a condition called endometriosis. In response to the hormonal cues to menstruate, the tissue bleeds, causing cramps. Endometriosis can make conception difficult, but curiously, once a woman with endometriosis has been pregnant, the cramps and bleeding usually subside.

Sometimes secretions in the vagina and cervix are hostile to sperm. If cervical mucus is unusually thick or sticky, as can happen during infection, sperm become entrapped and cannot move far enough to encounter an oocyte. Vaginal secretions may be so acidic or alkaline that they weaken or kill sperm. For example, some women with cystic fibrosis are unable to secrete bicarbonate from the cells lining their reproductive tracts, which is normally required to activate sperm. Douching daily with an acidic solution such as acetic acid (vinegar) or an alkaline solution, such as bicarbonate, can alter the pH of the vagina so that in some cases it is more receptive to sperm cells. Too little mucus is treated with low daily doses of oral estrogen. Sometimes mucus in a woman's body harbors antibodies that attack sperm. Infertility may also result if the oocyte fails to release sperm-attracting biochemicals.

One reason female infertility increases with age is that older women are more likely to produce oocytes with an abnormal chromosome number, which often causes spontaneous abortion. Losing very early embryos may appear to be infertility because the bleeding accompanying the aborted embryo resembles a heavy menstrual flow. The higher incidence of meiotic errors in older women may occur because their oocytes have been exposed longer to harmful chemicals, viruses, and radiation. In other cases, women are prone to producing oocytes with abnormal numbers of chromosomes. (See Applied Question 1 in chapter 13).

Infertility Tests

A number of medical tests can identify the cause or causes of infertility. The man is checked first, because it is easier, less costly, and certainly less painful to obtain sperm than oocytes.

Sperm are checked for number (sperm count), motility, and morphology (shape). An ejaculate containing up to 40 percent unusual forms is still considered normal, but many more than this can impair fertility. A urologist performs sperm tests. A genetic counselor may also be of help in identifying the cause of male infertility. He or she can interpret the results of a PCR analysis of the Y chromosome to detect deletions associated with lack of sperm. If a male cause of infertility is not apparent, the next step is for the woman to consult a gynecologist, who checks to see that the structures of the reproductive system are present and functioning.

Some cases of subfertility or infertility have no clear explanation. Psychological factors may be at play, or it may be that inability to conceive results from consistently poor timing. Sometimes a subfertile couple adopts a child, only to conceive one of their own shortly thereafter; many times, the couple's infertility remains a lifelong mystery.

Key Concepts

1. Male infertility is due to a low sperm count or sperm that cannot swim or are abnormal in structure.
2. Female infertility can be due to an irregular menstrual cycle or blocked uterine tubes. Fibroid tumors, endometriosis, or a misshapen uterus may prevent implantation of a fertilized ovum, and secretions in the vagina and cervix may inactivate or immobilize sperm. Oocytes may fail to release a sperm-attracting biochemical.
3. Early pregnancy loss due to abnormal chromosome number may be mistaken for infertility; this is more common among older women.
4. A variety of medical tests can pinpoint some causes of infertility.

21.2 Assisted Reproductive Technologies

A growing number of couples with fertility problems are turning to alternative ways to achieve pregnancy, many of which were perfected in nonhuman animals (see the Technology Timeline). In 2001 in the United States, the most recent year for which data have been analyzed, 1 percent of the approximately 4 million births in the United States were from ARTs (assisted reproductive technologies). The ART births account for 0.4 percent of single births and 16 percent of multiples, reflecting the fact that usually more than one fertilized ovum is implanted.

This section describes variations on the ART theme. The different procedures can be performed on material from the parents-to-be ("nondonor") or from donors, and may be "fresh" (collected just prior to the procedure) or "frozen" (preserved in liquid nitrogen). **Table 21.2** compares the success rates of the different ARTS for several hundred fertility clinics in the United States.

Donated Sperm— Intrauterine Insemination

The oldest assisted reproductive technology is **intrauterine insemination,** in which a doctor places donated sperm into a woman's reproductive tract, typically the cervix or uterus. The sperm are first washed free of seminal fluid, which can inflame female tissues. Her partner may be infertile or carry a gene for an inherited illness that the couple wishes to avoid passing to their child, or a woman may undergo intrauterine insemination if she desires to be a single parent without having sex.

The first intrauterine insemination in humans was done in 1790. For many years, physicians donated sperm, and this became a way for male medical students to earn a few extra dollars. By 1953, sperm could be frozen and stored and intrauterine insemination became much more commonplace. Today, donated sperm are frozen and stored in sperm banks, which provide the cells to obstetricians who perform the procedure. Intrauterine insemination costs about $125. However, if ovulation is induced to increase the chances, additional costs may exceed $3,000.

A couple who chooses intrauterine insemination can select sperm from a catalog that lists the personal characteristics of donors, such as blood type, hair and eye color, skin color, build, and even educational level and interests. Of course, not all of these traits are inherited. If a couple desires a child of one sex—such as a daughter to avoid passing on an X-linked disorder— sperm can be separated into fractions enriched for X-bearing or Y-bearing sperm.

Problems can arise in intrauterine insemination, as they can in any pregnancy.

Table 21.2

Comparison of Assisted Reproductive Technologies

Number of:	Technologies						
	Embryo transfer to host (surrogate)	IVF (fresh, nondonor)	IVF (frozen, nondonor)	IVF (fresh, donor)	IVF (frozen, donor)	GIFT	ZIFT
Treatments	1,210	73,406	13,083	7,581	2,721	549	763
Retrievals	1,114	62,881	NA	6,929	NA	489	683
Transfers	1,066	59,004	11,394	6,684	2,395	477	604
Pregnancies	459	22,567	2,906	3,413	704	162	242
Deliveries	382	18,793	2,324	2,920	563	121	204

Retrieval = removing oocyte or fertilized ovum from woman

NA = not applicable (frozen embryos are not retrieved)

(Data from a large sample of fertility clinics)

Landmarks in Reproductive Technology

	In Nonhuman Animals	In Humans
1782	Intrauterine insemination in dogs	
1790		Pregnancy reported from intrauterine insemination
1890s	Birth from embryo transplantation in rabbits	Intrauterine insemination by donor
1949	Cryoprotectant successfully freezes animal sperm	
1951	First calf born after embryo transplantation	
1952	Live calf born after insemination with frozen sperm	
1953		First reported pregnancy after insemination with frozen sperm
1959	Live rabbit offspring produced from *in vitro* ("test tube") fertilization (IVF)	
1972	Live offspring from frozen mouse embryos	
1976		First reported commercial surrogate motherhood arrangement in the United States
1978	Transplantation of ovaries from one cow to another	Baby born after IVF in United Kingdom
1980		Baby born after IVF in Australia
1981	Calf born after IVF	Baby born after IVF in United States
1982	Sexing of embryos in rabbits Cattle embryos split to produce genetically identical twins	
1983		Embryo transfer after uterine lavage
1984		Baby born in Australia from frozen and thawed embryo
1985		Baby born after gamete intrafallopian transfer (GIFT) First reported gestational surrogacy arrangement in the United States
1986		Baby born in the United States from frozen and thawed embryo
1989		First preimplantation genetic diagnosis (PGD)
1992		First pregnancies from sperm injected into oocytes
1994	Intracytoplasmic sperm injection (ICSI) in mouse and rabbit	62-year-old woman gives birth from donated oocyte
1995	Sheep cloned from embryo cell nuclei	Babies born following ICSI
1996	Sheep cloned from adult cell nucleus	
1998	Mice cloned from adult cell nuclei	Baby born 7 years after his twin
1999	Cattle cloned from adult cell nuclei	
2000	Pigs cloned from adult cell nuclei	
2001		Sibling born following PGD to treat sister for genetic disease Human preimplantation embryo cloned, survives to 6 cells
2003		3000$^+$ preimplantation genetic diagnoses performed to date
2004	Woman pays $50,000 to have her cat cloned	
2005	Dog cloned	Korean researchers perform somatic cell nuclear transfer to derive human embryonic stem cells for people with spinal cord injury or disease

For example, a man who donated sperm years ago developed a late-onset genetic disease, cerebellar ataxia (OMIM 608029). Eighteen children conceived using his sperm now face a 1 in 2 risk of having inherited the mutant gene. In 1983, the Sperm Bank of California became the first to ask donors if they wished to be contacted by their children years later. In 2002, the first such meeting occurred, evidently quite successfully.

A male's role in reproductive technologies is simpler than a woman's. A man can be a genetic parent, contributing half of his genetic self in his sperm, but a woman can be both a genetic parent (donating an oocyte) and a gestational parent (donating the uterus). Problems can crop up when a second female assists in conception and/or gestation.

A Donated Uterus— Surrogate Motherhood

If a man produces healthy sperm but his partner's uterus is absent or cannot maintain a pregnancy, a surrogate mother may help by being inseminated with the man's sperm. When the child is born, the surrogate mother gives the baby to the couple. In this variation of the technology, the surrogate is both the genetic and the gestational mother. Attorneys usually arrange surrogate relationships. The surrogate mother signs a statement signifying her intent to give up the baby, and she is paid for her nine-month job.

The problem with surrogate motherhood is that a woman may not be able to predict her responses to pregnancy and childbirth in the cold setting of a lawyer's office. When a surrogate mother changes her mind about giving up the baby, the results are wrenching for all. A prominent early case involved Mary Beth Whitehead, who carried the child of a married man for a fee and then changed her mind about relinquishing the baby. Whitehead's ties to "Baby M" were perhaps stronger because she was both the genetic and the gestational mother.

Another type of surrogate mother lends only her uterus, receiving a fertilized ovum conceived from a man and a woman who has healthy ovaries but lacks a functional uterus. This variation is called "embryo transfer to a host uterus." The gestational-only surrogate mother turns the child over to the donors of the genetic material.

In Vitro Fertilization

In *in vitro* **fertilization** (IVF), which means "fertilization in glass," sperm and oocyte join in a laboratory dish. Then the embryo is placed in the oocyte donor's uterus (or another woman's uterus), and, if all goes well, implants into the uterine lining.

Louise Joy Brown, the first "test-tube baby," was born in 1978. Initial media attention was great, with cartoons depicting a newborn with the word "Pyrex," a test-tube and glassware manufacturer, branded on her thigh. A prominent bioethicist said that IVF challenged "the idea of humanness and of our human life and the meaning of our embodiment and our relation to ancestors and descendants." Yet Louise is, despite her unusual beginnings, a rather ordinary young woman. The technology has since led to the births of more than a million children.

A woman might undergo IVF if her ovaries and uterus work but her uterine tubes are blocked. Using a laparoscope, a physician removes several of the largest oocytes and transfers them to a dish. Just before ovulation, 15 to 20 or so oocytes enlarge, and can be cultured in a dish. If left in the body, only one oocyte would pop out of the ovary, but in a culture, many of them can mature sufficiently to be fertilized *in vitro*. In the past, women took drugs to make them "superovulate" and produce many mature oocytes at once, but this is being phased out. Chemicals that mimic those in the female reproductive tract are added to the culture, and sperm are applied to the oocytes.

If the sperm cannot readily penetrate the oocyte, they may be sucked up into a tiny syringe and microinjected into the female cell. This is called **intracytoplasmic sperm injection** (ICSI), and it is more effective than IVF alone (**figure 21.4**). ICSI is very helpful for men who have low sperm counts or high percentages of abnormal sperm. The procedure even works with immature sperm, making fatherhood possible for men who cannot ejaculate, such as those who have suffered spinal cord injuries. ICSI has been very successful, performed on thousands of men with about a 30 percent success rate. The Bioethics: Choices for the Future on page 424 considers an unexpected problem with ICSI—transmitting infertility.

A day or so after sperm wash over the oocytes in the dish, or are injected into them, one or two of the embryos—balls of 8 or 16 cells—are transferred to the woman's uterus. If the hormone human chorionic gonadotropin appears in her blood a few days later, and its level rises, she is pregnant.

IVF costs from $6,500 to $15,000 per attempt, and the success rate is 29 percent. (The cover of a celebrity magazine featured a well-known actress who had finally had a

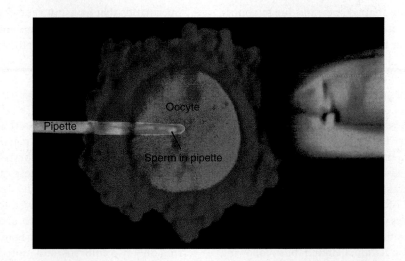

Figure 21.4 ICSI. Intracytoplasmic sperm injection (ICSI) enables some infertile men, men with spinal cord injuries, or men with certain illnesses to become fathers. A single sperm cell is injected into the cytoplasm of an oocyte.

Technology Too Soon? The Case of ICSI

Intracytoplasmic sperm injection (ICSI), available since 1995, has been extremely successful in enabling men with AIDS, paralysis, very low sperm counts, or abnormal sperm to become fathers. The birth defect rate is the same as for normal conceptions, although abnormal embryos and sex chromosome aberrations are slightly more common. But as more ICSI procedures are performed and tests on nonhuman animals continue, potential problems are emerging, based on the fact that ICSI bypasses what one researcher calls "natural sperm selection barriers."

ICSI is now commonly used on men who have azoospermia—lack of sperm—or oligospermia—very few sperm. The rare sperm are sampled from an ejaculate or taken from the testes with a needle, then injected into an oocyte. Sometimes these men produce spermatids but not mature sperm, and spermatids that have already elongated can also successfully fertilize oocytes using ICSI. However, a complication has developed that involves not science, but logic. About 10 percent of infertile men have microdeletions in the Y chromosome. When their sperm cells are used in ICSI, they pass on the infertility to their sons. This is true for other causes of infertility and subfertility, too. For example, if a man's above-average proportion of abnormally shaped sperm is due to abnormal apoptosis, he could pass a susceptibility to cancer to his offspring.

Bioethicists are debating whether it is right to intentionally conceive a male who is genetically destined to be infertile. On the positive side is the opportunity ICSI is providing to study that infertility. It is possible that adolescents with microdeleted Y chromosomes can produce viable sperm or spermatids, and if so, these cells can be sampled and stored for later use. In the past, these men would not have suspected that they had Y chromosome abnormalities until they had difficulty fathering children. Alternatives to transmitting deletions and other mutations include selecting and using only X-bearing sperm in ICSI, or selecting and implanting only female fertilized ova. Because of the transmission of Y-linked infertility with ICSI, men undergoing fertility testing and considering the procedure now have their Y chromosomes screened for deletions, and genetic counseling is provided.

Meanwhile, experiments on rhesus monkeys are pinpointing the sources of damage to those ICSI embryos that do not continue to develop:

- Injecting sperm at the site of the polar body on the oocyte can disrupt the meiotic spindle, leading to nondisjunction (an extra or missing chromosome).

- Injected sperm DNA does not always condense properly, also leading to nondisjunction.

- Spermatids that have not yet elongated are often unable to fertilize an oocyte. Culturing them in the laboratory until

they mature may help improve the odds of success.

- Spermatids may not be completely imprinted, leading to problems in gene expression.

- Mitotic cell cycle checkpoints are altered at the first division following ICSI.

- Injected sperm sometimes lack a protein that normally associates with the sex chromosomes. This may explain the elevation in sex chromosome anomalies.

- Injected sperm can include surface proteins normally left outside the oocyte, producing unanticipated effects. They may also include mitochondria from the male.

Despite these largely theoretical concerns, parents of children conceived with ICSI do not have any cause to worry, researchers insist. Tests on these children so far have not revealed any problems. Says one researcher, "In spite of its potential risks, ICSI still seems to be remarkably safe." Still, it would be comforting to some to have more research to back up this new way to start development. Ongoing studies are following the children of ICSI for longer times, and investigating any correlations between health problems in offspring and the cause of subfertility or infertility in the fathers.

---

baby after seven IVF attempts—a cost that clearly very few people can afford.) By contrast, two-thirds of embryos conceived through sexual intercourse implant. The lower success rate of IVF is due both to the difficulty of the procedure and to the fact that couples who choose it have a higher incidence of subfertility or infertility than the general population. Children born following IVF have twice the rate of birth defects (about 9 percent) compared to children con-

ceived naturally, which may also reflect the underlying medical problems of parents.

In the past, several embryos were implanted to increase the success rate, but this led to many multiple births. In many cases, embryos had to be removed to make room for others to survive. To avoid the multiples problem, clinicians began transferring only two embryos, and more recently, just one, after a study showed very similar birth rates for one or two.

Measures to improve the chance that IVF will culminate in a birth include:

1. Transferring embryos slightly later in development, at the blastocyst stage.

2. Culturing fertilized ova and early embryos with other cells that normally surround the oocyte in the ovary. These "helper" cells provide extra growth factors.

3. Screening early embryos for chromosome abnormalities, and implanting only those with apparently normal karyotypes.

Embryos resulting from IVF that are not soon implanted in the woman are frozen in liquid nitrogen, with cryoprotectant chemicals added to prevent salts from building up or ice crystals from damaging delicate cell parts. Freezing takes a few hours; thawing about a half hour. The longest an embryo has been frozen, stored, and then successfully revived is 13 years; the "oldest" pregnancy using a frozen embryo occurred 9 years after the freezing!

Gamete and Zygote Intrafallopian Transfer

IVF may fail because of the artificial environment for fertilization. In the late 1970s, a procedure called GIFT, which stands for **gamete intrafallopian transfer,** improves the setting. (Uterine tubes are also called fallopian tubes). Fertilization is assisted in GIFT, but it occurs in the woman's body rather than in glassware.

In GIFT, a woman has several of her largest oocytes removed. The man submits a sperm sample, and the most active cells are separated from it. The collected oocytes and sperm are deposited together in the woman's uterine tube, at a site past any obstruction that might otherwise block fertilization. GIFT is about 27 percent successful, and usually costs less than IVF.

A variation of GIFT is ZIFT, which stands for **zygote intrafallopian transfer.** In this procedure, an IVF ovum is introduced into the woman's uterine tube. Allowing the fertilized ovum to make its own way to the uterus seems to increase the chance that it will implant. ZIFT is 29 percent successful.

GIFT and ZIFT are done much less frequently than IVF. They often will not work for women who have scarred uterine tubes.

Oocyte Banking and Donation

Oocytes can be stored, as sperm are, but the procedure may introduce problems. An oocyte contains a large volume of water.

Freezing can cause ice crystals to form that can damage cell parts.

Candidates for preserving oocytes for later use include:

- Women wishing to have children later in life

- Women undergoing chemotherapy or other toxic or teratogenic treatments

- Women who work with toxins or teratogens

Oocytes are frozen in liquid nitrogen at −30 to −40 degrees Celsius, when they are at metaphase of the second meiotic division. At this time, the chromosomes are aligned along the spindle, which is sensitive to temperature extremes. If the spindle comes apart as the cell freezes, the oocyte may lose a chromosome, which would devastate development. Another problem with freezing oocytes is retention of a polar body, leading to a diploid oocyte. More than 100 babies have been born using frozen oocytes.

To avoid the difficulty of freezing oocytes, strips of ovarian tissue can be frozen, stored, thawed, and reimplanted at various sites, such as under the skin of the forearm or abdomen or in the pelvic cavity near the ovaries. The first child resulting from fertilization of an oocyte from reimplanted ovarian tissue was born in 2004. The mother, age 25, had been diagnosed with advanced Hodgkin's lymphoma in 1997. The harsh chemotherapy and radiation cured her cancer, but also destroyed her ovaries. Five strips of tissue from her left ovary were frozen and in 2003 several pieces of ovarian tissue were thawed and implanted in a pocket that surgeons crafted on one of her shriveled ovaries, very near the entrance to a uterine tube. Menstrual cycles resumed, and shortly thereafter, the woman became pregnant with her daughter, who is healthy. Freezing ovarian tissue is likely to become routine for cancer patients of childbearing age.

Women can also obtain oocytes from donors, typically younger women. Often these women are undergoing IVF and have "extra" harvested oocytes. The potential father's sperm and donor's oocytes are placed in the recipient's uterus or uterine tube, or fertilization occurs in the laboratory, and an 8- or 16-celled embryo is transferred to the woman's uterus.

The first baby to result from oocyte donation was born in 1984. The success rate ranges from 20 to 50 percent, and the procedure costs at least $10,000. The technique is useful for the reasons cited for freezing oocytes, as well as to avoid transmitting a disease-causing gene.

Embryo adoption is a variation on oocyte donation. A woman with malfunctioning ovaries but a healthy uterus carries an embryo that results when her partner's sperm is used in intrauterine insemination of a woman who produces healthy oocytes. If the woman conceives, the embryo is gently flushed out of her uterus a week later and inserted through the cervix and into the uterus of the woman with malfunctioning ovaries. The child is genetically that of the man and the woman who carries it for the first week, but is born from the woman who cannot produce healthy oocytes. "Embryo adoption" is also the term used to describe use of IVF "leftovers."

In another technology, cytoplasmic donation, older women have their oocytes injected with cytoplasm from the oocytes of younger women to "rejuvenate" the cells. Although resulting children conceived through IVF appear to be healthy, they are being monitored for a potential problem— heteroplasmy, or two sources of mitochondria in one cell. Researchers do not yet know the health consequences of having mitochondria from the donor cytoplasm plus mitochondria from the recipient's oocyte. These conceptions also have an elevated incidence of XO syndrome, which often causes spontaneous abortion. After the birth of the first child from this technique in 2001, the media made much of the unnatural situation of having three parents—the father, the mother, and the ooplasm donor. The U.S. government has banned this procedure unless it is part of a clinical trial.

Because oocytes are harder to obtain than sperm, oocyte donation technology has lagged behind that of sperm banks, but is catching up. One IVF facility that has run a donor oocyte program since 1988 has a patient brochure that describes 120 oocyte donors of various ethnic backgrounds, like a catalog of sperm donors. The oocyte

Preimplantation Genetic Diagnosis

Prenatal diagnostic tests such as amniocentesis, chorionic villus sampling, and fetal cell sorting can be used in pregnancies achieved with assisted reproductive technologies. A test called **preimplantation genetic diagnosis** (PGD) detects genetic and chromosomal abnormalities *before* pregnancy starts. The couple selects a very early embryo—termed a "preimplantation embryo" because it would not normally have yet arrived at the uterus for implantation—that has not inherited a specific detectable genetic condition. This was the technology used to select Adam Nash, whose umbilical cord stem cells cured his sister's Fanconi anemia (see figure 21.1). PGD has about a 29 percent success rate.

PGD is possible because one cell, or blastomere, can be removed for testing from an 8-celled embryo, and the remaining 7 cells can complete development normally in a uterus. Before the embryo is implanted into the woman, the single cell is karyotyped, or its DNA amplified and probed for particular genes that the parents carry. Healthy embryos are selected. At first, researchers implanted the remaining seven cells, but they have found that letting the embryo continue developing in the laboratory until day 5, when it is 80 to 120 cells, is more successful. Obtaining the cell to be tested is called "blastomere biopsy" (**figure 21.5**). Accuracy is about 97 percent. Errors are generally due to mosaics—when a somatic mutation occurs in a blastomere—or during amplification of blastomere DNA.

Although PGD is often reported as "news," the first children who had the procedure were born in 1989. In these first cases, probes for Y chromosome-specific DNA sequences were used to select females, who could not inherit X-linked conditions their mothers carried, such as Lesch-Nyhan syndrome (OMIM 300322; profound mental retardation with self-mutilative behavior) and adrenoleukodystrophy (OMIM

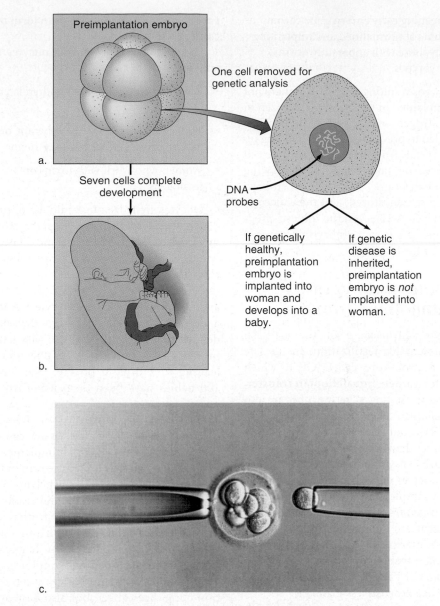

Figure 21.5   Preimplantation genetic diagnosis.   A blastomere biopsy provides material for assessing the health of an embryo. Preimplantation genetic diagnosis probes disease-causing genes or chromosome aberrations in an 8-celled preimplantation embryo. **(a)** A single cell is separated and tested to see if it contains a disease-causing gene combination or chromosome imbalance. **(b)** If it doesn't, the remaining seven cells, or the embryo after a few more cell divisions, is transferred to the oocyte donor to complete development. **(c)** This preimplantation embryo is held still by suction applied on the left. On the right, a pipette draws up a single blastomere. *In vitro* fertilization took place 45 hours previously.

300100), in which seizures and nervous system deterioration end in sudden death in early childhood. The alternative to PGD would have been to face the 25 percent chance of conceiving an affected male.

In March 1992, the first child was born who underwent PGD to avoid a specific inherited disease. Chloe O'Brien was checked as an 8-celled preimplantation embryo to see if she had escaped the cystic fibrosis that affected her brother. Since then, PGD has helped to select thousands of children free of several dozen types of inherited illnesses. It has been used for the better-known single-gene disorders as well as for many rare ones.

Today, preimplantation genetic diagnosis is increasingly being used to screen fertilized ova and early embryos derived from IVF for chromosome abnormalities before implanting them into women, which lowers the risk that the implanted embryos will be spontaneously aborted due to chromosomal abnormalities. PGD with IVF is primarily used for older women who have suffered several spontaneous abortions, because they are at higher risk of having oocytes with abnormal numbers of chromosomes. One woman who had undergone several failed IVF attempts had five preimplantation embryos karyotyped. Four were abnormal, but with different chromosomes affected. The fifth embryo became the couple's daughter. Obviously, one parent had very defective meiosis.

Like many technologies, preimplantation genetic diagnosis can introduce a bioethical "slippery slope" when it is used for reasons other than ensuring that a child is free of a certain disease, such as for gender selection. A couple with five sons might, for example, use PGD to select a daughter. But this use of technology might just be a new expression of age-old human nature, according to one physician who performs PGD. "From the dawn of time, people have tried to control the sex of offspring, whether that means making love with one partner wearing army boots, or using a fluorescence-activated cell sorter to separate X- and Y-bearing sperm. PGD represents a quantum leap in that ability—all you have to do is read the X and Y chromosome paints," he says.

While PGD used solely for family planning is certainly more civilized than placing baby girls outside the gates of ancient cities to perish, the American Society for Reproductive Medicine endorses the use of PGD for sex selection only to avoid passing on an X-linked disease. Yet even PGD to avoid disease can be controversial. A woman with early-onset familial Alzheimer disease used PGD to select a daughter free of the dominant mutant gene—but that child will have to experience life with her mother's illness.

Table 21.3 summarizes the assisted reproductive technologies.

RELATING THE CONCEPTS

Which ART did Gaby Vernoff undergo?

Key Concepts

1. In intrauterine insemination, donor sperm are placed in a woman's reproductive tract.
2. A genetic and gestational surrogate mother is intrauterinally inseminated, becomes pregnant, then gives the baby to the father and his partner. A gestational surrogate mother gestates a baby conceived *in vitro* with gametes from a man and a woman who cannot carry a fetus.
3. In IVF, sperm and oocyte unite outside the body, and the resulting embryo is transferred to the uterus. Early embryos can also be frozen and used later.
4. In GIFT, sperm and oocytes are placed in a uterine tube at a site past a blockage.
5. In ZIFT, an IVF embryo is placed in a uterine tube.
6. In embryo adoption, a woman who has undergone intrauterine insemination has an early embryo washed out of her uterus and transferred to a woman who lacks oocytes.
7. PGD removes cells from early embryos and screens them for genetic or chromosomal abnormalities.

Table 21.3

Some Assisted Reproductive Technologies

Technology	Procedures	Success/Cycle	Cost/Cycle
GIFT	Deposits collected oocytes and sperm in uterine tube.	27%	$8,000–$10,000
IVF	Mixes sperm and oocytes in a laboratory dish, with chemicals to simulate intrauterine environment to encourage fertilization.	29%	$6,500–$15,000
Intrauterine insemination	Places or injects washed sperm into the cervix or uterus.	5–25%	$125
ICSI	Injects immature or rare sperm into oocyte, before IVF.	28%	$10,000–$17,000
Ovulation induction	Drugs control timing of ovulation in order to perform a particular procedure.	28%	$3,000+
PGD	Searches for specific mutant allele in sampled cell of 8-celled embryo. Its absence indicates remaining 7-celled embryo can be nurtured and implanted in woman, and child will be free of genetic condition.	29%	$8,000–$15,000
Surrogate mother	Woman carries a pregnancy for a woman who cannot become or stay pregnant.		$10,000+
ZIFT	Places IVF ovum in uterine tube.	29%	$8,000–$13,000

21.3 Extra Embryos

Although the overall success rates of assisted reproductive technologies are not spectacular in terms of live births, at the early stages, ironically, they sometimes work too well. This leaves "extra" oocytes, fertilized ova, or very early embryos. Because human prenatal development cannot complete outside of a uterus, decisions must be made as to the fate of these biological materials. Clients of fertility facilities can either allow their oocytes, fertilized ova, or embryos to be stored indefinitely or discarded; donate them to other infertile couples; or donate them for use in research.

In the United States, nearly half a million embryos derived from IVF sit in freezers; some have been there for years. Most couples who donate embryos do so anonymously, with no intention of learning how their genetic offspring are raised. Scott and Glenda Lyons chose a different path when they learned that their attempt at IVF had yielded too many embryos.

In 2001, two of Glenda's 18 embryos were transferred to her uterus, and developed into twins Samantha and Mitchell. Through a website where couples chat about fertility issues, Scott and Glenda met and selected Bruce and Susan Lindeman to receive 14 remaining embryos. This second couple had tried IVF three times, with no luck. The Lyons' frozen embryos were shipped cross-country to a clinic where two were implanted in Susan's uterus. In July 2003, Chase and Jack Lindeman were born—genetic siblings of Samantha and Mitchell Lyons. But there were still embryos left. The Lyons gave permission for the Lindemans to send twelve embryos to a third couple, who used two to have twin daughters in August 2004. They are biological siblings of Samantha and Mitchell Lyons and Chase and Jack Lindeman (**figure 21.6**).

Donating fertilized ova and embryos for use in research is another alternative to disposing of them. The U.S. federal government does not fund research that uses human embryos, but certain state governments and private organizations do support such work. The results of these experiments sometimes challenge long-held ideas, indicating that we still have much to learn about early human prenatal development. This was the case for a study from Royal Victoria

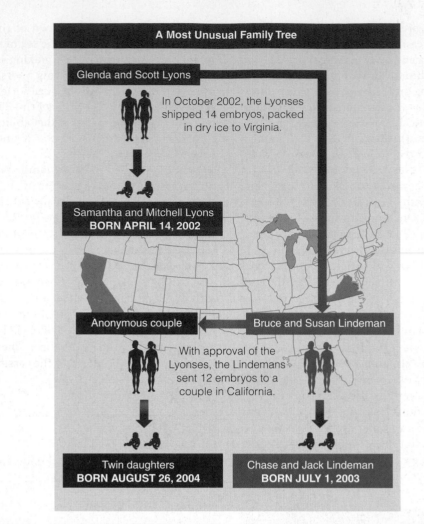

Figure 21.6 Using extra embryos. Six children resulted from Glenda and Scott Lyons' embryos. **(a)** The Lyons had a boy and a girl, then donated embryos to the Lindemans, who had twin boys. Finally, a couple in California used the Lyons' embryos to have twin daughters. **(b)** The Lyons and Lindemans have become friends.

Hospital at McGill University in Montreal. Researchers examined the chromosomes of sperm from a man with Klinefelter (XXY) syndrome. Many of the sperm would be expected to have an extra X chromosome, due to nondisjunction (see figure 13.12), which could lead to a preponderance of XXX and XXY offspring. Surprisingly, only 3.9 percent of the man's *sperm* had extra chromosomes, but five out of ten of his spare *embryos* had an abnormal X, Y, or chromosome 18. That is, even though most of the man's sperm were normal, his embryos weren't. The source of reproductive problems in Klinefelter disease, therefore, might not be in the sperm, but in early embryos—a finding that was previously unknown and not expected.

In another study, Australian researchers followed the fates of single blastomeres that had too many or too few chromosomes. They wanted to see whether the abnormal cells preferentially ended up in the inner cell mass, which develops into the embryo, or the trophectoderm, which becomes extra-embryonic membranes. The study showed that cells with extra or missing chromosomes become part of the inner cell mass much more frequently than one would expect by chance. This finding indicates that the ability of a blastomere sampled for PGD to predict health may depend on whether it is fated to be part of the inner cell mass.

Not everyone agrees that fertilized ova or embryos designated for discard should be used in research. With regulations not extending to privately funded research, ethically questionable experiments can happen. For example, researchers reported at a conference that they had mixed human cells from male embryos with cells from female embryos, to see if the normal male cells could "save" the female cells with a mutation. Sex was chosen as a marker because the Y chromosome is easy to detect. But the idea of human embryos with mixed sex parts caused a public outcry.

IVF is currently the major source of fertilized ova and early embryos for research, but this supply will diminish if yet another artificial reproductive technology becomes commonplace. **Polar body biopsy** is based on Mendel's first law, the segregation of alleles. In the technique, if a polar body resulting from the first meiotic division in a woman who is a carrier of an X-linked disorder has the mutant allele, then it is inferred that the oocyte to which it clings lacks that allele. Oocytes that pass this test can then be fertilized *in vitro* and the resulting embryo implanted in the woman. Polar body biopsy is possible because the polar body is attached to the much larger oocyte. A large pipette is used to hold the two cells in place, and a smaller pipette is used to separate the polar body. Then, DNA probes and FISH are used to look at genes and chromosomes in the polar body and infer the genotype of the oocyte (**figure 21.7**). Polar body biopsy followed by PGD is quite effective in avoiding conceptions with chromosome abnormalities or single-gene disorders.

ARTs introduce several bioethical issues. **Table 21.4** lists some complicated cases. Human genome information is providing more traits to track and perhaps control in coming generations. Already preimplantation genetic diagnosis is being used to screen out embryos destined to face a high risk of developing a particular disorder later in life, such as Huntington disease or cancer. It is easy to envision, instead of a quick Y chromosome FISH test, or a probe for a single gene that is mutant in a particular family, whole human genome DNA microarrays widely applied to gametes, fertilized ova, or preimplantation embryos. Who will decide which traits are worth living with, and which aren't?

ARTs operate on molecules and cells, but with repercussions for individuals and families. Ultimately, if it becomes widespread enough, it will affect the gene pool. Let us hope that regulations will evolve along with the technologies to assure that they are applied sensibly and humanely.

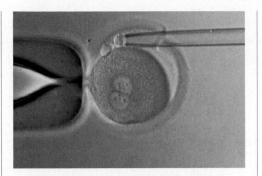

Figure 21.7 Polar body biopsy.
The fact that an oocyte shares a woman's divided genetic material with a much smaller companion, the polar body, allows physicians to screen oocytes for use in IVF. If a woman is a known carrier of a genetic disorder and the polar body contains the disease-causing gene variant, it is inferred that the oocyte has received the wild type allele. A polar body biopsy is possible because the polar body remains attached to the oocyte. In the procedure, a large pipette holds the oocyte still, and then a smaller pipette draws up the attached polar body. Genetic tests are then performed on the polar body.

Key Concepts

1. IVF produces extra fertilized ova and early embryos that may be used, frozen, donated, or discarded.
2. Used in basic research, such embryos are adding to our knowledge of early human development.
3. Polar body biopsy enables physicians to identify and select out defective oocytes.

Table 21.4
Assisted Reproductive Disasters

1. A physician in California used his own sperm to perform intrauterine insemination on 15 patients, telling them that he had used sperm from anonymous donors.

2. A plane crash killed the wealthy parents of two early embryos stored at –320°F (–195°C) in a hospital in Melbourne, Australia. Adult children of the couple were asked to share their estate with two 8-celled siblings-to-be.

3. Several couples in Chicago planning to marry discovered that they were half-siblings. Their mothers had been inseminated with sperm from the same donor.

4. Two Rhode Island couples sued a fertility clinic for misplacing several embryos.

5. Several couples in California sued a fertility clinic for implanting their oocytes or embryos in other women without donor consent. One woman requested partial custody of the resulting children if her oocytes were taken, and full custody if her embryos were used, even though the children were of school age and she had never met them.

6. A man sued his ex-wife for possession of their frozen fertilized ova. He won, and donated them for research. She had wanted to be pregnant.

Summary

21.1 Infertility and Subfertility

1. **Infertility** is the inability to conceive a child after a year of unprotected intercourse. Subfertility refers to individuals or couples who manufacture gametes, but may take longer than usual to conceive. **Assisted reproductive technologies** replace what is missing in reproduction.

2. Causes of infertility in the male include low sperm count, a malfunctioning immune system, a varicose vein in the scrotum, structural sperm defects, drug exposure, and abnormal hormone levels. In cases not associated with an obvious physical problem, a mutation is often the cause of the subfertility or infertility.

3. Female infertility can be caused by absent or irregular ovulation, blocked uterine tubes, an inhospitable or malshaped uterus, antisperm secretions, or lack of sperm-attracting biochemicals. Early pregnancy loss due to abnormal chromosome number is more common in older women and may appear to be infertility.

21.2 Assisted Reproductive Technologies

4. In **intrauterine insemination,** sperm are obtained from a donor and introduced into a woman's reproductive tract in a clinical setting.

5. A gestational and genetic surrogate mother provides her oocyte. Then intrauterine insemination is performed with sperm from a man whose partner cannot conceive or carry a fetus. The surrogate also provides her uterus for nine months. A gestational surrogate mother receives an *in vitro* fertilized ovum that belongs genetically to the couple who ask her to carry it.

6. In IVF, oocytes and sperm meet in a dish, fertilized ova divide a few times, and the resulting embryos are placed in the woman's body, circumventing blocked tubes or the inability of the sperm to penetrate the oocyte. **Intracytoplasmic sperm injection** introduces immature or nonmotile sperm into oocytes.

7. Embryos can be frozen and thawed and then complete development when placed in a woman's uterus.

8. **GIFT** introduces oocytes and sperm into a uterine tube past a blockage; fertilization occurs in the woman's body. **ZIFT** places an early embryo in a uterine tube.

9. Oocytes can be frozen and stored. In embryo adoption, a woman undergoes intrauterine insemination. A week later, the embryo is washed out of her uterus and introduced into the reproductive tract of the woman whose partner donated the sperm.

10. Seven-celled embryos can develop normally if a blastomere is removed at the 8-cell stage and cleared for abnormal chromosomes or genes. This is **preimplantation genetic diagnosis.**

21.3 Extra Embryos

11. Extra fertilized ova and early embryos generated in IVF are used, donated to couples, stored, donated for research, or discarded. They enable researchers to study aspects of early human development that they could not investigate in other ways.

12. **Polar body biopsy** enables physicians to perform genetic tests on polar bodies and to infer the genotype of the accompanying oocyte.

Review Questions

1. Which assisted reproductive technologies might help the following couples? (More than one answer may fit some situations.)

    a. A woman who is born without a uterus, but who manufactures healthy oocytes.

    b. A man whose cancer treatments greatly damage his sperm.

    c. A woman who undergoes a genetic test that reveals she will develop Huntington disease. She wants to have a child, but she does not want to pass on this presently untreatable illness.

    d. Two women who wish to have and raise a child together.

    e. A man and woman are each carriers of sickle cell disease. They do not want to have an affected child, but they also do not want to terminate a pregnancy.

    f. A woman's uterine tubes are scarred and blocked, so an oocyte cannot reach the uterus.

    g. A young woman must undergo abdominal radiation to treat ovarian cancer, but wishes to have a child in the future.

2. Why are men typically tested for infertility before women?

3. A man reads his medical chart and discovers that the results of his sperm analysis indicate that 22 percent of his sperm are shaped abnormally. He wonders why the physician said he had normal fertility if so many sperm are abnormally shaped. Has the doctor made an error?

4. Cite a situation in which both man and woman contribute to subfertility.

5. How does ZIFT differ from GIFT? How does it differ from IVF?

6. A Tennessee lower court, in ruling on the fate of seven frozen embryos in a divorce case, called them "children *in vitro.*" In what sense is this label incorrect?

7. Explain how preimplantation genetic diagnosis is similar to and different from CVS and amniocentesis.

8. What are some of the causes of infertility among older women?

9. How do each of the following assisted reproductive technologies deviate from the normal biological process?

    a. *in vitro* fertilization

    b. GIFT

    c. embryo adoption

    d. gestational surrogacy

    e. intrauterine insemination

    f. cytoplasmic donation

10. According to table 21.2, which ART has the highest success rate?

11. When Louise Joy Brown was born in 1978, many people were horrified and several government officials called for banning IVF. Today, more than a million people have been conceived using the technology.

Are there any experimental treatments or technologies today that people are wary of that might one day become routine?

Applied Questions

1. At the same time that 62- and 63-year-old women gave birth, actors Tony Randall and Anthony Quinn became fathers at ages 77 and 78—and didn't receive nearly as much criticism as the women. Do you think this is an unfair double standard, or a fair criticism based on valid biological information?

2. Many people spend thousands of dollars pursuing pregnancy. What might be an alternative solution to their quest for parenthood?

3. An Oregon man anonymously donated sperm that were used to conceive a child. The man later claimed, and won, rights to visit his child. Is this situation for the man more analogous to a genetic and gestational surrogate mother, or an oocyte donor who wishes to see the child she helped to bring into existence?

4. Big Tom is a bull with valuable genetic traits. His sperm are used to conceive 1,000 calves. Mist, a dairy cow with exceptional milk output, has many oocytes removed, fertilized *in vitro,* and implanted into surrogate mothers. With their help, Mist becomes the genetic mother of 100 calves—far more than she could give birth to naturally. Which two reproductive technologies performed on humans are based on these two agricultural examples?

5. State who the genetic parents are and who the gestational mother is in each of the following cases:

   a. A man was exposed to unknown burning chemicals and received several vaccines during the first Gulf war, and abused drugs for several years before and after that. Now he wants to become a father, but he is concerned that past exposures to toxins have damaged his sperm. His wife undergoes intrauterine insemination with sperm from the husband's brother, who has led a calmer and healthier life.

   b. A 26-year-old woman has her uterus removed because of cancer. However, her ovaries are intact and her oocytes are healthy. She has oocytes removed and fertilized *in vitro* with her husband's sperm. Two resulting embryos are implanted into the uterus of the woman's best friend.

   c. Max and Tina had a child by IVF in 1986. At that time, they had three extra embryos frozen. Two are thawed years later and implanted into the uterus of Tina's sister, Karen. Karen's uterus is healthy, but she has ovarian cysts that often prevent her from ovulating.

   d. Forty-year-old Christensen von Wormer wanted children, but not a partner. He donated sperm, which were used for intrauterine insemination of an Indiana mother of one. The woman carried the resulting fetus to term for a fee. On September 5, 1990, von Wormer held his newborn daughter, Kelsey, for the first time.

   e. Two men who live together want to raise a child. They go to a fertility clinic, have their sperm collected and mixed, and used to inseminate a friend, who nine months later turns the baby over to them.

6. Delaying childbirth until a woman is over age 35 is associated with certain physical risks, yet an older woman is often more mature and financially secure. Many women delay childbirth so that they can establish careers. Can you suggest societal changes, perhaps using a reproductive technology, that would allow women to more easily have both children and careers?

7. An IVF attempt yields 12 more embryos than the couple who conceived them can use. What could they do with the extras?

8. What do you think children born of an assisted reproductive technology should be told about their origins?

9. Wealthy couples could hire poor women as surrogates or oocyte donors simply because the adoptive mother does not want to be pregnant. Would you object to this practice? Why or why not?

10. Cloning could potentially help people who cannot make gametes. Madeline is fertile, but her partner Cliff had his testicles removed to treat cancer when he was a teenager. He did not bank testicular tissue. To have a child that is genetically theirs, the couple wishes to use a nucleus from one of Cliff's somatic cells, which would be transferred to an oocyte of Madeline's that has had its nucleus removed. The resulting cell would be cultured in the laboratory and then transferred to Madeline's uterus via standard IVF. Is the couple correct in assuming that the child would be genetically theirs? Cite a reason for your answer.

11. An IVF program in Bombay, India offers preimplantation genetic diagnosis to help couples who already have a daughter to conceive a son. The reasoning is that because having a male heir is of such great importance in this society, offering PGD can enable couples to avoid aborting second and subsequent female pregnancies. Do you agree or disagree that PGD should be used for sex selection in this sociological context?

Web Activities

Visit the Online Learning Center (OLC) at www.mhhe.com/lewisgenetics7. Select **Student Edition, chapter 21,** and **Web Activities** to find the website links needed to complete the following activities.

12. Go to the Centers for Disease Control and Prevention website. Click on ART Trends, and use the information to answer the following questions.

   a. Since 1996, to what extent has the use of assisted reproductive technologies (ART) in the U.S. increased?

   b. Which is more successfully implanted into the infertile woman's uterus, a fresh or frozen donor oocyte?

   c. Which is more successfully implanted into the infertile woman's uterus, a donated oocyte or one of her own?

   d. What are two factors that could complicate data collection on ART success rates?

13. A company called Extend Fertility provides oocyte freezing services, telling women to "set your own biological clock." The opening page states, "Today's women lead rich and busy lives—obtaining advanced degrees, pursuing successful careers, and taking better care of ourselves. As a result of this progress, many of us choose to have children later than our mothers did."

Look at the website. Discuss how it might be viewed by the following individuals:

a. A 73-year-old father of a healthy baby

b. A 26-year-old woman, married with no children but who wants them, told she must undergo chemotherapy for six months

c. An orphaned 10-year-old in Thailand

d. A healthy 28-year-old woman in the United States who wants to earn degrees in medicine, law, and business before becoming pregnant

e. A young mother in Mexico who is giving her son up for adoption because she cannot afford to raise him

Case Studies and Research Reports

14. Natallie Evans had to have her ovaries removed at a young age because they were precancerous, so she and her partner had IVF and froze their embryos for use at a later time. Under British law, both partners must consent for the continued storage of frozen embryos. Evans and her partner split, and he revoked his consent. She sued for the right to use the embryos. She told the court, "I am pleased to have the opportunity to ask the court to save my embryos and let me use them to have the child I so desperately want."

What information should the court consider in deciding this case? Whose rights do you think should be paramount?

15. The real case of Scott and Glenda Lyons, who donated extra embryos to two infertile couples, had a happy ending. A fictionalized case presented on the television program *Law and Order: Special Victims Unit* profiled a woman whose physician took her extra oocytes (without telling her) and fertilized them with sperm from men whose partners had attempted IVF without conceiving, to inflate his success rate. A few years after the woman's husband and daughter died in a car crash, she spotted a child in a playground who resembled her daughter, and stalked her. As the case unfolded, detectives discovered that the woman had a dozen children, all over the country, thanks to the doctor's unauthorized use of her oocytes. Suggest measures that can be taken at fertility clinics to minimize the chance of this type of crime occurring. (In the United States, the federal government does not oversee operation of such facilities.)

16. Colleen and Ellen were partners who had twin daughters using ART. Colleen was already in her forties when they decided to have children, and also had uterine fibroids. So they selected sperm from a sperm bank, which was used to perform intrauterine insemination on Ellen, but she did not conceive after several attempts. Next Ellen tried IVF, to no avail. Finally, the women decided to combine their contributions. Colleen had oocytes collected and fertilized and implanted in Ellen's uterus. The women signed legal documents declaring Ellen the sole parent, with the understanding that they would consider a more shared arrangement five years later. At the clinic, Colleen signed a form that waived her rights to her oocytes or children resulting from their being fertilized. She would later claim that she signed it because she thought she had to for the procedure to be done. Ellen's name was recorded on the birth certificates. Although both women were active parents, their relationship deteriorated, and when the girls were 6, Ellen moved them across the country and kept Colleen from visiting them. The case ended up in the courts, which ruled that Ellen is the mother. The Marin County Superior Court declared Colleen's relationship with the twins "legally irrelevant." Colleen has petitioned the Supreme Court of California to consider the case.

a. Who is the genetic mother and who is the gestational mother?

b. Do you agree with the Marin County Superior Court's decision that a genetic relationship is "legally irrelevant?"

c. What information would you need to decide whether Colleen was being discriminated against because she is gay?

Learn to apply the skills of a genetic counselor with these additional cases found in the *Case Workbook in Human Genetics:*

Charcot-Marie-Tooth disease

Male infertility

The Age of Genomics

CHAPTER 22

AN ALGA HELPS EXPLAIN A HUMAN DISEASE

Researchers are comparing genome information from an ever-expanding list of species. While revealing much about evolution, comparative genomics also sometimes illuminates a human disease. This is the case for the very rare Bardet-Beidl syndrome (BBS, OMIM 209900).

BBS causes learning disabilities, obesity, blindness, extra digits, and kidney problems. Causative genes have been linked to eight chromosomes. In 2003, researchers implicated impaired cilia, the organelles that fringe certain cells. Cilia wave, moving secretions across cell surfaces and the cells themselves, and they pick up signals. Abnormal or missing cilia can cause seemingly unrelated symptoms because so many cell types are affected. For example, early in embryonic development, waving cilia control how cells move to form tissues. When cilia do not form properly, organs may develop on the wrong side of the body! Cilia grow from structures called basal bodies.

To identify the genes behind BBS, researchers sought clues in genomes, looking for genes shared only by organisms that have ciliated cells. Researchers searched the genome sequences of a single-celled green alga that has basal bodies, humans, and a small plant in the mustard family whose cells lack basal bodies. The alga and humans share 4,348 genes, 3,660 of which are also in the plant genome. Subtracting the 3,660 genes that could not be involved in cilia formation left 688 genes. Researchers who work with BBS families consulted linkage data that pointed to 230 genes on part of human chromosome 2 as possible cilia genes. Only 2 of these 230 human genes were also in the alga, and one of these causes BBS. Researchers now have to figure out how the abnormal cilia cause the symptoms.

The ciliated green alga
Chlamydomonas reinhardtii.

Genetics is a young science, genomics younger still. As one field has evolved into another, milestones have come at oddly regular intervals. A century after Gregor Mendel announced and published his findings, the genetic code was deciphered; a century after his laws were rediscovered, the human genome was sequenced. This final chapter chronicles the rise of genomics.

22.1 How Genetics Became Genomics

The term *genome* was coined in 1920 by geneticist H. Winkler. A hybrid of "gene" and "chromosome," genome denotes a complete set of chromosomes and its genes. (The modern definition refers to all the DNA in a haploid set of chromosomes, accounting for non-protein-encoding sequences.) The term *genomics,* credited to T. H. Roderick, originated in 1986 to indicate the study of genomes. Thoughts of genomes in general, and of the human genome in particular, lay in the background throughout the twentieth century, as researchers defined and described the units of inheritance from various perspectives.

The human genome project began with deciphering signposts and discovering shortcuts to the huge undertaking. Many of the initial steps and tools grew from existing technology. Linkage maps from as long ago as the 1950s, and many family studies that associated chromosomal aberrations with syndromes, enabled researchers to assign some genes to their chromosomes. Then automated DNA sequencing took genetic analysis to a new level—information. **Figure 22.1** schematically illustrates the refinement and increasing resolution of different types of genetic maps.

Before and during the human genome project, many researchers investigated single genes that cause specific diseases, using an approach called **positional cloning.** The technique began with examining a particular phenotype corresponding to a Mendelian disorder in large families, then identifying parts of the genome that only affected relatives shared. Such sequences could serve as "markers" of the presumably tightly linked disease-causing gene. Instead of directly comparing DNA sequences, researchers con-

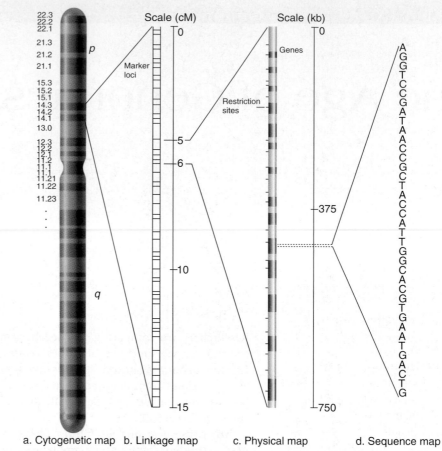

a. Cytogenetic map b. Linkage map   c. Physical map   d. Sequence map

Figure 22.1   Different levels of genetic maps.   (a) A cytogenetic map, based on associations between chromosome aberrations and syndromes, can distinguish DNA sequences that are at least 5,000 kilobases (kb) apart. **(b)** A linkage map derived from recombination data distinguishes genes hundreds of kb apart. **(c)** A physical map constructed from DNA pieces cloned in vectors and then overlapped distinguishes genes tens of kb apart. **(d)** A sequence map is the ultimate genetic map, consisting of all the nucleotide bases ordered as they are on the chromosomes.

sulted restriction fragment length polymorphisms (RFLPs). Recall from chapter 14 that RFLPs are sequence variations that alter the sizes of fragments that result when DNA is exposed to a type of restriction enzyme. DNA profiling detects larger fragments that result from short repeats. In positional cloning, single nucleotide polymorphisms (SNPs, or differences at particular sites) generate variant fragment sizes. If relatives with a disorder all have a SNP that removes or adds a cut site, their genomes, when cut with the corresponding restriction enzyme, will have pieces of a certain size that nonaffected relatives do not have.

Positional cloning experiments dominated medical genetics in the 1980s and 1990s, yielding a steady stream of discoveries of genes that cause such diseases

as Duchenne muscular dystrophy, cystic fibrosis, and Huntington disease (Reading 22.1). It was tedious and took years. Today, using the human genome sequence and tools to analyze it, many a graduate student has found a disease-causing gene in just weeks.

The entire way we look at gene discovery has changed in this new age of genomics. Today, gene discovery is largely an informational science, where researchers mine genome databases for DNA sequence similarities. Each time a new species has its genome sequenced, researchers can consult those done earlier and identify gene functions to fill in blanks. At first, the functions of nearly half of the genes uncovered in genomes were unknown. The overall focus has shifted from identifying single genes to considering hierarchies of multigene function.

Key Concepts

1. Human genome sequencing was built on linkage and cytogenetic information from decades of work.
2. Positional cloning located specific disease-causing genes in families using RFLP mapping of linked SNPs that affect fragment size.
3. Attention has shifted to understanding hierarchies of gene function.

RELATING THE CONCEPTS

Nicholas Katsanis is a geneticist at Johns Hopkins University who helped discover the gene that causes Bardet-Beidl syndrome. Said he, "It would have been nearly impossible to find this gene in any other way. It proves without a doubt the value of sequencing the genomes of a wide variety of species." Their technique used a computer to align the genome sequences of three species, and then used logic, based on knowledge of phenotypes, to deduce which of 230 candidate genes causes the disease. How is this approach faster than using positional cloning?

Technology Timeline

Evolution of the Human Genome Project

1985–1988	Idea to sequence human genome suggested at several scientific meetings.
1988	Congress authorizes the Department of Energy and the National Institutes of Health to fund the human genome project.
1989	Researchers at Stanford and Duke Universities invent DNA microarrays.
1990	Human genome project officially begins.
1991	Expressed sequence tag (EST) technology identifies protein-encoding sequences.
1992	First DNA microarrays available.
1993	Need to automate DNA sequencing recognized.
1994	U.S. and French researchers publish preliminary map of 6,000 genetic markers, one every 1 million bases along the chromosomes.
1995	Emphasis shifts from gene mapping to sequencing.
1996	Resolution passed to make all data public and update it daily at GenBank website.
1998	Public Consortium releases preliminary map of pieces covering 98 percent of human genome. Millions of sequences listed in GenBank. Directions for developing DNA microarrays posted on the Internet.
1999	Rate of filing of new sequences in GenBank triples.
	A full 30 percent of human genome sequenced by year's end.
	Public Consortium and two private companies race to complete sequencing.
2000	Ninety-nine percent of human genome sequenced privately.
	Microarray technology explodes as researchers investigate gene functions and interactions.
2001	Two versions of human genome sequence published in mid-February issues of *Science* and *Nature*.
2003	Finished version of human genome sequence announced to coincide with fiftieth anniversary of discovery of DNA structure.
	Entire protein-encoding part of human genome available on DNA microarrays.
2005 and beyond	Annotation of human genome sequence continues as genomics gradually impacts health care. (Number of species with sequenced genomes soars.)

22.2 The Human Genome Project Begins

The idea of sequencing an entire genome probably occurred to many researchers as soon as Watson and Crick determined the structure of DNA. In 1966, Francis Crick proposed sequencing all genes in *E. coli* to reveal how the organism works, writing that "this particular problem will keep very many scientists busy for a long time to come." Jacques Monod, who described the first genetic control system in bacteria, wrote at about the same time, "What is true for *Escherichia coli* is true for the elephant," meaning that to know genomes is to understand all life. It was a gross oversimplification.

For sequencing genomes to evolve from science fiction to science required the development of key tools and technologies (see the Technology Timeline). First was the need to obtain the nucleotide base sequence of pieces of DNA. Then, computer pro-

grams would have to align and detect overlaps in sequence in pieces cut from multiple copies of a genome, and assemble them to reconstruct each chromosome.

The Sanger Method of DNA Sequencing

Modern DNA sequencing instruments utilize a basic technique that Frederick Sanger developed in 1977. The goal is to generate a series of DNA fragments of identical sequence that are complementary to the sequence of interest. These fragments differ

in length from each other by one end base, as follows:

Sequence of interest: T A C G C A G T A C
Complementary
sequence:
Series of fragments:

A T G C G T C A T G
T G C G T C A T G
G C G T C A T G
C G T C A T G
G T C A T G
T C A T G
C A T G
A T G
T G
G

Discovering the Huntington Disease Gene

The story of the discovery of a marker for Huntington disease (HD) illustrates the gene-by-gene trajectory of research that immediately preceded the human genome project and has evolved with it. HD is an autosomal dominant neurodegenerative disorder with adult-onset symptoms of worsening gait, constant movement (except during sleep and in the final stage), and personality changes. Death usually comes 10 to 15 years after diagnosis. The HD gene is near the tip of the short arm of chromosome 4 and encodes a protein, huntingtin, that is synthesized in the brain. The mutation is an expanded triplet repeat that adds glutamines to the protein. In certain brain cells, the added material disrupts folding of huntingtin, interfering with its interactions with other proteins.

The search for the HD gene began with a large family in a remote village on the shores of Lake Maracaibo, Venezuela. Seven generations ago, in the 1800s, a local woman married a visiting Portuguese sailor who, according to folklore, walked as if intoxicated. Like many couples in the poor fishing village, the woman and her sailor had many children. Some grew up to walk in the same peculiar way as their father. Of their nearly 5,000 descendants, more than 250 living today have HD. This extended family presented a natural experiment to geneticists, and they began their quest for a marker by drafting a huge pedigree that eventually depicted more than 10,000 individuals and stretched to more than 100 feet in width (**figure 1**). The Venezuela family

Figure 1   **Nancy Wexler consults the enormous Venezuelan HD pedigree.**

was large enough to detect a marker-disease gene association. Studies at the DNA and chromosomal levels followed.

In 1981, Columbia University psychologist Nancy Wexler, whose mother would eventually die of HD, began annual visits to Lake Maracaibo. The people lived in huts perched on stilts, as their ancestors did. Wexler traded candy and blue jeans for blood samples and skin biopsies to bring back to a team of geneticists. Meanwhile, investigators at Massachusetts General Hospital were sampling tissue from an Iowa family of 41, in which 21 individuals had

HD. They extracted DNA from the samples, cut it with restriction enzymes, and tested the fragments with a set of labeled DNA probes. They were looking for a probe that bound only the DNA of sick people. This would reveal a RFLP unique to people with the disease. The team added the Venezuelan DNA. A group at Indiana University matched the probe data to pedigrees, seeking a pattern (**figure 2**).

With several hundred DNA probes and samples, the researchers expected the testing to take a long time, but luck was on their side. On a warm May night in 1983, the

Note that the entire complementary sequence appears in the sequence of end bases of each fragment. If the complement of the gene of interest can be cut into a collection of such pieces, and the end bases distinguished with a radioactive or fluorescent label, then polyacrylamide gel electrophoresis (see figure 14.7) can be used to separate the fragments by size. Then, once the areas of overlap are aligned, reading the labeled end bases in size order reveals the

sequence of the complement. Replacing A with T, G with C, T with A, and C with G establishes the sequence of the DNA in question.

Sanger invented a way to generate the DNA pieces. In a test tube, he included the unknown sequence and all of the biochemicals needed to replicate it, including supplies of the nucleotide bases. Some of each of the four types of bases were chemically modified at a specific location on the base

sugar to contain no oxygen atoms instead of one—in the language of chemistry, they were *di*-deoxyribonucleotides rather than deoxyribonucleotides. A radioactive nucleotide, usually C, was also included. DNA synthesis halted when DNA polymerase encountered a "dideoxy" base, leaving only a piece of the newly replicated strand.

Sanger repeated the experiment four times, using dideoxy A, T, C, then G. The

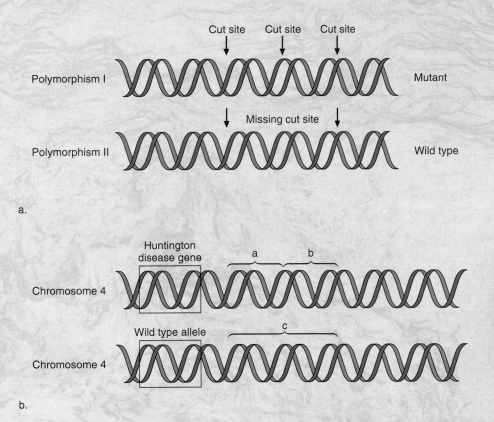

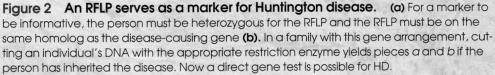

Figure 2 An RFLP serves as a marker for Huntington disease. (a) For a marker to be informative, the person must be heterozygous for the RFLP and the RFLP must be on the same homolog as the disease-causing gene **(b)**. In a family with this gene arrangement, cutting an individual's DNA with the appropriate restriction enzyme yields pieces *a* and *b* if the person has inherited the disease. Now a direct gene test is possible for HD.

twelfth probe tested, called G8, matched. In both families it bound only the DNA of the sick people. G8 was linked to, and inherited with, the HD gene. Until the gene itself was discovered in 1993, this marker was the basis of presymptomatic testing.

The next step was to localize G8 to a chromosome. The researchers used hybrid rodent/human cells that contain only one human chromosome each. The hybrid cell that included G8 had human chromosome 4. To extend the DNA sequence from the mark-er to the HD gene, the researchers located another probe that overlapped the first, then another probe that overlapped the second probe, and so on. This technique is called "chromosome walking." Finally, a computer aligned the probes according to their sequence overlaps, creating a contiguous, or "contig" map of the extended DNA sequence.

A hurdle was that the area surrounding a probe on a chromosome can harbor many genes. Researchers first narrowed down possible protein-encoding regions by looking for stretches of CGCGCG called "CpG islands," which precede genes. That analysis showed that the half-million-base-long contig map could include 100 genes! How would researchers know which were "candidate genes" for HD? They needed another clue, related to the phenotype. The researcher who had discovered that an expanding triplet repeat causes myotonic dystrophy (see figure 12.9) suggested that perhaps this type of mutation lay behind HD, too, since both disorders affect movement.

Looking for a triplet repeat was a long shot, but the researchers found it—a 210,000-base stretch of DNA in people who do not have HD is considerably longer in those who do. Next, the researchers looked for gene expression in cDNA libraries made from various differentiated cells. The expanding gene was indeed in brain cDNA libraries. It was the HD gene. Marker tests, which could only be done in certain families amid much uncertainty, were soon replaced with direct mutation tests, which any person can take.

four experiments were run in four lanes of a gel (**figure 22.2**). Today, fluorescent labels are used, one for each of the four base types, to reveal the sequence in a single experiment. The four types of pieces, ending in any of the four types of labeled bases, are collected in thin glass tubes called capillaries, rather than on cumbersome slabs of gel. The data appear as a sequential readout of the wavelengths of the fluorescence from the labels (**figures 22.3 and 22.4**). In the mid-1980s, Leroy Hood automated Sanger's method of DNA sequencing, which was essential to analyzing entire genomes.

DNA sequencing occurs on a vast scale today. Many individual laboratories or academic departments have one or more automated DNA sequencers, while companies devoted to genome sequencing may have hundreds. But the automated Sanger method of DNA sequencing is too slow to handle routine genome sequencing, either of human genomes in clinical applications, or of the many other species awaiting genome sequencing.

A newer method of DNA sequencing is called single-molecule detection. This technique unwinds a long, fluorescently labeled DNA molecule by passing it through a microfluidics environment, which is a small, fluid-filled chamber containing intricate nooks and crannies. Then

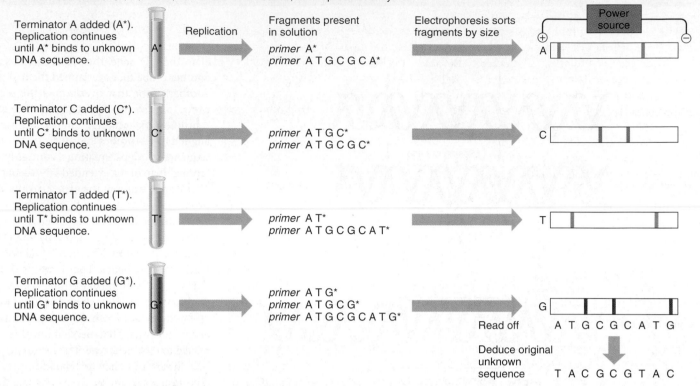

Four solutions contain unknown DNA sequence; primers (starting sequences); normal nucleotides A,T, C, and G; a radioactive nucleotide; and replication enzymes.

Terminator A added (A*). Replication continues until A* binds to unknown DNA sequence.

A*

Replication

Fragments present in solution

primer A*
primer A T G C G C A*

Electrophoresis sorts fragments by size

Power source

A

Terminator C added (C*). Replication continues until C* binds to unknown DNA sequence.

C*

primer A T G C*
primer A T G C G C*

C

Terminator T added (T*). Replication continues until T* binds to unknown DNA sequence.

T*

primer A T*
primer A T G C G C A T*

T

Terminator G added (G*). Replication continues until G* binds to unknown DNA sequence.

G*

primer A T G*
primer A T G C G*
primer A T G C G C A T G*

G

Read off   A T G C G C A T G

Deduce original unknown sequence   T A C G C G T A C

Figure 22.2   Determining the sequence of DNA.   In the Sanger method of DNA sequencing, complementary copies of an unknown DNA sequence are terminated early because of the incorporation of dideoxynucleotide "terminators." A computer deduces the sequence by placing the fragments in size order. Radioactive or fluorescent labels are used to visualize the sparse quantities of each fragment.

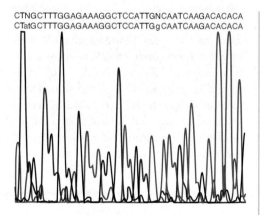

CTNGCTTTGGAGAAAGGCTCCATTGNCAATCAAGACACACA
CTatGCTTTGGAGAAAGGCTCCATTGgCAATCAAGACACACA

Figure 22.3   DNA sequence data.
In automated DNA sequencing, a readout of sequenced DNA is a series of wavelengths that represent the terminally labeled DNA base.

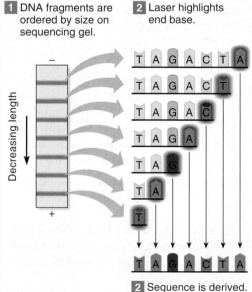

1 DNA fragments are ordered by size on sequencing gel.

2 Laser highlights end base.

Decreasing length

2 Sequence is derived.

Figure 22.4   Reading a DNA sequence.   A computer algorithm detects and records the end base from a series of size-ordered DNA fragments.

the unwound DNA passes a multilaser scanner that "reads" 10 to 30 million base pairs per second.

Many Goals

During the 1980s, the idea to sequence the human genome evolved for several reasons. It surfaced at a meeting held by the Department of Energy (DOE) in 1984 to discuss the long-term population genetic effects of exposure to low-level radiation. In 1985, at another gathering, Robert Sinsheimer, chancellor of the University of California, Santa Cruz, called for an institute to sequence the human genome, basically because it could be done. The next year, virologist Renato Dulbecco proposed that the key to understanding the origin of cancer lay in knowing the human genome sequence. In the summer of 1986, geneticists and molecular biologists convened at the Cold Spring Harbor Laboratory on New York's Long Island to discuss the feasibility of a human genome project. Worldwide planning soon began.

The decision to systematically sequence the human genome shifted the goals of life science research. A furious debate ensued, with detractors claiming that the project would be more gruntwork than a creative intellectual endeavor, comparing it to conquering Mt. Everest just because it is there. Some researchers feared that such an unprecedented "big science" project would divert government funds from basic research and AIDS. Finally, the National Academy of Sciences convened a committee of detractors and supporters to debate the feasibility, risks, and benefits of the project. The naysayers were swayed to the other side. So in 1988, Congress authorized the National Institutes of Health (NIH) and the DOE to fund the $3 billion, 15-year human genome project.

The government-sponsored human genome project officially began in 1990 with James Watson at the helm. Recognizing the profound effect that genetic information would have on public policy and on peoples' lives, the project set aside 3 percent of its budget for the ELSI program, which stands for ethical, legal, and social issues. ELSI helps ensure that genetic information is not misused to discriminate against people with particular genotypes.

The human genome project expanded in scope and participants. In addition to ten major sequencing centers, the largest in England, France, Germany, and Japan, many hundreds of other research laboratories and biotechnology companies contributed DNA sequence data.

22.3 Technology Drives the Sequencing Effort

The human genome project progressed in stages as sequencing technology and computer capability to align and derive continuous sequence improved. At first the project focused on developing tools and technologies to divide the genome into pieces small enough to sequence, and to improve the efficiency of sequencing techniques.

In 1991, two key inventions entered the picture. J. Craig Venter, then a government researcher at the NIH, introduced a very powerful shortcut called **expressed sequence tag** (EST) technology. ESTs are cDNAs that are ends of genes expressed in a particular cell type. Identifying ESTs enabled researchers to quickly pick out genes most likely to be implicated in disease, because they encode protein and are expressed in the cell types affected in a particular illness. ESTs also serve as landmarks among the chromosomes in genome sequencing. Venter left the NIH to develop ESTs at The Institute for Genomic Research (TIGR), which he founded. Also in 1991, Patrick Brown at Stanford University developed DNA microarrays, which are used to display cDNAs.

Sequencing and mapping techniques improved greatly from 1993 to 1998. At the start, researchers cut the genome into overlapping pieces of about 40,000 bases (40 kilobases), then randomly cut the pieces into small fragments. Researchers would cut up several genomes in a single experiment, so that many of the resulting fragments overlapped. By finding the overlaps, the pieces could be assembled to reveal the overall sequence. The greater the number of overlaps, the more likely that the final assembled sequence would not omit anything. Today, computers recognize the overlaps and align and derive the sequences, either beginning with larger sections or using many small pieces. It is important that the sites of overlap be unique sequences, found in only one place in the genome. Overlaps of repeated sequences present in several places could lead to more than one derived overall sequence. This is why the Y chromosome, with its many repeats, was extremely difficult to sequence, and had to be cut into very small pieces. Another problem is to determine which side of the double helix to sequence for each piece. **Figure 22.5** depicts how a DNA sequence might be reconstructed from overlapping pieces.

The human genome project examined thousands of pieces of DNA at a time. The sequences were cataloged continually in a public database called GenBank, and at some companies. Two general approaches tackled the large number of pieces of DNA (**figure 22.6**). The "clone-by-clone" technique aligned pieces one chromosome at a time, using a great deal of existing cytogenetic and linkage data to assign pieces to chromosomes. The "whole genome shotgun" approach shattered the entire genome,

Key Concepts

1. DNA sequencing and computer software to align DNA pieces were essential for genome projects to proceed.
2. In the Sanger method of DNA sequencing, complementary copies of an unknown DNA sequence are terminated early by incorporating dideoxynucleotides. A researcher or automated DNA sequencer deduces the sequence by labeling the end bases and placing fragments in size order.
3. The idea to sequence the human genome emerged in the mid-1980s with several goals. The project officially began in 1990.

Random fragments:	AGTCCT CTAG AGCTA
	CTACT TAGAGT CCTAGC
Alignment:	CTAG
	TAGAGT
	AGTCCT
	CCTAGC
	AGCTA
	CTACT
Sequence:	CTAGAGTCCTAGCTACT

Figure 22.5 Deriving a DNA sequence. Automated DNA sequencers first determine the sequences of short pieces of DNA, or sometimes of just the ends of short pieces. Then algorithms search for overlaps. By overlapping the pieces, the software derives the overall DNA sequence.

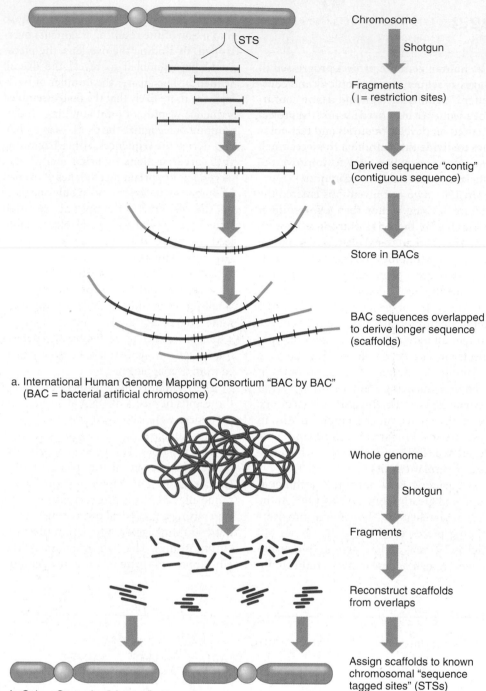

Chromosome

STS

Shotgun

Fragments
(| = restriction sites)

Derived sequence "contig"
(contiguous sequence)

Store in BACs

BAC sequences overlapped
to derive longer sequence
(scaffolds)

a. International Human Genome Mapping Consortium "BAC by BAC"
(BAC = bacterial artificial chromosome)

Whole genome

Shotgun

Fragments

Reconstruct scaffolds
from overlaps

Assign scaffolds to known
chromosomal "sequence
tagged sites" (STSs)

b. Celera Genomics "shotgun" approach

Figure 22.6 Two routes to the human genome sequence. **(a)** The International Consortium began with known chromosomal sites and overlapped large pieces, called contigs, that in turn were reconstructed from many small, overlapping pieces. **(b)** Celera Genomics shotgunned several copies of a genome into small pieces, overlapped them to form scaffolds, and then assigned scaffolds to known chromosomal sites. They used some Consortium data.

some sections (particularly duplicated sequences) that the clone-by-clone method detects, and it does not work well on large genomes.

One research group, the International Consortium, used the clone-by-clone approach, dividing several genomes into BAC clones. Recall from chapter 19 that a BAC clone is a bacterial artificial chromosome that contains a piece of foreign DNA. The BAC clones each housed about 100,000 bases of inserted human DNA known to correspond to specific places among the chromosomes called sequence tagged sites (STSs). An STS is an easily recognized short sequence found in only one place on a chromosome or in a genome. ESTs are a class of STSs. Several copies of each BAC were cut into up to 80 overlapping pieces, and the pieces sequenced. A powerful computer program then assembled the overlaps to derive the overall sequence for each chromosome. Finally, researchers filled in gaps and analyzed sequences, "annotating" each chromosome with functional descriptions of protein-encoding genes. More detailed views of the chromosomes continue to be published.

The second group to obtain the first draft sequence of the human genome was Celera Genomics Corporation of Rockville, Maryland, headed by J. Craig Venter. The Celera team shotgunned several copies of the genome into small pieces, and used a computer program to assemble the overlaps into larger pieces called scaffolds. They then used STSs to anchor the resulting 119,000 scaffolds to the chromosomes.

By many people's accounts, 1995 was a turning point in the human genome project, as "proteomics" entered the growing vocabulary of genome jargon. Recall from chapter 11 that the proteome is all the proteins that an organism can synthesize, and the field of proteomics also considers how the proteins are distributed among different cell types. Researchers began to see that proteomics would present an even greater challenge than sequencing the genome.

The year 1995 was also when efforts shifted from mapping genes to chromosomes to DNA sequencing. For the first time, the number of gene sequences deposited in GenBank outnumbered the number of papers in the scientific literature introducing disease genes identified by positional cloning. The reason: the automation of

then used a computer algorithm to identify overlaps and align them to derive a continual sequence. The task can be compared to cutting the binding off a large book, throwing it into the air, and reassembling the dispersed

pages in order. A "clone-by-clone" dismantling of the book would divide it into bound chapters. The whole genome shotgun approach would free every page. Whole genome shotgunning is faster, but it misses

DNA sequencing. Although similar to their 1980s prototypes in concept, devices could now produce lines of A, T, C, and G ten times as fast as in the early days.

The deadline for completion of the human genome project inched forward as optimism soared. In 1995, Venter's company, TIGR, developed software that could rapidly locate the unique sequence overlaps among many small pieces of DNA and assemble them into a continuous sequence, eliminating the preliminary step of gathering large guidepost pieces. Using this software, TIGR obtained the first complete genome sequence for an organism in under a year! Many more would follow.

The first organism to have its genome sequenced was *Haemophilus influenzae,* a bacterium that causes meningitis and ear infections. The researchers cut the bacterium's 1,830,137 bases into 24,000 fragments, sequenced them, and used the software to assemble the overlaps of several genomes worth of pieces. The human genome is 1,500 times larger than that of *H. influenzae.*

Meanwhile, the partners of the International Consortium established guidelines binding researchers to release new DNA sequences to the public database within 24 hours of determination. By 1998, Venter had founded Celera Genomics, and announced that they would sequence the human genome faster and at less cost than the Consortium, a feat possible because of the company's powerful assembler software. Unlike the Consortium's public posting of data, Celera planned to sell access to its sequence, even though they had to use some of the public data to get that sequence. And so began the fierce bioethical battle over public access to human genome information that continues today.

In 1999, the human genome project became intensely competitive, with Venter and Collins, director of the U.S. arm of the International Consortium since 1993, racing to finish and sacrificing a degree of accuracy for speed. Venter claimed privately to have reached the first draft goal for the human genome by March 2000. The Consortium's sequence was completed by June 22, following a month of frantic work by using a program called GigAssembler to put together thousands of preassembled pieces.

In the end, the battling factions called a truce. On June 26, 2000, Venter and Collins flanked President Clinton in the White House rose garden to unveil the "first draft" of the human genome sequence. The milestone capped a decade-long project involving thousands of researchers, which in turn was the culmination of a century of discovery. The historic June 26 date came about because it was the only opening in the White House calendar! In other words, the work was monumental; its announcement, staged.

Figure 22.7 is an overview of genome sequencing and one application—gene expression profiling using DNA microarrays, discussed in chapter 11. **Table 22.1** summarizes steps in genome sequencing and **Table 22.2** lists websites that have genome sequence information.

Table 22.1

Steps in Genome Sequencing and Analysis

1. Obtain chromosome maps with landmarks from classical linkage or cytogenetic studies, or RFLP sites.

2. Obtain chromosome pieces maintained in gene libraries, or shotgun entire sequence.

3. Sequence the pieces.

4. Overlap aligned sequences to extend the known sequence.

5. Compare the sequence to those in other species.

Table 22.2

Public Genome Databases

Organization	Website
GenBank	www.ncbi.nih.gov/Genbank
European Molecular Biology Laboratory (EMBL)	www.ebi.ac.uk/embl/index.html
DNA Data Bank of Japan	www.ddbj.nig.ac.jp
University of California, Santa Cruz Genome Browser	www.genome.ucsc.edu/
National Center for Biotechnology Information Map Viewer	www.ncbi.nlm.nih.gov/mapview/
National Human Genome Research Institute, NIH	www.genome.gov/
Ensemble Genome Browser	www.ensemble.org
U.S. Department of Energy Genomes to Life	http://doegenomestolife.org
Genomes OnLine Database (GOLD)	www.genomesonline.org/

22.4 Comparative Genomics

Sequencing the human genome has obvious practical applications in health care, forensics, and anthropology. However, most genome research has a much broader focus, going well beyond the genetic instructions for a single species. The human genome project, from the outset, planned to sequence the genomes of selected species that have been laboratory favorites, often serving as models of human conditions. But these species do not represent natural diversity. So as sequencing technology has improved, the list of organisms awaiting genome sequencing has grown rapidly.

The first genomes to be sequenced were microbial, because their smaller size makes them more amenable to whole genome

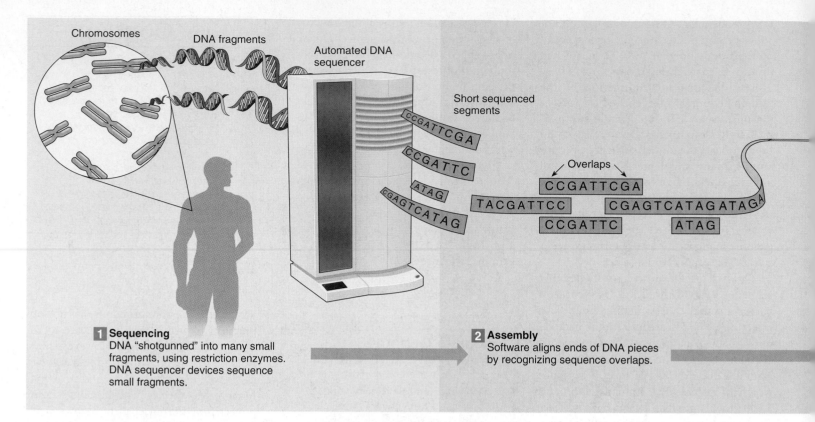

Figure 22.7 **Sequencing genomes.** Determining the DNA sequence of a genome is just a first step—albeit a huge one. After pieces are assembled, protein-encoding genes are identified, and then patterns of gene expression in different tissues are assessed.

shotgunning. The Genomes Online website keeps a running list of sequenced genomes, so it is easy to see that the shortest were completed first. Then our closest relatives were sequenced—mice, rats, chimps. Most important, however, have been the genomes of species at evolutionary crossroads, those that introduced a new trait or were the last to have an old one.

Comparing the DNA nucleotide base sequences of genomes and genome regions across different modern species helps us to better understand evolution. Researchers look for DNA sequences that are the same, or conserved, among species and therefore, by inference, through time. **Figure 22.8** shows one way of displaying sequence similarities, called a pictogram. DNA sequences from different species are aligned, and the bases at different points indicated. A large letter A, C, T or G indicates, for example, that all species examined have the same base at that site. A polymorphic site, in contrast, has different bases for different species.

The premise behind examining conserved sequences to reconstruct the past is

a. Not highly conserved

Figure 22.8   A pictogram indicates conservation of sequence.
Genomes from different species or individuals can be compared site by site. These pictograms are for short sequences in corresponding regions of the human, mouse, rat, and chicken genomes. A large letter means that all four species have the same base at that site. If four letters appear in one column, then the species are different. Pictogram **(a)** is not highly conserved; **(b)** is.

that any sequence that has withstood the test of time to be present in diverse species today must provide a vital function. Conserved sequences encode proteins or control transcription and/or translation. The field of **comparative genomics** uses conserved sequences to identify biologically important genome regions. It also considers differences among species.

Comparative genomics is establishing a framework to explain biodiversity and the evolutionary events that underlie it. Following are examples of the types of information revealed in conserved DNA sequences. **Figure 22.9** presents some organisms whose genomes have been or are being sequenced.

The minimum gene set required for life   The smallest microorganism known to be able to reproduce is *Mycoplasma genitalium*. It infects cabbage, citrus fruit, corn, broccoli, honeybees, and spiders, and causes respiratory illness in chickens, pigs, cows, and humans. Researchers call its tiny

Derived sequence

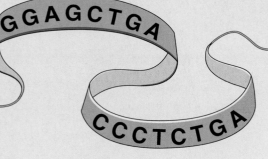

DNA microarrays

3 **Annotation**
Software searches for clues to locations of
protein-encoding genes. Databases from
other species' genomes searched for
similarities to identify gene functions.

4 **Gene expression
profiles and proteomics**
cDNAs made from mRNAs from differentiated
cells; displayed as DNA microarrays to create gene
expression profiles.

genome the "near-minimal set of genes for independent life." Of 480 protein-encoding genes, researchers think that 265 to 350 are essential for life. Considering how *Mycoplasma* uses its genes reveals the fundamental challenges of being alive. When the organism's genome was sequenced, nearly a quarter of the genes had no known counterpart in other organisms.

Fundamental distinctions among the three domains of life

Methanococcus jannaschii is a microorganism that lives at the bottoms of 2,600-meter-tall "white smoker" chimneys in the Pacific Ocean, at high temperature and pressure and without oxygen. As archaea, these cells lack nuclei, yet replicate DNA and synthesize proteins in ways similar to multicellular organisms. The genome sequence confirms that this organism represents a third form of life. Fewer than half of its 1,738 genes had known counterparts among the bacteria, other archaea, or eukaryotes when the genome was sequenced. Even the genome of *E. coli,* the organism that geneticists sup-

posedly knew the best, held surprises. The functions of more than a third of *E. coli*'s 4,288 genes remain a mystery.

The simplest organism with a nucleus

The paper in *Science* magazine that introduced the genome of the yeast *Saccharomyces cerevisiae* was entitled "Life with 6,000 Genes." But the unicellular yeast is more complex than this title implies. About a third of its 5,885 genes have counterparts among mammals, including at least 70 implicated in human diseases. Understanding what a gene does in yeast can provide clues to how it affects human health. For example, mutations in counterparts of yeast cell cycle control genes cause cancer in humans.

The basic blueprints of an animal

The genome of the tiny, transparent, 959-celled nematode worm *Caenorhabditis elegans* is packed with information on what it takes to be an animal. Thanks to researchers who, in the 1960s, meticulously tracked the movements of each cell as the animal developed, much of the biology of this organism

was already known before its 97 million DNA bases were revealed late in 1998. The worm's signal transduction pathways, cytoskeleton, immune system, apoptotic pathways, and even brain proteins are very similar to our own.

The sequencing of the fruit fly (*Drosophila melanogaster*) genome held a big surprise—it has 13,601 genes, fewer than the 18,425 in the much simpler worm. Discovering how genes control the development and functioning of an organism will be more complex than simply considering the contributions or numbers of single genes. Also, of 289 human disease-causing genes, 177 have counterparts in *Drosophila*. The fly might therefore serve as a model for humans in testing new treatments.

Life on land Before 450 million years ago, according to fossil evidence, life was confined to the seas, where it was abundant and diverse. Algae, microorganisms, and the first jawless fishes shared the depths. The first organisms to colonize land were the mosses, and for this reason, the genome

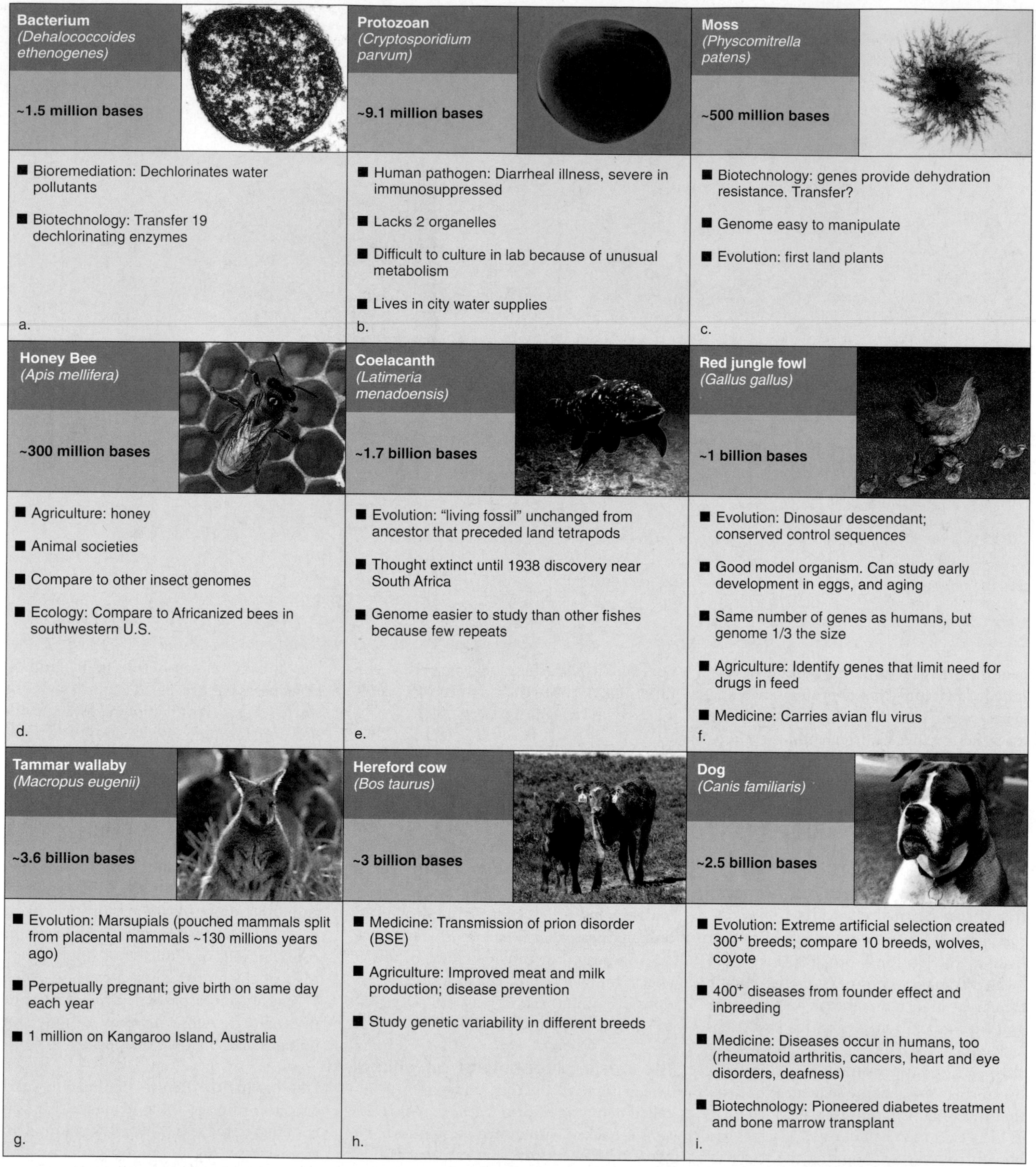

Figure 22.9 **A sampling of genomes, organized according to increasing evolutionary closeness to humans.** Listed are some items of interest about each.

of the modern moss *Physcomitrella patens* (figure 22.9c) is being sequenced. Mosses are highly streamlined plants, lacking familiar parts such as stems and leaves, and with only a few types of differentiated cells and tissues. They dominated landscapes until plants that had seeds and vessels evolved some 200 million years later.

By the time animals ventured onto land, plants had already taken root. Sea residents whose descendants were probably among those first land dwellers were the lobe-finned fishes, which have fleshy, strong fins that could have evolved into the first limbs. Two types of lobe-finned fishes persist today and remarkably resemble their fossilized forms—the lungfishes and two species of coelacanths. Because the lungfish genome is huge, researchers are tackling the smaller coelacanth genome (figure 22.9e). A coelacanth is about 5 feet long and weighs about 130 pounds. The huge fish was thought to be extinct until one was discovered near the coast of South Africa in 1938. Today they live in the Comoro Islands in the Indian Ocean. Information in the coelacanth genome may reveal the traits necessary for the evolution of the tetrapods—vertebrate animals with four limbs.

From Birds to Mammals The sequencing of the chicken genome (figure 22.9f) marked a number of milestones—the first agricultural animal, the first bird, and, as such, the first direct descendant of dinosaurs. The genome of the red jungle fowl *Gallus gallus* is remarkably like our own, minus many repeats, but its genome organization is intriguing. Like other birds, fishes, and reptiles, but not mammals, the chicken genome is distributed among very large macrochromosomes and tiny microchromosomes. Repeats may have been responsible for the larger sizes of mammalian chromosomes. However, because we do not share as many DNA sequences with the chicken as we do our fur-bearing relatives, those that we do have in common are likely to represent the long-sought control regions of the genome.

From Chimps to Humans Most comparisons of the human genome to those of other species seek similarities. But comparisons of our genome to the genome of the chimpanzee *Pan troglodytes,* our closest rel-

ative, aim to find the opposite—genetic *differences* may refine our knowledge about what makes humans unique, discussed in Reading 16.1. Our genomes differ by 1.2 percent, equaling about 40 million DNA base substitutions. Within those differences may lie the answers to compelling medical questions. Our cancer rates differ; humans are susceptible to malaria, but chimps are not; humans develop Alzheimer disease, and chimps do not; and the course of HIV is deadlier than the chimp version, SIV.

As researchers compare genomes, an international consortium is cataloging species according to selected short genome regions that vary among species but not much within species. The goal is to establish genetic "barcodes" for all species, much as barcodes are used to identify products in a store. The barcode for animals, for example, is a 648-base segment of the mitochondrial gene for the enzyme cytochrome c oxidase I. Genetic barcodes will be useful in identifying hard-to-culture microbes and hard-to-cultivate multicellular organisms.

Key Concepts

1. Comparative genomics usually seeks conserved sequences among species to correlate them with evolutionary transitions.
2. Comparisons between the chimp and human genomes focus on differences that help to distinguish our species.
3. Species can be distinguished using genetic "barcodes."

RELATING THE CONCEPTS

1. Did the researchers who searched the algal and plant genomes for sequences in common with the human genome look for conserved or nonconserved sequences?
2. The experiments that identified a gene that causes Bardet-Beidl syndrome looked at the genomes of two organisms that are common laboratory specimens—the alga *Chlamydomonas reinhardtii* and the plant *Arabidopsis thaliana.* Another research group published similar data at the same time, and they looked at eight model organisms. What are the benefits and limitations of doing comparative genomics on these species, compared to less familiar ones?

22.5 Into the Future

With the turn of the century, genetics was catapulted from a somewhat obscure science to an increasingly eclectic field (**figure 22.10**). The blossoming of genomics has provided a new view of humanity in the larger context of all life on earth.

On a personal level, human genome information will fuel an approach that is opposite that of classical genetics—focusing on the healthy many, rather than the ill few, to better understand disease. For example, researchers are assembling a "healthy cohort"—a group of exceptionally disease-free and vigorous

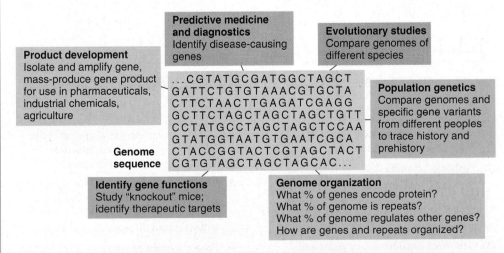

Figure 22.10 Some applications of human genome information.

people whose genomes can be searched to reveal shared gene variants that may keep them well. This project is not an attempt to define a "perfect" genetic blueprint, but one that might suggest new types of treatments based on how particular gene variants optimally function and interact. It is a broader version of the centenarian genome project, discussed in Reading 3.1. Comparing the lucky healthy cohort to groups of people who share the same molecularly defined disorder will help to pinpoint the exact deviations that underlie the pathology.

Another type of study made feasible by the sequencing of the human genome will scrutinize the genomes of people at high risk for an illness who do not have it. Why does one 40-year, two-pack-a-day smoker *not* develop lung cancer, while another does? Why do only some people who inherit *BRCA1* mutations develop breast cancer?

At the same time that human genome information reveals the underpinnings of health, the identification of disease-causing genes that has proceeded for half a century will continue, but with unprecedented precision. Breast cancer will no longer be considered "a" disease, but several distinct errors in cell cycling—sharing a phenotype, but not a genotype, and hence requiring different treatments. Each genetically influenced disease in an individual will be considered in the context of the entire genome.

Future treatments based on genome information will not be limited to correcting or circumventing genetic flaws, or supplying missing proteins. Discovering which gene variants contribute to which diseases, and how they do so, will enable us to identify risk factors that are easier to control—those from the environment. (This has, of course, been obvious for years for diseases such as lung cancer that have clear environmental triggers.) With the entire protein-encoding part of the human genome already available on a tiny chip, and fast, economical genome sequencing coming, the testing that Laurel and Mackenzie undergo in chapter 1 may soon represent a small fraction of what is possible.

Epilogue: Genome Information Will Affect You

Since the dawn of humanity, people have probably noted inherited traits, from height and body build, to hair and eye color, to talents, to behavioral quirks, to illnesses. Genetics provides the variety that makes life interesting.

The science of genetics grew out of questions surrounding plant and animal breeding, then became human-oriented in the mid-twentieth century with the recognition that certain characteristics and conditions run in families, sometimes recurring with predictable frequencies. Today, genetics and genomics are often in the news, and are beginning to impact many areas of clinical medicine. Once covered only in highly specialized scientific journals, genetic and genomic analyses of diseases are regularly featured in the top medical journals, as physicians embrace a field that was not so long ago barely touched upon in medical school.

You will likely encounter genetic technologies discussed in this book. In the near future, you might

- be offered a DNA microarray test to diagnose or treat a medical condition.

- serve on a jury and be asked to evaluate DNA profiling evidence.

- undergo a panel of genetic tests before trying to have a child.

- seek preimplantation genetic diagnosis to ensure that your child does not inherit a particular gene.

- help a parent or other loved one through chemotherapy, with some assurance, thanks to DNA testing, that the most effective drugs will be tried first.

- eat a genetically modified fruit or vegetable (you likely have already done this).

- receive a body part from a pig, with cell surfaces matched to your own.

- take medicine manufactured in a transgenic organism.

The list of applications of genetic technology is long and ever-expanding. I hope that this book has offered you glimpses of the future and prepared you to deal personally with the choices that genetic technology will present to you. Let me know your thoughts!

Ricki Lewis
ralewis@nycap.rr.com

Summary

22.1 How Genetics Became Genomics

1. Genetic maps have increased in detail and resolution, from cytogenetic and linkage maps to physical and sequence maps.

2. **Positional cloning** discovered individual genes by beginning with a phenotype and gradually identifying a causative gene, localizing it to a particular part of a chromosome.

3. Genomics considers many genes and compares genomes of different species.

22.2 The Human Genome Project Begins

4. Automated DNA sequencing and the ability of computers to align and derive long base sequences were vital to sequencing the human genome.

5. In the Sanger method of DNA sequencing, DNA fragments differing in size and with one labeled end base are aligned, and the sequence read off from the end bases.

6. People thought of sequencing the human genome for different reasons. The project officially began in 1990 under the direction of the DOE and NIH.

22.3 Technology Drives the Sequencing Effort

7. **Expressed sequenced tags** (ESTs) enable researchers to find protein-encoding genes and provide landmarks for genome sequencing.

8. In 1995, with better technology, the focus shifted from mapping to sequencing. Interest in proteomics began.

9. Several copies of a genome are cut and the pieces sequenced, overlapped, and aligned to derive the continuous sequence.

10. The International Consortium used BAC clones pinned to sequence tagged sites on chromosomes. Celera Genomics used a whole genome shotgun strategy.

11. The first draft of the genome sequence was announced in 2000.

22.4 Comparative Genomics

12. Identifying conserved regions among genomes of different species reveals genes with vital functions.

13. Distinctions between the chimp and human genome sequences can indicate how we differ from other primates.

22.5 Into the Future

14. Comparing the genomes of a healthy cohort to groups of people with certain diseases may reveal how gene combinations maintain health. Clues to health also lie in the genome sequences of people at high risk for certain disorders who do not become ill.

15. Understanding how genes contribute to disease will reveal more controllable risk factors.

Review Questions

1. How did the following technologies contribute to sequencing the human genome, or to the start of genomics?
   a. expressed sequence tags
   b. positional cloning
   c. automated DNA sequencing
   d. assembler computer programs

2. Why is a cytogenetic map less precise than an RFLP or a sequence map?

3. In 1966, Francis Crick suggested that knowing the genome of a simple bacterial cell would reveal how life works. What evidence indicates that he was mistaken?

4. Two difficulties in sequencing the human genome were the large number of repeat sequences, and the fact that DNA is double-stranded. Explain how these characteristics complicated sequencing efforts.

5. Why must several copies of a genome be cut up to sequence it?

6. If a researcher wanted to attempt to create a genome of a free-living life form, which organism could serve as a model? Cite a reason for your answer.

7. Why has availability of the human genome sequence made positional cloning obsolete?

8. Bacteria have only one DNA molecule (often called a chromosome). Which approach to genome sequencing would be used on them?

9. Many mammalian species have genomes that are about the same size as ours, but have different numbers of genes. How is this possible?

10. Why is it easier to detect conserved DNA sequences that control gene expression by comparing the human genome to a fish genome than to another primate genome?

Applied Questions

1. Celera Genomics actually sequenced several different genomes (one of which was J. Craig Venter's). Why would the sequences not be identical?

2. Suggest a few species you believe should have their genomes sequenced, and what information you think the sequences might reveal.

3. Which criteria do you think should be followed to decide the order in which genomes from different species are sequenced?

4. Restriction enzymes break a sequence of DNA bases into the following pieces:

   T T A A T A T C G

   C G T T A A T A T C G C T A G

   G C T T C G T T

   A A T A T C G C T A G C T G C A

   C T T C G T

   T A G C T G C A

   G T T A A T A T C G C T A G C T G C A

   How long is the original sequence? Reconstruct it.

5. One newly identified human gene has counterparts (homologs) in bacteria, yeast, roundworms, mustard weed, fruit flies, mice, and chimpanzees. A second gene has homologs in fruit flies, mice, and chimpanzees only. What does this information reveal about the functions of these two human genes with respect to each other?

6. It is possible now to measure your blood pressure using a machine available in a supermarket, or mail a blood sample to a laboratory to test for HIV infection or hereditary hemochromatosis. Do you think it would ever be useful, or safe, to be able to mail a plucked hair or cheek scraping to a laboratory and receive an analysis of your genome?

7. Suggest a study design to reveal why siblings with the same genetic disorder may experience different severities of the family's illness.

8. Headlines about sequencing genomes of such unusual organisms as sea squirts and pufferfish often serve as material for comedians. Why, scientifically, is it important to sequence the genomes of a variety of organisms?

9. If the "$1,000 genome" can be achieved, how should we decide who should have their genome sequenced, under what conditions, and who should have access to the resulting information?

Web Activities

Visit the Online Learning Center (OLC) at www.mhhe.com/lewisgenetics7. Select **Student Edition, chapter 22,** and **Web Activities** to find the website links needed to complete the following activities.

10. Go to the ELSI web pages. Discuss one societal concern arising from genomics, and how it might affect you.

Case Studies

11. After the tsunami that devastated countries bordering the Indian Ocean on December 26, 2004, many organisms never before seen washed up on shore. Presumably the undersea earthquake shook up their habitats and the waves took them to new places. Researchers collected specimens and sequenced DNA to try to classify the fishes. Consider the following

8-base segment known from cytogenetic analysis to be similar among the species:

a. Write the sequences for the two most closely related fishes based on this genome segment.

b. Which position(s) in the sequence are highly conserved?

c. Which position(s) in the sequence are the least conserved?

d. Which site is probably not essential, and how do you know this?

e. A coelacanth has the sequence C T A C T G G T for this section of the genome. According to the pictogram, which of the mystery fishes is the coelacanth's closest relative?

12. Reread chapter 1, look through this book again, and devise a plan for yourself. Which genes would you like to know about, and when? What will you do with the information? Will you tell anyone what you find out? Who?

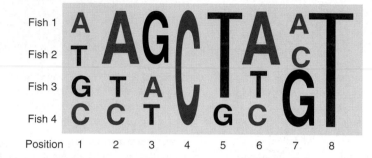

Fish 1	A	AG	C	TA	A	AT
Fish 2	T	A G	T	A	C	T
Fish 3	G	TA	C	T	T	G
Fish 4	C	CT	C	G	C	GT
Position	1	2 3	4	5	6 7	8

Learn to apply the skills of a genetic counselor with these additional cases found in the *Case Workbook in Human Genetics.*

Diffuse large B cell lymphoma

Muscle cell DNA microarray

Answers
to Relating the Concepts

Chapter 2

Page 22: Bone marrow cells, stem cells from cornea, nerve cells from retina

Page 38: Michael's brain had to receive signals from his eyes. In the mice the injected bone marrow stem cells homed in response to signals in the eyes.

Page 43: Michael's eye was injured, and so replacing the damaged tissue healed it. The mice, in contrast, had an inherited disease that could continue to affect the eye if mouse cells remained there.

Chapter 3

Page 63: The embryo

Page 64: The chicken experiments provide information on the critical period because the defect was caused at various times in prenatal development.

Chapter 4

Page 79:

1. bb
2. BB and Bb
3.

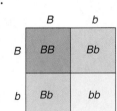

Page 90:

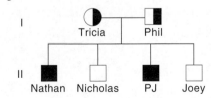

Chapter 5

Page 97: A gene can vary in many ways because it is made up of a long sequence of nucleotides.

Page 99: Variable expressivity

Page 100: Alkaptonuria is pleiotropic because all tissues that have melanin may be affected.

Chapter 6

Page 114: Ms. J's slightly masculine characteristics were not due to the absence of a gene *(SRY)*, but to a mutation in a different gene *(Wnt4)*.

Chapter 7

Page 136: The trait is very variable and has several genetic and environmental causes.

Page 141:

1. 2 Asians
2. Risk is 4% that condition will recur in a second child, but risk based on population (Native American) is .0036 × .0036. Counselor should state higher risk.

Page 146: Autism, anorexia nervosa, and bipolar disorder

Page 147: Identify a SNP pattern found only in people with the condition and the mutant gene.

Chapter 8

Page 160:

1. pedigree (if autosomal dominant)

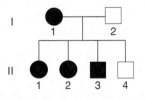

2. In this family the eating disorders can be inherited, but they could also be environmental, in which the children have learned to imitate their mother's behavior.

Chapter 9

Page 181: Mutations can occur in flu viruses as their genetic material replicates. Mutations enable the virus to infect new hosts, cause more severe symptoms, or spread more readily.

Chapter 10

Page 200:

1. Accumulating lipid blocks the part of the ER that untangles misfolded proteins.
2. The disorder is pleiotropic because any type of protein can misfold.

Chapter 11

Page 207: The hemoglobin molecule normally changes in form as time progresses and oxygen needs change. Similarly, cells in the same lineage can be looked at as stages in a continuum, from stem cell to progenitor cell to increasingly specialized daughter cells. The common theme is change over time.

Page 209: Gene expression profiling can predict which drugs will work without causing side effects, and how likely the cancer is to progress and spread.

Page 214: The ALL and MLL cells are probably more alike because they are both white blood cells and have the same function. Although the two types of pancreas cells derive from the same gland, they have very different functions and therefore very different sets of mRNAs.

Chapter 12

Page 219: The father's blood has both wild type and sickle hemoglobin. The mother's would be only wild type. (This assumes that the mother is not a carrier.)

Page 222: The mutation at a site other than the sixth amino acid is Juanita's second mutation. It is spontaneous because her father (and presumably her mother) do not have it.

Page 227: Missense

Chapter 13

Page 254: He is a mosaic.

Chapter 14

Page 279: The DNA analysis was able to localize the body geographically, but not specifically to a family. Other forensic evidence included the stomach contents and bone mineral content.

Chapter 15

Page 299:

1. When milk was introduced into the diet, people with lactose intolerance, the wild type condition, may have felt too ill to have many children. Gradually, the wild type allele's frequency would have diminished due to this selective pressure.

2. An example of negative selection is any condition that impairs an individual's fertility – perhaps a blood disorder that makes someone too weak to reproduce.

3. Lactose intolerance is wild type – that is, most common – when and where inability to digest milk does not affect an individual because the diet does not include this food.

Page 300: Geography played a role in the evolution of this gene because allele frequencies correspond to where agriculture brought dairy to a population.

Chapter 16

Page 311: *Australopithecus garhi* may have overlapped with *Homo habilis*, who overlapped with *H. erectus*.

Page 321: Either hypothesis could apply, depending upon who the Flores people really were and when their ancestors left Africa. Their DNA must be compared to that of modern peoples to distinguish between the hypotheses.

Page 322: mtDNA sequences would indicate where the founding females came from, and Y chromosome sequences would track the males.

Chapter 17

Page 346: Polio returned, despite the effectiveness of the vaccine, because for a vaccine to work on a population level, everyone must be vaccinated (or nearly everyone). When parents in Nigeria stopped vaccinating their children due to fear, the disease returned and spread.

Chapter 18

Page 356: My thyroid cancer arose in my thyroid, not in an oocyte. Therefore, it was not an inherited disease.

Page 374: Gene expression profiling enables physicians to subtype cancers that affect the same organ with much greater specificity, allowing choice of more effective therapies.

Chapter 19

Page 379: The technique is the same as the pig example, but the specific problem that is remedied is different. Therefore the use could be argued as being new and useful, although perhaps obvious.

Page 381: The original recombinant DNA experiments enabled bacteria to manufacture animal proteins. The pig example is an animal that produces a bacterial protein.

Chapter 20

Page 400: 1/4

Page 407:

1. Both

2. Somatic

Chapter 21

Page 427: IVF

Chapter 22

Page 435: Positional cloning must examine all of the genes in a particular chromosomal region. The comparative genomics approach, however, is based on function. It rules out possibilities much faster than does positional cloning.

Page 445:

1. Conserved.

2. The benefits of using well-studied organisms is that their genomes are well known. However, some diversity can be missed by focusing on just a few species.

Answers
to End-of-Chapter Questions

Chapter 1 Overview of Genetics
Answers to Review Questions

1. Gene pool, genome, chromosome, gene, DNA

2. **a.** An autosome does not carry genes that determine sex. A sex chromosome does.

   b. Genotype is the allele constitution in an individual for a particular gene. Phenotype is the physical expression of an allele combination.

   c. DNA is a double-stranded nucleic acid that includes deoxyribose and the nitrogenous bases adenine, guanine, cytosine, and thymine. DNA carries the genetic information. RNA is a single-stranded nucleic acid that includes ribose and the nitrogenous bases adenine, guanine, cytosine, and uracil. RNA carries out gene expression.

   d. A recessive allele determines phenotype in two copies. A dominant allele determines phenotype in one copy.

   e. Absolute risk refers to an individual's personal risk. Relative risk is in comparison to another group, and is less precise.

   f. A pedigree is a chart of family relationships and traits. A karyotype is a chart of chromosomes.

   g. A gene is a sequence of DNA that encodes a protein. A genome is all the DNA in a cell.

3. Inherited disease differs from other types of diseases in that recurrence risk is predictable for particular individuals in families; predictive testing detection is possible; and different populations have different characteristic frequencies of traits or disorders.

4. CF is caused by one malfunctioning gene. Height reflects the actions of several genes and environmental influences, such as diet.

5. A mutant is a variant, usually uncommon, but not necessarily harmful.

6. A genetic test can predict symptoms.

Answers to Applied Questions

1. The mutation in the German boy blocks the gene's effect, causing muscles to overgrow. In the poet, a mutation caused too much myostatin to be produced, decreasing muscle size.

2. Restricting publication of pathogen genome sequences might or might not prevent development of bioweapons, but would hamper research. An answer might balance the gains possible from microbiological research—such as new vaccines and treatments—against the possibility of abuse of that information.

3. 16

4. An individual shares more genes with a sibling (1/2) than with a first cousin (1/8).

5. The benefits: more convictions and exonerations would be obtained at the risk of invading the privacy of many people. Use of a population database is not foolproof. An adopted person might have an identical twin who is a suspect in a crime, and not know it. The database would implicate both twins. To be effective, DNA data should not be the only evidence considered.

6. "ATCG" on the http://gnn.tigr.org/articles/art_gallery.shtml website depicts James Watson's DNA.

7. An example is race and medicine. Correlating gene variants to a socially defined racial group may help in matching drug to patient, but could also foster discrimination and lead to medical errors because the association is not absolute.

8. It is useful to the many people who have been exonerated, but it is an invasion of privacy for the majority of people who are not criminals. However, people may sacrifice individual privacy protection if it eases law enforcement.

9. Many cartoons state that the genetic code is human and was recently cracked, that geneticists are evil or rich or both, and that genetically modified foods are known to be dangerous.

10. Oxidative stress profile for skin health and aging; obesity susceptibility profile; osteopenia

    Pro: If you have one, you can change your behavior to lower the risk.

    Con: If you think the problem is genetic, you may stop efforts to follow a healthy lifestyle. The tests are SNP based. Treat skin aging before wrinkles appear. Dangers unknown.

11. The wife

12. More likely than general population: coronary artery disease, kidney cancer, lung cancer, depression.

    Less likely than general population: addictive behaviors, diabetes.

13. **a.** DNA not included

    b. No DNA, government run

    c. Opinion. Important questions include informed consent of participants, whether they will benefit, privacy, and use of information.

14. Multiple sclerosis is multifactorial, and because of interacting, multiple causes, precise recurrence risks are not possible.

15. **a.** It is crucial to know what percentage of people with the carpal tunnel gene variant actually develop the condition, and whether an individual's job assignment would overwork the wrists.

    b. The test is not scientifically sound because a predisposition indicates increased probability of a future event, not an absolute prediction.

    c. Do not compel a person to take a genetic test, such as threatening job loss.

Chapter 2 Cells
Answers to Review Questions

1. **a.** 4 **b.** 6 **c.** 2 **d.** 1 **e.** 7 **f.** 3 **g.** 5

2. **a.** Tubulin forms microtubules and actin forms microfilaments, which comprise the cytoskeleton.

   b. Caspases carry out apoptosis.

   c. Changing levels of cyclins and kinases regulate the cell cycle.

   d. Checkpoint proteins provide choices during the cell cycle.

   e. Cellular adhesion molecules allow certain cell types to stick to each other.

3. Hormones, growth factors, cyclins, and kinases.

4. Specialized cells express different subsets of all the genes that are present in all cell types, except for red blood cells.

5. **a.** A bacterial cell is usually small and lacks a nucleus and other organelles. A eukaryotic cell contains membrane-bounded organelles, including a nucleus, that compartmentalize biochemical reactions.

  b. During interphase, cellular components are replicated. During mitosis, the cell divides, distributing its contents into two daughter cells.

  c. Mitosis increases cell number. Apoptosis eliminates cells.

  d. Rough ER is a labyrinth of membranous tubules, studded with ribosomes that synthesize proteins. Smooth ER is the site of lipid synthesis and lacks ribosomes.

  e. Microtubules are tubules of tubulin and microfilaments are rods of actin. Both form the cytoskeleton.

  f. A stem cell has greater developmental potential than a progenitor cell.

  g. A totipotent cell can differentiate as any cell type; a pluripotent cell's fate is more restricted.

6. **a.** Signal transduction: A cell surface receptor binds a first messenger, which is changed and binds a regulator within the cell, activating it. This activates an enzyme that catalyzes a reaction that produces a second messenger, which induces the cell's response. The body can't function if cells can't communicate.

  b. Apoptosis: A death receptor on a cell's surface receives a signal that activates caspases on the inside to cut up the cell and its DNA. Adhesion fails. Mitochondria make more caspases. Surface changes attract phagocytes. The cell blebs and bursts and is mopped up. Apoptosis prevents overgrowth.

7. Intermediate filaments have a similar shape and supportive function as part of the cytoskeleton, but consist of different proteins than microfilaments and microtubules.

8. Compartmentalization separates biochemicals that could harm certain cell constituents. It also organizes the cell so it can function more efficiently.

9. The plasma membrane is the backdrop that holds many of the molecules that intercept incoming signals. In response, these molecules contort in ways that amplify and spread the message.

10. The genome from a stem cell derived from an IVF embryo is that of the fertilized ovum; the genome in a stem cell obtained with SCNT is that of the nucleus donor, typically a patient.

Answers to Applied Questions

1. **a.** Lack of cell adhesion can speed the migration of cancer cells.

  b. Impaired signal transduction can block a message to cease dividing.

  c. Blocking apoptosis can cause excess mitosis, and an abnormal growth.

  d. Lack of cell cycle control can lead to too many mitoses.

  e. If telomerase is abnormal, a cell might not cease to divide when it normally would.

2. Because enzymes are proteins, genes encode them.

3. Nucleus and lysosomes

4. Stem cells maintain their populations because each mitosis produces a daughter cell that differentiates, as well as one that remains a stem cell.

5. Too frequent mitosis leads to an abnormal growth. Too little mitosis can limit growth or repair of damaged tissues. Too much apoptosis can kill healthy tissue. Too little apoptosis can lead to abnormal growths.

6. A cell in an embryo would not be in G_0 because it has to divide frequently to support the high growth rate.

7. Mitochondria

8. Disrupting microtubules can prevent the spindle apparatus from either forming or breaking down, either of which would halt mitosis. A drug that disables telomerase might enable chromosomes to shrink, which might stop mitosis. A drug can intercept an extracellular signal to divide.

9. Signals from outside the cell interact with receptors embedded in the plasma membrane, and the plasma membrane's interior face contacts the cytoskeleton.

10. **a.** Abnormal chloride channels in cell membranes of lung lining cells and pancreas.

  b. Lack of a transport protein in peroxisomes leads to buildup of long-chain fatty acids.

  c. Abnormal growth factors and signal transduction cause nerve overgrowth under skin.

  d. Lack of CAMs impairs wound healing.

  e. Syndactyly is a failure of apoptosis to fully separate digits.

11. Stem cells in the body may sense and respond to signals that indicate injury or a crisis. They may home to the affected site and divide to form cells that can heal the injury.

12. AIDS patient donates a cell with a nucleus, such as a skin fibroblast. Donor oocyte has its nucleus removed or destroyed. Patient's nucleus injected into oocyte. Oocyte with donor nucleus allowed to develop to blastocyst stage. Inner cell mass cells removed and cultured to become ES cells, then appropriate biochemicals applied to stimulate development to form a thymus gland. Tissue grafted into AIDS patient.

13. Senate bill 245 and House of Representatives bill 234 prohibit and criminalize reproductive cloning and SCNT to treat patients and in research. S303 and HR801 ban reproductive cloning but allow SCNT for research.

14. U.S., Germany, Norway = neither

  U.K., Israel, China = both

  Canada, France = IVF ok, no SCNT

  Belgium, Sweden = no IVF, SCNT ok

15. **a.** Anthony was more likely to have it, because his abnormal cells were not replaced immediately, as were Julia's.

  b. The stem cells came from an umbilical cord, not an embryo.

  c. Julia might have relapsed because not all of her affected cells were obliterated prior to the cord blood stem cell transplant.

Chapter 3 Development
Answers to Review Questions

1. **a.** 2 **b.** 2 **c.** 1 **d.** 2 **e.** 1 **f.** 1 **g.** 1

2. Male: Sperm are manufactured in seminiferous tubules packed into the testes, and mature and are stored in the epididymis. The epididymis is continuous with the vas deferens, which carries sperm to the urethra, which sends them out through the penis. The prostate gland and the seminal vesicles secrete into the vas deferens, and the bulbourethral glands secrete into the urethra, forming seminal fluid.

  Female: Oocytes develop within the ovaries and are released into the uterine tubes, which carry them to the uterus, which narrows to form the cervix, which opens to the vagina.

3. 2^{39}. This is an underestimate because it does not account for crossing over.

4. Mitosis divides somatic cells into two daughter cells with the same number of chromosomes as the diploid parent cell.

Meiosis forms gametes, in which the 4 daughter cells have half the number of chromosomes of the parent cell. Recombination occurs in meiosis.

5. Both produce gametes, but oogenesis takes years and spermatogenesis takes months.

6. In female gamete maturation, most of the cytoplasm concentrates in one huge cell. In male gamete maturation, 4 same-size sperm derive from an original cell undergoing meiosis.

7. Hundreds of millions of sperm are deposited in the vagina during intercourse. There, the sperm are chemically activated and the oocyte secretes a sperm attractant. Sperm tail movement and contraction of the woman's muscles aid sperm movement. When sperm contact follicle cells, their acrosomes release enzymes that penetrate the oocyte. When the plasma membranes of sperm and oocyte meet, a wave of electricity spreads physical and chemical changes over the oocyte surface, blocking other sperm. The nuclei approach, merge, and the chromosomes of the two cells meet. The fertilized ovum is a zygote.

8. Inner cell mass, zygote, morula, gastrula, notochord, fetus

9. Bone marrow cell becoming liver or muscle cell

10. Teratogens are more dangerous to an embryo than they are to a fetus because structures form in the first eight weeks.

11. If the birth defect is caused by a mutation, it can be passed to the next generation. If it is caused by a teratogen, it cannot.

12. Teratogens include thalidomide, alcohol, excess nutrients, cocaine, cigarettes, and infections. Exposure to teratogens during critical periods can have drastic effects.

13. Aging refers to the passage of time and we cannot reverse this process.

14. Accelerated aging disorders and studies on adopted individuals indicate an inherited component to longevity.

Answers to Applied Questions

1. Answers may vary. A commonly used cutoff point for experimentation is after the second week, when primary germ layers begin to form.

2. The chromosomes in a polar body resulting from the first meiotic division would be replicated, unlike those from the second meiotic division.

3. Yes

4. Stem or progenitor cells

5. A polar body does not have enough cytoplasm and organelles to support an embryo.

6. Some people believe that a pregnant woman should be held legally responsible for knowingly exposing an embryo or fetus to a teratogen.

7. Opinion

8. Prenatal environmental stress, such as lack of nutrients, can alter gene expression in ways that enhance utilization of nutrients, which is adaptive for the fetus but may set the stage for future diabetes or heart disease. Evidence is epidemiological and based on animal studies.

9. Don't: smoke, eat nitrites in cured meats, eat charred meats, eat a lot of fat, drink excess alcohol, breathe polluted air, drink excess coffee. Avoid excess sun, calories and risky behavior. Exercise.

10. Pigment patterns of cloned calves differ.

11. a. The basis of these tests is that the cells whose DNA is tested descend from the fertilized ovum, and should therefore be genetically identical to it.

   b. Cleavage embryo

   c. The structures in Diana and Max's fetus are more detailed than those in Anna and Peter's, but organs are present in both.

12. You must weigh the risk that one twin will die if they are separated versus the quality of their lives if they remain attached.

Chapter 4 Mendelian Inheritance

Answers to Review Questions

1. The law of segregation derives from the fact that during meiosis, alleles separate into different gametes. The law of independent assortment is based on the fact that distribution of alleles of two unlinked genes into gametes occurs at random.

2. The two laws of inheritance were derived from pea crosses. Without knowing about chromosomes, Mendel observed offspring phenotypes and predicted results of crosses.

3. a. An autosomal recessive trait is inherited from carriers and affects both sexes. An autosomal dominant trait can be inherited from one parent, who is affected. Autosomal recessive inheritance can skip generations; autosomal dominant inheritance cannot.

   b. Mendel's first law concerns inheritance of one trait. The second law follows inheritance of two genes on different chromosomes.

   c. A homozygote has identical alleles for a particular gene, and a heterozygote has different alleles.

   d. The parents of a monohybrid cross are heterozygotes for a single gene. Parents of a dihybrid cross are heterozygous for a pair of genes.

   e. A Punnett square tracks the distribution of alleles of genes on different chromosomes from parents to offspring. A pedigree depicts family members and their inherited traits.

4. If Mendel had chosen two traits on the same chromosome, he would have observed a higher percentage of offspring inheriting two alleles together rather than independently.

5. Blood relatives share alleles inherited from common ancestors.

6. The Egyptian pedigree only traces genealogy and not inheritance of traits. The pedigree a genetic counselor would use today would trace inherited traits too.

7. The parents of a person who is homozygous dominant for HD must be heterozygotes.

8. 100 percent

Answers to Applied Questions

1. a. Phenotype =all short   Genotype =all tt

   b. Phenotype =all tall   Genotype =all Tt

   c. Phenotype =all tall   Genotype =all TT

2. Heterozygotes

3. All white

4. Phenotype = all green inflated pods
Genotype = 9 VvGg, 3 V_gg, 3 vvG_, 1 vvgg

5. Gina and Spencer must be FfRr. Therefore their children have ratio of straight brown to frizzled brown to straight red to frizzled red of 9:3:3:1.

6. 0

7. Yes, because it isn't corrected in his sperm.

8. Autosomal dominant—it affects both sexes, and occurs every generation.

9. Pedigree

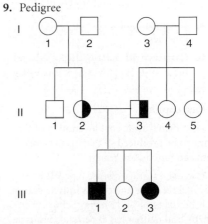

10. Carriers are I-3, I-4, II-4, III-3, III-4.

11. One in 2 chance the child has double eyelashes.

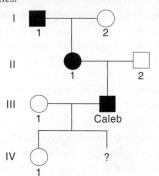

12. Consanguinity

13. Genotypic ratios:

4/32 = BbHhEe	2/32 = bbHhee
4/32 = bbHhEe	1/32 = BbHHEE
2/32 = BbHhEE	1/32 = BbhhEE
2/32 = BbHHEe	1/32 = BbHHee
2/32 = BbhhEe	1/32 = Bbhhee
2/32 = BbHhee	1/32 = bbHHEE
2/32 = bbHhEE	1/32 = bbhhEE
2/32 = bbHHEe	1/32 = bbHHee
2/32 = bbhhEe	1/32 = bbhhee

Phenotypic ratios:

9/32 = yellow urine, colored eyelids, short fingers

9/32 = red urine, colored eyelids, short fingers

3/32 = yellow urine, normal eyelids, short fingers

3/32 = red urine, normal eyelids, short fingers

3/32 = yellow urine, colored eyelids, long fingers

3/32 = red urine, colored eyelids, long fingers

1/32 = yellow urine, normal eyelids, long fingers

1/32 = red urine, normal eyelids, long fingers

14. II-1 and II-2 must be carriers. All of the individuals in generation I could be carriers.

15. a. Uncles and nieces having children together.

b. Carriers = III-5, III-6, III-9, IV-1 and niece, IV-6, IV-11, V-4, V-5, V-6, V-9, V-10

16. No

17. First cousins

18. His normal allele enables enough of the enzyme to be produced so that his nervous system can function normally.

19. Glycogen storage disease type VII is autosomal recessive and causes muscle cramps with exercise. Edna and Murray Schwartz are in their 70s, and neither has experienced muscle pain with exercise, although they are both sedentary, so would not know. Their son, Zeppo,

is a distance runner, as is his wife, Marsha. They are surprised when their daughter Kelly wants to try out for the gymnastics team, but becomes paralyzed with cramps upon exertion.

Macroglossia is autosomal dominant, and causes a large tongue, feeding problems, and snoring. The McDoofis family has a long history of babies with feeding problems, and loud snorers. In each generation, at least one individual has a very large tongue.

20. The retinoblastoma gene on chromosome 20 and the cystic fibrosis gene on chromosome 7 independently assort. The genes for xeroderma pigmentosum and brachydactyly type B1, each on chromosome 9, do not.

21. a. Autosomal recessive

b. Both sexes are affected, and it skips generations

c. Pedigree

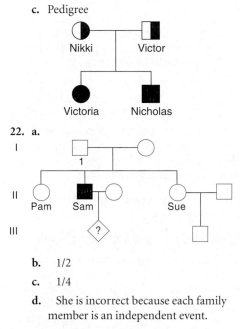

22. a.

b. 1/2

c. 1/4

d. She is incorrect because each family member is an independent event.

23. .10 × .002 = .0002 = 2 in 10,000

Chapter 5 Extensions and Exceptions to Mendel's Laws
Answers to Review Questions

1. a. Lethal alleles eliminate a progeny class that Mendel's laws predict should exist.

b. Multiple alleles create the possibility of more than two phenotypic classes.

c. Incomplete dominance introduces a third phenotype for a gene with two alleles.

d. Codominance introduces a third phenotype for a gene with two alleles.

e. Epistasis eliminates a progeny class when a gene masks another's expression.

f. Incomplete penetrance produces a phenotype that does not reveal the genotype.

g. Variable expressivity can make the same genotype appear to different degrees.

h. Pleiotropy can make the same genotype appear as more than one phenotype because subsets of effects are expressed.

i. A phenocopy mimics an inherited disorder, but is an environmental effect.

j. Conditions with the same symptoms but caused by different genes (genetic heterogeneity) will not recur with the frequency that they would if there was only one causative gene.

2. Dominant and recessive alleles are of the same gene, whereas epistasis is an interaction of alleles of different genes.

3. It can skip generations in terms of phenotype.

4. The I^A allele is codominant with the I^B allele; both are completely dominant to i.

5. Smoking

6. Maternal inheritance describes transmission of mitochondrial genes, which sperm do not usually contribute to oocytes and therefore these traits are always passed from mothers only. Linked genes are transmitted on the same chromosome. Mendel's second law applies to genes transmitted on different chromosomes.

7. Only females transmit maternally inherited traits. All of a woman's children inherit a mitochondrial trait, but a male does not pass the trait to his children.

8. Additional genes may affect expression of a particular gene, and therefore the same genotype for that gene may be associated with a different phenotype in different individuals.

9. Twenty-four linkage groups, including 22 pairs of autosomes, and the X and the Y.

Answers to Applied Questions

1. a. D **b.** A **c.** E **d.** C **e.** H **f.** F **g.** G **h.** G **i.** E

2. Haplotypes can reveal if people without symptoms have the haplotype associated with the condition. These people are non-penetrant. If the genotype is lethal, individuals with the haplotype containing the lethal allele should not exist.

3. Pleiotropic because each form of the disease has several symptoms. Genetically heterogeneic because they are caused by mutations in different genes.

4. a. 1/2 **b.** 1/2 **c.** 0 **d.** 0

5. 1/4

6. a. This alters the phenotype.

b. Bombay phenotype

7. It would be Mendelian because the gene that causes the condition when mutated is in the nucleus.

8. $.45 \times .05 = .0225$

9. The sperm are all of genotype *hhsese*. The oocytes are of genotype *HhSese*, with the alleles in cis:

$$
\begin{array}{c c}
H & \underline{\quad} & Se \\
\overline{h} & & se
\end{array}
$$

All sperm are *hse*.

Oocytes: *HSe* 49.5% *hse* 49.5%
  Hse 0.5% *hSe* 0.5%

Chance offspring like father =
  1 (*hse*) × 49.5 (*hse*) = 49.5

10. a. Male: *RrHhTt*
    Female: *rrhhtt*

  b. Parental progeny classes: Round eyeballs, hairy tail, 9 toes; square eyeballs, smooth tail, 11 toes

    Recombinant progeny classes: Round eyeballs, hairy tail, 11 toes; square eyeballs, smooth tail, 9 toes; round eyeballs, smooth tail, 11 toes; square eyeballs, hairy tail, 9 toes

  c. Crossover frequency between eyeball shape (*R*) and toe number (*T*): Round eyeballs, 11 toes = 6 + 4; square eyeballs, 9 toes = 6 + 4; crossover frequency = 20/100 = 20%

11. Ehlers-Danlos syndrome, FG syndrome, Kabuki syndrome, VATER association

12. Kearn-Sayre syndrome, heart failure, infections, anemia, hearing loss, pancreatic failure, visual loss

13. chromosome 3: genes for sodium channel, voltage-gated, type V, alpha and contactin 3 and chondroitin sulfate proteoglycan 5

14. Disorders that are caused by more than one mutation include muscular dystrophy, hemophilia, myotonic dystrophy, and epidermolysis bullosa.

15. a. a, b, c, d

  b. People with one copy have abnormalities so mild that they are not noticeable without a test. People with two copies can be severely affected.

  c. The parents are heterozygotes, so each of their offspring has a 1 in 4 probability of inheriting two wild type alleles, like Tina.

  d. Phenotype

Chapter 6 Matters of Sex

Answers to Review Questions

1. Sex is expressed at the chromosomal level as inheriting XX or XY; at the gonadal level by developing ovaries or testes; at the phenotypic level by developing male or female internal and external structures; and at the gender identity level by feelings.

2. Genes in the pseudoautosomal region are the same as certain X-linked genes, but X-Y genes share only sequence similarities.

3. a. Female **b.** Female **c.** Female

4. Absence of the *SRY* gene product and activation of other genes cause the Müllerian ducts to develop into ovaries. The ovaries produce female hormones, which influence the development of external and internal reproductive structures.

5. Feelings of very young children; twin studies in which identical twins are more likely to both be homosexual than are fraternal twins; fruit fly behavior; genes on the X chromosome that segregate with homosexuality.

6. A female homozygous dominant for an X-linked dominant allele would probably be so severely affected that she would not be alive or healthy enough to reproduce.

7. Coat color in cats is X-linked. In females, one X chromosome in each cell is inactivated, and the pattern of a calico cat's coat depends on which cells express which coat color allele. A male cat, with only one coat color allele, would have to inherit an extra X chromosome to be tortoiseshell or calico.

8. Inactivation of the gene in some cells but not others results in a patchy phenotype.

9. Each cell in a female's body contains only one active X chromosome, which makes females genetically equivalent (in terms of X-linked genes) to males.

10. An X-linked trait appears usually in males and may affect structures or functions not distinct to one sex. A sex-limited trait affects a structure or function distinct to one sex. A sex-influenced trait is inherited as a recessive in one sex and dominant in the other.

11. Mouse zygotes with two female pronuclei or two male pronuclei are abnormal. In humans, two male and one female genome in the same embryo yields placental tissue, while two female and one male genome yields a normal embryo with an abnormal placenta.

12. Genomic imprinting

Answers to Applied Questions

1. a. 1/2 **b.** 1/2 **c.** 1/2 **d.** A carrier might have symptoms if the mutated gene is expressed in tissues affected by the condition.

2. Girls may have milder cases because X inactivation might inactivate a more severe allele in some cells, and because hormonal differences from males may affect the phenotype.

3. The unevenness of the teeth of affected females may reflect expression of the defect in only some cells as the result of X inactivation.

4. Genomic imprinting

5. a. Autosomal recessive

  b. Sex-limited

  c. Males can pass it on.

6. Rett syndrome, OMIM 312750: autism, dementia, loss of hand use, jerky movements, seizures, poor growth

7. Cornelia de Lange syndrome, which causes growth retardation, characteristic facial features, and mental retardation, comes from the mother. Wolf-Hirschhorn syndrome, with severe growth retardation, microcephaly, "Greek helmet face," heart defects, and cleft lip or palate, comes from the father.

8. 1/4

9.

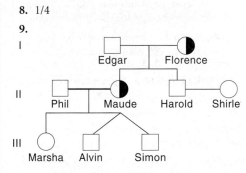

10. a. Because the phenotype is not severe.

  b. 1 in 2, because she is a carrier.

  c. 1 in 1

11. 1/2

12. a. Carriers = I-4, II-3, III-2, III-5 **b.** 1/2
  c. 1/2 **d.** Women would have to inherit kinky hair disease from an affected father and a carrier mother; affected males would not live long enough to have children.

Chapter 7 Multifactorial Traits

Answers to Review Questions

1. Body weight is more likely to be multi-factorial than eye color, because eating and exercise habits greatly influence weight. We can't change eye color.

2. FH, leptin deficiency, lipodystrophy, melanocortin-4 receptor deficiency

3. A Mendelian multifactorial trait is caused by one gene and environmental influences, whereas a polygenic multifactorial trait is caused by more than one gene and environmental influences.

4. Eye color has a greater heritability because the environment has a greater influence on height than on eye color.

5. Heritability for skin color changes because the duration and intensity of sun exposure contribute to skin color.

6. More genotypes specify a medium brown skin color than other colors.

7. **a.** 1/8 **b.** 1 /2 **c.** 1/4 **d.** 1/8

8. **a.** Empiric risk is based on observations of trait prevalence in a particular population or group of individuals.

   b. Twin studies approximate the degree of heritability by comparing trait prevalence among pairs of MZ twins to DZ twins. The greater the difference, the higher the heritability.

   c. Inherited traits may stand out in an adoptee's family where each member lives in the same environment, but the adopted individual has different genes.

   d. Association studies establish correlations between particular gene variants or sections of chromosomes (haplotypes) and inheritance of certain traits or susceptibilities.

9. Many people must be surveyed to determine whether a certain haplotype (and its SNPs) is exclusively associated with a particular phenotype. Enough DNA must be considered so that several SNPs are analyzed, and a case-control strategy must be employed to ensure that the same SNPs are not found among affected individuals as well as unaffected ones. Finally, if many SNPs are considered, many combinations must be correlated to phenotypes.

10. Cardiovascular: apolipoproteins, lipoprotein lipase and proteins that regulate blood pressure or homocysteine metabolism

Body weight: neuropeptide Y, genes that control leptin, melanocortin-4 receptor

11. During starvation, shrinking fat cells release less leptin, which stimulates neuropeptide Y and inhibits production of melanocortin-4 receptors. Appetite increases.

Answers to Applied Questions

1. Marbles = 1/8, Juice = 1/32, Angie = 1/16. Juice is the best choice.

2. General population risk is 0.1%, but risks are considerably higher if a blood relative is affected, suggesting inherited component.

3. **a.** The different affected sibs results and heritabilities from two populations suggest ADHD has a large environmental component. Increased relative risk among sibs and in adoptees of affected individuals also suggests a large environmental component.

   b. Differences in classroom restrictiveness.

   c. Attempt to identify specific genes and determine effects their protein products would have to exert to cause symptoms of ADHD.

   d. Drugs for ADHD are overprescribed, because diagnosis is based on symptoms, which might not always reflect illness. Knowing genetic susceptibility could avoid prescribing drugs to people in whom they would not work.

   e. A genetic test for ADHD might also identify highly creative, talented people! If there is a gene for ADHD, it is likely affected by other genes in ways that make the overall phenotype a talent, not a disorder.

4. A drug for obesity would block neuropeptide Y or stimulate production of the melanocortin-4 receptor.

5. A genetic change in a population would not occur in so short a time. Therefore, the cause for the increase in obesity is largely environmental, perhaps acting on inherited susceptibilities.

6. Heritability, 0.54, suggests the influence of genes and the environment is about equal.

7. Broad

8. http://www.bovineengineering.com/Heritability_Table.html traits with heritability <.5 in cattle: carcass loin eye area; efficiency of gain in feed lot; yearling weight; carcass tenderness; length of teats

9. 3 causes of death with high heritability: Alzheimer disease, cancer, asthma. These are multifactorial. 3 causes of death with low heritability: HIV infection, firearms injury, accidents. These are largely environmentally caused.

10. Yes. Because MZ twins are about 8 times as likely to both have anorexia as are DZ twins.

11. **a.** Topography of back of iris intensifies the color, or Tanya inherited dark pigment alleles from each parent

    b. Different allele combinations

    c. Height

    d. Height is genetically determined as evidenced by Jamal and Tanya's above average height since early childhood. Effect of diet is seen in their greater height compared to the parents, who did not eat as well.

    e. Eye color

Chapter 8 The Genetics of Behavior
Answers to Review Questions

1. Proteins whose products are involved in neurotransmission or signal transduction can affect behavior.

2. ADHD has a very distinctive phenotype; it is common; and linkage studies point to a candidate gene. In contrast, autism is rare, and linkage points to several candidate genes, suggesting genetic heterogeneity (different ways to inherit the same condition).

3. **a.** sleep disorders **b.** addiction **c.** addiction

4. We must understand how a gene variant affects behavior, and how other genes and the environment affect its expression.

5. Genetic heterogeneity; when a behavior is part of several disorders; behaviors that fall within the range of normal, although they are extreme; ability to imitate a behavior.

6. With age, the environment has had longer to affect gene expression.

Answers to Applied Questions

1. **a.** Environmental

   b. Better diagnosis, recognizing more forms

2. **a.** The same neurotransmitter controls different behaviors. The differences arise from which neurons are involved.

   b. Dopamine

3. Instead of demonstrating the existence and extent of inherited influences, DNA microarrays implicate specific genes, revealing clues to the biological basis of a behavior or behavioral disorder.

4. A SNP profile indicating increased risk could be helpful if the person can take action to lower the risk of developing the condition, but can be harmful if the person just gives up, thinking that fate cannot be avoided. SNP profiling might also be used to discriminate.

5. This is an opinion question. If drugs are outlawed based on their addiction potential, then alcohol and nicotine should be included with other drugs that are illegal. An alternate approach would be to outlaw drugs that impair one's ability to function. Such an approach might outlaw alcohol and cocaine, yet permit tobacco or possibly marijuana use.

6. Yes—a criminal cannot alter his or her inherited behavior. No—the environment can override inherited criminal tendencies.

7. Depression and bipolar disorder each might actually be several conditions that have similar symptoms.

8. Identify what is abnormal in the Utah family, develop a drug to treat the condition, and then try the drug on elderly people who are having difficulty sleeping.

9. The IQ test could be flawed or measure intelligence in a way that does not compensate for differences in educational opportunities.

10. Environmental component about equals genetic susceptibility

11. Research can isolate a chromosomal region where the DNA sequence is common to people with Wolfram syndrome, and eventually discover and describe the causative gene. The specific nature of the corresponding protein product may explain how a half normal gene dose causes the behavioral problems of relatives.

12. These data suggest a large inherited component to alcoholism.

13. Schizophrenia, bipolar disorder, unipolar depression

14. Advantages of genetic testing for an eating disorder: Alleviates guilt and can warn people at high risk so that they can take preventive action in themselves or in their children. Disadvantages: Genetic testing will not change the phenotype and may even discourage someone from trying to control the condition and stigmatization. Does not address underlying societal problem of pressure to be thin.

15. Non-genetic. Environmental cues can cause eating behavior, but metabolism is genetically controlled, so some people have more control over their weight than others.

16. a. Yes. Concordance is high and risk is elevated in first degree relatives.

    b. SSRI

    c. Do affected individuals have shared experiences that could explain development of OCD?

17. Both activate a pregnant woman's immune system, which may have affected the fetuses in similar ways.

18. How certain are they that the mouse phenotype represents the human phenotype? Are the doses similar? Do children with autism whose parents blame the condition on vaccines have genetic variants that indicate they are extra sensitive to mercury?

Chapter 9 DNA Structure and Replication
Answers to Review Questions

1. DNA is replicated so that it is not used up in directing the cell to manufacture protein.

2. Sugar, base, phosphate

3. Purines have a double ring structure, and pyrimidines have a single ring.

4. Width wouldn't be constant.

5. The nucleotide base sequence encodes information.

6. One end of a strand of nucleotides has a phosphate group attached to the 5′ carbon of deoxyribose. The other end has a hydroxyl group attached to the 3′ carbon. The opposite (complementary) strand has the reverse orientation.

7. 1E   2C   3D   4B   5A

8. Helicases, primase, DNA polymerase, ligase

9. An RNA primer temporarily starts each new DNA strand.

10. Histone protein, nucleosome, chromatin

11. It is highly folded.

12. Strands separate and are held apart. Primase makes RNA primer. DNA polymerase adds DNA bases to RNA primer. Proofreading, repair. Continuous on one strand only. RNA primers removed. Ligase seals sugar-phosphate backbone.

13. The strands of the DNA double helix run in opposite directions.

14. RNA forms a primer to which DNA polymerase can bind.

15. Hershey and Chase showed that DNA is the genetic material and protein is not. Meselson and Stahl showed that DNA replication is semiconservative, but not conservative or dispersive.

Answers to Applied Questions

1. The sugar-phosphate backbone of replicating DNA cannot attach.

2. Primase, DNA polymerase, helicase, binding proteins, and ligase are required for DNA replication.

3. Lacking DNA polymerase makes life impossible, because cells cannot divide.

4. a. A G C T C T T A G A G C T A A

   b. G G C A T A T C G G C C A T G

   c. T A G C C T A G C G A T G A C

5. Determining the structure of DNA led to discoveries of the mechanism of heredity, whereas sequencing the human genome provided information.

6. a. Freeze cells

   b. Thaw cells and activate development

   c. Only if there is a way to create organisms from stored DNA sequences.

7. CCACCCTTGGAGTTCACTCA GGTGGGAACCTCAAGTGAGT

8. Heteroplasmy of mtDNA

Chapter 10 Gene Action: From DNA to Protein
Answers to Review Questions

1. a. H bonds between A and T and G and C join the strands of the double helix.

   b. In DNA replication, a new strand is synthesized semiconservatively, with new bases inserted opposite their complementary bases to form a new strand.

   c. An mRNA is transcribed by aligning RNA nucleotides against their complements in one strand of the DNA.

   d. The sequence preceding the protein-encoding sequence of the mRNA base pairs with rRNA in the ribosome.

   e. A tRNA binds to mRNA by base pairing between the three bases of its anticodon and the three mRNA bases of a codon.

   f. The characteristic cloverleaf of tRNA is a consequence of H bonding between complementary bases.

2. The direction of the flow of genetic information in retroviruses is opposite that of the central dogma.

3. a. The start of a gene

   b. tRNA

   c. A pseudogene

   d. rRNA

   e. mRNA

4. a. Proteins with an incorrect sequence of amino acids may not function.

   b. If the initial amino acid is released, additional amino acids cannot add on.

   c. If rRNA cannot bind to the ribosome, then mRNAs cannot be translated into protein.

   d. If ribosomes cannot move, then a protein would not exceed two amino acids in length.

e. If a tRNA picks up the wrong amino acid, the protein's amino acid sequence will be abnormal.

5. Both respond to environmental conditions.

6. RNA contains ribose and uracil and is single-stranded. DNA contains deoxyribose and thymine and is double-stranded. DNA preserves and transmits genetic information; RNA expresses genetic information.

7. DNA replication occurs in the nucleus of eukaryotes. Prokaryotes have no nucleus so replication occurs in the cytoplasm. The same is true of transcription. In eukaryotes, translation occurs in the nucleus and cytoplasm, and some ribosomes are attached to the ER.

8. Transcription controls cell specialization by turning different sets of genes on and off in different cell types.

9. The same mRNA codon can be at the A site and the P site because the ribosome moves.

10. In transcription initiation, the DNA double helix unwinds locally, transcription factors bind near the promoter, and RNA polymerase binds to the promoter.

11. mRNA is the intermediate between DNA and protein, carrying the genetic information to ribosomes. tRNA connects mRNA and amino acids, and transfers amino acids to ribosomes for incorporation into protein. rRNA associates with proteins to form ribosomes.

12. Post-transcriptional changes to RNA include adding a cap and poly A tail, and removing introns.

13. Ribosomes consist of several types of proteins and rRNAs in two subunits of unequal size.

14. An overlapping code constrains protein structure because certain amino acids would always be followed by the same amino acids in every protein.

15. Transcription and translation recycle tRNAs and ribosomes.

16. Proinsulin is shortened to insulin after translation. RNA editing shortens apolipoprotein B post-transcriptionally.

17. Each of these types of abnormalities could affect more than one protein.

18. The amino acid sequence determines a protein's conformation by causing attractions and repulsions between different parts of the molecule.

19. A two-nucleotide genetic code would only specify 16 types of amino acids. There are 20 amino acids in biological proteins.

20. To create a mature mRNA for translation, snurps (small RNAs and proteins) excise introns in pre-mRNAs. Ribozymes assist in peptide bond formation. Proteins make up part of the structure of ribosomes, and enzymes are involved in protein synthesis.

21. Primary is amino acid sequence, secondary is close interactions, and tertiary is far apart interactions. Primary is most important in determining structure.

22. Protein folding starts right away and is very fast. Delay the process and the protein won't fold.

Answers to Applied Questions

1. a. 46
   b. 44

2. a. A A U G U G A A C G A A C U C U C A G
   b. U G A A C C C G A U A C G A G U A A T
   c. C C G A C G U U A U C G G C A U C U A
   d. C C U U A U G C A G A U C G A U C G U

3. a. C G A T A G A C A G T A T T T T C T C C T
   b. C A C C G C A T A A G A A A A G G C C C A T C C
   c. C T C C C T T A A G A A A G A G T T C G T T C A
   d. T C C T T T T G G G G A G A A T A A T A T C T A

4. Many answers are possible, using combinations of *his* (CAU or CAC), *ala* (CGU, GCC, GCA, GCG), *arg* (CGU, CGC, CGA, CGG), *ser* (AGU, AGC, AGA, AGG), *leu* (CUU, CUC, CUA, CUG), *val* (GUC, GUG), and *cys* (UGU, UGC).

5. Several answers are possible because of the redundancy of the genetic code. One answer is: C A T A C C T T T G G G A A A T G G.

6. There is only one genetic code.

7. Use ACA with any triplet other than CAA, and see whether threonine or histidine occurs. Whichever occurs, ACA encodes.

8. 26,927

9. 125, but this doesn't account for synonymous codons.

10. 1

11. 3,777 DNA bases (plus promoter sequence)

12. Cystic fibrosis. Chaperones would ensure that the CFTR ion channel folds correctly so that it can take its normal place in the plasma membrane.

13. a. *Candida cylindrical* (fungus) CTG = serine, normal CUG = leucine
    b. *Mycoplasma* (microorganism) CGG is not used, normal = arginine
    c. *Acetabularia* (alga) UAA = glutamine, normal = STOP

14. http://medweb.bham.ac.uk/http/depts/clin_neuro/teaching/tutorials/parkinsons1.html
Parkinson disease and Lewy body dementia brains accumulate neurofilaments that ubiquitin binds, indicating they are destined for proteasomes.

http://www.lougehrigsdisease.net/als_causes_of_als.htm
"The aggregates may be a byproduct from overwhelmed cells attempting to repair incorrectly folded proteins."

http://www.the-scientist.com/yr2003/may/hot_030519.html
Huntingtin protein plugs up proteasomes.

15. a. Shortens   b. No change   c. Shortens
    d. Lengthens   e. No protein made

Chapter 11 Control of Gene Expression
Answers to Review Questions

1. Transcription and translation must be controlled so that the appropriate amounts of proteins are present for specific cell types.

2. Adipose cells contain mostly fat. Muscle cells are filled with contractile proteins. Liver cells have many detoxification enzymes.

3. After the genetic code was elucidated, the steps of transcription and translation needed to be determined. After sequencing the genome, genes and their functions had to be identified, and the functions of non-protein encoding sequences discovered.

4. Oxygen level

5. Progenitor cells in the pancreas divide to give rise to daughter cells that differentiate as either exocrine or endocrine cells.

6. A genetic change is a change to DNA. An epigenetic change is not to DNA, but to something that affects it.

7. Histones are proteins, so genes encode them. Yet histones control which genes are transcribed, and under what conditions.

8. Methylation, RNA interference

9. Pattern of binding of methyl, acetyl, and phosphate groups

10. The nucleic acid sequence of the viral gene to be silenced.

11. It combines different parts of a gene.

12. Most of the genome doesn't encode protein. DNA encodes rRNAs and tRNAs, both vital for protein synthesis. Just because we do

not know what a DNA sequence does doesn't mean it does not have a function.

13. Genes come in pieces. Genes can move. The same sequence of DNA can encode more than one protein because of exon shuffling.

14. Introns interrupt genes. Exons from different genes can contribute to the same protein. An intron on one strand may be an exon on the other.

15. One protein is degraded faster than the other.

16. Encoding rRNA and tRNA; introns; repeats; promoters and other control sequences

Answers to Applied Questions

1. Dermatomics = genes that affect the skin; musculomics = genes that affect muscle; imaginomics = genes that affect creativity

2. Acetylase

3. Parts of the protein-encoding genes mix and match. The genes contain many large introns.

4. Because histones control expression of many if not all genes, altering histones' methylation patterns could affect several genes, thereby causing multiple symptoms.

5. Make an siRNA that blocks the mRNA transcribed from the fusion gene.

6. RNAi is more specific in blocking expression of a particular gene.

7. 3

8. Alternate splicing

9. Transcription factors are most needed before birth.

10. SARS inhibition

11. The promoter for the adjacent gene activates the aromatase gene.

12. Gene expression patterns and the alleles differ. The man has mutations in oncogenes or tumor suppressor genes and those that control blood pressure. The woman has wild type alleles of these genes.

Chapter 12 Gene Mutation
Answers to Review Questions

1. A germinal mutation occurs in a gamete or a fertilized ovum, and therefore affects all the cells of an individual and is more serious than a somatic mutation, which affects only some tissues and is not transmitted to future generations.

2. The gene for collagen is prone to mutation because it is very symmetrical.

3. A spontaneous mutation can arise if a DNA base is in its rare tautomeric form at the instant

when the replication fork arrives. A wrong base inserts opposite the rare one.

4. Mutational hot spots are direct repeats or symmetrical regions of DNA.

5. Gaucher disease is caused by an insertion, a missense mutation, or a crossover with a pseudogene.

6. Mutations in the third codon position can be silent. Mutations in the second position may replace an amino acid with a similarly shaped one. 61 codons specify 20 amino acids.

7. (1) A degenerate codon, (2) a mutation that replaces an amino acid with a structurally similar one, (3) a mutation that replaces an amino acid in a nonessential part of the protein, and (4) a mutation in an intron

8. A conditional mutation is only expressed under certain conditions, such as increased temperature or exposure to particular drugs or chemicals.

9. Frameshift, deletion, duplication, insertion, transposable element

10. Retention of an intron and expanding triplet repeats may provide a new function for a gene, which may cause disease.

11. A jumping gene can disrupt gene function by altering the reading frame or shutting off transcription.

12. A new recessive mutation will not become obvious until two heterozygotes produce a homozygous recessive individual with a phenotype.

13. The gene is expanding.

14. Short repeats can cause mispairing during meiosis. Long triplet repeats add amino acids, which can disrupt the encoded protein's function, often adding a function. Repeated genes can cause mispairing in meiosis and have dosage-related effects.

15. Whether the symptoms differ in type rather than severity.

16. Excision repair corrects ultraviolet-induced pyrimidine dimers. Mismatch repair corrects replication errors. They use different enzymes.

17. Conditional

18. Division and distribution of newly replicated and parental DNA strands are skewed so that strands most likely to have new mutations go to the most mature cell types.

19. **a.** CAU (his) to CAC (his)

    b. GGU (gly) to UGU (cys)

Answers to Applied Questions

1. Environment because the mutation is the same in the U.S. and Africa.

2. The frameshift mutation can create a stop codon, leading to a shortened polypeptide.

3. Any change that produces UAA, UAG, or UGA.

4. A transcription factor controls the expression of several genes, in a time and tissue-specific manner. Therefore a mutation in it affects several genes, producing multiple symptoms.

5. GAU to AAU or GAC to AAC

6. The second boy's second mutation, further in the gene, restores the reading frame so that part of the dystrophin protein has a normal structure, providing some function.

7. *asn* to *lys:* AAU to AAA AAC to AAG *ile* to *thr:* AUU to ACU AUC to ACC AUA to ACA

8. Nonsense

9. *Arg* to *his:* CGU to CAU CGC to CAC

10. GAU to GUU or GAC to GUU

11. Nonsense

12. The repair enzymes could correct UV-induced pyrimidine dimers.

13. **a.** *de novo,* because not in parents

    b. The mutation creates an intron splice site.

    c. Emery-Dreifuss muscular dystrophy, dilated cardiomyopathy, Dunnigan-type familial partial lipodystrophy, limb girdle muscular dystrophy, obesity, Charcot-Marie-Tooth disease

14. Marcia's case is more severe because she lacks CFTR protein, whereas Jan is missing part of the protein.

15. They shouldn't be concerned, because they have mutations in two different genes.

16. GGA (gly) to AGA (arg) GGG (gly) to AGG (arg)

Chapter 13 Chromosomes
Answers to Review Questions

1. Essential parts of a chromosome are telomeres, the centromere, and origin of replication sites. A centromere includes repeats of alpha satellites; centromere-associated proteins; and centromere protein A.

2. Protein-encoding genes become denser from the telomeres inward toward the centromere.

3. Centromeres and telomeres contribute to chromosome stability and have many repeats.

4. **a.** Homologs do not separate in meiosis I or II, leading to a gamete with an extra or missing chromosome.

    b. DNA replicates, but is not apportioned into daughter cells, forming a diploid gamete.

c. Increased tendency for nondisjunction in the chromosome 21 pair.

d. Crossing over in the male yields unbalanced gametes, which can fertilize oocytes, but too much or too little genetic material halts development.

e. A gamete including just one translocated chromosome will have too much of part of the chromosome, and too little of other parts. Excess chromosome 21 material causes Down syndrome.

5. Patches of octaploid cells in liver tissue may arise as a result of abnormal mitosis in a few liver cells early in development.

6. a. A XXX individual has no symptoms, but she may conceive sons with Klinefelter syndrome by producing XX oocytes.

b. A female with XO Turner syndrome has wide-set nipples, flaps of skin on the neck, and no secondary sexual development.

c. A female with trisomy 21 Down syndrome. Phenotype includes short, sparse, straight hair, wide-set eyes with epicanthal folds, a broad nose, protruding tongue, mental retardation, and increased risk of a heart defect, suppressed immunity, and leukemia.

7. Basketball players may have an extra Y chromosome that makes them tall.

8. Triploids are very severely abnormal. Trisomy 21 is the least severe trisomy, and involves the smallest chromosome. Klinefelter syndrome symptoms are worse if there is more than one extra X chromosome.

9. A balanced translocation causes duplications or deletions when a gamete contains one translocated chromosome, plus has extra or is missing genes from one of the chromosomes involved in the translocation. A paracentric or pericentric inversion can cause duplications or deletions if a crossover occurs between the inverted chromosome and its homolog. Isochromosomes result from centromere splitting in the wrong plane, duplicating one chromosome arm but deleting the other.

10. Chromosomes would not contort during meiosis because their genes are aligned.

11. a. FISH: Fluorescently-labeled DNA probes bind homologous regions on chromosomes.

b. Amniocentesis: Fetal cells and fluid are removed from around a fetus. Cells are cultured and their chromosomes stained or exposed to DNA probes, and karyotyped.

c. Chorionic villus sampling: Chromosomes in chorionic villus cells are directly karyotyped.

d. Fetal cell sorting: A fluorescence activated cell sorter separates fetal from maternal cells, and fetal chromosomes are karyotyped.

12. Trisomies 5 and 16 involve chromosomes with higher gene densities than 13 and 18. Therefore, trisomies 5 and 16 stop development earlier and are less likely to be seen in spontaneous abortions and births than trisomies 13 and 18.

13. 48

14. A female cannot have Klinefelter syndrome because she does not have a Y chromosome, and a male cannot have Turner syndrome because he has a Y chromosome.

15. Nondisjunction in oocyte. Nondisjunction in sperm. Large deletion in X chromosome.

Answers to Applied Questions

1. a. 35

b. Karen Martini

c. The risks in the right hand column are higher because they include many conditions.

d. One in 66 is greater than 1 percent.

e. Age 40 trisomy 21 risk = 1/106. Age 45 risk = 1/30. Her risk approximately triples in the five years.

2. The person is a girl missing part of the short arm of chromosome 5. This is cri du chat syndrome, and she will be mentally retarded with a catlike cry.

3. FISH

4. a. Reciprocal translocation

b. She doesn't have extra or missing genes.

c. She might have a child with translocation Down syndrome.

5. At the second mitotic division, replicated chromosomes failed to separate, yielding one of four cells with an extra two sets of chromosomes.

6. Down syndrome caused by aneuploidy produces an extra chromosome 21 in each cell. In mosaic Down syndrome, the extra chromosome is only in some cells. In translocation Down syndrome, unbalanced gametes lead to an individual with extra chromosome 21 material in each cell.

7. The XXY son could have gotten two X chromosomes from his mother, or an XY-bearing sperm from his father.

8. a. A translocation carrier can produce an unbalanced gamete that lacks chromosome 22 material.

b. The microdeletion may be more extensive than the deleted region in the translocation individuals.

c. Translocation family members might be infertile or have offspring with birth defects.

9. a. 4 + 5, 9 + 10, 11 + 12   The lower chromosome # should have more bases.

b. 19

c. Y

d. 6.2 times

10. Student should create a karyotype.

11. Chromosome 13:

cataract-clouded lens	OMIM	601885
deafness		220290
leukemia/lymphoma—blood cancer		602221
ectodermal dysplasia—no hair or nails, dark skin		129500

12. a. Uniparental disomy

b. The first condition might arise from an oocyte that has two copies of the long arm of chromosome 14 being fertilized by a sperm that lacks this segment. The situation would be the reverse for the second condition.

c. A deletion mutation could remove the copy of the gene that is expressed.

13. One of the Watkins probably has a balanced translocation, because there is more than one Down syndrome case. The two spontaneous abortions were the result of unbalanced gametes. Their problems are likely to repeat with a predictable and high frequency, because the translocated chromosome is in half of the carrier parents' gametes. In contrast, the Phelps' child with Down syndrome is more likely the result of nondisjunction, which is unlikely to repeat. The Phelps child has trisomy 21 Down syndrome; the Watkins' child may only have a partial extra copy of chromosome 21.

Chapter 14 When Allele Frequencies Stay Constant
Answers to Review Questions

1. A gene pool refers to a population.

2. Evolution is not occurring.

3. Knowing the incidence of the homozygous recessive class makes it possible to derive the "q" part of the Hardy-Weinberg equation.

4. The possibility of two unrelated Caucasians without a family history of CF having an

affected child is $1/4 \times 1/23 \times 1/23 = 1/2{,}116$. If one person knows he or she is a carrier, the risk is $1/4 \times 1/23 = 1/92$.

    5. a. Protein-encoding genes could affect the phenotype.

      b. VNTRs are longer than STRs so are more informative because they are rarer.

      c. Rare. Any rare trait is more helpful in narrowing down a list of suspects than a more common trait.

6. Population databases are necessary to interpret DNA fingerprints because alleles occur with different frequencies in different populations.

7. For females use the standard formula. For males, gene frequency equals phenotypic frequency.

Answers to Applied Questions

1. $q^2 = .25$ $q = .5$ Carrier percentage $= 2pq = (2)(.5)(.5) = .5$

2. $q^2 = .001$ $q = .0316$ $p = .968$
$2pq =$ carrier frequency $= (2)(.0316)(.968) = .061$

3. a. $q^2 = 1/190 = 0.005$. Square root of $0.005 = 0.071 =$ frequency of mutant allele q

    b. Frequency of wild type allele $= p = 1 - 0.071 = 0.929$

    c. Carriers $= 2pq = 2 \times 0.071 \times 0.929 = 0.132$

    d. Nonrandom mating

4. $0.1 \times 0.1 = 0.01 =$ chance both are carriers. If both are carriers, $0.25 \times 0.01 = 0.0025 = 0.25\%$ chance child will be affected.

5. $q^2 = 1/8{,}000 = 0.000125$ $q = 0.011$ $p = 1 - 0.011 = 0.989$

Carrier frequency $= 2pq = 2 \times 0.011 \times 0.989 = 0.022$

6. $4/177 = 0.0225 = 2.25\%$ are carriers

7. Students genotypes: 6 TT 4 Tt 10 tt

Parents genotypes:
of 6 TT students $= 9\ TT\ 2\ Tt\ 1\ tt$
of 4 Tt students $= 2\ TT\ 4\ Tt\ 2\ tt$
of 10 tt students $= 8\ Tt\ 12\ tt$

Students allele frequencies: $T = p = 1/2Tt + TT = 1/2\ (4) + 6 = 8/20$ students $= .4$

$t = q = .6$

Parents allele frequencies: $T = p = 1/2Tt + TT = 1/2\ (14) + 11 = 18/40$ parents $= .45$

$t = q = .55$

The gene is evolving because the allele frequencies change between generations.

8. It is a good idea for catching criminals and in deterring crime, but will compromise privacy for the majority of the population, who are not criminals.

9. a. In 1990, DNA profiling did not consider as many sites in the genome as it does now, and juries and judges did not understand the outlandish statistics.

    b. Sperm cells in victim, any somatic cells of suspects and victim

    c. Their ethnic/population group

    d. It is unfair to convict and punish suspects without using an available technology to evaluate their guilt or innocence (opinion).

    e. Eddie Joe Lloyd was exonerated in 2002, after serving 17 years in prison for a rape and murder of a 16-year-old girl in Detroit in 1984. Police tricked him into confessing by feeding him details, then using his repeating the details as evidence of knowledge of the crime. Semen stain on underwear used to strangle her was used as evidence, but not tested for DNA.

10. Autosomal recessive class $= 900/10{,}000 = .09 = q^2$ $q = .3$ $p = .7$

Normal lashes $= p^2 + 2pq = .49 + .42$

Homozygote with normal lashes $= .49/.49 + .42 = .538$

11. a. Genes 2 and 4 have 2 alleles because they generate 2 bands on the gel

    b. $(.2)^2 + (2)(.3)(.7) + (.10)^2 + (2)(.4)(.2) = .04 \times .42 \times .01 \times .16 = .0000268$

    c. The man might have close male relatives who cannot be ruled out as suspects.

    d. It makes his guilt even more likely.

Chapter 15 Changing Allele Frequencies

Answers to Review Questions

1. a. Agriculture

    b. Cities, as groups of immigrants arrive and mix in

    c. Endangered species, survivors of massacres or natural disasters and their descendants

    d. Introducing an inherited disease into a population

2. a. Highly virulent TB bacteria are selected against because they kill hosts quickly. Resistant hosts are selected for because they survive infection and live to reproduce. In this way, TB evolved from an acute systemic infection into a chronic lung infection. In the 1980s, antibiotic-resistant TB strains led to re-emergence of the disease.

    b. Bacteria that become resistant to antibiotics by mutation or by acquiring resistance factors from other bacteria selectively survive in the presence of antibiotics.

    c. Viral diversity is low at the start of infection because the immune system is vulnerable—viral variants aren't necessary. As symptoms ebb, viral diversity increases, then it decreases again as the immune system becomes overwhelmed.

    d. CF may be maintained in populations where heterozygosity protects against diarrheal disease.

3. Increasing homozygosity increases the chance that homozygous recessives will arise, who may be too unhealthy to reproduce. The population may decline.

4. A genome sequence can reveal new parts of a pathogen or biochemical reactions that a new drug can target.

5. Natural selection

6. Sickle cell disease protects against malaria.

7. The incidence of the phenotype is less than that predicted by the genotype and mode of inheritance.

8. Decreased variation in mitochondrial and Y chromosome DNA sequences indicate the genetic uniformity of a population bottleneck.

9. In the Dunker population, the frequencies of blood type genotypes are quite different from those of the surrounding U.S. population and the ancestral population in Europe.

10. The most common CF allele is more common in France (70 percent) than in Finland (45 percent). Therefore the test would be more likely to detect a carrier in France than in Finland.

11. The same mutation in a population group and linkage disequilibrium (shared haplotypes) suggest a founder effect.

12. The causes of founder effects and population bottlenecks differ. A founder effect reflects a small group moving to start a new population, whereas a population bottleneck results from removal of individuals with certain genotypes from the population.

13. A gradual cline might reflect migration over many years. An abrupt cline could be due to a cataclysmic geological event that separates two populations, such as an earthquake or flood.

14. Genetic drift is the chance sampling of some genotypes from a population, and this may lead to nonrandom mating, in which the most fit individuals reproduce more successfully. Environmental conditions help determine which genotypes reproduce and are thus selected for.

15. a. A founder effect is a type of genetic drift that occurs when a few individuals found a settlement and their alleles constitute a new gene pool, amplifying the alleles they introduce and eliminating others.

   b. Linkage disequilibrium is the inheritance of certain combinations of alleles of different genes at a higher frequency than can be accounted for by their individual frequencies.

   c. Balanced polymorphism is the persistence of a disease-causing allele in a population because the heterozygote enjoys a health or reproductive advantage.

   d. Genetic load is the proportion of deleterious alleles in a population.

16. Shifts in allele frequency should parallel people's movements, which are often described in historical records, and explained by sociology and anthropology.

Answers to Applied Questions

1. Not balanced polymorphism because being homozygous recessive for glycophorin C deficiency does not cause symptoms.

2. Tasting bitter is harmful—see whether people who can taste bitter substances are overrepresented among people with cancer, because they may have avoided protective vegetables.

   Tasting bitter is protective—see whether people who cannot taste bitter are overrepresented among those who have been poisoned.

3. Microevolution refers to changes in allele frequencies in populations. Examples: changing virulence of infectious diseases, antibiotic resistance, infectious diseases that jump species.

4. Balanced polymorphism as an explanation requires 12 rare events; drift requires one, the sequestering of the population.

5. Natural selection

6. a. All modern Afrikaners with porphyria variegata descend from the same person in whom the disorder originated.

   b. One person, who had the dental disorder, contributed disproportionately to future generations.

   c. Heterozygotes for cystic fibrosis and sickle cell disease resist certain infectious diseases, maintaining the disease-causing allele in populations.

   d. The Pima Indian population has a high incidence of an allele that predisposes to develop type 1 diabetes mellitus, but it wasn't expressed until certain dietary and lifestyle changes became popular.

   e. The Amish and Pakistani groups have high incidences of certain inherited diseases because of consanguinity.

   f. Migration patterns are responsible for the different frequencies of the galactokinase deficiency allele across Europe.

   g. The same haplotype for CJD in different populations reflects descent from a shared ancestor followed by migration.

   h. Varying mtDNA sequences along the Nile river valley are due to migration.

   i. The alleles responsible for BRCA1 breast cancer originated in different ancestors for Ashkenazim and African-Americans.

7. a. Founder effect

   b. Geographical barriers and natural selection, acting over time, make two populations have different variants of inherited characteristics.

   c. Nonrandom mating

   d. Natural selection

   e. Population bottleneck

8. Treating PKU has increased the proportion of mutant alleles in the population because without treatment, affected individuals would not have been able to reproduce.

9. a. The mutations arose independently.

   b. If other groups with many smart people have high carrier rates for these disorders. If the smart people are the carriers.

10. The high incidence is due to extreme consanguinity—nearly everyone is related to nearly everyone else. Tracking the incidence of this condition is difficult because the symptoms exist independently, and may be caused by environmental factors rather than genes.

11. Balanced polymorphism

12. a. Balanced polymorphism **b.** Founder effect **c.** Migration

13. West Nile virus-associated illness is evolving. It arrived in the U.S. in New York City in 1999 and has since spread nearly everywhere in the nation.

14. European males settled in India. Later, Asian immigrants arrived. The male European Indians had children with Asian Indian women.

Chapter 16 Human Origins and Evolution
Answers to Review Questions

1. Hominoids are ancestral to apes and humans, whereas hominids are ancestral to humans only. Therefore, hominoids are more ancient.

2. Physically, chimpanzees are not as similar to us as were the australopithecines, yet the australopithecines are in a different genus from us.

3. A single gene can control the rates of development of specific structures, causing enormous differences in the relative sizes of organs in two species.

4. The hunting behavior of *A. garhi* and the time and place where it lived suggests that it could have been a bridge between *Australopithecus* and *Homo*.

5. Fossil evidence and mtDNA evidence support an "out of Africa" emergence of modern humans. Anthropological evidence indicates three waves of migration of native Americans from Siberia, but molecular evidence suggests one migration.

6. Y chromosome and mitochondrial DNA sequences enable researchers to trace the genetic contributions of fathers and mothers, respectively.

7. Bipedalism, larger brain, improved fine coordination

8. Exons are highly conserved because they affect the phenotype by encoding protein and are therefore subject to natural selection.

9. Small insertions and deletions might alter the regulation of the same protein in chimps and humans, but not introduce a new structure or function or drastically alter an existing one.

10. Estimating passage of time requires a steady interval, such as a minute. Mutation rates vary.

11. a. Differences in genes that control rates of development of particular structures

   b. Variants of single genes with obvious effects, such as body hair

   In general, differences in gene expression can account for phenotypic distinctions between humans and chimps despite similarity in genome sequence.

12. Gene sequences are more specific because different codons can encode the same amino acids.

13. Mutation rates are not the same across genomes. Genes mutate at different rates.

14. One gene encodes one polypeptide, and so comparing the evolution of a gene tracks a tiny part of the biology of the organism. DNA hybridization assesses relationships among many genes, and thus means more. Also, much of the genome does not encode protein.

15. A chromosome band may contain many genes, and so comparing them is not specific.

16. Knowledge of a DNA, RNA, or polypeptide sequence and the mutation rate is necessary to construct an evolutionary tree diagram. An assumption is that mutation rate is constant. A limitation is that only one biochemical is considered, and not large scale characteristics such as behavior and anatomy.

17. Human and chimp genomes differ by numbers of copies of many genes and DNA sequences.

18. Keratin gene mutation accounts for minimal body hair on humans. Other possibly key single gene traits of humanity are a larger brain, longer fetal development, and spoken language. Controversial, though.

19. Sterilizing people with mental retardation. Encouraging poor people to limit family size. Avoiding marriage to a person who carries a disease-causing allele.

20. Positive and negative natural selection do not have intent to alter a population in a particular way. Positive selection increases the prevalence of an adaptive trait and negative selection acts against a harmful trait. Positive eugenics encourages reproduction and negative eugenics discourages it.

21. The hypothesis that modern humans and Neanderthals interbred was disproved by the degree of DNA sequence differences.

22. Tools from 50,000 years ago from South Carolina suggest migration to the Atlantic coast before migration from Mongolia to Pacific coast.

Answers to Applied Questions

1. a. A small duplication occurred in human chromosome 11 to give rise to the Betazoid karyotype. The Klingon and Romulan karyotypes could have arisen from fusion of human chromosomes 15 and 17.

   b. The Betazoids are our closest relatives because of greater similarity in chromosome bands and chromosome arrangement. Cytochrome *c* sequences and intron pattern in the collagen gene are identical between humans and Betazoids.

   c. They are not distinct species because they can interbreed.

   d. (3)

2. Negative eugenics:
   —sterilizing people who are mentally retarded
   —restricting certain groups from immigrating
   —encouraging people with inherited disease or carriers not to reproduce

Positive eugenics:
   —a sperm bank where Nobel prizewinners make deposits
   —people seeking very smart people as mates
   — governments paying people with the best jobs and education to have larger families

3. People in developing nations might be alarmed by white-coated strangers wielding hypodermic needles seeking blood samples. A compromise would be to collect hair instead of blood, and use PCR to amplify the genes. Another compromise is to offer vaccines in exchange for tissue samples.

4. The action was eugenic because it indirectly selected against nonwhite students. It was not eugenic in that, superficially, all applicants were evaluated using the same criteria.

5. The child would be a modern human.

6. A law can be passed stating that no laws can be passed or restrictions imposed based upon genetic data.

7. Spinal muscular atrophy is on human chromosome 5 and might be on horse chromosome 21

8. Ease of cultivation might not be a biologically sound way to select a model organism, because that species might not suffer the same inherited disorders as humans, or have the same mutant phenotypes.

9. The *BAZ1B* gene encodes a bromodomain next to a zinc finger domain. It is deleted in Williams-Beuren syndrome, which is a developmental disorder, OMIM 605681. Chimp and baboon have similar sequences.

10. Correct genetic conditions: woolly hair, dwarfism, albinism Incorrect: feeblemindedness

11. Assumes Y chromosome represents surnames. This isn't true for Jewish immigrants to the US early in the 20th century, for example. Sometimes names are changed. Y leaves out females, mtDNA leaves out males, sometimes data don't agree. These techniques do not account for mixing of populations, especially in recent time, which can be considerable. Ethnicity and race are not biological distinctions.

12. Opinion. The guidelines are eugenic if the goal is to decrease the incidence of the conditions, to improve society.

13. a. Interpretation depends upon which data are considered. One conclusion is that the known *Homo* brains are larger than the other three, supporting correlation of brain size to intelligence. However, inclusion of *Homo floresiensis,* if she really is *Homo,* contradicts this hypothesis.

    b. A lot of evidence is missing.

14. a. 6 mya = most recent common ancestor of human and chimp
       1.8 mya = *Homo* appearing
       0.5 mya = Neanderthal split from modern human lineage

    b. Lice live on us.

    c. Lice spread as *Homo* spread geographically. Need to analyze more genes.

    d. Need fossil evidence too.

15. a. Forced sterilization

    b. If a trait in one family is deemed inherited, it might be assumed to be inherited in another family, when that might not be the case.

    c. Wealthy people seeking partners among other wealthy people

    d. How common the negative traits were in the general population. Whether the family members were really related biologically. Environmental influences on the traits considered. Evidence of family members without the negative traits who have beneficial ones.

16. Males with the E-M81 subhaplotype left Africa from Morocco to go to Spain and Portugal.

Chapter 17 Genetics of Immunity
Answers to Review Questions

1. 1. d   2. e, g   3. a   4. b   5. h   6. b

2. a. T = cellular immunity, make cytokines
      B = humoral immunity, make antibodies

   b. Innate immunity is broad and inborn and adaptive immunity is specific and acquired.

   c. Primary immune response occurs on first exposure to foreign antigen and secondary response is to subsequent exposures.

   d. Cellular is T cells and humoral is B cells.

   e. Autoimmunity is attack against the body; allergy is attack against a harmless foreign antigen.

   f. Inherited immunity from genes; acquired from infection.

3. Blood type is determined by specific glycoprotein antigens on red blood cell surfaces. Incompatibility results when a person's immune system manufactures antibodies that attack red blood cells bearing antigens of other blood types.

4. a. Collapse of entire immune system

   b. Increase in viral infections and cancer

   c. Increase in bacterial infections

   d. Total collapse of immune system

5. a. Destruction of bacteria and viruses, stimulation of inflammation

   b. Destruction of bacteria, yeasts, some viruses

   c. Bind to and stimulate destruction of bacteria

   d. Cause changes in body that are inhospitable to pathogens

6. Different viruses elicit production of the same cytokines, which cause the same symptoms.

7. HIV shuts down immunity by attacking T cells, macrophages, and then other immune system cells.

8. If too different, the bone marrow will be rejected. If too similar, it may not fight the condition that the original, recipient's bone marrow could not fight.

9. HIV mutates rapidly, replicates rapidly, the site where it binds to T cells is shielded, and resistance alleles are very rare.

10. Thymocytes are selected as others die, rather than undergoing an "education."

11. Allergens stimulate production of IgE antibodies that bind mast cells, causing them to release allergy mediators. In an autoimmune disorder, antibodies attack the body's cells and tissues.

12. a. Transplanted bone marrow cells (the graft) recognize cells in the recipient (the host) as foreign. The donor's cells attack the host's cells.

    b. ADA deficiency may result in severe combined immune deficiency, in which a lack of ADA poisons T cells, which then cannot activate B cells.

    c. Autoantibodies attack cells that line joints.

    d. In AIDS, HIV infects helper T cells and reproduces, eventually killing enough T cells to overcome cellular immunity. Opportunistic infections occur.

    e. Grass antigens (allergens) induce production of IgE that causes mast cells to release histamines, causing allergy symptoms.

13. T cells control B cell function.

14. Memory and plasma B cells respond specifically to one antigen following a cytokine cue from a T cell. Plasma B cells secrete antigen-specific antibodies and are in the circulation for only a few days. Memory B cells remain, providing a fast response the next time the antigen is encountered.

15. A polyclonal antibody response attacks an invader at several points simultaneously, hastening recovery. MAbs are useful as a diagnostic tool because of their specificity.

Answers to Applied Questions

1. O negative blood lacks ABO cell surface antigens and the Rh antigens, so it is less likely to evoke a rejection reaction than other types of blood.

2. Autoimmune, because of presence of autoantibodies

3. Autoimmune

4. It is working. Lowered IgE signifies less of an allergic reaction, and increased levels of IgG and IgA indicate protection.

5. Antibiotics treat the bacterial infection, but not the inflammation in the joints, which is an autoimmune response.

6. Innate antibodies. A mutation in the host that prevents the pathogen from infecting.

7. Memory cells alert the immune system of the first exposure, and ensure that a secondary immune response occurs on subsequent exposure to the Coxsackie virus.

8. Colony stimulating factor

9. The bonobo's cells would bear antigens specific to a particular human.

10. Stockpile antibiotics locally, but control their use. Better vaccine coverage. Publicize what people should look out for, and how to disinfect themselves or suspicious materials.

11. SCID heritability is higher because it is caused by a mutation that usually removes a gene's function. An allergy in contrast, reflects an overactive, misdirected immune response, which is more acquired than inherited.

12. http://allonhealth.com/human-immune-system-booster.htm

    This website claims that "Barleygreen" ingredients "give your body the nutrients it needs to rebuild your immune system." Assumes your immune system is broken down. Product claims to prevent immune system diseases, slow aging, fight bacteria and viruses, destroy cancer cells, and distinguish "live" and "dead" nutrients. A nutrient is a chemical and cannot be alive. This product is vitamins, minerals, and enzymes. The enzymes are digested by the stomach and small intestine, before they can

function. Taking a multivitamin, or eating a balanced diet, would have the same effect as this product.

13. a. Rh

    b. At first dominant and recessive designations were used to indicate alleles within blood groups, but then numbers were used to account for the multiple alleles. Blood type antigens can interact.

    c. Blood group antigen genes will only independently assort if they are on different chromosomes. If they are linked, blood types may be inherited together within families.

14. a. Might b. Will not c. Will d. Might

15. a. African ancestry correlates to inherited gene variants that are associated with a heightened inflammatory response.

    b. Environment, particularly stress, and access to health care services

    c. People choose partners within the population or social group, which keeps alleles prevalent.

    d. Compare disease rates in populations of African Americans at both ends of the economic spectrum in a particular country.

16. a. Allograft b. Isograft c. Xenograft d. Autograft

17. There will not be an Rh incompatibility because the female would have to be Rh⁻.

Chapter 18 The Genetics of Cancer
Answers to Review Questions

1. One inherits a susceptibility to cancer, not the cancer itself. Cancer may affect only somatic tissue. Many mutations may be necessary for cancer to develop.

2. Genetic tests can subtype cancers, enabling more specific diagnoses and treatments.

3. Different cancers at the genetic level can affect the same cell type.

4. Growth factors, hormones from outside. Cyclins, kinases, telomere length from within.

5. Cancer cells divide faster than cells from which they derive. Some normal cells divide faster than cancer cells.

6. No

7. Only an inherited cancer susceptibility can pass to future generations.

8. Cancer cells divide continuously and indefinitely; they are heritable, transplantable, dedifferentiated, and lack contact inhibition.

9. Cancer stem cells; dedifferentiation; shift in percentage of tissue that consists of dividing cells; uncontrolled tissue repair.

10. Cancer is a consequence of disruption of the cell cycle. Deletion of a tumor suppressor gene and translocation of an oncogene next to a highly active gene would have the same effect of uncontrolled division.

11. a. An overexpressed transcription factor could function as an oncogene, causing too frequent cell division.

   b. Mutations in the *p53* gene lift tumor suppression, allowing too many cell divisions.

   c. Mutations in the retinoblastoma gene lift tumor suppression.

   d. The *myc* oncogene allows too frequent cell division.

   e. DNA repair fixes errors that would otherwise lead to cancer (turning on an oncogene or turning off a tumor suppressor).

   f. Mutations in the *APC* gene lift tumor suppression by making the DNA more prone to replication errors.

   g. *CHEK2* controls *BRCA1*, so a mutation in it would lift tumor suppression, causing cancer.

12. A cancer cell could have a point mutation.

13. a. A population study examines disease incidence in different populations, at any time.

   b. A case-control study analyzes pairs of people who differ only in the characteristic of interest.

   c. A prospective study is any study that evaluates results as they occur, rather than relying on recall.

14. Inducing apoptosis halts runaway cell division. Inducing differentiation enables cancer cells to specialize, which would slow down their division rate. Blocking hormone receptors prevents cancer cells from receiving signals to divide. Inhibiting angiogenesis prevents tumors from building a blood supply. Blocking telomerase stops cell division.

Answers to Applied Questions

1. Retinoblastoma

2. DNA expression microarrays could predict which women would respond to chemotherapy and/or radiation and which would not.

3. a. Missing both *p53* alleles is lethal.

   b. A person with two mutant *p53* alleles could be conceived if both parents have one mutant allele.

4. It is an oncogene because it caused overexpression of a transcription factor.

5. Tumor suppressor. The normal function prevents aneuploidy, which prevents cancer. Caretaker, because of the widespread effect on the genome.

6. The cells from the original tumor are not genetically identical.

7. Treatment for leukemia is more complex than removal of a solid organ because bone marrow must be replaced.

8. No diet can guarantee that an individual will not develop cancer.

9. Are certain *p53* mutations more prevalent among exposed workers, compared to people not exposed to PAHs? If so, the PAHs may cause the mutation.

10. Advantage – will identify patients at highest risk of cancer metastasis. Disadvantage – will miss some patients whose cancers spread by another route or mechanism.

11. A new drug might alter expression of stemness genes, but this might affect non-cancer stem cells too – perhaps even causing a cancer.

12. Tumor suppressors. These genes cause cancer when inactivated or removed.

13. If the receptor is blocked, the cell cannot receive the signal to divide.

14. a. *hTERT* keeps telomeres long by encoding telomerase. When overexpressed it is an oncogene.

   b. *p53* is a tumor suppressor gene that when underexpressed due to a germinal mutation and a somatic mutation causes Li-Fraumeni family cancer syndrome.

15. The woman could have a familial cancer syndrome.

16. a. The counselor should ask about ethnic background, and other types of cancer in the family.

   b. This woman would probably not benefit from a *BRCA1* test, because her two affected blood relatives were older when affected.

   c. A complication of *BRCA1* testing is that if no mutation is found, a person might assume that she cannot develop breast cancer. We do not know all of the ways that this disease can develop, and in fact it may be several different disorders.

17. a. Oncogene because it is overexpressed in cancer.

   b. Overexpression of this gene gets rid of p53, needed to repair DNA.

   c. The p53 protein is normal, but the extra Pirh2 will take it away.

   d. Targeting proteasomes could compromise their normal functions, allowing misfolded proteins to accumulate.

18. Are there cancer-causing mutations in the germline? If the mutation is in affected as well as non-affected tissues, it would likely also be in the germline and be transmissible to the next generation.

Chapter 19 Genetic Technologies: Amplifying, Modifying and Monitoring DNA

Answers to Review Questions

1. A disease-causing gene variant used as a diagnostic test. A transgenic pig whose parts are accepted in a human body as a transplant. A transgenic crop plant that is more nutritious than the natural version.

2. PCR and recombinant DNA technology are both done on DNA, but PCR amplifies DNA whereas recombinant DNA adds a gene.

3. a. Restriction enzymes cut DNA at specific sequences. They can be used to create DNA fragments for constructing recombinant DNA molecules.

   b. In gene targeting, genes of interest are added to embryonic stem cells where they exchange places with a homologous gene. After construction of chimeric embryos and crosses of mosaic animals, heterozygotes are bred to yield homozygous transgenic animals with a particular gene knocked out.

   c. Cloning vectors carry DNA molecules into cells.

4. Antibiotics are used to set up a system where only cells that have taken up foreign DNA can survive.

5. Human insulin DNA cut with a restriction enzyme, vector DNA cut with the same restriction enzyme, *E. coli*, DNA ligase, selection mechanism (such as antibiotic)

6. Bacteria couldn't manufacture human proteins if the genetic code was not universal.

7. A drug obtained using recombinant DNA technology does not contain contaminants found in proteins extracted from organisms.

8. Foreign DNA can be inserted on a virus, carried across plasma membranes in liposomes, microinjected, electroporated, or sent in with particle bombardment.

9. Transgenic technology is not precise because introduced DNA is not directed to a particular chromosomal locus, as it is in gene targeting.

10. All decrease gene expression.

11. A gene expression profile measures transcription in a particular cell type, whereas a DNA sequence variation analysis indicates inherited alleles, present in every cell.

12. A DNA microarray displays gene expression. A haplotype displays inherited gene variants.

Answers to Applied Questions

1. 20 cycles × 2 minutes/cycle = 40 minutes

2. PCR can directly and quickly detect HIV rRNA, rather than a delayed sign that the human body is responding to the presence of the virus.

3. A goat produces human EPO in its semen.

A mouse produces jellyfish GFP in its plasma.

A chicken produces human clotting factor in its egg whites.

4. Human collagen produced in transgenic mice is less likely to include infectious agents than collagen obtained from hooves and hides. It is also the human protein, which is less likely to stimulate an immune response than the cow type.

5. a and b are opinions. c. Perhaps people do not object to genetic modification of bacteria because bacteria are not easily visible and the drug is used to treat an illness. GM foods are more familiar, and do not improve upon an existing product, as Humulin does.

6. Mice can express human genes because they use the same genetic code.

7. Another gene specifies the same enzyme; another gene specifies a different enzyme with the same or a similar function; the enzyme isn't vital.

8. Gene targeting

9. Gene targeting shows the result of no CFTR protein. A transgenic model shows effects of an abnormal CFTR protein. Gene targeting models the most extreme expression of CF.

10. Unlikely genes could be overlooked.

11. Select protein hormones that affect emotion and measure the expression levels of their genes in blood before and after watching an emotional film. Or scan whole genome for genes expressed only during emotional situations to identify non-obvious candidate genes.

12.

Insulin	type 1 diabetes mellitus
Beta interferon	multiple sclerosis
EPO	kidney failure, anemia
TPA	stroke, coronary heart disease
CSF	restores white blood cell count

13. 20,000$^+$ patents for DNA submitted; 1,300$^+$ for human DNA sequences.

14. Object: A novel idea in 1989 is now obvious.

Support: He thought of using the noncoding DNA first.

15. Give a ewe hormones to make her superovulate. Apply ram sperm to ewe's reproductive tract. Flush out fertilized ova, and inject them with human *AAT* gene linked to sheep gene promoter that will control its expression. Place GM fertilized ova into sheep surrogate mothers. Select heterozygous offspring, breed them to obtain homozygotes. Milk female homozygotes and extract human drug.

16. **a.** A2, D4, E2, F7, G2

  b. B3, C2, C7, F1

  c. any open circle

  d. Compare samples from same people taken at different times.

  e. Compare samples from same people before and after sunburn.

Chapter 20 Gene Testing, Genetic Counseling, and Gene Therapy

Answers to Review Questions

1. It is economically feasible if it prevents births of infants who would require very expensive, lifelong health care.

2. Newborn = sickle cell disease; young adult = amniocentesis; middle age = inherited forms of Alzheimer disease

3. A genetic counselor meets with families to discuss inherited conditions, suggests and interprets DNA and chromosome and maternal serum marker tests, arranges referrals to support groups or other health care professionals.

4. Advantage of virtual counseling is that information is more widely and quickly available. Disadvantage is lack of psychological support at stressful time.

5. How common is the disorder? How invasive is the test and can it be administered with others already being done? How much money and suffering does testing save?

6. (a) Replace protein; (b) use recombinant DNA technology to obtain pure, human protein; (c) gene therapy.

7. Blood removal doesn't change the gene variant that causes the disease.

8. *Ex vivo* gene therapy alters cells outside the body, then injects or implants them. SCID is treated this way. *In situ* gene therapy is a localized procedure on accessible tissue, such as treating melanoma. *In vivo* gene therapy occurs inside the body, such as a nasal spray to deliver the CFTR gene to a person with cystic fibrosis.

9. Germline gene therapy can affect evolution because the changes are heritable.

10. Researchers should consider the amount of DNA the virus can carry, the types of cells the virus normally infects, how stable the incorporated vector is in the human genome, if toxic effects are associated with use of the virus, and whether or not the virus stimulates a strong immune response.

11. AV and AAV are viral vectors, and the dystrophin gene is stitched into them using recombinant DNA technology. A liposome is a fatty bubble that can encase the gene and introduce it into a cell across the lipid-rich plasma membrane.

12. A liver is in only one place in the body. Muscle is more difficult to treat because it comprises much of the body's bulk. It is nearly everywhere.

13. A bone marrow transplant is a cell implant; genes are not altered. Removing bone marrow and adding a functional gene is gene therapy because the DNA is altered.

14. Gene therapy makes cancer cells more "visible" to the immune system.

15. Gene therapy for SCID due to ADA deficiency provides the wild type gene to T cell progenitors.

16. Hydroxyurea reactivates production of fetal hemoglobin, which dilutes the sickled hemoglobin so that cells can reach the lungs and pick up oxygen faster, which prevents sickling.

Answers to Applied Questions

1. Testing for untreatable disorders helps doctors rule them out in diagnosing other conditions.

2. To what degree will violating the confidentiality harm the patient? To what degree will not violating the confidence harm someone else?

3. HIV would have to have the genes that make it destroy the immune system deleted.

4. Small amounts of nitric oxide could be used to prevent sickling. The cells sickle only in a low-oxygen environment.

5. There is no dopamine gene to manipulate—a gene therapy would be applied to an enzyme necessary to synthesize dopamine. It is difficult to deliver a gene therapy to the brain, because of the blood-brain barrier.

6. To treat Duchenne muscular dystrophy, the gene for dystrophin is delivered via a retrovirus into immature muscle cells.

To treat sickle cell disease, the beta globin gene is delivered as naked DNA into immature red blood cells.

To treat glioma, mouse fibroblasts are given retroviruses with a herpes gene that encodes thymidine kinase. The fibroblasts are implanted into the tumor, and an anti-herpes drug given, which selectively destroys dividing cells, including the cancer cells.

7. *In situ*, because only the scalp need be treated.

8. Jesse Gelsinger was relatively healthy. Children with Canavan disease have no other treatment options.

9. Any disease where the affected tissue can be reached and an overexpressed gene silenced or a deficient gene's activity boosted or replaced. RNA interference to squelch mRNA for huntingtin in Huntington disease. For cystic fibrosis, a viral vector to introduce *CFTR* gene into airway lining cells.

10. PKU, maple syrup urine disease, medium chain acyl-CoA dehydrogenase deficiency

11. a. The counselor should explain to the couple that even though the fetus has an extra X chromosome, the individual will most likely not have any related symptoms other than perhaps great height.

b. The counselor might suggest a technique to separate X-bearing sperm, to increase the chances of conceiving a female. Or, the couple could test the fetus for the muscular dystrophy gene and terminate an affected fetus.

c. Explain that risk increases with age, but a young woman can still conceive a child with trisomy 21—it is just rare.

d. If the parents are of normal height, then their child with achondroplasia is a new mutation, and there should be no elevated risk to other children.

e. Genetic counselors tell patients that amniocentesis can rule out certain chromosomal and biochemical disorders, but it cannot guarantee a healthy child.

12. Which patient is sicker, closer to death, more likely to be cured, did not contribute to her condition by smoking, or has more to contribute to society.

13. OMIM 259420 describes osteogenesis imperfecta type III, which, unlike the other forms, is recessive. It is autosomal. In addition to easily broken bones, the whites of the eyes may appear bluish and the teeth rotted. The genetic counselor should take a family history to determine the mode of inheritance. One or both parents may be affected if the disorder is dominant. The counselor should point out symptoms other than broken bones, describe how collagen contributes to bone structure and integrity (the phenotype), and how the disorder is transmitted (the genotype).

14. a. Phenotype

b. Hannah's blood cells do not make the missing enzyme.

c. Have prenatal diagnosis for the disorder, or preimplantation genetic diagnosis (discussed in the next chapter) to select an unaffected embryo.

15. A genetic counselor.

Chapter 21 Reproductive Technologies
Answers to Review Questions

1. a. Surrogate mother

b. Intrauterine insemination

c. Oocyte donation, preimplantation genetic diagnosis

d. Intrauterine insemination

e. Preimplantation genetic diagnosis or intrauterine insemination

f. IVF, ZIFT, or GIFT

g. Preservation of her ovarian tissue in her arm

2. It is easier, less costly, and less painful to detect infertility in men than in women.

3. A man can have up to 40 percent abnormally shaped sperm and still be considered fertile.

4. A man with a low sperm count and a woman with an irregular menstrual cycle.

5. ZIFT and GIFT occur in the uterine tube, whereas IVF takes place in the uterus. ZIFT and IVF transfer a zygote, whereas GIFT transfers gametes. In IVF, fertilization occurs outside the body.

6. They are "embryos *in vitro*" because a uterus is required for the embryos to develop—and there is no guarantee that this will happen.

7. Preimplantation genetic diagnosis is similar to amniocentesis and CVS in that it allows prenatal detection of disease-causing genes. It is different in that it takes place much earlier in gestation.

8. Endometriosis, scarred uterine tubes, irregular ovulation, nondisjunction

9. a. Fertilization occurs outside of the body.

b. Oocytes and sperm are collected and placed in the uterine tubes.

c. Conception occurs in a woman other than the one who gives birth.

d. Conception occurs outside the body, and a woman other than the genetic mother carries the fetus.

e. Conception does not occur as a result of sexual intercourse, but in a Petri dish.

f. The nucleus and cytoplasm in a cell come from different individuals.

10. Embryo transfer to host (surrogate)

11. Somatic cell nuclear transfer to yield stem cells for regenerative medicine

Answers to Applied Questions

1. An older man fathering a child does not have to alter his physiology the way a postmenopausal woman must to conceive.

2. Adoption

3. Oocyte donor

4. Big Tom illustrates intrauterine insemination. Mist illustrates surrogate motherhood.

5. a. The genetic parents are the sperm donor and the woman, who is also the gestational mother.

b. The genetic parents are the woman whose uterus is gone, and her husband. The gestational mother is the woman's friend.

c. The genetic parents are Max and Tina; the gestational mother is Karen.

d. The genetic parents are von Wormer and the Indiana woman, who was also the gestational mother.

e. The genetic and gestational mother is the woman who is the friend of the men. The genetic father can be determined if DNA profiles of the child are compared to those of the sperm donors.

6. Younger women can freeze oocytes or early embryos, to be fertilized or implanted years later.

7. Extra preimplantation embryos can be donated to infertile couples.

8. People will vary in when they think children born from assisted reproductive technologies should be told of their origins.

9. Paying for reproductive services because one is lazy is not the same as seeking assistance because one has a fertility problem.

10. The child would be a clone of Cliff.

11. Agree: it prevents abortion

    Disagree: sex selection is abhorrent

12. **a.** For year 2000 data, use of ART has increased 54% since 1996.

    b. Fresh

    c. Her own

    d. If she becomes naturally pregnant; couple who have had ART more than once: couple who have had more than one type of ART; several causes of infertility

13. **a.** "It's great that older women can have babies too!"

    b. Relief – she can freeze oocytes to use after her cancer treatment.

    c. "Why won't one of those older women adopt me?"

    d. Relief – "I can complete my education before having kids."

    e. Anger that women in another country can keep their kids and can delay having them to pursue education or a career.

14. The court might investigate why the man revoked his consent, and whether he has children. One might argue that the woman's rights should be paramount, because she is the one who would be pregnant.

15. Find some way to tag donated material so that identity mix-ups can't occur. Require several people to handle each sample, checking each other. Develop a DNA test, perhaps of SNP patterns, to uniquely identify each sample.

16. **a.** Colleen is the genetic mother and Ellen the gestational mother.

    b. Opinion

    c. Are there precedents in which the egg donor is denied custody and a surrogate awarded custody and a man is part of the picture too?

Chapter 22 The Age of Genomics

Answers to Review Questions

1. **a.** Expressed sequence tags sped the discovery of protein-encoding genes by working with cDNAs reverse transcribed from mRNAs.

   b. Positional cloning identified many important disease-causing genes.

   c. Automated DNA sequencing sped determination of the base sequence of the genome.

   d. Assembler computer programs overlapped genome pieces.

2. A cytogenetic map provides only a few landmarks per chromosome, whereas a sequence map consists of the DNA bases, and an RFLP map reflects sequence differences.

3. We still do not know what all of the genes of *E. coli* do. Having a list of genes doesn't reveal how an organism functions.

4. Repeated sequences could be mapped to several chromosomal sites. Not knowing which strand a sequence comes from could lead to duplicating some parts of the genome sequence and missing others.

5. The pieces must be overlapped to derive the sequence, so several genome copies must be used.

6. *Mycoplasma genitalium*, because it is the smallest genome sequenced.

7. If the affected protein is known its DNA sequence can be used to identify the corresponding part of the genome sequence.

8. Whole genome sequencing

9. Genomes differ widely in the percent that encodes protein.

10. Demonstrating conservation requires going farther back in time than the split of humans and chimps from a shared ancestor. The human and chimp genomes are so alike that the conserved genes cannot be distinguished from more recent ones.

Answers to Applied Questions

1. SNPs differ among individuals.

2. Species that represent leaps in evolution or that humans use

3. Criteria: size, ease, importance to agriculture, importance to human health care, importance to understanding biodiversity or evolution, economics.

4. G C T T C G T T A A T A T C G C T A G C T G C A

5. The first gene is more ancient than the second.

6. A genome analysis without someone to interpret the results could lead to decisions or actions based on erroneous information. Genes interact with each other and the environment, and many are not fully penetrant, so that genotype does not reflect phenotype. This isn't so for a blood pressure reading or cholesterol measurement, which is accurate when taken.

7. List all of the symptoms of the family's illness, and devise tests for other genes that may cause them directly, or by interacting with the product of the recognized causative gene. Also investigate different environmental exposures of the family members that might account for different symptoms or symptom severity.

8. Comparative genomics reveals evolutionary relatedness and may identify useful model organisms. For example, a sea squirt's primitive kidney is used to test new drugs.

9. If genome sequencing remains costly, access might be decided in a way similar to distribution of organs for transplant—by who could benefit most in terms of survival.

10. Psychological impact and stigma: learning that you have a high risk of developing a certain disease, or that your child-to-be has inherited an illness.

11. **a.** #1 (AAGCTAAT) and #2 (TAGCTACT)

    b. 4 and 8

    c. 1

    d. 1, because it differs in all 4 fishes

    e. 3

Glossary

A

absolute risk The probability that an individual will develop a particular condition, based on family history and/or test results.

acrocentric chromosome A chromosome in which the centromere is located close to one end.

adaptive immunity A slow, specific immune response that develops after exposure to a foreign antigen.

adenine One of two purine nitrogenous bases in DNA and RNA.

allele An alternate form of a gene.

alternate splicing Building different proteins by combining exons of a gene in different ways.

amino acid A small organic molecule that is a protein building block.

amniocentesis A prenatal diagnostic procedure performed on a small sample of amniotic fluid, which contains fetal cells and biochemicals. A chromosome chart is constructed from cultured fetal cells, and tests for certain inborn errors of metabolism are conducted on fetal biochemicals.

anaphase The stage of mitosis when the centromeres of replicated chromosomes part.

aneuploid A cell with one or more extra or missing chromosomes.

antibody A multisubunit protein, produced by B cells, that binds a specific foreign antigen at one end, alerting other components of the immune system or directly destroying the antigen.

anticodon A three-base sequence on one loop of a transfer RNA molecule that is complementary to an mRNA codon, and brings together the appropriate amino acid and its mRNA.

antigen A molecule that elicits an immune response.

antigen-presenting cell A cell displaying a foreign antigen.

antiparallel The head-to-tail arrangement of the two entwined chains of the DNA double helix.

B

apoptosis A form of cell death that is a normal part of growth and development.

assisted reproductive technologies Procedures that replace a gamete or the uterus to help people with fertility problems have children.

association study A case-control study in which genetic variation, often measured as SNPs that form haplotypes, is compared between people with a particular condition and unaffected individuals.

autoantibodies Antibodies that attack the body's own cells.

autoimmunity An immune attack against one's own body.

autosomal dominant The inheritance pattern of a dominant allele on an autosome. The phenotype can affect males and females and does not skip generations.

autosomal recessive The inheritance pattern of a recessive allele on an autosome. The phenotype can affect males and females and can skip generations.

autosome A chromosome that does not have a gene that determines sex.

balanced polymorphism Maintenance of a harmful recessive allele in a population because the heterozygote has a reproductive advantage.

Barr body A dark-staining, inactivated X chromosome in a cell.

base excision repair Removal of up to five bases to correct damage due to reactive oxygen species.

B cell A type of lymphocyte that secretes antibody proteins in response to nonself antigens displayed on other immune system cells.

biotechnology The alteration of cells or biochemicals with a specific application.

blastocyst A hollow ball of cells descended from a fertilized ovum.

blastomere A cell of a blastocyst.

C

cancer A group of disorders resulting from loss of cell cycle control.

carcinogen A substance that causes cancer.

case-control study An epidemiological method in which people with a particular condition are compared to individuals as much like them as possible, but without the disease.

cDNA library A collection of DNA molecules reverse transcribed from the mRNAs in a particular cell type.

cell The fundamental unit of life.

cell cycle A cycle of events describing a cell's preparation for division and division itself.

cellular immune response T cells release cytokines to stimulate and coordinate an immune response.

centriole A structure in cells that organizes microtubules to form the mitotic spindle apparatus.

centromere The largest constriction in a chromosome, located at a specific site in each chromosome type.

chaperone protein A protein that binds a polypeptide as it begins to fold, directing the folding.

chorionic villus sampling (CVS) A prenatal diagnostic technique that analyzes chromosomes in chorionic villus cells, which, like the fetus, descend from the fertilized ovum.

chromatid A single, very long DNA molecule and its associated proteins, forming half of a replicated chromosome.

chromatin DNA and its associated histone proteins.

chromatin remodeling Adding or removing chemical groups to or from histones, which can alter gene expression.

chromosome A structure within a cell's nucleus that carries genes. A chromosome consists of a continuous molecule of DNA and proteins wrapped around it.

cleavage A series of rapid mitotic cell divisions after fertilization.

clines Allele frequencies that change from one area to another.

cloning vector A piece of DNA used to transfer DNA from a cell of one organism into the cell of another.

coding strand The side of the double helix for a particular gene from which RNA is not transcribed.

codominant A heterozygote in which both alleles are fully expressed.

codon A continuous triplet of mRNA that specifies a particular amino acid.

comparative genomics Identifying conserved DNA sequences among genomes of different species.

complementary base pairs The pairs of DNA bases that hydrogen bond together; adenine hydrogen bonds to thymine and guanine to cytosine in the DNA double helix.

complementary DNA (cDNA) A DNA molecule that is the complement of an mRNA, copied using reverse transcriptase.

concordance A measure indicating the degree to which a trait is inherited; expressed as the percentage of twin pairs in which both members express a trait.

conditional mutation A genotype that is expressed only under certain environmental conditions.

conformation The three-dimensional shape of a molecule.

consanguinity Blood relatives having children together.

critical period The time during prenatal development when a structure is sensitive to damage from a mutation or an environmental intervention.

crossing over An event during prophase I when homologs exchange parts.

cytogenetics A discipline that matches phenotypes to detectable chromosomal abnormalities.

cytokine A biochemical that a T cell secretes that controls immune function.

cytoplasm Cellular contents other than organelles.

cytosine One of the two pyrimidine nitrogenous bases in DNA and RNA.

cytoskeleton A framework composed of protein tubules and rods that supports the cell and gives it a distinctive form.

D

dedifferentiated A cell less specialized than the cell it descends from, such as a cancer cell.

deletion mutation A missing sequence of DNA or part of a chromosome.

deoxyribonucleic acid (DNA) The genetic material; the biochemical that forms genes.

deoxyribose The 5-carbon sugar in a DNA nucleotide.

differentiation Cell specialization, reflecting differential gene expression.

dihybrid cross Breeding individuals heterozygous for two traits.

diploid A cell containing two sets of chromosomes.

dizygotic (DZ) twins Twins that originate as two different fertilized ova.

DNA *See* **deoxyribonucleic acid.**

DNA microarray *See* **microarray.**

DNA polymerase (DNAP) An enzyme that inserts into replicating DNA and corrects mismatched base pairs.

DNA probe A labeled short sequence of DNA that, when applied to a biological sample, binds its complement, revealing its locus.

DNA profiling A biotechnology that detects differences in the number of copies of certain DNA repeats between individuals. Used to rule out or establish identity.

DNA replication Construction of a new DNA double helix using the information in parental strands as a template.

DNA variation screening Use of DNA microarrays to screen a sample for SNPs and mutations in a particular gene or genome segment.

dominant A gene variant expressed when present in even one copy.

duplication An extra copy of a gene or DNA sequence, usually caused by misaligned pairing in meiosis.

E

ectoderm The outermost primary germ layer.

embryo In humans, prenatal development until the end of the eighth week. Embryo cells can be distinguished from each other, but all basic structures are not yet present.

embryonic stem (ES) cell A cell from a preimplantation embryo and then cultured that can give rise to all differentiated cell types.

empiric risk Probability that a trait will recur based upon its incidence in a population.

endoderm The innermost primary germ layer of the primordial embryo.

endoplasmic reticulum (ER) A labyrinth of membranous tubules on which proteins, lipids, and sugars are synthesized.

enzyme A type of protein that speeds the rate of a specific biochemical reaction, making it fast enough to be compatible with life.

epigenetic A layer of information placed on a gene that is a modification other than a change in DNA sequence.

epistasis One gene masking expression of another.

equational division The second meiotic division, producing four cells from two.

euchromatin Parts of chromosomes that do not stain and that contain active genes.

eugenics The control of individual reproductive choices to achieve a societal goal.

eukaryotic cell A complex cell containing organelles, including a nucleus.

euploid A somatic cell with the normal number of chromosomes for that species.

excision repair Enzyme-catalyzed removal of pyrimidine dimers in DNA.

exon Part of a gene that encodes amino acids.

expressed sequence tags (ESTs) Short cDNAs used to find protein-encoding genes.

expressivity Degree of expression of a phenotype.

***ex vivo* gene therapy** Genetic alteration of cells removed from a patient, then reinfused or implanted back.

F

fetus The prenatal human after the eighth week of development, when structures grow and specialize.

fluorescence *in situ* hybridization (FISH) A technique that binds fluorescently labeled DNA probes to complementary sequences on a chromosome.

founder effect A type of genetic drift in which a few individuals leave to found a new settlement, perpetuating a subset of the alleles from the original population.

frameshift mutation A mutation that alters a gene's reading frame.

fusion protein A protein that forms from transcription of two genes as a unit and then translation. Can cause cancer.

G

gamete   A sex cell.

gamete intrafallopian transfer (GIFT)   An infertility treatment in which sperm and oocytes are placed in a woman's uterine tube.

gastrula   A three-layered embryo.

gene   A sequence of DNA that instructs a cell to produce a particular protein.

gene expression   Transcription of a gene.

gene expression profiling   Use of DNA microarrays to detect the types and amounts of cDNAs reverse transcribed from the mRNAs in a particular cell source.

gene pool   All the genes in a population.

gene targeting   A biotechnology in which an introduced gene exchanges places with its counterpart on a host cell's chromosome by homologous recombination.

gene therapy   Replacing a malfunctioning gene to alleviate symptoms.

genetic code   The correspondence between specific RNA triplets and the amino acids they specify.

genetic counselor   A medical specialist who calculates risk of recurrence of inherited disorders in families, applying the laws of inheritance to pedigrees.

genetic drift   Changes in gene frequencies that occur when small groups of individuals are reproductively separated from a larger population.

genetic heterogeneity   A phenotype that can be caused by variants of any of several genes.

genetic load   The collection of deleterious recessive alleles in a population.

genetics   The study of inherited variation.

genome   The complete set of genetic instructions in the cells of a particular type of organism.

genomic imprinting   Differing of the phenotype depending upon which parent transmits a particular allele.

genomic library   A collection of DNA pieces representing the genome of an individual, including introns.

genomics   The study of the functions and interactions of many genes at a time, or comparing genomes.

genotype   The allele combinations in an individual that cause particular traits or disorders.

genotypic ratio   The ratio of genotype classes expected in the progeny of a particular cross.

germline gene therapy   Genetic alterations of gametes or fertilized ova, which perpetuate the change throughout the organism and transmit it to future generations.

germline mutation   A mutation in every cell in an individual.

glycolipid   A molecule that consists of a sugar bonded to a lipid.

glycoprotein   A molecule that consists of a sugar bonded to a protein.

Golgi apparatus   An organelle, consisting of flattened, membranous sacs, that packages secretion components.

gonads   Paired structures in the reproductive system where sperm or oocytes are manufactured.

growth factor   A protein that stimulates mitosis.

guanine   One of the two purine nitrogenous bases in DNA and RNA.

H

haploid   A cell containing one set of chromosomes.

haplotype   A series of known DNA sequences or single nucleotide polymorphisms linked on a chromosome.

Hardy-Weinberg equilibrium   An idealized state in which gene frequencies in a population do not change from generation to generation.

hemizygous   The sex that has half as many X-linked genes as the other; a human male.

heritability   An estimate of the proportion of phenotypic variation in a group due to genes.

heterochromatin   Dark-staining genetic material that is inactive but that maintains the chromosome's structural integrity.

heterogametic sex   The sex with two different sex chromosomes; a human male.

heteroplasmy   Mitochondria within the same cell having different alleles of a particular gene.

heterozygous   Having two different alleles of a gene.

histone   A type of protein around which DNA entwines.

hominids   Animals ancestral to humans only.

hominoids   Animals ancestral to apes and humans only.

homogametic sex   The sex with identical types of sex chromosomes; the human female.

homologous pairs   Chromosomes with the same gene sequence.

homologous recombination   A naturally occurring process in which a piece of DNA exchanges places with its counterpart on a chromosome.

homozygous   Having two identical alleles of a gene.

human leukocyte antigen (HLA) complex   Polymorphic genes closely linked on the short arm of chromosome 6 that encode cell surface proteins important in immune system function.

humoral immune response   Process in which B cells secrete antibodies into the bloodstream.

I

incomplete dominance   A heterozygote intermediate in phenotype between either homozygote.

independent assortment   The random arrangement of homologous chromosome pairs, in terms of maternal or paternal origin, down the center of a cell in metaphase I. Inheritance of a gene on one chromosome does not influence inheritance of a gene on a different chromosome. (Mendel's second law).

infertility   The inability to conceive a child after a year of unprotected intercourse.

inflammation   Part of the innate immune response that causes an infected or injured area to swell with fluid, turn red, and attract phagocytes.

innate immunity   Components of immune response that are present at birth and do not require exposure to an environmental stimulus.

inner cell mass   A clump of cells on the inside of the blastocyst that will continue developing into an embryo.

insertion mutation   A mutation that adds DNA bases.

in situ gene therapy   Localized gene therapy in an easily accessible body part.

intermediate filament   A type of cytoskeletal component made of different proteins in different cell types.

interphase   Stage when a cell is not dividing.

intracytoplasmic sperm injection (ICSI)   Injection of a sperm cell nucleus into an oocyte, to overcome lack of sperm motility.

intrauterine insemination   An infertility treatment in which donor sperm are placed in the cervix or uterus.

introns   Base sequences within a gene that are transcribed but are excised from the mRNA before translation into protein.

in vitro fertilization (IVF)   Placing oocytes and sperm in a laboratory dish with appropriate biochemicals so that fertilization occurs, then, after a few cell divisions, transferring the embryos to a woman's uterus.

in vitro gene therapy   Direct genetic manipulation of cells in the body.

K

karyotype   A size-order chromosome chart.

L

law of independent assortment   *See* **independent assortment.**

law of segregation   *See* **segregation.**

lethal allele   An allele that causes death before reproductive maturity or halts prenatal development.

ligand   A molecule that binds to a receptor.

ligase   An enzyme that catalyzes the formation of covalent bonds in the sugar-phosphate backbone of a nucleic acid.

linkage   Genes on the same chromosome.

linkage disequilibrium   Extremely tight linkage between DNA sequences.

linkage maps   Maps that show how genes are ordered on chromosomes, determined from crossover frequencies between pairs of genes.

lipid   A type of organic molecule that has more carbon and hydrogen atoms than oxygen atoms. Includes fats and oils.

lysosome   A saclike organelle containing enzymes that degrade debris.

M

macroevolution   Genetic change sufficient to form a new species.

major histocompatibility complex (MHC)   A gene cluster, on chromosome 6 in humans, that includes many genes that encode components of the immune system.

manifesting heterozygote   A female carrier of an X-linked recessive gene who expresses the phenotype because the normal allele is inactivated in some affected tissues.

meiosis   A type of cell division that halves the number of chromosomes to form haploid gametes.

mesoderm   The middle primary germ layer.

messenger RNA (mRNA)   A molecule of RNA complementary in sequence to the template strand of a gene. Messenger RNA carries the information that specifies a particular protein product.

metacentric chromosome   A chromosome with the centromere approximately in the center.

metaphase   The stage of mitosis when chromosomes align along the center of the cell.

metastasis   Spread of cancer from its site of origin to other parts of the body.

microarray   Also called a DNA chip. A set of target genes embedded in a glass chip, to which labeled cDNAs from a sample bind and fluoresce. Microarrays show patterns of gene expression.

microevolution   Change of allele frequency in a population.

microfilament   A solid rod of actin protein that forms part of the cytoskeleton.

microtubule   A hollow structure built of tubulin protein that forms part of the cytoskeleton.

mismatch repair   Proofreading of DNA for misalignment of short, repeated segments.

missense   A single base change mutation that alters an amino acid in the gene product.

mitochondrion   An organelle consisting of a double membrane that houses enzymes that catalyze reactions that extract energy from nutrients.

mitosis   Division of somatic (nonsex) cells.

mode of inheritance   The pattern in which a gene variant passes from generation to generation, dominant or recessive, autosomal or sex-linked.

molecular evolution   Changes in protein and DNA sequences over time used to estimate how recently species diverged from a common ancestor.

monoclonal antibody (MAb)   A single antibody type, produced from a B cell fused to a cancer cell (a hybridoma).

monohybrid cross   A cross of two individuals who are heterozygous for a single trait.

monosomy   A human cell with 45 (one missing) chromosomes.

monozygotic (MZ) twins   Twins that originate as a single fertilized ovum; identical twins.

morula   The very early prenatal stage that resembles a mulberry.

multifactorial trait   A trait or illness determined by several genes and the environment.

mutagen   A substance that changes, adds, or deletes a DNA base.

mutant   An allele that differs from the normal or most common allele, altering the phenotype.

mutation   A change in a protein-encoding gene that affects the phenotype.

N

natural selection   Differential survival and reproduction of individuals with particular phenotypes in particular environments, which may alter allele frequencies in subsequent generations.

neural tube   A structure in the embryo that develops into the brain and spinal cord.

nitrogenous base   A nitrogen-containing base that is part of a nucleotide.

nondisjunction   The unequal partitioning of chromosomes into gametes during meiosis.

nonsense mutation   A point mutation that changes an amino-acid-coding codon into a stop codon, prematurely terminating synthesis of the encoded protein.

nucleic acid   DNA or RNA.

nucleolus   A structure within the nucleus where ribosomes are assembled from ribosomal RNA and protein.

nucleosome   A unit of chromatin structure.

nucleotide   The building block of a nucleic acid, consisting of a phosphate group, a nitrogenous base, and a 5-carbon sugar.

nucleotide excision repair   Replacement of up to 30 nucleotides to correct DNA damage of several types.

nucleus   A large, membrane-bounded region of a eukaryotic cell that houses DNA.

O

oncogene   A dominant gene that promotes cell division. An oncogene normally controls the cell cycle, but causes cancer when overexpressed.

oocyte   The female gamete (sex cell).

oogenesis   Oocyte development.

organelle   A specialized structure in a eukaryotic cell that carries out a specific function.

P

paracentric inversion   An inverted chromosome that does not include the centromere.

PCR   *See* **polymerase chain reaction.**

pedigree   A chart consisting of symbols connected by lines that depict the genetic relationships and transmission of inherited traits in related individuals.

penetrance   Percentage of individuals with a genotype who have an associated phenotype.

pericentric inversion   An inverted chromosome that includes the centromere.

peroxisome   An organelle consisting of a double membrane that houses enzymes with various functions.

phenocopy   An environmentally caused trait that occurs in a familial pattern, mimicking inheritance.

phenotype   The expression of a gene in traits or symptoms.

plasma membrane The selective barrier around a cell, consisting of proteins, glycolipids, glycoproteins, and lipid rafts on or in a phospholipid bilayer.

plasmid A small circle of double-stranded DNA found in some bacteria. Used as a vector in recombinant DNA technology.

pleiotropic A Mendelian disorder with several symptoms. Different symptom subsets may occur in different individuals.

point mutation A single base change in DNA.

polar body A product of female meiosis that contains little cytoplasm and does not continue to develop into an oocyte.

polar body biopsy A genetic test performed on a polar body to infer the genotype of the attached oocyte.

polygenic traits Traits determined by more than one gene.

polymerase chain reaction (PCR) A nucleic acid amplification technique in which a specific DNA sequence is replicated in a test tube to rapidly produce many copies.

polymorphism A DNA base or sequence at a certain chromosomal locus that varies in at least 1 percent of individuals in a population.

polyploid (cell) A cell with one or more extra sets of chromosomes.

population A group of interbreeding individuals.

population bottleneck Decrease in allele diversity resulting from an event that kills many members of a population, followed by restoration of population numbers.

population genetics The study of allele frequencies in different groups of individuals.

population study Comparison of disease incidence in different groups of people.

positional cloning Identifying a gene by beginning with a phenotype within a large family and narrowing down the segment of a chromosome that includes the gene.

preimplantation genetic diagnosis (PGD) Removing a cell from an 8-celled embryo and testing it for a disease-causing gene or chromosomal imbalance to deduce the genotype of the embryo.

primary germ layers The three layers of an embryo.

primary (1°) structure The amino acid sequence of a protein.

progenitor cell A cell whose descendants can follow any of several developmental pathways.

prokaryote cell A cell that does not have a nucleus or other organelles. One of the three domains of life.

promoter A control sequence near the start of a gene.

pronuclei DNA packets in the fertilized ovum.

prophase The first stage of mitosis or meiosis, when chromatin condenses.

prospective study A study that follows two or more groups.

proteasome A multiprotein structure shaped like a barrel through which misfolded proteins pass and are dismantled.

protein A type of macromolecule that is the direct product of genetic information; a chain of amino acids.

proteome The set of proteins a cell produces.

proteomics Study of the proteins produced in a particular cell type under particular conditions.

proto-oncogene A gene that normally controls the cell cycle. When overexpressed, it functions as an oncogene, causing cancer.

pseudoautosomal region Genes on the tips of the Y chromosome that have counterparts on the X chromosome.

pseudogene A gene that does not encode protein, but whose sequence very closely resembles that of a coding gene.

Punnett square A diagram used to follow parental gene contributions to offspring.

purine A type of organic molecule with a two-ring structure; adenine and guanine are purines.

pyrimidine A type of organic molecule with a single-ring structure; cytosine, thymine, and uracil are pyrimidines.

Q

quantitative trait loci Genes that determine polygenic traits.

quaternary (4°) structure A protein that has more than one polypeptide subunit.

R

reading frame The grouping of DNA base triplets encoding an amino acid sequence.

receptor A structure on a cell that binds a specific molecule.

recessive An allele whose expression is masked by another allele.

reciprocal translocation A chromosome aberration in which two nonhomologous chromosomes exchange parts, conserving genetic balance but rearranging genes.

recombinant A series of alleles on a chromosome that differs from the series of either parent.

recombinant DNA technology Transferring genes between species.

reduction division The first meiotic division, which halves the chromosome number.

relative risk Probability that an individual from a population will develop a particular condition in comparison to another group, usually the general population.

replication fork Locally opened portion of a replicating DNA double helix.

ribonucleic acid (RNA) A nucleic acid whose bases are A, C, U, and G.

ribose A 5-carbon sugar in RNA.

ribosomal RNA (rRNA) RNA that, with proteins, comprises ribosomes.

ribosome An organelle consisting of RNA and protein that is a scaffold for protein synthesis.

risk factor A characteristic associated with increased likelihood of developing a particular medical condition.

RNA *See* **ribonucleic acid.**

RNA interference A natural process that destroys specific mRNA molecules using small interfering RNAs that result from transcribing short sequences on both DNA strands.

RNA polymerase (RNAP) An enzyme that adds RNA nucleotides to a growing RNA chain.

Robertsonian translocation A chromosome aberration in which two short arms of nonhomologous chromosomes break and the long arms fuse, forming one unusual, large chromosome.

S

secondary (2°) structure Folds in a polypeptide caused by attractions between amino acids close together in the primary structure.

segregation The distribution of alleles of a gene into separate gametes during meiosis. (Mendel's first law).

self-renewal Defining property of a stem cell; the ability to yield a daughter cell like itself in terms of developmental potential.

semiconservative replication DNA synthesis along each half of the double helix.

sex chromosome A chromosome containing genes that specify sex.

sex-influenced trait Phenotype caused when an allele is recessive in one sex but dominant in the other.

sex-limited trait A trait that affects a structure or function present in only one sex.

short tandem repeats (STRs) Repeats of 2 to 10 DNA bases that are compared in DNA profiling.

signal transduction A series of biochemical reactions and interactions that pass information from outside a cell to inside, triggering a cellular response.

single nucleotide polymorphism (SNP) Single base sites that differ among individuals. A SNP is present in at least 1 percent of a population.

somatic cell A nonsex cell, with 23 pairs of chromosomes in humans.

somatic cell nuclear transfer Transfer of a somatic cell's nucleus to an enucleated egg, and growth to the blastocyst stage to obtain inner cell mass cells, which are cultured to yield embryonic stem (ES) cells. Given appropriate stimulation, the ES cells divide to produce needed cells.

somatic gene therapy Genetic alteration of a specific somatic tissue, not transmitted to future generations.

somatic mutation A mutation occurring in only a subset of somatic (nonsex) cells.

sperm *See* **spermatozoon.**

spermatogenesis Sperm cell development.

spermatozoon (sperm) A mature male reproductive cell (meiotic product).

S phase The stage of interphase when DNA replicates.

spindle A structure composed of microtubules that pulls sets of chromosomes apart in a dividing cell.

spontaneous mutation A genetic change that results from mispairing when the replication machinery encounters a base in its rare tautomeric form.

SRY gene The sex-determining region of the Y. This gene controls whether the unspecialized embryonic gonad will develop as testis or ovary. If the gene is activated, the gonad becomes a testis; if not, an ovary forms.

stem cells Cells that give rise to other stem cells that retain the potential to differentiate (specialize) further, as well as to cells that differentiate or give rise to cells that differentiate.

submetacentric chromosome A chromosome in which the centromere establishes a long arm and a short arm.

sugar-phosphate backbone The "rails" of a DNA double helix, consisting of linked, alternating deoxyribose and phosphate groups, oriented opposite one another.

synonymous codons DNA triplets that specify the same amino acid.

synteny The correspondence of genes located on the same chromosome in several species.

T

tandem duplication A duplicated sequence of DNA located right next to the original sequence on a chromosome.

T cell A type of lymphocyte that produces cytokines and coordinates the immune response.

telomerase An enzyme, including a sequence of RNA, that adds DNA to chromosome tips.

telomere A chromosome tip.

telophase The stage of mitosis or meiosis when daughter cells separate.

template strand The DNA strand carrying the information to be transcribed.

teratogen A substance that causes a birth defect.

tertiary (3°) structure Folds in a polypeptide caused by interactions between amino acids and water. This draws together amino acids that are far apart in the primary structure.

thymine One of the two pyrimidine bases in DNA.

transcription Manufacturing RNA from DNA.

transcription factor A protein that activates the transcription of other genes.

transfer RNA (tRNA) A type of RNA that connects mRNA to amino acids during protein synthesis.

transgenic organism An individual with a genetic modification in every cell.

transition A point mutation altering a purine to a purine or a pyrimidine to a pyrimidine.

translation Assembly of an amino acid chain according to the sequence of base triplets in a molecule of mRNA.

translocation Exchange between nonhomologous chromosomes.

translocation carrier An individual with exchanged chromosomes, but no signs or symptoms. The person has the usual amount of genetic material, but it is rearranged.

transposon A gene or DNA segment that moves to another chromosome.

transversion A point mutation altering a purine to a pyrimidine or vice versa.

trisomy A human cell with 47 chromosomes (one extra).

tumor suppressor gene A recessive gene whose normal function is to limit the number of divisions a cell undergoes.

U

uniparental disomy Inheriting two copies of the same gene from one parent.

uracil One of the four types of bases in RNA; a pyrimidine.

V

variable number of tandem repeats (VNTRs) Repeats of 10 to 80 DNA bases that are compared in DNA profiles.

virus An infectious particle built of nucleic acid in a protein coat.

W

wild type The most common phenotype in a population for a particular gene.

X

X inactivation The inactivation of one X chromosome in each cell of a female mammal, occurring early in embryonic development.

X-linked Genes on an X chromosome.

X-Y homologs Y-linked genes that are similar to genes on the X chromosome.

Y

Y-linked Genes on a Y chromosome.

Z

zygote A prenatal human from the fertilized ovum stage until formation of the primordial embryo, at about two weeks.

zygote intrafallopian transfer (ZIFT) An assisted reproductive technology in which an ovum fertilized *in vitro* is placed in a woman's uterine tube.

Credits

Line Art/Text

Chapter 2

Figure 2.26 From Piero Anversa: *New England Journal of Medicine*, January 1, 2002 Copyright ©2002 Massachusetts Medical Society, Waltham, MA. Reprinted by permission.

Chapter 3

Box 3.2 Reprinted by permission of TMS Reprints

Chapter 4

Battling Batten Disease, Nathan's Story by Phil Milto

Chapter 6

Box 6.1 Courtesy Professor Jennifer A. Marshall-Graves, Australian National University. **Box 6.3** Courtesy David Page, Massachusetts Institute of Technology; Howard Hughes Medical Institute Investigator. **Box 6.5** Plate No. 16 of the 38-Plate Edition. The above has been reproduced from Ishihara's Tests for Colour Deficiency published by Kanehara Trading Inc., located at Tokyo in Japan. But tests for color deficiency cannot be conducted with this material. For accurate testing, the original plates should be used.

Chapter 7

Figure 7.1 Data and print from Gordon Mendenhall, Thomas Mertens, and Jon Hendrix, "Fingerprint Ridge Count" in *The American Biology Teacher,* vol. 51, no. 4, April 1898, pp. 204-6. **Figure 7.11** Graph from www.unitedhealthfoundation.org. Reprinted by permission.

Chapter 8

Table 8.1 Source: Psychiatric Genomics Inc., Gaithersburg, MD. The information was collated form the Surgeon General's 1999 Report on Mental Health. **Figure 8.9** ©Robert Gilliam. **Figure 8.10** Source: Maher, Brandan A., "The Infection Connection in Schizophrenia". *The Scientist*, November 2003, p. 30.

Chapter 13

Box 13.1 Modified from Willard, 1998. *Curr Opin Genet Dev* 8:219-25. **Figure 13.12** From *Color Atlas of Genetics* by Eberhard Passarge, p. 401. Copyright ©2001. Reprinted by permission of Thieme Medial Publishers, Inc.

Chapter 15

Box, Chapter 15 "Rape as a Weapon or War" from Amnesty International Publications. ©Amnesty International Publications. Reprinted by permission of Amnesty International, http://www..amnesty.org. **Figure 15.15** Source: Data from Smadar Avigad, et al., A single origin of phenylketonuria in Yemenite Jews, *Nature* 344:170, March 8, 1990.

Chapter 16

Box 16.2 Excerpt by Blaine Deatherage-Newsom, "If we could eliminate disabilities from the population, should we? Results of a survey on the Internet." Reprinted by permission. **Figure 16.10** From "Primate

shadow play" by Richard A. Gibbs and David L. Mellon in *Science* Vol. 299, p. 2442, February 28, 2003. Copyright ©2003 AAAS. Reprinted by permission of AAAS. **Figure 16.13** Human Chromosome Colour Index by Bhanu Chowdhary, Texas A & M University. Reprinted by permission of Bhanu Chowdhary. **Figure 16.19** From "The African diaspora: mitochondrial DNA and the Atlantic Slave Trade" by Antonio Salas et al from *American Journal of Human Genetics*, March 2004, p. 458, figure 1. Copyright ©2004. Reprinted by permission of The University of Chicago Press.

Chapter 18

Figure 18.8 Source: Clarke, Michael F. "At the root of brain cancer", *Nature*, 432:281, November 2004. **Figure 18.11** Source: Beachy, Philip A., et al, Tissue repair and stem cell renewal In carcinogenesis", *Nature*, 432:324-331, November 2004. **Figure 18.13** Adapted from "Mechanism of Action of BCR-ABL" by John Goldman and Juania Melo: *New England Journal of Medicine* 344:1084. Reprinted by permission of Massachusetts Medical Society.

Photos

Chapter 1

Opener: ©Vol. 19/Corbis; **Figure 1.1:** ©Susan McCartney/Photo Researchers; **Figure 1.3(4):** ©CNRI/Photo Researchers; **Figure 1.4(5-6):** ©PhotoDisc/Getty R-F; **Figure 1.4(7):** ©Richard Laird/Getty Images; **Figure 1.5(man):** ©Corbis R-F; **Figure 1.5(chimp):** ©PhotoDisc/Getty R-F; **Figure 1.5(fish):** ©Byrappa Venkatesh, IMCB, Singapore; **Figure 1.5(mouse):** ©PhotoDisc/Vol.# 8; **Figure 1.5(sea squirt):** ©PhotoDisc/Getty R-F; **Figure 1.5(fruit fly):** ©David M. Phillips/Photo Researchers; **Figure 1.5(yeast):** ©Dr. Stanley Flegler/Visuals Unlimited; **Figure 1.6a:** ©Lester Bergman/ProjectMasters; **Figure 1.6b:** ©Sunstar/Photo Researchers; **Figure 1.8:** Roger Berry, Portrait of a DNA Sequence, 1998, dichroic glass and steel, 18" X 51'. Photo: Joseph Coulombe; **Figure 1.9:** Courtesy of the Massachusetts Historical Society; **Figure 1.10:** ©Jay Sand; **Figure 1.11a:** Courtesy, Thierry LaCombe & Jean Pierre Bruno, I.N.R.A. , France; **Figure 1.11b:** ©Alexander Lowry/Photo Researchers; **Figure 1.12:** Courtesy Ingo Potrykus, Swiss Federal Institute of Technology; **Figure 1.13:** ©Alexis Rockman, 2000. Courtesy Gorney Bravin + Lee, New York.

Chapter 2

Opener: ©Corbis Images/Warren Morgan; **Figure 2.1:** ©Muscular Dystrophy Association; **Figure 2.2:** ©Manfred Kage/Peter Arnold; **Figure 2.3(top right):** ©David M. Phillips/The Population Council/Science Source/Photo Researchers; **Figure 2.3(left):** ©K.R. Porter/Photo Researchers; **Figure 2.3(right):** ©Biophoto Associates/Science Source/Photo Researchers; **Reading 2.1, Fig. 1:** ©SPL/Photo Researchers; **Figure 2.6:** ©Prof. P. Motta & T. Naguro/SPL/Photo Researchers; **Figure 2.7:** ©D. Friend-D. Fawcett/Visuals Unlimited; **Figure 2.8:** ©Bill Longcore/Photo Researchers; **Figure 2.9b:** ©Gordon Leedale/BioPhoto Associates; **Reading 2.2, fig. 2:**

Courtesy, Cystic Fibrosis Foundation; **Figure 2.11:** ©Visuals Unlimited; **Figure 2.20(top left):** ©David McCarthy/Photo Researchers; **Figure 2.13:** ©P. Motta/SPL/Photo Researchers; **Figure 2.14b:** ©Bart's Medical Library/Phototake; **Figure 2.16b:** ©From Dr. A.T. Sumner, "Mammalian Chromosomes from Prophase to Telophase," *Chromosoma*, 100:410-418, 1991. ©Springer-Verlag; **Figure 2.17(all):** ©Ed Reschke; **Figure 2.19:** From L. Chong, et al. 1995. "A Human Telomeric Protein." *Science*, 270:1663-1667. ©1995 American Association for the Advancement of Science. Photo courtesy, Dr. Titia DeLange; **Figure 2.20(right):** ©Peter Skinner/Photo Researchers; **Figure 2.25(left):** ©AP/Wide World Photos; **Figure 2.25(center, right):** Courtesy of Advanced Cell Technologies, Inc.

Chapter 3

Opener: ©Peter Willi/Superstock; **Figure 3.8:** ©Ed Reschke/Peter Arnold; **Figure 3.9b:** ©David M. Phillips/Visuals Unlimited; **Figure 3.10:** Courtesy R.J. Blandau; **Figure 3.13b:** ©Francis LeRoy/BioCosmos/SPL/Photo Researchers; **Figure 3.14(1, 3):** ©Petit Format/Nestle/Science Source/Photo Researchers; **Figure 3.14(2):** ©P.M. Motta & J. Van Blerkom/SPL/Photo Researchers; **p. 58:** ©AP/Wide World Photos; **Figure 3.17:** ©Steve Wewerka; **Figure 3.18a:** ©Petit Format/Nestle/Photo Researchers; **Figure 3.18b:** ©CBSC/Phototake; **Figure 3.18c:** ©Donald Yaeger/Camera M.D. Studios; **Figure 3.19:** ©Richard Nowitz/Phototake; **Figure 3.21b-d:** From Streissguth, A.P., Landesman-Dwyer, S., Martin, J.C., & Smith, D.W. July 1980. "Teratogenic effects of alcohol in human and laboratory animals." *Science*, 209(18):353-361. ©1980 American Association for the Advancement of Science; **Figure 3.22a:** Courtesy Dr. Francis Collins; **Figure 3.22b:** ©AP/Reagan Presidential/Wide World Photos; **Figure 3.23:** ©Xinhua-Chine Nouvelle/Gamma Presse; **Reading 3.1, Fig. 1:** ©Mitch Wojnarowicz/Image Works.

Chapter 4

Opener: ©PhotoDisc/Vol. 2; **Figure 4.1:** ©Sands Steven/Corbis Sygma; **Figure 4.15:** ©Nancy Hamilton/Photo Researchers.

Chapter 5

Opener: From G. Pierard, A. Nikkels. April 5, 2001. "A Medical Mystery." *New England Journal of Medicine*, 344: p. 1057. ©2001 Massachusetts Medical Society. All rights reserved; **Figure 5.1a:** ©Porterfield-Chickering/Photo Researchers; **Figure 5.2:** From Genest, Jacques, Jr., Lavoie, Marc-Andre. August 12, 1999. "Images in Clinical Medicine." *New England Journal of Medicine,* pp 490. ©1999, Massachusetts Medical Society. All Rights Reserved; **Figure 5.5a:** ©North Wind Picture Archives.

Chapter 6

Opener: ©Medical Research Council/Photo Researchers; **Figure 6.2:** ©Biophoto Associates/Photo Researchers; **p. 117:** ©Dr. Walter Just; **Figure 6.5:** ©Ward Odenwald, National Institute of Neurological Disease and Stroke; **Figure 6.7:** Courtesy, Dr. Mark A.

Index

Page numbers followed by an "f" indicate figures; numbers followed by a "t" indicate tables.

staining, 249
  swelling, squashing, and untangling, 248–50
Chromosome 1, 317
Chromosome 2, 316–17
Chromosome 3, 250f, 317
Chromosome 5, 243t, 257
Chromosome 12, 316
Chromosome 13, 242, 245
Chromosome 14, 245, 316
Chromosome 15, 257, 259f, 263–64
Chromosome 16, 243t
Chromosome 18, 242, 316
Chromosome 19, 243t
Chromosome 20, 316
Chromosome 21, 243t, 245, 317. *See also* Down syndrome
Chromosome 22, 243t, 245, 316
Chromosome abnormality
  abnormal structure, 242, 257–62, 263t
  in cancer cells, 364
  definition of, 242
  deletion, 251t, 257–58, 257f, 263t
  duplication, 251t, 257–58, 257f, 259f, 263t
  inversion, 251t, 257f, 261–62, 261–62f, 263t
  translocation. *See* Translocation
Chromosome number, 49, 248–49
  abnormal, 242, 251–57, 251t, 263t
Chromosome walking, 437
Chronic beryllium disease, 226
Chronic granulomatous disease, 68t, 123t, 341, 342t
Chronic myeloid leukemia (CML), 243t, 359t, 364–65, 365f
Cilia, 30
  abnormal or missing, 433
Ciliated protozoa, genetic code in, 193
Circadian pacemaker, 161
Circumcision, 119, 403
"Cis" configuration, of linked genes, 104, 105f
CJD. *See* Creutzfeldt-Jakob disease
Cleavage, 56–58, 57f, 60t
Cleft chin, 81f
Cleft lip, 135, 135f, 140t, 144t
Cleft palate, 100, 135, 144t
Cleopatra, 89f
Cline, 287
Clinical trial, of gene therapy, 405t
Clitoris, 49, 49f, 119
Cloacal exstrophy, 119
Clomiphene, 420
Cloning
  bioethics of, 58–59
  positional, 434
  reproductive, 42, 58–59
  therapeutic, 42, 58
Cloning vector, 381–82, 382f, 382t
Clotting disorder, 100
Clotting factors, 148, 148t, 227
Clotting factor therapy, 402
Clovis point, 322
Clubfoot, 141t
CML. *See* Chronic myeloid leukemia
Coat color
  in cats, 127, 128f
  in cattle, 59
Cocaine, 65, 163, 163f
Cockayne syndrome, 68t
Coding strand, of DNA, 186, 187f, 189, 190f, 198f
CODIS, 277f
Codominance, 97–98, 98f, 101t
Codon, 187, 192f, 196, 196f
  start, 193
  stop, 193, 196, 196f, 227
  synonymous, 193–95, 233–34
Codon usage bias, 195
Coefficient of relatedness, 141, 141t, 147t
Coelacanth, 445
Cognition, evolution in primates, 315
Cohanim, 14, 14f

Cohesin, 214
Colchicine, 248
Collagen, 176, 186t
  conformation of, 220, 220f
  disorders of, 219–20, 219t
  production by transgenic silkworms, 388, 388f
  type VI, 254t
  type X, 390
Collectin, 337
Colon cancer, 4, 4f, 236, 243t, 347t, 369f, 370–72, 371f, 373t
Colonoscopy, 371
Colony stimulating factor, 186t, 340, 340t, 347, 373, 384t
Colorblindness, 68t, 88, 121, 124–25, 124f
Colorectal cancer. *See* Colon cancer
Combined DNA Index System (CODIS), 277f
Common ancestor, 8, 8f, 245, 307f, 313
Common cold, 333
Comparative genomic hybridization, 392, 393t
Comparative genomics, 441–45, 444f
Complementary base pairing, 178, 178f, 181, 181f, 186, 187f, 189, 313, 313f, 380
Complementary DNA (cDNA) library, 382–83, 383f
Complement system, 337, 339
Complete penetrance, 99
Complex trait. *See* Multifactorial trait
Concordance, 144–45, 144t, 147t
Conditional mutation, 234
Conditional probability, 90–91
Cone cells, 124–25
Congenital generalized hypertrichosis (CGH), 124, 126f
Congenital rubella syndrome, 66
Conjoined twins, 61–62, 62f
  Chang and Eng, 61
  Hensel sisters, 62, 62f
Connective tissue, 22, 40f
Conquistadors, 351
Consanguinity, 83, 89f, 99, 285–86
Conservative replication, 180–81, 180f
Constant region, of antibody, 339, 339f
Contact inhibition, 36
  in cancer cells, 360, 361t
Contig map, 437
Control sequence, 213, 213t
Cooley, Thomas, 219
Cooley's anemia, 219
Corn
  bt, 386
  genetically modified, 387t
Corneal transplant, 348
Corona radiata, 56, 57f
Cortés, Hernán, 351
Cosmic rays, 225, 225t
Cotton, genetically modified, 378, 387t
Coumadin derivatives, 65t
Cousins, genetic relationships between, 142, 142f, 143t
Cowpox, 345–46
CpG island, 437
Creutzfeldt-Jakob disease (CJD), 233, 286
  variant, 200f, 201, 297
Crick, Francis, 174, 175f, 175t, 192, 194f, 435
Cri-du-chat syndrome (5p⁻), 257–58
Crigler-Najjar syndrome, 271f, 289t, 290
Crime scene investigation. *See* Forensic science
Crohn disease, 147, 243t, 347t
Cro-Magnons, 307f, 310, 320
Crossing over, 50f, 52, 52f, 85, 104–5, 105f, 107, 291
  in inversion loop, 261, 261–62f
Croticaciduria, 250f
Crowd disease, 350–51
Cruciferous vegetable, 372, 372f
Cryoprecipitate, 403
Crystallin, 254t
Culture, 314
Curly hair, 83–85

CXCR4 receptor, 343
Cyanosis, 232–33
Cyclic AMP, 38f
Cyclin, 36, 186t
Cyclopamine, 47
Cyclopia, 47
Cystathione beta synthase, 254t
Cystic fibrosis (CF), 2, 4, 4f, 31, 31f, 68t, 82, 97, 434
  carrier of, 268, 270–71, 271t, 285
  defect in, 199
  ΔF508 allele, 268, 300, 392, 392f
  diarrheal disease and, 297
  enzyme replacement therapy in, 402
  expressivity, 99
  frequency of CF gene, 268
  gene sequences in nonhuman species, 312
  gene therapy in, 404, 405t, 406f, 407, 413
  genetic test for, 97
  haplotypes in pedigree in, 108, 108f
  historical, archeological, and linguistic correlations of, 300
  microarray analysis of, 392, 392f
  mode of inheritance of, 80t
  multiple alleles for, 97, 101t
  mutation in, 221, 221t, 227
  phenocopy of, 100
  prenatal diagnosis of, 426
  treatment of, 31, 31f
  uniparental disomy in, 263
Cytochrome b5 reductase, 233
Cytochrome *c*, 317, 318f
Cytochrome *c* oxidase I, 445
Cytogenetic map, 434f
Cytogenetics, 106, 242, 248
  technology timeline, 249
Cytokine, 186t, 333, 337–38, 337f, 340, 340t, 344
  therapeutic uses of, 341, 347–48
Cytokine receptor, 410
Cytokinesis, 33–34f, 35, 50–51f, 369
Cytoplasm, 24, 25f
Cytoplasmic donation, 425
Cytosine, 3, 5, 174, 176, 176f, 178, 178f
Cytoskeleton, 29–30f, 30–33, 32f
Cytotoxic T cells, 336–37f, 340, 341f, 341t, 343

Daka (fossil), 308
Dalton, John, 124
Damage tolerance, 235, 236f
Danish Adoption Register, 144
Darfur region (Sudan), genocide by rape in, 285
Davenport, Charles, 323
Deatherage-Newsom, Blaine, 325, 325f
"Death receptor," 36, 37f
deCODE Genetics, 13, 13t
Dedifferentiation, of cancer cells, 360–61, 361t, 363f
Deford family, 82
Deinococcus radiodurans, 234
Deletion, 222–24, 223f, 227, 227t, 229, 251t, 257–58, 257f, 263t
  in cancer cells, 365
  microdeletion, 258
δ cells, pancreatic, 208f
Dementia, 156
Density shift experiment, 180–81, 180f
Dentinogenesis imperfecta, 212, 212f, 212t
Dentin phosphoprotein, 212, 212f
Dentin sialophosphoprotein, 212, 212f
Dentin sialoprotein, 212, 212f
Deoxyribonucleic acid. *See* DNA
Deoxyribose, 173, 186, 187f, 188t
Depression, clinical, 159t, 164, 165f, 167t
Dermal ridges, 136–37, 137f
Dermatitis herpetiformia, 336t
Dermatoglyphics, 136–37, 137f
DES. *See* Diethylstilbestrol
DeSilva, Ashanthi, 408, 409f
Detoxification, 27–28, 28f